Heat and Thermodynamics

Heat and Thermodynamics

Theory, Problems, and Solutions

Upendranath Nandi

CAMBRIDGE
UNIVERSITY PRESS

CAMBRIDGE
UNIVERSITY PRESS

Shaftesbury Road, Cambridge CB2 8EA, United Kingdom

One Liberty Plaza, 20th Floor, New York, NY 10006, USA

477 Williamstown Road, Port Melbourne, VIC 3207, Australia

314–321, 3rd Floor, Plot No. 3, Splendor Forum, Jasola District Centre, New Delhi – 110025, India

103 Penang Road, #05–06/07, Visioncrest Commercial, Singapore 238467

Cambridge University Press is part of Cambridge University Press & Assessment,
a department of the University of Cambridge.

We share the University's mission to contribute to society through the pursuit of
education, learning and research at the highest international levels of excellence.

www.cambridge.org
Information on this title: www.cambridge.org/9781009263924

First published 2025

Printed in India by Nutech Print Services, New Delhi 110020

A catalogue record for this publication is available from the British Library

ISBN 978-1-009-26392-4 Paperback

Cambridge University Press & Assessment has no responsibility for the persistence
or accuracy of URLs for external or third-party internet websites referred to in this
publication and does not guarantee that any content on such websites is, or will
remain, accurate or appropriate

For EU product safety concerns, contact us at Calle de José Abascal, 56, 1°, 28003 Madrid, Spain,
or email eugpsr@cambridge.org.

To my brilliant son, Sukalpan Nandi,
and wife, Alpana Nandi, without whom I would be nothing.
They always comfort and console, never complain
or interfere, ask for nothing, and endure all.

Contents

Foreword

Thermodynamics has been an empirical science since it started in the mid-17th century. The laws of thermodynamics are based on observation alone, largely on the argument that the reverse has never been observed—such as why heat flows from a hot body to a cold one. It is no wonder, then, that thermodynamics was developed as an important and most faithful tool for physicists and engineers to deal with the working principle of fluids and the design of pressure cookers, steam engines, and steamships. A major driving force behind thermodynamics has been the heat engines that convert heat into work, and in an efficient way, as shown in the works of Sadi Carnot.

It took the genius of Ludwig Boltzmann to give thermodynamics a mathematical and statistical framework, including the all-important concept of entropy as the measure of possible number of microstates. Thermodynamics, including heat, has been a pillar of classical physics for a very long time. Its range of applicability has increased from biology to geology, and from cosmology to economics, which are beyond the imagination of those who formulated the subject.

Books on thermodynamics are generally of two types: first, those based on its engineering aspects, and second, those focusing on its more physical concepts. For undergraduate physics students, it is desirable to have a textbook that discusses the physical concepts clearly, without getting abstruse. There should be enough experiments and applications to keep the students interested, and at the same time, there should be a clear exposition of the theoretical ideas to prepare a strong base for them, leading, ultimately, to statistical mechanics.

In our system of education, greater emphasis is given to performance in examinations than on the assimilation of, and the ability to apply, the actual knowledge. Gradually but irreversibly the in-depth analytical questions are being replaced by multiple choice type questions, possibly for logistic reasons. This trend will definitely continue in the new National Education Policy of India. And that is where the classical textbooks on thermodynamics may fall out of favor to the present generation of students.

Fortunately, this book has done a nice balancing job. It has sacrificed neither the rigour and thorough, in-depth exposition of ideas, nor the interesting applications and thought-provoking problems. These problems, strewn throughout the book, should be a big help to the students. The book covers topics beyond a standard thermodynamics syllabus, such as Planck's theory of radiation. I applaud my old friend, Dr Upendranath Nandi, a very experienced teacher and researcher at one of the top-level undergraduate colleges in Kolkata, for producing such a textbook.

I believe that this book will be of help to all the students taking physics as undergraduate major, not only for the university syllabus but also for the competitive examinations.

Anirban Kundu
University of Calcutta

Preface

The book *Heat and Thermodynamics: Theory, Problems, and Solutions* is an informal, readable introduction to the basic ideas of thermal physics. It is aimed at making the reader comfortable with this text as a first course in **Heat and Thermodynamics**. The basic principles and phenomenological aspects required for the development of the subject are discussed at length. In particular, the extremum principles of entropy and free energies are presented elaborately to make the content of the book comprehensive. The book provides a succinct presentation of the material so that the student can more easily determine the major objective of each section of a particular chapter. In fact, thermal physics is not the subject in physics that starts with its epigrammatic equations—Newton's, Maxwell's, or Schrodinger's, which provide accessibility and direction. Instead, it (thermodynamics) can be regarded as a subject formed by the set of rules and constraints governing interconversion and dissipation of energy in macroscopic systems. Further, the syllabus of statistical mechanics for graduate students has changed significantly with the introduction of **National Education Policy 2020**.

Thermal physics has established the principles and procedures needed to understand and explain the properties of systems consisting of macroscopically large numbers of particles, typically of the order of 10^{23} or so. Examples of such collections of systems include the molecules in a closed vessel, the air in a balloon, the water in a lake, the electrons in a piece of metal, and the photons (electromagnetic wave packets) emitted by the Sun. By developing the macroscopic classical thermodynamic descriptions, the book *Heat and Thermodynamics: Theory, Problems, and Solutions* provides insights into basic concepts and relationships at an advanced undergraduate level. This book is updated throughout, providing a highly detailed, profoundly thorough, and comprehensive introduction to the subject. The laws of probability are used to predict the bulk properties like stiffness, heat capacity, and the physics of phase transition, and magnetization of such systems.

The book is divided into eleven chapters. An introduction to the subject is presented in Chapter 1. The laws of thermodynamics are elaborately discussed in subsequent chapters. The concept of thermodynamic potentials and the Maxwell relations are comprehensively described in Chapter 10. Applications of thermodynamics to gases, condensed matter, and phase transitions and critical phenomena are dealt with in detail. The basic thermodynamic principles are liberally illustrated with numerous solved examples that demonstrate how the principles are applied to the general laws of energy and entropy to engines, refrigerators, chemical reactions, phase transformations, and mixtures. The topics covered in various chapters of the book are systematically presented in a lucid and simple language, and developed from a basic level understandable to the majority of the students. Most of the national level examinations are now based on multiple choice questions (MCQ) and numerals. Keeping this in view, questions and exercises are interspersed with the text to help students consolidate the learning, and a set of solved numerical problems, and MCQs with answers have been included at the end of each chapter. These worked out examples explore applications not just within physics but also to engineering, chemistry, biology, geology, atmospheric science, astrophysics, cosmology, and everyday life. I hope such problems would trigger the intellect of the students.

Both professors and students are encouraged to email me at unphys@scottishchurch.ac.in if they have comments/corrections/questions or just want to opine.

Upendranath Nandi

Acknowledgments

First and foremost, I would like to thank my wife, Alpana, and son, Sukalpan, for standing beside me throughout the period of writing the book. They have been my constant inspiration and motivation for continuing to improve my knowledge and move my career forward. Simply, they are my rock, and I dedicate this book to them. I express my love to my wonderful son, Sukalpan, for always making me smile and for understanding those late evenings when I was writing this book instead of enjoying time driving with him. I hope that one day he will read this book and understand why I spent so much time in front of my computer. I would like to thank my parents for allowing me to follow my ambitions throughout my childhood. My family has always supported me throughout my career and in authoring this book for which I really appreciate much.

I am grateful to the brilliant teachers who helped me teach the subject **Heat and Thermodynamics**. I am indebted to my outstanding teachers for the strength they gave to make my dreams come true, for the academic world they showed, for the challenges they have made me face, and for the courage to be leading every race. They inspired me to find my goal, and made me capable of realizing it to write this book. It is whole-heartedly expressed that the advice from these great teachers proved to be a landmark effort toward the success of writing this book. Whoever I am today is because of these teachers. I am thankful to my old friends, Prof. Anirban Kundu, Prof. Debnarayan Jana, and Prof. Tanusri Ghosh, for their constant support, amenable suggestions, and appropriate corrections that have improved the quality of the book.

Having an idea and turning it into a book is as hard as it sounds. The experience is both internally challenging and rewarding. I especially want to thank my colleagues for their wonderful collaboration and support. It was always a pleasure coming to work every day with such lovely and engaging people. I thank them for being an excellent source of inspiration, and especially for creating an environment in which textbook writing is valued and encouraged. I would like to thank all the members of the publishing team whose assistance proved to be a milestone in the accomplishment of my end goal. These personalities are Mr Ankush Kumar, Mr Vikash Tiwari, and Mrs Vaishali Thapliyal of Cambridge University Press and Assessment, whose confidence in this project has always exceeded my own, which finally provides the book in this shape.

I would like to express my sincere thanks to my beloved students who provided support, discussed, read, wrote, offered comments, allowed me to quote their remarks and assisted in the editing, proofreading, and design. I would feel highly satisfied and amply rewarded in case the student community is benefited from the book to any substantial extent. I would be thankful for any type of suggestion for the improvement of the content of the book.

Upendranath Nandi

CHAPTER 1

Introduction to Heat and Thermodynamics

Thermodynamics is a funny subject. The first time you go through it, you don't understand it at all. The second time you go through it, you think you understand it, except for one or two small points. The third time you go through it, you know you don't understand it, but by that time you are so used to it, it doesn't bother you anymore.

—Arnold Sommerfeld

1.1 Introduction

"Thermodynamics" is the branch of science that deals with the macroscopic properties of matter. In this branch of physics, concepts about heat and work and their inter-conversion, energy and energy conversion, and working principle of heat engines with their efficiency are mainly discussed. The name "thermodynamics" was originated from two Greek words: "therme" means "heat" and "dynamics" means "power" or "energy". Thus, matter related to heat and energy is primarily paid attention in this subject. Further, it is believed that the term "thermodynamics" arises from the fact that the macroscopic thermodynamic variables used to describe a thermodynamic system depend on **the temperature of the system**.

Thermodynamics is the branch of physics in which the system under investigation consists of a large number of atoms and molecules contributing to the macroscopic matter of the system. The average physical properties of such a thermodynamic system are determined by applying suitable conservation equations such as conservation of mass, conservation of energy, and the laws of thermodynamics in equilibrium. The equilibrium state of a macroscopic system is achieved when the average physical properties of the system do not change with time and the system is not driven by any external driving force during the course of investigation. The interrelationships among the various physical properties are established with the help of associated thermodynamic relations derived from the laws of thermodynamics. These average (macroscopic) properties of thermodynamic systems are determined from the macroscopic parameters such as volume V, pressure P, and temperature T, which do not depend on the detailed positions and velocities of the atoms and molecules of the macroscopic matter in the system. These macroscopic quantities are called **thermodynamic coordinates, variables or parameters**. Further, these macroscopic properties depend on each other. Therefore, from the measurements of a subset of these properties, the rest of them can be calculated using the associated thermodynamic relations.

There are **four laws in thermodynamics** that are explained in terms of the microscopic constituents within the framework of statistical mechanics. These laws can be successfully applied to a wide variety of physical systems in science and engineering, especially in physical sciences, physical chemistry, chemical engineering, and mechanical engineering for extracting physical properties of the systems. This branch of physics not only provides the exact description of the states of a thermodynamic system in equilibrium but also provides an approximate description (to a very high degree of accuracy!) of a huge number of relatively **slow processes** occurring in physical sciences. A process is considered to be "slow" when the dynamics of evolution of the process is slow compared to the rate of atomic and molecular relaxation processes. It should be noted that most of the time, these atomic and molecular relaxation processes are fast on the human time scale of seconds and milliseconds. On the other hand, the thermodynamic processes have to be slow enough so that an initially homogeneous system remains homogeneous during the evolution of the process. During such slow processes, systems pass through a sequence of nearly equilibrium states and remain in quasi-equilibrium state. This facilitates the successful applications of the laws of thermodynamics in determining the physical properties of such systems.

According to Callen, "Thermodynamics is the study of the restrictions on the possible properties of matter that follow from the symmetry properties of the fundamental laws of physics". With the development of the subject, it was subsequently realized that this statement seems to link thermodynamics with statistical physics and modernizes the term "thermodynamics" in true sense. The branch "thermodynamics" deals with the relationship between thermal, mechanical, and chemical interactions and equilibrium properties (and small fluctuations about equilibrium) on a macroscopic scale. Though the macroscopic behavior of a system is described in thermodynamics, the mechanism behind that behavior is not always described in it (thermodynamics).

In thermodynamics, **system** and **surrounding** play a vital role in describing the evolution of thermodynamic processes. Systems in thermodynamics are generally classified as **open, closed,** and **isolated** and serve the purpose of actual realization of physical systems for the description of different thermodynamic processes. The surroundings are defined as the rest part of the universe except the chosen system. An **open thermodynamic system** can exchange both energy and matter with the surroundings, whereas an **isolated thermodynamic system** will not allow the flow of matter as well as energy to the surroundings. A **closed thermodynamic system** is defined as the system that allows only the flow of energy but not the quantity of matter to the environment. In the case of a closed thermodynamic system, a suitable thermodynamic partition is used so that it can allow only the flow of thermal energy but not matter. Such a closed system has a set of equilibrium states characterized by the thermodynamic variables and is generally described by the laws of thermodynamics in equilibrium. These equilibrium states are the basic elements of the theory in thermodynamics. During the evolution of a thermodynamic system following a certain process, transition from one such equilibrium state to another takes place. The intermediate stages between two such equilibrium states are considered quasi-static such that the laws of thermodynamics can be applied successfully to each of these intermediate states. Further, the path followed by the thermodynamic parameters during such evolution of the system may be reversible or irreversible. The laws of thermodynamics deal with such possible transitions between equilibrium states and exchange of energy between the system and the surroundings during these transitions. When a thermodynamic system makes a transition from one state to another, it may pass through a number of nonequilibrium states as well. In such cases, the relation between the end states and the total effect of the transition is primarily considered for the description of the thermodynamic system. To deal with such nonequilibrium states between the end states and for the derivation of physical properties, advanced theoretical concepts in physics have to be adopted.

1.2   What is thermodynamics?

Thermodynamics is the branch of physical sciences that deals with the macroscopic properties of a system, and the relation between heat and other forms of energy such as mechanical, chemical, or electrical. This branch of physics was formulated in the 18th and 19th centuries, long before the acceptance of atomistic view of material particle by the scientific community. This branch is mainly related to the thermal motion of the constituent particles of a system. In classical thermodynamics, physical properties of a thermodynamic system are described in detail in equilibrium condition. This branch of physics is based on certain empirical laws, that is, there is no way to prove such laws and is therefore phenomenological; nevertheless, it is powerful, and its description is exact. Each of these empirical laws introduces a new concept, for example, the zeroth law of thermodynamics introduces temperature, the first law of thermodynamics introduces internal energy, and the second law of thermodynamics introduces entropy. These concepts provide a definite meaning to physically measurable quantities and indicate useful correlations between them. These physically observable macroscopic parameters of a thermodynamic system are determined in equilibrium situation. This branch occupies a special position in physics for any form of energy on its transformation finally changes into energy of thermal motion.

The laws of thermodynamics are very general and can be successfully used to describe a wide variety of physical and chemical systems in equilibrium. The scope of thermodynamics is so general that it would work even if matter did not consist of atoms and molecules! The main disadvantage of thermodynamics is that one has to understand a lot of abstract concepts before applying them to interesting applications. It is to be noted that more general theories are more abstract. Thermodynamics is in line with it.

1.3   Why one should study thermodynamics?

A natural physical system contains a huge number of atoms or molecules. For example, a system of an ideal gas of volume one cm^3 under normal conditions contains $N_L = 2.69 \times 10^{19}$ atoms. Here, N_L is the so-called **Loschmidt number**. If we want to study the dynamics of such a system, we have to use Newton's equations to describe the motion of the atoms, which will lead to a large number of differential equations. Direct solutions of such a huge number of differential equations with necessary boundary conditions are really impossible, even using a superfast computer. It should be mentioned that in order to obtain the macroscopic features of a system, one does not need too detailed information about the motion of the individual particles, that is, the microscopic behavior of the system. One has to rely upon the average properties of such a system. These average physical properties are characterized by macroscopic thermodynamic quantities such as volume V, pressure P, temperature T, and various thermodynamic potentials. It is to be noted that these macroscopic quantities arise and make sense only in systems with a large number of atoms or molecules. They do not make any physical sense for a system with five or ten atoms or molecules. In order to find out the macroscopic physical properties of such a system consisting of a huge number of atoms or molecules and the relations between them, the branch of physics "thermodynamics" should be used. Other advantages of using thermodynamics are the following:

1. It uses the phenomenological approach to build up the foundation of the subject. The description at the micro-level is completely ignored in this branch of physics.

2. It focuses on the directly or indirectly observed macroscopic quantities in the experiment.

3. It is based on some simple empirical laws formulated on the basis of a number of principles taken from observations. For example, energy is conserved, and two bodies are in thermal equilibrium when they are at the same temperature, etc.

4. It provides an indication of whether a thermodynamic process would occur spontaneously or not.

5. It sets the achievable limit of theoretical efficiency of a thermodynamic heat engine.

6. It forms the basis of understanding steam engines, refrigerators, and other machines. Thermodynamics is a discipline with an exceptionally wide range of applicability.

7. It can successfully explain natural phenomena such as spontaneous flow of heat from a hotter body to a colder one, blue of the sky, spherical shape of raindrops, thinner and colder atmosphere of the earth at higher altitude, red color of the colder stars, bluish-white color of the hotter stars, yellow appearance of the Sun, and ultimate collapse of high mass star to form black holes. These observations indicate that the range of applications of thermodynamics is versatile.

8. It forms the basis of low temperature in physics.

These advantages indicate the importance of studying thermodynamics.

1.4 Thermodynamic description of a system

Heat flows from hotter to colder bodies. The flow of heat in the reverse direction is not realized in nature unless some input energy is supplied. One can then ask the question: Is it possible to explain such irreversibility of heat flow from the exact knowledge of the positions and velocities of the constituent particles or molecules of a system? But daily experience indicates that it is not required to know the exact positions and velocities of the particles of a metal bar in order to explain why heat spreads out from the hot end to the cold one. The kinetic theory explains heat to be a result of random motions and collisions of molecules. The laws of mechanics, no doubt, govern collisions between molecules and may give us a complete microscopic picture, but they do not agree with the concept of irreversibility. In order to get rid of this difficulty, **can we not investigate a many-particle system without going into its internal structure?** This technique has been adopted in thermodynamics. This subject is very simple and leads to very accurate results. But the method used in thermodynamics is practically a phenomenological one. The primary objective of this subject is to establish a relationship between directly observable quantities such as pressure, volume, and temperature. It does not deal with the internal mechanisms of the processes determining the behavior of the system. The laws of thermodynamics govern the behavior of a large molecular population and are not concerned with the fate of individual molecules. The internal mechanisms remain unknown in the formalism of thermodynamics. For example, it cannot explain why a copper wire cools upon rapid expansion while a rubber, under similar conditions, becomes hot.

Thus, the subject **thermodynamics** deals with the phenomenological description of macroscopic properties of a system consisting of a large number of atoms or molecules or particles in equilibrium. Such a phenomenological description is based on a number of empirical observations that are expressed in the form of laws in thermodynamics. Using these empirical observations, a coherent and logical description of the subject is then constructed. These mathematical descriptions lead to some useful physical concepts and a variety of testable relationships among various thermodynamic quantities. These concepts provide useful guidelines to describe the macroscopic properties of a thermodynamic system. It should be mentioned further that some more fundamental physical concepts (microscopic) are needed to justify the empirical laws of thermodynamics. For example, attempts are made in statistical mechanics to obtain these empirical laws starting from classical or quantum mechanical equations for the evolution of collections of particles. Applying the laws of thermodynamics, it becomes much easier to study the macroscopic properties of a thermodynamic system in equilibrium. The condition of equilibrium of a system demands that the physical properties of the system should not be changed appreciably with time during the course of investigation. This concept of equilibrium becomes subjective if it depends on the observation time. For example, the glass of a window remains in a state of equilibrium as a solid over many decades, but it is experimentally observed to flow like a fluid over the time scales of millennia. On the other hand, the equilibrium between matter and radiation in the early universe was known to exist for the first few minutes of the Big Bang. Thus, the time scales play an important role in determining the state of a thermodynamic system and, hence, its physical properties.

1.5 Historical development of kinetic theory of gases and thermodynamics

In thermodynamics, a gaseous system is characterized by thermodynamic variables such as pressure P, temperature T, volume V, internal energy U, and entropy S. The physical properties of such a system are expressed in terms of these variables in the form of certain empirical laws. Boyle's law, discovered in 1661, is one such law. According to this law, the pressure P exerted by a gas of given mass is inversely proportional to the volume V of the enclosed gas at a particular temperature T. Boyle proposed two alternative atomistic explanations for this pressure P:

1. The first alternative is: air is composed of particles that repel each other. This repulsion is very similar to that of the coiled-up pieces of wool or springs.

2. The second alternative is: air is composed of whirling particles that push each other away by impacts.

Sir Isaac Newton took up the first hypothesis and proved mathematically the origin of pressure considering the repulsion of neighboring particles. He further showed that this force of repulsion must be inversely proportional to the interparticle distance. On the other hand, Boyle tried to associate the second hypothesis with Descartes' ethereal vortices, but it lacked a quantitative foundation in the 17th century. However, the second hypothesis gained qualitative support from the well-known concept that heat is related to the motion of atoms or molecules or constituent particles of the medium. Further, the second hypothesis was found to be consistent with the experimental observation that the pressure of a medium, say air, increases with the increase in temperature.

Boyle, Newton, and other scientists assumed that gases are made of tiny atomic particles, and based on this assumption, they tried to explain the physical properties of gaseous systems. The actual atomic theory of gases got its foundation more than 150 years later. It was assumed in the KTG that a gas consists of a large number of identical, discrete particles called atoms or molecules moving rapidly in all directions, and the consequences obtained from such rapid motions of the atoms were used to explain the physical behavior of the gaseous systems. As the interatomic forces in gases are very small, such random motion of the molecules is possible, and often, these forces are neglected in gases. These interatomic forces are short-range forces and play an important role in solids and liquids. The elements of KTG were developed by Maxwell, Boltzmann, Clausius, and other scientists between 1860s and 1880s. Its success is remarkable. It provides a molecular interpretation of the temperature of a gas, and its interpretations are consistent with the gas laws and Avogadro's hypothesis. The specific heat capacities of many gases are correctly explained by the KTG. It can also explain various measurable properties of gases such as viscosity, thermal conductivity, and diffusion of material particles. It correctly correlates these physical properties with molecular parameters such as sizes and masses.

The Swiss mathematical physicist Daniel Bernoulli formulated the KTG quantitatively and derived Boyle's law by calculating the force exerted on a movable piston by the impacts of gas molecules moving with speed v in a closed vessel of volume V. Bernoulli assumed that the frequency of impacts is proportional to the speed v of the molecules and the force of each impact is proportional to the momentum $m\,v$ of the molecules, and showed that the pressure exerted by the gas would be proportional to the kinetic energy of the particles. Further, he explained that at a fixed temperature, pressure is proportional to the density of the medium and suggested that at a standard density, the temperature of a gaseous system itself could be defined in terms of pressure. Thus, a beautiful idea was introduced in Bernoulli's theory that heat or temperature could be identified with the kinetic motion of the molecules in a gaseous system.

It should be mentioned that in the 18th century, the kinetic theory was not widely accepted by the scientific community. Most of the scientists preferred the repulsion theory due to Newton. This theory was compatible with the idea that heat is a fluid, known as the "caloric", rather than the kinetic energy associated with the motion of the molecules. This caloric was assumed to be composed of particles that repel each other and are attracted to the atoms of ordinary matter. The caloric theory explained the increase in gas pressure with a corresponding increase in temperature because at higher temperature, the gas acquires more of the self-repelling caloric fluid. In this theory, temperature itself was assumed to be related to the density of caloric, that is, the amount of the caloric fluid divided by the volume.

Using this definition of temperature, the increase in temperature of a gas due to compression (this is accompanied by accumulation of the same amount of caloric over a smaller volume) can be explained by the caloric theory even though heat is not added to the system from outside, or the decrease in temperature during expansion (this is accompanied by spreading out of the same amount of caloric over a larger volume) even though no heat is lost in the process. However, an anomalous observation was experimentally realized by Joule whose importance was not recognized until much later. In that experiment, it was found that no change in temperature practically occurs during the free expansion of a gas into a vacuum. There were also a few interesting results explained by the caloric theory. For example, the latent heat of phase transitions and the heat absorbed or released in chemical reactions were successfully explained by the caloric theory assuming that some caloric is "attached" to the individual atoms or compounds. According to the caloric theory, the relation between pressure and volume of a gas was determined by the "unbound" or "free" caloric that filled the space between the atoms or molecules. No plausible explanation of these phenomena was offered by the KTG. According to the KTG, it was assumed that the atoms or molecules move at constant speeds in between collisions. This movement of the atoms seems incompatible with the generally accepted concept that an ethereal fluid fills all space uniformly. In the early 19th century, the caloric theory finally gained credibility when Laplace used it to calculate the speed of sound in gases and resolved a long-standing mismatch between the theoretically calculated value and the experimentally observed value. This theory gained indirect support from the acceptance of the particle nature of light. This is because light and heat were widely regarded as qualitatively similar phenomena in physical sciences.

According to the modern KTG, atoms or molecules of a gas occupy a very small part of the volume of the enclosure, and are thus able to move freely in straight lines as enough empty space is available for their movement. This motion is not affected by other atoms or molecules or by any resistance offered by the ethereal fluid. The modern model of the kinetic theory of gas is radically different from the one proposed by the English chemist John Dalton at the beginning of the 19th century. Dalton explicitly adopted Newton's hypothesis that the pressure exerted by a gas is due to the short-range repulsive forces between the molecules and considered these molecules as being surrounded by caloric in contact with those of their neighbors. Further, the molecules of different elements have different sizes, and equal number of them for different gases would occupy different amounts of volume in the position space. This would provide a pressure–volume relation at a constant temperature.

Dalton used the supplementary idea of Newton's hypothesis that atoms or molecules of different elements do not repel each other, but they may even attract. He used this hypothesis to explain why air did not separate into its components under the influence of gravity. Because of self-repulsion, each kind of atom would diffuse among the others rather than clustering with their own kind. Another consequence of this idea is Dalton's "law of partial pressures". This states that the total pressure of a gas mixture is simply the sum of the pressures each kind of gas would exert if it were occupying the entire space by itself. It is to be noted that this law can also be derived from the KTG without considering the differential forces between similar and dissimilar atoms or molecules.

Dalton assumed the simplest possible formula for each chemical compound and prepared his famous table of "atomic weights". For example, he assumed that a water molecule contains one atom of oxygen and one of hydrogen. If that were true, then using modern data, since one gram of hydrogen reacts with 8 grams of oxygen to form 9 grams of water, the atomic weight of oxygen relative to hydrogen would be 8. This idea of Dalton was not in agreement with Gay-Lussac's discovery of the "law of combining volumes", which states that **in gaseous reactions, the volumes of the reactants and products are related to each other by ratios of small integers**. For example, two liters of hydrogen combine with one liter of oxygen to form two liters of water. The most plausible atomistic interpretation of this law is that the volume of each gas is proportional to the number of particles it contains. Then, two particles of hydrogen would combine with one of oxygen to form two particles of water. That would imply that a "particle" of oxygen can split into two parts, one for each of the resulting water particles; thus, gases may be composed of particles, each of which is made of two or more atoms. It would also imply that the atomic weight of oxygen is 16, not 8.

Dalton criticized the accuracy of Gay-Lussac's measurements and raised objections to the experimental results as they contradicted his own theory. It was argued by him that since the particles of different elements have different sizes and are in contact with each other, the volume of a gas cannot be proportional to the number of Daltonian particles. Further, the idea that an oxygen molecule is composed of two oxygen atoms violates the principle that atoms of the same kind repel each other. The Italian physicist Amedeo Avogadro first explored the consequences of the results of Gay-Lussac's experiments and put forward the chemical atomic theory into its modern form. Avogadro conjectured that the ultimate particles dealt in kinetic theory are not the atoms but the molecules, each of which may contain one or more atoms of the same or different kinds.

"Avogadro's hypothesis" provides a basic guideline to build up a foundation of chemical physics and states that **every kind of gas contains the same number of molecules in a given volume under the same conditions**. After 1860, the chemists calculated the atomic weights of elements forming gaseous compounds using this hypothesis. It was found that the gaseous forms of many common elements, such as hydrogen, nitrogen, oxygen, and chlorine, are composed of diatomic molecules, though others, like mercury, were later found to be composed of one molecule. This hypothesis was found to be in favor of the KTG insofar as it implied that the volume occupied by a certain number of molecules is independent of their size and shape, which suggests that these molecules are not ordinarily in contact with each other. In 1859, Maxwell showed that the hypothesis could be derived from the knowledge of the KTG. In 1860, James Clerk Maxwell derived the expected distribution of molecular speeds in a gas starting from the mechanics of individual molecular collisions. Over the next several years, the KTG developed rapidly, and many macroscopic properties of gases were computed in equilibrium.

The kinetic theory of gases together with the concept of mean free path relates some physical properties like viscosity, thermal conduction of diffusion of matter and provides some insight into the behavior of ideal gases. Brownian motion is one of the important characteristic features of the motion of the molecules. It provided a different dependency of mean square displacement on time. Later, the random walk was discovered to map the motion of molecules and simulate the real-life problems.

It was accepted in the 1850s that heat is a form of energy, and there is a close relationship between heat and energy. Sadi Carnot, in 1824, used the relationship between heat and energy and developed the foundation of steam engines by calculating and comparing their efficiencies. Rudolf Clausius and William Thomson (Kelvin) formulated the statements of both the first law of thermodynamics (the total energy is conserved) and the second law of thermodynamics (heat does not flow spontaneously from a colder body to a hotter) around 1850. Other formulations of the second law of thermodynamics were quickly realized, and in particular, Kelvin understood some of the general implications of the law.

In 1857, Clausius revived the idea that gases consist of molecules in motion. Ludwig Boltzmann, in 1872, described the time evolution of a gas in detail considering the gas in equilibrium state as well as in nonequilibrium state. In the 1860s, Clausius introduced the concept of entropy as a ratio of heat to temperature and stated the second law of thermodynamics in terms of the increase of this quantity, entropy. Using H-Theorem, Boltzmann showed that the entropy in equilibrium must always increase with time. Hence, it was assumed that Boltzmann had successfully proved the second law. But it was then noticed that his derivation could be run in reverse since molecular collisions were assumed to be reversible, and would then imply the opposite of the second law. It was realized later that Boltzmann' s original equation implicitly assumed that before each collision, molecules are uncorrelated, but not afterward. This introduced a fundamental asymmetry in time. Early in the 1870s, Maxwell and Kelvin thought that the second law of thermodynamics cannot be formally derived from microscopic physics because of human inability to track large numbers of molecules. In responding to objections related to reversibility, Boltzmann realized around 1876 that there are many more states in a gas and these states are random rather than seem orderly. He then argued that as entropy always tends to a maximum and a system always loves a macrostate with maximum number of microstates, entropy must be proportional to the logarithm of the number of possible states of the system. He also formulated ideas about ergodicity.

In the early 1900s, many interesting as well as fundamental discoveries took place in quantum theory. These incidents largely overshadowed the developments of thermodynamics and little fundamental work was done on this subject. Nevertheless, the second law of thermodynamics, by that time, was considered to be regarded as a basic principle of physics. The ergodic theory received not much importance in physics; however, it became an active area of pure mathematics. Various ergodic properties were established for many kinds of simple systems during the period from the 1920s to the 1960s.

In the present century, computers are being routinely used to perform impressive works in almost every branch of physics. It has become highly essential to deal with thermodynamic systems in which molecules are extremely complex, interacting in a very complex manner. These complex systems are modeled with a suitable pattern, and simulations are performed with the help of computers to extract the physical properties. Some of these complex systems include amorphous systems, polymer systems, aqueous systems and mixtures at extreme conditions, and mixtures with very complex molecules. There are even more complex systems, such as paint, various types of foods, and cosmetics. It is very difficult to obtain experimentally the equilibrium state in such complex systems; hence, it becomes difficult to analyze such systems for the determination of composition as well as physical properties with greater accuracy. It should be mentioned that this overview of the development of thermodynamics reflects the background, knowledge, and ignorance of the author. Although the four laws and some related standard thermodynamic prescriptions are fixed points, there is plenty of scope for different viewpoints of what should be included and what not in such a short story of such a rich topic.

1.6 Contents of the book

This book contains 11 chapters, which are organized in the following way: An introduction on the subject is presented in **Chapter 1**. The historical background and chronological development of the subject are described in detail in this chapter. A thermodynamic system consists of a large number of atoms or molecules or particles that are in motion and interact among themselves with some force. This makes a challenging task to study the macroscopic properties of such a system by applying Newton's laws of motion. As a result, a number of assumptions are made by neglecting volume of the molecules as well as interactions among the molecules. Various aspects of the KTG, the pressure exerted by the molecules of such a gaseous system, the well-known gaseous laws, and the specific heat are discussed in detail in **Chapter 2**. The molecules of a gaseous system, in general, are identical and distinguishable at normal temperature and pressure and follow classical statistics. These molecules move randomly with all possible values of velocity in space. Description of the dynamical behavior of such a system includes the distribution of velocity (Maxwell–Boltzmann law), energy, and momentum in various dimensions. Average values of these quantities play a major role in characterizing a gaseous system. These characteristic features are elaborated in **Chapter 3**. As an experimental verification of the distributions of velocities, experiments due to Stern, Zartman, and Ko, and the Doppler broadening of spectral lines are described in detail in this chapter. Further, the law of equipartition of energy is derived, and the anomaly in the specific heat problem is resolved by introducing the concept of degree of freedom in this chapter.

Collisions between molecules in a gaseous system indicate that the molecules have a finite size. These collisions serve as examples of random events. Collision probabilities among the molecules, distributions of free paths and mean free path, and estimation of such mean free path following Clausius' and Maxwell's methods are presented in detail in **Chapter 4**. A gaseous system exhibits various physical properties like viscosity, thermal conductivity, and diffusion. These properties of ideal gaseous systems are explained physically from the kinetic aspects of the molecules/atoms. Brownian motion is an important characteristic feature exhibited by the molecules of a fluid medium. This Brownian motion with its theoretical explanations due to Einstein and Langevin is highlighted in this chapter.

Several physicists, including Regnault, Andrews, and Amagat, conducted experiments on a variety of gases like CO_2, N_2, H_2, He, and O_2 and measured volume V as a function of temperature T and pressure P. These experimental results indicate that the ideal gas equations are incapable of explaining the behavior of such gases, particularly at low temperatures and at high pressures. This necessitated other equations of state for the description of physical behavior of these gases. As a result, a number of equations of state due to Van der Waals, Clausius, Berthelot, Dieterici, Saha and Bose, and Redlich–Kwong were proposed to explain these experimental results on real gases. A brief note about these equations of state with primary emphasis on the Van der Waals equation of state is presented in **Chapter 5**. Considering the volume of the molecules and some sort of interaction between the molecules, Van der Waals equation of state is derived mathematically. The reduced form of this equation of state, law of corresponding state, and other implications associated with this equation of state are discussed in detail in this chapter. Heat is the outcome of molecular motion in a fluid medium, whereas it is the result of thermal vibrations of the molecules about their equilibrium positions in solids. The flow of heat occurs via the processes of conduction, convection, and radiation in various media. The process of heat flow using conduction in solid medium is discussed elaborately in **Chapter 6**. One-dimensional heat flow equation due to Fourier is derived under the condition of variable state, and this equation is used to solve the problems of heat flow through conducting bar, spherical and cylindrical conductors in steady state condition. For the determination of thermal conductivity K, some experiments are also discussed in this chapter.

There are two ways of studying the physical behavior of a thermodynamic system. One is a microscopic behavior, which deals with the study of atoms and molecules of the system. The second way is to study the macroscopic (average) behavior of extremely large number of atoms/molecules constituting the system and is the subject matter of thermodynamics in which we deal with quantities, such as temperature T, pressure P, volume V, and internal energy U and interrelationship among these variables. The chronological development of the subject and basic terminologies such as surroundings, open and closed systems, intensive and extensive thermodynamic variables, thermodynamic equilibrium, indicator diagram, state function, reversibility and irreversibility, and various thermodynamic processes are discussed in detail in **Chapter 7**. The zeroth law of thermodynamics and the concept of temperature are highlighted in this chapter. **Chapter 8** deals with the concepts of heat and work, internal energy, and the first law of thermodynamics. Using the formalism of the first law of thermodynamics, expressions for various physical parameters like specific heat (C_P and C_V), isothermal compressibility (β_T), and volume expansion coefficient (α) are derived for ideal as well as real gases. Application of the first law of thermodynamics to a magnetic system is also described in detail in **Chapter 8**.

Chapter 9 provides a detailed discussion of the second law of thermodynamics, including its various statements, the concept of entropy and its properties, the principle of entropy increase, and the entropy of a perfect gas. Topics such as the Gibbs paradox and its resolution, Carnot's theorem, heat engines and their efficiencies, illustrated with both pressure–volume and entropy–temperature diagrams, are also thoroughly covered. Additionally, the thermodynamic temperature scale and the unattainability of absolute zero are discussed in this chapter.

Thermodynamic potentials such as internal energy U, enthalpy H, Helmholtz free energy F, and Gibbs free energy G with their formulations, properties, and applications are presented in **Chapter 10**. Concepts of thermodynamic potentials are utilized to study the properties of surface films, the variation of surface tension of such films with temperature, and the calculation of magnetic work. The $T-S$ diagram of a paramagnetic material and the principle of adiabatic demagnetization are used to produce low temperature. The first and second-order phase transitions and the related equations due to Clausius–Clapeyron and Ehrenfest are also detailed in **Chapter 10**. Maxwell's thermodynamic relations and well-known thermodynamic properties of ideal and Van der Waals gas are derived. Joule's experiment, Joule–Thomson porous plug experiment and its application to ideal and real gas, and inversion temperature are also covered in this chapter.

Chapter 11 presents the properties of thermal radiation, the blackbody radiation spectrum, and their explanations using classical theories such as Wien's distribution law, the Rayleigh–Jeans law, and the ultraviolet catastrophe. It also discusses Planck's quantum postulates, Planck's law of blackbody radiation, and its experimental verification. Furthermore, the Rayleigh–Jeans law, Stefan–Boltzmann law, and Wien's displacement law are derived from Planck's radiation law. Additionally, using the formulations of Bose and Einstein, Planck's law of blackbody radiation is derived. The thermodynamic properties of a photon gas are also explored in this chapter.

A substantial number of solved numerical problems, tailored to topics covered in exams like JAM and JEST, as well as other similar examinations, are included at the end of each chapter. Additionally, multiple choice questions with solutions are provided to further enhance the readers' problem-solving skills.

CHAPTER 2

Kinetic Theory of Ideal Gases

I think a strong claim can be made that the process of scientific discovery may be regarded as a form of art. This is best seen in the theoretical aspects of Physical Science. The mathematical theorist builds up on certain assumptions and according to well understood logical rules, step by step, a stately edifice, while his imaginative power brings out clearly the hidden relations between its parts. A well-constructed theory is in some respects undoubtedly an artistic production. A fine example is the famous Kinetic Theory of Maxwell, ... The theory of relativity by Einstein, quite apart from any question of its validity, cannot but be regarded as a magnificent work of art.

—Sir Ernest Rutherford

Learning Outcomes

After reading this chapter, the reader will be able to

- State the assumptions of kinetic theory of gases (KTG)
- Explain the concept of pressure and calculate the expression for it
- Demonstrate mathematically the gas laws using the expression for pressure derived from KTG
- Present the kinetic interpretation of temperature
- Derive the expression for specific heat at constant volume C_V and constant pressure C_P
- Explain the concept of degree of freedom
- Solve numerical problems and multiple choice questions on KTG

2.1 Introduction

The kinetic theory of gases (KTG) is a theoretical model that describes the physical properties of a gaseous system in terms of a large number of submicroscopic particles, such as atoms, molecules, and small particles. These constituent elements are in random motion and collide constantly with each other and also with the walls of the container. Considering the molecular composition and characteristic features of such random motion of the molecules, various macroscopic properties of the gaseous system, such as pressure, temperature, viscosity, thermal conductivity, and mass diffusivity can be explained with the help of KTG. In this theory, it is postulated that the pressure exerted by a gas is due to the collision of atoms or molecules moving at different velocities on the walls of a container. It basically attempts to explain the macroscopic properties that are related to the microscopic phenomenon. The physical properties of solids and liquids, in general, are described by their shape, size, mass, volume, etc. Gases, however, have no definite shape, and size. Furthermore, their mass and volume are not directly measurable. In such cases, the KTG can be successfully applied to extract the physical properties of the gaseous system.

In the KTG, the atoms or molecules of a gas are considered as very tiny particles like a point (dot) on a plain paper with a small mass attached to it. The interparticle distance, in this case, is much larger than the size of the particle, and they occupy a very small portion of volume of the container, that is, there is a huge amount of available space in the container to be occupied by the particles. In other words, the volume of the particles is considered negligible (almost zero volume) compared to the volume of the container.

In the KTG, it is assumed that a gaseous system contains a large number of atoms or molecules. These atoms or molecules are in constant motion and undergo perfectly elastic collisions in which the energy or momentum of the particles does not change during collision. These atoms or molecules also collide with the walls of the container and exert force on them. This force when calculated per unit area gives rise to pressure. This pressure is proportional to the number of colliding particles per unit time per unit area on the wall of the container. Since these particles are always in motion, the average kinetic energy per particle is proportional to the temperature of the gas. In the formalism of KTG, these atoms or molecules are further treated as independent particles with no force of interactions (attractive or repulsive) operating between them. Due to the lack of such interactions, atoms or molecules move randomly in all directions in a straight line between two collisions in the available space. The average distance between two such collisions is defined as the mean free path which is used to determine the size of the molecules.

Using the assumption that a gas consists of a large number of identical and distinguishable discrete molecules called atoms or molecules, attempts are made in KTG to explain the macroscopic properties of a gaseous system in terms of the motion of such molecules. The basis of KTG was developed by Maxwell, Boltzmann, and Clausius between 1860s and 1880s, and using this basis, attempts were made to explain various characteristic features of gaseous, solid as well as liquid states of matter. However, in this chapter, only the KTG is dealt exhaustively. This theory has its importance in developing a correlation between the microscopic phenomena and the macroscopic properties of a gaseous system. The main characteristic features of the KTG are that the physical properties of any gaseous system can be determined in terms of three measurable macroscopic variables: pressure P, volume V, and temperature T. In simple terms, it also helps to compute various physical parameters related to the random motion of the molecules and collisions among the molecules. In addition, the model of KTG helps in understanding the physical facts such as transport phenomena, Brownian motion, and the problem of random walk.

2.2   Various states of matter

Matter can exist in several distinct forms, known as states of matter. Commonly encountered in everyday life are four states: solid, liquid, gas, and plasma. Beyond these, many other states of matter, such as Bose–Einstein condensates and neutron-degenerate matter, are observed under extreme conditions. For instance, Bose–Einstein condensates form at ultracold temperatures, while neutron-degenerate matter occurs in ultradense environments. A brief introduction to these four fundamental states of matter is provided below.

2.2.1   Solid states of matter

The following are the properties of matter in the solid state:

1. A solid is a closely packed structure of ions, atoms, or molecules. The interparticle distance is very short in this case.

2. A solid material has a definite volume and a definite shape. Solids can only change their shape by the application of a large external force. It has a very high value of isothermal bulk modulus.

3. The forces acting between the particles in a solid are very strong so that the particles cannot move freely but can only vibrate about their equilibrium positions.

4. The coefficient of thermal expansion of a solid is very low.

5. It is nearly incompressible, exhibiting a very low isothermal coefficient of compression.

6. The particles in a crystalline solid are packed following a definite pattern that repeats in all directions. A solid, in general, has short-range as well as long-range order.

7. Solid materials possess a well-defined melting point.

8. The physical properties of solids are generally anisotropic.

9. Amorphous solids have no long-range order and ground states at thermal equilibrium.

10. By melting, solids can be transformed into liquids, and similarly by freezing, liquids can be transformed into solids. Furthermore, through the process of sublimation, solids can also be changed directly into gases.

11. Different solids exhibit varying behaviors, making it impossible to define a universal equation of state for them.

2.2.2   Liquid states of matter

The following are the properties of matter in the liquid state:

1. The interparticle separation in liquids small, though slightly larger than that in solids.

2. In the case of liquids, intermolecular forces are strong. The molecules have enough energy so that they move relative to each other. Therefore, a liquid has no definite shape.

3. The density of a liquid is generally smaller than that of the corresponding solid. But the best-known exception to this behavior is water.

4. In a liquid, short-range order (arrangement of atoms or molecules) is preserved, but long-range order is not maintained.

5. A liquid is nearly incompressible, meaning its volume remains almost unaffected by pressure.

6. At an external pressure higher than the pressure corresponding to the triple point of a solid substance, the solid becomes the corresponding liquid if it is heated above its melting point.

7. Different liquids behave differently; hence, they cannot have a common equation of state.

2.2.3   Gaseous states of matter

The following are the properties of matter in the gaseous state:

1. A gas occupies the entire volume of the container in which it is confined; it has no definite shape or volume.

2. Due to thermal motion, gas molecules are in random and chaotic motion at ordinary temperature and hence, they have no order even at short range.

3. The interparticle distance is very large in this state of matter.

4. The interparticle or intermolecule interaction is very weak (negligible). Thermal energy is much larger than the intermolecular attraction even at ordinary temperature.

5. Due to the large intermolecular separation, its density is very low.

6. The gas is highly compressible, that is, the isothermal coefficient of compression is very high.

7. The isothermal bulk modulus, which is the reciprocal of the isothermal coefficient of compression, is very low for the gaseous state of matter.

8. It has very high thermal volume coefficient of expansion. This indicates a gaseous state that expands significantly with an increase in temperature at constant pressure.

9. A liquid can be converted into a gas at constant pressure by heating it to its boiling point or by reducing the pressure while maintaining a constant temperature.

10. A gas is called a vapor below the critical temperature, and the gas can only be liquefied by compression without cooling. A vapor can exist in equilibrium with a liquid (or a solid), when the pressure of the gas equals the vapor pressure of the liquid or the solid.

11. If the gas is above the critical temperature, it cannot be liquefied only by increasing pressure.

12. Above the critical temperature and critical pressure, a gas behaves as a supercritical fluid. In this state, liquid and gas cannot be distinguished, that is, the difference between them disappears. A supercritical fluid has the properties similar to that of a gas, but its density is higher than that of a gas. In some cases, this attributes solvent-like properties of the supercritical fluid and leads to useful practical applications. For example, supercritical carbon dioxide is applied to extract caffeine to manufacture decaffeinated coffee.

2.3   The kinetic theory of gases

The KTG is a theoretical model that describes the macroscopic properties of a gaseous system. It is based on the random motion of a large number of sub-microscopic particles that include atoms and molecules. Furthermore, the pressure exerted by the gas arising due to collisions among themselves and with the walls of the container is explained by the KTG. This theory can also explain the empirical laws relating to the macroscopic parameters of the gaseous systems. Various physical properties such as temperature, viscosity, thermal conductivity, and diffusion of molecules can be explained within the framework of KTG. At the same time, these macroscopic properties can also be microscopically connected with each other with the help of KTG.

In this kinetic model of a gaseous system, it is assumed that the atoms or molecules exhibit random and chaotic motion. Such random motion of the molecules of a gaseous system is schematically shown in Figure 2.1. These molecules of gas collide with each other and also with the walls of the container. There are ample experimental evidences in favor of such motion of the molecules, which are responsible for various physical phenomena. Some of the evidences in favor of the kinetic model of the molecules are mentioned below.

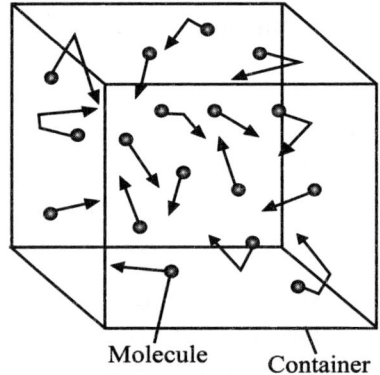

Figure 2.1 Random motion of the molecules of an ideal gas.

2.3.1 Evidence in favor of the kinetic theory of gases

There are some direct and indirect evidences in favor of the KTG. These evidences are briefly mentioned below:

1. A gas always occupies the whole volume of the container. This shows that the gas molecules are always in random motion, and they try to expand.

2. All gases spontaneously diffuse into one another. This spontaneous diffusion of gases in arbitrary directions implies that the gases are in random and chaotic motion.

3. Gases are highly compressible. If a very high pressure is applied, the volume of a gas can be reduced to a very small value. This implies that most parts of the container are vacant. That is, the actual volume of the gas molecules is very small in comparison to the volume of the container.

4. The phenomenon of effusion is also an indication of the motion of the molecules of a gas.

5. In 1911, L. Dunoyer[1] proved experimentally that in between two successive collisions, gas molecules move in straight lines. This is also an indication of the motion of gas molecules.

6. Brownian motion, a ceaseless, random, or chaotic motion is observed in pollen grains suspended in a liquid, colloidal suspension under an ultramicroscope. This type of motion is observed in all types of media.

---

[1]L. Dunoyer, Sur la réalisation d'un rayonnement matériel d'origine purement thermique. Cinétique expérimentale, *Le Radium* 8, no. 4 (1911): 142–146.

A. Einstein[2] and M. Smoluchowski[3] developed the theory of Brownian motion on the basis of KTG and the existence of Brownian motion of the particles in the fluid was experimentally verified by Perrin.[4]

2.3.2 Postulates of the kinetic theory of gases

In general, a gas consists of a large number of randomly moving point particles (atoms or molecules) interacting each other with a complicated force. This force of interaction depends on the distance between the particles in a complicated manner and is also influenced by external factors such as temperature and pressure. These parameters determine the physical properties of a gaseous system and have certain relations among them. Experiments are carried out to find these relations among the variables: pressure, temperature, volume, etc., and are expressed in the form of laws. In the case of a real gas, these laws are complicated. On the other hand, all sorts of interactions except interaction through elastic collisions are neglected in the case of an ideal gas. Also, the volume of the particles (atoms or molecules) is neglected in this case. Thus, an ideal gas is a theoretical concept. A simple theoretical model known as the **kinetic theory of gases** is developed to explain the experimental observations about the behavior of such ideal gases and attempts are made to compare these experimental results with those found in real gases. This theoretical model of KTG is based on the following assumptions or postulates:

1. Gases are composed of a large number of atoms or molecules that behave like hard, spherical point-like objects in a state of constant, random motion. These atoms or molecules are treated as very small particles like a point (dot) on a plain paper with a small mass.

2. These point particles move in a straight line in between two collisions among themselves or before colliding with the walls of the container.

3. The interparticle distance among these particles is generally much larger than the particle size. As a result, there is large free unoccupied space in the container. The volume of the particle is negligible (zero volume) compared to the volume of the container. As these particles are in constant, random motion, they occupy the total volume of the container and therefore, the volume of the container is to be treated as the volume of the gases.

4. There is no force of interaction between the atoms or molecules of the gas or between the atoms or molecules of the gas and the walls of the container. Any

---

[2] A. Einstein, On the Movement of Small Particles Suspended in Stationary Liquids Required by the Molecular-kinetic Theory of Heat, *Ann. Phys.* 322, no. 8 (1905): 549–560.

[3] M. Von Smoluchowski, Zur kinetischen Theorie der Brownschen Molekularbewegung und der Suspensionen, *Ann. Phys.* 21, no. 14 (1906): 756–780.

[4] J. Perrin, Brownian Movement and Molecular Reality, *Ann. Chim. Phys. Eighth Series* 78, no. 12 (1909): 1278.

type of attractive or repulsive interactions among atoms or molecules are not considered, and hence, they are treated as independent particles.

5. Due to the absence of interactions among the particles, and as the total volume of the container is available for movement, these particles move randomly in all directions but in straight lines in between two collisions.

6. Collisions among the particles with each other or collisions of the particles with the walls of the container are assumed to be perfectly elastic, that is, no change in energy or momentum of the particles takes place in such collisions.

7. The average kinetic energy of such a collection of gas particles depends only on the temperature of the gas and is not influenced by any other external factors present.

8. When the particles of the gas collide with the walls of the container, a change in momentum takes place whose time rate of change provides force. This force when calculated per unit area gives rise to pressure on the walls of the container. This indicates that the pressure of the gas is proportional to the number of particles colliding in unit time per unit area on the wall of the container, that is, the frequency of collisions plays an important role in determining the pressure on the wall.

Depending upon these assumptions, the model of KTG is established. Using this model, some physical variables are provided kinetic interpretation, expressions for various quantities, and different laws of gases are derived. The model generates some new insights about the behavior of gases, and at the same time refers to its limitations.

2.3.3   Pressure exerted by an ideal gas

We consider an ideal gas contained within a volume V, at a pressure P, and a temperature T. Such a gaseous system with its molecules is shown in Figure 2.1. The gas contains N number of molecules, and the mass of each molecule is m so that the mass of the gas is $M = m \times N$. These molecules are in a state of random motion in the isotropic and homogeneous space of the container and collide continuously against the walls of the container. Momenta of the molecules are transferred to the walls of the container during such collisions. Such momenta computed per second gives rise to force exerted by the gas on the walls of the container. Due to such frequent collisions, a continuous force is applied on the walls. This force is equal to the total momentum imparted to the walls per second. The pressure exerted by the gas on the walls is determined by dividing this experienced force by the surface area of the walls.

In order to derive the expression for pressure exerted by an ideal gas, the following assumptions are made:

1. Any small sample of the ideal gas consists of an enormous number of molecules N. The walls of the container are assumed to be very smooth and the collisions

of the molecules with the walls are assumed to be perfectly elastic. Only the perpendicular component of velocity is changed upon collision, which gives rise to pressure.

2. In the absence of any external field of force, the molecules are distributed uniformly throughout the available space inside the container of volume V.

3. The molecular density $n = \frac{N}{V}$ is assumed to be constant. This indicates that the number of molecules dN in a small element of volume dV will be given by $dN = n \, dV = \frac{N}{V} \, dV$.

4. As the molecules move in all directions with every possible values of velocity, no direction is preferred with a higher probability for the velocity of any molecule. This clearly shows that there would be as many molecules moving in one direction as in another at any moment.

5. All molecules will not have the same speed. Speeds may be assumed to vary continuously over the range from 0 to ∞. If $N(v)dv$ represents the number of molecules with speeds between v and $v + dv$, it is assumed that $N(v)dv$ remains constant at equilibrium even though the molecules are perpetually colliding and changing their speeds.

Since the velocity vectors of the gaseous molecules have no preferred direction in space, we are at first to find out the number of molecules having their speeds within the range v and $v + dv$ and in the directions between θ and $\theta + d\theta$, and ϕ and $\phi + d\phi$. Let $d^3N_{v,\theta,\phi}$ represents this number of molecules. Obviously, this number must be proportional to the solid angle subtended by the area at the origin. This number of molecules $d^3N_{v,\theta,\phi}$ must be proportional to the solid angle $d\Omega$, that is, we must have

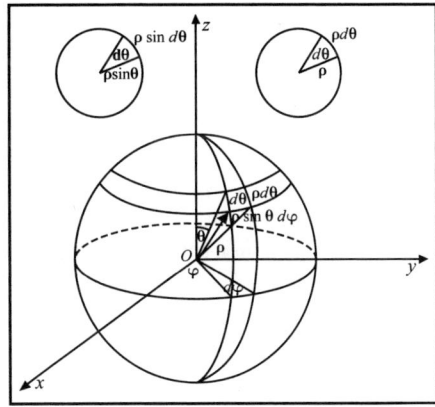

Figure 2.2 Solid angle and elementary area in spherical polar coordinates.

$$d^3N_{v,\theta,\phi} \propto d\Omega; \quad \Rightarrow \quad d^3N_{v,\theta,\phi} = K d\Omega = K \frac{dA}{v^2},$$

where K is the constant of proportionality. It should be noted that solid angle is measured by the area intercepted by the cone-like structure by a sphere of unit radius with the apex as the center. The elementary area dA is given by $dA = v^2 \sin\theta d\theta d\phi$ and is shown in Figure 2.2.

Using this definition of solid angle and the expression for elementary area dA, we get

$$d^3N_{v,\theta,\phi} = K\frac{dA}{v^2} = K\frac{v^2\sin\theta d\theta d\phi}{v^2} = K\sin\theta d\theta d\phi. \qquad (2.1)$$

In order to get the number of molecules $N(v)dv$ having speeds between v and $v + dv$, we have to integrate equation (2.1) over all possible directions of space, that is, θ from 0 to π and ϕ from 0 to 2π. This gives rise to

$$N(v)dv = \int\limits_{\theta=0}^{\pi}\int\limits_{\phi=0}^{2\pi} K\sin\theta d\theta d\phi = K.2.2\pi. \qquad \Rightarrow \qquad K = \frac{N(v)dv}{4\pi}. \qquad (2.2)$$

Using equation (2.2) in equation (2.1), we get

$$d^3N_{v,\theta,\phi} = \frac{N(v)dv}{4\pi}\sin\theta d\theta d\phi. \qquad (2.3)$$

The number of molecules $d^3n_{v,\theta,\phi}$ having their speeds within the range v and $v+dv$ and directions θ and $\theta+d\theta$, and ϕ and $\phi+d\phi$ per unit volume will be given by

$$d^3n_{v,\theta,\phi} = \frac{n(v)dv}{4\pi}\sin\theta d\theta d\phi. \qquad (2.4)$$

where $n(v)dv$ is the number of molecules per unit volume having speeds between v and $v + dv$.

Now, let us consider the group of molecules approaching the small area dA of the portion of the wall of the container. Many of these molecules will undergo collisions along the way. But we consider only those molecules that lie within the cylinder whose side is of length $v\,dt$, where dt is such a small time interval that practically no collisions are made during this time. The number of molecules striking the area dA in time dt will be given by

$$d^3n_{v,\theta,\phi} \times \text{volume of the cylinder of length } v\,dt \text{ and base area } dA\,\cos\theta$$
$$= d^3n_{v,\theta,\phi} \times dA\cos\theta \times v\,dt.$$

So the momentum transferred to the wall by these molecules will be given by

$$= d^3n_{v,\theta,\phi} \times dA\cos\theta \times v\,dt \times (2mv\cos\theta) = d^3n_{v,\theta,\phi} \times (2mv^2\cos^2\theta) \times dA \times dt.$$

We know that the time rate of change of momentum is the force and the force per unit area is the pressure. So the pressure exerted by the molecules of the ideal gas on the wall is given by

$$P = \int\limits_{v=0}^{\infty} \int\limits_{\theta=0}^{\frac{\pi}{2}} \int\limits_{\phi=0}^{2\pi} d^3 n_{v,\theta,\phi} \times (2mv^2 \cos^2 \theta)$$

$$= \int\limits_{v=0}^{\infty} \int\limits_{\theta=0}^{\frac{\pi}{2}} \int\limits_{\phi=0}^{2\pi} \frac{n(v)dv}{4\pi} \sin\theta d\theta d\phi \times (2mv^2 \cos^2 \theta)$$

$$= \frac{m}{2\pi} \int\limits_{v=0}^{\infty} v^2 n(v)dv \int\limits_{\theta=0}^{\frac{\pi}{2}} \sin\theta \cos^2\theta d\theta \int\limits_{\phi=0}^{2\pi} d\phi = \frac{m}{2\pi} n \langle v^2 \rangle \times \frac{1}{3} \times 2\pi. \qquad (2.5)$$

This leads to the expression for pressure as

$$P = \frac{1}{3} m\, n\, \langle v^2 \rangle = \frac{1}{3} m \frac{N}{V} \langle v^2 \rangle = \frac{1}{3} \frac{M}{V} \langle v^2 \rangle = \frac{1}{3} \rho \langle v^2 \rangle, \qquad (2.6)$$

where ρ is the volume density of the ideal gas.

It is to be noted that the limit of integration of the variable θ in equation (2.5) has been considered from 0 to $\frac{\pi}{2}$ because as θ varies from $0 \rightarrow \frac{\pi}{2}$ and ϕ from $0 \rightarrow 2\pi$, all the points in the hemisphere above $X-Y$ plane are covered. Since $X-Y$ plane is the wall of the container, all the molecules above it are considered in the calculation of pressure and the molecules below it are not considered.

2.3.4 Some physical facts and the kinetic theory of gases

In this section, we use the basic kinetic features of the motion of gas molecules and the mathematical expression for pressure exerted by the gas to explain the following physical phenomena related to the behavior of an ideal gas:

1. The kinetic interpretation of temperature

2. Gas laws

   (a) Boyle's law
   (b) Pressure law
   (c) Charles' law
   (d) Graham's law of diffusion
   (e) Avogadro's hypothesis
   (f) Dalton's law of partial pressure
   (g) Clapeyron's equation

3. Specific heat of gases

4. Degrees of freedom

Before presenting the kinetic interpretation of the above facts, the concept of root mean square (r.m.s.) is discussed below.

Root mean square velocity in terms of molecular weight and temperature

The r.m.s. velocity of the molecules of a gas is defined as the square root of the average of the velocity-square of the molecules. This r.m.s. velocity also has a unit of velocity. As the particles are moving in all directions, the average velocity is zero for a typical sample of a gas. Therefore, r.m.s. velocity instead of the average is defined and used in the interpretation of various physical quantities.

The expression for r.m.s. velocity of the molecules of an ideal gas is derived below in terms of the molecular weight M_o and the absolute temperature T from the expression for pressure P. From the KTG, this expression for pressure is given by

$$P = \frac{1}{3}m\,n\,\langle v^2 \rangle; \quad \Rightarrow \quad P = \frac{1}{3}m\,\frac{N}{V}\,\langle v^2 \rangle \quad \Rightarrow \quad PV = \frac{1}{3}m\,N\,\langle v^2 \rangle \qquad (2.7)$$

and

$$PV = NK_BT, \qquad (2.8)$$

where N is the total number of molecules in the volume V of the gas. From equations (2.7) and (2.8), we get

$$\frac{1}{3}m\,N\,\langle v^2 \rangle = NK_BT. \qquad (2.9)$$

This gives rise to

$$\langle v^2 \rangle = \frac{3K_BT}{m}; \quad \Rightarrow \quad \sqrt{\langle v^2 \rangle} = \sqrt{\frac{3K_BT}{m}}. \qquad (2.10)$$

Equation (2.10) gives the r.m.s. velocity as a function of temperature in absolute scale.

We know that Boltzmann constant K_B is $\frac{R}{N_A}$ and the molecular weight M_o is $m \times N_A$. Hence, the above equation reduces to

$$\sqrt{\langle v^2 \rangle} = \sqrt{\frac{3\,R\,T}{N_A\,m}} = \sqrt{\frac{3\,R\,T}{M_o}}. \qquad (2.11)$$

Equation (2.11) demonstrates that the r.m.s. velocity increases with the increase in temperature T and decreases with the increase in molecular weight M_o. This explains the rare occurrence of hydrogen gas in the atmosphere.

Energy per degree of freedom

Using equation (2.10), the average kinetic energy of a molecule of an ideal gas can be expressed in the following way:

$$\frac{1}{2}m\left\langle v^2 \right\rangle = \frac{1}{2}m\frac{3K_BT}{m} = \frac{3}{2}K_BT. \tag{2.12}$$

A molecule in an ideal gas has three degrees of freedom as it has only kinetic energy, and no potential energy. According to the law of equipartition of energy, this energy will be equally shared among these three degrees of freedom. Hence, the energy per degree of freedom of a molecule can be written as

$$E_K = \frac{1}{2}K_BT. \tag{2.13}$$

It should be noted that this result will be proved later in a rigorous way.

The kinetic interpretation of temperature

Degree of hotness or coldness of a body is determined by its temperature. It is a well-established fact that a body at a high temperature rejects heat to a body at a low temperature when these two bodies are in contact with each other. The definition of temperature is introduced in **zeroth law of thermodynamics** and the thermal state of a body is indicated by it, which ultimately determines the direction of flow of heat. In thermodynamics, temperature is considered as one of the fundamental thermodynamic variables. Equation (2.12) shows that the average kinetic energy E_K of the molecules of an ideal gas is directly proportional to the temperature T of the gas. This E_K is independent of the volume, pressure, and nature of the gas. Equation (2.12) can be utilized to provide kinetic interpretation of temperature.

Using equations (2.6) and (2.12), pressure P and the average kinetic energy E_K can be correlated in the following way:

$$P = \frac{1}{3}mn\langle v^2\rangle = \frac{2}{3}\times n\times\left(\frac{1}{2}m\langle v^2\rangle\right) = \frac{2}{3}\times n\times E_K = \frac{2}{3}\times\frac{N}{V}\times E_K; \quad \Rightarrow \quad PV = \frac{2}{3}NE_K. \tag{2.14}$$

Comparing this equation (2.14) with that of the ideal gas equation $PV = n_mRT$, we get

$$\frac{2}{3}n_mN_AE_K = n_mRT; \quad \Rightarrow \quad E_K = \frac{3}{2}\frac{R}{N_A}T = \frac{3}{2}K_BT. \tag{2.15}$$

Here $n_m = \frac{N}{N_A}$ is the number of moles of the gas. The factor K_B is the Boltzmann constant. Its value is 1.38×10^{-23} JK^{-1}.

Equation (2.15) indicates the fact that the average kinetic energy E_K per molecule of an ideal gas is proportional to the absolute temperature T. At $T = 0$, equation (2.15) implies that $E_K = 0$. This indicates that at absolute zero of temperature, there will be no motion of the molecules of the ideal gas. This is the kinetic or molecular interpretation of absolute zero. The motion of the molecules will exist only at a finite temperature. It should be mentioned that in practice, E_K is finite at absolute zero of temperature indicating a deviation from the ideal behavior of the gas molecules. Equation (2.15) thus shows that the average kinetic energy of a molecule of an ideal gas is independent of the volume, pressure, or the type of molecules. Hence, the mean kinetic energy of all molecules will have the same value at a given temperature in spite of the dissimilarities in their masses and chemical nature.

Graham's law of diffusion or effusion

Graham's law, popularly known as **Graham's law of effusion**, was formulated by Thomas Graham in the year 1848. Thomas Graham carried out experiments to characterize the effusion process and discovered a very important characteristic feature about the molecules of a gas. This characteristic feature is that lighter molecules will travel faster than the heavier molecules of a gas. Before the formal description of Graham's law of diffusion or effusion, brief ideas about diffusion and effusion are presented below.

Diffusion: Diffusion is defined as the phenomenon in which molecules or particles of a fluid from higher concentration region move to the lower concentration region. In this process, particles or molecules of a fluid spread out through the medium at a certain rate. This rate of diffusion depends on temperature and density of the particles or molecules. For example, if perfume is sprayed at one end of a room, its smell can be sensed at the other end. This occurs because of the diffusion phenomenon.

Effusion: Effusion is a special case of diffusion. In this process, particles or molecules escape or leak through a small hole whose diameter is much smaller than the mean free path of the particles or molecules. Under this situation, all the molecules reaching the hole will pass through, as collisions between molecules in these places will be negligibly small.

Graham's law of diffusion or effusion states that "when two gases at the same temperature and pressure are allowed to diffuse into each other, the rate of diffusion or of effusion of each gas is inversely proportional to the square root of the molecular weight of the gas". This law can also be stated as "when two gases at the same temperature and pressure are allowed to diffuse into each other, the rate of diffusion or of effusion of each gas is inversely proportional to the square root of the density of the gas".

We consider the case of diffusion between two gases of densities ρ_1 and ρ_2, molecular weights M_1 and M_2 at the same temperature T and pressure P. It is reasonable to assume that the rate of diffusion r is proportional to the r.m.s. speed

$\sqrt{\langle v^2 \rangle}$ of the molecules of the gas. If r_1 and r_2 be the rates of diffusion of the two gases and $\langle v_1^2 \rangle$ and $\langle v_2^2 \rangle$ be the mean square speeds, we can write

$$\frac{r_1}{r_2} = \frac{\sqrt{\langle v_1^2 \rangle}}{\sqrt{\langle v_2^2 \rangle}}. \tag{2.16}$$

From the expression for pressure P given by equation (2.6), we have

$$P = \frac{1}{3} \rho \langle v^2 \rangle. \quad \Rightarrow \quad \sqrt{\langle v^2 \rangle} = \sqrt{\frac{3P}{\rho}}. \tag{2.17}$$

If the pressures of the two gases are the same, we can write from equation (2.17)

$$\frac{\sqrt{\langle v_1^2 \rangle}}{\sqrt{\langle v_2^2 \rangle}} = \sqrt{\frac{\rho_2}{\rho_1}}. \tag{2.18}$$

Combination of equations (2.16) and (2.18) leads to

$$\frac{r_1}{r_2} = \sqrt{\frac{\rho_2}{\rho_1}}. \tag{2.19}$$

This equation (2.19) shows that at the same temperature and pressure, the rate of diffusion of the molecules of two gases is inversely proportional to the square root of the density of the gases. This is known as **Graham's law of diffusion or effusion.**
Furthermore, from equation (2.10), we have

$$\langle v^2 \rangle = \frac{3K_B T}{m}; \quad \Rightarrow \quad \sqrt{\langle v^2 \rangle} = \sqrt{\frac{3K_B T}{m}} = \sqrt{\frac{3RT}{M_o}}, \tag{2.20}$$

where M_o is the molecular weight of the gas. When a gas is confined in a container of fixed volume V with a small aperture, the rate of effusion must be proportional to the average speed as well as r.m.s. speed of the molecules. When the temperature is kept constant, both average and r.m.s. speed will be inversely proportional to the square root of molar mass of the molecules of the gas. Hence, it can be concluded that at constant temperature and pressure, the rate of effusion is inversely proportional to the square root of molar mass of the gas. This is known as **Graham's law of effusion**.

Avogadro's hypothesis

Avogadro's hypothesis states that "at the same temperature and pressure, equal volumes of all gases contain equal number of molecules".

We consider two gases 1 and 2. Let m_1 and N_1 be, respectively, the mass of each molecule and the number of molecules of the first gas and m_2 and N_2 be, respectively, the mass of each molecule and the number of molecules of the second gas. Furthermore, let us assume that V and P be the common values of volume and pressure of the two gases.

From the expression for pressure given by equation (2.6), we have

$$PV = \frac{1}{3}m\ N\ \langle v^2 \rangle. \tag{2.21}$$

For the two gases, this equation (2.21) can be written as

$$PV = \frac{1}{3}m_1\ N_1\ \langle v_1^2 \rangle \tag{2.22}$$

and

$$PV = \frac{1}{3}m_2\ N_2\ \langle v_2^2 \rangle. \tag{2.23}$$

Here, $\langle v_1^2 \rangle$ and $\langle v_2^2 \rangle$ are the mean square speeds of the molecules of the first and second gas, respectively. Thus, from equations (2.22) and (2.23), we have

$$\frac{1}{3}m_1\ N_1\ \langle v_1^2 \rangle = \frac{1}{3}m_2\ N_2\ \langle v_2^2 \rangle. \qquad \Rightarrow \qquad m_1\ N_1\ \langle v_1^2 \rangle = m_2\ N_2\ \langle v_2^2 \rangle. \tag{2.24}$$

As the temperatures of the two gases are the same, the average kinetic energy of the molecules of the two gases will also be the same, that is,

$$\frac{1}{2}m_1\ \langle v_1^2 \rangle = \frac{1}{2}m_2\ \langle v_2^2 \rangle; \qquad \Rightarrow \qquad m_1\ \langle v_1^2 \rangle = m_2\ \langle v_2^2 \rangle. \tag{2.25}$$

From equations (2.24) and (2.25), we can conclude that

$$N_1 = N_2. \tag{2.26}$$

Thus, equation (2.26) shows that at the same temperature and pressure, the equal volume of all gases contains the equal number of molecules. This statement is known as the **Avogadro's hypothesis**.

Dalton's law of partial pressure

Dalton's law of partial pressure states that **the pressure exerted by a mixture of several gases at a temperature equals the sum of the pressure exerted by each gas occupying the same volume as that of the mixture at the same temperature**.

We consider a container of volume V in which a number of gases $1, 2, 3, \cdots$ are kept at a temperature T. Mass of one molecule of the gases are $m_1, m_2, m_3, \cdots$ and the number of molecules of the gases are $N_1, N_2, N_3, \cdots$. The total pressure will be due to all the different types of molecules, assuming that there is no interaction among the molecules and the gaseous systems are in a state of thermal equilibrium at a temperature T.

We know that the expression for the pressure of an ideal gas is given by

$$P = \frac{1}{3}m\,n\,\langle v^2 \rangle = \frac{1}{2}m\,\langle v^2 \rangle \frac{2}{3}n = \frac{3}{2}K_B T \frac{2}{3}n = nK_B T. \qquad (2.27)$$

This equation (2.27) can be written as

$$n = \frac{P}{K_B T}. \qquad (2.28)$$

When all the different gases are mixed together in the same volume V and at the same temperature T to form the mixture, we must have (in the absence of chemical reaction)

$$N = N_1 + N_2 + N_3 + \cdots ; \qquad \Rightarrow \quad \frac{N}{V} = \frac{N_1}{V} + \frac{N_2}{V} + \frac{N_3}{V} + \cdots .$$

$$\Rightarrow \quad n = n_1 + n_2 + n_3 + \cdots .$$

Using equation (2.28) for various gases, we get

$$\Rightarrow \quad \frac{P}{K_B T} = \frac{P_1}{K_B T} + \frac{P_2}{K_B T} + \frac{P_3}{K_B T} + \cdots .$$

This leads to the fact that

$$P = P_1 + P_2 + P_3 + \cdots \qquad (2.29)$$

where $P_1, P_2, P_3, \cdots$ are, respectively, the partial pressures of the gases $1, 2, 3, \cdots$. Thus, the total pressure P is the sum of the partial pressure of the individual components of the gas mixture if the temperature and volume of the container remain the same. **This is known as Dalton's law of partial pressure**.

It should be noted here that the partial pressure P_i of a particular component in a gas mixture is given by

$$P_i = x_i P \qquad (2.30)$$

where $x_i = \frac{n_i}{\sum_i n_i}$ is the mole fraction of the ith gas.

Clapeyron's equation from the kinetic theory of gases

The Clapeyron equation is an equation of state of an ideal gas, which combines the Boyle–Mariotte law (relation between P and V at constant temperature), Gay-Lussac's law (relation between V and T at constant pressure), and Avogadro's hypothesis.

The Clapeyron equation is the simplest equation of state that is applicable with a certain degree of accuracy to real gases at low pressures and high temperatures. When the gases, such as atmospheric air and combustion products in gas engines are close in their properties to an ideal gas, this equation can be successfully applied to them.

We know that the expression for pressure of an ideal gas is given by

$$P = \frac{1}{3} m \, n \, \langle v^2 \rangle, \tag{2.31}$$

where m is the mass of one molecule of the gas, $\langle v^2 \rangle$ is the mean square speed, and n is the number of molecules of the gas per unit volume.

Equation (2.31) can be written as

$$P = \frac{1}{3} \left(\frac{n}{N_A} \right) m \, N_A \, \langle v^2 \rangle = \frac{1}{3} \left(\frac{n}{N_A} \right) M_O \, \langle v^2 \rangle,$$

where $M_O = m N_A$ is the molecular weight of the gas.

Using $\langle v^2 \rangle = \frac{3RT}{M_O}$ in the above expression, we get

$$P = \frac{1}{3} \left(\frac{n}{N_A} \right) M_O \frac{3RT}{M_O} = n \frac{R}{N_A} T = n K_B T.$$

Hence, we have $P = n \, K_B \, T$. This is Clapeyron's equation. This equation helps us to determine the number of molecules per unit volume at a particular pressure and temperature. This equation is also utilized to prove Dalton's law of partial pressure and to find the pressure difference in the diffusion problems.

2.3.5   Calculation of specific heat from kinetic theory of gases

We consider one mole of an ideal gas at a temperature T, at a pressure P, and having volume V. It contains N_A number of molecules, and the mass of each molecule is m. Specific heat of a system is defined as the amount of energy (in the form of heat) required to increase its temperature through 1° per unit mass of the system.

The total kinetic energy of the system of one mole of an ideal gas is

$$E = N_A \epsilon = N_A \times \frac{3}{2} K_B T = \frac{3}{2} N_A K_B T = \frac{3}{2} RT.$$

The molar heat capacity C_V is defined as the energy required to raise the temperature of one mole of an ideal gas by one kelvin at constant volume. Hence,

$$C_V = \left(\frac{\delta Q}{dT} \right)_V = \left(\frac{dE}{dT} \right)_V = \left[\frac{d}{dT} \left(\frac{3}{2} RT \right) \right]_V = \frac{3}{2} R = 2.98 \text{ cal mol}^{-1} \text{ K}^{-1}.$$

Again, using Mayer's relation $C_P - C_V = R$, we get the specific heat at constant pressure C_P as

$$C_P = C_V + R = \frac{3}{2} R + R = \frac{5}{2} R = 4.87 \text{ cal mol}^{-1} \text{ K}^{-1}.$$

The ratio of specific heats $\gamma = \frac{C_P}{C_V}$ is then given by

$$\gamma = \frac{C_P}{C_V} = \frac{\frac{5}{2}R}{\frac{3}{2}R} = \frac{5}{3} = 1.67.$$

Thus, the following consequences about the specific heat of gases can be made using the KTG:

1. Specific heats C_V and C_P are the same at all temperatures, that is, specific heats are independent of temperature. They do not depend upon temperature.

2. The ratio of specific heats $\gamma = \frac{C_P}{C_V}$ is also the same for all ideal gases at all temperatures.

3. Specific heats are independent of the nature of the gases.

The following observations are in order about the values of specific heats calculated from KTG:

1. For monoatomic gases, values of C_V and γ agree well with the experimental results.

2. Further, according to the KTG, the specific heats are found to be practically independent of temperature for monoatomic gases. This prediction has to be tested experimentally.

3. Regarding the numerical estimates of specific heats, there is a clear deviation between the results of the KTG and the experimental values for diatomic and polyatomic gases. Values of C_V and γ for some common gases at room temperature are shown in Table 2.1.

Table 2.1 Molar heat capacities for common gases at room temperature

Gas	C_V (cal mol^{-1} K^{-1})	$\gamma = \frac{C_P}{C_V}$
Ar	2.98	1.67
He	2.98	1.66
H_2	4.88	1.41
O_2	5.03	1.401
N_2	4.96	1.404
Cl_2	6.15	1.360
CO_2	6.80	1.304
NH_3	6.65	1.310

These apparent disagreements can be understood from the concept of the degree of freedom of the molecules of these gases and the law of equipartition of energy.

2.3.6 Degree of freedom

If a particle moves along a straight line, say, along the X-axis, its position is explicitly described by one coordinate only. The particle is then said to have one degree of freedom. Such a particle is shown schematically in Figure 2.3.

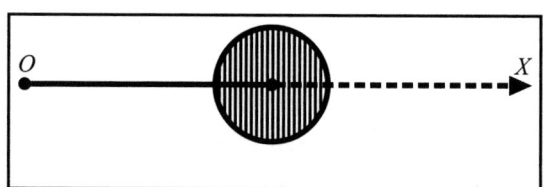

Figure 2.3 Motion of a particle in one dimension.

If the particle moves over a plane, we need two coordinates to specify the position of the particle. Its degree of freedom is then 2. Such a particle is shown schematically in Figure 2.4. Similarly, if the particle moves in a three-dimensional space, we need three coordinates to specify the position of the particle. Its degree of freedom is then 3.

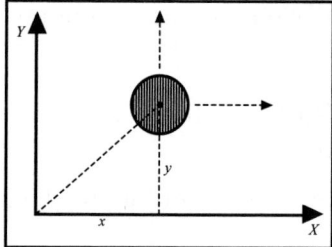

Figure 2.4 Motion of a particle in two dimensions.

Molecules, however, are not geometric mass points but are of finite size. They possess not only mass but also the moment of inertia and, therefore, have rotational kinetic energy, apart from that of the translational kinetic energy. So, molecules are expected to have rotational degrees of freedom.

In case, the molecules do not have perfectly rigid structures, they could also vibrate, giving rise to still more degrees of freedom. Spectroscopic analysis of light emitted or absorbed by molecules in the infrared demonstrates convincingly the rotations and vibrations of molecules.

Hence, we can define degrees of freedom in the following way:

The degrees of freedom of a dynamical system are defined as the number of independent coordinates required to specify its position and configuration.

Sometimes, we can define the degrees of freedom using the number of squared terms present in the energy expression. Hence, degrees of freedom can be defined as:

The number of independent quadratic variables (squared terms) by which the energy of a system is determined.

Degree of freedom in terms of the number of constraints N_C

The number of constraints plays an important role in determining the degrees of freedom of a system. If there are N_C number of constraints and N number of particles in a system, the degrees of freedom will be given by

$$N_{df} = 3N - N_C.$$

For a single atom, $N = 1$ and $N_C = 0$, so $N_{df} = 3 \times 1 - 0 = 3$.

For a diatomic molecule, $N = 2$ and $N_C = 1$ (bond length of the diatomic molecule), so $N_{df} = 3 \times 2 - 1 = 5$.

Solution of the specific heat problem

We consider an ideal gas consisting of Avogadro's number of molecules N_A. Each molecule has N_{df} degrees of freedom. The amount of energy contained in a single degree of freedom can be obtained from the principle of equipartition of energy, which states that: **For a system in thermal equilibrium at a temperature T, the total energy is equally partitioned among several degrees of freedom and for each degree of freedom, it is equal to $\frac{1}{2}K_BT$, where K_B is the Boltzmann constant.**

So, the system under consideration will have energy

$$E = N_A \, N_{df} \, \frac{1}{2}K_BT = \frac{1}{2} \, N_{df} \, N_A \, K_BT = \frac{1}{2} \, N_{df} \, R \, T. \qquad (2.32)$$

Differentiating equation (2.32) with respect to T, we get

$$C_V = \left(\frac{dE}{dT}\right)_V = \left[\frac{d}{dT}\left(\frac{1}{2} \, N_{df} \, R \, T\right)\right]_V = \frac{1}{2} \, N_{df} \, R. \qquad (2.33)$$

Hence,

$$C_P = C_V + R = \frac{1}{2} \, N_{df} \, R + R = R\left(1 + \frac{N_{df}}{2}\right). \qquad (2.34)$$

The ratio of specific heat, also known as the adiabatic constant γ, is given by

$$\gamma = \frac{C_P}{C_V} = \frac{R\left(1 + \frac{N_{df}}{2}\right)}{\frac{1}{2} \, N_{df} \, R} = \frac{\left(1 + \frac{N_{df}}{2}\right)}{\frac{N_{df}}{2}} = 1 + \frac{2}{N_{df}}. \qquad (2.35)$$

This is an important general expression between γ and N_{df}. While the principle of thermodynamics gives only an expression for the difference of two specific heats, the kinetic theory of the molecules of a gas, with the law of equipartition, predicts their actual magnitudes and their ratio γ in terms of the degrees of freedom N_{df} and the universal gas constant R. It is worth mentioning that according to the molecular theory C_P, C_V, and γ are all constants and independent of temperature.

Values of γ for various types of gases

1. **Monoatomic gas:** For a monoatomic gas, each molecule has degrees of freedom $N_{df} = 3$. From equations (2.33), (2.34), and (2.35), we have

$$C_V = \frac{1}{2} N_{df} R = \frac{3}{2}R, \qquad C_P = R\left(1 + \frac{N_{df}}{2}\right) = \frac{5}{2}R, \qquad \text{and}$$

$$\gamma = 1 + \frac{2}{N_{df}} = 1 + \frac{2}{3} = 1.67.$$

This is in good agreement with the values of C_V and γ for monoatomic gases, and their specific heats are also found to be practically independent of temperature, as predicted.

2. **Diatomic gas:** For a diatomic gas, each molecule has degrees of freedom $N_{df} = 5$. From equations (2.33), (2.34), and (2.35), we have

$$C_V = \frac{1}{2} N_{df} R = \frac{5}{2}R, \qquad C_P = R\left(1 + \frac{N_{df}}{2}\right) = \frac{7}{2}R, \qquad \text{and}$$

$$\gamma = 1 + \frac{2}{N_{df}} = 1 + \frac{2}{5} = 1.40.$$

These are almost exactly the values of H_2, N_2, etc.

But, Cl_2 is an interesting exception to the group of diatomic molecules at room temperature. It indicates something more. The atomic bonds are not perfectly rigid; they may also vibrate along the line joining them, and thereby introduce two vibrational degrees of freedom since the vibrational energy is partly kinetic and partly potential. Then, $N_{df} = 7$.

$$C_V = \frac{7}{2}R, \qquad C_P = R\left(1 + \frac{N_{df}}{2}\right) = \frac{9}{2}R, \qquad \text{and}$$

$$\gamma = 1 + \frac{2}{N_{df}} = 1 + \frac{2}{7} = 1.285.$$

3. **Triatomic gas:** In a gas of triatomic molecules, if the atoms are arranged at the vertices of a triangle, the number of degrees of freedom would be $N_{df} = 6$: three translational and three rotational.

From equations (2.33), (2.34), and (2.35), we have

$$C_V = \frac{1}{2} N_{df} R = \frac{6}{2}R = 3R, \qquad C_P = R\left(1 + \frac{N_{df}}{2}\right) = \frac{8}{2}R = 4R, \qquad \text{and}$$

$$\gamma = 1 + \frac{2}{N_{df}} = 1 + \frac{2}{6} = 1.33.$$

In a gas of triatomic molecules, if the atoms are linearly arranged, the number of degrees of freedom would be $N_{df} = 3 \times N_P - N_C = 3 \times 3 - 2 = 7$.

From equations (2.33), (2.34), and (2.35), we have

$$C_V = \frac{1}{2} N_{df} R = \frac{7}{2}R, \qquad C_P = R\left(1 + \frac{N_{df}}{2}\right) = \frac{9}{2}R, \qquad \text{and}$$

$$\gamma = 1 + \frac{2}{N_{df}} = 1 + \frac{2}{7} = 1.285.$$

4. **Polyatomic gases:** A polyatomic molecule has more than two atoms in a molecule. Like any other molecule, it has three degrees of freedom for translational motion. Each molecule can also rotate about its center of mass with an angular velocity with components along the three axes: X, Y, and Z. Hence, each molecule has three degrees of freedom for rotational motion. In addition to this, at high enough temperatures, the molecule (if it is soft) can vibrate easily with different modes of different frequencies, say, f. This is due to the presence of many interatomic bonds in the molecule. Hence, it has f degrees of freedom for vibrational motion. According to the law of equipartition of energy, for each degree of freedom for the translational and rotational motion, the molecule has an average energy of $\frac{1}{2}K_B T$, but for each mode of vibration with a certain frequency, the average energy is $K_B T$, since each vibrational mode involves potential as well as kinetic energy. Hence, the total energy of one mole of such a polyatomic gas can be written as

$$E = \left[\frac{3}{2}K_B T + \frac{3}{2}K_B T + f K_B T\right] \times N_A = (3 + f)\, RT.$$

This gives specific heat at constant volume C_V as $C_V = (3 + f)\, R$ and $C_P = (4 + f)\, R$. So, the ratio of specific heats γ is given by

$$\gamma = \frac{C_P}{C_V} = \frac{4 + f}{3 + f}.$$

It should be noted that the relation $C_P - C_V = R$ is true for any ideal gas, whether mono, di, or polyatomic.

As the number of atoms in a molecule increases, the degrees of freedom may also be expected to increase. The theory thus predicts a decreasing ratio of specific heats and this feature is in general in agreement with the experiment.

1. γ is never greater than 1.67.

2. γ is never less than 1.

3. In some cases, using the values of the ratio of specific heats γ, we get fractional degrees of freedom N_{df}, which is meaningless. This clearly indicates that the simple concept of equipartition of energy is obviously not the full story behind the problem.

4. The divergence between the theory and the experiment becomes more prominent when we take into account the variation in temperature.

γ for a mixture of two ideal gases

Suppose n_1 moles of an ideal gas with N_{df}^1 degrees of freedom per molecule are mixed with n_2 moles of another ideal gas with N_{df}^2 degrees of freedom per molecule at a temperature T. The ratio of specific heats γ for the mixture can be calculated in the following way:

The total internal energy of the mixture system is given by

$$E = N_A\, n_1\, N_{df}^1\, \frac{1}{2}K_BT + N_A\, n_2\, N_{df}^2\, \frac{1}{2}K_BT$$

$$= \frac{1}{2}K_BTN_A\left(n_1\, N_{df}^1 + n_2\, N_{df}^2\right) = \frac{1}{2}RT\left(n_1\, N_{df}^1 + n_2\, N_{df}^2\right).$$

The specific heat at constant volume C_V is given by

$$C_V = \left(\frac{dE}{dT}\right)_V = \left[\frac{d}{dT}\frac{1}{2}RT\left(n_1\, N_{df}^1 + n_2\, N_{df}^2\right)\right]_V = \frac{1}{2}R\left(n_1\, N_{df}^1 + n_2\, N_{df}^2\right).$$
(2.36)

As $C_P - C_V = (n_1 + n_2)\,R$, we have

$$C_P = C_V + (n_1 + n_2)\,R = \frac{1}{2}R\left(n_1\, N_{df}^1 + n_2\, N_{df}^2\right) + (n_1 + n_2)\,R. \qquad (2.37)$$

From equations (2.36) and (2.37), we have

$$\gamma = \frac{C_P}{C_V} = \frac{n_1(2 + N_{df}^1) + n_2(2 + N_{df}^2)}{n_1\, N_{df}^1 + n_2\, N_{df}^2}. \qquad (2.38)$$

This provides the required value of the adiabatic constant γ for the mixture of two ideal gases.

Example: Calculate the adiabatic constant γ for a mixture of one mole ($n_1 = 1$) of an ideal monoatomic gas with degrees of freedom per molecule $N_{df}^1 = 3$ and one mole ($n_2 = 1$) of another ideal diatomic gas with degrees of freedom per molecule $N_{df}^2 = 5$.

According to equation (2.38), the adiabatic constant γ is given by

$$\gamma = \frac{C_P}{C_V} = \frac{n_1(2 + N_{df}^1) + n_2(2 + N_{df}^2)}{n_1\, N_{df}^1 + n_2\, N_{df}^2}.$$

Using the values given in the problem, we get

$$\gamma = \frac{1(2+3) + 1(2+5)}{1 \times 3 + 1 \times 5} = \frac{3}{2}.$$

2.4 Concluding remarks about the kinetic theory of gases

2.4.1 Success of the kinetic theory of gases

The KTG was developed by several scientists, including Rudolf Clausius and James Clerk Maxwell over a period of nearly four hundred years. This theory describes how the molecular properties are related to the macroscopic behaviors of an ideal gas. The assumptions used in KTG about molecular behavior and their subsequent improvements can be used both as the basis for other theories about molecules and to solve real-world problems. The KTG has the following main advantages:

1. The assumptions used in the KTG about the motion of the molecules of an ideal gas are the following:

   (a) Molecules are constantly moving;

   (b) Molecules of an ideal gas have negligible volume;

   (c) Molecules have negligible intermolecular forces;

   (d) Molecules undergo perfectly elastic collisions.

2. Using the assumptions mentioned in (1), the KTG provides the basis that the kinetic energy of the molecules of an ideal gas depends upon temperature.

3. The kinetic theory of heat was developed by Rudolf Clausius. According to his prescription, the energy of a gaseous system in the form of heat is related to the kinetic energy of the molecules of the gas.

4. The KTG describes the macroscopic behaviors of an ideal gas in terms of the properties at the molecular level.

5. The KTG forms the foundation of equation of the state for an ideal gas. This equation helps us to explain successfully the empirical gas laws.

6. The KTG provides the kinetic interpretation of temperature.

7. The expression for pressure exerted by an ideal gas can be successfully derived with the help of the KTG.

8. The KTG provides quantitative estimates of the specific heats of a gaseous system.

9. It paves the way of degree of freedom to be introduced in resolving the anomaly of specific heat.

2.4.2   Limitations of the kinetic theory of gases

It has been established that the KTG is a useful tool for describing the physical behaviors of the atoms and molecules of an ideal gas. In spite of this great success for an ideal gas, it has certain limitations which lead to the differences in physical behavior between real and ideal gases. Some of these limitations are briefly described below.

1. It has been assumed in the KTG that the force of interaction between the molecules is negligible, and the volume of the molecules is negligible compared to the volume of the container. These assumptions are not valid in the case of a real gas. Intermolecular forces are present in real gases. These intermolecular forces become dominant when the molecules of a gas move more slowly, particularly at low temperatures.

2. It is assumed in KTG that molecules of an ideal gas have no volume, but this is not true for the molecules of a real gas. The molecules of a real gas do have a certain volume. This provides greater volume for a real gas at high pressure than would be predicted from KTG. Furthermore, with the compression of a real gas, the mean free path of its molecules decreases, and the molecules suffer collisions more frequently. This increases the pressure exerted by a real gas compared to the prediction of KTG.

3. Assumptions used in KTG are valid only under certain conditions. These conditions are found to be realized at low pressure, where molecules have lots of empty space to move in, and the volume of the molecules becomes very small compared to the total volume available. These conditions may also occur at high temperatures. Under this situation, the molecules possess a very high speed, and hence, a very high kinetic energy. As a result, molecules overcome easily the attractive forces operating between them.

4. This theory cannot explain why scents, such as those from perfume or baking cookies, take much longer to travel across a room.

5. The physical properties of liquids cannot be fully explained by KTG. The reason for this is assigned to the fact that molecules are very much close to each other in liquids, but still, a certain degree of constant random motion is retained with the molecules. As a result, liquids have no definite shape, and can diffuse.

6. The molecules are arranged in fixed positions in a solid, that is, the molecules are unable to move freely to assume new positions. Only they can vibrate about their equilibrium positions. Hence, the KTG cannot be applied to explain the properties of solids. Solids have fixed volumes and definite shapes and cannot diffuse, in general.

2.5   Solved numerical problems

1. Show that the pressure exerted by a gas is equal to two-thirds of the mean kinetic energy of translation of the molecules per unit volume.

   Answer: The mean kinetic energy of the molecules of an ideal gas is given by

$$E = \frac{1}{2} M \langle v^2 \rangle,$$

   where M is the mass of all the molecules of the gaseous system having N number of molecules in a volume V. So the kinetic energy of the gaseous system per unit volume is given by

$$E_V = \frac{E}{V} = \frac{1}{2} \frac{M}{V} \langle v^2 \rangle = \frac{1}{2} \rho \langle v^2 \rangle. \quad \Rightarrow \quad \langle v^2 \rangle = \frac{2E_V}{\rho}.$$

   Again, we have

$$P = \frac{1}{3} m n \langle v^2 \rangle = \frac{1}{3} \frac{M}{V} \langle v^2 \rangle = \frac{1}{3} \rho \langle v^2 \rangle.$$

   Using the value of $\langle v^2 \rangle$ from the above expression, we get the expression for pressure P as

$$P = \frac{1}{3} \rho \langle v^2 \rangle = \frac{1}{3} \rho \frac{2E_V}{\rho} = \frac{2}{3} E_V.$$

2. The mean kinetic energy of translation of a molecule of hydrogen at 0°C is 5.64×10^{-14} ergs, and the universal gas constant R is given by $R = 8.32 \times 10^7$ erg mol^{-1}K^{-1}. Calculate Avogadro's number N_A.

   Answer: The expression for pressure P for one mole of an ideal gas is given by

$$P = \frac{1}{3} m n \langle v^2 \rangle = \frac{1}{3} m \frac{N_A}{V} \langle v^2 \rangle; \quad \Rightarrow \quad N_A = \frac{3PV}{m \langle v^2 \rangle} = \frac{3RT}{m \langle v^2 \rangle}.$$

   Here, the ideal gas equation for one mole $PV = RT$ has been used. Inserting the values of the parameters, we get Avogadro's number N_A as

$$N_A = \frac{3RT}{m \langle v^2 \rangle} = \frac{3 \times 8.32 \times 10^7 \times 273}{2 \times 5.64 \times 10^{-14}} = 6.023 \times 10^{23}.$$

3. Assume that the fresh air is composed of nitrogen N_2 (78%) and oxygen O_2 (21%). Find the r.m.s. speed of N_2 and O_2 at 20°C.

   Answer: For nitrogen, the molar mass is $M_{N_2} = 0.0280$ kg mol^{-1}. Here, temperature is $T = 20°C = 293$ K and $R = 8.314$ Jmol^{-1}K^{-1}.

The r.m.s. speed for nitrogen is given by

$$V_{\text{r.m.s.}}^{N_2} = \sqrt{\frac{3RT}{M_{N_2}}} = \sqrt{\frac{3 \times 8.314 \times 293}{0.0280}} = 511 \text{ ms}^{-1}.$$

For oxygen, the molar mass is $M_{O_2} = 0.0320$ kg mol^{-1}. The r.m.s. speed for oxygen is given by

$$V_{\text{r.m.s.}}^{O_2} = \sqrt{\frac{3RT}{M_{O_2}}} = \sqrt{\frac{3 \times 8.314 \times 293}{0.0320}} = 478 \text{ ms}^{-1}.$$

4. If the r.m.s. speed of methane gas in the atmosphere of Jupiter is 471.8 ms^{-1}, show that the temperature of the surface of Jupiter is well below 0°C.

   Answer: The r.m.s. speed for methane gas is given by

   $$V_{\text{r.m.s.}} = 471.8 \text{ ms}^{-1} \text{ and the molar mass of methane gas is}$$
   $$M = 16.04 \times 10^{-3} \text{ kg mol}^{-1}.$$

   The value of the universal gas constant is $R = 8.314$ Jmol^{-1}K^{-1}. We have to calculate the temperature on the surface of Jupiter.

   The expression for r.m.s. speed is given by

   $$V_{\text{r.m.s.}} = \sqrt{\frac{3RT}{M}}; \quad \Rightarrow \quad T = \frac{MV_{\text{r.m.s.}}^2}{3R}$$
   $$= \frac{(471.8)^2 \times 16.04 \times 10^{-3}}{3 \times 8.314} = 143 \text{ K} = -130°C.$$

   Hence, the temperature on the surface of Jupiter is well below 0°C.

5. The temperature and pressure of a gas are respectively given by 80°C and 5×10^{-10} Nm^{-2}. Calculate the number of molecules per m^3 of the gas if Boltzmann constant is $K_B = 1.38 \times 10^{-23}$ JK^{-1}.

   Answer: The temperature of the gas is $T = 80°C = 353$ K, and Boltzmann constant is $K_B = 1.38 \times 10^{-23}$ JK^{-1}. Here, the pressure of the gas is $P = 5 \times 10^{-10}$ Nm^{-2} and the volume of the gas is $V = 1$ m^3.

   The number of molecules in the gas is given by

   $$n = \frac{PV}{K_B T} = \frac{5 \times 10^{-10} \times 1}{1.38 \times 10^{-23} \times 353} = 1.02 \times 10^{11}.$$

6. A beam of 10^{20} oxygen molecules moving with a speed of 2×10^3 ms^{-1} strikes an area of 4 cm^2 at an angle of 30° with the normal of a wall per second. Find the pressure exerted by the gas on the wall. The mass of one atom of the gas is $= 2.657 \times 10^{-26}$ kg.

Answer: Mass of one O_2-molecule is $= (2 \times 2.657 \times 10^{-26})$ kg $= 5.314 \times 10^{-26}$ kg. Therefore, the mass of 10^{20} molecules is $(10^{20} \times 5.312 \times 10^{-26})$ kg $= 5.314 \times 10^{-6}$ kg.

As the speed of an oxygen molecule is 2×10^3 ms^{-1}, the momentum of 10^{20} oxygen molecules is

$$P = mv = 5.314 \times 10^{-6} \times 2 \times 10^3 = 10.628 \times 10^{-3} \text{ kg ms}^{-1}.$$

The component of momentum normal to the wall is

$$10.628 \times 10^{-3} \times \cos 30°.$$

So the pressure exerted by these oxygen molecules is given by

$$P = \frac{F}{A} = \frac{10.628 \times 10^{-3} \times \cos 30°}{4 \times 10^{-4}} \text{ Nm}^{-2} = 23.009 \text{ Nm}^{-2}.$$

7. Calculate the average thermal energy of a helium atom

   (a) at a temperature of 27°C.

   (b) on the surface of the Sun with a temperature of 6000 K.

   (c) inside the core of a star with a temperature of 10^7 K.

   Answer:

   (a) At a temperature of 27°C, the average thermal energy of a helium atom is given by

   $$E_K = \frac{3}{2}K_BT = \frac{3}{2} \times (1.38 \times 10^{-23} \text{ JK}^{-1}) \times 300 \text{ K} = 6.21 \times 10^{-21}\text{J}$$
   $$= 0.03876 \text{ eV}.$$

   Hence, the average thermal energy of a helium atom at the given temperature is 6.21×10^{-21} J $= 0.03876$ eV.

   (b) On the surface of the Sun, the temperature is 6000 K. Using the formula, the average thermal energy corresponding to this temperature is given by

   $$E_K = \frac{3}{2}K_BT = \frac{3}{2} \times (1.38 \times 10^{-23} \text{ JK}^{-1}) \times 6000 \text{ K} = 1.241 \times 10^{-19}\text{J}$$
   $$= 0.775 \text{ eV}.$$

   Hence, the average thermal energy on the surface of the Sun with a temperature of 6000 K is 1.241×10^{-19} J $= 0.775$ eV.

(c) Inside the core of a star, the temperature is 10^7 K. Using the formula, the average thermal energy at this temperature is given by

$$E_K = \frac{3}{2}K_B T = \frac{3}{2} \times \left(1.38 \times 10^{-23} \text{ JK}^{-1}\right) \times 10^7 \text{ K} = 2.07 \times 10^{-16}\text{J}$$
$$= 1292.13 \text{ eV}.$$

Hence, the average thermal energy inside the core of a star at a temperature of 10^7 K is 2.07×10^{-16} J $= 1292.13$ eV.

It should be mentioned here that helium has a fully filled shell, that is, $_1s^2$. The ionization energy for the first electron of the helium atom is 25.6 eV and that for the second electron is 54.44 eV. Thus, inside the core of a star with a temperature of 10^7 K, all helium atoms are in the ionized state.

8. The mass of an oxygen molecule is 5.28×10^{-26} kg, and its mean $\langle v \rangle$ at N.T.P. is 4.25×10^2 m/s. Find the average kinetic energy of an oxygen molecule at 0°C.

Answer: The expressions for average and r.m.s. speed are respectively given by

$$\langle v \rangle = \sqrt{\frac{8K_B T}{m\pi}} \quad \text{and} \quad v_{\text{r.m.s.}} = \sqrt{\frac{3K_B T}{m}}.$$

Taking ratio of these two, we get

$$\frac{v_{\text{r.m.s.}}}{\langle v \rangle} = \sqrt{\frac{\frac{3K_B T}{m}}{\frac{8K_B T}{m\pi}}} = \sqrt{\frac{3\pi}{8}} \quad \Rightarrow \quad v^2_{\text{r.m.s.}} = \frac{3\pi}{8}(\langle v \rangle)^2.$$

Hence, the average kinetic energy of an oxygen molecule is given by

$$= \frac{1}{2}mv^2_{\text{r.m.s.}} = \frac{1}{2}m\frac{3\pi}{8}(\langle v \rangle)^2 = \frac{3\pi m}{16}(\langle v \rangle)^2$$
$$= \frac{3}{16} \times 3.14 \times 5.28 \times 10^{-26} \times (4.25 \times 10^2)^2 = \mathbf{5.614 \times 10^{-21} \text{ J}}.$$

9. One mole of a gas is enclosed in a cubic volume of side 0.2 m. If the translational speed of these molecules is 483 ms^{-1}, find out the pressure exerted by the gas on the side of the cubic enclosure. Given: the mass of each molecule is 5×10^{-26} kg.

Answer: Time interval Δt between two successive collisions on the wall of the cubic volume is given by

$$\Delta t = \frac{2L}{v_x} = \frac{2 \times 0.2 \text{ m}}{483 \text{ ms}^{-1}} = 8.3 \times 10^{-4} \text{ s}.$$

The change in momentum in the collision with the wall is $\Delta p_x = 2mv_x = 2 \times (5 \times 10^{-26}$ kg$) \times (483$ ms$^{-1}) = 4.83 \times 10^{-23}$ N s.

The force exerted by one molecule is $F_x = \frac{\Delta p_x}{\Delta t} = \frac{4.83 \times 10^{-23} \text{ N-s}}{8.3 \times 10^{-4} \text{ s}} = 0.582 \times 10^{-19}$ N.

The total force exerted by one mole of the gas is $(0.582 \times 10^{-19}$ N$) \times (6 \times 10^{23}) = 3.49 \times 10^4$ N.

So, the average pressure exerted by the gas is $\frac{3.49 \times 10^4 \text{ N}}{3 \times 4 \times 10^{-2} \text{ m}^2} = 2.9 \times 10^5$ N m^{-2}.

10. Dust particles suspended in a monoatomic gas are in equilibrium with the gas. If the gas is at 300 K, calculate the mean kinetic energy of translation of a dust particle. If the mass of a particular suspended particle is 10^{-27} kg, calculate the root mean square speed $\sqrt{\langle v^2 \rangle}$.

Answer: The mean translational kinetic energy per molecule is given by

$$E_K = \frac{3}{2} K_B T = \frac{3}{2} \times \left(1.38 \times 10^{-23} \text{ JK}^{-1}\right) \times 300 \text{ K} = 6.2 \times 10^{-21} \text{ J} = 0.04 \text{ eV}.$$

Again we know that

$$v_{\text{r.m.s.}} = \sqrt{\langle v^2 \rangle} = \sqrt{\frac{3RT}{M_o}} = \sqrt{\frac{3RT}{mN_A}} = \sqrt{\frac{3K_B T}{m}},$$

where M_o is the molecular weight. Inserting the numerical values of the parameters, we get

$$v_{\text{r.m.s.}} = \sqrt{\frac{3 \times \left(1.38 \times 10^{-23} \text{ JK}^{-1}\right) \times 300 \text{ K}}{10^{-27}}} = \sqrt{3 \times 1.38 \times 3} \times 10^3 \text{ ms}^{-1}$$

$$= 3.5 \times 10^3 \text{ ms}^{-1}.$$

11. Find the number of molecules in an ideal gas at 27°C and pressure of 20 mm Hg. Mean kinetic energy of a molecule at 27°C is 4×10^{-14} ergs, density of $Hg = 13.6$ g/c.c.

Answer: Here, the pressure is given by

$$P = 20 \text{ mm Hg} = 2 \text{ cm Hg} = 2 \times 13.6 \times 980 \text{ dyne cm}^{-2}.$$

We know that the mean kinetic energy is $E_K = \frac{3}{2} K_B T$.

From Clapeyron's equation, we have

$$P = nK_B T; \quad \Rightarrow \quad K_B = \frac{P}{nT}.$$

Using this value of K_B, we get the expression for mean kinetic energy as

$$E_K = \frac{3}{2}\frac{P}{nT}T; \quad \Rightarrow \quad n = \frac{3}{2}P \times \frac{1}{E_K} = \frac{3}{2} \times 2 \times 13.6 \times 980 \times \frac{1}{4 \times 10^{-14}}$$

$$= 9.996 \times 10^{17}.$$

12. Show that the number of molecules striking the unit area of a wall per second is given by

$$dn = \frac{P}{\sqrt{2\pi m K_B T}},$$

where symbols have their usual meaning.

Answer: We know that the number of molecules striking the unit area of a wall per second is given by $\frac{n\langle v \rangle}{4}$, where n is the number of molecules per unit volume and $\langle v \rangle$ is average velocity given by $\langle v \rangle = \sqrt{\frac{8K_B T}{\pi m}}$.

From Clapeyron's equation, we have

$$P = nK_B T; \quad \Rightarrow \quad n = \frac{P}{K_B T}.$$

Using these two values, we get the number of molecules striking the unit area of a wall per second as

$$dn = \frac{\frac{P}{K_B T}\sqrt{\frac{8K_B T}{\pi m}}}{4} = \frac{1}{4}\frac{P}{\sqrt{K_B T}}\frac{\sqrt{8}}{\sqrt{\pi m}} = \frac{P}{\sqrt{2\pi m K_B T}}.$$

13. The number of molecules per c.c. of a gas is 2.7×10^{19} at N.T.P. Calculate the number of molecules per c.c. of the gas at a temperature of 39°C and at a pressure of 10^{-6} mm Hg.

Answer: We know that $P = \frac{1}{3}\, m\, n\, \langle v^2 \rangle$. Let P_1 and P_2 be the pressures, respectively, at N.T.P. and at the given temperature and pressure.

Then we have

$$\frac{P_1}{P_2} = \frac{n_1}{n_2}\frac{\langle v_1^2 \rangle}{\langle v_2^2 \rangle} = \frac{n_1}{n_2}\frac{T_1}{T_2} \quad \text{as } v \propto \sqrt{T}.$$

Hence, we have

$$n_2 = \frac{n_1 T_1 P_2}{P_1 T_2} = \frac{2.7 \times 10^{19} \times 10^{-7} \times 13.6 \times 981 \times 273}{76 \times 13.6 \times 981 \times (39 + 273)} = 3.1 \times 10^{10} \text{ cm}^{-3}.$$

14. At standard temperature and pressure, a cylinder of fixed capacity 44.8 L contains helium gas. Calculate the amount of heat needed to raise the temperature of the gas in the cylinder by 15°C. Given that $R = 8.31$ J mol^{-1} K^{-1}.

Answer: At N.T.P., the volume of the gas is 44.8 L. So, the enclosure contains two moles of the gas.

Being monoatomic, the molar specific heat at constant volume of helium is $C_V = \frac{3}{2}R$. Since the volume of the cylinder is fixed, the heat required to raise its temperature ΔT is determined by C_V. Hence, the required heat is given by

$$\delta Q = n \times C_V \times \Delta T = 2 \times 1.5R \times 15.0 = 45R = 45 \times 8.31 \text{ J} = 374 \text{ J}.$$

15. Determine the number of degrees of freedom for

(**a**) a particle moving on the circumference of a circle.

(**b**) a rigid body moving in space with one point fixed.

(**c**) the bob of the conical pendulum.

(**d**) a dumbbell moving freely is space.

Answer:

(**a**) In this case, the particle moves in a plane, so it has two degrees of freedom. However, the two coordinates that define the position of the particle at any instant of time are related by the constraint equation

$$x^2 + y^2 = R^2; \quad \Rightarrow \quad x^2 + y^2 - R^2 = 0, \text{ where } R \text{ is the radius of}$$
$$\text{the circle.}$$

Clearly, there exists one constraint that is holonomic, and hence, the number of degrees of freedom is $N_{df} = 2 - 1 = 1$.

(**b**) Let us consider the point on the rigid body, which is fixed to be a particle at the origin. Clearly, for this particle, we have $x_1 = 0, y_1 = 0$, and $z_1 = 0$. Consider two other particles having coordinates x_2, y_2, z_2 and x_3, y_3, z_3. The three constraint equations for these particles are (distance between any pair of particles is a constant):

$$(x_1 - x_2)^2 + (y_1 - y_2)^2 + (z_1 - z_2)^2 = r_{12}^2 = r_2^2 = \text{constant},$$
$$(x_1 - x_3)^2 + (y_1 - y_3)^2 + (z_1 - z_3)^2 = r_{13}^2 = r_3^2 = \text{constant},$$
$$\text{and } (x_2 - x_3)^2 + (y_2 - y_3)^2 + (z_2 - z_3)^2 = r_{23}^2 = \text{constant}.$$

Hence, the number of degrees of freedom for the rigid body is $N_{df} = 3 \times 2 - 3 = 3$.

(c) If the distance between the bob and the point of suspension of the pendulum is l, we have the equation of constraint

$$x^2 + y^2 + z^2 = l^2$$

x, y, and z are the coordinates of the bob at any instant of time.

The constraint imposed on the bob is holonomic, and hence, the degrees of freedom for the bob of the conical pendulum are $N_{df} = 3 - 1 = 2$.

(d) A dumbbell consists of two massive particles connected to a rigid rod of constant length, say, l. If at any time t, x_1, y_1, z_1 and x_2, y_2, z_2 be the coordinates of the two particles of the dumbbell, we have the equation of constraint as

$$(x_2 - x_1)^2 + (y_2 - y_1)^2 + (z_2 - z_1)^2 = l^2.$$

The constraint is holonomic, and hence, the degrees of freedom are $N_{df} = 6 - 1 = 5$.

16. Consider the linear model of the hydrogen cyanide (HCN) molecule. Hence, show that the molar-specific heat of HCN gas at constant pressure must be $5.5R$.

    Answer: A molecule that is composed of n atoms has three translational degrees of freedom and two or three rotational degrees of freedom if it is linear or nonlinear, respectively. If it is linear, it has $3n - 5$ vibrational degrees of freedom. If it is nonlinear, it has $3n - 6$ vibrational degrees of freedom.

    HCN molecule is linear. So, it has three translational degrees of freedom, two rotational degrees of freedom, and four vibrational degrees of freedom.

    So, the total degrees of freedom is $f = 3 + 2 + 4 = 9$. Hence, the molar specific heat at constant volume of HCN molecule is $C_V = \frac{fR}{2} = \frac{9R}{2}$.

    So, the molar-specific heat at constant pressure C_P is given by

    $$C_P = C_V + R = \frac{9R}{2} + R = \frac{11R}{2} = 5.5R.$$

17. Calculate the average kinetic energy of an oxygen molecule at a temperature of $0°C$. Given that $R = 8.314 \text{ Jmol}^{-1}\text{K}^{-1}$ and $N_A = 6.023 \times 10^{23}$.

    Answer: Oxygen is a diatomic molecule; therefore, it has 5 degrees of freedom (3 translational and 2 rotational).

    So the kinetic energy is $E_K = 5 \times \frac{1}{2}K_B T = \frac{5}{2} \times \frac{RT}{N_A} = \frac{5}{2} \times \frac{8.314 \times 273}{6.023 \times 10^{23}} = 9.4 \times 10^{-21} \text{J}$.

2.6   Multiple choice questions with answers

1. The average kinetic energy of gas molecules $\langle E_K \rangle$ is

   (a) $K_B T$ (b) $\frac{1}{2} K_B T$ (c) $\frac{3}{4} K_B T$ (d) $\frac{3}{2} K_B T$

   Answer: The correct choice is (d).

2. The average kinetic energy of gas molecules $\langle E_K \rangle$ in d-dimension is

   (a) $d K_B T$ (c) $\frac{3d}{4} K_B T$

   (b) $\frac{d}{2} K_B T$ (d) $\frac{3d}{2} K_B T$

   Answer: The correct choice is (b).

3. At a temperature $T = 300$ K, two nitrogen molecules are at a distance of 1×10^{-10} m. The gravitational potential energy of one nitrogen molecule in the field of an other is $V(r)$, and the mean thermal translational kinetic energy per nitrogen molecule is E_K. We then have

   (a) $V(r) = E_K$ (c) $V(r) << E_K$

   (b) $V(r) < E_K$ (d) $V(r) >> E_K$

   Answer: The correct choice is (c).

   Solution: The gravitational potential energy of one nitrogen molecule is

   $$V(r) = -\frac{Gm_1 m_2}{r} = -\frac{6.67 \times 10^{-11} \text{ N m}^2 \text{ kg}^{-2} \left(28 \times 1.67 \times 10^{-27} \text{ kg} \right)^2}{1 \times 10^{-10} \text{ m}}$$
   $$= -1.458 \times 10^{-51} \text{ J}.$$

   On the other hand, the mean thermal translational kinetic energy per nitrogen molecule is

   $$E_K = \frac{3}{2} K_B T = \frac{3}{2} \times 1.38 \times 10^{-23} \text{ JK}^{-1} \times 300 \text{ K} = 6.21 \times 10^{-21} \text{ J}.$$

   Hence, we have $V(r) << E_K$.

4. In terms of the number of degrees of freedom n of a molecule in a gas, the ratio of specific heats is then given by

   (a) $1 + \frac{1}{n}$ (b) $1 + \frac{1}{2n}$ (c) $1 + \frac{2}{n}$ (d) $\frac{2n}{2n-1}$

   Answer: The correct choice is (c).

   Solution: The energy associated with each degree of freedom of one mole of gas is given by $\frac{1}{2} RT$, where R is the universal gas constant. Hence, if there is n degree of freedom, the energy is then given by $U = \frac{n}{2} RT$.

We know that the specific heat at constant volume is given by $C_V = \frac{dU}{dT}$. We then have

$$C_V = \frac{d}{dT}\left(\frac{n}{2}RT\right) = \frac{n}{2}R.$$

Using the relation $C_P - C_V = R$ (for one mole of an ideal gas), we get

$$C_P = C_V + R = \frac{n}{2}R + R = \left(\frac{n}{2} + 1\right)R.$$

Then, the ratio of specific heats is given by

$$\gamma = \frac{C_P}{C_V} = \frac{\left(\frac{n}{2} + 1\right)R}{\frac{n}{2}R} = 1 + \frac{2}{n}.$$

5. For which gas, the ratio of specific heats $\left(\frac{C_P}{C_V}\right)$ will be the largest? [JEST 2014]

    (a) Monoatomic   (c) Triatomic

    (b) Diatomic   (d) Hexaatomic

    Answer: The correct choice is (a).

    Solution: $C_P/C_V = \gamma = (1 + \frac{2}{f})$, where f is the degree of freedom.

    For monoatomic gas: $f = 3$, for diatomic gas: $f = 6$, for triatomic gas: $f = 9$, and for hexagonal: $f = 18$.

6. Consider a one-dimensional gas of N noninteracting particles of mass m with the Hamiltonian for a single particle is given by $H = \frac{p^2}{2m} + \frac{1}{2}m\omega^2(x^2 + 2x)$. The high temperature-specific heat in units of $R = NK_B$ (K_B is the Boltzmann constant) is   [GATE 2019]

    (a) 1   (b) 1.5   (c) 2   (d) 2.5

    Answer: The correct choice is (a).

    Solution: $< H > = < \frac{p^2}{2m} > + \frac{1}{2}m\omega^2 < x^2 > + \frac{1}{2}m\omega^2 < 2x > = \frac{NK_BT}{2} + \frac{NK_BT}{2} + U_0$.

    $\Rightarrow < H > = N K_B T$

    $\Rightarrow C_v = \frac{\partial H}{\partial T} = NK_B$.

7. A rigid triangular molecule consists of three non-co-linear atoms joined by rigid rods. The constant pressure molar specific heat C_p of an ideal gas consisting of such molecules is   [JAM 2015]

    (a) $6R$ (b) $5R$ (c) $4R$ (d) $3R$

    Answer: The correct choice is (c).

Solution: The degrees of freedom of the rigid triangular molecule is $= 6$.

The internal energy is given by $U = \frac{6RT}{2}$. So, the specific heat at constant volume is $\Rightarrow C_V = (\frac{\partial U}{\partial T})_V = 3R \Rightarrow C_P = C_V + R = 4R$.

8. The temperature of an ideal gas is increased from 120 to 480 K. The r.m.s. velocity of the gas molecules at 480 K is $v_{\text{r.m.s.}}$. Then, the r.m.s. velocity of the gas molecule at 120 K is

   (a) $2 v_{\text{r.m.s.}}$ (b) $\frac{1}{2}v_{\text{r.m.s.}}$ (c) $v_{\text{r.m.s.}}$ (d) $\frac{1}{4}v_{\text{r.m.s.}}$

   Answer: The correct choice is (b).

9. One mole of a monatomic gas $(\gamma = \frac{5}{3})$ is mixed with one mole of a diatomic gas $(\gamma = \frac{7}{5})$. The value of γ for the mixture is then given by

   (a) 1   (b) 2   (c) 2.5   (d) 1.5

   Answer: The correct choice is (d).

10. If r is the number of existing phases, f the degrees of freedom, and n the number of components in a system, then Gibbs phase rule is

    (a) $r + f = n$ (c) $r + f = n + 2$
    (b) $r + f = n + 1$ (d) $r + f = n - 1$

    Answer: The correct choice is (c).

2.7   Exercise

2.7.1   Short answer type questions

1. State some experimental evidences in favor of the kinetic aspect of the molecules of a gas.

2. What do you mean by homogeneity and isotropy of space?

3. Calculate the pressure exerted by a gas having 2.5×10^{19} molecules per cubic cm at a temperature of $27°C$.

4. Briefly explain the concept of temperature according to the KTG.

   [Delhi University 2021]

5. What do you mean by r.m.s. velocity of the molecules of a gas? Calculate the r.m.s. velocity of hydrogen at N.T.P. Compare the r.m.s. velocity of hydrogen at 300 K with that of oxygen at 1200 K.

6. What do you mean by degree of freedom? Find out the relation between the ratio of specific heat at constant pressure and that of at constant volume with the degree of freedom.

7. State the law of equipartition of energy. Show that the average translational kinetic energy per degree of freedom is $\frac{1}{2}K_B T$.

8. State Graham's law of diffusion and Dalton's law of partial pressure. Prove these laws with the help of the KTG.

9. Prove Clapeyron's equation $P = n\,K_B\,T$ from the KTG.

10. What do you mean by "degrees of freedom" of a dynamical system? What are the degrees of freedom of a diatomic molecule? [WBSU 2019]

11. Find the ratio of specific heats for a (i) monoatomic, (ii) diatomic, and (iii) triatomic gas.

12. How is the atomicity of gas molecules related to the ratio of two specific heats? [WBSU 2019]

13. State the principle of equipartition of energy.

14. Give the kinetic interpretation of pressure and calculate the pressure exerted by a perfect gas. How is the pressure expected to vary with height in an isothermal vertical column of a gas?

15. Assume that the molecules of a hydrogen gas are dumbbell shaped. Calculate the total molecular translational and rotational energy of 1 gm of such a hydrogen gas at temperatures 100 and 300 K. Justify whether the specific heat at constant volume in this case will be the same or not at these two temperatures.

2.7.2 Long answer type questions

1. What is an ideal gas? Under what conditions of pressures and temperatures can a real gas be assumed to behave as an ideal gas?

2. State the postulates of the KTG. Using the basis of these postulates, derive the expression for pressure exerted by an ideal gas.
   [Christ (Deemed to be University), Bengaluru 2018, ST. Joseph's College (Autonomous), Bengaluru 2019]

3. Using the concepts of the KTG, prove the following gas laws: (i) Boyle's law, (ii) Charles' law, (iii) Pressure law, (iv) Dalton's law of partial pressure, (v) Avogadro's hypothesis, and (vi) Graham's law of partial pressure.

4. Using the relation $P = \frac{1}{3}\rho\langle v^2 \rangle$, prove that the average kinetic energy of a molecule of an ideal gas is directly proportional to absolute temperature of the gas.

5. Derive an expression for solid angle.

6. Two vessels A and B are identical. A has 1 gm hydrogen at $0°C$ and B has 1 gm nitrogen at $0°C$.

   (a) Which vessel does contain more molecules and how much?

   (b) In which vessel is the pressure of the gas higher and how much?

   (c) In which vessel is the average speed of molecules larger and how much?

7. Calculate the energy per degree of freedom of a molecule of a gas.

8. In a triatomic molecule, the atoms are linearly arranged. Find the molar-specific heat at constant volume of such a gas, assuming the temperature to be such that the vibrational levels are not excited.

9. What is the success of the KTG?

10. What are the limitations of the KTG?

11. On what factors does the average kinetic energy of gas molecules depend? What will be the kinetic energy of the molecules at the absolute zero?

2.7.3   Numerical problems

1. Calculate the average kinetic energy of thermal neutrons at temperature $27°C$.
   [0.0258 eV] [Delhi University 2021]

2. Calculate the r.m.s. speed of hydrogen at standard temperature and pressure. Given that 1 L of hydrogen weighs 0.08987 g at S.T.P.   [1839.14 ms^{-1}]

3. Calculate the temperature at which the r.m.s. velocity of a gas will be half of its value at $0°C$ at constant pressure.   [$-204.75°C$]

4. Calculate the average translational kinetic energy of 1 gm of helium at standard temperature and pressure. Given $R = 8.3$ J(gm $-$ mol $-$ K)$^{-1}$.
   [5.65×10^{-21} J]

5. The average kinetic energy of a molecule of hydrogen gas at $0°C$ is 5.64×10^{-14} ergs, and the molar gas constant R is 8.31×10^{7} ergs/$°C$. Calculate Avogadro's number.   [6.04×10^{23}]

6. If one gm mole of an ideal monoatomic gas at temperature T_1 is mixed with one gm mole of another diatomic gas at temperature T_2 without the loss of any energy, then find out the final temperature of the mixture.
   [$\frac{3T_1 + 5T_2}{8}$] [Burdwan University 2020]

7. Two moles of H_2 at $30°C$ is mixed with one mole of helium at $60°C$. Calculate the final temperature of the mixture.   [$36.9°C$]

8. The mean kinetic energy of a molecule at $27°C$ is 8.0×10^{-14} ergs. Calculate the number of molecules per cubic cm of a perfect gas at $27°C$ and at a pressure of 40 mm of mercury.   [9.9×10^{17} molecules/cc]

9. Show that the r.m.s. velocity of oxygen is $\sqrt{2}$ times that of sulfur dioxide.

10. Calculate the average kinetic energy of an oxygen molecule (in ergs), if it occupies a volume of 2×10^4 c.c. at a pressure of 10^5 dyne-cm^{-2}. Use the value of Avogadro's number $N_A = 6.023 \times 10^{23}$.   [$4.9 \times 10^{-14}$ ergs]

11. A vessel contains two non-reacting gases: neon (monoatomic) and oxygen (diatomic). The ratio of their partial pressures is 3:2. Find the ratio of (a) the number of molecules and (b) the mass density of Ne and O_2 in the vessel. The atomic mass of Ne is 20.2 and the molecular mass of O_2 is 32.0.

[(a) 3/2; (b) 0.947]

CHAPTER 3

Dynamics of Ideal Gases

All the mathematical sciences are founded on the relations between physical laws and laws of numbers.

—James Clerk Maxwell

Learning Outcomes

After reading this chapter, the reader will be able to

- Learn the basic concept of the theory of probability

- List the assumptions used in the derivation of Maxwell's speed distribution law

- Derive Maxwell's speed distribution law and test its validity experimentally

- Calculate average, root mean square and most probable speed, energy, and momentum in one, two, and three dimensions, respectively

- State and prove the law of equipartition of energy

- Calculate the specific heat of gases

- Solve numerical problems and multiple choice questions on the distribution of molecular speed, energy, and momentum

3.1   Introduction

In Chapter 2, various characteristic features of a gaseous system based on the model of the kinetic theory of gases (KTG) have been discussed elaborately. Macroscopic properties and various relations among the thermodynamic variables have been explained in terms of this kinetic model. According to the assumptions used in this model, a gaseous system is composed of a large number of particles (atoms or molecules) with practically no volume occupied by them. Most of the times, these molecules move randomly through empty space at temperatures above absolute zero, and such motions remain unaffected by the presence of other particles.

This motion of the molecules is extremely chaotic and is characterized by straight-line trajectories interrupted by collisions with other molecules or with a physical boundary. In such a collision, the transfer of kinetic energy with a change in direction takes place depending on the nature of the relative kinetic energies of the particles. Any individual molecule collides with others at a huge rate, typically of the order of a billion times per second. This chapter is focused to present a comprehensive and quantitative discussion on the distributions of velocities, energies, and momenta of these molecules in various dimensions.

Measurement of the velocities of the molecules at a given time leads to a large distribution of values; some molecules may move very slowly and others very quickly. As these molecules move constantly in different directions, the velocity could be momentarily equal to zero. For a quantitative presentation of such velocity, one has to find out the **distribution function for the velocities**. This function provides the number of molecules having velocity within a specified range and indicates the dependency of this number of molecules on the velocity itself, temperature, and mass of the molecules. It is to be noted that the molecules in a sample of gas have average kinetic energy and average speed collectively, but they move at different speeds individually. As there are a large number of molecules and collisions involved, the distribution of molecular speed and the average speed of this large number of molecules are constant. This distribution of the speed of the molecules of a gas is known as a **Maxwell–Boltzmann distribution**, and it represents the fractional numbers of molecules in a bulk gaseous sample as a function of the speed of molecules at a given temperature. The distribution shows interesting characteristic features to be discussed later.

The objective of this chapter is to find out this distribution function considering the probabilistic approach proposed by Maxwell. It is demonstrated that this distribution of molecular speeds depends both on the mass of the molecules and the temperature. The speed distribution law for gaseous molecules enunciated by Maxwell–Boltzmann is very similar to that of liquids, even though the speeds are much larger in gases than in liquids. In solids, atoms are arranged in their equilibrium positions, and hence, they do not have translational energy anymore. But these atoms vibrate about their equilibrium positions. The only exception is the solid helium, in which atoms can move around. This solid helium is known to be a "quantum solid". To characterize this motion of the molecules at a given temperature and pressure, the following points are addressed in detail in this chapter:

1. Using the concept of probability theory, the distribution function for the molecular speeds of gaseous molecules is derived.

2. Average speed, root mean square (r.m.s.) speed, and most probable speed of the molecules are calculated.

3. Starting from the speed distribution law, distribution laws for energy and momentum of the molecules are derived. The expressions for average energy, r.m.s. energy, most probable energy, average momentum, r.m.s. momentum, and most probable momentum are also derived.

4. Experimental evidences in favor of Maxwell's speed distribution law are presented.

5. The concept of degrees of freedom is introduced, and the law of equipartition of energy is derived.

6. As an application of the law of equipartition of energy, the classical theory of specific heat problem in solids is thoroughly discussed.

3.2   Distribution of velocities: Maxwell–Boltzmann law of distribution of velocities in an ideal gas

It appears from this proposition that the velocities are distributed among the particles, according to the same law as errors are distributed among the observations in the theory of the method of least squares.

—James Clerk Maxwell

Let us first study the essence of the problem of distribution and its concept. At first sight, the distribution of molecules by velocities may apparently mean the determination of the number of molecules having a specified velocity. But, frankly speaking, the probable number of molecules with exactly – mathematically exactly – the given velocity is zero. Clearly, while the number of different values of the velocity is infinite, the number of molecules is falling to the share of each arbitrarily given velocity equals zero. Hence, the meaningful question should be: how many molecules have velocities within a certain interval near the present value of the velocity?

Statistical problems are always posed exactly in this fashion. For example, the distribution of population in a country by age means the determination of the probable number of people having an age within a definite interval of values, and not the determination of the probable number of people having a mathematically exact age, for the latter is devoid of any sense. When we make a statement that Mr X is 58 years old, we do not mean that he is exactly 58 years, 0 days, 0 hours, 0 minutes, and 0 seconds old. The statement only conveys or intends to convey the idea that Mr X has an age between 58 and 59 years.

3.2.1   An idea about the velocity space

Velocity space plays an important role in describing the motion of the molecules of an ideal gas. An idea of this velocity space can be illustrated in the following way. At a particular instant of time t, let us consider a gas having N molecules enclosed in a vessel of volume V. These molecules move randomly. Let us attach a vector to each molecule that represents its velocity in magnitude and direction, as shown in Figure 3.1(a). These vectors (not the molecules) are then transferred to a common origin, as shown in Figure 3.1(b). In this process of transfer, the property that a vector does not change when it is translated parallel to itself is used. The lengths of these velocity vectors are then removed, and only the end points are kept in the considered space. This provides a number of points equal to the number of molecules in the space known as **velocity space**.

In this way, all the molecules of a gaseous system are represented by points on the velocity space, and hence, such a gaseous system can be mathematically described on such a velocity diagram with OX, OY, and OZ as

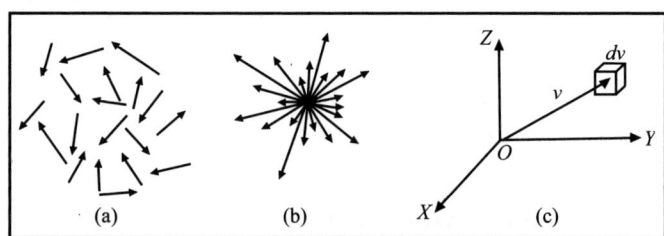

Figure 3.1 Idea of velocity space.

the coordinate axes shown in Figure 3.1(c). In this velocity space, an infinitesimal volume element dv is generally considered for the calculation of macroscopic physical parameters of the gaseous system.

3.2.2   The distribution function for the velocity of the molecules

In order to find the distribution function for the velocity of the molecules of a gas, the following assumptions are made:

1. The laws of classical mechanics can be used to describe the motion of atoms and molecules.

2. The number of molecules in a macroscopic volume is very large. In fact, there are about 10^{19} molecules in 1 cm^3 at normal conditions. This statement holds good up to a very high value of vacuum. Because of this large number of molecules, the impacts of individual molecule on the walls merge with each other and give rise to time-independent pressure. Hence, the expression for pressure does not contain time.

3. The average separation between the molecules is much larger than the molecular size and the typical range of intermolecular forces. This assumption allows us to neglect the interaction among the molecules, and the gas under consideration is assumed to behave as ideal, with the internal energy dominated by the kinetic energy of the molecules. In describing the macroscopic properties of the ideal gas at equilibrium, collisions between the molecules can be neglected. However, this assumption is violated if pressure is increased and/or temperature is decreased, and the gas behaves as nonideal

one. Different types of equations of state are then used to describe the behavior of such systems.

4. It is assumed that the molecules are distributed uniformly within a container. However, this uniform distribution of molecules is violated when the gas under consideration is subject to strong enough potential fields. For example, the density (concentration) of molecules in the atmosphere decreases with the increase in height. However, this variation in a laboratory-size container would be so slow that the distribution of the molecules within the gas container can still be assumed to be approximately uniform.

5. The molecular space is assumed to be isotropic such that there is a uniform distribution of the directions of velocities of the molecules. This is the hypothesis of molecular chaos and is found to be always true.

Considering the effects of these assumptions, the idea of distribution function for the velocities of the molecules is developed below. In practice, it is observed that the number of molecules $n(v)dv$ per unit volume with speed between v and $v + dv$ depends on the following factors:

1. The number of molecules $n(v)dv$ per unit volume with speed between v and $v + dv$ is directly proportional to the speed interval dv. So,

$$n(v)dv \propto dv. \qquad \Rightarrow \qquad n(v)dv = \alpha dv,$$

where α is the constant of proportionality.

2. $n(v)dv$ depends on the velocity itself. Even for identical intervals, it would be different at different values of the velocity. This means that the constant of proportionality factor α will be a function of velocity. This provides

$$\alpha = f(v).$$

3. Finally, $n(v)dv$ should be proportional to the total number of molecules n per unit volume, that is,

$$n(v)dv \propto n.$$

Combining these factors, we get the number of molecules $n(v)dv$ per unit volume with speed between v and $v + dv$ as

$$n(v)dv = nf(v)dv.$$

The function $f(v)$ is called the distribution function. If $dv = 1$, $f(v) = \frac{n(v)\ dv}{n}$. This indicates that the distribution function is the fraction of molecules whose velocities are within a unit interval of velocities at v. The quantity $\frac{n(v)\ dv}{n}$ has the meaning of probability. It is the probability of any molecule in unit volume having a velocity that is within a unit interval near the velocity v. Therefore, the distribution function is also called probability density.

In an assembly of gaseous molecules, the molecules are all identical, but the velocities of the molecules are not exactly the same. The velocities of some of the molecules may be quite high, while others may have low velocities at any instant of time and in the next instant, the picture of the distribution of the velocities may be altered. But on the average, the overall picture will remain the same. In fact, velocity may range from zero to infinity. Even if the velocities of all the molecules be identical at any instant that is highly improbable, they will be greatly disturbed due to collision, and the equality will be destroyed in no time. Most of the molecules may have the velocities within a certain range determined by the conditions of temperature and pressure. The law that expresses the number of molecules having a certain velocity in terms of the velocity and other known quantities is called the "law of distribution of velocities" of the ideal gas molecules due to Maxwell–Boltzmann.

An assembly of gas molecules is a statistical system, and the problem of finding the distribution of velocities among the molecules is essentially a statistical problem. According to the prescription of Maxwell, the law of distribution of molecular velocity will be deduced by using the **theory of probability**. Maxwell first enunciated the theorem that **"the components of the molecular velocities are distributed amongst the molecules according to the same law as errors are distributed amongst the observations"**. He then applied the important theorem of the **theory of probability** that the probability of a composite event is equal to the product of the probability of the individual events provided they are independent and derived the law of distribution function for the velocity of the molecules in an ideal gas.

3.3 Maxwell–Boltzmann distribution law of velocities of an ideal gas in three dimensions

Molecules in an ideal gas do not travel at the same speed but at speeds with a large variation in magnitudes. The r.m.s. speed is one kind of average of the speed, but many molecules move faster than this speed, and many even move at slower speed. The actual distribution of speeds of the molecules has several interesting implications not only in thermodynamics but also in other branches of physics. For individual molecules, the motion of the molecules is random in magnitude and direction, but for a gas of many molecules, there is a predictable distribution of molecular speeds. This predictable distribution of molecular speeds, after its originators, is known as the Maxwell–Boltzmann distribution. This distribution function is calculated based on the assumptions of the KTG and the theory of probability. This distribution law of molecular speeds has also been verified by several physicists experimentally.

The Maxwell–Boltzmann distribution function for speeds of the molecules in an ideal gas at a temperature T is given by the expression

$$f(v) = 4\pi \left(\frac{m}{2\pi K_B T} \right)^{3/2} v^2 \, e^{-\left(\frac{mv^2}{2K_B T} \right)}, \tag{3.1}$$

where m is the mass of one molecule of the gas, K_B is the Boltzmann constant, and v is the velocity of the molecule. This distribution function is derived below. Its average value, r.m.s. value, and most probable value are also calculated.

James Clerk Maxwell

James Clerk Maxwell was born on June 13, 1831 at 14 India Street, Edinburgh, Scotland, and was a scientist in the field of mathematical physics. By formulating the classical theory of electromagnetic radiation, he brought together electricity, magnetism, and light as manifestations of the same phenomenon for the first time. This was his most notable achievement in life. The equations for electromagnetism developed by Maxwell have been called the "second great unification in physics". In 1865, Maxwell theoretically derived in a research publication entitled **"A Dynamical Theory of the Electromagnetic Field"** that both electric and magnetic fields propagate through space as electromagnetic waves moving at the speed of light. Maxwell proposed that light is an undulation in the same medium that is the cause of electric and magnetic phenomena. He assisted in developing the distribution law of molecular speeds known as **Maxwell–Boltzmann distribution law**. This law is treated as a statistical means of describing various aspects of the KTG. He is also famous for presenting the first durable color photograph in 1861. He paid attention to resolve the issue related to the nature of Saturn's rings. It should be mentioned that this problem eluded scientists for two hundred years.

Maxwell made discoveries that laid the foundation for fields such as the special theory of relativity and quantum mechanics. Many physicists regard Maxwell as the greatest 19th-century scientist who has made profound influence on 20th-century physics and believe that Maxwell's contributions to science are of the same magnitude as those of Isaac Newton and Albert Einstein. Maxwell was voted the third greatest physicist of all time, behind only Newton and Einstein, in a millennium poll – a survey of the 100 most prominent physicists. On the centenary of Maxwell's birthday, Einstein described Maxwell's work as the "most profound and the most fruitful that physics has experienced since the time of Newton". Maxwell died on November 5, 1879 (aged 48) at Cambridge, England.

3.3.1   Derivation of the Maxwell–Boltzmann law of distribution of velocities in an ideal gas in three dimensions

Consider an ideal gas in a state of thermal equilibrium at a temperature T. Further, assume that the molecular collisions do not disturb the density of the gas so that, on average, it remains constant. The gas molecules are regarded to have identical masses. Let us calculate the number of molecules per unit volume that have their velocities lying between a small range v and $v + dv$, where v may have any direction in space. This number of molecules varies according to Maxwell's speed distribution law. In order to derive this distribution law, the following assumptions are made:

1. Equilibrium exists with a large number of molecules that are considered to be point masses.

2. The molecular density of the gas remains constant throughout the equilibrium condition.

3. Molecules are colliding elastically with one another, with all possible speeds lying in the range 0 to ∞.

4. There is no intermolecular force of attraction, that is, the molecules are assumed to be ideal.

5. In the body of the gas, there is no mass motion or convection current.

6. For a particular velocity, all directions are equally probable. The isotropy of the gas makes it immaterial in which coordinate system the results are expressed in.

7. The velocity component along the three mutually perpendicular directions is independent of one another so that the distribution of one component is independent of the other.[1]

8. The change of velocity to lie within a certain limit is a function of velocity and the limits only. It is independent of the other influences.

9. The gas molecules have no vibrational or rotational energies.

---

[1] J. C. Maxwell, Illustrations of the Dynamical Theory of Gases - Part I: On the Motions and Collisions of Perfectly Elastic Spheres, *Phil. Mag.* 19 (1860): 29–31.

We shall now consider the small range of velocities v and $v + dv$ that will have components between v_x and $v_x + dv_x$, v_y and $v_y + dv_y$, and v_z and $v_z + dv_z$ along the three axes. According to the prescription, the number of molecules per unit volume having the velocities lying between v_x and $v_x + dv_x$ can be denoted as $n(v_x)dv_x$. Obviously, $n(v_x)dv_x$ must be some function of n and v_x, say $nf(v_x)$, where $f(v_x)$ is a function of v_x to be determined. Then the probability that any molecule selected at random will have velocity components lying between v_x and $v_x + dv_x$ is $f(v_x)dv_x$, since the probability depends on the velocity and the range of velocity alike. Maxwell assumed that, as the velocity components v_x, v_y, and v_z of a molecule are perpendicular to each other, the distribution of one of these components among the molecules will not depend upon the values of the other components. Hence, $f(v_x)$ is independent of v_y and v_z. Similarly, the probabilities along Y and Z directions are, respectively, given by $f(v_y)dv_y$ and $f(v_z)dv_z$. It is to be noted that the function f is the same along the three directions as the molecular velocity space is assumed to be isotropic in nature. Since the probabilities for a composite event are multiplied, the probability that the velocity components of a molecule will lie between v_x and $v_x + dv_x$, v_y and $v_y + dv_y$, and v_z and $v_z + dv_z$ is given by

$$f(v_x)dv_x f(v_y)dv_y f(v_z)dv_z = f(v_x)f(v_y)f(v_z)dv_x dv_y dv_z.$$

So, the probability that n such molecules per unit volume will have velocities between v_x and $v_x + dv_x$, v_y and $v_y + dv_y$, and v_z and $v_z + dv_z$ is given by

$$nf(v_x)dv_x f(v_y)dv_y f(v_z)dv_z = nf(v_x)f(v_y)f(v_z)dv_x dv_y dv_z.$$

We now represent all the molecules in a velocity diagram with three coordinate axes OX, OY, and OZ, respectively, as shown in Figure 3.2. In this velocity diagram, a molecule having the velocity components v_x, v_y, and v_z will be represented by a point whose coordinates are v_x, v_y, and v_z. Such a point is called a velocity point. All the molecules whose velocity components lie in the range v_x and $v_x + dv_x$, v_y, $v_y + dv_y$, and v_z and $v_z + dv_z$ will be contained in the volume element $dv_x dv_y dv_z$ and the number of such molecules is

$$nf(v_x)f(v_y)f(v_z)dv_x dv_y dv_z.$$

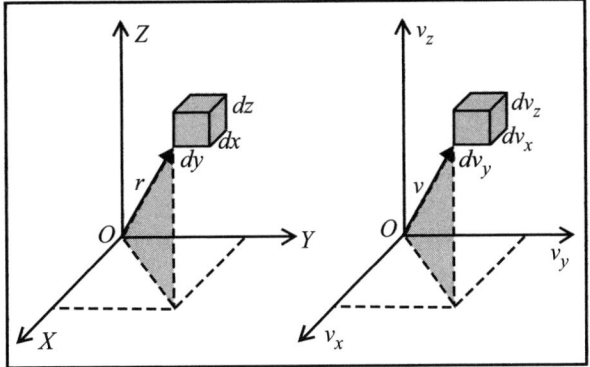

Figure 3.2 A configuration space volume element dr = dxdydz at spatial position $\vec{r}$ (left). The equivalent velocity space element. Together these two elements constitute a volume element $dV_{vol} = drdv$ at position (r, v) in phase space (right).

Now these molecules have the resultant velocity v, where $v^2 = v_x^2 + v_y^2 + v_z^2$; hence, their number must also be equal to

$$nF(v)dv_x dv_y dv_z,$$

that is, product of some function of v and the interval $dv_x dv_y dv_z$. Further, this number of molecules will not depend on the inclination of v to the axis as the directions of the axes are quite arbitrary, and there is noting to distinguish one direction from another in a gas at rest. Thus,

$$nf(v_x)f(v_y)f(v_z)dv_x dv_y dv_z = nF(v)dv_x dv_y dv_z = n\phi(v^2)dv_x dv_y dv_z, \qquad (3.2)$$

where $\phi(v^2)$ is some function of v^2. In writing this, we make use of the fact that after a large number of collisions, the distribution will be isotropic and depend upon v^2 only.

To solve this equation (3.2), it is to be noted that for a fixed value of v, $\phi(v^2)$ is a constant, and hence, $d\left[\phi(v^2)\right] = 0$. This leads to

$$d\left[\phi(v^2)\right] = d\left[f(v_x)f(v_y)f(v_z)\right] = 0;$$
$$\Rightarrow \quad f'(v_x)dv_x f(v_y)f(v_z) + f(v_x)f'(v_y)dv_y f(v_z) + f(v_x)f(v_y)f'(v_z)dv_z = 0.$$

Dividing throughout by $f(v_x)f(v_y)f(v_z)$, we get

$$\frac{f'(v_x)}{f(v_x)}dv_x + \frac{f'(v_y)}{f(v_y)}dv_y + \frac{f'(v_z)}{f(v_z)}dv_z = 0. \qquad (3.3)$$

Since v may have any direction, its components may change, but v may remain constant in magnitude. Hence, we have

$$v^2 = v_x^2 + v_y^2 + v_z^2 = \text{constant}; \qquad \Rightarrow \qquad v_x dv_x + v_y dv_y + v_z dv_z = 0. \qquad (3.4)$$

It is clear from equation (3.4) that the differentials dv_x, dv_y, and dv_z are not mutually independent, but these differentials can take any value that will necessarily satisfy equation (3.4). The method of Lagrange of undetermined multipliers is used to relax this constraint. In this method of Lagrange undetermined multiplier, the constraining relation is multiplied by a constant, and the resulting expression is added to the constrained equation. In this present problem, let us choose α as the undetermined multiplier. Multiplying equation (3.4) by α, we get

$$\alpha v_x dv_x + \alpha v_y dv_y + \alpha v_z dv_z = 0. \qquad (3.5)$$

Adding equations (3.3) and (3.5), we get

$$\left[\frac{f'(v_x)}{f(v_x)} + \alpha v_x\right] dv_x + \left[\frac{f'(v_y)}{f(v_y)} + \alpha v_y\right] dv_y + \left[\frac{f'(v_z)}{f(v_z)} + \alpha v_z\right] dv_z = 0. \qquad (3.6)$$

Since dv_x, dv_y, and dv_z are independent of one another, each of the three terms in equation (3.6) must be individually zero. Hence, we must have

$$\left[\frac{f'(v_x)}{f(v_x)} + \alpha v_x\right] dv_x = 0; \quad \Rightarrow \quad \frac{f'(v_x)}{f(v_x)} dv_x = -\alpha v_x dv_x.$$

$$\Rightarrow \quad \ln[f(v_x)] = -\frac{\alpha v_x^2}{2} + \ln a.$$

$$\Rightarrow \quad \ln\left[\frac{f(v_x)}{a}\right] = -\frac{\alpha v_x^2}{2}; \quad \Rightarrow \quad \frac{f(v_x)}{a} = e^{-\frac{\alpha v_x^2}{2}}.$$

This leads to

$$f(v_x) = ae^{-\frac{\alpha v_x^2}{2}} = ae^{-bv_x^2}, \qquad (3.7)$$

where a and $b\left(= \frac{\alpha}{2}\right)$ are two constants. This should be mentioned here that the method of Lagrange undetermined multipliers helps one to discover the form of the function $f(v)$, and it is observed that $f(v)$ is a decaying exponential function of velocity. Following the similar procedure for y and z components, we have

$$f(v_y) = ae^{-bv_y^2}, \quad \text{and} \quad f(v_z) = ae^{-bv_z^2}. \qquad (3.8)$$

So, the probability that a molecule will have velocities within v and $v + dv$ is given by

$$f(v_x)f(v_y)f(v_z)dv_x dv_y dv_z = a^3\, e^{-b(v_x^2+v_y^2+v_z^2)} dv_x dv_y dv_z. \qquad (3.9)$$

Let n be the total number of molecules per unit volume and $n(v)dv$ be the number of molecules out of n whose velocities lie within v and $v + dv$. Hence, the probability that a molecule will have velocity within v and $v + dv$ is $\frac{n(v)dv}{n}$. Equating this probability with that of equation (3.9), we get

$$\frac{n(v)dv}{n} = a^3\, e^{-b(v_x^2+v_y^2+v_z^2)} dv_x dv_y dv_z.$$

This leads to

$$n(v)dv = na^3\, e^{-b(v_x^2+v_y^2+v_z^2)} dv_x dv_y dv_z. \qquad (3.10)$$

This gives the number of molecules per unit volume with velocity between v and $v + dv$.

Evaluation of the constant a in terms of b

We know from equation (3.10) that the number of molecules $n(v)dv$ per unit volume with velocity between v and $v + dv$ is given by

$$n(v)dv = na^3 \, e^{-b(v_x^2+v_y^2+v_z^2)}dv_x dv_y dv_z. \qquad (3.11)$$

Integrating equation (3.11) over all velocity range from $-\infty$ to ∞, we get the total number of molecules per unit volume as

$$n = \int_{-\infty}^{\infty} n(v)dv = na^3 \int_{-\infty}^{\infty}\int_{-\infty}^{\infty}\int_{-\infty}^{\infty} e^{-b(v_x^2+v_y^2+v_z^2)}dv_x dv_y dv_z.$$

$$n = na^3 \int_{-\infty}^{\infty} e^{-bv_x^2}dv_x \int_{-\infty}^{\infty} e^{-bv_y^2}dv_y \int_{-\infty}^{\infty} e^{-bv_z^2}dv_z. \quad \Rightarrow \quad n = na^3 \sqrt{\frac{\pi}{b}}\sqrt{\frac{\pi}{b}}\sqrt{\frac{\pi}{b}}.$$

This leads to

$$a = \sqrt{\frac{b}{\pi}}. \qquad (3.12)$$

Using this value of a in equation (3.11), the number of molecules $n(v)dv$ per unit volume with velocity between v and $v + dv$ comes out to be

$$n(v)dv = n\left(\frac{b}{\pi}\right)^{\frac{3}{2}} e^{-b(v_x^2+v_y^2+v_z^2)}dv_x dv_y dv_z. \qquad (3.13)$$

The Maxwell–Boltzmann speed distribution law in spherical polar coordinates

In spherical polar coordinates (v, θ, ϕ), the Cartesian volume $dv_{vol} = dv_x dv_y dv_z$ becomes $dv_{polar} = v^2 sin\theta d\theta d\phi dv$. Using this elementary volume dv_{polar} and the relation $v^2 = v_x^2 + v_y^2 + v_z^2$, the number of molecules $n'(v)dv$ per unit volume having speeds between v and $v + dv$ in the direction between θ and $\theta + d\theta$ and ϕ and $\phi + d\phi$ can be written as

$$n'(v)dv = n\left(\frac{b}{\pi}\right)^{\frac{3}{2}} e^{-bv^2} v^2 sin\theta d\theta d\phi dv. \qquad (3.14)$$

So, the total number of molecules $n(v)dv$ per unit volume having speeds between v and $v + dv$ in all possible directions is given by

$$n(v)dv = n\left(\frac{b}{\pi}\right)^{\frac{3}{2}} e^{-bv^2} v^2 \, dv \int_0^{\pi} sin\theta d\theta \int_0^{2\pi} d\phi = n\left(\frac{b}{\pi}\right)^{\frac{3}{2}} e^{-bv^2} v^2 \, dv \times 2 \times 2\pi.$$

This leads to

$$n(v)dv = 4\pi n \left(\frac{b}{\pi}\right)^{\frac{3}{2}} e^{-bv^2} v^2 \, dv. \tag{3.15}$$

Equation (3.15) gives the number of molecules per unit volume having speed between v and $v + dv$ moving in all possible directions in velocity space.

Evaluation of the constant b

In order to evaluate the value of the constant b, we compute below the total energy E_T of an ideal gas containing N number of molecules.

According to Maxwell's speed distribution law, the number of molecules $N(v)dv$ with speed lying between v and $v + dv$ in a volume V in spherical polar coordinates is given by

$$N(v)dv = 4\pi N \left(\frac{b}{\pi}\right)^{\frac{3}{2}} e^{-bv^2} v^2 dv. \tag{3.16}$$

Now, a free molecule of mass m moving with velocity v will have kinetic energy E as

$$E = \frac{1}{2}mv^2 \quad \text{so that} \quad \Rightarrow \quad dv = \frac{dE}{\sqrt{2mE}}. \tag{3.17}$$

Using these values in equation (3.16), we get the number of molecules with energy lying between E and $E + dE$ in the volume V as

$$N(E)dE = 2\pi N \left(\frac{\beta}{\pi}\right)^{3/2} e^{-\beta E} E^{\frac{1}{2}} dE, \tag{3.18}$$

where the factor β is given by $\beta = \frac{2b}{m}$.

Using this number $N(E)dE$ from equation (3.18), the total energy E_T of the gaseous system can be calculated as

$$E_T = \int_0^\infty E N(E) dE = \int_0^\infty E \, 2\pi N \left(\frac{\beta}{\pi}\right)^{3/2} e^{-\beta E} E^{\frac{1}{2}} dE$$

$$= 2\pi N \left(\frac{\beta}{\pi}\right)^{3/2} \int_0^\infty e^{-\beta E} E^{\frac{3}{2}} dE. \tag{3.19}$$

Using the value of the definite integral $\int_0^\infty e^{-\beta E} E^{\frac{3}{2}} dE = \frac{3}{4}\sqrt{\frac{\pi}{\beta^5}}$, we have

$$E_T = 2\pi N \left(\frac{\beta}{\pi}\right)^{\frac{3}{2}} \cdot \frac{3}{4}\sqrt{\frac{\pi}{\beta^5}} = \frac{3}{2}\frac{N}{\beta}. \tag{3.20}$$

We know from the KTG that the total energy of a system with N number of ideal gas molecules is given by

$$E_T = \frac{3}{2} N K_B T, \tag{3.21}$$

where K_B is the Boltzmann constant and T is the temperature in absolute scale. Comparing these two equations (3.20) and (3.21), we get

$$\frac{3}{2}\frac{N}{\beta} = \frac{3}{2} N K_B T; \qquad \Rightarrow \qquad \beta = \frac{1}{K_B T}. \tag{3.22}$$

Substituting this value of $\beta = \dfrac{1}{K_B T}$ in equation (3.18), we get the number of molecules with energy lying between E and $E + dE$ in the volume V as

$$N(E)dE = 2\pi N \left(\frac{1}{\pi K_B T} \right)^{\frac{3}{2}} e^{-\left(\frac{E}{K_B T} \right)} E^{\frac{1}{2}} dE. \tag{3.23}$$

Equation (3.23) represents the number of molecules with energy lying between E and $E + dE$ in the same assembly and is known as the Maxwell–Boltzmann law for the **distribution of energy** among the molecules of a noninteracting gaseous system.

For small values of E, the term $e^{-\left(\frac{E}{K_B T} \right)}$ is nearly unity, and hence, $N(E)$, that is, the number of molecules with energy E is proportional to $E^{\frac{1}{2}}$ and so for $E = 0$, $N(E) = 0$.

Now, using the value of β from equation (3.22), the constant b comes out to be

$$b = \frac{m\beta}{2} = \frac{m}{2 K_B T}. \tag{3.24}$$

Using this value of b in equation (3.16), the number of molecules with speed between v and $v + dv$ in the volume V comes out to be

$$N(v)dv = 4\pi N \left(\frac{m}{2\pi K_B T} \right)^{\frac{3}{2}} e^{-\left(\frac{mv^2}{2 K_B T} \right)} v^2 dv. \tag{3.25}$$

This equation (3.25) gives the number of molecules in the volume V with speed between v and $v + dv$ moving in all possible directions in velocity space. This is known as the **Maxwell–Boltzmann speed distribution law**. Equation (3.25) shows that this number $N(v)dv$ depends on the following three important factors:

1. Velocity v of the molecules through the exponential term $e^{-\left(\frac{mv^2}{2 K_B T} \right)}$ and v^2 at constant temperature T.

2. Temperature T through the exponential term $e^{-\left(\frac{mv^2}{2K_BT}\right)}$ and $T^{-\frac{3}{2}}$ at constant velocity v, and

3. Mass m of the molecules through the exponential term $e^{-\left(\frac{mv^2}{2K_BT}\right)}$ and $m^{+\frac{3}{2}}$.

Variation of the number of molecules $N(v)dv$ as a function of velocity v following Maxwell–Boltzmann distribution law at four different temperatures is shown in Figure 3.3. This variation will be discussed quantitatively later.

Important features of Maxwell's speed distribution law

The following important features are to be noted about Maxwell's speed distribution law:

1. The curve for Maxwell's speed distribution law is a Gaussian one, and Maxwell's velocity distribution in one dimension is a Gaussian distribution or normal distribution.

2. With the increase in temperature, the height of the peak of the distribution curve decreases, and the probability of finding molecules with higher velocity increases. However, the area under the curve that represents the total probability remains the same, and it is equal to 1. The average kinetic energy also increases with the increase in temperature, and consequently, molecules move with higher and higher velocity.

3. When the temperature tends to zero Kelvin, the distribution curve will be along the probability axis, that is, $F(v)$-axis. But in the limit $T \to \infty$, probability distribution will be along the velocity axis.

4. The height of the peak of the distribution curve increases with the increase in the molar mass of the gas molecules, whereas the probability of finding molecules having higher and higher velocities decreases.

5. This distribution curve is symmetric about $v = 0$. Hence, the average velocity in one dimension is zero. This does not mean that all the molecules are at rest. As there is no intermolecular interaction, both directions $+x$ and $-x$ are equally probable. This means that the probability of finding molecules having velocity in the range between $+v$ and $+(v+dv)$ and $-v$ and $-(v+dv)$ are equal. This makes the average velocity to be zero. In one dimension, this statement can be justified mathematically in the following way:

$$\langle v \rangle = \int\limits_{-\infty}^{\infty} vf(v)dv = \int\limits_{-\infty}^{\infty} v\left(\frac{m}{2\pi K_BT}\right)^{\frac{1}{2}} e^{-\left(\frac{mv^2}{2K_BT}\right)}dv = 0,$$

as the integrand is an odd function of v.

The average speed of an ideal gas in three dimensions

The average value of a physical quantity with a probability distribution is determined by multiplying the distribution function by the physical quantity to be averaged and integrating the product over all possible values. This is analogous to calculating averages of discrete distributions, where each value is multiplied by the number of times it occurs, the results are added, and finally is divided by the number of values. The integral is analogous to the first two steps, and the normalization is analogous to division by the number of values. As the number of points in the velocity space is distributed almost continuously, we can use integration instead of summation. Following these prescriptions, the average speed of the molecules of an ideal gas obeying Maxwell's law of molecular speed distribution is calculated below.

We consider a system of N identical, independent, and distinguishable molecules of an ideal gas occupying a volume V at a temperature T. Each molecule has a mass m. These identical molecules obey Maxwell's law of molecular speed distribution. According to equation (3.25), the number of molecules $N(v)dv$ with speed in the range between v and $v + dv$ in the given volume V is given by

$$N(v)dv = 4\pi N \left(\frac{m}{2\pi K_B T} \right)^{\frac{3}{2}} e^{-\left(\frac{mv^2}{2K_B T} \right)} v^2 \, dv, \tag{3.26}$$

where symbols have their usual meaning. The average speed $\langle v \rangle$ of these molecules is then given by

$$\langle v \rangle = \frac{\displaystyle\int_0^\infty v \, N(v) \, dv}{\displaystyle\int_0^\infty N(v) \, dv} = \frac{\displaystyle\int_0^\infty v \times 4\pi N \left(\frac{m}{2\pi K_B T} \right)^{3/2} e^{-\left(\frac{mv^2}{2K_B T} \right)} v^2 \, dv}{\displaystyle\int_0^\infty 4\pi N \left(\frac{m}{2\pi K_B T} \right)^{3/2} e^{-\left(\frac{mv^2}{2K_B T} \right)} v^2 \, dv}. \tag{3.27}$$

Equation (3.26) has been used here. It should be noted that the integral

$$\int_0^\infty 4\pi N \left(\frac{m}{2\pi K_B T} \right)^{3/2} e^{-\left(\frac{mv^2}{2K_B T} \right)} v^2 \, dv$$

in the denominator of equation (3.27) gives the total number of molecules N. Using this result, equation (3.27) can be written as

$$\langle v \rangle = \frac{1}{N} \int_0^\infty v 4\pi N \left(\frac{m}{2\pi K_B T} \right)^{3/2} e^{-\left(\frac{mv^2}{2K_B T} \right)} v^2 \, dv$$

$$= \frac{1}{N} 4\pi N \left(\frac{m}{2\pi K_B T} \right)^{3/2} \int_0^\infty e^{-\left(\frac{mv^2}{2K_B T} \right)} v^3 \, dv,$$

In order to evaluate the integral, let us put $\frac{mv^2}{2K_BT} = x$. This gives rise to

$$2vdv = \frac{2K_BT}{m}dx. \text{ So, } dv = \frac{K_BT}{m}\frac{dx}{\sqrt{x\frac{2K_BT}{m}}} = \sqrt{\frac{K_BT}{2m}}x^{-1/2}dx.$$

Substituting these values in the above expression, we get the average speed $\langle v \rangle$ as

$$\langle v \rangle = 4\pi \left(\frac{m}{2\pi K_BT}\right)^{3/2} \int_0^\infty e^{-x} \left(\frac{2K_BT}{m}x\right)^{\frac{3}{2}} \sqrt{\frac{K_BT}{2m}}x^{-\frac{1}{2}}dx.$$

$$= 4\pi \left(\frac{m}{2\pi K_BT}\right)^{\frac{3}{2}} \left(\frac{2K_BT}{m}\right)^{\frac{3}{2}} \sqrt{\frac{K_BT}{2m}} \int_0^\infty e^{-x}xdx$$

$$= \frac{4}{\pi^{1/2}} \sqrt{\frac{K_BT}{2m}}\Gamma(2) = \sqrt{\frac{8K_BT}{\pi m}}. \tag{3.28}$$

Equation (3.28) shows that the average speed $\langle v \rangle = \sqrt{\frac{8K_BT}{\pi m}}$ is directly proportional to the square root of the absolute temperature T and is inversely proportional to the square root of mass m of one molecule of the ideal gas. With the increase in T, average speed $\langle v \rangle$ increases, and the distribution curve shown in Figure 3.3 becomes more flat at the higher temperature. This indicates that a larger number of molecules shift toward higher speeds with the increase in temperature T.

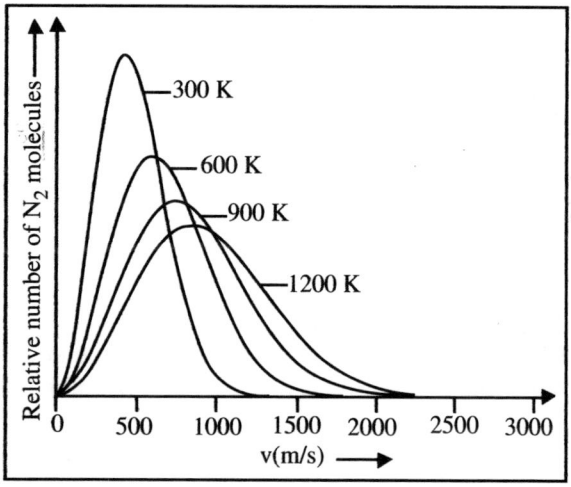

Figure 3.3 Maxwell–Boltzmann speed distribution law of molecules at four different temperatures.

The most probable speed of an ideal gas in three dimensions

The most probable speed is defined as the speed possessed by the maximum number of molecules. The number of molecules with velocity in the range v and $v + dv$ is given by equation (3.26) that is recast as

$$N(v)dv = 4\pi N \left(\frac{m}{2\pi K_BT}\right)^{3/2} v^2 e^{-\left(\frac{mv^2}{2K_BT}\right)} dv. \tag{3.29}$$

Then, the number of molecules per unit velocity interval at v is given by

$$N(v) = 4\pi N \left(\frac{m}{2\pi K_B T}\right)^{3/2} v^2 e^{-\left(\frac{mv^2}{2K_B T}\right)}. \tag{3.30}$$

Here, the symbols have their usual meaning.

The most probable speed v_m is defined as that speed at which the number of molecules per unit speed interval is maximum. This can be achieved by differentiating equation (3.30) with respect to speed v and equating the result to zero. These steps are followed in the sequence below:

$$\frac{d}{dv}[N(v)]_{v=v_m} = \frac{d}{dv}\left[4\pi N \left(\frac{m}{2\pi K_B T}\right)^{3/2} v^2 e^{-\left(\frac{mv^2}{2K_B T}\right)}\right]_{v=v_m} = 0$$

$$\Rightarrow \left[4\pi N \left(\frac{m}{2\pi K_B T}\right)^{3/2}\right]\left[(2v)e^{-\left(\frac{mv^2}{2K_B T}\right)} + v^2 e^{-\left(\frac{mv^2}{2K_B T}\right)}\left(-\frac{mv}{K_B T}\right)\right]_{v=v_m} = 0$$

$$\Rightarrow \left[4\pi N \left(\frac{m}{2\pi K_B T}\right)^{3/2}\right] e^{-\left(\frac{mv^2}{2K_B T}\right)}\left[(2v) + v^2\left(-\frac{mv}{K_B T}\right)\right]_{v=v_m} = 0.$$

This provides the most probable speed v_m as

$$v_m = \sqrt{\frac{2K_B T}{m}}. \tag{3.31}$$

This equation (3.31) shows that the most probable speed v_m is proportional to the square root of temperature T of the gas, is inversely proportional to the square root of mass m of one molecule and is independent of the total number of molecules N. With the increase in T, the most probable speed v_m increases, and the distribution curve shown in Figure 3.3 becomes more flat. This indicates that a smaller number of molecules move with this most probable speed v_m at higher temperature.

The number of molecules moving at this most probable speed v_m can be calculated in the following way: Inserting v_m in place of v into equation (3.30), we get

$$N(v_m) = 4\pi N \left(\frac{m}{2\pi K_B T}\right)^{3/2} v_m^2 \; e^{-\left(\frac{mv_m^2}{2K_B T}\right)}.$$

Using $v_m = \sqrt{\frac{2K_B T}{m}}$ from equation (3.31), the expression for the number $N(v_m)$ becomes

$$N(v_m) = 4\pi N \left(\frac{m}{2\pi K_B T}\right)^{3/2}\left[\sqrt{\frac{2K_B T}{m}}\right]^2 e^{-m\left[\sqrt{\frac{2K_B T}{m}}\right]^2/2K_B T} = \frac{4N}{e}\sqrt{\frac{m}{2\pi K_B T}}. \tag{3.32}$$

Equation (3.32) shows that the number of molecules corresponding to the most probable speed v_m is proportional to the square root of mass m of one molecule at a fixed temperature T and for a fixed total number of molecules N, is inversely proportional to the square root of temperature T of the gas for fixed m and N, and is directly proportional to the total number of molecules N of the gas.

The following important facts are observed from equations (3.31) and (3.32):

1. With the increase in temperature T, the most probable speed v_m increases, but the number of molecules $N(v_m)$ moving at this speed v_m decreases.

2. Another interesting feature is noted from the multiplicative results of equations (3.31) and (3.32). This multiplication gives rise to

$$v_m \times N(v_m) = \sqrt{\frac{2K_BT}{m}} \times \frac{4N}{e}\sqrt{\frac{m}{2\pi K_BT}} = \frac{4N}{e\sqrt{\pi}}. \qquad (3.33)$$

This result shows that the product of the most probable speed v_m and the number of molecules $N(v_m)$ moving at this speed v_m depends only on the total number of molecules N of the gas; it is not influenced by any other parameters.

3. In an ideal gas, the speed of sound is given by

$$v_s = \sqrt{\frac{\gamma P}{\rho}} = \sqrt{\frac{\gamma K_BT}{m}},$$

where we have used $P = nK_BT$ (Clapeyron's equation) and $\rho = n\,m$. It is clear that the various average speeds that we have just calculated are all of the order of the speed of sound, that is, a few hundred meters per second at room temperature. In ordinary air ($\gamma = 1.4$), the sound speed is about 84% of the most probable speed and about 74% of the average speed. Because sound waves ultimately propagate via molecular motion, it makes sense that they travel at slightly less than the most probable and average molecular speeds.

Reduced form of Maxwell's speed distribution law in three dimensions

The number of molecules $N(v)dv$ with speed in the range between v and $v + dv$ is given by equation (3.26). We recast the equation as

$$N(v)dv = 4\pi N \left(\frac{m}{2\pi K_BT}\right)^{3/2} v^2\, e^{-\left(\frac{mv^2}{2K_BT}\right)}dv. \qquad (3.34)$$

We scale the velocity v of the molecules by the most probable velocity v_m and we call this as the so-called reduced velocity v_r, that is, $v_r = \frac{v}{v_m}$. When Maxwell's speed distribution law is expressed in terms of v_r, we get its reduced form.

Using the value of v_m from equation (3.31), we get the reduced velocity as

$$v_r = \frac{v}{v_m} = \frac{v}{\sqrt{\frac{2K_BT}{m}}}.$$

This provides

$$v = v_m \times v_r = \sqrt{\frac{2K_BT}{m}}v_r; \quad \text{and} \quad \Rightarrow \quad dv = \sqrt{\frac{2K_BT}{m}}dv_r.$$

Substituting these values in equation (3.34), we get

$$N(v_r)dv_r = 4\pi N \left(\frac{m}{2\pi K_BT}\right)^{3/2} \times \left[e^{-v_r^2 \times \left(\frac{2K_BT}{m}\right) \times \frac{m}{2K_BT}}\right]$$

$$\times \left(v_r^2\frac{2K_BT}{m}\right) \times \left(\sqrt{\frac{2K_BT}{m}}dv_r\right).$$

After simplification, we get

$$N(v_r)dv_r = \frac{4}{\sqrt{\pi}}\, N\, e^{-v_r^2}\, v_r^2 dv_r. \tag{3.35}$$

Equation (3.35) provides the reduced form of Maxwell's speed distribution law. It is to be noted that the reduced form is independent of temperature T.

Root mean square speed of the molecules of an ideal gas in three dimensions

At a certain temperature T, the instantaneous velocities of the molecules in an ideal gas would have a wide range of values, theoretically from zero to infinity. Some of these values would be zero, and some of them would have very high values, but the majority of these values would fall into a more or less well-defined range of velocities. Calculation of average velocity for such a collection of ideal gas molecules will lead to zero value, because the molecules in the gas are in random thermal motion. On average, there will be an almost equal number of molecules moving in one direction as in the opposite direction. Hence, the velocity vectors of opposite signs would all cancel out, and the average velocity would come out to be zero. In physics, this zero average velocity is not very useful. Therefore, in such a case, the averaging of velocity is calculated in a slightly different way by considering the idea of r.m.s. velocity. The mathematical expression for such r.m.s. velocity of a Maxwellian gas is calculated below.

Consider a system of N identical, independent, and distinguishable molecules of a gas occupying a volume V at a temperature T. Each molecule of the gas has mass m. These identical particles, in classical statistical mechanics, are described by Maxwell–Boltzmann statistics, and the distribution of molecular speeds is in accordance with Maxwell's law. The mean square speed of the molecules of such a gas is given by

$$\langle v^2 \rangle = \frac{\displaystyle\int_0^\infty v^2\, N(v)\, dv}{\displaystyle\int_0^\infty N(v)\, dv} = \frac{\displaystyle\int_0^\infty v^2 4\pi N \left(\frac{m}{2\pi K_B T}\right)^{3/2} v^2 e^{-\left(\frac{mv^2}{2K_B T}\right)} dv}{\displaystyle\int_0^\infty 4\pi N \left(\frac{m}{2\pi K_B T}\right)^{3/2} v^2 e^{-\left(\frac{mv^2}{2K_B T}\right)} dv}$$

$$= \frac{1}{N}\int_0^\infty v^2 4\pi N \left(\frac{m}{2\pi K_B T}\right)^{3/2} v^2 e^{-\left(\frac{mv^2}{2K_B T}\right)} dv.$$

Equation (3.26) for $N(v)dv$ has been used in the above expression. It should be noted further that the integral

$$\int_0^\infty 4\pi N \left(\frac{m}{2\pi K_B T}\right)^{3/2} v^2 e^{-\left(\frac{mv^2}{2K_B T}\right)} dv$$

gives the total number of molecules N. Using this result, the expression for mean square speed comes out to be

$$\langle v^2 \rangle = \frac{1}{N} 4\pi N \left(\frac{m}{2\pi K_B T}\right)^{3/2} \int_0^\infty v^4 e^{-\left(\frac{mv^2}{2K_B T}\right)} dv$$

$$= 4\pi \left(\frac{m}{2\pi K_B T}\right)^{3/2} \frac{3}{8}\left(\frac{2K_B T}{m}\right)^2 \sqrt{\frac{2\pi K_B T}{m}} = \frac{3K_B T}{m}.$$

Hence, the r.m.s. speed is given by $v_{\text{r.m.s.}} = \sqrt{\langle v^2 \rangle} = \sqrt{\frac{3K_B T}{m}}$. This result $v_{\text{r.m.s.}} = \sqrt{\frac{3K_B T}{m}}$ shows that the r.m.s. speed $v_{\text{r.m.s.}}$ is directly proportional to the square root of the absolute temperature T and is inversely proportional to the square root of mass m of one molecule. This indicates that the molecules with lighter mass will have larger $v_{\text{r.m.s.}}$ comparable to or even greater than the escape velocity for those molecules. Hence, gases with lighter mass molecules are rarely observed in the atmosphere.

Temperature dependence of the Maxwell–Boltzmann speed distribution law

The distribution law of molecular speeds of an ideal gas is given by Maxwell–Boltzmann and is recast as

$$N(v)dv = 4\pi N \left(\frac{m}{2\pi K_B T}\right)^{\frac{3}{2}} e^{-\left(\frac{mv^2}{2K_B T}\right)} v^2\, dv. \tag{3.36}$$

This Maxwell–Boltzmann distribution law of molecular speeds is shown in Figure 3.3 at four selective temperatures that are mentioned in the graph against each curve.

In order to describe the temperature dependence of this distribution law, we have to focus on two important factors:

1. Normalization term $\left(\frac{m}{2\pi K_B T}\right)^{\frac{3}{2}}$ and

2. The exponential term $e^{-\left(\frac{mv^2}{2K_B T}\right)}$.

The overall characteristic features of the temperature dependence of the distribution law are primarily determined by these two factors. The salient features of this dependence are briefly described below.

1. At lower temperature, say 300 K, the molecules have less kinetic energy. Therefore, the speeds of the molecules are lower, and the distribution has a smaller range as shown in Figure 3.3. The area under such a curve represents the total number of molecules of the gas. Hence, in order to maintain the same area under the curve, the height of the graph has to increase as the graph shifts to the left at a lower temperature.

2. As the temperature of the molecules increases, say 1200 K, a larger fraction of the molecules achieves greater amounts of kinetic energy. This causes the Boltzmann plots to spread out, that is, the Maxwell–Boltzmann distribution shifts to the higher speeds. This broadening at a higher temperature (corresponding to 1200 K) is clearly observed in Figure 3.3. The height of the graph decreases with increase in temperature T and the graph shifts to the right. Thus, the total area under the curve, and hence, the total number of molecules remain the same.

3. With the increase in temperature, the most probable speed v_m increases, but the number of molecules at the most probable speed decreases with the increase in temperature T.

4. With the increase in temperature, the r.m.s. speed $v_{\text{r.m.s.}}$ increases, but the number of molecules at the root mean square speed decreases with the increase in temperature T.

5. In the limit $T \to 0$, the normalization term approaches ∞, but the exponential term approaches 0. The product of these terms results 0. Hence, the function approaches zero.

6. In the limit $T \to \infty$, the normalization term approaches 0, but the exponential term approaches 1. Again, the product of these terms results 0. Hence, the function approaches zero.

7. The total area under the curve would increase if molecules from an external source are allowed to enter the sample of the gas. On the other hand, the total area under the curve would decrease if molecules are allowed to leave the sample of the gas.

Discussion on the variation of probability distribution of the molecules of an ideal gas with molecular speed

The Maxwell–Boltmann distribution function of molecular speeds, that is, the number of molecules $N(v)dv$ within the range of speed between v and $v + dv$ is given by equation (3.26). This distribution function for O_2 molecules is shown in Figure 3.4 at room temperature. The most probable speed v_m, average speed v_{av}, and r.m.s. speed $v_{\text{r.m.s.}}$ are shown in the figure by vertical dotted lines. It is observed from the figure that according to their magnitudes, these speeds follow the sequence: $v_m < v_{av} < v_{\text{r.m.s.}}$. Further, all these speeds increase with the increase in temperature.

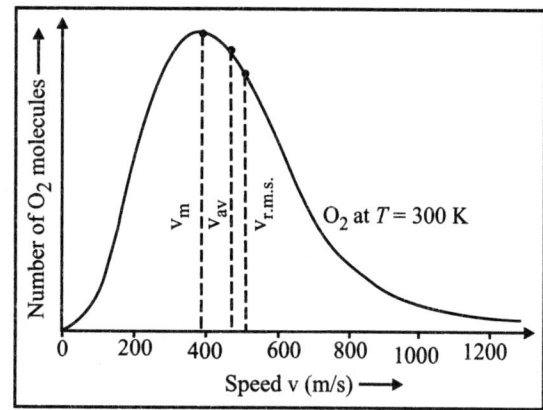

Figure 3.4 Maxwell's speed distribution for O_2 molecules at room temperature.

At a constant temperature, the following characteristic features of the speed dependence of the distribution function $f(v)$ from equation (3.26) are observed:

1. The factor $4\pi\left(\frac{m}{2\pi K_B T}\right)^{3/2}$ plays the role of a normalization constant. This normalization constant assures that $N(0, \infty) = N$ with $\int_0^\infty f(v)dv = 1$.

2. The factor of v^2 in equation (3.26) means that $f(0) = 0$, that is, there are practically no molecules with zero speed. All the molecules are in random motion. For small v, the factor v^2 dominates, and the curve looks like a parabola, that is, the number of molecules increases following a parabola.

3. The exponential factor $e^{-\left(\frac{mv^2}{2K_B T}\right)}$ in equation (3.26) shows that $\lim_{v\to\infty} f(v) = 0$ and the graph has an exponential tail. This indicates that there are only a few molecules moving at several times the r.m.s. speed.

4. It is well understood that the exponential function is a strongly varying one rather than the square function of the same variable. Hence, in the limit of larger speed, the probability distribution function decreases due to the exponential factor. The superposition of these two factors v^2 and $e^{(-((mv^2)/(2K_B T)))}$ provides the probability distribution function with the single-peaked shape shown in Figure 3.4.

5. At a particular temperature, this peak occurs at the most probable speed v_m that increases with temperature T.

What does the area under a Maxwell–Boltzmann distribution represent?

The X- and Y-axes of the Maxwell–Boltzmann speed distribution graph give, respectively, the speed and the number of molecules per unit speed interval. The total area under the entire curve is obtained by carrying out the integration from $v = 0$ to $v = \infty$, and this area represents the total number of molecules in the gas.

Why are hydrogen and helium so rare in the atmosphere of the earth?

Hydrogen and helium are the common elements in the universe, hydrogen being the most common and helium the second-most common. It should be mentioned that helium is constantly produced on the earth by radioactive decay. But these two elements are rarely observed in the atmosphere of the earth. This created a puzzle about the composition of the earth's atmosphere. A qualitative understanding of this puzzle is achieved from the values of the r.m.s. speeds of these elements. The molecules of these elements have lower mass and their r.m.s. speeds are higher than other gas molecules such as oxygen and nitrogen. Even these speeds can reach speeds above the earth's escape velocity, that is about 11.2 km s^{-1}. As a result, these elements can escape from the atmosphere into space. This successfully explains the reason that the lighter mass elements have escaped from the atmosphere than other heavier elements over the billions of years that the earth has existed. These lighter mass elements are now hardly found in the atmosphere.

Important consequences of Maxwell's velocity distribution law

There are some important consequences of Maxwell's law of velocity distribution. These are stated below.

1. Maxwell's law of velocity distribution can be used to derive the expression for pressure P of an ideal gas from the first principles and can also explain the equation of state $PV = NK_BT$ for an ideal gas that was known empirically for a long time. Hence, the microscopic basis of the perfect gas equation was really established from Maxwell's law.

2. This law can also be used to calculate the number of molecules crossing a unit area (from one side) per unit time. This number is found to be $\frac{n\langle v^2 \rangle^{1/2}}{\sqrt{6\pi}} \sim \frac{P}{(mT)^{1/2}}$. By investigating the evaporation from a liquid surface, Irving Langmuir verified the above equation experimentally.

3. It can successfully explain the specific heat behavior of solids at high temperatures (Dulong–Petit's law). In this high-temperature regime, the contribution of phonon to the specific heat can be ignored, that is, the role of quantum statistics can be ignored. The application of classical statistics provides satisfactory results.

3.3.2   The momentum distribution function in three dimensions

According to Maxwell–Boltzmann law of molecular speed distribution, the number of molecules $N(v)dv$ within speeds in the range between v and $v + dv$ is given by

$$N(v)dv = 4\pi N \left(\frac{m}{2\pi K_B T}\right)^{3/2} e^{-\left(\frac{mv^2}{2K_B T}\right)} v^2 dv.$$

From this expression, the momentum distribution law for the molecules of an ideal gas can be obtained in the following way:

The momentum p of a molecule is given by $p = m\,v$. Hence, we have

$$v = \frac{p}{m}; \quad \text{and} \Rightarrow \quad dv = \frac{dp}{m}.$$

With these substitutions, the number of molecules $N(p)dp$ within the momentum range between p and $p + dp$ is given by

$$N(p)dp = 4\pi N \left(\frac{m}{2\pi K_B T}\right)^{3/2} e^{-\left(\frac{m}{2K_B T} \times \frac{p^2}{m^2}\right)} \left(\frac{p^2}{m^2}\right) \frac{dp}{m}$$

$$= \frac{4\pi N}{m^3} \left(\frac{m}{2\pi K_B T}\right)^{3/2} e^{-\left(\frac{p^2}{2m K_B T}\right)} p^2 dp$$

$$= 4\pi N \left(\frac{1}{2\pi m K_B T}\right)^{3/2} e^{-\left(\frac{p^2}{2m K_B T}\right)} p^2 dp. \tag{3.37}$$

Equation (3.37) gives the number of molecules of an ideal gas within the momentum range between p and $p + dp$ at a particular temperature T. This is the momentum distribution law in three dimensions. Similar to speed distribution law, this momentum distribution law will also have average, most probable, and r.m.s. value. These parameters in three dimensions are calculated below.

Average momentum in three dimensions

We consider a system of N identical, independent, and distinguishable molecules of an ideal gas occupying a volume V at a temperature T. Each molecule has a mass m. The average momentum $\langle p \rangle$ of the molecules in such an ideal gas can be derived as follows:

$$\langle p \rangle = \frac{\displaystyle\int_0^\infty p\, N(p)\, dp}{\displaystyle\int_0^\infty N(p)\, dp} = \frac{\displaystyle\int_0^\infty p \times 4\pi N \left(\frac{1}{2\pi m K_B T}\right)^{3/2} e^{-\left(\frac{p^2}{2m K_B T}\right)} p^2 dp}{\displaystyle\int_0^\infty 4\pi N \left(\frac{1}{2\pi m K_B T}\right)^{3/2} e^{-\left(\frac{p^2}{2m K_B T}\right)} p^2 dp}. \tag{3.38}$$

Equation (3.37) for $N(p)dp$ has been used in the above expression. It should be noted further that the integral

$$\int_0^\infty 4\pi N \left(\frac{1}{2\pi m K_B T}\right)^{3/2} e^{-\left(\frac{p^2}{2m K_B T}\right)} p^2 dp$$

in the denominator of equation (3.38) gives the total number of molecules N. Using this result, the expression for average momentum $\langle p \rangle$ comes out to be

$$\langle p \rangle = \frac{1}{N} \int_0^\infty p \, 4\pi N \left(\frac{1}{2\pi m K_B T}\right)^{3/2} e^{-\left(\frac{p^2}{2m K_B T}\right)} p^2 dp$$

$$= \frac{1}{N} \times 4\pi N \left(\frac{1}{2\pi m K_B T}\right)^{3/2} \int_0^\infty e^{-\left(\frac{p^2}{2m K_B T}\right)} p^3 dp.$$

In order to find the value of the above integral, let us put

$$\frac{p^2}{2m K_B T} = x, \quad \Rightarrow 2p dp = 2m K_B T dx. \quad \text{So,} \, dp = \sqrt{\frac{m K_B T}{2}} x^{-\frac{1}{2}} dx.$$

Again, when $p \to 0$, $x \to 0$, and as $p \to \infty$, then $x \to \infty$.

Substituting these values in the above expression, we get the average momentum $\langle p \rangle$ as

$$\langle p \rangle = 4\pi \left(\frac{1}{2\pi m K_B T}\right)^{3/2} \int_0^\infty (2\pi m K_B T)^{3/2} x^{3/2} e^{-x} \sqrt{\frac{m K_B T}{2}} x^{-\frac{1}{2}} dx.$$

$$= 4\pi \left(\frac{1}{2\pi m K_B T}\right)^{3/2} (2\pi m K_B T)^{3/2} \sqrt{\frac{m K_B T}{2}} \int_0^\infty e^{-x} x dx$$

$$= \sqrt{\frac{8m K_B T}{\pi}} \times \Gamma(2) = \sqrt{\frac{8m K_B T}{\pi}}. \tag{3.39}$$

Equation (3.39) shows that the average momentum $\langle p \rangle = \sqrt{\frac{8m K_B T}{\pi}}$ is directly proportional to the square root of the absolute temperature T and is directly proportional to the square root of mass m of the ideal gas molecule. Further, with an increase in T, average momentum $\langle p \rangle$ increases. Alternatively, equation (3.39) can be obtained from the expression for average speed in the following way:

$$\langle p \rangle = m \times \langle v \rangle = m \times \sqrt{\frac{8 K_B T}{m\pi}} = \sqrt{\frac{8m K_B T}{\pi}}. \tag{3.40}$$

Hence, the result obtained is identical to that provided by equation (3.39).

Most probable momentum in three dimensions

Using equation (3.37), the expression for the most probable momentum p_m of the molecules in an ideal gas can be derived as follows:

$$\frac{d}{dp}\left[f(p)\right] = \frac{d}{dp}\left[4\pi\left(\frac{1}{2\pi m K_B T}\right)^{3/2} e^{-\left(\frac{p^2}{2m K_B T}\right)} p^2\right] = 0;$$

$$\Rightarrow 2p\, e^{-\left(\frac{p^2}{2m K_B T}\right)} + p^2\left(-\frac{p}{m K_B T}\right) e^{-\left(\frac{p^2}{2m K_B T}\right)} = 0;$$

$$\Rightarrow p\, e^{-\left(\frac{p^2}{2m K_B T}\right)}\left[2 - \frac{p^2}{m K_B T}\right] = 0.$$

This shows that either $p = 0$, and/or $e^{-\left(\frac{p^2}{2m K_B T}\right)} = 0$, that is, $p \to \infty$ and/or $\left[2 - \frac{p^2}{m K_B T}\right] = 0$, so $p = \pm\sqrt{2m K_B T}$. As we have considered the magnitude of momentum only, p cannot be negative, and hence, the most probable momentum is

$$p_m = \sqrt{2m K_B T}.$$

Root mean square momentum in three dimensions

To calculate the root mean square (r.m.s.) momentum of the molecules in an ideal gas, we proceed as follows:

The mean square momentum $\langle p^2 \rangle$ is given by

$$\langle p^2 \rangle = \frac{\displaystyle\int_0^\infty p^2\, N(p)\, dp}{\displaystyle\int_0^\infty N(p)\, dp} = \frac{\displaystyle\int_0^\infty p^2 \times 4\pi N\left(\frac{1}{2\pi m K_B T}\right)^{3/2} e^{-\left(\frac{p^2}{2m K_B T}\right)} p^2 dp}{\displaystyle\int_0^\infty 4\pi N\left(\frac{1}{2\pi m K_B T}\right)^{3/2} e^{-\left(\frac{p^2}{2m K_B T}\right)} p^2 dp}. \qquad (3.41)$$

Note that equation (3.37) has been used in the above expression. It should be mentioned further that the integral

$$\int_0^\infty 4\pi N\left(\frac{1}{2\pi m K_B T}\right)^{3/2} e^{-\left(\frac{p^2}{2m K_B T}\right)} p^2 dp$$

in the denominator of equation (3.41) gives the total number of molecules N. Using this result, the expression for the mean square momentum $\langle p^2 \rangle$ comes out to be

$$\langle p^2 \rangle = \frac{1}{N} \int_0^\infty p^2 \, 4\pi N \left(\frac{1}{2\pi m K_B T} \right)^{3/2} e^{-\left(\frac{p^2}{2m K_B T} \right)} p^2 \, dp$$

$$= \frac{1}{N} \times 4\pi N \left(\frac{1}{2\pi m K_B T} \right)^{3/2} \int_0^\infty e^{-\left(\frac{p^2}{2m K_B T} \right)} p^4 \, dp.$$

In order to evaluate the above integral, let us put

$$\frac{p^2}{2m K_B T} = x, \quad \Rightarrow \quad 2p \, dp = 2m K_B T \, dx \quad \text{and} \quad dp = \sqrt{\frac{m K_B T}{2}} x^{-\frac{1}{2}} dx.$$

Again, when $p \to 0$, $x \to 0$ and as $p \to \infty$, then $x \to \infty$.

Substituting these values in the above expression, we get the expression for mean square momentum $\langle p^2 \rangle$ as

$$\langle p^2 \rangle = 4\pi \left(\frac{1}{2\pi m K_B T} \right)^{3/2} \int_0^\infty (2\pi m K_B T)^2 \, x^2 \, e^{-x} \sqrt{\frac{m K_B T}{2}} \, x^{-\frac{1}{2}} \, dx.$$

$$= \frac{2\pi}{\pi^{3/2}} (2m K_B T) \int_0^\infty e^{-x} x^{3/2} \, dx = \frac{4}{\sqrt{\pi}} m K_B T \, \Gamma \left(\frac{5}{2} \right)$$

$$= \frac{4}{\sqrt{\pi}} m K_B T \times \frac{3}{2} \times \frac{1}{2} \sqrt{\pi} = 3m K_B T.$$

Hence, the r.m.s. momentum $p_{\text{r.m.s.}}$ is given by

$$p_{\text{r.m.s.}} = \sqrt{\langle p^2 \rangle} = \sqrt{3m K_B T}. \tag{3.42}$$

Equation (3.42) shows that the r.m.s. momentum $p_{\text{r.m.s.}} = \sqrt{3m K_B T}$ is directly proportional to the square root of the absolute temperature T and is directly proportional to the square root of mass m of the ideal gas molecule. With the increase in T, the r.m.s. momentum $p_{\text{r.m.s.}}$ also increases.

Variance in momentum in three dimensions

The variance in momentum σ_p^2 is given by

$$\sigma_p^2 = \langle p^2 \rangle - \langle p \rangle^2 = 3m K_B T - \frac{8m K_B T}{\pi} = m K_B T \left(3 - \frac{8}{\pi} \right) = 0.45 \, m K_B T.$$

As T increases, the variance σ_p^2 in the momentum distribution also increases.

Standard deviation in momentum in three dimensions

The standard deviation in the momentum distribution σ_p is given by

$$\sigma_p = \sqrt{\sigma_p^2} = \sqrt{\langle p^2 \rangle - \langle p \rangle^2} = \sqrt{3mK_BT - \frac{8mK_BT}{\pi}}$$

$$= \sqrt{mK_BT \left(3 - \frac{8}{\pi}\right)} = 0.67 \sqrt{mK_BT}.$$

With an increase in T, the standard deviation σ_p in the momentum distribution also increases.

Average energy per molecule from the momentum in three dimensions

The average kinetic energy of an ideal gas molecule in three dimensions can be obtained from the momentum distribution law in the following way:

$$\langle E \rangle = \left\langle \frac{p^2}{2m} \right\rangle = \frac{1}{2m} \langle p^2 \rangle = \frac{1}{2m} \times 3mK_BT = \frac{3}{2}K_BT.$$

This is the same result as obtained from the equipartition theorem.

3.3.3   Maxwell–Boltzmann energy distribution law from the velocity distribution law in three dimensions

The law of distribution of energy can be derived from the law of distribution of velocity given by equation (3.26), which is recast as

$$N(v)dv = 4\pi N \left(\frac{m}{2\pi K_BT}\right)^{\frac{3}{2}} e^{-\left(\frac{mv^2}{2K_BT}\right)} v^2 \, dv. \tag{3.43}$$

For a particle of mass m moving with velocity v, the energy (totally kinetic) is given by $E = \frac{1}{2}mv^2$. This gives $v^2 = \frac{2E}{m}$ and $dv = \frac{dE}{\sqrt{2mE}}$.

Using these values in the above equation, the number of particles $N(E)dE$ having energy between E and $E + dE$ can be written as

$$N(E)dE = 4\pi N \left(\frac{m}{2\pi K_BT}\right)^{\frac{3}{2}} e^{-\left(\frac{E}{K_BT}\right)} \left(\frac{2E}{m}\right) \frac{dE}{\sqrt{2mE}}.$$

After simplification, this reduces to

$$N(E)dE = \frac{2\pi N}{(\pi K_BT)^{\frac{3}{2}}} e^{-\left(\frac{E}{K_BT}\right)} E^{\frac{1}{2}} dE. \tag{3.44}$$

This is known as the Maxwell–Boltzmann law of distribution of energy among the ideal gas molecules. Variation of the number of molecules with energy between E and $E+dE$ as a function of energy E is shown in Figure 3.5 at two different temperatures T_1 (dotted line) and T_2 (solid line). The area under the kinetic energy distribution curve indicates the total number of molecules within a given range of kinetic energies. If the area under the curve is small, it represents a smaller number of molecules, and if the area under the curve is large, it indicates a larger number of molecules. Further, if temperature increases from T_1 to T_2, the kinetic energy of the molecules increases, and the distribution curve for kinetic energy shifts toward higher kinetic energy.

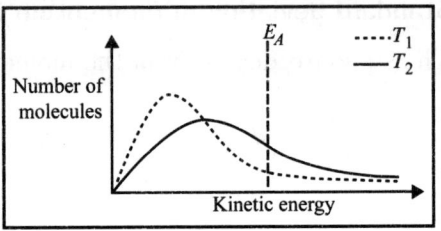

Figure 3.5 Variation of number of molecules as a function of energy at two different temperatures.

The activation energy E_A is defined as the minimum amount of energy required by the reactant molecules in order to collide successfully with the molecules of the products. This is shown by the dotted vertical line in Figure 3.5. This activation energy E_A is obtained from the Arrhenius equation: $K = A\, e^{-\left(\frac{E_A}{K_B T}\right)}$, where A is a constant. This is a mathematical function that gives the rate of a reaction as a function of the activation energy E_A and the temperature T.

At higher temperature T_2, the following characteristic features of the curve shown in Figure 3.5 are observed:

1. The area under the energy distribution curve to the right of E_A corresponding to the temperature T_2 is greater than the area under the curve to the right of E_A corresponding to the temperature T_1.

2. At the temperature T_2, a larger number of molecules possess energy greater than E_A compare to that of the curve at T_1.

3. As there are larger number of molecules above E_A at T_2 than at T_1, the reaction is, therefore, faster at higher temperature T_2 than at lower temperature T_1.

Calculation of the average energy $\langle E \rangle$ in three dimensions

Consider a system of N identical, independent, and distinguishable molecules of a gas occupying a volume V at a temperature T. Each molecule has a mass m. In classical statistical mechanics, these identical particles can be described by Maxwell–Boltzmann statistics. The average energy $\langle E \rangle$ of these particles is given by

$$\langle E \rangle = \frac{\int\limits_0^\infty E N(E)\,dE}{\int\limits_0^\infty N(E)\,dE}. \tag{3.45}$$

Using equation (3.44), we get

$$\langle E \rangle = \frac{\int_0^\infty E \frac{2\pi N}{(\pi K_B T)^{3/2}} e^{-\left(\frac{E}{K_B T}\right)} E^{\frac{1}{2}} dE}{\int_0^\infty \frac{2\pi N}{(\pi K_B T)^{3/2}} e^{-\left(\frac{E}{K_B T}\right)} E^{\frac{1}{2}} dE} = \frac{1}{N} \int_0^\infty E \frac{2\pi N}{(\pi K_B T)^{3/2}} e^{-\left(\frac{E}{K_B T}\right)} E^{\frac{1}{2}} dE.$$

This leads to

$$\langle E \rangle = \frac{1}{N} \frac{2\pi N}{(\pi K_B T)^{3/2}} \int_0^\infty e^{-\left(\frac{E}{K_B T}\right)} E^{3/2} dE = \frac{2\pi}{(\pi K_B T)^{3/2}} \frac{3\sqrt{\pi}}{4} (K_B T)^{5/2} = \frac{3}{2} K_B T.$$

(3.46)

Here, it should be noted that the integral

$$\int_0^\infty \frac{2\pi N}{(\pi K_B T)^{3/2}} e^{-\left(\frac{E}{K_B T}\right)} E^{1/2} dE$$

gives the total number of molecules N. Thus, it is observed from equation (3.46) that the average energy per molecule is $\frac{3}{2} K_B T$. A free particle has three degrees of freedom. This also signifies that the average energy per degree of freedom is $\frac{1}{2} K_B T$. This is the law of the equipartition of energy.

Calculation of the most probable energy $\langle E_m \rangle$ in three dimensions

Most probable energy is defined as the energy, which the maximum number of molecules possess. The number of molecules with energy in the range E and $E + dE$ is given by equation (3.44), which is recast as $N(E) dE = \frac{2\pi N}{(\pi K_B T)^{3/2}} e^{-\left(\frac{E}{K_B T}\right)} E^{1/2} dE$. From this expression, the number of molecules per unit energy interval comes out to be

$$N(E) = \frac{2\pi N}{(\pi K_B T)^{3/2}} e^{-\left(\frac{E}{K_B T}\right)} E^{1/2}. \qquad (3.47)$$

Here, the symbols have their usual meanings.

It is to be noted that the most probable energy is that energy at which the number of particles per unit energy interval is maximum. This can be achieved by differentiating equation (3.47) with respect to energy E and equating the result to zero.

$$\frac{d}{dE}\left[N(E)\right]_{E=E_m} = \frac{d}{dE}\left[\frac{2\pi N}{(\pi K_B T)^{3/2}} \, e^{-\left(\frac{E}{K_B T}\right)} E^{\frac{1}{2}}\right]_{E=E_m} = 0$$

$$\Rightarrow \left[\frac{2\pi N}{(\pi K_B T)^{3/2}}\right]\left[\left(-\frac{1}{K_B T}\right) e^{-\left(\frac{E}{K_B T}\right)} E^{\frac{1}{2}} + e^{-\left(\frac{E}{K_B T}\right)} \frac{1}{2} E^{-(1/2)}\right]_{E=E_m} = 0$$

$$\Rightarrow \left[\frac{2\pi N}{(\pi K_B T)^{3/2}}\right] e^{-\left(\frac{E}{K_B T}\right)} \left[\left(-\frac{1}{K_B T}\right) E^{(1/2)} + \frac{1}{2}\frac{1}{E^{1/2}}\right]_{E=E_m} = 0.$$

$$\Rightarrow \left[\frac{2\pi N}{(\pi K_B T)^{3/2}}\right] E^{1/2} e^{-\left(\frac{E}{K_B T}\right)} \left[\frac{1}{2} - \frac{E}{K_B T}\right]_{E=E_m} = 0.$$

This gives $E = 0$ and/or $E \to \infty$ and/or $E = \frac{1}{2}K_B T$. The first two solutions represent minima, and hence, the most probable energy E_m is given by

$$E_m = \frac{1}{2}K_B T. \tag{3.48}$$

The number of particles at this most probable energy E_m is given by

$$N(E_m) = \frac{2\pi N}{(\pi K_B T)^{3/2}} \, e^{-\left(\frac{E_m}{K_B T}\right)} E_m^{1/2}.$$

Using $E_m = \frac{1}{2}K_B T$, this number of particle $N(E_m)$ becomes

$$N(E_m) = \frac{2\pi N}{(\pi K_B T)^{3/2}} \, e^{-\frac{K_B T}{2 K_B T}} \left(\frac{1}{2}K_B T\right)^{1/2} = \sqrt{\frac{2}{\pi e}}\frac{N}{K_B T}. \tag{3.49}$$

Equations (3.48) and (3.49) provide interesting physical information that requires a "stop and think." The most probable energy E_m is proportional to the temperature of the gas and is independent of the number of particles. Thus, with the increase in T, the most probable energy E_m increases, but the number of particles at this energy decreases (see equation 3.49). Another interesting feature of equation (3.49) is the following: multiplying both sides of this equation with the most probable energy E_m, we get the following result:

$$E_m \times N(E_m) = \sqrt{\frac{2}{\pi e}}\frac{N}{K_B T} \times \frac{1}{2}K_B T = \frac{N}{\sqrt{2\pi e}}. \tag{3.50}$$

Equation (3.50) indicates that the product of the most probable energy E_m and the number of particles $N(E_m)$ at this energy depends only on the number of particles N of the gas. This is a very interesting result.

Some comments on the energy distribution law in three dimensions

The following points are to be noted for the energy distribution law in three dimensions:

1. This energy distribution law is independent of the mass of the molecules, and it is a function of temperature only.

2. When $\frac{E}{K_B T}$ is low, $N(E)dE$ varies with $\sqrt{\frac{E}{K_B T}}$, that is, parabolic about the horizontal energy axis. As $\frac{E}{K_B T}$ increases, it reaches to a maximum value corresponding to the given temperature, and then decreases exponentially.

3. The most probable energy is given by $E_m = \frac{1}{2}K_B T$.

4. The height of the peak is given by the value of $\frac{N(E)dE}{N}$ at $E = \frac{1}{2}K_B T$. This gives $\frac{N(E)dE}{N}$ at $E = E_m = \frac{1}{K_B T}\sqrt{\frac{2}{\pi e}}$.

5. With the increase in temperature, the position of the peak increases, whereas the height of the peak decreases. But the area under the curve remains the same. Thus, it is clear that with the increase in temperature, the probability of finding molecules having higher temperature increases. This is expected as average kinetic energy is proportional to temperature.

6. When temperature tends to infinity, the energy distribution will be along the horizontal axis, that is, along the energy axis. If the temperature is zero kelvin, then the distribution curve will be along the vertical axis, that is, along $\frac{N(E)dE}{N}$-axis.

7. The average energy of a molecule at a temperature T is $\langle E \rangle = \frac{3}{2}K_B T$.

8. The average of the square of energy is $\langle E^2 \rangle = \frac{15}{4}K_B^2 T^2$.

9. Variance in energy is given by $\sigma_E^2 = \langle E^2 \rangle - \langle E \rangle^2 = \frac{3}{2}K_B^2 T^2$.

10. Standard deviation in energy is given by $\sigma_E = \sqrt{\langle E^2 \rangle - \langle E \rangle^2} = \sqrt{\frac{3}{2}}K_B T$.

3.4 Distribution laws in two dimensions

In this section, the distribution laws of velocity, energy, and momentum of the molecules of an ideal gas in two dimensions are discussed in detail.

3.4.1 The velocity distribution law in two dimensions

We consider an ideal gas whose molecules are constrained to move in a two-dimensional space bounded by a curvilinear closed curve (rectangle, square, circle, ellipse, etc.). The velocity v of a molecule can be represented by a point in a two-dimensional velocity space ($v_x - v_y$ space). In terms of its components, this velocity v is expressed as $v^2 = v_x^2 + v_y^2$.

Following Maxwell's method of probabilities, as discussed earlier, we have

$$f(v_x) = a\, e^{-bv_x^2}; \quad \text{and} \quad f(v_y) = a\, e^{-bv_y^2},$$

where $f(v_x)dv_x$ and $f(v_y)dv_y$ signify the probabilities that a molecule has velocity components in the range v_x and $v_x + dv_x$ and v_y and $v_y + dv_y$, respectively. Then the probability that a molecule will have velocity components between v_x and $v_x + dv_x$ and v_y and $v_y + dv_y$ is given by

$$f(v_x)dv_x f(v_y)dv_y = f(v_x)f(v_y)dv_x dv_y = a\, e^{-bv_x^2} \times a\, e^{-bv_y^2} dv_x dv_y$$
$$= a^2\, e^{-b(v_x^2+v_y^2)} dv_x dv_y.$$

The constant "a" can be calculated in terms of the constant "b" in the following way: As the total probability is 1, we have

$$\int_{-\infty}^{+\infty} \int_{-\infty}^{+\infty} f(v_x)f(v_y)dv_x dv_y = 1; \quad \Rightarrow \quad a^2 \int_{-\infty}^{+\infty} e^{-bv_x^2} dv_x \int_{-\infty}^{+\infty} e^{-bv_y^2} dv_y = 1.$$

$$\Rightarrow \quad a^2 \left(\sqrt{\frac{\pi}{b}}\right)^2 = 1; \quad \Rightarrow \quad a = \sqrt{\frac{b}{\pi}}.$$

In terms of "b", the distribution function can then be expressed as

$$f(v_x)f(v_y)dv_x dv_y = \left(\frac{b}{\pi}\right) e^{-b(v_x^2+v_y^2)} dv_x dv_y. \tag{3.51}$$

The constant "b" can be evaluated from the expression for pressure P in the following way: Pressure P, defined as force per unit length of the bounding curve, is given by

$$P = 2mn \sum_0^\infty n_{v_x} v_x^2 dv_x; \quad \Rightarrow \quad P = 2m \int_0^\infty n f(v_x) v_x^2 dv_x;$$

$$P = 2mn \int_0^\infty \sqrt{\frac{b}{\pi}} e^{-bv_x^2} v_x^2 dv_x; \quad \Rightarrow \quad P = 2mn \sqrt{\frac{b}{\pi}} \frac{1}{4} \sqrt{\frac{\pi}{b^3}} = nK_BT,$$

where m is the mass of a molecule, and n is the number of molecules per unit area.

On simplification, we have

$$\frac{2K_BT}{m} = \frac{1}{b}; \quad \Rightarrow \quad b = \frac{m}{2K_BT}.$$

Using this value of b, we get the value of the constant a as

$$a = \sqrt{\frac{b}{\pi}} = \sqrt{\frac{m}{2\pi K_BT}}.$$

Using these values of "a" and "b", the velocity distribution law due to Maxwell in two dimensions is given by

$$f(v_x)f(v_y)dv_xdv_y = a^2 \, e^{-\left(\frac{mv^2}{2K_BT}\right)} \, dv_xdv_y = \frac{m}{2\pi K_BT} \, e^{-\left(\frac{mv^2}{2K_BT}\right)} \, dv_xdv_y.$$

In view of the **isotropy** of the distribution, we have to integrate over the polar angle θ in two-dimensional space. Integrating over the angle θ, ranging from 0 to 2π, we have

$$\int_0^{2\pi} \frac{m}{2\pi K_BT}e^{-\left(\frac{mv^2}{2K_BT}\right)}vdvd\theta = \frac{m}{K_BT}e^{-\left(\frac{mv^2}{2K_BT}\right)}vdv = f(v)dv, \qquad (3.52)$$

where $dv_xdv_y = vdvd\theta$ and $f(v)dv$ may be interpreted as the probability that a molecule chosen at random has a speed between v and $v + dv$. Equation (3.52) provides Maxwell's velocity distribution law in two dimensions.

Some comments about velocity distribution law in two dimensions

The following points are to be noted about the velocity distribution law of the molecules of an ideal gas in two dimensions.

1. For a particular gas and at a constant temperature, this distribution law varies with speed as $\sim v \, e^{-\left(\frac{mv^2}{2K_BT}\right)}$. At low speeds, the polynomial part dominates, and at high speeds, the exponential part dominates. Hence, $\frac{N(v)dv}{N}$ first increases linearly, it reaches a maximum value, and then decreases exponentially.

2. With the increase in temperature, the height of the peak decreases, and the probability of finding molecules with higher speed increases. But the area under the curve that represents the total probability remains the same and it is equal to 1.

3. When temperature tends to infinity, the distribution will be along the horizontal axis, that is, along the velocity axis. If the temperature is zero kelvin, then the distribution curve will be along the vertical axis, that is, along $\frac{N(v)dv}{N}$-axis.

4. With the increase in molar mass of the gas molecules, the height of the peak increases, and the probability of finding molecules having higher and higher speed decreases.

5. The average speed of a molecule is $\langle v \rangle = \frac{\pi K_B T}{m}$.

6. The average of the square of speed is $\langle v^2 \rangle = \frac{2K_B T}{m}$.

7. The most probable speed is given by $v_m = \sqrt{\frac{K_B T}{m}}$.

8. Variance in speed is given by $\sigma_v^2 = \langle v^2 \rangle - \langle v \rangle^2 = 0.43\frac{K_B T}{m}$.

9. Standard deviation in speed is given by $\sigma_v = \sqrt{\langle v^2 \rangle - \langle v \rangle^2} = 0.656\sqrt{\frac{K_B T}{m}}$.

3.4.2 The energy distribution law in two dimensions

Maxwell's speed distribution law in two dimensions is given by

$$f(v)dv = \frac{N(v)dv}{N} = \frac{m}{K_B T}\, v\, e^{-\left(\frac{mv^2}{2K_B T}\right)} dv, \qquad (3.53)$$

where $f(v)dv = \frac{N(v)dv}{N}$ is the probability or fraction of molecules having speed within the range v and $v + dv$, N is the total number of molecules, and m is the mass of a single molecule.

For the molecule of mass m moving with velocity v, the energy (totally kinetic) is given by $E = \frac{1}{2}mv^2$. This gives $v^2 = \frac{2E}{m}$ and $dv = \frac{dE}{\sqrt{2mE}}$.

Using these values in equation (3.53), the number of molecules $N(E)dE$ having energy between E and $E + dE$ in two dimensions can be written as

$$f(E)dE = \frac{N(E)dE}{N} = \left(\frac{1}{K_B T}\right) e^{-\left(\frac{E}{K_B T}\right)} dE. \qquad (3.54)$$

This equation (3.54) gives the Maxwell–Boltzmann law of distribution of energy among the ideal gas molecules in two dimensions.

Some comments about the energy distribution law in two dimensions

The following points are to be noted about the energy distribution law of the molecules of an ideal gas in two dimensions.

1. This energy distribution law is independent of the mass of the molecules, and is a function of temperature only.

2. When $E = 0$, $f(E) = \left(\frac{1}{K_B T}\right)$. With the increase in E, $f(E)$ decreases exponentially. As it does not pass through a maximum, there is nothing called most probable energy, but it is maximum at $E = 0$.

3. With the increase in temperature, value of $f(E)$ at $E = 0$ decreases, and the exponential decay becomes slower, but the area under the curve remains the same. Further, with the increase in temperature, probability of finding molecules having higher energy increases. This is expected as average kinetic energy is proportional to temperature.

4. When temperature tends to infinity, the distribution will be along the horizontal axis, that is, along the energy axis. If the temperature is zero kelvin, then the distribution curve will be along the vertical axis, that is, along the $f(E)$-axis.

5. The average energy of a molecule is $\langle E \rangle = K_B T$.

6. The average of the square of energy is $\langle E^2 \rangle = 2\, K_B^2 T^2$.

7. Variance in energy is given by $\sigma_E^2 = \langle E^2 \rangle - \langle E \rangle^2 = K_B^2 T^2$.

8. Standard deviation in energy is given by $\sigma_E = \sqrt{\langle E^2 \rangle - \langle E \rangle^2} = K_B T$.

3.4.3   The momentum distribution law in two dimensions

For a molecule of mass m moving with velocity v, momentum is given by $p = mv$. This gives $dv = \frac{dp}{m}$.

Using these values in equation (3.53), the number of molecules $N(p)dp$ having momentum between p and $p + dp$ in two dimensions can be written as

$$N(p)dp = N \left(\frac{p}{mK_B T}\right) e^{-\left(\frac{p^2}{2mK_B T}\right)} dp. \qquad (3.55)$$

This equation (3.55) gives the Maxwell–Boltzmann law of distribution of momentum among the ideal gas molecules in two dimensions.

Some comments about the momentum distribution law in two dimensions

The following points are to be noted about the momentum distribution law of the molecules of an ideal gas in two dimensions.

1. For a particular gas and at a particular temperature, if we plot $\frac{N(p)dp}{N}$ versus p, it gives a Gaussian distribution.

2. The distribution is peaked at $p = 0$, and the height of the peak is $\frac{1}{\sqrt{2\pi m K_B T}}$.

3. For a particular gas, with the increase in temperature, the peak height decreases, and the distribution spreads out.

4. When temperature tends to infinity, the distribution will be along the horizontal axis, that is, along the momentum axis. If the temperature is zero kelvin, then the distribution curve will be along the vertical axis, that is, along $\frac{N(p)dp}{N}$-axis.

5. The average momentum is $\langle p \rangle = 0$, as the distribution is symmetric.

6. The average of the square of momentum is $\langle p^2 \rangle = m K_B T$.

7. Variance in momentum is given by $\sigma_p^2 = \langle p^2 \rangle - \langle p \rangle^2 = m K_B T$.

8. Standard deviation in momentum is given by $\sigma_p = \sqrt{\langle p^2 \rangle - \langle p \rangle^2} = \sqrt{m K_B T}$.

3.5 Distribution laws in one dimension

We consider a hypothetical collection of molecules of a gas that moves along a particular direction, say, along x-direction. As the three dimensions are independent to each other, this is equivalent to consider the x-component of the velocity of a molecule that moves in arbitrary directions. For such a distribution, we would like to calculate the distribution functions for speed, momentum, and energy.

3.5.1 The velocity distribution law in one dimension

Maxwell's speed distribution law in one dimension is given by

$$f(v_x)dv_x = \frac{N(v_x)dv_x}{N} = \sqrt{\frac{m}{2\pi K_B T}}\ e^{-\left(\frac{mv_x^2}{2K_B T}\right)}dv_x, \qquad (3.56)$$

where $f(v_x)dv_x = \frac{N(v_x)dv_x}{N}$ is the probability or fraction of molecules having speed within the range v_x and $v_x + dv_x$, N is the total number of molecules, m is the mass of a single molecule, K_B is the Boltzmann constant, and T is the temperature in absolute scale.

Some comments about the velocity distribution law in one dimension

The following points are to be noted about the velocity distribution law of the molecules of an ideal gas in one dimension.

1. Maxwell's velocity distribution in one dimension is a Gaussian distribution or normal distribution.

2. With the increase in temperature, the height of the peak decreases, and the probability of finding molecules having higher velocity increases. But the area under the curve that represents the total probability remains the same, and it is equal to 1.

3. When temperature tends to infinity, the distribution will be along the horizontal axis, that is, along the velocity axis. If temperature is zero kelvin, then the distribution curve will be along the vertical axis, that is, along $\frac{N(v_x)dv_x}{N}$-axis.

4. With the increase in the molar mass of the gas molecules, the height of the peak increases, and the probability of finding molecules having higher and higher velocity decreases.

5. The distribution curve is symmetric about $v_x = 0$. And hence, the average velocity is $\langle v_x \rangle = 0$.

6. The average velocity along a particular direction, say, along $+x$ axis, is $\langle v_x \rangle = \frac{1}{4}\sqrt{\frac{8K_BT}{\pi m}} = \frac{1}{4}\langle v \rangle$, where $\langle v \rangle$ is the average speed of the molecules in three dimensions.

7. The average of the square of velocity is $\langle v_x^2 \rangle = \frac{K_BT}{m}$.

8. Variance in speed is given by $\sigma_{v_x}^2 = \langle v_x^2 \rangle - \langle v_x \rangle^2 = \frac{K_BT}{m}$.

9. Standard deviation in velocity is given by $\sigma_{v_x} = \sqrt{\langle v_x^2 \rangle - \langle v_x \rangle^2} = \sqrt{\frac{K_BT}{m}}$.

3.5.2 The energy distribution law in one dimension

For a molecule of mass m moving with velocity v_x, the energy (totally kinetic) is given by $E = \frac{1}{2}mv_x^2$. This gives $v_x^2 = \frac{2E}{m}$ and $dv_x = \frac{dE}{\sqrt{2mE}}$.

Using these values in equation (3.56), the number of molecules $N(E)dE$ having energy between E and $E + dE$ in one dimension can be written as

$$f(E)dE = \frac{N(E)dE}{N} = \left(\frac{1}{\pi K_BT}\right)^{\frac{1}{2}} E^{-\frac{1}{2}} e^{-\left(\frac{E}{K_BT}\right)} dE. \qquad (3.57)$$

This equation (3.57) gives the Maxwell–Boltzmann law of distribution of energy among the ideal gas molecules in one dimension.

Some comments about the energy distribution law in one dimension

The following points are to be noted about the energy distribution law of the molecules of an ideal gas in one dimension.

1. This energy distribution law is independent of the mass of the molecules, and it is a function of temperature only.

2. When E is very small, $f(E)$ varies with $= \frac{1}{\sqrt{E}}$. When E is much greater compared to the thermal energy $K_B T$, $f(E)$ decreases exponentially. As it does not pass through a maximum, there is nothing called most probable energy.

3. With the increase in temperature, the value of $f(E)$ at $E = 0$ decreases, and the exponential decay becomes slower, but the area under the curve remains the same. Further, with the increase in temperature, the probability of finding molecules having higher energy increases. This is expected as average kinetic energy is proportional to temperature.

4. When temperature tends to infinity, the distribution will be along the horizontal axis, that is, along the energy axis. If the temperature is zero kelvin, then the distribution curve will be along the vertical axis, that is, along $f(E)$-axis.

5. The average energy is $\langle E \rangle = \frac{1}{2} K_B T$.

6. The average of the square of energy is $\langle E^2 \rangle = \frac{3}{4} K_B^2 T^2$.

7. Variance in energy is given by $\sigma_E^2 = \langle E^2 \rangle - \langle E \rangle^2 = \frac{1}{2} K_B^2 T^2$.

8. Standard deviation in energy is given by $\sigma_E = \sqrt{\langle E^2 \rangle - \langle E \rangle^2} = \frac{1}{\sqrt{2}} K_B T$.

3.5.3   The momentum distribution law in one dimension

For a molecule of mass m moving with velocity v_x, momentum is given by $p = m \, v_x$. This gives $dv_x = \frac{dp}{m}$.

Using these values in equation (3.56), the number of molecules $N(p)dp$ having momentum between p and $p + dp$ in one dimension can be written as

$$N(p)dp = N \left(\frac{1}{2\pi m K_B T} \right)^{\frac{1}{2}} e^{-\left(\frac{p^2}{2m K_B T} \right)} dp. \qquad (3.58)$$

This equation (3.58) gives the Maxwell–Boltzmann law of distribution of momentum among the ideal gas molecules in one dimension.

Some comments about the momentum distribution law in one dimension

The following points are to be noted about the momentum distribution law of the molecules of an ideal gas in one dimension.

1. For a particular gas and at a particular temperature, if we plot $\frac{N(p)dp}{N}$ versus p, it gives a Gaussian distribution.

2. The distribution is peaked at $p = 0$ and the height of the peak is $\frac{1}{\sqrt{2\pi m K_B T}}$.

3. For a particular gas, with the increase in temperature, the peak height decreases, and the distribution spreads out.

4. When temperature tends to infinity, the distribution will be along the horizontal axis, that is, along the momentum axis. If the temperature is zero kelvin, then the distribution curve will be along the vertical axis, that is, along $\frac{N(p)dp}{N}$-axis.

5. The average momentum is $\langle p \rangle = 0$, as the distribution is symmetric.

6. The average of the square of momentum is $\langle p^2 \rangle = m K_B T$.

7. Variance in momentum is given by $\sigma_p^2 = \langle p^2 \rangle - \langle p \rangle^2 = m K_B T$.

8. Standard deviation in momentum is given by $\sigma_p = \sqrt{\langle p^2 \rangle - \langle p \rangle^2} = \sqrt{m K_B T}$.

3.6 Experimental verification of the Maxwell–Boltzmann law of distribution of velocities in an ideal gas

Up to now, Maxwell's speed distribution function for the molecules of an ideal gas has been theoretically derived and utilized to obtain expressions for average speed, most probable speed, and r.m.s. speed, average momentum, most probable momentum and r.m.s. momentum and average energy, most probable energy and r.m.s. energy in one, two, and three dimensions. The first direct experimental verification of Maxwell's speed distribution law was carried out by Stern in 1920. His experimental technique was subsequently modified by Zartman and Ko. There are some indirect pieces of evidence available in the literature as well. In this section, some of the direct or indirect experimental techniques with necessary theory are briefly outlined in favor of the distribution law of molecular speeds in an ideal gas.

3.6.1 Doppler broadening of spectral lines

We know that in an excited atom, electrons in the outermost orbit jump to still higher orbits. If the energy supplied from an external source is not sufficient to knock these electrons out, they fall back into their original positions, and the difference in energy is emitted in the form of radiation. If an atom at rest is excited, we expect that there will be no Doppler broadening. The spectral line will have an exceedingly small width. However, in actual practice, we observe a finite width of the spectral line. The width of the spectral line at which the intensity is half of its maximum value is known as **half-width** and is experimentally found to depend inversely on the square root of the molecular weight of the gaseous substance. This experimental result can be explained on the basis of Maxwell's speed distribution law for the electrons taking part in the transition from one state to another, and hence, this experimental result is considered as one of the direct experimental evidences of Maxwell's distribution law of molecular speeds.

Using this speed distribution law, the mathematical expression for half-width b is derived below, and is found to be consistent with the experimental findings.

The frequency of radiation emitted from an atom at rest is given by

$$\nu_0 = \frac{c}{\lambda_0}. \tag{3.59}$$

If we consider an assembly of electrons moving freely and behaving as the molecules of an ideal gas, the change in frequency will be given by Doppler's effect. To compute this change in frequency, we proceed as follows:

To make the calculation simple, we consider the motion of an electron along the X-axis. If the electron moves toward an observer with a velocity v_x, its frequency will be given by

$$\nu_0' = \nu_0 \left(1 + \frac{v_x}{c}\right), \tag{3.60}$$

where c is the velocity of light. On the other hand, if the electron moves away from the observer, its frequency will correspond to

$$\nu_0' = \nu_0 \left(1 - \frac{v_x}{c}\right). \tag{3.61}$$

Since v_x can have all possible values from 0 to ∞, all frequencies about ν_0 are possible. That is, the spectral line will have an infinite width. However, since the number of electrons having large velocity is very small, the intensity falls off rapidly about a central maximum.

To understand it better, we recall that the frequency is inversely proportional to wavelength. So, the changed frequency

$$\nu_0 \left(1 \pm \frac{v_x}{c}\right).$$

will correspond to the changed wavelength

$$\lambda_0 \mp x.$$

The value of x can be calculated in the following way. In the case of an electron moving toward the observer, we will have

$$\lambda_0 - x = \frac{c}{\nu_0 \left(1 + \frac{v_x}{c}\right)} = \frac{c}{\nu_0} \left(1 + \frac{v_x}{c}\right)^{-1}. \tag{3.62}$$

Using binomial expansion and retaining terms up to first order in $\frac{v_x}{c}$, we find that

$$\lambda_0 - x \approx \frac{c}{\nu_0}\left[1 - \frac{v_x}{c}\right] = \frac{c}{\nu_0} - \frac{v_x}{\nu_0} = \lambda_0 - \frac{v_x}{\nu_0}.$$

This leads to the value of x as

$$x = \frac{v_x}{\nu_0}. \tag{3.63}$$

This equation (3.63) suggests that the spread of a spectral line, taken care of by x, is related to the velocity v_x of the electrons. We also know that the intensity of a spectral line is proportional to the number of electrons taking part in the transition for the particular spectral line, that is, $I \propto N$. In order to determine the intensity of the spectral line, we should know the number of electrons $N(v_x)dv_x$ having velocity components between v_x and $v_x + dv_x$. This is obtained from Maxwell's speed distribution law. Hence, we have

$$N(v_x)dv_x = N\,a\,e^{-\left(\frac{mv_x^2}{2K_BT}\right)}dv_x.$$

Hence, the corresponding intensity due to these electrons is given by

$$I_x = I_0 e^{-\left(\frac{mv_x^2}{2K_BT}\right)} = I_0 e^{-\left(\frac{mx^2\nu_0^2}{2K_BT}\right)}.$$

The value of v_x from equation (3.63) has been used in the above expression. Let "b" be the value of x at which intensity becomes half of its maximum value, that is, $I_b = \frac{I_0}{2}$. This leads to

$$I_b = \frac{I_0}{2} = I_0 e^{-\left(\frac{mb^2\nu_0^2}{2K_BT}\right)}; \quad \Rightarrow \quad 2 = e^{\left(\frac{mb^2\nu_0^2}{2K_BT}\right)}.$$

Taking the natural logarithm of both sides of the above expression, we find that

$$\ln 2 = \frac{mb^2\nu_0^2}{2K_BT}; \quad \Rightarrow \quad b = \frac{1}{\nu_0}\sqrt{\frac{2K_BT\ln 2}{m}} = \frac{\lambda_0}{c}\sqrt{\frac{2RT\ln 2}{M_o}}, \tag{3.64}$$

where $M_o = mN_A$ is the molecular weight. Thus, it is observed from equation (3.64) that the half-width b of a spectral line is inversely proportional to the square root of the molecular weight M_o of the substance emitting it. This indicates that the intensity distribution of the spectral lines for hydrogen atoms (smaller molecular weight) should be diffused, whereas those for the lines of cadmium and mercury (larger molecular weight) should be sharp. This physical observation clearly provides the reason why heavier nuclei should be used for precision work in experimental techniques. This is in conformity with the experimental observation and lends support to Maxwell's speed distribution law.

3.6.2   Thermionic emission of electrons from metallic filaments

When a metallic wire is heated continuously by passing a current through it, electrons are emitted from the wire with different velocities. By considering these electrons as a Maxwellian gas, Richardson–Dushman obtained an expression for the current density J as a function of temperature T of the filament made of the metallic wire. Such current density J is given by

$$J = AT^2 e^{-\left(\frac{\phi}{K_B T}\right)}, \tag{3.65}$$

where A is an arbitrary constant and ϕ is the work function. This equation (3.65) qualitatively explains the phenomenon of thermionic emission and provides indirect evidence in favor of Maxwell's molecular speed distribution law.

3.6.3   Thermal neutrons in a nuclear reactor

Nuclear reactors provide a copious source of thermal neutrons. The thermal neutrons are produced by a process known as **nuclear fission** and moderated within the reactor to form a gas of thermal neutrons. To a reasonable approximation, this gas of thermal neutrons obeys the Maxwell–Boltzmann speed distribution law at a temperature equal to that of the moderating material. Typically, light or heavy water at a temperature above 300 K serves as the moderating material. A part of the moderator can be locally heated or cooled to produce, respectively, a hot or a cold source of thermal neutrons. As a result, highly energetic or less energetic thermal neutrons are produced within the reactor. The motion of these thermal neutrons provides indirect evidence in favor of Maxwell's molecular speed distribution law.

3.6.4   Stern's experiment

Maxwell's speed distribution (sometimes known as Maxwell–Boltzmann distribution law) is an equilibrium distribution law. In 1920, Stern carried out an experiment for verifying the Maxwell's law of speed distribution.[2] The experimental arrangement due to Stern is shown in Figure 3.6. The apparatus consists of an open vessel having a hot gas of cesium with a narrow horizontal opening hole. Cesium is taken as the source of atoms, and is heated in the oven placed in

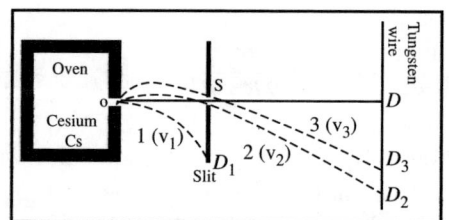

Figure 3.6 Stern's experimental arrangement to verify Maxwell's speed distribution law.

---

[2]O. Stern, Eine direkte Messung der thermischen Molekulargeschwindigkeit, *Z. Physik* 2 (1920): 49–56.

a long highly evacuated chamber ($P \sim 10^{-8}$ mm of Hg) so that the cesium atoms do not suffer any collision among themselves in the space. A nozzle slit S is placed at a distance of 1 meter from it. A thin tungsten wire is placed at a distance of 1 M from the slit S. It serves as a target. The entire arrangement is placed along one strictly horizontal line and is enclosed in a highly evacuated chamber.

The cesium atoms flow out of the oven through the nozzle "O". In the absence of gravitational field, only those atoms that emerge horizontally from the oven would be able to pass through the slit and would strike the collector (the tungsten wire) at D, irrespective of their velocities. However, due to gravitational field, the path of each cesium atom would be a parabola. The atoms (represented by path 1) emerging from the nozzle with a small velocity horizontally along the X-axis will not be able to pass through the narrow slit S and will not be able to reach the target D. It strikes the slit S at point "D_1". The atoms (represented by path 2) emerging

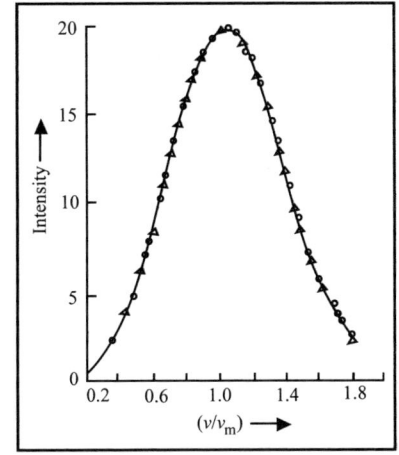

Figure 3.7 Variation of ionization current with $\frac{v}{v_m}$ in Stern's experiment.

from the nozzle at a small angle θ will be able to pass through the slit S and will strike the tungsten wire target at a point D_2. Similarly, the atoms, represented by path 3, will strike the tungsten wire target at the point D_3.

The tungsten wire target is heated by an electric current passing through it. When cesium atoms strike the wire target, they get ionized positively. These positively charged ions, leaving the target, are collected by the negatively charged cylinder surrounding the target (not shown in the diagram). Thus, an electric current of ions passes between the tungsten wire and the cylinder that can be measured with sufficient accuracy. The ionic current in the collecting cylinder gives the number of atoms hitting the target per unit time. Moving the target in a vertical direction at different positions, the ionic current, and hence, the number of atoms hitting the target is measured at different heights. It should be noted that different heights correspond to different velocities of the atoms. Hence, we get the number of atoms with different velocities by measuring currents at different heights. The measurement of the ion current as a function of velocity normalized to the most probable velocity is shown in Figure 3.7. This gives us the distribution of atoms with different values of velocities. This characteristic feature is in complete agreement with Maxwell's velocity distribution law.

3.6.5 Stern–Zartman experiment

The primary objective of this experiment was to determine the speeds of the molecules of a gas in thermal motion. This experiment was first carried out by O. Stern in 1920. Later, it was modified by Stern and Zartman and is known as the Stern–Zartman experiment. One of the achievements of this experiment was that it confirmed the validity of the foundations of the KTG. The experimental arrangement is shown in Figure 3.8.

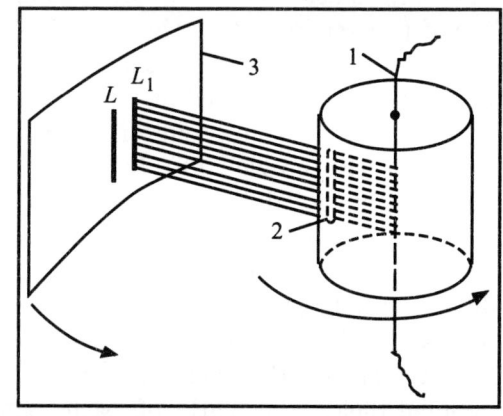

Figure 3.8 Schematic of the Stern–Zartman experiment: (1) Silver-coated platinum wire, (2) Slit that forms a beam of silver atoms, (3) Brass plate on which the silver atoms were deposited, (L) Denotes the location of the band of silver atoms deposited on the brass plate when the apparatus was stationary, (L_1) Denotes the location of the band of silver atoms deposited on the brass plate when the apparatus was rotated.

In this experiment, a rarefied silver vapor was used as a source of gas. This gas of silver vapor was produced by vaporizing the silver coating from the platinum wire heated by an electric current. This is shown by (1) in Figure 3.8. This platinum wire was placed in the middle of an air-evacuated vessel. This provided free escape of the silver atoms from the wire in all directions. A barrier with a slit (shown by 2 in Figure 3.8) was set up in the path of the escaping atoms and a narrow beam of escaping silver atoms was obtained. This narrow beam of silver atoms then impinged on a brass plate (shown by 3 in Figure 3.8) maintained at room temperature. The silver atoms escaping from the slit were deposited on the brass plate as a narrow band, and a silver image of the slit (shown by L in Figure 3.8) was formed on the brass plate. The entire apparatus was set in rapid rotation with the help of a special device about an axis parallel to the plane of the plate. Due to the rotation of the apparatus, the silver atoms impinged on another site (shown by L_1 in Figure 3.8) on the brass plate. The reason behind is that due to rotation, the plate shifted as the silver atoms traversed the distance from the slit to the plate (see Figure 3.8).

As the angular velocity ω of the apparatus increases, the shift of the plate increases. But the shift decreases with the increase in speed v of the silver atoms. By measuring the distance between L and L_1, the speed v of the silver atoms can be determined. Since the silver atoms move at different speeds, the band formed at the brass plate spreads out and becomes broader as the apparatus rotates. Further, it is known that the density of the silver atoms deposited at a given point of the band is proportional to the number of atoms moving with a given speed. Hence, the maximum density of the deposit corresponds to the most probable velocity of the atoms. Mapping of the densities at different points provided the number of silver atoms with their corresponding velocities. The values obtained in the Stern–Zartman experiment for the most probable and other velocities of the silver atoms were in good agreement with the theoretical values obtained from Maxwell's speed distribution law for the molecules.

3.6.6   Zartman–Ko experiment

In 1930, I. F. Zartman and C. C. Ko modified Stern's experimental technique to study the distribution of molecular velocities in a gas. The apparatus used by them is shown in Figure 3.9. This experimental arrangement provides a direct test of the Maxwell–Boltzmann velocity distribution law.

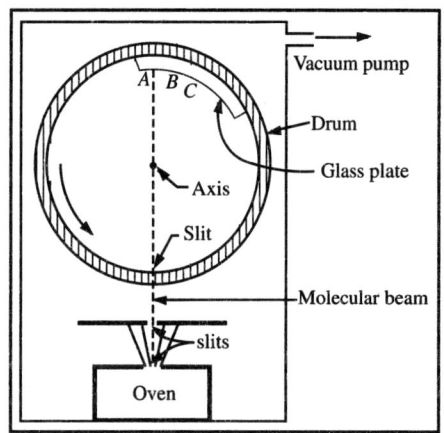

The experimental set-up used by Zartman and Ko consists of an oven that produces Cs (cesium) vapor at about 800°C. Some of these vaporized cesium atoms escape through a slit in the wall of the oven and are collimated by another slit S placed a short distance away from the oven.

Figure 3.9 Experimental arrangement due to Zartman–Ko to verify Maxwell's speed distribution law.

A drum is placed above the second slit S, and the drum is rotated about a horizontal axis nearly at 6000 rpm. A slit is also attached with the drum, and a glass plate is fitted on the inner surface just opposite to the slit. At some instant of time, all the slits will be in a same straight line, and a burst of Cs molecules enters the drum. These molecules reach the opposite face of the drum, and get deposited on the glass plate at various times depending upon their speeds.

When the drum is stationary, the beam of the Cs molecules entering into it through the slits will go straight and strike the glass plate at the same point (A in Figure 3.9). But when the drum rotates, the faster and slower molecules strike at different parts of the plate (B and C in Figure 3.9). Some molecules moving with still slower velocity strike at the slit S. From the resulting distribution of deposited

Cs molecules on the plate, it is possible to infer about the distribution of speeds of the molecules in the beam. This distribution is found to be in good agreement with the prediction of Maxwell–Boltzmann speed distribution law.

3.6.7 Limitations of the Maxwell–Boltzmann statistics

There are certain limitations on the Maxwell–Boltzmann statistics. These are listed below:

1. The Maxwell–Boltzmann statistics is only applicable to an isolated gas-molecular system in equilibrium with negligible mutual interaction and in dilute form. The separation between the molecules is large such that the individual molecule can be distinguished.

2. This distribution function cannot directly resolve the Gibbs paradox. To remove the Gibbs paradox, the expression has to be divided by N! to satisfy the additive property of entropy.

3. This statistics cannot be applied to sub-atomic particles, such as electrons or protons.

4. This statistics cannot be applied to photon. If applied to a photon gas, it predicts a continuously increasing number of photon as the frequency increases that does not satisfy Planck's law.

5. The problems encountered by applying the Maxwell–Boltzmann statistics are solved by the quantum statistics such as Bose–Einstein and Fermi–Dirac statistics. When the density of phase points in phase space is very small, all statistics merge to Maxwell–Boltzmann one at high temperatures.

6. In quantum mechanics, the particles considered are identical and indistinguishable, whereas in classical mechanics, the considered particles are identical and distinguishable. Actually, when the number of particles is much less than the number of available states, the indistinguishability of the particles is lost (as in the case of high temperature), then all quantum statistics reduce to Maxwell–Boltzmann statistics.

3.7 Law of equipartition of energy

The equipartition theorem states that *the total kinetic energy of a dynamical system consisting of a large number of particles in thermal equilibrium is equally divided among its all degrees of freedom, and the average energy associated with each degree of freedom is $\frac{1}{2}K_B T$, where T is the temperature in absolute scale and K_B is the Boltzmann constant.*

Let us discuss the equipartition theorem with the help of an example. Consider a box containing a number of ping-pong balls. Suppose that these balls are initially stationary. Now, some energy is randomly thrown into the box. This energy will be shared amongst the ping-pong balls such that they start to move about. These motions of the balls are irregular and completely random. This result is exactly the same as predicted by the equipartition theorem – the energy is shared evenly amongst the translational degrees of freedom X, Y, and Z. The equipartition theorem predicts that amongst the accessible modes of motion, the available energy will be shared evenly. This theorem can also make quantitative predictions about how much energy will appear in each degree of freedom. In particular, it states that on average, each quadratic degree of freedom will possess an amount of energy equal to $\frac{1}{2}K_B T$.

A "quadratic degree of freedom" is one for which the energy depends on the square of some property. For example, the kinetic and potential energies are associated with translational, rotational, and vibrational energy. Each translational degree of freedom has energy equal to $\frac{1}{2}mv^2$, each rotational degree of freedom has energy equal to $\frac{1}{2}I\omega^2$, and each vibrational degree of freedom has energy equal to $\frac{1}{2}mv^2$. These three types of degrees of freedom have a quadratic dependence on the velocity (or angular velocity in the case of rotation) and therefore, all follow the equipartition theorem. It is to be noted that when a harmonic oscillator vibrates in potential, say V, we consider both kinetic and potential energy, that is, the potential energy is counted as an additional degree of freedom. Further, we know that in a molecular system, vibrational motion is highly quantized, and most of the molecules are in their ground vibrational state at room temperature, and higher energy levels are not thermally accessible. Therefore, for such a system, the equipartition theorem has to be applied at very high temperatures.

3.7.1   Derivation of the law of equipartition of energy

In order to derive the law of equipartition of energy, the following assumptions are made:

1. The system consists of free particles only.

2. The energy of the system is the sum of the square of the position or momentum coordinates whose number is equal to the number of degrees of freedom.

3. The energy is not discrete and continuous in nature.

We consider a dynamical system with f degrees of freedom. Classically, the system is described by f position coordinates $(q_1,\ q_2,\ q_3, \cdots, q_f)$ and the corresponding f momenta coordinates $(p_1,\ p_2,\ p_3, \cdots, p_f)$. The total energy of the dynamical system can be written as a function of these position and momenta coordinates, that is,
$$E = E(q_1,\ q_2,\ q_3, \cdots, q_f, p_1,\ p_2,\ p_3, \cdots, p_f).$$

Let p_i be any particular momentum, and then the total energy can be expressed as

$$E = E_i(p_i) + E'(q_1, \ q_2, \ q_3, \cdots, q_f, p_1, \ p_2, \ p_3, \cdots, p_f), \qquad (3.66)$$

where $E_i(p_i)$ is the function of momentum p_i and the second term in equation (3.66) is the function of all the positions and the momenta coordinates excluding the momentum p_i.

Under this situation, the average energy of the particle with momentum p_i can be obtained classically as

$$\langle E_i \rangle = \frac{E}{N} = \frac{\displaystyle\int E_i(p_i) e^{-\frac{E}{K_B T}} dq_1, \ dq_2, \ dq_3, \cdots, dq_f, dp_1, \ dp_2, \ dp_3, \cdots, dp_f}{\displaystyle\int e^{-\frac{E}{K_B T}} dq_1, \ dq_2, \ dq_3, \cdots, dq_f, dp_1, \ dp_2, \ dp_3, \cdots, dp_f}. \qquad (3.67)$$

Since $E = E_i(p_i) + E'(q_1, \ q_2, \ q_3, \cdots, q_f, p_1, \ p_2, \ p_3, \cdots, p_f)$, we have

$$\langle E_i \rangle = \frac{\displaystyle\int E_i(p_i) e^{-\frac{E_i(p_i)+E'}{K_B T}} dq_1, \ dq_2, \ dq_3, \cdots, dq_f, dp_1, \ dp_2, \ dp_3, \cdots, dp_f}{\displaystyle\int e^{-\frac{E_i(p_i)+E'}{K_B T}} dq_1, \ dq_2, \ dq_3, \cdots, dq_f dp_1, \ dp_2, \ dp_3, \cdots, dp_f}. \qquad (3.68)$$

Equation (3.69) can be written as

$$\langle E_i \rangle = \frac{\displaystyle\int_{-\infty}^{\infty} E_i(p_i) e^{-\frac{E_i(p_i)}{K_B T}} dp_i}{\displaystyle\int_{-\infty}^{\infty} e^{-\frac{E_i(p_i)}{K_B T}} dp_i} \times \frac{\displaystyle\int e^{-\left(\frac{E'}{K_B T}\right)} dq_1, \ dq_2, \ dq_3, \cdots, dq_f dp_1, \ dp_2, \ dp_3, \cdots, dp_f}{\displaystyle\int e^{-\left(\frac{E'}{K_B T}\right)} dq_1, \ dq_2, \ dq_3, \cdots, dq_f dp_1, \ dp_2, \ dp_3, \cdots, dp_f}.$$

$$(3.69)$$

Here the integrals containing $e^{-\left(\frac{E'}{K_B T}\right)}$ extend overall $q's$ and $p's$ except p_i. Such a separation of integrals is possible because $E'(q_1, \ q_2, \ q_3, \cdots, q_f, p_1, \ p_2, \ p_3, \cdots, p_f)$ is independent of p_i. It is obvious that the integrals containing $e^{-\left(\frac{E'}{K_B T}\right)}$ are equal and cancel out. We therefore have

$$\langle E_i \rangle = \frac{\displaystyle\int_{-\infty}^{\infty} E_i(p_i) e^{-\frac{E_i(p_i)}{K_B T}} dp_i}{\displaystyle\int_{-\infty}^{\infty} e^{-\frac{E_i(p_i)}{K_B T}} dp_i}.$$

Again, E_i is quadratic in p_i, so we can write $E_i = \frac{p_i^2}{2\,m}$. Therefore, the above integral takes the form

$$\langle E_i \rangle = \frac{\displaystyle\int_{-\infty}^{\infty} \frac{p_i^2}{2\,m}\, e^{-\frac{p_i^2}{2\,m}\over K_B T}\, dp_i}{\displaystyle\int_{-\infty}^{\infty} e^{-\frac{p_i^2}{2\,m}\over K_B T}\, dp_i}.$$

The integrals in the numerator and denominator can be evaluated using gamma function. On simplification, we get

$$\langle E_i \rangle = \frac{1}{2}\, K_B T. \tag{3.70}$$

This expression shows that the average kinetic energy associated with single degree of freedom is $\frac{1}{2}\, K_B T$ per particle. This represents the law of equipartition of energy in classical statistical mechanics. For example, a free particle in three dimensions has three degrees of freedom, so its kinetic energy will be given by $E = \frac{3}{2}\, K_B T$. Thus, the equipartition theorem plays an important role in finding energy of a system if the number of degrees of freedom of the system is known.

3.7.2 Application of the law of equipartition of energy to specific heat

The equipartition theorem has been applied in different physical problems. The problem of specific heat is one of them. In Figure 3.10, variation of specific heat of hydrogen gas is shown as a function of temperature. The different regions, such as translational, rotational, and vibrational at different ranges of temperatures are indicated in the figure. It should be noted that the values of specific heat also change with the change of these regions. This problem of specific heat is discussed below within the framework of classical statistical mechanics.

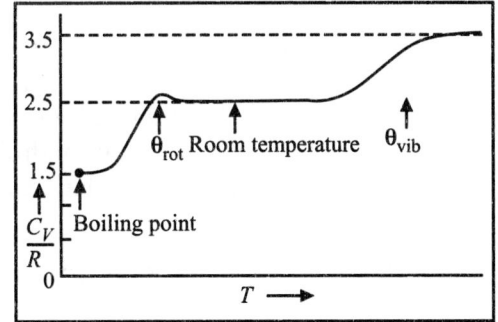

Figure 3.10 Variation of molar specific heat of hydrogen as a function of temperature T. Temperature is plotted along X-axis in a logarithmic scale. It is to be noted that hydrogen boils at 20 K.

Classical theory of specific heat (Dulong and Petit law)

We consider a crystal (solid) consisting of N_A (Avogadro's number) atoms at a particular temperature T. The specific heat at constant volume C_V (here, it is molar specific heat as we have considered N_A number of molecules) of this solid involves the calculation of the total energy at the temperature T. This energy is mostly due to the vibration of the atoms about their equilibrium positions. This vibrational energy of the crystal containing N_A atoms is equivalent with the energy of a system of $3N_A$ harmonic oscillators. The classical theory of specific heat of solid is based on the following assumptions:

1. Each solid consists of a large number of atomic particles executing harmonic motion about their equilibrium positions. They are called atomic oscillators.

2. The atomic oscillators vibrate with the same frequency, but their energies are different because they vibrate with different amplitudes.

3. The internal energy is largely contributed due to vibrational energy of the atomic oscillators.

4. There is no restriction on the energy values of the atomic oscillators, and it can take values ranging from zero up to infinity continuously.

5. The total energy E_T of the crystal at a given temperature T is N_A times the average energy $\langle E \rangle$ of the oscillator.

The classical calculation of specific heat assumes that each atom is a classical three-dimensional harmonic oscillator that executes simple harmonic motions independently about the mean position. Under these circumstances, one can calculate the total internal thermal energy of the crystal consisting of Avogadro's number of atoms N_A by finding out the average energy of a single classical oscillator and multiplying it by Avogadro's number N_A. The energy of a classical harmonic oscillator of natural angular frequency ω is given by

$$E = \frac{p_x^2}{2\,m} + \frac{1}{2}\,m\,\omega^2\,x^2. \tag{3.71}$$

Here, the first term on the right-hand side of equation (3.71) represents the kinetic energy (p_x is the momentum), and the second term represents the potential energy (x is the displacement from the equilibrium position, and m is the mass of the classical oscillator).

The average energy $\langle E \rangle$ of the oscillators can then be calculated using the basic concepts of classical statistical mechanics. This average energy $\langle E \rangle$ is given by

$$\langle E \rangle = \frac{\sum\limits_{0}^{\infty} E \, dN}{\sum\limits_{0}^{\infty} dN}, \tag{3.72}$$

where dN is the number of atomic oscillators having their energies lying between E and $E + dE$ at a given temperature T, and this number, according to classical statistical mechanics, is proportional to the Boltzmann factor $e^{-\left(\frac{E}{K_B T}\right)}$. Then, from equation (3.72), we can write

$$\langle E \rangle = \frac{\sum\limits_{0}^{\infty} E \, dN}{\sum\limits_{0}^{\infty} dN} = \frac{\sum\limits_{0}^{\infty} E \, e^{-\left(\frac{E}{K_B T}\right)}}{\sum\limits_{0}^{\infty} e^{-\left(\frac{E}{K_B T}\right)}}. \tag{3.73}$$

Using equation (3.71) into equation (3.73), we get

$$\langle E \rangle = \frac{\sum\limits_{0}^{\infty} \left(\frac{p_x^2}{2m} + \frac{1}{2} m \omega^2 x^2\right) e^{-\left(\frac{\left(\frac{p_x^2}{2m} + \frac{1}{2} m \omega^2 x^2\right)}{K_B T}\right)}}{\sum\limits_{0}^{\infty} e^{-\left(\frac{\left(\frac{p_x^2}{2m} + \frac{1}{2} m \omega^2 x^2\right)}{K_B T}\right)}}.$$

$$= \frac{\sum\limits_{0}^{\infty} \left(\frac{p_x^2}{2m}\right) e^{-\left(\frac{p_x^2}{2m K_B T}\right)} e^{-\left(\frac{\left(\frac{1}{2} m \omega^2 x^2\right)}{K_B T}\right)}}{\sum\limits_{0}^{\infty} e^{-\left(\frac{\left(\frac{p_x^2}{2m}\right)}{K_B T}\right)} e^{-\left(\frac{\left(\frac{1}{2} m \omega^2 x^2\right)}{K_B T}\right)}} + \frac{\sum\limits_{0}^{\infty} \left(\frac{1}{2} m \omega^2 x^2\right) e^{-\left(\frac{\left(\frac{1}{2} m \omega^2 x^2\right)}{K_B T}\right)} e^{-\left(\frac{p_x^2}{2m K_B T}\right)}}{\sum\limits_{0}^{\infty} e^{-\left(\frac{\left(\frac{1}{2} m \omega^2 x^2\right)}{K_B T}\right)} e^{-\left(\frac{p_x^2}{2m K_B T}\right)}}.$$

Replacing the summation by integration, we get

$$\langle E \rangle = \frac{\int\limits_{-\infty}^{\infty} \left(\frac{p_x^2}{2m}\right) e^{-\left(\frac{p_x^2}{2m K_B T}\right)} dp_x}{\int\limits_{-\infty}^{\infty} e^{-\left(\frac{\left(\frac{p_x^2}{2m}\right)}{K_B T}\right)} dp_x} + \frac{\int\limits_{-\infty}^{\infty} \left(\frac{1}{2} m \omega^2 x^2\right) e^{-\left(\frac{\left(\frac{1}{2} m \omega^2 x^2\right)}{K_B T}\right)} dx}{\int\limits_{-\infty}^{\infty} e^{-\left(\frac{\left(\frac{1}{2} m \omega^2 x^2\right)}{K_B T}\right)} dx}.$$

$$= \frac{K_B T}{2} + \frac{K_B T}{2} = K_B T.$$

Since each oscillator has three degrees of freedom, the classical value of the vibrational energy of the crystal containing N_A atoms is given by

$$E_T = 3 \, N_A \, \langle E \rangle = 3 \, N_A \, K_B T. \tag{3.74}$$

So, the specific heat at constant volume C_V is given by

$$C_V = \left(\frac{dE_T}{dT} \right)_V = 3 \, N_A \, K_B = 3 \, R = 3 \times 6.023 \times 10^{26} \times 1.38 \times 10^{-23}$$

$$= 25 \text{ kJ(k–mol)}^{-1} \text{K}^{-1}.$$

The expression $C_V = 3 \, R$ shows that the specific heat at constant volume is independent of temperature and is valid for most of the solids at high temperature. This is known as the Dulong–Petit law. Similarly, if the solid consists of N_A atoms of type A and N_A atoms of type B, then the specific heat of the system per mole would be given by $6 \, R$.

In the derivation of energy corresponding to each degree of freedom, the internal structure of the particles in the gaseous system has not been considered. Only the "translational" kinetic energy associated with the velocity of the molecules has been considered. This model is valid for monoatomic gases, such as helium and neon. But the molecules of most of the gases are composed of two or more atoms, and their arrangements are very important to form the molecules. Hence, the internal structure in such a diatomic molecule or triatomic molecule provides additional energy states for the molecule. These additional energy states are associated with the rotation and vibration of the molecule, and their energy can also be calculated using equipartition theorem. Further, we have to take into account the possibility of electronic excitation of one or both of the atoms in the molecule to calculate the total energy of the system.

Some comments about Dulong–Petit's law

Some important comments about Dulong–Petit's law are in order:

1. This law shows that the specific heat of solids is constant and independent of temperature.

2. The prediction of this law is in quantitative agreement with the experimental results at high temperatures only if sources of specific heat other than the lattice vibration are subtracted.

3. But experimentally, it is observed for most of the solids that the specific heat starts decreasing with the decrease in temperature at a particular temperature (Debye temperature) and approaches 0 as temperature tends to absolute zero. This experimental result could not be explained by Dulong–Petit law.

4. Dulong–Petit law also fails for light elements such as boron, beryllium, and carbon (as diamond), for which $C_V = 14$, 16, and 6.1 kJ(k–mol)$^{-1}$ K^{-1}, respectively, at room temperature. This is because the Debye temperature of these elements is much above 300 K.

These inadequacies indicate that something fundamental is missed out in the classical theory of specific heat as prescribed by Dulong–Petit.

3.7.3   Limitations of the law of equipartition of energy

The prediction of the equipartition theorem does not agree with the measured heat capacity of gases. Other drawbacks of the equipartition theorem are listed below:

1. It cannot predict the temperature variation of specific heat.

2. In 1907, Einstein pointed out that the anomaly in the specific heat was due to quantum effects. At low temperatures, some vibrational modes become "frozen" and do not absorb energy, and hence, do not contribute to the specific heat.

3. With the increase in temperature, the rotational motion of the molecules is activated first, and then the vibrational motion gets activated at still higher temperature. Thus, when the rotational modes are active, the vibrational modes of the molecules are not so active. Hence, almost no energy is used to increase the vibrational energy of the molecules, and the vibrational modes do not contribute to the heat capacity at low temperature.

4. If a system is isolated from the rest of the world (this happens in the case of a micro-canonical ensemble), the energy in each normal mode of such a system is constant. Energy is not transferred from one mode to another. For such a system, the equipartition theorem does not hold. The amount of energy in each normal mode remains fixed at its initial value.

It is recognized that all states with the same energy in an ergodic system are equally likely populated in thermal equilibrium. For such an ergodic system, the law of equipartition theorem holds good. Hence, for an ergodic system, it is possible to exchange energy among all its various forms within the system, or with an external heat bath in the canonical ensemble. Unfortunately, there are a very small number of physical systems that are rigorously proven to be ergodic. The hard-sphere system of Yakov Sinai is a famous example of ergodic system.

Why is the specific heat of a diatomic gas greater than that of a monoatomic gas? Discuss qualitatively.

Answer: The number of independent ways in which a gas molecule can move in three directions gives the degree of freedom. A monoatomic gas molecule can have translational motion in three directions. Thus, a monoatomic gas molecule has three degrees of freedom, whereas the diatomic gas molecule has five degrees of freedom, because in addition to translational motion, it has rotational motion also (clockwise and anticlockwise). Further, at higher temperatures it can vibrate about the axis joining the two atoms; thus, it has an additional degree of freedom. Hence, in total, it has six degrees of freedom at higher temperatures. If the same amount of heat is given to monoatomic and diatomic gas molecules, monoatomic gas molecules will spend all the heat energy in their translational motions while the diatomic molecules will spend a part of heat energy in their translational motions and a part of it in their rotational motions.

The temperature rise of a gas molecule results from translational kinetic energy of gas molecules. Thus, comparing monoatomic and diatomic gas molecules, for the same amount of heat given, the diatomic molecules get lesser amount of energy for translational motion. So, to have the same amount of temperature rise in diatomic gases, we have to provide more heat to it. Thus, a diatomic gas has more specific heat.

3.8   Solved numerical problems

1. Ten particles are moving with the speed of 2, 3, 4, 5, 5, 5, 6, 6, 7, and 9 m s^{-1}. Calculate the average speed, r.m.s. speed, and most probable speed.

   Answer: The average speed $\langle v \rangle$ of the ten particles is given by

   $$\langle v \rangle = \frac{2+3+4+5+5+5+6+6+7+9}{10} \ \text{ms}^{-1} = 5.2 \ \text{ms}^{-1}.$$

   To calculate the r.m.s. speed $v_{\text{r.m.s.}}$, let us first find out the mean square speed $\langle v^2 \rangle$ in the following way:

   $$\langle v^2 \rangle = \frac{2^2+3^2+4^2+5^2+5^2+5^2+6^2+6^2+7^2+9^2}{10} \ \text{ms}^{-1} = 30.6 \ \text{m}^2\text{s}^{-2}.$$

   So, the r.m.s. speed is $v_{\text{r.m.s.}} = \sqrt{30.6 \ \text{m}^2\text{s}^{-2}} = 5.53 \ \text{ms}^{-1}$.

   The most probable speed is 5 ms^{-1} because three of the ten particles have that speed.

2. Calculate the mean speed of a nitrogen molecule of mass m in a nitrogen gas at temperature of $T = 273$ K.

Answer: The mean speed of a gas molecule of mass m at a temperature T Kelvin is given by

$$\langle v \rangle = \sqrt{\frac{8K_BT}{\pi m}}.$$

Here, m is the mass of a nitrogen molecule and is given by $m = 28 \times (1.67 \times 10^{-24}$ g$) = 28 \times 1.67 \times 10^{-27}$ kg, where 28 is the molecular weight of nitrogen and $T = 273$ K. So, the mean speed of nitrogen molecule is

$$\langle v \rangle = \sqrt{\frac{8 \times (1.38 \times 10^{-23}) \times 273}{3.14 \times (28 \times 1.67 \times 10^{-27})}} = 4.53 \times 10^2 \text{ m s}^{-1}.$$

3. The mass of an oxygen molecule is 5.28×10^{-26} kg, and its mean velocity $\langle v \rangle$ at N.T.P. is 4.25×10^2 ms^{-1}. Calculate the average kinetic energy of an oxygen molecule at a temperature of 0°C.

Answer: Using Maxwell's speed distribution law, the expression for average speed and the r.m.s. speed of the molecules of an ideal gas are, respectively, given by

$$\langle v \rangle = \sqrt{\frac{8KT}{m\pi}} \quad \text{and} \quad v_{\text{r.m.s.}} = \sqrt{\frac{3KT}{m}}.$$

Taking the ratio of these two, we get

$$\frac{v_{\text{r.m.s}}}{\langle v \rangle} = \sqrt{\frac{\frac{3KT}{m}}{\frac{8KT}{m\pi}}} = \sqrt{\frac{3\pi}{8}}. \quad \text{This leads to} \quad v_{\text{r.m.s.}}^2 = \frac{3\pi}{8}(\langle v \rangle)^2.$$

So, the average kinetic energy of oxygen molecules

$$\frac{1}{2}mv_{\text{r.m.s.}}^2 = \frac{1}{2}m\frac{3\pi}{8}(\langle v \rangle)^2 = \frac{3m\pi}{16}(\langle v \rangle)^2$$
$$= \frac{3}{16} \times 3.14 \times 5.28 \times 10^{-26} \times (4.25 \times 10^2)^2 = 5.614 \times 10^{-21} \text{ J}.$$

4. In a sample of nitrogen gas, the molar mass of N_2 is 28.0 g mol^{-1}. Find the ratio of the number of N_2 molecules with a speed very close to 300 ms^{-1} to the number with a speed very close to 100 ms^{-1} at a temperature of 27°C.

Answer: Here, we have temperature $T = 300$ K, Boltzmann constant $K_B = 1.38 \times 10^{-23}$ JK^{-1}, and molecular weight of N_2, $M = 0.0280$ kg mol^{-1} so that the mass of one nitrogen molecule is $m = 4.65 \times 10^{-26}$ kg.

Substituting these values, we get the required ratio as

$$\frac{f(300 \text{ ms}^{-1})}{f(100 \text{ ms}^{-1})} = \frac{4\pi N\left(\frac{m}{2\pi K_B T}\right)^{3/2} (300 \text{ ms}^{-1})^2 \exp\left[-\frac{m(300 \text{ ms}^{-1})^2}{2K_B T}\right]}{4\pi N\left(\frac{m}{2\pi K_B T}\right)^{3/2} (100 \text{ ms}^{-1})^2 \exp\left[-\frac{m(100 \text{ ms}^{-1})^2}{2K_B T}\right]}$$

$$= 9 \times \frac{\exp\left[-\frac{4.65\times10^{-26} \text{ kg}\times(300 \text{ ms}^{-1})^2}{2\times1.38\times10^{-23} \text{ JK}^{-1}\times300 \text{ K}}\right]}{\exp\left[-\frac{4.65\times10^{-26} \text{ kg}\times(100 \text{ ms}^{-1})^2}{2\times1.38\times10^{-23} \text{ JK}^{-1}\times300 \text{ K}}\right]} = 5.74.$$

5. At the same temperature, two containers are filled with molecules of ideal gases have masses m_1 and m_2, respectively. The mean speed of the molecules of the second gas is 10 times the r.m.s. speed of the molecules of the first gas. Find the ratio of $\frac{m_1}{m_2}$ to the nearest integer.

Answer: The mean speed of the molecules of the second gas is $\langle v_2 \rangle = \sqrt{\frac{8K_B T}{\pi m_2}}$ and the r.m.s. speed of the molecules of the first gas is $\sqrt{\langle v_1^2 \rangle} = \sqrt{\frac{3K_B T}{m_1}}$.

According to the problem, we have

$$\sqrt{\frac{8K_B T}{\pi m_2}} = 10 \times \sqrt{\frac{3K_B T}{m_1}}; \quad \Rightarrow \quad \sqrt{\frac{m_1}{m_2}} = 10 \times \sqrt{\frac{3\pi}{8}};$$

$$\Rightarrow \quad \frac{m_1}{m_2} = \frac{300\pi}{8} = 117.75 \approx 118.$$

6. Show that Maxwell's speed distribution law is normalized, that is,

$$\int_0^\infty f(v)\, dv = 1.$$

Answer: According to Maxwell's speed distribution law, the number of molecules with speed between v and $v + dv$ is given by

$$N(v)dv = 4\pi N\left(\frac{m}{2\pi K_B T}\right)^{3/2} e^{-\left(\frac{mv^2}{2K_B T}\right)} v^2\, dv.$$

So, the fraction of molecules, that is, the probability distribution function for the velocity is given by

$$f(v)dv = \frac{N(v)dv}{N} = 4\pi \left(\frac{m}{2\pi K_B T}\right)^{3/2} e^{-\left(\frac{mv^2}{2K_B T}\right)} v^2\, dv.$$

Let us calculate the integral: $\displaystyle\int_0^\infty f(v)dv = \int_0^\infty 4\pi\left(\frac{m}{2\pi K_BT}\right)^{3/2} e^{-\left(\frac{mv^2}{2K_BT}\right)} v^2 \, dv.$

Let us put $\dfrac{mv^2}{2K_BT} = x;$ $\Longrightarrow 2vdv = \dfrac{2K_BT}{m}dx.$ So, $dv = \dfrac{K_BT}{m}\dfrac{dx}{\sqrt{x\frac{2K_BT}{m}}} =$

$\sqrt{\dfrac{K_BT}{2m}}\, x^{-1/2}dx.$

Using these substitutions, the integral becomes

$$\int_0^\infty f(v)dv = \int_0^\infty 4\pi\left(\frac{m}{2\pi K_BT}\right)^{3/2} e^{-x}\left(\frac{2K_BT}{m}x\right)\sqrt{\frac{K_BT}{2m}}\,x^{-1/2}dx.$$

$$= 4\pi\left(\frac{m}{2\pi K_BT}\right)^{3/2}\left(\frac{2K_BT}{m}\right)\sqrt{\frac{K_BT}{2m}}\int_0^\infty e^{-x}\,x^{1/2}\,dx$$

$$= 2\pi\,\frac{1}{\pi^{3/2}}\,\Gamma\left(\frac{3}{2}\right) = 2\pi\,\frac{1}{\pi^{3/2}}\,\frac{\sqrt{\pi}}{2} = 1.$$

This shows that Maxwell's speed distribution law is normalized.

7. For a Maxwellian gas, show that $\langle v\rangle \times \langle\frac{1}{v}\rangle = \frac{4}{\pi}.$

   Answer: According to Maxwell's speed distribution law, the number of molecules with speed between v and $v + dv$ is given by

   $$N(v)dv = 4\pi N\left(\frac{m}{2\pi K_BT}\right)^{3/2} e^{-\left(\frac{mv^2}{2K_BT}\right)}v^2\,dv.$$

   So the expression for $\langle\frac{1}{v}\rangle$ for such a Maxwellian gas is given by

   $$\left\langle\frac{1}{v}\right\rangle = \frac{1}{N}\int_0^\infty \frac{1}{v}N(v)dv = \frac{1}{N}\int_0^\infty 4\pi N\left(\frac{m}{2\pi K_BT}\right)^{3/2} e^{-\left(\frac{mv^2}{2K_BT}\right)}v^2 dv$$

   $$= 4\pi\left(\frac{m}{2\pi K_BT}\right)^{3/2}\int_0^\infty e^{-\left(\frac{mv^2}{2K_BT}\right)}v\,dv$$

   $$= 4\pi\left(\frac{m}{2\pi K_BT}\right)^{3/2}\frac{1}{2\left(\frac{m}{2K_BT}\right)} = \left(\frac{2m}{\pi K_BT}\right)^{1/2}.$$

   Again, the average speed $\langle v\rangle$ for a Maxwellian gas is given by $\langle v\rangle = \sqrt{\dfrac{8K_BT}{\pi m}}.$

Using these two values for $\langle v \rangle$ and $\langle \frac{1}{v} \rangle$, we get

$$\langle v \rangle \times \left\langle \frac{1}{v} \right\rangle = \sqrt{\frac{8K_B T}{\pi m}} \times \left(\frac{2\,m}{\pi K_B T} \right)^{1/2} = \frac{4}{\pi}.$$

8. Using Maxwell's speed distribution law, show that the fraction of molecules within the momentum range between p and $p + dp$ is given by

$$f(p)dp = 4\,\pi \left(\frac{1}{2\pi m K_B T} \right)^{3/2} e^{-\left(\frac{p^2}{2m K_B T} \right)} p^2 dp.$$

Answer: According to Maxwell's speed distribution law, the number of molecules within the range of speed between v and $v + dv$ is given by

$$N(v)dv = 4\pi N \left(\frac{m}{2\pi K_B T} \right)^{3/2} e^{-\left(\frac{mv^2}{2K_B T} \right)} v^2 dv. \implies \frac{N(v)dv}{N}$$

$$= 4\pi \left(\frac{m}{2\pi K_B T} \right)^{3/2} e^{-\left(\frac{mv^2}{2K_B T} \right)} v^2 dv.$$

The momentum p is given by $p = m\,v; \implies v = \frac{p}{m}$. So, $dv = \frac{dp}{m}$.

Hence, the fraction of number of molecules $f(p)dp$ within the momentum range between p and $p + dp$ is given by

$$f(p)dp = \frac{N(p)dp}{N} = 4\pi \left(\frac{m}{2\pi K_B T} \right)^{3/2} e^{(-)\frac{m}{2K_B T}\left(\frac{p^2}{m^2} \right)} \left(\frac{p^2}{m^2} \right) \frac{dp}{m}$$

$$= \frac{4\pi}{m^3} \left(\frac{m}{2\pi K_B T} \right)^{3/2} e^{-\left(\frac{p^2}{2m K_B T} \right)} p^2 dp$$

$$= 4\pi \left(\frac{1}{2\pi m K_B T} \right)^{3/2} e^{-\left(\frac{p^2}{2m K_B T} \right)} p^2 dp.$$

9. Using Maxwell's speed distribution law, show that the fraction of molecules within the energy range between E and $E + dE$ is given by

$$f(E)dE = \frac{N(E)dE}{N} = 2\,\pi \left(\frac{1}{\pi K_B T} \right)^{3/2} e^{-\left(\frac{E}{K_B T} \right)} E^{\frac{1}{2}} dE.$$

Answer: According to Maxwell's speed distribution law, the number of molecules within the speed range between v and $v + dv$ is given by

$$N(v)dv = 4\pi N\left(\frac{m}{2\pi K_B T}\right)^{3/2} e^{-\frac{mv^2}{2K_B T}} v^2 dv.$$

Within this formalism, the molecules of the gaseous system are treated as ideal. Hence, these molecules do not possess any potential energy. So, the energy possessed by the molecules is totally kinetic. The kinetic energy of a molecule is given by $E = \frac{1}{2}mv^2$.

Thus, $v^2 = \left(\frac{2E}{m}\right)$; $\implies v = \sqrt{\frac{2E}{m}}$. This gives $dv = \frac{1}{2}\left(\frac{2E}{m}\right)^{-1/2} 2\frac{dE}{m} = \left(\frac{2E}{m}\right)^{-1/2} \frac{dE}{m}$.

So, the number of molecules $N(E)dE$ within the energy range between E and $E + dE$ is given by

$$N(E)dE = 4\pi N\left(\frac{m}{2\pi K_B T}\right)^{3/2} e^{(-)\frac{m}{2K_B T}\left(\frac{2E}{m}\right)} \left(\frac{2E}{m}\right) \left(\frac{2E}{m}\right)^{-1/2} \frac{dE}{m}$$

$$= 4\pi N\left(\frac{m}{2\pi K_B T}\right)^{3/2} e^{(-)\frac{m}{2K_B T}\left(\frac{2E}{m}\right)} \left(\frac{2E}{m}\right)^{1/2} \frac{dE}{m}$$

$$= 2\pi N\left(\frac{1}{\pi K_B T}\right)^{3/2} e^{-\left(\frac{E}{K_B T}\right)} (E)^{1/2} \, dE.$$

Hence, the fraction of number of molecules within the energy range between E and $E + dE$ is given by

$$f(E)dE = \frac{N(E)dE}{N} = 2\pi\left(\frac{1}{\pi K_B T}\right)^{3/2} e^{-\left(\frac{E}{K_B T}\right)} E^{\frac{1}{2}} dE.$$

10. The Doppler broadening of spectral line increases with r.m.s. speed of the atoms in the source of light. Which one should give narrower spectral line: (i) a mercury 198 lamp at 300 K or (ii) a krypton 86 lamp at 77 K?

Answer: Mass of one Hg-molecule is given by $m_1 = \frac{198}{N_A}$ and its r.m.s. speed is $v_1 = \sqrt{\frac{3K_B T_1}{m_1}}$. Here, N_A is Avogadro's number.

Similarly, mass of one Kr-molecule is given by $m_2 = \frac{86}{N_A}$ and its r.m.s. speed is $v_2 = \sqrt{\frac{3K_B T_2}{m_2}}$. Hence, we get

$$\frac{v_1}{v_2} = \sqrt{\frac{T_1}{m_1} \times \frac{m_2}{T_2}} = \sqrt{\frac{300}{198} \times \frac{86}{77}} = 1.30.$$

Thus, $v_1 > v_2$. So, the Krypton lamp at 77 K will give narrower spectral line.

11. The de-Broglie wavelength λ of a particle of mass m moving with velocity v is given by $\lambda = \frac{h}{mv}$, where h is Planck's constant. Suppose that Maxwell's speed distribution law, in terms of this wavelength λ, is given as

$$F(\lambda)d\lambda = a\lambda^{-b}\exp\left(-\frac{c}{\lambda^2}\right)d\lambda,$$

then find the values of the constants a, b, and c.

Answer: Maxwell's speed distribution law is given by

$$f(v)dv = 4\pi \left(\frac{m}{2\pi K_B T}\right)^{3/2} e^{-\frac{mv^2}{2K_B T}} v^2 dv,$$

where symbols have their usual meaning. The de-Broglie wavelength λ is given by $\lambda = \frac{h}{mv}$ so that $v = \frac{h}{m\lambda}$. This gives rise to $dv = |\frac{h}{m\lambda^2}d\lambda|$. Using these values in the above expression, we get

$$F(\lambda)d\lambda = 4\pi \left(\frac{m}{2\pi K_B T}\right)^{3/2} e^{-\frac{m\left(\frac{h}{m\lambda}\right)^2}{2K_B T}} \left(\frac{h}{m\lambda}\right)^2 \frac{h}{m\lambda^2}d\lambda$$

$$= 4\pi \left(\frac{m}{2\pi K_B T}\right)^{3/2} \lambda^{-4} e^{-\frac{h^2}{2mK_B T\lambda^2}} d\lambda.$$

Comparing this expression with $F(\lambda)d\lambda = a\lambda^{-b}exp\left(-\frac{c}{\lambda^2}\right)d\lambda$, we get

$$a = 4\pi\left(\frac{h^2}{2\pi m K_B T}\right)^{3/2}; \quad b = 4 \quad \text{and } c = \frac{h^2}{2mK_B T}.$$

12. Calculate the fraction of number of the oxygen molecule with velocities between 199 ms^{-1} and 201 ms^{-1} at temperature 27°C.

Answer: From Maxwell's speed distribution law, we know that the fractional number of molecules within the speed range between v and $v + dv$ is given by

$$f(v)dv = \frac{N(v)dv}{N} = 4\pi\left(\frac{m}{2\pi K_B T}\right)^{\frac{3}{2}} e^{-\left(\frac{mv^2}{2K_B T}\right)}v^2 dv.$$

In this problem, we have the average velocity $v = \frac{199+201}{2}$ ms^{-1} = 200 ms^{-1} and $dv = (201 - 199)$ ms^{-1} = 2 ms^{-1}.

Mass of one oxygen molecule is $m = 32 \times 1.67 \times 10^{-27}$ kg, Boltzmann constant $K_B = 1.38 \times 10^{-23}$ JK^{-1}, and the temperature is $T = (27 + 273)$ K = 300 K.

Using these values in the above expression, we get

$$f(v)dv = 4 \times 3.14 \times \left(\frac{32 \times 1.67 \times 10^{-27}}{2 \times 3.14 \times 1.38 \times 10^{-23} \times 300} \right)^{\frac{3}{2}} e^{-\left(\frac{32 \times 1.67 \times 10^{-27} \times (200)^2}{2 \times 1.38 \times 10^{-23} \times 300} \right)}$$
$$\times (200)^2 \times 2 = 2.29 \times 10^{-3}.$$

13. Maxwell's speed distribution function $f(v)$ has the same value at $v_1 = 500$ ms^{-1} and $v_2 = 600$ ms^{-1} for nitrogen molecules. Determine the temperature at which this happens. Given: mass of one nitrogen molecule is $m = 28 \times 1.67 \times 10^{-27}$ kg and Boltzmann constant $K_B = 1.38 \times 10^{-23}$ JK^{-1}.

Answer: Let T be the temperature at which Maxwell's velocity distribution functions at these two velocities v_1 and v_2 have equal values, that is, $f(v_1) = f(v_2)$.

Using the expression for Maxwell's velocity distribution law, we have

$$4\pi \left(\frac{m}{2\pi K_B T} \right)^{3/2} e^{-\left(\frac{mv_1^2}{2K_B T} \right)} v_1^2 = 4\pi \left(\frac{m}{2\pi K_B T} \right)^{3/2} e^{-\left(\frac{mv_2^2}{2K_B T} \right)} v_2^2.$$
$$\Rightarrow e^{-\left(\frac{mv_1^2}{2K_B T} \right)} v_1^2 = e^{-\left(\frac{mv_2^2}{2K_B T} \right)} v_2^2.$$

This gives $\Rightarrow \frac{v_2^2}{v_1^2} = e^{-\left(\frac{mv_1^2}{2K_B T} \right)} e^{\frac{mv_2^2}{2K_B T}}. \quad \Rightarrow 2\ln\left(\frac{v_2}{v_1} \right) = \left(\frac{m}{2K_B T} \right) [v_2^2 - v_1^2];$

$$\Rightarrow T = \left(\frac{m}{4K_B} \right) \frac{v_2^2 - v_1^2}{\ln\left(\frac{v_2}{v_1} \right)}.$$

Substituting the values of the parameters, we get

$$T = \frac{28 \times 1.67 \times 10^{-27} \times (36 - 25) \times 10^4}{4 \times 1.38 \times 10^{-23} \ \ln\left(\frac{600}{500} \right)} = 511.08 \text{ K}.$$

14. The velocity and the energy distribution curves become maximum, respectively, at most probable speed v_m and most probable energy E_m. The energy corresponding to v_m is $\frac{1}{2} m v_m^2$. Explain whether E_m and $\frac{1}{2} m v_m^2$ are equal or not.

    Answer: For energy to be maximum, we must have

    $$\frac{d}{dE} \left[2\pi N \left(\frac{1}{\pi K_B T} \right)^{3/2} e^{-\left(\frac{E}{K_B T} \right)} (E)^{1/2} \right]_{E=E_m} = 0;$$

    $$\Rightarrow \frac{d}{dE} \left[e^{-\left(\frac{E}{K_B T} \right)} (E)^{1/2} \right]_{E=E_m} = 0.$$

    $$\Rightarrow \left[e^{-\left(\frac{E}{K_B T} \right)} \left[\frac{1}{2} (E)^{-1/2} - \frac{E^{1/2}}{K_B T} \right] \right]_{E=E_m} = 0; \Rightarrow E_m = \frac{1}{2} K_B T.$$

    This is the value of energy corresponding to the most probable energy E_m. But the energy corresponding to most probable speed v_m is given by

    $$E_{v_m} = \frac{1}{2} m v_m^2 = \frac{1}{2} m \left(\sqrt{\frac{2K_B T}{m}} \right)^2 = K_B T.$$

    Hence, it is observed that $E_m \neq E_{v_m}$.

    The reason behind this discrepancy can be stated in the following way: According to Maxwell–Boltzmann speed distribution law, the number of molecules within a certain range of velocity or energy is determined by the probabilities, and it should be independent of how we choose to represent them. Therefore, we must have $f(v)dv = f(E)dE$, not $f(v) = f(E)$. This is important because the relationship between velocity v and energy E is nonlinear; its sizes get distorted as we carry out the transformation. In fact, we have $v \propto E^{\frac{1}{2}}$ and this leads to $dv \propto \frac{1}{2}dE$. Hence, $dv \neq dE$. This means that at the corresponding energies and velocities, we have $f(v) \neq f(E)$. The relation $dv \propto \frac{1}{2}dE$ exactly leads to the observed discrepancy.

15. Assuming the energy distribution law for an ideal gas obeying Maxwell–Boltzmann distribution, calculate the total internal energy of the given gas.

    Answer: The energy distribution law for an ideal gas obeying Maxwell–Boltzmann distribution is given by

    $$N(E)dE = \frac{2N}{\sqrt{\pi}} \beta^{\frac{3}{2}} E^{\frac{1}{2}} e^{-\beta E} dE,$$

    where $\beta = \frac{1}{K_B T}$ and other symbols have their usual meaning.

The total internal energy E_T of the given gas is given by

$$E_T = \int\limits_0^\infty E N(E) dE = \int\limits_0^\infty E \frac{2N}{\sqrt{\pi}} \left(\frac{1}{K_B T}\right)^{\frac{3}{2}} E^{\frac{1}{2}} e^{-\frac{E}{K_B T}} dE$$

$$= \frac{2N}{\sqrt{\pi}} \left(\frac{1}{K_B T}\right)^{\frac{3}{2}} \int\limits_0^\infty E^{\frac{3}{2}} e^{-\frac{E}{K_B T}} dE = \frac{2N}{\sqrt{\pi}} \left(\frac{1}{K_B T}\right)^{\frac{3}{2}} \frac{3(K_B T)^2}{4} = \frac{3}{2} N K_B T.$$

Here, the value of the standard integral $\int\limits_0^\infty x^3 e^{-\alpha x} dx = \frac{3}{4\alpha} \sqrt{\frac{\pi}{\alpha}}$ has been used.

Hence, the total internal energy of the ideal gas obeying Maxwell–Boltzmann distribution is $E_T = \frac{3}{2} N K_B T$.

16. Using Maxwell's speed distribution function, calculate the mean value of the projection of velocity on X-axis $\langle v_x \rangle$ and the mean value of the modulus of this projection $\langle |v_x| \rangle$. Assume that the mass of each molecule is equal to m and the gas is at a temperature T.

Answer: The number of molecules with x-component of velocity between v_x and $v_x + dv_x$ is given by

$$N(v_x) dv_x = N \left(\frac{m}{2\pi K_B T}\right)^{1/2} e^{-\left(\frac{m v_x^2}{2 K_B T}\right)} dv_x.$$

So, the average of the x-component of velocity of a Maxwellian gas is given by

$$\langle v_x \rangle = \frac{1}{N} \int\limits_{-\infty}^\infty v_x N(v_x) dv_x = \frac{1}{N} \int\limits_{-\infty}^\infty v_x N \left(\frac{m}{2\pi K_B T}\right)^{1/2} e^{-\left(\frac{m v_x^2}{2 K_B T}\right)} dv_x = 0.$$

The average of the modulus of x-component of velocity of a Maxwellian gas is given by

$$\langle |v_x| \rangle = \frac{1}{N} \int\limits_{-\infty}^\infty |v_x| dN_{v_x} = 2 \times \frac{1}{N} \int\limits_0^\infty v_x N \left(\frac{m}{2\pi K_B T}\right)^{1/2} e^{-\left(\frac{m v_x^2}{2 K_B T}\right)} dv_x$$

$$= \sqrt{\frac{2 K_B T}{m \pi}}.$$

17. If v_x and v_y represent the two Cartesian components of velocity of a molecule in a gas, find the average value of $(av_x + bv_y)^2$ in terms of K_B, T, and m.

Answer: The molecules of an ideal gas obey Maxwell's law of velocity distribution at an equilibrium temperature T. The average value of $(\alpha v_x + \beta v_y)^2$ can be calculated in the following way:

$$\langle (\alpha v_x + \beta v_y)^2 \rangle = \langle (\alpha^2 v_x{}^2) \rangle + \langle (\beta^2 v_y{}^2) \rangle + 2 \langle (\alpha \beta v_x v_y) \rangle$$
$$= \alpha^2 \langle (v_x{}^2) \rangle + \beta^2 \langle (v_y{}^2) \rangle + 2\alpha\beta \langle (v_x v_y) \rangle .$$

Again, $\langle v_x \rangle = \int\limits_{-\infty}^{\infty}\int\limits_{-\infty}^{\infty}\int\limits_{-\infty}^{\infty} \left(\frac{m}{2\pi K_B T}\right)^{3/2} e^{-\left(\frac{m(v_x^2+v_y^2+v_z^2)}{2K_B T}\right)} v_x dv_x dv_y dv_z = 0.$

Again, $\langle v_x^2 \rangle = \int\limits_{-\infty}^{\infty}\int\limits_{-\infty}^{\infty}\int\limits_{-\infty}^{\infty} \left(\frac{m}{2\pi K_B T}\right)^{3/2} e^{-\left(\frac{m(v_x^2+v_y^2+v_z^2)}{2K_B T}\right)} v_x^2 dv_x dv_y dv_z = \frac{K_B T}{m}.$

Similarly, $\langle v_y \rangle = 0$ and $\langle v_y^2 \rangle = \frac{K_B T}{m}$.

So,

$$\langle (\alpha v_x + \beta v_y)^2 \rangle = \alpha^2 \langle v_x{}^2 \rangle + \beta^2 \langle v_y{}^2 \rangle + 2\alpha\beta \langle (v_x v_y) \rangle$$
$$= \alpha^2 \frac{K_B T}{m} + \beta^2 \frac{K_B T}{m} + 0 = (\alpha^2 + \beta^2) \frac{K_B T}{m}.$$

18. If v_x, v_y, and v_z represent the three Cartesian components of velocity of a molecule in a gas, find the average values of the following quantities in terms of K_B, T, and m.

(a) $\langle v_x \rangle$

(b) $\langle v_x^2 \rangle$

(c) $\langle v_x v_z \rangle$

(d) $\langle (v_x + bv_y)^2 \rangle$, where b is a constant.

Answer:

(a) From Maxwell's speed distribution law, we have

$$\langle v_x \rangle = \int_{-\infty}^{\infty} v_x f(v_x) dv_x = \int_{-\infty}^{\infty} v_x \left(\frac{m}{2\pi K_B T} \right)^{\frac{1}{2}} e^{-\left(\frac{mv_x^2}{2K_B T} \right)} dv_x$$

$$= \left(\frac{m}{2\pi K_B T} \right)^{\frac{1}{2}} \int_{-\infty}^{\infty} v_x e^{-\left(\frac{mv_x^2}{2K_B T} \right)} dv_x = 0. \text{ [This is an odd function.]}$$

(b) From Maxwell's speed distribution law, we have

$$\langle v_x^2 \rangle = \int_{-\infty}^{\infty} v_x^2 f(v_x) dv_x = \int_{-\infty}^{\infty} v_x^2 \left(\frac{m}{2\pi K_B T} \right)^{\frac{1}{2}} e^{-\left(\frac{mv_x^2}{2K_B T} \right)} dv_x$$

$$= \left(\frac{m}{2\pi K_B T} \right)^{\frac{1}{2}} \int_{-\infty}^{\infty} v_x^2 e^{-\left(\frac{mv_x^2}{2K_B T} \right)} dv_x$$

$$= \left(\frac{m}{2\pi K_B T} \right)^{\frac{1}{2}} \times \frac{1}{2} \times \sqrt{\frac{\pi}{\left(\frac{m}{2K_B T} \right)^3}} = \frac{K_B T}{m}.$$

(c) From Maxwell's speed distribution law, we have

$$\langle v_x v_z \rangle = \int_{-\infty}^{\infty} \int_{-\infty}^{\infty} v_x v_z f(v_x v_z) dv_x dv_z$$

$$= \int_{-\infty}^{\infty} \int_{-\infty}^{\infty} v_x v_z f(v_x) f(v_z) dv_x dv_z$$

$$= \int_{-\infty}^{\infty} v_x f(v_x) dv_x \times \int_{-\infty}^{\infty} v_z f(v_z) dv_z.$$

$$= \left(\frac{m}{2\pi K_B T} \right)^{\frac{1}{2}} \int_{-\infty}^{\infty} v_x e^{-\left(\frac{mv_x^2}{2K_B T} \right)} dv_x$$

$$\times \left(\frac{m}{2\pi K_B T} \right)^{\frac{1}{2}} \int_{-\infty}^{\infty} v_z e^{-\left(\frac{mv_z^2}{2K_B T} \right)} dv_z = 0 \times 0 = 0.$$

(d) Now, $\langle (v_x + bv_y)^2 \rangle = \langle v_x^2 \rangle + b^2 \langle v_y^2 \rangle + 2b\langle v_x \rangle \langle v_y \rangle$.

Using Maxwell's speed distribution law, if we calculate $\langle v_x \rangle$ and $\langle v_y \rangle$, we get the value 0 and for $\langle v_x^2 \rangle$ and $\langle v_y^2 \rangle$, we get the value $\frac{K_B T}{m}$, the where T is the temperature of the gas.

So, $\langle (v_x + bv_y)^2 \rangle = \frac{K_B T}{m} + b^2 \frac{K_B T}{m} + 0 = \frac{K_B T}{m}(1 + b^2)$.

19. Write down the expression for energy distribution function of a Maxwellian gas. Hence, find $\langle E \rangle$ and $\langle E^2 \rangle$.

    Answer: The kinetic energy of a free particle is given by

    $$E = \frac{1}{2}mv^2. \qquad \text{This provides} \qquad dv = \frac{dE}{(2mE)^{\frac{1}{2}}}.$$

    The expression for the velocity distribution of a Maxwellian gas is given by

    $$f(v)dv = 4\pi \left(\frac{m}{2\pi K_B T} \right)^{\frac{3}{2}} e^{-\frac{mv^2}{2K_B T}} v^2 dv.$$

    Using values of v and dv in terms of E and dE in the above expression, we get

    $$f(E)dE = \frac{2}{\sqrt{\pi}} \frac{1}{(K_B T)^{\frac{3}{2}}} e^{-\left(\frac{E}{K_B T} \right)} E^{\frac{1}{2}} dE.$$

    Hence, the average value of energy E can be written as

    $$\langle E \rangle = \int_0^\infty E f(E) dE = \int_0^\infty E \frac{2}{\sqrt{\pi}} \frac{1}{(K_B T)^{\frac{3}{2}}} e^{-\left(\frac{E}{K_B T} \right)} E^{\frac{1}{2}} dE.$$

    The integration leads to $\langle E \rangle = \frac{3}{2} K_B T$.

    Similarly, we get the average value of square of energy as

    $$\langle E^2 \rangle = \int_0^\infty E^2 f(E) dE = \int_0^\infty E^2 \frac{2}{\sqrt{\pi}} \frac{1}{(K_B T)^{\frac{3}{2}}} e^{-\left(\frac{E}{K_B T} \right)} E^{\frac{1}{2}} dE$$

    $$= \frac{2}{\sqrt{\pi}} (K_B T)^2 \times \frac{5}{2} \times \frac{3}{2} \times \frac{1}{2} \sqrt{\pi} = \frac{15}{4} (K_B T)^2.$$

20. Calculate the variance and standard deviation in energy in three dimensions.

    Answer: The variance in energy σ_E^2 in three dimensions is given by

    $$\sigma_E^2 = \langle E^2 \rangle - \langle E \rangle^2.$$

We know that $\langle E^2 \rangle = \frac{15}{4}(K_B T)^2$ and $\langle E \rangle = \frac{3}{2} K_B T$. Using these two values in the above expression, we get the variance in energy as

$$\sigma_E^2 = \langle E^2 \rangle - \langle E \rangle^2 = \frac{15}{4}(K_B T)^2 - \left(\frac{3}{2} K_B T\right)^2$$

$$= \frac{15}{4}(K_B T)^2 - \frac{9}{4}(K_B T)^2$$

$$= \frac{3}{2}(K_B T)^2.$$

Standard deviation in energy σ_E in three dimension is given by

$$\sigma_E = \sqrt{\sigma_E^2} = \sqrt{\frac{3}{2}(K_B T)^2} = \sqrt{\frac{3}{2}} K_B T.$$

21. What is the difference between the energy distribution curve and velocity distribution curve at the same equilibrium temperature?

Answer: The graphical variation of energy and velocity distribution curves are similar to each other. At a fixed temperature, the number of molecules in both cases increases, becomes maximum and decreases further with the increase in velocity.

Again, we know that $E = \frac{1}{2}mv^2; \implies \frac{dE}{E} = 2\frac{dv}{v}.$

This equation indicates that the relative change in E is double to that of v at the same equilibrium temperature. This implies that the spread of the energy distribution curve is greater indicating its flatter characteristics compared to the velocity distribution curve.

22. Consider Maxwell's distribution for velocities v of molecules at equilibrium temperature T in two dimensions.

(a) Find the average value of v at equilibrium temperature T.
(b) Find the average value of v^2 at equilibrium temperature T.
(c) Find the average value of $\frac{1}{v}$ at equilibrium temperature T.

Answer: Maxwell's distribution for velocities v of molecules at equilibrium temperature T in two dimensions is given by

$$f(v) = 2\pi \left(\frac{m}{2\pi K_B T}\right) e^{-\frac{mv^2}{2K_B T}} v \, dv.$$

(a) At equilibrium temperature T, the average value of velocity v of this two-dimensional gas is given by

$$\langle v \rangle = \int_0^\infty v f(v) dv = 2\pi \frac{m}{2\pi K_B T} \int_0^\infty v^2 \exp\left[-\frac{mv^2}{2K_B T}\right] dv$$

$$= \frac{m}{K_B T} \frac{1}{2} \frac{1}{\left(\frac{m}{2K_B T}\right)^{\frac{3}{2}}} \frac{3}{2} = \sqrt{\frac{\pi K_B T}{2m}}.$$

(b) At equilibrium temperature T, the average value of v^2 of this two-dimensional gas is given by

$$\langle v^2 \rangle = \int_0^\infty v^2 f(v) dv = 2\pi \frac{m}{2\pi K_B T} \int_0^\infty v^3 \exp\left[-\frac{mv^2}{2K_B T}\right] dv$$

$$\doteq \frac{m}{K_B T} \frac{1}{2} \frac{1}{\left(\frac{m}{2K_B T}\right)^2} = \frac{2K_B T}{m}.$$

(c) At equilibrium temperature T, the average value of $\frac{1}{v}$ of this two-dimensional gas is given by

$$\left\langle \frac{1}{v} \right\rangle = \int_0^\infty \frac{1}{v} f(v) dv = 2\pi \frac{m}{2\pi K_B T} \int_0^\infty \exp\left[-\frac{mv^2}{2K_B T}\right] dv$$

$$= \frac{m}{K_B T} \frac{1}{2} \frac{1}{\left(\frac{m}{2K_B T}\right)^{\frac{1}{2}}} \frac{1}{2} = \left(\frac{m\pi}{2K_B T}\right)^{\frac{1}{2}}.$$

23. Consider a gas of atoms obeying Maxwell–Boltzmann statistics. Calculate the average value of $e^{\vec{a}\cdot\vec{p}}$ over all the momenta $\vec{p}$ of each of the particles, where $\vec{a}$ is a constant vector, and a is the magnitude, m is the mass of each atom, T is the temperature in absolute scale, and K_B is the Boltzmann constant.

Answer: The average value of $\langle e^{\vec{a}\cdot\vec{p}} \rangle$ is given by

$$\langle e^{\vec{p}\cdot\vec{a}} \rangle = \int_{-\infty}^\infty \int_{-\infty}^\infty \int_{-\infty}^\infty f(p_x, p_y, p_z) e^{\vec{p}\cdot\vec{a}} dp_x dp_y dp_z,$$

where $f(p_x, p_y, p_z)$ is Maxwell probability distribution at a temperature T.

So, we have

$$\langle e^{\vec{p}.\vec{a}} \rangle = \int\limits_{-\infty}^{\infty} A_x e^{-\frac{p_x^2}{2mK_BT}} e^{p_x a_x} dp_x \int\limits_{-\infty}^{\infty} A_y e^{-\frac{p_y^2}{2mK_BT}} e^{p_y a_y} dp_y$$

$$\int\limits_{-\infty}^{\infty} A_z e^{-\frac{p_z^2}{2mK_BT}} e^{p_z a_z} dp_z. = e^{\frac{-(a_x^2+a_y^2+a_z^2)mK_BT}{2}} \int\limits_{-\infty}^{\infty} A_x e^{-\frac{(P_x-mK_BTa_x)^2}{2mK_BT}} dp_x$$

$$\times \int\limits_{-\infty}^{\infty} A_y e^{-\frac{(P_y-mK_BTa_y)^2}{2mK_BT}} dp_y \int\limits_{-\infty}^{\infty} A_z e^{-\frac{(P_z-mK_BTa_z)^2}{2mK_BT}} dp_z$$

$$= e^{\frac{-(a_x^2+a_y^2+a_z^2)mK_BT}{2}} .1.1.1 = e^{-\frac{1}{2}a^2 mK_BT}.$$

24. A bottle of volume 1 L is filled with hydrogen at N.T.P. Find the number of molecules in the energy range 0.0235 to 0.0236 eV.

Answer: The number of molecules in the energy range E to $E + dE$ is given by

$$N(E)dE = \frac{2N}{\sqrt{\pi}(K_BT)^{3/2}} E^{1/2} e^{-E/K_BT} dE.$$

There are 6×10^{23} molecules in 23.4 L at N.T.P. Thus, the number of molecules in 1 L is $N = \frac{6 \times 10^{23}}{22.4}$.

Also,

$$K_BT = \left(8.617 \times 10^{-5} \text{eV/K}\right) \times (273\text{K}) = 0.0235 \text{ eV}.$$

The energy range given is from 0.0235 to 0.0236 eV, so $dE = 0.0001$ eV. Writing this energy range as E to $E + dE$, so $E = K_BT$, or $E/(K_BT) = 1$. Also, $\frac{dE}{K_BT} = \frac{0.0001eV}{0.0235} = \frac{1}{235}$. Thus, we have

$$N(E)dE = \frac{2N}{\sqrt{\pi}} \left(\frac{E}{K_BT}\right)^{1/2} e^{-(E/K_BT)} \left(\frac{dE}{K_BT}\right)$$

$$= \frac{2 \times 6 \times 10^{23}}{22.4 \times \sqrt{3.14}} \times \frac{1}{2.718} \times \frac{1}{235} \approx 4.5 \times 10^{19}.$$

3.9   Multiple choice questions and answers

1. At thermal equilibrium, a gas satisfies Maxwell's velocity distribution law with the following parameters: V_{av} is the average velocity of the molecules; V_{mp} is the most probable velocity, and $V_{r.m.s.}$ is the r.m.s. velocity. Select the correct sequence for V_{av}, V_{mp}, and $V_{r.m.s.}$:

   (a) $V_{av} > V_{r.m.s.} > V_{mp}$

   (b) $V_{r.m.s.} > V_{mp} > V_{av}$

   (c) $V_{av} > V_{mp} > V_{r.m.s.}$

   (d) $V_{r.m.s.} > V_{av} > V_{mp}$

   Answer: The correct choice is (a).

2. The most probable velocity of the molecules of an ideal gas increases by a factor of 4 when its temperature is increased by some factor. By what factor will the r.m.s. velocity increase due to the same increase in temperature?

   (a) $\sqrt{3}/2$ (c) 4

   (b) 2   (d) 16

   Answer: The correct choice is (c).

   Solution: We know that the r.m.s. speed and the most probable speed are, respectively, given by $V_{r.m.s.} = \sqrt{\frac{3K_BT}{m}}$ and $V_{mp} = \sqrt{\frac{2K_BT}{m}}$. Thus, we see that both V_{mp} and $V_{r.m.s.}$ are proportional to $\sqrt{T}$. Hence, $V_{r.m.s.}$ will also increase by a factor of 4.

3. The most probable velocity and the r.m.s. velocity of an ideal gas obeying Maxwellian distribution of the velocity are, respectively, V_{mp} and $V_{r.m.s.}$. The magnitude of the ratio $\frac{V_{r.m.s.}}{V_{mp}}$ is

   (a) $\sqrt{3/2}$

   (b) $3/2$

   (c) $\sqrt{2/3}$

   (d) $2/3$

   Answer: The correct choice is (a).

   Solution: For a Maxwellian distribution of velocity, we know

   $$V_{mp} = \sqrt{\frac{2K_BT}{m}} \text{ and } V_{r.m.s.} = \sqrt{\frac{3K_BT}{m}}. \text{ So, } \frac{V_{r.m.s.}}{V_{mp}} = \sqrt{\frac{3}{2}}.$$

4. The speed v of the molecules of mass m of an ideal gas obeys Maxwell's velocity distribution law at an equilibrium temperature T. Let $(V_x, V_y,$ and $V_z)$ be the components of the velocity and K_B the Boltzmann constant. Here, α and β are constants. The average value of $(\alpha V_x - \beta V_y)^2$ is

(a) $(\alpha^2 - \beta^2)\left(\frac{K_B T}{m}\right)$

(b) $(\alpha^2 + \beta^2)\left(\frac{K_B T}{m}\right)$

(c) $(\alpha + \beta)^2 \left(\frac{K_B T}{m}\right)$

(d) $(\alpha - \beta)^2 \left(\frac{K_B T}{m}\right)$

Answer: The correct choice is (b).

Solution: Ideal gas obeys Maxwell's velocity distribution law at equilibrium temperature. Then average value of $(\alpha V_x - \beta V_y)^2$ can be calculated in the following way:

The average of the quantity is given by

$$\left\langle \left(\alpha V_x - \beta V_y\right)^2 \right\rangle = \left\langle \left(\alpha^2 V_x^{\,2}\right)\right\rangle + \left\langle \left(\beta^2 V_y^{\,2}\right)\right\rangle - 2\left\langle \left(\alpha\beta V_x V_y\right)\right\rangle.$$

But, we know that

$$\langle (V_x)\rangle = 0, \ \langle (V_y)\rangle = 0 \text{ and } \left\langle (V_x^{\,2})\right\rangle = \left\langle (V_y^{\,2})\right\rangle = \frac{K_B T}{m}.$$

The above average can then be calculated as:

$$\left\langle (\alpha V_x - \beta V_y)^2 \right\rangle = \alpha^2 \left\langle \left(V_x^{\,2}\right)\right\rangle + \beta^2 \left\langle \left(V_y^{\,2}\right)\right\rangle - 2\alpha\beta \, \langle (V_x)\rangle \, \langle (V_y)\rangle$$

$$= \alpha^2 \frac{K_B T}{m} + \beta^2 \left(\frac{K_B T}{m}\right) 2\alpha\beta \times 0 \times 0 = (\alpha^2 + \beta^2)\left(\frac{K_B T}{m}\right).$$

5. The speed v of the molecules of mass m of an ideal gas obeys Maxwell's velocity distribution law at an equilibrium temperature T. Let $(V_x, V_y,$ and $V_z)$ be the components of the velocity and K_B the Boltzmann constant. Here, α and β are constants. The average value of $(\alpha V_x V_y)^2$ is

(a) 0

(b) $(\alpha^2)\left(\frac{K_B T}{m}\right)^2$

(c) $(\alpha^2)\left(\frac{K_B T}{2m}\right)^2$

(d) $(\alpha^2)\left(\frac{2K_B T}{m}\right)^2$

Answer: The correct choice is (b).

Solution: We know that an ideal gas obeys Maxwell's velocity distribution law at equilibrium temperature. Then average value of $(\alpha V_x V_y)^2$ can be calculated in the following way:

We have $\left\langle (\alpha V_x V_y)^2)^2 \right\rangle = \alpha^2 \left\langle \left(V_x{}^2 V_y{}^2 \right) \right\rangle$ and $\left\langle (V_x{}^2) \right\rangle = \left\langle (V_y{}^2) \right\rangle = \left\langle (V_z{}^2) \right\rangle = \frac{K_B T}{m}$.

Then, the average value of the above quantity becomes

$$\left\langle \left((\alpha V_x V_y)^2 \right)^2 \right\rangle = \alpha^2 \left\langle (V_x{}^2) \right\rangle \left\langle (V_y{}^2) \right\rangle = \alpha^2 \left(\frac{K_B T}{m} \right) \left(\frac{K_B T}{m} \right)$$

$$= \alpha^2 \left(\frac{K_B T}{m} \right)^2.$$

6. A gas molecule having mass m is in thermal equilibrium at a temperature T. Let v_x, v_y, v_z be the Cartesian components of velocity $\vec{v}$ of a molecule. The mean value of $(v_x - \alpha v_y + \beta v_z)^2$ is   [JAM 2010]

(a) $(1 + \alpha^2 + \beta^2) \frac{K_B T}{m}$

(b) $(1 - \alpha^2 + \beta^2) \frac{K_B T}{m}$

(c) $(\alpha^2 + \beta^2) \frac{K_B T}{m}$

(d) $(\beta^2 - \alpha^2) \frac{K_B T}{m}$

Answer: The correct choice is (a).

$$(v_x - \alpha v_y + \beta v_z)^2 = v_x^2 + \alpha^2 v_y^2 + \beta^2 v_y^2 - 2\alpha v_x v_y + 2\beta v_z v_x - 2\beta v_y v_z$$

$$< (v_x - \alpha v_y + \beta v_z)^2 > = < v_x^2 > + \alpha^2 < v_y^2 > + \beta^2 < v_y^2 >$$

$$- 2\alpha < v_x >< v_y > + 2\beta < v_z >< v_x > - 2\beta\alpha < v_y >< v_z >;$$

$$< v_x^2 > = < v_y^2 > = < v_z^2 > = \frac{K_B T}{m} \text{ and } < v_x > = < v_y > = < v_z > = 0$$

$$< (v_x - \alpha v_y + \beta v_z)^2 > = (1 + \alpha^2 + \beta^2) \frac{K_B T}{m}.$$

7. A hypothetical speed distribution for a sample of a gas with N number of particles is shown in Figure 3.11. The distribution function $P(V)$ is given by

(a) $P(V) = \frac{a}{V_0} V$ for $0 \leq V \leq V_0$.

(b) $P(V) = a$ for $V_0 \leq V \leq 2V_0$.

(c) $P(V) = 0$ for $V \geq 2V_0$.

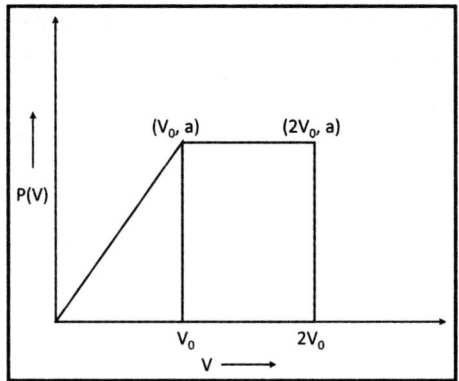

Figure 3.11 Figure for MCQ problem 7.

The number of particles having speeds between $1.2V_0$ and $1.9V_0$ is given by

(a) $\frac{N}{5}$

(c) $\frac{2N}{21}$

(b) $\frac{7N}{15}$

(d) None of the above

Answer: The correct choice is (b).

Solution: Since the total probability is one, the area under the curve should be one. This indicates that

$$\implies \frac{1}{2}aV_0 + a\left(2V_0 - V_0\right) = 1. \implies \frac{3}{2}aV_0 = 1. \implies a = \frac{2}{3V_0}.$$

Now, the area between $1.2V_0$ and $1.9V_0$ is given by $\frac{2}{3V_0} \times (1.9V_0 - 1.2V_0)$. So, the number of molecules having speeds between $1.2V_0$ and $1.9V_0$ is given by $N \times \frac{2}{3} \times (0.7) = \frac{7N}{15}$.

8. A rigid triangular molecule consists of three non-collinear atoms joined by rigid rods. The constant pressure molar specific heat (C_P) of an ideal gas consisting of such molecules is   [JAM 2015]

(a) $6R$

(b) $5R$

(c) $4R$

(d) $3R$

Answer: The correct choice is (c).

Degrees of freedom = 6. So the internal energy is $\Rightarrow U = \frac{6RT}{2}$.

This gives rise to $\Rightarrow C_V = (\frac{\partial U}{\partial T})_V = 3R \Rightarrow C_p = C_V + R = 4R$.

9. A gas of molecular mass m is at temperature T. If the gas obeys Maxwell–Boltzmann velocity distribution, the average speed of molecules is given by   [JAM 2011]

(a) $\sqrt{\frac{K_B T}{m}}$

(b) $\sqrt{\frac{2K_B T}{m}}$

(c) $\sqrt{\frac{2K_B T}{\pi m}}$

(d) $\sqrt{\frac{8K_B T}{\pi m}}$

Answer: The correct choice is (d).

10. A hypothetical speed distribution for a sample of N gas particles is the following: $P(V) = 0$ for $V > 2V_0$. How many particles have speeds between $1.2V_0$ and $1.9V_0$?

(a) $\frac{N}{5}$ (c) $\frac{2N}{21}$

(b) $\frac{7N}{15}$ (d) None of the above

Answer: The correct choice is (b).

Solution: Since the total probability is one, the area of the figure should be one. This indicates that

$$\implies \frac{1}{2}aV_0 + a(2V_0 - V_0) = 1. \implies \frac{3}{2}aV_0 = 1. \implies a = \frac{2}{3V_0}.$$

Now, the area between $1.2V_0$ and $1.9V_0$ is given by $\frac{2}{3V_0} \times (1.9V_0 - 1.2V_0)$. So, the number of molecules having speeds between $1.2V_0$ and $1.9V_0$ is given by $N \times \frac{2}{3} \times (0.7) = \frac{7N}{15}$.

11. An ideal gas consists of three-dimensional poly-atomic molecules. The temperature is such that one vibrational mode is excited. If R denotes the gas constant, then the specific heat at constant volume of one mole of the gas at this temperature is   [JAM 2018]

(a) $3R$ (c) $4R$

(b) $\frac{7}{2}R$ (d) $\frac{9}{2}R$

Answer: The correct choice is (c).

Solution: For a polyatomic gas, $C_P = (4 + f)R$; $C_V = (3 + f)R$, as $f = 1, C_V = 4R$.

12. A vessel of volume V contains a mixture of 1 mole of hydrogen and 1 mole of oxygen (both considered as ideal). Let $f_1(v)dv$ denote the fraction of hydrogen molecules with speed between v and $(v+dv)$, and $f_2(v)dv$ denote the same for oxygen molecules. Then

(a) $f_1(v) + f_2(v) = f(v)$ obeys Maxwell's distribution law.

(b) $f_1(v)$, $f_2(v)$ will obey Maxwell's distribution law separately.

(c) Neither $f_1(v)$ nor $f_2(v)$ will obey Maxwell's distribution law.

(d) $f_2(v)$ and $f_1(v)$ will be the same.

Answer: The correct choice is (b).

13. Let V_{av}, $V_{r.m.s.}$, and V_{mp}, respectively, denote the mean speed, r.m.s. speed, and most probable speed of the molecule in an ideal monatomic gas at absolute temperature T. The mass of the molecule is m then?

(a) No molecule can have a speed greater than $\sqrt{2}V_{r.m.s.}$

(b) No molecule can have a speed less than $\frac{V_{mp}}{\sqrt{2}}$

(c) $v_{mp} < v_{av} < v_{r.m.s.}$

(d) The average kinetic energy of a molecule is $\frac{3}{4}mv_{mp}^2$

Answer: The correct choices are (c) and (d).

14. Suppose the temperature of N_2 gas is tripled and the molecules of N_2 gas dissociate into atom. The r.m.s. speed of atoms, in terms of $v_0 = \sqrt{\frac{3K_BT}{M}}$, will then be

(a) $\sqrt{6}v_0$

(b) $\sqrt{6v_0}$

(c) $\sqrt{3}v_0$

(d) $\sqrt{3v_0}$

Answer: The correct choice is (a).

15. The following four gases are at the same temperature. In which gas do the molecules have the maximum r.m.s. speed?

(a) H_2

(b) O_2

(c) N_2

(d) CO_2

Answer: The correct choice is (a).

16. E_{0_2} and E_{H_2}, respectively, represent the average kinetic energy of a molecule of oxygen and hydrogen. If the two gases are at the same temperature T, which of the following statements is true?

(a) $E_{0_2} > E_{H_2}$

(b) $E_{0_2} = E_{H_2}$

(c) $E_{0_2} < E_{H_2}$

(d) Nothing can be said about the magnitude of E_{0_2} and E_{H_2} as the information given is insufficient.

Answer: The correct choice is (b).

17. The r.m.s. speed of the molecules of an enclosed gas is v. What will be the r.m.s. speed if the pressure is doubled and the temperature remaining the same?

(a) $\frac{v}{2}$

(b) v

(c) $2v$

(d) $4v$

Answer: The correct choice is (b).

3.10 Exercise

3.10.1 Short answer type questions

1. What do you mean by velocity distribution function?

2. Explain the idea of velocity space.

3. Write down the assumptions of Maxwell's speed distribution law.
[Calcutta University 2022]

4. Calculate the constants a and b.

5. Why is isotropic distribution of particles needed to derive Maxwell's velocity distribution law? [Calcutta University 2021]

6. What do you mean by average and r.m.s. velocity of the molecules of a gas?

7. What is the significance of most probable velocity?

8. Discuss the temperature dependence of Maxwell's speed distribution law.

9. Compare the average speed in different dimensions.

10. Compare the most probable and r.m.s. speed in different dimensions.

11. Why are hydrogen and helium rare in the earth's atmosphere?

12. Write down momentum distribution law in three dimensions.

13. What do you mean by Doppler broadening of spectral lines?
[Burdwan University 2021]

14. Define degree of freedom of a dynamical system.

15. How is the atomicity of gas molecules related to the ratio of specific heats?

16. Write down Maxwell's speed distribution law at a particular temperature T.

17. Write down Maxwell's distribution law in terms of energy at a particular temperature T.

18. Plot Maxwell's speed distribution law at two different temperatures and explain the variation.

19. State the law of equipartition of energy.

3.10.2   Long answer type questions

1. Find out the relation between the ratio of specific heats at constant pressure and at constant volume with a degree of freedom.

2. Show that the average translational kinetic energy of a free particle per degree of freedom is $\frac{1}{2}K_BT$.

3. Write down Maxwell's speed distribution law at a particular temperature T. Plot this speed distribution law as a function of velocity at two different temperatures and comment on the variation.

4. Define average speed, r.m.s. speed, and most probable speed. Find out expressions for these speeds using Maxwell's speed distribution law. Also find out the ratio of these speeds.

5. Write down the characteristic features of Maxwell's speed distribution law.

6. Give the graphical interpretation of Maxwell's velocity distribution law.
   [St. Joseph's College (Autonomous), Bengaluru 2019]

7. Show that Maxwell's speed distribution law is normalized.
   [Calcutta University 2022]

8. Calculate average momentum, most probable momentum, and r.m.s. momentum of a gaseous system obeying Maxwell's speed distribution law in three dimensions.

9. Calculate average momentum, most probable momentum, and r.m.s. momentum of a gaseous system obeying Maxwell's speed distribution law in two dimensions.

10. Calculate average momentum, most probable momentum, and r.m.s. momentum of a gaseous system obeying Maxwell's speed distribution law in one dimension.

11. Write down Maxwell–Boltzmann law of distribution of velocities for molecules of a gas. Hence, obtain the relation between most probable velocity v_m, average velocity v_{av}, and r.m.s. velocity $v_{\text{r.m.s.}}$ for molecules of the gas. Show that $v_{\text{r.m.s.}} > v_{av} > v_m$.   [Calcutta University 2020, Delhi University 2021]

12. Write down the Maxwell–Boltzmann distribution of molecular velocities in an ideal gas. Why does the peak of the curve showing Maxwell's velocity distribution move toward the higher velocity at higher temperature?
        [Burdwan University 2020]

13. Write some comments on the energy distribution law in three dimensions.

14. Describe an experiment to validate Maxwell's speed distribution law.

15. What are the degrees of freedom of a diatomic molecule?

16. Find out the relation between the degrees of freedom and specific heats of a gas.

17. Find the ratios of specific heats for a (i) monatomic, (ii) diatomic, and (iii) triatomic gas.

18. Deduce the law of equipartition of energy. How can Dulong and Petit's law be explained with the help of this law?

19. State the principle of equipartition of energy. How one can estimate about the atomicity of the gas molecules from their ratio of specific heats?
        [Burdwan University 2020]

20. A perfect gas of molecules of mass m each is at absolute temperature T. The velocity of a molecule is denoted by $\vec{v}$. Stating the assumptions clearly, find an expression for the average number of molecules per unit volume with x component of velocity between v_x and $v_x + dv_x$ irrespective of the values of the other two components. Find also the expression for the average number of molecules per unit volume having speed between v and $v + dv$.

21. Using the model of KTG explain why peaks in Maxwell's distribution curves of molecular speeds move toward higher speed at higher temperature.

22. Stating the assumptions clearly deduce Maxwell's distribution law of molecular speed, and hence, find the energy distribution law. Find an expression for the most probable speed of the molecules.

23. Use Maxwell–Boltzmann velocity distribution law to obtain the momentum distribution law for identical and noninteracting molecules.

24. Obtain the distribution law of molecular speed for a two-dimensional gas.

25. According to Maxwell's speed distribution law, the number of molecules per unit volume with speed between v and $v + dv$ is given by

$$n(v)dv = na^3 e^{-b(v_x^2+v_y^2+v_z^2)} dv_x dv_y dv_z,$$

where symbols have their usual meaning. Calculate the constant "a" in terms of the constant "b". [Calcutta University 2022]

26. A vessel contains N molecules of an ideal gas at temperature T. Using the equation $f(v_x)dv_x = ae^{-bv_x^2} dv_x$, show that the number of molecules having speed lying between v and $v + dv$ is given by

$$dN_v = N(v)dv = 4\pi N \; a^3 \; e^{-bv^2} v^2 dv,$$

where v_x is the x-component of the velocity v and $v^2 = v_x^2 + v_y^2 + v_z^2$. (i) Plot $N(v)$ versus v for two different temperature T and $4T$ on the same graph. (ii) Find out an expression for the most probable v_m and the number of molecules $N(v_m)$ having speed v_m. (iii) Estimate the percentage of the total molecules whose speed lies between $v_m \pm 0.01 v_m$.

27. U is a Cartesian component of the molecular velocity (say the x-component). Show that the ratio of the square of the velocity of sound in an ideal gas to the mean square value of u equals the ratio of the specific heat at constant pressure to the specific heat at constant volume.

3.10.3 Numerical problems

1. Show that the probability of having speed between v and $v + dv$ becomes independent of temperature if the velocity of a molecule is normalized by its average value. [Calcutta University 2020]

2. $dn_v = 4\pi na^3 e^{-bv^2} v^2 dv$ is the number of molecules having lying between v and $v + dv$, where $a = \sqrt{\frac{m}{2\pi K_B T}}$ and $b = \frac{m}{2K_B T}$. Plot dn_v versus v for two different temperatures T_1 and $T_2 (T_2 > T_1)$. What is the significance of the total area under the curve and v-axis. [Vidyasagar University 2019]

3. Find the fraction of gas molecules whose velocities are greater than the r.m.s. value by at least 2%. [0.203] [Calcutta University 2020]

4. Calculate the r.m.s. velocity of hydrogen at N.T.P. Compare the r.m.s. velocity of hydrogen at 300 K with that of oxygen at 1200 K.
[V_{rms}(H2) $= 1933.82$ ms^{-1}; V_{rms}(H2)$/V_{rms}$(O2) $= 2$]

5. Calculate the probability that the speed of an O_2 molecule will lie between 200 and 201 ms^{-1} at 300 K. Mass of oxygen molecule is 32 units. [0.00114] [Calcutta University 2020]

6. At what temperature, pressure remaining constant, will the r.m.s. velocity of a gas be half its value at 0°C? [68.25 K]

7. Calculate the average translational kinetic energy of 1 gm of helium at N.T.P. Given $R = 8.3$ J $(gm - mol - K)^{-1}$. [849.71 J]

8. The average kinetic energy of a molecule of hydrogen at 0°C is 5.64×10^{-14} ergs, and the molar gas constant $R = 8.31 \times 10^7$ ergs. Calculate Avogadro's number N_A. [6.034×10^{23}]

9. Two moles of H_2 at 30°C is mixed with one mole of helium at 60°C. Find the final temperature of the mixture. [309.9 K]

10. The mean kinetic energy of a molecule at 27°C is 8.0×10^{-14} ergs. Calculate the number of molecules per cubic cm of a perfect gas at 27°C and at a pressure of 40 mm of mercury. [9.8×10^{20} molecules/cc]

11. Calculate the total molecular translational and rotational energy of 1 gm of hydrogen gas at 100 K and at 300 K, assuming the molecules to be dumbbell shaped. Will the specific heat at constant volume be the same at these two temperatures? [2077.59 J and 6232.77 J]

12. Using Maxwell's speed distribution law, calculate the fraction of argon gas molecules with a speed of 305 ms^{-1} at 500 K. [0.00141]

13. Find out the probability with which speed of the oxygen molecule will lie between 100 ms^{-1} and 101 ms^{-1} at 200 K temperature. Given: Molecular weight of oxygen is 32, Avogadro's number is 6.023×10^{23} molecules/mol, and Boltzmann constant is 1.38×10^{-23} J K^{-1}. [5.35×10^{-6}]
[Burdwan University 2020]

14. Calculate the fraction of molecules of gas within 1% of the most probable speed at normal temperature and pressure. Will it be the same for all gases at all temperatures? [0.0165]

15. 0.46 moles of argon gas is kept at a temperature 500 K. How many molecules have the speed of 305 ms^{-1}? [3.9×10^{20}]

16. Calculate the values of most probable speed v_{mp}, mean speed v_m, and r.m.s. speed $v_{\text{r.m.s.}}$ for xenon gas at 298 K.
[$V_{mp} = 1933.82$ ms^{-1}; $V_m = 219.182$ ms^{-1}; and $V_{rms} = 237.89$ ms^{-1}]

17. Show that the ratio of the speed of sound in an ideal gas to the average speed of its molecules is $\sqrt{\frac{\gamma \pi}{8}}$, where $\gamma = \frac{C_P}{C_V}$.

18. Show that the average value of v^3 of a Maxwellian gas is $\frac{4}{\sqrt{\pi}} \left(\frac{2K_B T}{m} \right)^{3/2}$.

19. In a sample of hydrogen sulfide H_2S $\left(M = 34.1 \text{ g mol}^{-1}\right)$ at a temperature of 300 K, estimate the ratio of the number of molecules that have speeds very close to $v_{\text{r.m.s.}}$ to the number that have speeds very close to $2v_{\text{r.m.s.}}$.   [22.5]

20. Calculate the time taken by a nitrogen molecule moving at the r.m.s. speed to travel a room of length 10 m at room temperature. Assume that the molecule does not suffer any collision. $[t \sim 0.02 \text{ s}]$

21. Calculate the temperature at which the r.m.s. speed of hydrogen and oxygen molecules will be equal to their escape velocities from the earth's gravitational field. The radius of the earth is 6400 km.
$$[T_{H_2} = 10.1 \times 10^3 \text{ K}, \ T_{O_2} = 16 \times T_{H_2}]$$

22. The quantity $(v - \langle v \rangle)^2 = v^2 - 2v\langle v \rangle + \langle v \rangle^2$ is the square of the deviation of the speed of a molecule from the average or the mean speed. Find the mean value of this quantity using Maxwell's speed distribution law and then take the square root of the final result to prove that the r.m.s. deviation of the distribution is $\sqrt{\left(3 - \frac{8}{\pi}\right)\left(\frac{K_B T}{m}\right)}$.

CHAPTER 4

Journey toward Real Gas and Transport Phenomena

These motions [Brownian motion] were such as to satisfy me, after frequently repeated observation, that they arose neither from currents in the fluid, nor from its gradual evaporation, but belonged to the particle itself.

—Robert Brown

Learning Outcomes

After reading this chapter, the reader will be able to

- Express the meaning of sphere of influence and collision frequency
- Derive the distribution function for the free paths among the molecules and demonstrate the concept of mean free path
- Calculate the expression for mean free path following Clausius and Maxwell
- Derive the expression for pressure exerted by a gas using the survival equation
- Calculate the expressions for viscosity, thermal conductivity, and diffusion coefficient of a gaseous system
- Demonstrate Brownian motion with its characteristics and calculate the mean square displacement of a particle executing Brownian motion
- State the idea of a random walk problem
- Solve numerical problems and multiple choice questions on the mean free path, viscosity, thermal conduction, diffusion, Brownian motion, and random walk

4.1   Introduction

Gases are distinguished from other forms of matter, not only by their power of indefinite expansion so as to fill any vessel, however large, and by the great effect heat has in dilating them, but by the uniformity and simplicity of the laws which regulate these changes.

—James Clerk Maxwell

The molecules of an ideal gas are considered as randomly moving point particles. From the concept of kinetic theory of gases (KTG), it is well established that even at room temperature, such point molecules of the ideal gas move at very large speeds. The average value of this speed can be determined assuming that the molecules obey Maxwell's speed distribution law and is given by the following expression

$$\langle v \rangle = \sqrt{\frac{8K_B T}{m\pi}}, \tag{4.1}$$

where the symbols have their usual meanings. This equation (4.1) can be used to estimate the average speed of **air particles**. Air mainly consists of nitrogen (almost 78%) and oxygen (almost 21%) particles. At a temperature of 20°C (293 K), the mean speed of the nitrogen molecules is found to be about 470 ms^{-1}. In this calculation, the mass of a nitrogen molecule is taken to be $m = 4.65 \times 10^{-26}$ kg. From such enormous speeds of the molecules, one should expect that the molecules should disappear from a given vessel in no time. But in reality, this does not happen, and it leads to an apparent paradox. Clausius suggested a simple way out of this paradox. According to him, the gas molecules have **finite size** and move randomly colliding with each other. In between two collisions, molecules follow a straight-line path in an ideal gas, as described by the theory of KTG. During this random motion, they collide with one another and take much larger time to diffuse through the medium or the enclosing vessel. This resolves the paradox. One may then ask the following questions:

1. What is the average distance traveled by a molecule between such successive collisions?

2. What is the physical reason of such collisions?

3. What is the number of collisions per second suffered by the molecules?

Answers to these questions are provided from the knowledge of the mean free path. In this chapter, the distribution law of free paths and expressions for mean free paths following various methods are derived. Relevant experimental techniques in determining the mean free path are also discussed.

Transport phenomenon in physics involves the transfer of various entities, such as mass, momentum, or energy through a medium, fluid, or solid. This movement is due to the nonuniform conditions existing within the medium. For example, differences in the concentration of various chemical species in a medium lead to the relative motion among themselves, and this mass transport is generally referred to as **diffusion**. The existence of the velocity gradient between different layers within a fluid results in the transport of momentum. This is generally referred to as **viscous flow**. Similarly, differences in temperature in various parts of the medium result in the transport of energy and this process is usually known as **conduction of heat**. These three processes often occur physically simultaneously. Mathematical descriptions of these three phenomena show that there are many similarities in deriving the final expressions for these processes. Combustion is such an example where a flowing, viscous, fluid mixture undergoes chemical reactions that produce heat and various chemical species. This heat is conducted away, and the various chemical species interdiffuse with one another.

In this chapter, a variety of applications of the KTG are discussed to explain many important features of an ideal gases (sometimes of real gases also) from a molecular point of view. Some of these applications include

1. Two-body collision frequency

2. Distribution of free paths and mean free path

3. Wall collision frequency

4. The transport of various physical entities of the gaseous system leading to physical properties, such as viscosity, diffusion, and conduction of heat, and

5. Problems of Brownian motion and random walk

4.2   Molecular collisions

The molecules of an ideal gas move randomly in all directions with all possible values of velocity. Such a random distribution of velocities of the ideal gas molecules is shown in Figure 4.1 in the velocity space. On an average, particles in the air move at supersonic speeds that are distributed statistically in the velocity space. This statistical distribution of the velocities indicates that some of the molecules possess significantly higher speeds in the velocity space. About 1% of the molecules have a speed of more than $1000 \, \text{ms}^{-1}$. Out of one billion, one molecule

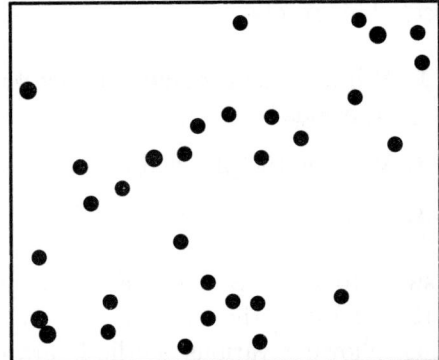

Figure 4.1 Random distribution of velocities of the molecules of an ideal gas in velocity space.

even reaches a speed of 2000 ms^{-1}. Thus, the molecules of a gas generally have a distribution for their velocities. One may then ask: Why is the smell of an open perfume bottle not immediately sense at the other end of a room, as one would expect that the molecules of the constituting elements of perfume will move at speeds of several hundred meters per second? Day-to-day experience shows that it obviously takes some time for the fragrance to be noticed. It thus makes an apparent contradiction.

The apparent contradiction mentioned above lies in the fact that the gas particles of the perfume do not have a free path all along the entire room during their motion. The molecules will suffer frequent collisions with other molecules and change their direction of motion in a random way. These frequent collisions among the molecules result in a series of zigzag paths of unequal length for each molecule. These zigzag paths are called free paths. The trajectory of such a molecule randomly moving with average speed $\langle v \rangle$ is shown in Figure 4.2. The distance a molecule

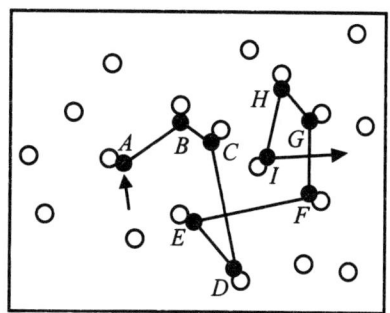

Figure 4.2 The trajectory of a molecule moving in a gas in velocity space.

can travel on an average without colliding with other molecules is called the mean free path. In the case of opening a perfume bottle at one corner of a room, the relatively small mean free path of the fragrance particles prevents the perfume from being sensed immediately on the other corner of the room.

In this section, collisions suffered by the molecules per second, the concept of mean free path and the related phenomena are presented in detail. Attempts are also made to answer the questions:

1. What is the number of collisions per unit area per unit time?

2. How are the free paths distributed among the molecules?

3. What is a mean free path? How is it calculated?

4. What is meant by the collision frequency? How is it calculated?

4.2.1 The number of collisions per unit area per second

The molecules of an ideal gas exhibit random motion and move in all directions with all possible values of velocity. During the course of motion, molecules continually hit the walls of the container and generate pressure. The number of molecules of a gas that strikes or cross a given surface per unit time can be calculated in the following way:

Let us consider an ideal gas consisting of N number of molecules at a temperature T. Further, we consider an elementary area dA on the surface of one of the walls of the container. It is known that an element of volume in spherical coordinates in velocity space is given by

$$dV = v^2 \sin\theta d\theta d\phi dv.$$

Let dn_v be the number of molecules per unit volume having velocity between v and $v + dv$. Then, the number of molecules N_0 in the elementary volume dV will be given by

$$N_0 = dn_v \, dV.$$

Now, the number of molecules that will move toward the elementary area dA will be given by

$$dn_{v\theta\phi} = \frac{N_0}{4\pi} d\omega.$$

The factor 4π arises due to the total solid angle, and the factor $d\omega$ is the solid angle subtended by dA at dV. This solid angle $d\omega$ is given by

$$d\omega = \frac{dA \, \cos\theta}{v^2}.$$

Using the values of N_0 and $d\omega$, the number of molecules moving toward dA is given by

$$dn_{v\theta\phi} = \frac{dn_v \, dV}{4\pi} \times \frac{dA \, \cos\theta}{v^2}; \quad \Rightarrow \quad dn_{v\theta\phi} = \frac{dn_v \, v^2 \sin\theta d\theta d\phi dv}{4\pi} \times \frac{dA \, \cos\theta}{v^2}$$

$$= \frac{dA \, dn_v \, \sin\theta \cos\theta d\theta d\phi dv}{4\pi}.$$

Hence, the number of molecules striking the elementary area dA in time dt coming from all possible directions and all possible distances is given by

$$N = \int \int \int dn_{v\theta\phi} \, dt = \frac{dA \, dt}{4\pi} \int_0^{\frac{\pi}{2}} \sin\theta \cos\theta d\theta \int_0^{2\pi} d\phi \int_0^{\infty} v \, dn_v$$

$$= \frac{dA \, dt}{4\pi} \frac{1}{2} \, 2\pi \int_0^{\infty} v \, dn_v = \frac{dA \, dt}{4} \, n \, \langle v \rangle,$$

where $\langle v \rangle$ is the average velocity of the molecules of the gas. So, the number of molecules striking a given surface of unit area per unit time is

$$n_{\text{strike}} = \frac{N}{dA \, dt} = \frac{1}{4} \, n \, \langle v \rangle. \tag{4.2}$$

This is the expression for the number of molecules that strike the wall of unit cross-section per unit time. Using the expression for average velocity $\langle v \rangle = \sqrt{\frac{8K_B T}{\pi m}}$, equation (4.2) can be expressed as

$$n_{\text{strike}} = \frac{1}{4} n \sqrt{\frac{8K_B T}{\pi m}} = n\sqrt{\frac{K_B T}{2\pi m}}. \tag{4.3}$$

This equation (4.3) shows that the number of molecules that strike the wall of unit cross-section per unit time depends on the number density n, temperature T, and mass of the molecule m.

4.2.2 Mass of gas striking a wall of unit cross-sectional area per unit time

We consider a wall of unit cross-sectional area. The molecules coming from a particular direction perpendicular to the wall will strike this wall. If $\langle v_x \rangle$ be the average velocity in that direction and n be the number density of gas molecules, then the number of collisions with the wall of unit cross-sectional area per unit time is given by $\langle v_x \rangle \times n$. Hence, the mass of the gas m_{strike} striking the wall of unit cross-sectional area per unit time is given by

$$m_{\text{strike}} = m \times \langle v_x \rangle \times n, \tag{4.4}$$

where m is the mass of a single molecule of the gas.

Now the average velocity $\langle v_x \rangle$ along a particular direction, say, along the positive x-direction, is related to the average speed $\langle v \rangle$ of the molecules by

$$\langle v_x \rangle = \frac{1}{4}\langle v \rangle. \tag{4.5}$$

Using the equation $P = nK_B T$ and the value of $\langle v_x \rangle$, the expression for m_{strike} from equation (4.4) can be written as

$$m_{\text{strike}} = m \times \frac{1}{4}\langle v \rangle \times \frac{P}{K_B T} = m \times \frac{1}{4}\sqrt{\frac{8K_B T}{m\pi}} \times \frac{P}{K_B T}$$

$$= \sqrt{\frac{m}{2\pi K_B T}} \times P = \sqrt{\frac{M}{2\pi RT}} \times P, \tag{4.6}$$

where M is the molar mass of the gas. At temperatures, where the vapor pressure of a substance is very low, even less than a tenth of a mm Hg, the rate of evaporation of a substance may be considered to be independent of vapor around it. Further, under a high vacuum, the rate of condensation would be same as the rate of evaporation. Equation (4.6) has a very important implication. **Langmuir used**

this equation (4.6) to find out the vapor pressure of metals. According to this equation, the mass of gas condensing, and hence, the evaporation per unit time can be written as

$$m_{\text{strike}} = \sqrt{\frac{M}{2\pi RT}} \times P = 43.74 \times 10^{-6} \, P \, \sqrt{\frac{M}{T}}, \qquad (4.7)$$

where P is in the unit of mm Hg or torr, and T is in Kelvin. From this equation (4.7), the expression for vapor pressure P comes out to be

$$P = \frac{m_{\text{strike}}}{43.74 \times 10^{-6}} \sqrt{\frac{T}{M}}. \qquad (4.8)$$

For Tungsten ($M = 183.86$ g mol^{-1}) at $T = 2800$ K, the loss of weight from a heated filament per square cm per second is found to be 0.43×10^{-6} g. Hence, the vapor pressure P for tungsten filament, from equation (4.8), is found to be

$$P = \frac{0.43 \times 10^{-6} \text{ g} \times \sqrt{2800}}{43.74 \times 10^{-6} \times \sqrt{183.86}} \text{ mm Hg} = 29 \times 10^{-6} \text{ mm Hg}. \qquad (4.9)$$

This simple example provides an interesting insight about a particular physical property of a metal, that is, the determination of vapor pressure using the concept of KTG. This is a success of KTG.

4.3 Mean free path

Molecules in a gas collide with one another. We assume that in these collisions, momentum and energy remain conserved, and the ideal gas laws are applicable. The paths followed by the molecules in between collisions are of different lengths. The mean free path λ is the average distance a particle travels between collisions. It (λ) depends on the size of the particles and the density of the gas. With the increase in size of the particles or the density of the gas, collisions become more frequent, and hence, λ decreases. If the particle were all by itself, then the mean free path would be infinite. In this chapter, the crucial importance of collisions between particles is emphasized within the framework of the KTG. The motions of individual particles are quantified in more detail. The key ideas covered are:

1. The concept of the mean free path of the molecules of an ideal gas

2. Survival probabilities of the molecules

3. Brownian motion of the molecules and

4. The random walk problem

A close inspection of Figure 4.2 indicates that some of the free paths are short while others are long. The average length of these free paths is defined as the mean free path. Thus, a mean free path is the mean distance traveled by a molecule between two successive collisions. This is denoted by the symbol λ. Such a mean free path is schematically represented in Figure 4.3.

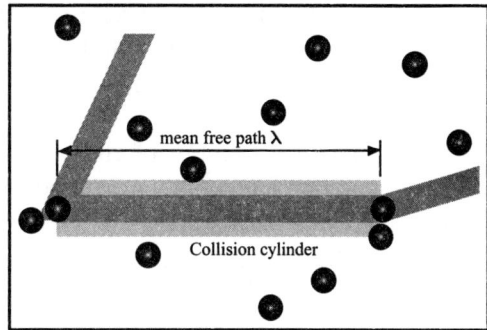

Figure 4.3 Demonstration of mean free path.

If λ_1, λ_2, λ_3, $\cdots$, λ_N are the successive free paths traversed by a molecule in a total time t and N is the total number of collisions suffered by this molecule during this time interval t, then the mean free path λ is given by

$$\lambda = \frac{\lambda_1 + \lambda_2 + \lambda_3 + \cdots + \lambda_N}{N}. \tag{4.10}$$

In terms of the average speed $\langle v \rangle$ of the molecules, equation (4.10) can be written as

$$\lambda = \frac{\langle v \rangle \times t}{N}. \tag{4.11}$$

Let τ be the mean free time between two successive collisions. In this case, τ is given by $\tau(= \frac{t}{N})$. Using this mean free time τ, the mean free path λ comes out to be

$$\lambda = \frac{\langle v \rangle \times t}{N} = \langle v \rangle \times \tau = \frac{\langle v \rangle}{p_c}, \tag{4.12}$$

where $p_c = \tau^{-1}$ denotes the collision frequency. It measures the average number of collisions suffered by the molecules per second. The mean free path of oxygen molecules, under normal conditions, is about $\lambda \approx 2 \times 10^{-7}$ m. This mean free path is slightly smaller than the wavelength of visible light, which is in the range from 4×10^{-7} to 6×10^{-7} m. However, it should be mentioned that the mean free path λ is larger (by two orders of magnitude) than the average intermolecular separation of 3×10^{-9} m.

Using the expressions for average speed $\langle v \rangle$ and the mean free path λ [derived in the later section, $\lambda = \frac{1}{\sqrt{2}\pi d^2 n} = \frac{K_B T}{\sqrt{2}\pi d^2 P}$, where d is the diameter of the molecule], the collision frequency p_c can be expressed in the following form:

$$p_c = \frac{\langle v \rangle}{\lambda} = \frac{\sqrt{\frac{8K_B T}{m\pi}}}{\frac{K_B T}{\sqrt{2}\pi d^2 P}} = \sqrt{\frac{16\pi d^4 P^2}{m K_B T}}. \tag{4.13}$$

At a temperature of 293 K and a pressure of 1 bar, a collision frequency of 7×10^9 s^{-1} results for the nitrogen molecule, that is, on an average, a single nitrogen molecule will collide with 7 billion other molecules within one second! Think of this horrible situation!!

To obtain the total number of the collisions per unit volume per unit time, the collision frequency p_c has to be multiplied by the particle density n only. It must be noted that each collision involves two particles, so that a factor of $\frac{1}{2}$ must be taken into account. So, the total number of collisions n' per unit volume per unit time will be determined in the following way:

$$n' = \frac{1}{2} \times n \times p_c = \frac{1}{2} \times \frac{P}{K_B T} \times \sqrt{\frac{16\pi d^4 P^2}{m K_B T}} = \sqrt{\frac{4\pi d^4 P^4}{m K_B^3 T^3}}. \tag{4.14}$$

This equation (4.14) shows that for the molecules of a given gas, the total number of collisions per unit volume per unit time n' is directly proportional to the square of pressure P and is inversely proportional to one and half power of temperature T. For example, at normal pressure and temperature, for the molecules of nitrogen gas, n' is found to be $n' = 8.7 \times 10^{34}$ m^{-3} s^{-1}, that is, 8.7×10^{34} the number of collisions occurs within one second in a volume of one cubic meter. **If we think of this number, we will be simply puzzled out!!** Thus, atoms and molecules are scattered many times per second as they move through a gas under normal atmospheric conditions.

4.3.1 Sphere of influence and collision cross-section

It is assumed in the kinetic theory of ideal gases that the molecules of an ideal gas are identical and noninteracting and behave as perfectly elastic spheres moving randomly with all possible values of velocities in all possible directions. In executing such random motions, these molecules collide continuously against each other.

To develop a quantitative understanding of random collisions, it is essential to first introduce the concepts of the sphere of influence and collision cross-section. These two quantities can be illustrated as follows: A gas molecule is assumed to have a spherical shape, with d representing the diameter of a single

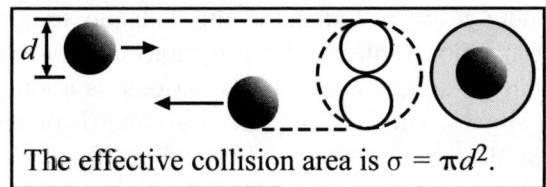

The effective collision area is $\sigma = \pi d^2$.

Figure 4.4 Sphere of influence and collision cross-section.

molecule. Taking the center of the molecule as a center, a sphere of radius equal to the diameter of the molecule d is drawn. Then, this molecule will collide with all those molecules whose centers lie within this sphere. Such a sphere is called a "sphere of influence" and is schematically shown in Figure 4.4.

We consider a hypothetical situation that, at a certain instant of time, all the molecules of a gas are at rest, and only one molecule is moving. This molecule moves with the average velocity $\langle v \rangle$ against the static background formed by the "frozen" molecules. We assume further that the molecules are perfectly elastic spheres of diameter d. At the instant of collision, the center-to-center distance of the colliding molecules would be d. This center-to-center distance d would thus be the same if the radius of the moving molecule was increased to d, and the molecules at rest were all shrunk to mere geometric points. So, the effective cross-sectional area of the mobile molecule, that is, *collision cross-section* σ will be

$$\sigma = \pi d^2. \tag{4.15}$$

If we take a cross section along the diameter of the sphere of influence, a collision cross-section would be obtained. The area $\pi d^2 = \pi(2r)^2$ would be the collision cross-section σ. Such a collision cross-section is shown in Figure 4.5. It is the effective area which determines the probability that a molecule of diameter d will collide with another molecule of the same diameter.

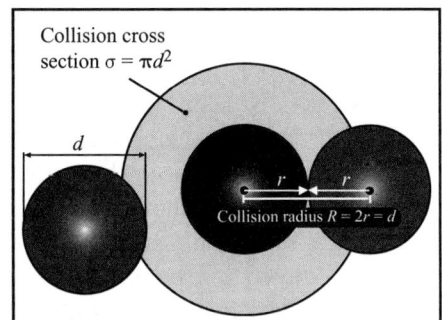

Figure 4.5 Collision cross-sectional area.

4.3.2 Concept of two- and three-body collisions

In a gaseous system, two-body collisions occur very frequently, but the occurrence of three-body problem is rare. In order to occur a three-body problem, a two-body problem must occur, and that must have to spent a certain period of time during, which a third molecule must have to hit the given pair from a distance of mean free path. The lifetime of a binary collision can be considered as the time required by a molecule to traverse the distance equal to its molecular diameter is

$$= \frac{d}{\sqrt{2}\,\langle v \rangle}. \tag{4.16}$$

The third molecule takes time to traverse the distance of the mean free path λ is

$$= \frac{\lambda}{\sqrt{2}\,\langle v \rangle}. \tag{4.17}$$

So, the number of three-body collisions within the lifetime of a binary collision is given by

$$= \frac{\frac{d}{\sqrt{2}\,\langle v \rangle}}{\frac{\lambda}{\sqrt{2}\,\langle v \rangle}} = \frac{d}{\lambda}. \tag{4.18}$$

Hence, the ratio of the frequency of the three-body collisions to two collisions is

$$= \frac{d}{\lambda}. \tag{4.19}$$

At ordinary conditions, $\lambda \approx 10^{-7}$ m and $d \approx 10^{-10}$ m. Hence, $\frac{d}{\lambda} = \frac{1}{1000}$. For this reason, the trimolecular reactions in the gas phase involving three-body collisions occur less rapidly than bimolecular reactions.

4.3.3   Distribution of free paths and the collision probability

To illustrate the concept of the distribution of free paths for the molecules of an ideal gas, let us consider the following example. Consider a man shooting aimlessly with an Avtomat Kalashnikova Model 1947 (AK-47) in a thick forest. Every bullet coming out of the gun eventually hits a tree, but some travel farther than others. This situation is analogous to the flights of molecules of a gas. A moving molecule will collide with another molecule, but when this collision would take place is uncertain. It is a probabilistic event. In between two collisions, a molecule moves in a straight line, which is known as free path. We now aim to determine how these free paths are distributed among the molecules in a given gas.

Survival probability and the distribution of free paths

The concept of **mean free path** provides an estimate about the mean distance traveled by a molecule between collisions, but it does not tell us anything about the *distribution* of free paths of the molecules. In other words, if we want to know - what is the likelihood of the molecule having a collision after traveling a certain distance? - cannot be answered from the knowledge of mean free path.

In order to have an answer of this question, we proceed in the following way: We consider an ideal gas consisting of N_0 number of molecules that make collisions among themselves, and the distance between two successive collisions may be anything ranging from zero to a very large value. Let N be the number of molecules having free paths x. Due to collisions, some of these molecules will be removed from the group. Let dN be the number of molecules removed from the group when the free path changes from x to $x + dx$. This number dN will be proportional to the number of molecules N in the group and also to the length dx of the path. Therefore, we have

$$dN \propto N \times dx; \quad \Rightarrow \quad dN = -\alpha \, N \, dx, \tag{4.20}$$

where α is the proportionality constant and is known as the collision probability. This constant α is independent of both N and dx. The negative sign in equation (4.20) indicates that a collision removes a molecule from the group. Rearranging equation (4.20), we get

$$\frac{dN}{N} = -\alpha \; dx.$$

Integrating this expression between proper limits yields

$$\int_{N_0}^{N} \frac{dN}{N} = -\alpha \int_{x=0}^{x} dx; \quad \Rightarrow \quad \ln\left(\frac{N}{N_0}\right) = -\alpha \; x.$$

This leads to

$$N = N_0 \; e^{-\alpha \; x}. \tag{4.21}$$

Thus, equation (4.21) shows that the number of molecules N remaining in the group falls off exponentially with the free path length x. Using equations (4.20) and (4.21), we get

$$dN = -\alpha \; N_0 \; e^{-\alpha \; x} \; dx. \tag{4.22}$$

In order to calculate the expression for mean free path λ, let us proceed in the following way: Let dN_1 be the number of molecules, each having free path x_1, dN_2 be the number of molecules each having free path x_2, dN_3 be the number of molecules each having free path x_3, and so on. Hence, the mean free path λ of these molecules can be mathematically expressed as

$$\lambda = \frac{x_1 \; dN_1 + x_2 \; dN_2 + x_3 \; dN_3 + \cdots}{dN_1 + dN_2 + dN_3 + \cdots} = \frac{\int_0^\infty x \; dN}{\int_0^\infty dN}$$

$$= \frac{1}{N_0} \int_0^\infty x \; dN = \frac{1}{N_0} \int_0^\infty x \; \alpha \; N_0 \; e^{-\alpha \; x} \; dx = \alpha \int_0^\infty x \; e^{-\alpha \; x} \; dx.$$

Here, we have used equation (4.22), and the negative sign has been neglected as it only implies that collisions remove molecules from the group of the given free path.

In order to evaluate the above integral, let us put $\alpha \; x = y$. Then, we get $dx = \frac{dy}{\alpha}$. Again, when $x \to 0, y \to 0$ and $x \to \infty, y \to \infty$. Hence, we get the expression for λ as

$$\lambda = \alpha \int_0^\infty \frac{y}{\alpha} \; e^{-y} \; \frac{dy}{\alpha} = \frac{1}{\alpha} \int_0^\infty y^{2-1} \; e^{-y} dy = \frac{1}{\alpha} \; \Gamma(2) = \frac{1}{\alpha}.$$

This shows that **the collision probability α is equal to the reciprocal of the mean free path λ**. Using this value of α in equation (4.21), we get

$$N = N_0\, e^{-\frac{x}{\lambda}}. \tag{4.23}$$

This is known as the **survival equation**. This equation (4.23) indicates that if we have a sample of N_0 molecules at the beginning, only $N_0 e^{-\frac{x}{\lambda}}$ of these molecules will survive a collision in traversing a distance x. The following comments about equation (4.23) are in order:

1. Similar equations are also found in other areas of physics. For example, radioactive decay obeys a similar law. Also, Biots' law in optics follows a similar trend. This law describes the exponential decay character of the intensity of an incident beam after it has traversed a distance x in a medium.

2. The distribution of free paths gives the probability that a molecule may describe a distance x without suffering collision. The probability of such collision, that is, the collision probability is the reciprocal of mean free path λ.

3. This equation (4.23) shows that the fraction of molecules with free paths larger than λ is only $e^{-1} = 0.37$.

4. If we take into account Maxwell's speed distribution law, the calculation of the distribution function for free paths becomes slightly more difficult. This calculation was done by Jeans, and the result can be expressed as

$$f(x) = e^{-\left(\frac{1.04 \times x}{\lambda}\right)}. \tag{4.24}$$

Figure 4.6 shows the plot of $\frac{N}{N_0}$ as a function of $\frac{x}{\lambda}$. The ordinate gives the fractional number of molecules with free paths greater than any fraction of the mean free path λ. Further, it is observed from the figure that the fractional number of molecules with free paths greater than λ is 37%, while that with free paths shorter than λ is 63%.

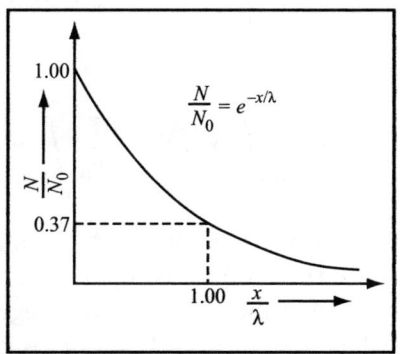

Figure 4.6 Plot of fractional number of molecules $\frac{N}{N_0}$ as a function of $\frac{x}{\lambda}$.

Figure 4.7 shows the plot of $\left|\frac{dN}{dx}\right|$ as a function of x. The ordinate gives the number of molecules per unit path length with free paths between x and $x + dx$. The actual number dN having free paths between x and $x + dx$ is, however, represented by the area of a narrow vertical strip, as shown in Figure 4.7.

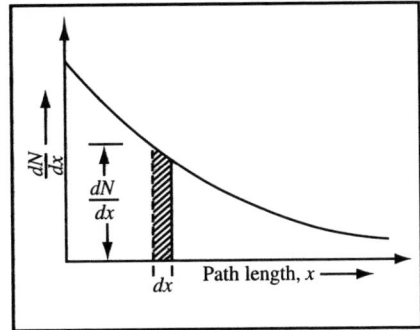

Figure 4.7 Plot of $\left|\frac{dN}{dx}\right|$ as a function of x.

4.3.4 Calculation of mean free path

In this section, expressions for the mean free path of the molecules of an ideal gas are derived using various methods. These methods include elementary approach, Clausius approach, and Maxwell's approach. To calculate the mean free path, we make the following assumptions:

1. The gas molecules behave like elastic hard sphere.

2. The molecules have definite size.

3. The molecules move in a force-free region, that is, there is no force of repulsion or attraction between the molecules.

Calculation of mean free path: an elementary approach

We consider a gaseous system consisting of N number of molecules in a volume V, and the system is in thermal equilibrium at a temperature T. The number of molecules per unit volume is then given by $n = \frac{N}{V}$. Each molecule has a mass m and a diameter d. We assume that the molecules of the gas undergo random collisions. We can say mathematically that the probability of a molecule suffering a collision in a small interval

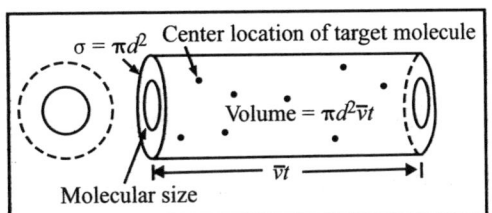

Figure 4.8 Cylindrical volume with base area equal to collision cross-section to calculate mean free path.

of time dt is independent of the history of past collisions made by it. As a simple analogy, we can consider that if we throw a dice, the probability of getting a **two** does not depend on the preceding throw, where a two may or may not have appeared. Similarly, losing or winning **a toss** before a match is independent of the earlier results.

These molecules are assumed to be identical and perfectly elastic sphere. Further, it is assumed that the molecules do not exert any force of repulsion or attraction on each other. In order to calculate the expression for the mean free path of the molecules in the elementary method, the following assumptions are made

1. Only one molecule chosen at random moves through the gas with an average velocity $\langle v \rangle$.

2. Except this molecule, all other molecules are at rest.

3. The sphere of influence of the molecules has a radius d, that is, equal to the diameter of the molecule.

The selected molecule will collide per second with all the molecules whose centers lie within a cylinder of radius d and length $\langle v \rangle$. Such a cylinder of defined cross section is schematically shown in Figure 4.8. The number of molecules within this cylinder is then given by $= \pi d^2 \langle v \rangle \times n$, where n is the number of molecules per unit volume.

Number of collisions made by the selected molecule in 1 second is then given by

$$= \pi \, d^2 \, \langle v \rangle \, \times n.$$

Hence, the mean time taken in one collision is $\frac{1}{\pi d^2 \langle v \rangle \, \times n}$ seconds. This is the time interval between two successive collisions. So, the distance traveled by a molecule during this time interval is given by

$$= \text{average speed} \times \text{average time} = \langle v \rangle \times \frac{1}{\pi d^2 \langle v \rangle \, \times n} = \frac{1}{\pi d^2 \, \times n}.$$

This is the expression for mean free path λ and is given by

$$\lambda = \frac{1}{\pi d^2 \, \times n}. \tag{4.25}$$

Let us pause for a moment and ask the following questions about equation (4.25):

1. What is the implication of equation (4.25)?

2. What is the information obtained from equation (4.25)?

To get answers to these questions, let us have a look at equation (4.25). It provides us that the mean free path λ is inversely proportional to the macroscopic collision cross-section σ and is inversely proportional to the number density n. Through σ, λ is inversely proportional to the second power of diameter d of the molecule. This indicates that the mean free path λ will be less for a denser and/or a heavier gas. These facts are in good agreement with common observations and lend support to the basic features of KTG. **Hence, the aesthetic beauty of kinetic theory is that it has the ability to relate experimentally the measurable macroscopic**

quantity, such as the mean free path to a microscopic quantity like the size of the molecule.

It is to be noted that the above calculation of λ has a number of unsatisfactory characteristics. The assumption that all the molecules except one are at rest – is not true. In fact, the molecules are in constant motion in all possible directions. Hence, it is a crude approximation. More complete calculations for the mean free path must take into account the motion of all molecules. When the idea of motion of the molecules with a well-defined functional form is taken into account, the expression for cross section is slightly changed, and this becomes $\sigma = \sqrt{2}\pi d^2$ for a gas with Maxwellian distribution of molecular speeds.

The free time between two collisions is an important parameter for a gaseous system. Its mean, that is, *mean free time* τ can be defined as $\tau = \frac{\lambda}{\langle v \rangle}$, where $\langle v \rangle$ is the average speed of the molecules, and λ is the mean free path. Values for mean speeds, mean free paths, and mean free times for some natural gases at normal temperature and pressure are listed in Table 4.1.

Table 4.1 Mean speed, mean free path, and mean free time of some gases

Species	Mean speed ($\langle v \rangle$ ms^{-1})	Mean free path ($\lambda(nm)$)	Mean free time ($\tau(ns)$)
Ar	380	63	0.165
CO_2	362	39	0.108
N_2	454	59	0.130
O_2	425	63	0.149

Relation between mean free path λ, temperature T, and pressure P

For one mole of an ideal gas, we have the equation of state

$$P V = R T.$$

Dividing both sides by Avogadro's number N_A, we get

$$\frac{P V}{N_A} = \frac{R T}{N_A}; \quad \Rightarrow \frac{P}{n} = K_B T; \quad \Rightarrow n = \frac{P}{K_B T}. \tag{4.26}$$

The relation for the number density in one mole, $n = \frac{N_A}{V}$, and the expression for Boltzmann factor $K_B = R/N_A$ have been used in equation (4.26). Using equation (4.26) into equation (4.25), we get

$$\lambda = \frac{1}{\pi d^2 n} = \frac{1}{\pi d^2} \times \frac{K_B T}{P}. \tag{4.27}$$

Thus, equation (4.27) shows that the mean free path λ varies directly as the absolute temperature T and inversely as the pressure P. Further, equation (4.28) points toward a very interesting concept. Suppose at a constant temperature T, pressure P of the system is reduced to a very small value using a vacuum pump. Under this situation, λ will approach to a very high value. Now the question is: Will the value of λ approach ∞? It should be noted that λ cannot increase indefinitely. At the most the value of λ will be equal to the dimensions of the container. This finds an interesting application in getting well-directed molecular beams for research purposes.

It should be noted further that for point molecules, molecular diameter d approaches zero. Hence, the collision cross-section σ also approaches zero. Under this situation, the mean free path approaches infinity, that is, $\lambda \to \infty$.

Size of the particles

In order to calculate the expression for the mean free path λ, molecules are considered to be *hard spheres* with a finite diameter d, whereas real molecules do not behave as hard spheres. This "hard sphere" assumption makes the calculation of the mean free path much simpler. It should be noted that at N.T.P., the mean free path of the molecules of an ideal gas is nearly 310 times the nominal atomic diameter (which is of the order of 0.3 nm) and nearly 28 times the average molecular separation (which is of the order of 3.3 nm). The value of the diameter of a molecule can be calculated from the expression for the mean free path λ. Some typical values of diameter d of the molecules are given in Table 4.2.

Table 4.2 Diameter of some molecules.

Atom or molecule	Symbol	d (in nm)
Argon	Ar	0.340
Oxygen	O_2	0.354
Nitrogen	N_2	0.375
Carbon dioxide	CO_2	0.390

Calculation of the mean free path: Clausius approach

In the elementary method for the calculation of the mean free path λ, it was assumed that only one molecule moves with average velocity $\langle v \rangle$, and the rest of the molecules are considered to be at rest, that is, other molecules are "frozen". In real situation, this assumption is not valid. Further, if the target molecules move, the collision probability goes up. Clausius introduced the concept of relative velocity to make a correction to the expression for the mean free path obtained from the elementary method. He argued that if the relative velocity of one molecule with respect to all other molecules could be found, essentially, the molecule would move (with relative velocity) while others would be at rest. Hence, Clausius improved the calculation for the mean free path slightly better by introducing the idea of relative velocity. In the method introduced by Clausius, the following steps are followed to calculate the mean free path:

1. One molecule is chosen at random from the gaseous system.

2. The relative velocity of this molecule with respect to all other molecules is calculated. This technique makes all other molecules at rest with respect to the molecule under consideration.

3. All molecules of the gas move with the same average velocity $\langle v \rangle$.

We consider a gaseous system consisting of N number of molecules in a volume V and at a temperature T. The number of molecules per unit volume is then given by $n = \frac{N}{V}$. It is assumed that these molecules are identical and perfectly elastic sphere, and the diameter of each molecule is d. We consider two molecules A and B, moving with velocities v_1 and v_2, respectively. Further, we assume that the molecule B is moving at an angle θ with respect to A. The relative velocity of A with respect to B is then given by

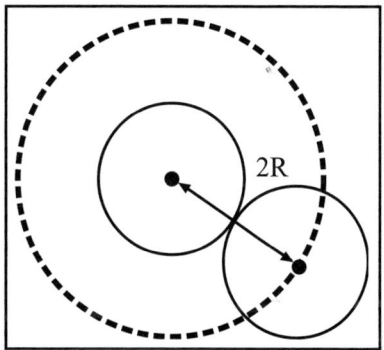

Figure 4.9 Effective radius to calculate the collision cross-section of a molecule.

$$v_{\text{rel}} = \sqrt{v_1^2 + v_2^2 - 2v_1 v_2 \cos\theta}. \qquad (4.28)$$

To find the mean relative velocity $\langle v_{\text{rel}} \rangle$ of A with respect to all other B-type molecules, we proceed in the following way: Let $dn_{\theta,\phi}$ be the number of molecules per unit volume moving between θ and $\theta + d\theta$ and ϕ and $\phi + d\phi$. Then the mean relative velocity $\langle v_{\text{rel}} \rangle$ of A with respect to B type molecules in velocity space is given by

$$\langle v_{\text{rel}} \rangle = \frac{\int v_{\text{rel}} \, dn_{\theta,\phi}}{\int dn_{\theta,\phi}}. \qquad (4.29)$$

But, the number of molecules in the directions between θ and $\theta + d\theta$ and ϕ and $\phi + d\phi$ is given by

$$dn_{\theta,\phi} = \frac{n}{4\pi} d\omega = \frac{n}{4\pi} \frac{dA}{v^2} = \frac{n}{4\pi} \frac{v^2 \sin\theta \, d\theta \, d\phi}{v^2} = \frac{n}{4\pi} \sin\theta \, d\theta \, d\phi, \qquad (4.30)$$

where n is the molecular density.

Using equations (4.28) and (4.30), the mean relative velocity $\langle v_{rel} \rangle$ can be written as

$$\langle v_{rel} \rangle = \frac{\frac{n}{4\pi} \int\limits_{0}^{\pi} (v_1^2 + v_2^2 - 2v_1 v_2 \cos\theta)^{1/2} \sin\theta\, d\theta \int\limits_{0}^{2\pi} d\phi}{\frac{n}{4\pi} \int\limits_{0}^{\pi} \sin\theta\, d\theta \int\limits_{0}^{2\pi} d\phi}. \tag{4.31}$$

The limits of integration are such that all molecules in velocity space are included. Therefore, the above integral can be written as

$$\langle v_{rel} \rangle = \frac{1}{2} \int\limits_{0}^{\pi} (v_1^2 + v_2^2 - 2v_1 v_2 \cos\theta)^{1/2} \sin\theta\, d\theta \qquad \left[\int\limits_{0}^{\pi} \sin\theta\, d\theta = 2 \right]. \tag{4.32}$$

In order to evaluate the integral, let us substitute $v_1^2 + v_2^2 - 2v_1 v_2 \cos\theta = z$. This provides $2v_1 v_2 \sin\theta\, d\theta = dz$.

Therefore, equation (4.32) can be written as

$$\langle v_{rel} \rangle = \frac{1}{2} \times \frac{1}{2v_1 v_2} \int\limits_{(v_1 \sim v_2)^2}^{(v_1+v_2)^2} z^{1/2}\, dz = \frac{1}{2} \times \frac{1}{3v_1 v_2} \times [(v_1+v_2)^3 - (v_1 \sim v_2)^3]. \tag{4.33}$$

Clausius assumed that $v_1 = v_2 = \langle v \rangle$, that is, all molecules move with the same average velocity $\langle v \rangle$. From equation (4.33), we then have

$$\langle v_{rel} \rangle = \frac{1}{6\langle v \rangle^2} \times 8\langle v \rangle^3 = \frac{4}{3} \langle v \rangle. \tag{4.34}$$

So, in moving over a distance $\langle v \rangle$, the number of collisions made by the molecule moving with relative velocity $\langle v_{rel} \rangle$, is $\pi d^2 n \langle v_{rel} \rangle$.

Hence, the mean free path is given by

$$\lambda = \frac{\langle v \rangle}{\pi d^2 \langle v_{rel} \rangle\, n} = \frac{\langle v \rangle}{\pi d^2 \frac{4}{3}\langle v \rangle n} = \frac{3}{4\pi d^2 n} = \frac{3}{4} \frac{1}{\pi\, d^2\, n}. \tag{4.35}$$

Equation (4.35) provides the expression for the mean free path due to Clausius. Although this approach is a step better than the elementary one, it is also open to criticism as the assumption $v_1 = v_2 = \langle v \rangle$ is not valid. From the concept of KTG, it is known that the velocities of the molecules are really distributed according to Maxwell's distribution law. Hence, this distribution law has to be taken into account for the calculation of mean free path λ. This velocity distribution function is specifically considered in the calculation of the mean free path using Maxwell's approach.

Calculation of the mean free path: Maxwell approach

In order to calculate the expression for the mean free path, Maxwell also followed the technique of relative velocity introduced by Clausius. He first calculated the mean relative velocity of a molecule A, moving with velocity v_1, relative to all B-type molecules moving with velocity v_2 and then averaging over all possible values of v_2. He then averaged this value over all possible v_2 values, ranging from 0 to ∞. He next multiplied this mean velocity by the number of molecules moving with velocity v_1 and averaging over all possible values of v_1 from 0 to ∞. Thus, the mean velocity of any molecules with respect to all others in velocity space was obtained. Thus, the following steps are adopted in Maxwell's method for the calculation of the mean free path:

1. The relative velocity v_{rel} of one A-type molecule w.r.t another B-type molecule is calculated first.

2. The mean relative velocity $\langle v_1(\text{rel})\rangle$ of one A-type molecule w.r.t all other B-type molecules is then calculated assuming that the velocities of B-type molecules are distributed according to Maxwell speed distribution law.

3. The mean relative velocity $\langle v_{\text{all}}(\text{rel})\rangle$ of all A-type molecules w.r.t all other B-type molecules is calculated next assuming that the velocities of A-type molecules are distributed according to Maxwell speed distribution law.

4. The mean free path λ is then calculated using the expression:

$$\lambda = \frac{\langle v\rangle}{\pi\, d^2\,\langle v_{\text{all}}(\text{rel})\rangle\, n} = \frac{\langle v\rangle}{\pi\, d^2\,\sqrt{2}\langle v\rangle\, n} = \frac{1}{\sqrt{2}} \times \frac{1}{\pi\, d^2\, n}.$$

In order to prove that the mean relative velocity $\langle v_{\text{all}}(\text{rel})\rangle$ of all A-type molecules w.r.t all other B-type molecules is $\sqrt{2}\langle v\rangle$, we proceed in the following way:

From equation (4.33), the mean relative velocity $\langle v_{\text{rel}}\rangle$ of a molecule moving with velocity v_1 with respect to molecules having velocity v_2, is,

$$\langle v_{\text{rel}}\rangle = \frac{1}{2}\cdot\frac{1}{3v_1 v_2}\left[(v_1 + v_2)^3 - (v_1 \sim v_2)^3\right]. \tag{4.36}$$

This gives rise to

$$\langle v_{\text{rel}}\rangle = \frac{3v_1^2 + v_2^2}{3v_1} \qquad \text{when } v_1 > v_2 \tag{4.37}$$

and

$$\langle v_{\text{rel}}\rangle = \frac{3v_2^2 + v_1^2}{3v_2} \qquad \text{when } v_1 < v_2. \tag{4.38}$$

Therefore, the mean relative velocity $\langle v_1(\text{rel})\rangle$ of A-type molecule with respect to all other B-type molecules is

$$\langle v_1(\text{rel})\rangle = \frac{\displaystyle\int_0^\infty \langle v_{\text{rel}}\rangle . 4\pi n \left(\frac{m}{2\pi K_B T}\right)^{3/2} \exp\left(-\frac{mv_2^2}{2K_B T}\right) v_2^2 dv_2}{\displaystyle\int_0^\infty 4\pi n \left(\frac{m}{2\pi K_B T}\right)^{3/2} \exp\left(-\frac{mv_2^2}{2K_B T}\right) v_2^2 dv_2}.$$

The above expression leads to

$$\langle v_1(\text{rel})\rangle = \int_0^\infty \langle v_{\text{rel}}\rangle . 4\pi \left(\frac{m}{2\pi K_B T}\right)^{3/2} \exp\left(-\frac{mv_2^2}{2K_B T}\right) v_2^2 dv_2. \tag{4.39}$$

The integral in the denominator gives the number of molecules per unit volume n. Hence, we get

$$\langle v_1(\text{rel})\rangle = \int_0^{v_1} 4\pi \left(\frac{m}{2\pi K_B T}\right)^{3/2} . \frac{3v_1^2 + v_2^2}{3v_1} \exp\left(-\frac{mv_2^2}{2K_B T}\right) v_2^2 dv_2$$

$$+ \int_{v_1}^\infty 4\pi \left(\frac{m}{2\pi K_B T}\right)^{3/2} . \frac{3v_2^2 + v_1^2}{3v_2} \exp\left(-\frac{mv_2^2}{2K_B T}\right) v_2^2 dv_2. \tag{4.40}$$

The appropriate values of $\langle v_{\text{rel}}\rangle$ in appropriate ranges from equations (4.37) and (4.38) have been used in equation (4.40).

Now, molecule A again may move according to Maxwell speed distribution law with all possible values of v_1 ranging from 0 to ∞. So, the mean relative velocity of any molecule with respect to all other molecules moving in all possible manners is

$$\langle v_{\text{all}}(\text{rel})\rangle = \frac{\displaystyle\int_0^\infty \langle v_1(\text{rel})\rangle dn_{v_1}}{\displaystyle\int_0^\infty dn_{v_1}} = \frac{1}{n}\int_0^\infty \langle v_1(\text{rel})\rangle dn_{v_1}$$

$$= \frac{1}{n}\int_0^\infty 4\pi n \left(\frac{m}{2\pi K_B T}\right)^{3/2} \langle v_1(rel)\rangle e^{-\frac{mv_1^2}{2K_B T}} v_1^2 dv_1.$$

This leads to

$$\langle v_{\text{all}}(\text{rel})\rangle = \int\limits_0^\infty 4\pi \left(\frac{m}{2\pi K_B T}\right)^{3/2} \langle v_1(\text{rel})\rangle e^{-\frac{mv_1^2}{2K_B T}} v_1^2 dv_1. \tag{4.41}$$

Substituting values of $\langle r_1(\text{rel})\rangle$ from (4.40) into equation (4.41), we obtain

$$\langle v_{\text{all}}(\text{rel})\rangle = 16\pi^2 \left(\frac{m}{2\pi K_B T}\right)^3 \left[\int\limits_0^\infty v_1^2 e^{-\frac{mv_1^2}{2K_B T}} dv_1 \int\limits_0^{v_1} \frac{3v_1^2 + v_2^2}{3v_1} e^{-\frac{mv_2^2}{2K_B T}} v_2^2 dv_2\right] \tag{4.42}$$

$$+16\pi^2 \left(\frac{m}{2\pi K_B T}\right)^3 \left[\int\limits_0^\infty v_1^2 e^{-\frac{mv_1^2}{2K_B T}} dv_1 \int\limits_{v_1}^\infty \frac{3v_2^2 + v_1^2}{3v_2} e^{-\frac{mv_2^2}{2K_B T}} v_2^2 dv_2\right]. \tag{4.43}$$

Let us denote the term in the first third bracket [ ] by I_1 and the second third bracket [ ] by I_2. Hence, we get

$$I_2 = \int\limits_0^\infty v_1^2 e^{-\frac{mv_1^2}{2K_B T}} dv_1 \int\limits_{v_1}^\infty \frac{3v_2^2 + v_1^2}{3v_2} e^{-\frac{mv_2^2}{2K_B T}} v_2^2 dv_2 \quad \left(\text{where } b = \frac{m}{2K_B T}\right).$$

Now,

$$\int\limits_{v_1}^\infty \frac{3v_2^2 + v_1^2}{3v_2} e^{-bv_2^2} v_2^2 dv_2 = \int\limits_{v_1}^\infty v_2^3 e^{-bv_2^2} dv_2 + \frac{v_1^2}{3}\int\limits_{v_1}^\infty v_2 e^{-bv_2^2} dv_2$$

$$= \frac{1}{2b}\int\limits_{v_1}^\infty v_2^2 e^{-bv_2^2} d(bv_2^2) + \frac{v_1^2}{6b}\int\limits_{v_1}^\infty e^{-bv_2^2} d(bv_2^2)$$

$$= \frac{1}{2b^2}\int\limits_{bv_1^2}^\infty z e^{-z} dz + \frac{v_1^2}{6b}\left[-e^{-bv_2^2}\right]_{v_1}^\infty.$$

Using $bv_2^2 = z$, the above expression can be written as

$$= \frac{1}{2b^2}\left[-ze^{-z} - e^{-z}\right]_{bv_1^2}^\infty + \frac{v_1^2}{6b}\left[-e^{-bv_2^2}\right]_{v_1}^\infty = \frac{1}{2b^2}\left(bv_1^2 e^{-bv_1^2} + e^{-bv_1^2}\right) + \frac{v_1^2}{6b} e^{-bv_1^2}.$$

Therefore,

$$I_2 = \frac{1}{2b}\int\limits_0^\infty v_1^4 e^{-2bv_1^2}dv_1 + \frac{1}{2b^2}\int\limits_0^\infty v_1^2 e^{-2bv_1^2}dv_1 + \frac{1}{6b}\int\limits_0^\infty v_1^4 e^{-bv_1^2}dv_1$$

$$= \frac{1}{2b}\cdot\frac{3}{8}\sqrt{\frac{\pi}{(2b)^5}} + \frac{1}{2b^2}\cdot\frac{1}{4}\sqrt{\frac{\pi}{(2b)^3}} + \frac{1}{6b}\cdot\frac{3}{8}\sqrt{\frac{\pi}{(2b)^5}} = \frac{1}{8\sqrt{2}}\sqrt{\frac{\pi}{(b)^7}}.$$

It can be shown that the term in the first [ ] bracket of equation (4.43), that is, I_1 is also equal to I_2. The integral given by I_1 can be evaluated in the following way:

$$I_1 = \int\limits_0^\infty v_1^2 e^{-bv_1^2}dv_1 \int\limits_0^{v_1} \frac{3v_1^2 + v_2^2}{3v_1} e^{-bv_2^2} v_2^2 dv_2, \qquad \text{where } b = \frac{m}{2K_BT}.$$

In the integral I_1, integration with respect to v_2 is first performed for given values of v_1 from 0 to ∞, followed by integration with respect to v_1 from 0 to ∞. The inequality for the variables v_1 and v_2 is $v_1 > v_2 > 0$, and this can be equivalently written as $v_2 < v_1 < \infty$.

To evaluate I_1, we can invert the order or integration so that the integration with respect to v_1 is first carried out from v_2 to ∞, and then we integrate with respect to v_2 from 0 to ∞.

$$I_1 = \int\limits_0^\infty v_2^2 e^{-bv_2^2}dv_2 \int\limits_{v_2}^\infty \left(v_1^3 + \frac{v_2^2}{3}v_1\right) e^{-bv_1^2}dv_1$$

$$= \int\limits_0^\infty v_2^2 e^{-bv_2^2}dv_2 \left[\int\limits_0^\infty v_2 e^{-bv_1^2}dv_1 + \int\limits_{v_2}^\infty \frac{v_2^3}{3}v_1 e^{-bv_1^2}dv_1\right]$$

$$= \int\limits_0^\infty v_2^2 e^{-bv_2^2}dv_2 \left[2\left(\frac{1}{2b}\right)^2 e^{-bv_2^2} + \frac{4}{3}v_2^2 e^{-bv_2^2}\left(\frac{1}{2b}\right)\right]$$

$$= \int\limits_0^\infty 2\left(\frac{1}{2b}\right)^2 e^{-bv_2^2} v_2^2 dv_2 e^{-bv_2^2} + \int\limits_0^\infty \frac{4}{3}\left(\frac{1}{2b}\right) e^{-2bv_2^2} v_2^4 dv_2. \qquad (4.44)$$

This gives rise to

$$I_1 = \left(\frac{1}{2b}\right)\frac{4}{3}\cdot\frac{3}{8}\sqrt{\frac{\pi}{(2b)^5}} + 2\left(\frac{1}{2b}\right)^2\cdot\frac{1}{4}\sqrt{\frac{\pi}{(2b)^3}} = \frac{1}{2}\sqrt{\frac{\pi}{2b}}\left(\frac{1}{2b}\right)^3 + \frac{1}{2}\sqrt{\frac{\pi}{2b}}\left(\frac{1}{2b}\right)^3.$$

$$= \sqrt{\frac{\pi}{2b}}\left(\frac{1}{2b}\right)^3 = \frac{1}{8\sqrt{2}}\sqrt{\frac{\pi}{(b)^7}} = I_2.$$

Hence, we get

$$\langle v_{\text{all}}(\text{rel})\rangle = 16\pi^2 \left(\frac{m}{2\pi K_B T}\right)^3 \left[\frac{1}{8\sqrt{2}}\sqrt{\frac{\pi}{b^7}} + \frac{1}{8\sqrt{2}}\sqrt{\frac{\pi}{b^7}}\right] = 16\pi^2 \left(\frac{b}{\pi}\right)^3 \frac{1}{4\sqrt{2}}\sqrt{\frac{\pi}{b^7}}$$

$$= \frac{2\sqrt{2}}{\sqrt{\pi b}} = 2\sqrt{2}\frac{1}{\sqrt{\frac{m\pi}{2K_B T}}} = \sqrt{2}\sqrt{\frac{8K_B T}{m\pi}} = \sqrt{2}\,\langle v\rangle,$$

where $\langle v\rangle$ is the average velocity of the molecules given by $\langle v\rangle = \sqrt{\frac{8K_B T}{m\pi}}$.

Therefore, the mean free path due to Maxwell is given by

$$\lambda = \frac{\langle v\rangle}{\pi d^2\, n\, \langle v_{\text{all}}(\text{rel})\rangle} = \frac{\langle v\rangle}{\pi d^2\, n\, \sqrt{2}\langle v\rangle} = \frac{1}{\sqrt{2}\,\pi d^2 n}. \tag{4.45}$$

This is the expression for the mean free path of the molecules of a gaseous system due to Maxwell. It should be noted that in this approach for the calculation of mean free path, the molecules are treated as hard spheres, whereas the real molecules are not really hard spheres; they possess different shapes. For noble gases, the collisions are probably close to being perfectly **elastic**, so the hard sphere approximation is probably a good approximation. But, the real molecules may have a **dipole moment** and hence, have a significant electrical interaction as they approach each other. This has been taken into consideration by using an electrical potential for the molecules to refine the calculation and also by using the measured **viscosity of the gas** as a parameter to refine the estimate of the mean free path of molecules in a real gas.

Effect of vacuum on the mean free path

To discuss the effect of vacuum on the mean free path, we take the help of the following example. Consider an oxygen molecule with a diameter of 4.0×10^{-10} m at room temperature. Let the pressure be $P = 10^{-x}$ Torr $= 10^{-x}$ mm of Hg $= \frac{(10^{-x})}{760}$ atm $= \frac{(10^{-x} \times 1.01 \times 10^5)}{760}$ Pa, where x is an integer.

According to Maxwell's treatment, the expression for the mean free path λ is given by

$$\lambda = \frac{1}{\sqrt{2}\pi d^2 n} = \frac{RT}{\sqrt{2}\pi d^2 N_A P}$$

$$= \frac{8.31\ \text{JK}^{-1} \times 273\ \text{K}}{\sqrt{2}\pi(4.0 \times 10^{-10})^2 \times (6.02 \times 10^{23})(1.01 \times 10^5\ \text{Pa}) \times \frac{10^{-x}}{760}}$$

$$= 3.992 \times 10^{x-5}\ \text{m}.$$

For $P = 10^{-5}$ Torr (in the laboratory), we have

$$\lambda = 3.992\ \text{m}.$$

For $P = 10^{-9}$ Torr (ultrahigh vacuum), we have

$$\lambda = 39920 \text{ m}.$$

> **From an experimental point of view, these results are very interesting. At a very high vacuum, the mean free path of the molecules will be extremely large. Hence, there will be negligible scattering of the molecules, and we will get a highly collimated beam of molecules. Therefore, scattering experiments are performed in the laboratory at a very high vacuum.**

4.3.5 Concept of pressure from mean free path

In this section, we will explore the concept of pressure through the lens of the mean free path.

Equation for pressure from the mean free path

In the gas assembly, let dV be an elemental volume at a distance r from an elemental area dA, on the container wall XY, in a direction making an angle θ with the normal to dA as shown in Figure 4.10. The total number of molecules in dV is $n\ dV$, where n is the molecular density. If p_c be the collision frequency of any molecule, the total number of collisions within this elementary volume dV in time dt is given by $\frac{1}{2}\ p_c\ n\ dV\ dt$, the factor $\frac{1}{2}$ is introduced to avoid counting each collision twice, since the collision between molecules 1 and 2 and that between 2 and 1 is a single collision.

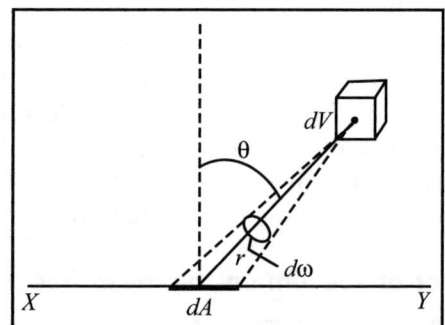

Figure 4.10 Pressure using the concept of mean free path.

Now, each collision generates two new free paths. So, the number of free paths originating from the elementary volume dV in time dt due to these collisions is

$$2 \times \frac{1}{2}\ p_c\ n\ dV\ dt = p_c\ n\ dV\ dt.$$

Since these free paths start off uniformly in all directions, the number of molecules heading toward the elemental area dA is given by

$$N_0 = \frac{p_c n dV dt}{4\pi} \times d\omega, \tag{4.46}$$

where $d\omega$ is the solid angle subtended by dA at dV. This solid angle $d\omega$ is given by $d\omega = \frac{dA\cos\theta}{r^2}$ and, in spherical polar coordinates (r, θ, ϕ), the elementary volume dV is given by $dV = r^2 \sin\theta d\theta d\phi dr$. Using these expressions in equation (4.46), we get the number of molecules N_0 as

$$N_0 = \frac{1}{4\pi} \times r^2 \sin\theta d\theta d\phi dr \times p_c n dt \times \frac{dA\cos\theta}{r^2} = \frac{1}{4\pi} p_c n \, dAdt \, \sin\theta \cos\theta d\theta d\phi dr.$$

(4.47)

Out of these number N_0, the number of molecules N reaching the elementary area dA without any collision in a very small interval of time dt can be obtained from the survival equation. Hence, N will be given by

$$N = N_0 \, e^{-\frac{r}{\lambda}} = \frac{1}{4\pi} p_c n dAdt \sin\theta \cos\theta d\theta e^{-\frac{r}{\lambda}} d\phi dr.$$

These N molecules will make collisions with the area dA, suffer a change in momentum, and generate pressure. The change in momentum of a molecule per collision is given by

$$mv\cos\theta - (-mv\cos\theta) = 2mv\cos\theta.$$

So, the total change in momentum of all the molecules (from all the directions and distances) striking the area dA in time dt is given by

$$\frac{1}{4\pi} \, p_c \, n \, dAdt \, 2mv \int_0^{\frac{\pi}{2}} \cos^2\theta \sin\theta d\theta \int_0^{\infty} e^{-\frac{r}{\lambda}} dr \int_0^{2\pi} d\phi$$

$$= \frac{1}{4\pi} \, p_c \, n \, dAdt \times 2mv \times \frac{1}{3} \times \lambda \times 2\pi$$

$$= \frac{1}{3} mnv \, p_c \, \lambda dAdt = \frac{1}{3} mnv^2 dAdt. \quad (\text{since } p_c = \frac{v}{\lambda})$$

Let $\langle dF \rangle$ be the average force exerted by the molecules in area dA due to this change in momentum. Then, this average force $\langle dF \rangle$ per unit time will be given by

$$\langle dF \rangle = \frac{1}{3} mn \langle v^2 \rangle dA.$$

So, the expression for pressure P is given by

$$P = \frac{\langle dF \rangle}{dA} = \frac{1}{3} mn \langle v^2 \rangle = \frac{1}{3} mn \langle v^2 \rangle,$$

which is the required expression for pressure exerted by the molecules of an ideal gas.

Expression for pressure of a two-dimensional ideal gas

Consider a two-dimensional ideal gas whose molecules are constrained to move in a plane (two-dimensional box), and let dl be an element of the boundary. Consider molecules striking the boundary at an angle θ (shown in Figure 4.11) with the normal to the boundary and let $f(v)$ be the velocity distribution (normalized) for the two-dimensional gas. The number of molecules having speeds between v and $v + dv$ is $nf(v)dv$, n being the number of molecules per unit area.

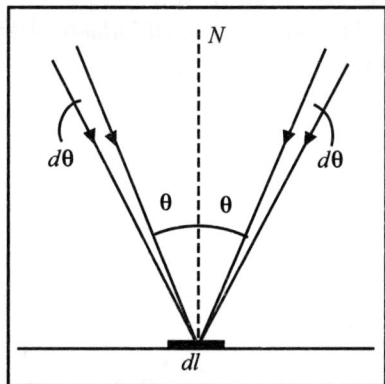

Figure 4.11 Length element dl to calculate pressure of an ideal gas in two dimensions.

The number of molecules with speed v striking a length dl at an angle between θ and $\theta + d\theta$ per unit time is given by

$$dN = nf(v)dv \times v\cos\theta dl \times \left(\frac{d\omega}{2\pi}\right),$$

where 2π is the total solid angle in two dimensions and $d\omega = 2d\theta$.

So, the number of molecules striking per unit length per unit time is given by

$$N = n \int\limits_{0}^{\infty} vf(v)dv \int\limits_{0}^{\frac{\pi}{2}} \frac{2d\theta}{2\pi}\cos\theta = n\frac{\langle v \rangle}{\pi} \quad (dl = 1). \qquad (4.48)$$

With reference to Figure 4.11, the expression for the force exerted by the molecules can be written as

$$F = \int nf(v)dv \, \frac{2d\theta}{2\pi} \, v\cos\theta \times 2mv\cos\theta dl = 2\frac{mn\langle v^2 \rangle}{\pi} \int\limits_{0}^{\frac{\pi}{2}} \cos^2\theta d\theta \times dl.$$

$$= \frac{2mn\langle v^2 \rangle}{\pi}\left[\int\limits_{0}^{\frac{\pi}{2}} \frac{1}{2}(1 + \cos 2\theta)\right]dl = \frac{2mn\langle v^2 \rangle}{\pi} \times \frac{1}{2}\left[\frac{\pi}{2} + \left(\frac{\sin 2\theta}{2}\right)_{0}^{\frac{\pi}{2}}\right]dl.$$

Hence, the expression for **pressure** exerted by the molecules of a two-dimensional gas is given by

$$p = \frac{F}{dl} = \frac{2mn\langle v^2 \rangle}{2\pi} \times \frac{\pi}{2} = \frac{1}{2}mn\langle v^2 \rangle = \frac{1}{2}\rho\langle v^2 \rangle.$$

For a two-dimensional ideal gas, the function $f(v)$ is the normalized isotropic distribution of velocities of a two-dimensional gas.

4.4 Transport phenomena

The KTG establishes the fact that each molecule of a gas has a finite mass and is characterized by random molecular velocity. Therefore, a moving molecule possesses mass, momentum, and energy. Thus, as a molecule moves from one region of a container to another, it can be regarded as a carrier of these three physical quantities. When a gas is in equilibrium, there is no net transfer of mass, momentum, or energy since the rate of transfer across a given plane is exactly balanced by an equal movement in the opposite direction. However, when the gas is endowed with macroscopic velocity, that is, the entire gas as a whole or a part of it is moving in a particular direction, the following three cases may occur singly or jointly:

1. When different regions of a gas move at varying velocities, relative motion is established between adjacent layers. In this scenario, the faster-moving layers transfer momentum to the slower-moving ones. As a result, there is a net transport of momentum across an imaginary plane, aligned with the preferential direction of motion. This phenomenon is known as viscosity. Notably, unlike in liquids, where viscosity arises from frictional forces between layers, in gases, it is driven by the random thermal motion of molecules.

2. When various regions of a gas have different temperatures, the molecules in the higher-temperature areas transfer thermal energy to the molecules in the lower-temperature regions. This process, driven by the need to achieve thermal equilibrium, results in **the phenomenon of thermal conduction**.

3. When different regions of a gas have varying concentrations, molecules from areas of higher concentration move toward areas of lower concentration. This movement results in the transport of matter, leading to **the phenomenon of diffusion**.

Thus, it is to be noted that the physical phenomena such as viscosity, thermal conduction, and diffusion, respectively, represent the transport of momentum, energy, and mass. For this reason, these physical processes are categorized under the title of **transport phenomena**. Such physical processes are of vital importance and frequently appear in various branches of physical sciences.

4.4.1 Connection between viscosity and thermal conductivity of a gas from the concept of mean free path

The above mentioned three transport phenomena involve the transfer of three different physical entities: *momentum*, *thermal energy*, and *mass*. According to KTG, the molecules of a gas are in a state of thermal agitation and attain an equilibrium state by transporting momentum, heat (thermal energy), and mass from one layer of the gas to another layer, giving rise to viscosity, thermal conductivity, and diffusion, respectively. In this section, the relation between the coefficient of viscosity η and the thermal conductivity K of a gaseous system will be established from the concept of mean free path λ.

In order to establish the relation between the coefficient of viscosity η and the thermal conductivity K, we consider a gaseous system in a volume V with temperature T and pressure P. Further, we denote the value of the physical quantity momentum or thermal energy by H, and this quantity has different values in different layers of the gaseous system. Let dV be an elementary volume at $P(r,\theta)$ from the origin O so that it is situated on a layer at a height $z = r \cos\theta$ from the XY-plane. Such an elementary volume dV is shown in Figure 4.12. If H be the value of the physical entity at XY-plane, that is, at $z=0$, it would have a value $H + \frac{dH}{dz}$ r $\cos\theta$ at the layer on which the elementary volume dV is situated. Further, it is *assumed* that H *increases upwards.*

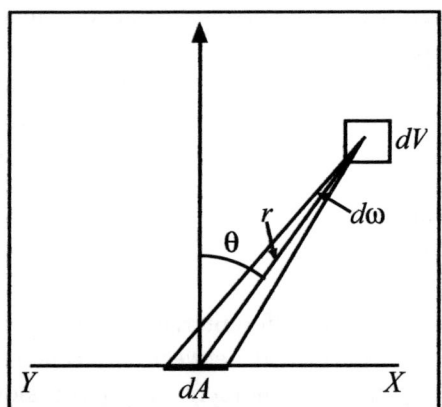

Figure 4.12 A schematic diagram illustrating the calculation of the average number of molecules crossing an elementary area dA in the XY-plane.

At an identical layer below the reference plane XY, the value of H would then be $H - \frac{dH}{dz}$ r $\cos\theta$.

We assume further that dn_v is the number of molecules per unit volume with velocity between v and $v + dv$. Hence, the number of molecules in the elementary volume dV is dn_v dV. The molecules inside this elementary volume dV will suffer collisions and would come out from the volume in all possible directions.

If p_c be the collision frequency of the molecules, the number of molecules suffering collisions in the elementary volume dV per second will be given by

$$p_c \, dn_v \, dV.$$

Since each collision involves two molecules, then the number of collisions in time dt will be

$$\left(\frac{1}{2}\right) (p_c \, dn_v \, dV \, dt).$$

Each collision produces *two* new paths following which two molecules come out. The total number of new paths for the molecules to escape from dV in time dt by way of collisions is therefore given by

$$2 \times \frac{1}{2} \, (p_c \, dn_v \, dV \, dt) = p_c \, dn_v \, dV \, dt.$$

These molecules $p_c\, dn_v\, dV\, dt$ will come out of the elementary volume dV in all possible directions. So, the number of molecules moving toward an elementary area dA of the reference plane are those contained in the solid angle subtended by dA at dV. This solid angle is given by $d\Omega = \frac{dA\cos\theta}{r^2}$.

So, the number of molecules, say N_0, moving toward the area dA in time dt will be given by

$$N_0 = \left(\frac{p_c\, dn_v\, dV\, dt}{4\pi}\right) d\Omega = \left(\frac{p_c\, dn_v\, dV\, dt}{4\pi}\right)\left(\frac{dA\cos\theta}{r^2}\right).$$

Some of these molecules, however, will go away from their paths due to collisions during the journey. The number of molecules N actually reaching the area dA in time dt will be obtained from the **survival equation** and is given by

$$N = N_0\, e^{-\frac{r}{\lambda}} = \left(\frac{p_c\, dn_v\, dV\, dt}{4\pi}\right)\left(\frac{dA\cos\theta}{r^2}\right)e^{-\frac{r}{\lambda}}$$

$$= \frac{dA\cos\theta}{4\pi r^2}\left(\frac{v}{\lambda}\right)dn_v\,(r^2\sin\theta\, d\theta\, d\phi\, dr)\,e^{-\frac{r}{\lambda}}dt;$$

we have used $(dV = r^2\sin\theta\, d\theta\, d\phi\, dr)$

$$= \frac{dA\, dt}{4\pi}v\, dn_v\cos\theta\sin\theta d\theta d\phi\,\frac{e^{-\frac{r}{\lambda}}}{\lambda}\, dr.$$

Here, we have used the relation $p_c = \frac{v}{\lambda}$.

The total transport of the physical entity H in the downward direction (assuming that the molecules from a certain layer carry with them the entity characteristic of the layer) through dA in time dt is given by

$$H_{\text{down}} = \frac{dA\, dt}{4\pi}\int_{v=0}^{\infty}\int_{\theta=0}^{\frac{\pi}{2}}\int_{\phi=0}^{2\pi}\int_{r=0}^{\infty}v\, dn_v\cos\theta\sin\theta d\theta d\phi\frac{e^{-\frac{r}{\lambda}}}{\lambda}\left(H + \frac{dH}{dz}r\cos\theta\right)dr.$$

Similarly, the total transport of H through dA in time dt in the upward direction is given by

$$H_{\text{up}} = \frac{dAdt}{4\pi}\int_{v=0}^{\infty}\int_{\theta=0}^{\frac{\pi}{2}}\int_{\phi=0}^{2\pi}\int_{r=0}^{\infty}v\, dn_v\cos\theta\sin\theta d\theta d\phi\frac{e^{-\frac{r}{\lambda}}}{\lambda}\left(H - \frac{dH}{dz}r\cos\theta\right)dr.$$

So, the net transfer of H from the top layer to the bottom layer toward the reference plane through the area dA in time dt is given by

$$H_{\text{net}} = H_{\text{down}} - H_{\text{up}} = \frac{2dAdt}{4\pi}\frac{dH}{dz}\int\limits_{v=0}^{\infty} v\,dn_v \int\limits_{0}^{\frac{\pi}{2}} \cos^2\theta\sin\theta d\theta \int\limits_{0}^{2\pi} d\phi \int\limits_{0}^{\infty} \frac{re^{-\frac{r}{\lambda}}}{\lambda}dr$$

$$= \frac{dAdt}{4\pi}\frac{dH}{dz}\int\limits_{0}^{\infty} v\,dn_v \times \frac{1}{3} \times 2\pi \times \lambda\Gamma(2) = \frac{1}{3}dAdt\frac{dH}{dz}\lambda\int\limits_{0}^{\infty} v\,dn_v$$

$$= \frac{1}{3}dAdt\frac{dH}{dz}\lambda n\langle v\rangle, \qquad (\Gamma(2) = 1).$$

This expression leads to the net transfer of H as

$$H_{\text{net}} = \frac{1}{3}\,dA\,dt\,\frac{dH}{dz}\,\lambda\,n\,\langle v\rangle. \tag{4.49}$$

We shall now identify momentum and thermal energy with H in turn and obtain the expressions for the coefficient of viscosity η and thermal conductivity K of the gaseous system, respectively.

The coefficient of viscosity of a gas

We know that in the case of viscosity, the concerned physical entity H stands for momentum mv. Hence, we have

$$H \equiv \text{momentum} = m\,v; \quad \Rightarrow \quad \frac{dH}{dz} = m\,\frac{dv}{dz}.$$

Using this expression in equation (4.49), we get the net transfer of momentum per second, that is, the force F as

$$F = \frac{H_{\text{net}}}{dt} = \frac{1}{3}\,dA\,n\,\langle v\rangle\,\lambda\,\frac{dH}{dz} = \frac{1}{3}\,dA\,n\,\langle v\rangle\,\lambda\,m\,\frac{dv}{dz}. \tag{4.50}$$

Now, according to the definition of the coefficient of viscosity η, we have the expression for viscous force F_v as

$$F_v = \eta\,dA\,\frac{dv}{dz}. \tag{4.51}$$

Equating equations (4.50) and (4.51), we get

$$\eta\,dA\,\frac{dv}{dz} = \frac{1}{3}\,dA\,n\,\langle v\rangle\,\lambda\,m\,\frac{dv}{dz}.$$

This leads to the expression for the coefficient of viscosity η as

$$\eta = \frac{1}{3}\,m\,n\,\langle v\rangle\,\lambda = \frac{1}{3}\,\rho\,\langle v\rangle\,\lambda, \tag{4.52}$$

where $mn = \rho$ is the mass density of molecules of the gaseous system. This is the expression for the coefficient of viscosity obtained from the KTG.

It should be mentioned that the expression $\eta = \frac{1}{3}\, \rho\, \langle v \rangle\, \lambda$ is an approximate one and requires corrections for the following reasons:

1. The mean free path used is not Maxwell's mean free path,

2. The velocity distribution and the free path distribution have been disregarded in writing some of the expressions, etc.

Equation (4.52) provides us a remarkable result based on the elementary kinetic theory. According to this equation, the coefficient of viscosity η of a gaseous system is found to be directly proportional to the average velocity $\langle v \rangle$. Further, it enables us to determine the molecular diameter d, since η is a directly measurable physical quantity. It is noted from equation (4.52) further that the viscosity coefficient η is independent of pressure. However, it has a $T^{\frac{1}{2}}$ dependence on temperature through the average velocity $\langle v \rangle$. Both these conclusions are in conformity with experimental results and are presented in detail below.

Effect of pressure P on the coefficient of viscosity η

For instance, for pressures from a few mm of mercury up to several atmospheres, the coefficient of viscosity of a gas is found to be independent of pressure. However, at very low or very high pressures, this does not hold. At very low pressure, intermolecular collisions are rare, and the mean free path becomes comparable with the dimensions of the apparatus. But the number density decreases continuously as pressure is lowered. Consequently, the coefficient of viscosity decreases as pressure decreases. This fact was experimentally verified by Crookes.

Warburg and von Babo showed that at very high pressures, the coefficient of viscosity η increases as pressure P increases. This could be understood in the following way: With the increase in pressure, the intermolecular distance decreases. As a result, the intermolecular force increases. Hence, the relative velocity between two adjacent layers decreases. Further, at such high pressures, the mean free path becomes comparable with the molecular size and leads to frequent collisions among the molecules. Due to these factors, at high pressure, the coefficient of viscosity increases with the increase in pressure.

Effect of temperature T on the coefficient of viscosity η

In general, the viscous property of a fluid depends strongly on temperature. Using the expressions for average speed $\langle v \rangle = \sqrt{\frac{8 K_B T}{m \pi}}$ and mean free path $\lambda = \frac{1}{\sqrt{2}\pi d^2 n}$ into equation (4.52), the mathematical form for the coefficient of viscosity η of a gaseous system can be written as

$$\eta = \frac{1}{3} \, m \, n \, \langle v \rangle \, \lambda = \frac{1}{3} \, m \, n \, \sqrt{\frac{8K_B T}{m\pi}} \, \frac{1}{\sqrt{2}\pi d^2 n} = \frac{m}{3\sqrt{2}\pi d^2} \, \sqrt{\frac{8K_B}{m\pi}} \, \sqrt{T}. \qquad (4.53)$$

This equation (4.53) shows that the viscosity of a gas increases with temperature. This could be understood as follows: With the increase in temperature T, the velocity of the gaseous molecules increases that increases the momentum flux. Since the transfer of momentum is caused by the free motion of gas molecules between collisions, increasing the thermal velocity of the molecules results in a larger viscosity. Thus, the viscosity of a gaseous system increases with temperature.

In the gas phase, the intermolecular forces can be neglected to a good approximation, but in a liquid phase, these forces have a major effect on the viscous property. In liquids, the coefficient of viscosity usually decreases with the increase in temperature. The reason for the reduction in viscosity with increasing temperature in a liquid is that most of the energy goes into overcoming intermolecular forces, thereby making it easier for the molecules to move past each other.

The thermal conductivity K

In the case of thermal conduction, the physical entity H represents the thermal energy E, so that

$$\frac{dH}{dz} = \frac{dE}{dz}.$$

Using equation (4.49) and the above expression, we get the flow of thermal energy per second as

$$F = \frac{H_{\text{net}}}{dt} = \frac{1}{3} \, dA \, n \, \langle v \rangle \, \lambda \frac{dH}{dz} = \frac{1}{3} \, dA \, n \, \langle v \rangle \, \lambda \frac{dE}{dz}. \qquad (4.54)$$

We know from the law of conduction of heat that the quantity of heat Q flowing per second across an area dA is given by

$$Q = K \, dA \, \frac{dT}{dz}, \qquad (4.55)$$

where K is the coefficient of thermal conduction, and $\frac{dT}{dz}$ is the temperature gradient. Equating equations (4.54) and (4.55), we get

$$K \, dA \, \frac{dT}{dz} = \frac{1}{3} \, dA \, n \, \langle v \rangle \, \lambda \frac{dE}{dz}; \quad \Rightarrow \quad K = \frac{1}{3} \, n \, \langle v \rangle \, \lambda \frac{dE}{dT}.$$

Let m be the mass, c_v the specific heat at constant volume, and T the temperature of the gaseous system, we then have $\frac{dE}{dT} = m\,c_v$. Using this value of $\frac{dE}{dT}$, the expression for K comes out to be

$$K = \frac{1}{3}\,n\,\langle v \rangle\,\lambda\,m\,c_v = \frac{1}{3}\,m\,n\,\langle v \rangle\,\lambda\,c_v = \frac{1}{3}\,\rho\,\langle v \rangle\,\lambda\,c_v. \tag{4.56}$$

This equation (4.56) gives the expression for thermal conductivity K of a gaseous system.

The following points are to be noted from equation (4.56):

1. The mean free path λ is inversely proportional to the pressure P, that is, $\lambda \propto \frac{1}{P}$. Again, the molar concentration c_m of a gas is directly proportional to the pressure P, that is, $[c_m] \propto P$. Combining these two results, it is observed that the thermal conductivity K becomes independent of pressure P.

2. Equation (4.56) indicates that the thermal conductivity K is directly proportional to average molecular speed $\langle v \rangle$. Therefore, the theory predicts that the thermal conductivity K is directly proportional to the square root of temperature T. However, in actual practice, the thermal conductivity increases more rapidly than the prediction. This indicates that when the intermolecular forces come into play, they begin to influence the transport of energy.

Relation between the coefficient of viscosity η and the thermal conductivity K

Using the value of coefficient of viscosity $\eta = \frac{1}{3}\,m\,n\,\langle v \rangle\,\lambda$ from equation (4.52) into equation (4.56), we get

$$K = \frac{1}{3}\,\rho\,\langle v \rangle\,\lambda\,c_v = \eta\,c_v. \tag{4.57}$$

This is the relation between the thermal conductivity K and the coefficient of viscosity η for an ideal gaseous system as obtained from the KTG. The relation $K = \eta\,c_v$ incidentally establishes a relationship between a purely *mechanical* (coefficient of viscosity) and a purely *thermal* (thermal conductivity) phenomenon in a gaseous system. The following comments about the ratio $\frac{K}{\eta\,c_v}$ are in order:

1. It is to be noted from equation (4.57) that the ratio $\frac{K}{\eta\,c_v}$ is constant, equal to 1, and this ratio is the same for all gases.

2. Experimental results show that the ratio $\frac{K}{\eta\,c_v}$ is greater than one. A more rigorous calculation shows that this ratio varies between 1.00 and 2.57 and is different for different gases, that is, this ratio depends on the nature of the gas.

3. For polyatomic molecules, the transfer of energy takes place via both translational and rotational degrees of freedom. This consideration leads to the value of the ratio as

$$\frac{K}{\eta\, c_v} = \frac{9\gamma - 5}{4},$$

where γ is the ratio of specific heats. This relation shows that the ratio decreases with the increase in atomicity.

The process of diffusion

Let us consider two gases, hydrogen and oxygen, enclosed in two jars made of glass. The jar containing hydrogen is inverted over the jar containing oxygen, and the lids are removed. As such, in either direction, there will be no large scale movement of these gases that are at the

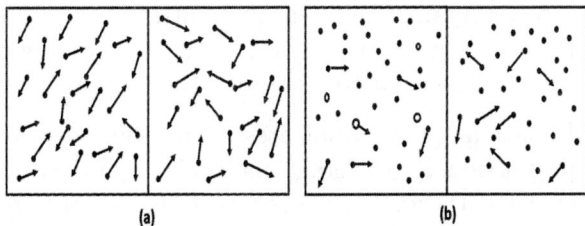

(a) (b)

Figure 4.13 (a) Self-diffusion: Two identical samples of a gas in a container are separated by a barrier. As soon as the barrier is removed, they diffuse into one another. (b) Diffusion of unlike molecules.

same temperature and pressure. But after some time, one finds that the two gases mix with one another. Heavier oxygen molecules in the lower jar move against gravity. This gradual intermixing of gases is called diffusion.

The process of diffusion is responsible for spreading the smell of flowers in all directions. Similarly, due to the diffusion process only, we can sense, sitting in the study room, what is being cooked in the kitchen. Diffusion is a direct consequence of random molecular motion when there are inequalities in concentration between different parts of the system. Molecules diffuse from higher concentration regions toward lower concentration regions. Figure 4.13(a) schematically shows the diffusion of like molecules. Diffusion of unlike molecules is shown in Figure 4.13(b). Thus, diffusion is the phenomenon arising out of the transference of mass due to difference in concentration and results finally in equilibrium distribution in concentration. Diffusion may be of two different types: one is *self-diffusion*, when it occurs in a single gaseous system and the other interdiffusion when two different gases communicate with each other.

For finding out an expression for diffusion coefficient D, we shall confine ourselves to the case of *self-diffusion* only. Let n be the concentration on the reference plane XY (Figure 4.12) and those above and below the XY-plane be $n + \frac{dn}{dz}$ and $n - \frac{dn}{dz}$, respectively. Here $\frac{dn}{dz}$ is the concentration gradient.

We consider an elemental volume dV at (r, θ) from an area dA of the reference plane XY as shown in Figure 4.12. The vertical height of dV from the reference plane is $z = r \cos \theta$. Let dn_v be the number of molecules per unit volume having velocity lying between v and $v + dv$ obeying Maxwell's speed distribution law. Then, the number of molecules contained in the elementary volume dV having velocity between v and $v + dv$ is given by

$$4\pi \left(n + r \cos \theta \frac{dn}{dz} \right) A^3 \, e^{-bv^2} v^2 dv \, dV.$$

Because of collisions, the number of new paths dn' arising out of the elemental volume dV in time dt is given by

$$dn' = \left(\frac{v}{\lambda} \right) 4\pi \left(n + r \cos \theta \frac{dn}{dz} \right) A^3 \, e^{-bv^2} \, v^2 dv \, dV \, dt.$$

Of these, the number of molecules moving toward the area dA in time dt is given by

$$dn = \frac{dn'}{4\pi} \frac{dA \, \cos \theta}{r^2} = \frac{dA \, \cos \theta}{4\pi r^2} \frac{v^3}{\lambda} \left(n + r \cos \theta \frac{dn}{dz} \right) 4\pi A^3 \, e^{-bv^2} dv \, dV \, dt$$

$$= A^3 dA \, dt \, \sin \theta \cos \theta \left(n + r \cos \theta \frac{dn}{dz} \right) d\theta d\phi$$

$$\times v^3 \, e^{-bv^2} dv \, \frac{dr}{\lambda}. \quad (dV = r^2 \sin \theta d\theta d\phi dr).$$

Out of these again, the number of molecules that actually cross the area dA in time dt is obtained from survival equation and is given by $dn \times e^{-\frac{r}{\lambda}}$. Hence, the total number of molecules crossing dA in time dt in the downward direction is given by

$$N \downarrow = A^3 \, dA \, dt \int \int \int \int \left(n + r \cos \theta \frac{dn}{dz} \right) \sin \theta \, \cos \theta \, d\theta d\phi \frac{e^{-\frac{r}{\lambda}}}{\lambda} dr \, v^3 \, e^{-bv^2} dv.$$

Similarly, the total number of molecules crossing the area dA in time dt in the upward direction is given by

$$N \uparrow = A^3 \, dA \, dt \int \int \int \int \left(n - r \cos \theta \frac{dn}{dz} \right) \sin \theta \, \cos \theta \, d\theta d\phi \frac{e^{-\frac{r}{\lambda}}}{\lambda} dr \, v^3 \, e^{-bv^2} dv.$$

Hence, the net number of molecules N_{net} crossing dA in time dt in the downward direction is given by

$$N_{\text{net}} = N \downarrow - N \uparrow = 2A^3 \, dA \, dt \frac{dn}{dz} \int_0^\infty v^3 \, e^{-bv^2} \, dv \int_0^{\frac{\pi}{2}} \sin\theta \cos^2\theta d\theta \int_0^{2\pi} d\phi \int_0^\infty \frac{re^{-\frac{r}{\lambda}}}{\lambda} dr$$

$$= 2A^3 \, dA \, dt \frac{dn}{dz} \times \frac{1}{2b^2}\Gamma(2) \times \frac{1}{3} \times 2\pi \times \lambda\Gamma(2)$$

$$= \frac{1}{3} dA \, dt \frac{dn}{dz} \times \frac{A^3}{b^2} \times 2\pi \times \lambda \quad (as \ \Gamma(2) = 1)$$

$$= \frac{1}{3} dA \, dt \frac{dn}{dz} \sqrt{\frac{8K_BT}{\pi m}} \left(as \ A = \sqrt{\frac{m}{2\pi K_BT}}, \ b = \frac{m}{2K_BT} \right) = \frac{1}{3}\lambda\langle v\rangle \ dA \ dt \frac{dn}{dz}.$$

Here, the expression for average velocity $\langle v\rangle = \sqrt{\frac{8K_BT}{\pi m}}$ has been used.

The **coefficient of diffusion** D is defined as the *number of molecules that cross unit area per unit time for unit concentration gradient*, that is, $\frac{dn}{dz} = 1$. Using these conditions, we get the expression for the diffusion coefficient D as

$$D = \frac{1}{3} \lambda \langle v\rangle. \tag{4.58}$$

Using value of the coefficient of viscosity η from equation (4.52) into equation (4.58), we get

$$D = \frac{1}{3} \lambda \langle v\rangle = \frac{1}{3} \rho \langle v\rangle \lambda \times \frac{1}{\rho} = \frac{\eta}{\rho}. \tag{4.59}$$

Equation (4.59) provides the relation between the coefficient of self-diffusion D and the coefficient of viscosity η, as obtained from the KTG.

Some comments about the self-diffusion coefficient D

According to equation (4.58), the expression for self-diffusion coefficient D is given by

$$D = \frac{1}{3} \langle v\rangle \lambda. \tag{4.60}$$

Using the expressions for average speed $\langle v\rangle = \sqrt{\frac{8K_BT}{m\pi}}$ and the mean free path $\lambda = \frac{1}{\sqrt{2}\pi d^2 n}$ into equation (4.60), the mathematical expression for the self-diffusion coefficient D of a gaseous system can be written as

$$D = \frac{1}{3} \times \sqrt{\frac{8K_BT}{m\pi}} \times \frac{1}{\sqrt{2}\ \pi d^2 n} = \frac{1}{3\sqrt{2}\pi d^2 n} \sqrt{\frac{8K_B}{m\pi}} \sqrt{T}$$

$$= \frac{1}{3\sqrt{2}\pi d^2} \sqrt{\frac{8K_B^3}{m\pi}} \frac{T^{\frac{3}{2}}}{P}. \tag{4.61}$$

The following points about equation (4.61) are in order:

1. This equation shows that the self-diffusion coefficient D is directly proportional to three and half powers of temperature T and is inversely proportional to pressure P.

2. Experimentally, the inverse dependence of D on P is observed in the gaseous system, but the temperature dependence is not found to be in accordance with $\propto T^{1.5}$ but with the value of 1.75 to 2. This change is observed due to the presence of attractive forces between molecules.

3. Equation (4.61) shows that $D \propto \frac{1}{\sqrt{m}}$. This means that the amount of diffusion per second (diffusion rate) is inversely proportional to $\sqrt{m}$. This is in agreement with Graham's law of diffusion.

4.4.2   Brownian motion and its significance

In KTG, it was assumed that the molecules of a gas are in a state of constant random motion, and based on this assumption, several physical properties of a gaseous system were explained providing approximate (at least) numerical estimates. This assumption was directly verified when English Botanist Robert Brown observed the motion of very small pollen grains suspended in an aqueous solution. The particles were seen to exhibit a completely erratic and perpetual movement. This motion is termed as **Brownian Motion**. Such an erratic motion of a fine particle suspended in an aqueous solution is schematically shown in Figure 4.14. This Brownian motion of a particle is influenced by several physical factors, such as temperature and size of the particles and has a number of characteristics features. Some of these characteristic features are mentioned below:

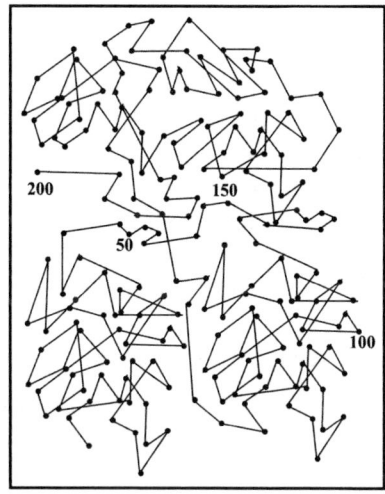

Figure 4.14 Brownian motion of a fine particle suspended in an aqueous solution.

1. The Brownian motion of particles is continuous, completely random, and irregular.

2. No two particles execute the same type of motion. (This implies that it cannot be caused by fluid currents.)

3. Smaller particles execute faster motion, and hence, are more noticeable in experimental situation.

4. The motion becomes more vigorous and lively when the temperature is increased, or a less viscous fluid is taken. For example, it is just perceptible in glycerin and very active in gas.

5. The Brownian motion of the particles is about the same in all directions.

6. This motion is independent of external influences.

The discovery of Brownian motion puzzled scientists for a long time. At first, it was regarded as a property of **organic matter**, but it was established that small particles of **inorganic matter** suspended in both liquids and gases also executed Brownian motion. Thereafter, the origin of this Brownian motion was attributed to (i) surface tension, (ii) non-homogeneity in temperature, (iii) chemical or electrical action, and so on. However, none of these explanations were found to be adequate. After a systematic study, Wiener and Guy proposed that Brownian motion is perhaps due to the bombardment of suspended particles by the molecules of the surrounding fluid. For bigger particles, the forces due to molecular impacts almost completely balance, whereas for smaller particles, the unbalanced force makes them move in a completely random manner.

The phenomenon of Brownian motion provides a very useful picture of the gaseous state of matter. One can suppose that like gas molecules, Brownian particles are also in random motion and frequently collide with each other. In fact, Brownian motion is readily observed in gases because intermolecular forces are negligibly small. This phenomenon provides us a way of visualizing a microstate in a physical system. It is due to this reason that the works of a botanist gained importance in the subject matter of study in physics.

Einstein's theory of Brownian motion

We cannot solve our problems with the same thinking we used when we created them.

—Albert Einstein

In 1905, Einstein gave a mathematical theory on Brownian motion. His arguments were based on physical processes that take place inside the medium. In 1908, Langevin re-derived Einstein's formula by considering the equation of motion of suspended particles. We will, however, first discuss Einstein's theory. Einstein gave an exact description of Brownian motion in terms of the effects of the random collisions between the molecules of the liquid and the suspended particles. Although the effect of each impact is small, the net result of a large number of them–nearly 10^{20} per second under normal collisions–gives rise to "drunken man's walk". Einstein quantified the problem by relating the diffusion coefficient D of the suspended particles to the properties of the molecules responsible for the collision. That is, *he calculated the diffusion coefficient from the erratic motion of particles arising from molecular bombardment.*

We also know that molecules of the solute dissolved in a dilute solution exert pressure, called *osmotic pressure*. (This is known as van't Hoff's law and states that the osmotic pressure is numerically equal to the pressure, which the dissolved substance would exert, if it were assumed to behave like a gas having the same volume and temperature as the given solution.) If a concentration gradient exists between different parts of the solution, *suspended particles will diffuse under the osmotic pressure difference*. This can also be used to calculate the diffusion coefficient D. Einstein calculated the diffusion coefficient D from the random motion of suspended particles as well as the osmotic pressure difference between different parts of the fluid medium caused by the difference in concentration of the suspended particles. He then equated these two expressions for the diffusion coefficient D to calculate the mean square displacement of a Brownian particle.

Calculation of the diffusion coefficient D from random molecular motion

The random motion of the molecules causes Brownian particles to diffuse, and their motion is totally erratic. For simplicity, we confine ourselves to one-dimensional Brownian motion and assume that, on an average, each particle is displaced through a distance s in time τ. Let us consider a cylinder of cross-sectional area

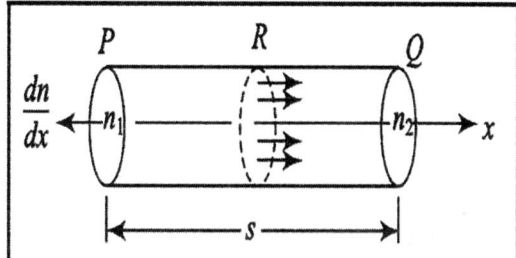

Figure 4.15 Calculation of diffusion coefficient for one-dimensional Brownian motion.

A and length s within the medium with its axis parallel to the X-axis, as shown in Figure 4.15. Its end faces are denoted by P and Q. Let the molecular concentration of the Brownian particles at P be n_1, and that at Q be n_2, such that $n_1 > n_2$. Let n be the mean concentration along the length of the cylinder, and there exists a molecular concentration gradient $\frac{dn}{dx}$ along the axis of the cylinder. This $\frac{dn}{dx}$ makes the suspended particles to diffuse.

The number of particles Γ_R crossing P to the right in time τ is given by

$$\Gamma_R = \frac{n_1}{2} s\, A. \tag{4.62}$$

Factor $\frac{1}{2}$ has been introduced because only half of the particles contained in a cylinder of volume $s\, A$ situated to the left of P will enter the cylinder PQ at P during this time. Similarly, the number of particles Γ_L entering Q and moving along the negative X-direction in time τ is given by

$$\Gamma_L = \frac{n_2}{2} s\, A. \tag{4.63}$$

So, the excess number of particles crossing the vertical plane R at the center of the cylinder in the positive x-direction in time τ is given by

$$\Gamma_{\text{net}} = \Gamma_R - \Gamma_L = (n_1 - n_2)\frac{s\,A}{2}. \tag{4.64}$$

Hence, the number of Brownian particles diffusing through R to the right per unit area per unit time will be given by

$$\frac{\Gamma_{\text{net}}}{A\,\tau} = \frac{(n_1 - n_2)\frac{s\,A}{2}}{A\,\tau} = (n_1 - n_2)\frac{s}{2\,\tau} = -\frac{s^2}{2\,\tau}\frac{dn}{dx}. \tag{4.65}$$

Here, we have used the relation $n_1 - n_2 = -s\frac{dn}{dx}$ since $n_2 = n_1 + \frac{dn}{dx}\,s$.

As the diffusion coefficient D is defined as the number of particles crossing per unit area per unit time when the concentration gradient $\frac{dn}{dx}$ is unity, it readily follows from equation (4.65) that

$$D = \frac{s^2}{2\tau}. \tag{4.66}$$

This is the expression for diffusion coefficient D obtained from the random molecular motion of the suspended particles.

Calculation of the diffusion coefficient D from the osmotic pressure difference

Let us now calculate the expression for diffusion coefficient D from the osmotic pressure difference. Einstein argued that van't Hoff's law should hold for dilute solutions as well as for dilute suspensions. Let the osmotic pressures exerted by the Brownian particles on the faces P and Q of the cylinder be, respectively, p_1 and p_2. If we treat Brownian particles like the molecules of an ideal gas, we can write

$$p_1 = n_1 K_B T \qquad \text{and} \qquad p_2 = n_2 K_B T,$$

where K_B is Boltzmann constant, and T is temperature of the solution. Since $n_1 > n_2$ and T remains the same, the pressure p_1 will be greater than the pressure p_2 and the osmotic pressure difference $(p_1 - p_2)$ will be positive. This pressure difference will give rise to a force F given by

$$F = (p_1 - p_2)A = (n_1 - n_2)K_B T A. \tag{4.67}$$

This force F tends to push the molecules toward the right in the chosen cylinder and is experienced by all $(n\,A\,s)$ particles contained therein. Hence, the magnitude of the force experienced by any one of the Brownian particles is given by

$$f = \frac{F}{n\,A\,s} = \frac{(n_1 - n_2)K_B T A}{n\,A\,s} = \frac{(n_1 - n_2)}{n}\frac{K_B T}{s} = -\frac{K_B T}{n}\frac{dn}{dx}. \tag{4.68}$$

When such a particle moves under the influence of this force due to the osmotic pressure difference, it will be subjected to viscous drag. Assuming all Brownian particles to be of spherical shape of radius r, the viscous force, according to Stokes' law, is given by $f = 6\pi\eta rv$. Equating this viscous force to that given by equation (4.68), we get

$$f = 6\pi\eta rv = -\frac{K_B T}{n}\frac{dn}{dx}; \qquad \Rightarrow \qquad nv = -\frac{K_B T}{6\pi\eta r}\frac{dn}{dx}. \tag{4.69}$$

The product $(n\,v)$ defines the number of particles moving to the right per unit area per second and is, by definition, equal to $-D\,\frac{dn}{dx}$. Using this definition and equation (4.69), it readily follows that

$$D = \frac{K_B T}{6\pi\eta r}. \tag{4.70}$$

This is the expression for diffusion coefficient D as obtained from the osmotic pressure difference. On comparing equations (4.66) and (4.70), we get

$$\frac{s^2}{2\tau} = \frac{K_B T}{6\pi\eta r}; \qquad \Rightarrow \qquad s^2 = \frac{K_B T}{3\pi\eta r}\tau = \frac{RT}{N_A}\frac{1}{3\pi\eta r}\tau. \tag{4.71}$$

This result is known as **Einstein's equation** for the Brownian motion. It is clear from equation (4.71) that the observed diffusion of the Brownian particles is directly related to the molecular properties of gases. The ultramicroscope was introduced in 1903, and this aided quantitative studies of Brownian motion by making visible small colloidal particles whose greater activity could be measured more easily. During the period from 1905 to 1911, several important measurements of this kind were made. During this span of time, the French physicist Jean Baptiste Perrin was successful in verifying Einstein's analysis of Brownian motion, and for this beautiful work, Perrin was awarded the Nobel Prize for Physics in 1926. Perrin's work established the physical theory of Brownian motion and ended the skepticism about the existence of atoms and molecules as actual physical entities.

Vertical distribution of Brownian particles (sedimentation equilibrium)

When a suspension is left undisturbed for a period of time, sedimentation equilibrium is established. The forces governing this type of equilibrium, specifically the vertical distribution of Brownian particles, are:

1. The force due to gravity mg that acts on the particles, if unopposed, would tend to deposit them on the bottom of the container, where the potential energy due to gravity would be a minimum, and

2. The forces due to the collision of the fluid molecules with the particles, if unopposed, would distribute the particles uniformly throughout the fluid.

Under the action of these two types of forces, sedimentation equilibrium is established, and the vertical distribution of the Brownian particles is achieved. To determine this vertical distribution of the Brownian particles, we consider first a vertical column of the gas particles at a temperature T in dynamic equilibrium. Further, we consider the forces acting on a horizontal slice of thickness dZ of the gas bounded by surfaces at heights Z and $Z + dZ$ (Z-axis is taken vertically upward), where the pressures are, respectively, P and $P + dP$. This is schematically shown in Figure 4.16.

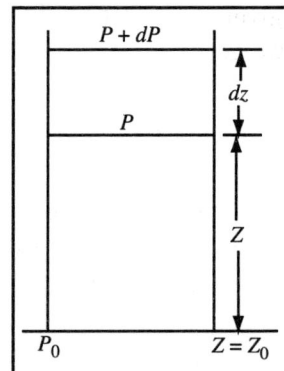

Figure 4.16 Vertical distribution of Brownian particles.

The force acting in the downward direction is the weight and is given by

$$A \, n \, m \, g \, dZ,$$

where n is number of particles per unit volume at height Z, and m is the mass of a particle. The upward force arises from the pressure difference acting on the lower and upper faces of the slice. This force is given by

$$A \, dP.$$

Since the slice is in equilibrium, we must have

$$-A \, n \, m \, g \, dZ = A \, dP. \tag{4.72}$$

The negative sign arises because $\frac{dP}{dZ}$ is obviously negative. Using the ideal gas equation and Boltzmann constant $K_B = \frac{R}{N_A}$, we get

$$P \, V = R \, T; \quad \Rightarrow \quad P = \frac{N_A \, K_B T}{V} = n \, K_B T. \tag{4.73}$$

For an isothermal change, we have from equation (4.73)

$$dP = K_B \, T \, dn. \tag{4.74}$$

Combining equations (4.72) and (4.74), we get

$$\frac{dn}{n} = -\frac{m \, g}{K_B \, T} \, dZ.$$

Integrating this expression between heights $Z = 0$ and $Z = Z$, where the densities of Brownian particles are n_0 and n, respectively, we get

$$\int_{n_0}^{n} \frac{dn}{n} = -\frac{m\,g}{K_B\,T} \int_0^Z dZ; \quad \Rightarrow \quad \ln n - \ln n_0 = -\frac{m\,g}{K_B\,T}\,Z.$$

This leads to

$$n = n_0\,e^{-\frac{m\,g}{K_B\,T}\,Z}. \tag{4.75}$$

Equation (4.75) gives the variation of the number of particles per unit volume as a function of height. This equation can also be applied to Brownian particles of uniform size if the weight mg is replaced by the effective weight of a particle in a fluid taking buoyancy into account. Assuming the spherical shape of the particles with radius a, the effective weight of the particle can be written as

$$\frac{4}{3}\,\pi\,a^3\,\rho\,g - \frac{4}{3}\,\pi\,a^3\,\rho'\,g = \frac{4}{3}\,\pi\,a^3\,(\rho - \rho')\,g,$$

where ρ and ρ' are, respectively, the densities of the material of the particles and the fluid medium. Using this expression, equation (4.75) can be written as

$$n = n_0\,e^{-\frac{\frac{4}{3}\,\pi\,a^3\,(\rho - \rho')\,g\,N_A\,Z}{R\,T}}, \tag{4.76}$$

where K_B has been replaced by $\frac{R}{N_A}$.

Between 1900 and 1912, Jean Baptiste Perrin conducted a groundbreaking series of experiments to verify equation (4.76) and determined the value of Avogadro's number (N_A). For this remarkable achievement, he was awarded the Nobel Prize in Physics in 1926. A schematic overview of his experimental work is outlined below.

Perrin's work: Determination of Avogadro's number

Perrin carried out an extensive series of researches in Brownian motion and was able to verify both the predicted form of the distribution function and obtain a value for Avogadro's number N_A of the correct order of magnitude. Perrin's experimental research work for the determination of N_A is presented elaborately below.

Preparation of emulsion: Perrin selected emulsions of gambose (gum) and mastic (resin) obtained by alcoholic solutions of these substances to excess water. For a uniform emulsion, that is, an emulsion containing homogeneous grain, he subjected the initial emulsions to ultracentrifuging when the granules separated out as sediments. The sediments were brought to suspension again, and grains of approximately equal size were collected by fractional centrifuging. Perrin carried out his experiments with such emulsions.

Experiment: A schematic diagram of the experimental setup used by Perrin is shown in Figure 4.17. In this experimental setup, a drop of the emulsion is put in a cell made out of a hollow slide about 1 mm in height. The cell is kept surrounded by a water bath at a constant temperature. The suspension is viewed with a

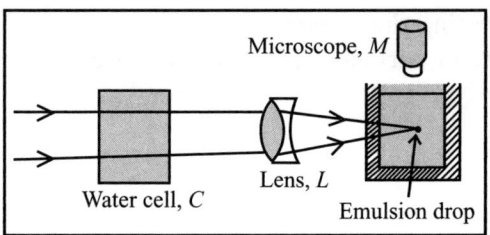

Figure 4.17 Schematic diagram of Perrin's experimental setup to observe Brownian motion.

high-power vertical microscope M. Because of the small depth of focus only a thin layer ($\sim$ 1 micron) of the suspension will be in focus at a time, and the grains in that layer will be sharply defined. By raising or lowering the microscope by means of the focus controls, one can bring layers at different depths into focus. There is a coarse and a fine focus control on the microscope. The fine focus control is marked off in divisions, and this is used to measure the difference in height between different layers in the suspension. To facilitate observation of the particles, the field is illuminated by a strong light source, filtering out the heat rays by a water cell C, and using a condensing lens L. To cut down the number of particles to be counted, a diaphragm with a small central hole is used to limit the field of view. The observation in a particular layer is taken only when the distribution in different layers is steady.

Determination of $\frac{n_0}{n}$ and $Z - Z_0$: As the granules move, it is difficult to count them. Instead, therefore, instantaneous photomicrographs are taken, and the number is counted conveniently at leisure. Since the granules are strongly illuminated, they appear as distinct as possible. The number of particles in a certain layer is calculated as the mean of several observations at regular intervals. In this way, n_0/n is obtained, and $z - z_0$ is known directly from the vertical shift of the microscope.

Determination of density of the granules: Perrin determined the density of the granules ρ in three different ways. The simplest one consists of finding masses m_1 and m_2 of water and emulsion, respectively, required to fill a specific gravity bottle-like vessel (pycnometer). Next, by desiccation in an oven, the mass m_3 of the granules contained in the emulsion is determined.

If ρ_w is the density of water, the volume of the specific gravity bottle will be = $\frac{m_1}{\rho_w}$ and the volume of water in emulsion will be $\frac{m_2 - m_3}{\rho_w}$. The volume of the granules will be

$$\frac{m_1}{\rho_w} - \frac{m_2 - m_3}{\rho_w} = \frac{m_1 - m_2 + m_3}{\rho_w},$$

and the density of granules will be

$$\rho = \frac{m_3\,\rho_w}{m_1 - m_2 + m_3}.$$

Determination of radius: Perrin determined the radius of the granules also by three different ways. The most convenient one is by using Stokes' law which states that the viscous drag experienced by a sphere of radius a in an infinitely extended fluid of viscosity η is $6\pi\eta a v$, where v is the terminal velocity of the sphere. If the sphere falls under its own weight, we get

$$6\pi\eta a v = \frac{4}{3}\pi a^3(\rho - \rho')g; \quad \Rightarrow \quad a^2 = \frac{9}{2}\frac{\eta v}{(\rho - \rho')g}.$$

The emulsion is contained in a capillary tube and the steady rate of fall v of the cloud of granules noted. This expression is then used to evaluate a.

We thus have all the quantities in the right side of equation as known. So, N_A can be readily determined.

Value of Avogadro's number N_A from Perrin's work: From series of experiments, Perrin obtained the value of Avogadro's number N_A as 6.82×10^{23}. This value is, however, higher than the recently accepted value $N_A = 6.0251 \times 10^{23}$. This discrepancy was mainly due to the nonavailability of the exact value of coefficient of viscosity η. Further, the conditions for the application of Stokes' law were not fully met in the experimental situations, neither the particles were perfectly spherical nor did they fall through an infinite ocean of the fluid.

Avogadro's number is a fundamental physical constant giving the number of molecules in one mole of a substance. Combining it with values of atomic weight, **atomic masses** can be determined. Further, combining it with Faraday's constant, a value for the **fundamental unit of charge** can be obtained.

Langevin's theory of Brownian motion

On inspection of the characteristics features of Brownian motion, Langevin made a conclusion that each Brownian particle collides nearly 10^{21} times per second with the molecules of the liquid and in one second, its velocity changes (in magnitude as well as direction) nearly 10^7 times. Thus, it is almost impossible to trace the path followed by an individual molecule during collision. It is also impossible to predict the exact position of the molecule at a given time. Under this situation, Langevin considered the existence of average force acting on the molecule. He assumed that this average force acting on a suspended particle due to molecular bombardment consists of two terms: (i) a frictional force and (ii) a fluctuating component. In terms of these two forces, the equation of motion of a free Brownian particle can be written as

$$m\ddot{x} = F + F_x, \tag{4.77}$$

where m is the mass of the suspended particle, $\ddot{x}$ denotes its second-order derivative with respect to time t, F is the frictional force, and F_x is X-component of the fluctuating force due to molecular bombardment.

According to the observation of Langevin, it is known that, on an average, a suspended particle suffers one collision with the molecules of the liquid in about 10^{-21} s. Therefore, the mean free path of the fluid molecules is very small compared with the size of the suspended particles. Hence, the fluid can be considered as a continuous medium. Langevin further assumed that the suspended particles are spherical in shape and the frictional force due to Stokes law is given by

$$F = -6\pi\eta r\dot{x} = -C\dot{x}, \tag{4.78}$$

where $C = 6\pi\eta r$, and r is the radius of the suspended particle. Combining equations (4.77) and (4.78), we can write

$$m\ddot{x} = -C\dot{x} + F_x. \tag{4.79}$$

At each collision, the direction of motion of each suspended particle changes; hence, the fluctuating force component F_x will have quite an unpredicted value. It may be positive or even negative. Therefore, the expected displacement will be zero if the motion of suspended particles is considered over a time $t >> 10^{-21}$ s. This is because the probabilities of positive and negative displacement are equal. This problem is resolved by evaluating the mean square displacement and working with x^2 rather than x. This indicates that the equation of motion should be written in terms of x^2. This is obtained by simply multiplying equation (4.79) throughout by x. This gives rise to

$$mx\ddot{x} = -Cx\dot{x} + xF_x. \tag{4.80}$$

We note further that

$$\frac{d}{dt}(x^2) = 2x\dot{x}; \qquad \text{and} \qquad \frac{d^2}{dt^2}(x^2) = 2x\dot{x} + 2(\dot{x}^2).$$

Using these results in equation (4.80), we get

$$\frac{m}{2}\frac{d^2}{dt^2}(x^2) - m(\dot{x}^2) = -\frac{C}{2}\frac{d}{dt}(x^2) + xF_x. \tag{4.81}$$

Equation (4.81) is valid for each suspended Brownian particle in the fluid medium. Taking an average of this expression over a large number of particles, we get

$$\frac{m}{2}\left\langle\frac{d^2}{dt^2}(x^2)\right\rangle - m\left\langle(\dot{x}^2)\right\rangle = -\frac{C}{2}\left\langle\frac{d}{dt}(x^2)\right\rangle + \left\langle xF_x\right\rangle. \tag{4.82}$$

The symbol $\langle\rangle$ denotes the average over all the Brownian particles. It should be mentioned that the term $\left\langle xF_x\right\rangle$ in equation (4.82) will become zero as both x and F vary randomly. The Brownian particles are in thermal equilibrium with the molecules of the fluid, and the mean kinetic energy associated with each degree of freedom is $\frac{1}{2}K_BT$. This shows that

$$m\left\langle(\dot{x}^2)\right\rangle = K_BT.$$

Using these results in equation (4.82) and rearranging, we get

$$\frac{m}{2}\left\langle\frac{d^2}{dt^2}(x^2)\right\rangle + \frac{C}{2}\left\langle\frac{d}{dt}(x^2)\right\rangle = K_BT.$$

$$\Rightarrow \quad \left\langle\frac{d^2}{dt^2}(x^2)\right\rangle + \omega^2\left\langle\frac{d}{dt}(x^2)\right\rangle = \frac{2K_BT}{m}, \tag{4.83}$$

where $\omega^2 = \frac{C}{m}$. Note that ω has the dimension of inverse of time and signifies the frequency of collisions.

To solve this equation, we assume that averaging over space and differentiation with respect to time are commutative, that is,

$$\left\langle\frac{d^2}{dt^2}(x^2)\right\rangle = \frac{d}{dt}\left\langle(x^2)\right\rangle,$$

and introduce a change of variable by defining

$$\frac{d}{dt}\left\langle(x^2)\right\rangle = u.$$

Using these results, equation (4.83) can be expressed as

$$\dot{u} + \omega^2 u = \frac{2K_BT}{m}. \tag{4.84}$$

Equation (4.84) is a first-order nonhomogeneous differential equation. The general solution of this equation is given by

$$u = \frac{2K_BT}{C} + A\ exp\left(-\omega^2 t\right). \tag{4.85}$$

We know that $\omega^2 = \frac{C}{m}$. As mass m of the suspended Brownian particle is very small, ω^2 is very large. As a result, the exponential term decays very rapidly and can be neglected safely. Under this condition, equation (4.85) takes a very compact form given by

$$\frac{d}{dt}\left\langle(x^2)\right\rangle = u = \frac{2K_BT}{C}. \tag{4.86}$$

This equation (4.86) can be readily integrated, say from $t = 0$ to $t = \tau$. This gives rise to

$$\int\limits_0^\tau \frac{d}{dt}\left\langle (x^2) \right\rangle dt = \frac{2K_B T}{C} \int\limits_0^\tau dt.$$

The desired result is

$$\left\langle (x_\tau^2) \right\rangle - \left\langle (x_0^2) \right\rangle = \Delta \left\langle (x^2) \right\rangle = \frac{2K_B T}{C}\tau. \tag{4.87}$$

On substituting the value of $C = 6\pi\eta r$, we get

$$\Delta \left\langle (x^2) \right\rangle = \frac{K_B T}{3\pi\eta r}\tau. \tag{4.88}$$

This result is the same as given by equation (4.71). Perrin used this equation (4.88) to measure Avogadro's number, N_A, and established beyond doubt the existence of molecules as well as collisions among the molecules. In a way, these studies put kinetic theory on a sound basis.

It should be noted that $\Delta \left\langle (x^2) \right\rangle$ is not the actual displacement of Brownian particles. It is the mean of the squares of the projections of actual displacement on the x-axis. The factor $\Delta \left\langle (x^2) \right\rangle$ is obtained in the following way:

1. Take a snapshot of the suspension at $t = 0$ and $t = \tau$, and measure the component of displacement along any arbitrarily chosen direction, say x-axis.

2. Determine $\delta(x^2)$ for each particle and take a sum of these $\delta(x^2)$.

3. Divide this sum by the number of particles to obtain $\Delta \left\langle (x^2) \right\rangle$.

Perrin worked with 100 different particles of known size in his experiments. It should be pointed out that the Brownian motion is so complex that it is quite inconvenient to work with a large number of particles. Therefore, the motion of one particle for N successive intervals of time (when N is a large number) is almost equivalent to the motion of N particles during a single interval of time. This corresponds to the assumption that averaging and differentiation are commutative.

4.5   Random walk

Many processes can be discretized into individual steps. What happens in the next step may depend on only the current state or also on what happened in earlier steps. If it depends only on the current state, the process is memoryless and fits the definition of a Markov chain. A Markov chain where the events are analogous steps in some parameter space can be modeled as a random walk. A random walk is a mathematically formalized succession of random steps. A random walk on a lattice, where each step can only lead from a lattice point to a directly neighboring lattice point is a particularly simple model.

A number of physical situations arise in physics, where a system develops in time or space through a finite number of discrete steps. Sometimes these steps are random in direction and independent of the preceding or succeeding ones. The study of the net motion of a particle following such a random sequence of steps is referred to as the problem of **random walk**. The tossing of a coin can be considered as a simple example of a random walk problem. Suppose a coin is tossed at regular intervals of time t_c (say). Further, we set the criterion: If the coin shows a head, we move a step to the right, and if the coin shows a tail, we move a step to the left. Then after a certain time $t = N \times t_c$, our position $x(t)$ will depend upon the sequence of heads and tails. Both space and time parameters are discrete in this case. This is a simple example of a random walk problem.

Further, consider the net motion of a highly intoxicated person who begins to stroll from a single light post. He is so intoxicated that he may put each step in any random direction with a range of different lengths. Such a motion may be visualized as

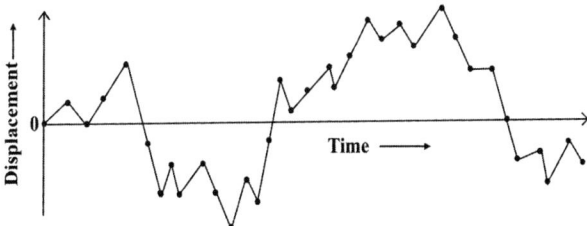

Figure 4.18 Schematic of one-dimensional random walk. The position of the person is marked after each step.

another common example of a random walk. This is shown in Figure 4.18. A number of methods (models) such as conditioning, generating functions, difference equations, the theory for Markov chains, the theory of branching processes, martingales, counting paths, mirroring, and time reversal are generally used in the literature to analyze the random walk problems.

The models used to describe random walk problems are also utilized to numerous applications such as (i) Brownian motion, (ii) neutron diffusion, (iii) turbulence in fluids, (iv) motion of electrons through a metal, (v) motion of holes in a semiconductor, and (vi) motion of defects in crystals. Therefore, a thorough understanding of the random walk problem is essential to the readers in physics. Further, it is known that the motion in more than one dimension can be studied by breaking up into its individual components. Therefore, it is sufficient to consider the random walk problem only in one dimension.

A *random walk* is a stochastic sequence $\{S_n\}$, with $S_0 = 0$, defined by

$$S_n = \sum_{k=1}^{n} X_k,$$

where $\{X_k\}$ are independent and identically distributed random variables. The random walk is simple if $X_k = \pm 1$, with $P(X_k = 1) = p$ and $P(X_k = -1) =$

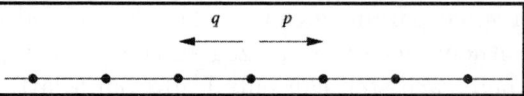

Figure 4.19 A simple random walk.

$1 - p = q$. Let us consider a particle performing a random walk on the integer points of the real line, where it moves to one of its neighboring points in each step. This is pictorially shown in Figure 4.19.

A random walk can also be studied in higher dimensions. In two dimensions, each point has four neighbors, whereas in three dimensions, each point has six neighbors. A simple random walk is **symmetric** if the particle has the same probability for each of the neighbors. In this case, p and q will be equal to each other.

In this chapter, we will only study simple random walks in one dimension. We are interested in answering the following questions:

1. What is the probability that the particle will ever reach a particular point a? (The case $a = 1$ is often called "The monkey at the cliff".)

2. What time does it take to reach a?

3. What is the probability of reaching $a > 0$ before $-b < 0$? ("The gambler's ruin".)

4. If the particle after n steps is at $a > 0$, what is the probability that (i) it has been on the positive side since the first step? and (ii) it has never been on the negative side? ("The Ballot problem".)

5. How far away from 0 will the particle get in n steps?

4.5.1   Calculation of the occupation probability

The random walk problem is of considerable interest in statistical physics. This problem was originally solved by Markoff. This random walk problem can be thought of as an application of binomial distribution and can be formulated on the basis of a simple example: the motion of a drunkard that can be described by the following steps:

1. The drunkard starts from a lamp post on a street.

2. He takes some steps forward and some steps backward.

3. Each forward step has a certain probability, and each backward step has also a certain probability.

4. Further, each step is of the same length and is completely independent of the preceding step.

A particle undergoing random displacements, as in Brownian motion, forms a physical situation similar to the drunkard's tottering. We consider such a particle undergoing successive displacements in one dimension. The length of each step is, say, l_s. We address the following question: **what is the probability that the particle, after N such displacements, will be at a distance $x = m \, l_s$, where m is an integer?**

Let n_1 be the number of displacements to the right and n_2 to the left. Since the total number of steps is N, we will have

$$N = n_1 + n_2. \tag{4.89}$$

The net displacement, say, to the right is (assuming $n_1 > n_2$)

$$m \, l_s = (n_1 - n_2) \, l_s = [n_1 - (N - n_1)] \, l_s = (2n_1 - N) \, l_s. \tag{4.90}$$

Solving these two equations (4.89) and (4.90), we get

$$n_1 = \frac{N + m}{2} \qquad \text{and} \qquad n_2 = \frac{N - m}{2}. \tag{4.91}$$

If the probability that the displacements to the right are p, and that of to the left is q, we have then $p + q = 1$. We are assuming that these probabilities are guided by the binomial distribution. Hence, the probability of a sequence of n_1 steps to the right and n_2 steps to the left is given by $= p^{n_1} q^{n_2}$. However, there are various ways in which a particle may undergo n_1 displacements to the right and n_2 displacements to the left. For example, suppose a particle undergoes 3 displacements, two of which are to the right and one to the left. This may happen in 3 different ways given by $\frac{3!}{2! \, 1!} = 3$. These ways are shown below.

Combination of steps for the problem

First step	Second step	Third step
$\longrightarrow$	$\longrightarrow$	$\longleftarrow$
$\longrightarrow$	$\longleftarrow$	$\longrightarrow$
$\longleftarrow$	$\longrightarrow$	$\longrightarrow$

In the problem out of N steps, the number of distinct possibilities leading to n_1 steps to the right and n_2 steps to the left is given by

$$\frac{N!}{n_1! \, n_2!}. \tag{4.92}$$

Therefore, the probability of n_1 displacements to the right and n_2 to the left in any order is given by

$$\frac{N!}{n_1! \, n_2!} \, p^{n_1} \, q^{n_2}. \tag{4.93}$$

Since the probability of a displacement in either direction is $\frac{1}{2}$; we have, in the case of binomial distribution, $p = q = \frac{1}{2}$. Therefore, the probability that the particle will be at $x = m \, l_s$ after N displacements is given by

$$P(m, N) = \frac{N!}{\left(\frac{N+m}{2}\right)! \, \left(\frac{N-m}{2}\right)!} \left(p^{\frac{1}{2}(N+m)}\right) \left(q^{\frac{1}{2}(N-m)}\right) = \frac{N!}{\left(\frac{N+m}{2}\right)! \, \left(\frac{N-m}{2}\right)!} \left(\frac{1}{2}\right)^N. \tag{4.94}$$

Taking natural logarithm, we get

$$\ln P(m, N) = \ln N! - N \ln 2 - \ln \left(\frac{N+m}{2}\right)! - \ln \left(\frac{N-m}{2}\right)!. \tag{4.95}$$

For large N, the exact form of Stirling approximation is given by

$$N! = N^N \, e^{-N} \, (2\pi \, N)^{\frac{1}{2}}.$$

Taking natural logarithm, we get

$$\ln N! = N \ln N - N + \frac{1}{2}\ln(2\pi N) = \left(N + \frac{1}{2}\right) \ln N - N + \frac{1}{2}\ln(2\pi). \tag{4.96}$$

Using Stirling's approximation written in equation (4.96), we can write an equation (4.95) as

$$\ln P(m, N) = \left(N + \frac{1}{2}\right) \ln N - N + \frac{1}{2}\ln(2\pi) - N \ln 2$$

$$- \left(\frac{N+m}{2} + \frac{1}{2}\right) \ln \left(\frac{N+m}{2}\right) + \left(\frac{N+m}{2}\right) - \frac{1}{2}\ln(2\pi)$$

$$- \left(\frac{N-m}{2} + \frac{1}{2}\right) \ln \left(\frac{N-m}{2}\right) + \left(\frac{N-m}{2}\right) - \frac{1}{2}\ln(2\pi)$$

$$= \left(N + \frac{1}{2}\right) \ln N - \frac{1}{2}\ln(2\pi) - N \ln 2 - \left(\frac{N+m}{2} + \frac{1}{2}\right) \ln \left(\frac{N+m}{2}\right)$$

$$- \left(\frac{N-m}{2} + \frac{1}{2}\right) \ln \left(\frac{N-m}{2}\right)$$

$$= \left(N + \frac{1}{2}\right) \ln N - \frac{1}{2}\ln(2\pi) - N \ln 2$$

$$- \left(\frac{N+m}{2} + \frac{1}{2}\right) \ln \left[\frac{N}{2}\left(1 + \frac{m}{N}\right)\right] - \left(\frac{N-m}{2} + \frac{1}{2}\right) \ln \left[\frac{N}{2}\left(1 - \frac{m}{N}\right)\right]$$

$$= \left(N + \frac{1}{2}\right) \ln N - \frac{1}{2}\ln(2\,\pi) - N\,\ln 2$$

$$- \left[\left(\frac{N+m}{2} + \frac{1}{2}\right) + \left(\frac{N-m}{2} + \frac{1}{2}\right)\right] \ln \frac{N}{2}$$

$$- \left(\frac{N+m}{2} + \frac{1}{2}\right) \ln \left(1 + \frac{m}{N}\right) - \left(\frac{N-m}{2} + \frac{1}{2}\right) \ln \left(1 - \frac{m}{N}\right).$$

If $m << N$, which is generally true, we have

$$\ln \left(1 + \frac{m}{N}\right) = \frac{m}{N} - \frac{m^2}{2N^2} + \frac{m^3}{3N^3} - \cdots \approx \frac{m}{N} - \frac{m^2}{2N^2}$$

and

$$\ln \left(1 - \frac{m}{N}\right) = -\frac{m}{N} - \frac{m^2}{2N^2} - \frac{m^3}{3N^3} - \cdots \approx -\frac{m}{N} - \frac{m^2}{2N^2}.$$

Using these approximations, the above expression can be written as

$$\ln P(m, N) = \left(N + \frac{1}{2}\right) \ln N - \frac{1}{2}\ln(2\,\pi) - N\,\ln 2 - (N+1)\ln\frac{N}{2}$$

$$- \left(\frac{N+m}{2} + \frac{1}{2}\right)\left(\frac{m}{N} - \frac{m^2}{2N^2}\right) - \left(\frac{N-m}{2} + \frac{1}{2}\right)\left(-\frac{m}{N} - \frac{m^2}{2N^2}\right),$$

$$= \ln 2 - \left(\frac{1}{2}\right)\ln N - \frac{1}{2}\ln(2\,\pi) - \left(\frac{N+m}{2} + \frac{1}{2}\right)\left(\frac{m}{N} - \frac{m^2}{2N^2}\right)$$

$$+ \left(\frac{N-m}{2} + \frac{1}{2}\right)\left(\frac{m}{N} + \frac{m^2}{2N^2}\right).$$

$$= \ln 2 - \left(\frac{1}{2}\right)\ln N - \frac{1}{2}\ln(2\,\pi) - \frac{m^2}{2N^2}(N-1).$$

As N is very large, we can approximately take $(N-1) \approx N$. Under this situation, the above expression takes the form

$$\ln P(m, N) = \ln 2 - \left(\frac{1}{2}\right)\ln N - \frac{1}{2}\ln(2\,\pi) - \frac{m^2}{2N}. \qquad (4.97)$$

This gives rise to

$$P(m, N) = \left(\frac{2}{\pi N}\right)^{\frac{1}{2}} e^{-\frac{m^2}{2N}}. \qquad (4.98)$$

From this, we note that although $P(m, N)$ is a discontinuous function, the terminal points of a random walk form the outline of a smooth bell-shaped curve. Now, let us suppose that the length of each step l_s is very small and $x = m\,l_s$. One can then approximate $P(m, N)$ by a continuous probability density function $P_C(x, N)$ such that

$$P(m, N) = CP_C(x, N), \qquad (4.99)$$

where C is the constant of proportionality. As the probability of finding the random walker is a certainty, C can be calculated from the following normalization condition.

$$\int_{-\infty}^{\infty} P_C(x, N)dx = \frac{1}{C} \int_{-\infty}^{\infty} P(m, N)dx = 1.$$

Using equation (4.98), we get

$$\frac{1}{C} \left(\frac{2}{\pi N} \right)^{\frac{1}{2}} \int_{-\infty}^{\infty} e^{-\frac{m^2}{2N}} dx = 1; \quad \Rightarrow \quad \frac{1}{C} \left(\frac{2}{\pi N} \right)^{\frac{1}{2}} \int_{-\infty}^{\infty} e^{-\frac{x^2}{2Nl_s^2}} dx = 1,$$

$$\Rightarrow \quad \frac{1}{C} \left(\frac{2}{\pi N} \right)^{\frac{1}{2}} 2 \int_{0}^{\infty} e^{-\frac{x^2}{2Nl_s^2}} dx = 1.$$

Let $\frac{x^2}{2Nl_s^2} = z$. This leads to $\Rightarrow dx = \sqrt{\frac{Nl_s^2}{2}} z^{-\frac{1}{2}} dz$. Using these substitutions, we get the value of C as

$$\Rightarrow \quad \frac{1}{C} \times 2l_s \times \frac{1}{\sqrt{\pi}} \times \sqrt{\pi} = 1 \qquad \Rightarrow \quad C = 2l_s.$$

So, the continuous probability density function $P_C(x, N)$ takes the form

$$P_C(x, N) = \left(\frac{1}{2\pi N l_s^2} \right)^{\frac{1}{2}} e^{-\frac{x^2}{2Nl_s^2}}. \qquad (4.100)$$

This is the probability function for $x(= m\, l_s)$ excess displacements to the right.

Equation (4.100) can be rewritten in a slightly different form by noting that if the time taken in N jumps is t and $\nu = \tau^{-1}$ denotes the number of jumps per unit time, then $N = \nu t = \frac{t}{\tau}$. This gives

$$P_C(x, N) = \frac{1}{\sqrt{4\pi Dt}} e^{-\left(\frac{x^2}{4D\,t} \right)}, \qquad (4.101)$$

where $D = \frac{t^2}{2\tau}$ is known as the diffusion coefficient. This equation (4.101) gives the desired probability that the particle is at a distance x at time t. It represents a Gaussian distribution curve.

We can find out the probability that the drunkard is between x and $x + dx$ after N steps. We have $x = m\, l_s$ and $m = n_1 - n_2 = n_1 - (N - n_1) = 2n_1 - N$. Let $P(x, N)\, dx$ be the probability that the drunkard is between x and $x + dx$ after N steps and is given by

$$P(x, N)dx = P(m, N)dm = P(m, N)\frac{dx}{2l_s}.$$

Here, we have written $dx = 2l_s\, dm$ because m can take only integral values separated by an amount $\Delta m = 2$. So, the probability that the drunkard is at a distance x after N steps is

$$P(x, N) = \left(\frac{1}{2\pi l_s^2\, N}\right)^{\frac{1}{2}} e^{-\left(\frac{x^2}{2\,N\,l_s^2}\right)}. \tag{4.102}$$

This is the normal or Gaussian distribution.

4.5.2 Calculation of the mean square displacement of a Brownian particle

The mean square displacement of the particle is defined as

$$\langle x^2 \rangle = \int_{-\infty}^{\infty} x^2 P_C(x, N)dx. \tag{4.103}$$

Using equation (4.101) in equation (4.103), we get

$$\langle x^2 \rangle = \int_{-\infty}^{\infty} x^2 \frac{1}{\sqrt{4\pi Dt}}\, e^{-\left(\frac{x^2}{4D\,t}\right)}dx = \frac{1}{\sqrt{4\pi Dt}} \int_{-\infty}^{\infty} x^2\, e^{-\left(\frac{x^2}{4D\,t}\right)}dx. \tag{4.104}$$

Since the integrand is symmetric in x, this equation can be written as

$$\langle x^2 \rangle = \frac{1}{\sqrt{\pi Dt}} \int_{0}^{\infty} x^2\, e^{-\left(\frac{x^2}{4D\,t}\right)}dx. \tag{4.105}$$

Now, we introduce a new variable by defining $\frac{x^2}{4Dt} = \alpha$ so that $x\, dx = 2Dt\, d\alpha$ and $x^2\, dx = 4(Dt)^{\frac{3}{2}}\alpha^{\frac{1}{2}}d\alpha$. Then equation (4.105) takes a compact form:

$$\langle x^2 \rangle = \frac{4Dt}{\sqrt{\pi}} \int_{0}^{\infty} \alpha^{\frac{1}{2}}\, e^{-\alpha}d\alpha = \frac{4Dt}{\sqrt{\pi}}\, \Gamma\left(\frac{3}{2}\right). \tag{4.106}$$

Since $\Gamma(\frac{3}{2}) = \frac{\sqrt{\pi}}{2}$, we obtain

$$\langle x^2 \rangle = 2\,Dt. \tag{4.107}$$

This gives the same result as that obtained by Einstein for mean square displacement by considering the molecular concentration gradient.

4.5.3   Random walk in three dimensions

The continuous probability density function $P_C(x, N)$ of a random walk problem in one dimension is given by equation (4.101), which is recast as

$$P_C(x, N) = \frac{1}{\sqrt{4\pi Dt}}\, e^{-\left(\frac{x^2}{4D\,t}\right)}, \tag{4.108}$$

with $D = \frac{t^2}{2\tau}$ as the diffusion coefficient. Using the expression for mean square displacement of such a random walk problem in one dimension from equation (4.107) in equation (4.108), we get

$$P_C(x, N) = \frac{1}{\sqrt{2\pi \langle x^2 \rangle}}\, e^{-\left(\frac{x^2}{2\langle x^2 \rangle}\right)}. \tag{4.109}$$

This result does no longer depend on step size, not even implicitly because we have removed the dependence on step number N. Therefore, it can be generalized to three dimensions. Since the random walks along the three pairwise orthogonal directions in Cartesian space are independent of each other, we have

$$P_{3D}(x, y, z)dxdydz = P_{1D}(x)dx P_{1D}(y)dy P_{1D}(z)dz. \tag{4.110}$$

We pose the question of the distribution of mean square end-to-end distances $\left\langle \vec{R}^2 \right\rangle$ with the Cartesian components of the end-to-end vector $\vec{R}$ being $x = R_x$, $y = R_y$, and $z = R_z$. So we have

$$\left\langle \vec{R}^2 \right\rangle = \left\langle R_x{}^2 \right\rangle + \left\langle R_y{}^2 \right\rangle + \left\langle R_z{}^2 \right\rangle = NK_L^2, \tag{4.111}$$

where K_L is the Kuhn length. For symmetry reasons, we have

$$\left\langle R_x{}^2 \right\rangle = \left\langle R_y{}^2 \right\rangle = \left\langle R_z{}^2 \right\rangle = \frac{NK_L^2}{3}. \tag{4.112}$$

Using the result of equation (4.112), the continuous probability density function $P_{1D}(x, N)$ in one dimension can be written as

$$P_{1D}(x, N) = \sqrt{\frac{3}{2\pi NK_L^2}}\, e^{-\left(\frac{3R_x^2}{2NK_L^2}\right)} \tag{4.113}$$

and analogous expressions for $P_{1D}(y)$ and $P_{1D}(z)$. We have reintroduced parameter N, which is now the number of Kuhn segments. Yet, by discussing a continuous probability density distribution, we have removed dependence on a lattice model. This is necessary since the steps along dimensions x, y, and z differ for each Kuhn segment. We find

$$P_{3D}(N, \vec{R}) = \left(\frac{3}{2\pi N K_L^2}\right)^{\frac{3}{2}} e^{-\left(\frac{3R^2}{2NK_L^2}\right)}. \tag{4.114}$$

This is the expression for continuous probability density distribution function for a random walk in three dimensions.

4.6   Solved numerical problems

1. The average speed and the radius of hydrogen molecules are, respectively, 1840 ms^{-1} and 1.37×10^{-10} m. Calculate (i) collision cross-section, (ii) collision frequency, and (iii) mean free path of hydrogen molecule. The volume density of the number of molecules is 3×10^{25} m^{-3}.

   Answer: According to the problem, we have the radius of a hydrogen molecule as 1.37×10^{-10} m and the number density n as 3×10^{25} m^{-3}. With the help of these data, collision cross-section, collision frequency, and mean free path can be easily calculated using respective formulas.

   (i) The collision cross-section is given by

   $$\sigma = \pi d^2 = 4\pi \left(1.37 \times 10^{-10} \text{ m}\right)^2 = 23.6 \times 10^{-20} \text{ m}^2.$$

   (ii) The collision frequency is given by

   $$p_c = \pi d^2 n \langle v \rangle = \sigma n \langle v \rangle.$$

   Using the values of the parameters, we get the collision frequency as

   $$p_c = \left(23.6 \times 10^{-20} \text{ m}^2\right) \times \left(3 \times 10^{25} \text{ m}^{-3}\right) \times \left(1.84 \times 10^3 \text{ ms}^{-1}\right) = 1.3 \times 10^{10} \text{ s}^{-1}.$$

   (iii) The mean free path λ is given by

   $$\lambda = \frac{1}{\pi d^2 n} = \frac{1}{\sigma n} = \frac{1}{\left(23.6 \times 10^{-20} \text{ m}^2\right) \times \left(3 \times 10^{25} \text{ m}^{-3}\right)}$$
   $$= 141 \times 10^{-9} \text{ m} = 141 \text{ nm}.$$

   It is to be noted that the number of collisions per second is very large, of the order of 10^{10}. For this reason, it is almost impossible to follow the trajectory of a molecule because the path of a molecule is made up of so many kinks and zigzags. Further, it is noted that the mean free path λ is much larger than the intermolecular distance, which is only a few nanometer (≈ 3 nm).

2. The mean free path of molecules of a gas is 10^{-8} cm. If the number density of the gas is 10^{19} cm^{-3}, calculate the diameter of the molecule.

   Answer: From the expression of mean free path λ, the expression for square of diameter d of a molecule is given by

   $$d^2 = \left(\frac{1}{\sqrt{2}}\right)\left(\frac{1}{\pi \lambda n}\right) = \left(\frac{1}{\sqrt{2}}\right)\left(\frac{1}{22/7 \times 10^{-8} \times 10^{19}}\right)$$
   $$= 0.22502 \times 10^{-11} \text{ cm}^2.$$

   Taking the square root of the above expression, we get the diameter d of the molecule as

   $$d = 0.1500 \times 10^{-5} \text{ cm} = 15 \times 10^{-7} \text{ cm} = 15 \times 10^{-9} \text{ m} = 15 \text{ nm}.$$

3. Calculate the ratio of the mean free paths of the molecules of two gases if the ratio of the number density per cm^3 of the gases is $5 : 3$ and the ratio of the diameters of the molecules of the gases is $4 : 3$.

   Answer: We know that the mean free path λ is given by

   $$\lambda = \left(\frac{1}{\sqrt{2}}\right)\left(\frac{1}{\pi d^2 n}\right),$$

   where d is the diameter of a molecule of a gas and n is the number density. So, for the two gases, we have $\lambda_1 = \left(\frac{1}{\sqrt{2}}\right)\left(\frac{1}{\pi d_1^2 n_1}\right)$ and $\lambda_2 = \left(\frac{1}{\sqrt{2}}\right)\left(\frac{1}{\pi d_2^2 n_2}\right)$.

   Taking the ratio of these two expressions, we get

   $$\frac{\lambda_1}{\lambda_2} = \left(\frac{d_2^2 n_2}{d_1^2 n_1}\right) = \left(\frac{3}{4}\right)^2 \times \left(\frac{3}{5}\right) = \frac{27}{80}.$$

4. Clausius assumed that all molecules move with average velocity $\langle v \rangle$ with respect to the container. Prove that the average of relative velocity, that is, velocity of one molecule with respect to another is $\frac{4}{3}\langle v \rangle$.

   Answer: The relative velocity v_{rel} of one molecule moving with $\langle v \rangle$ w.r.t another molecule moving with $\langle v \rangle$ is given by

   $$v_{\text{rel}} = \sqrt{(\langle v \rangle^2 + \langle v \rangle^2 - 2 \times \langle v \rangle \times \langle v \rangle \times \cos \theta)} = 2 \times \langle v \rangle \times \sin \frac{\theta}{2}.$$

Now, all the directions of velocity are equally probable. So, the probability that $\langle v \rangle$ lying within the solid angle between θ and $\theta + d\theta$ is

$$f(\theta) = \frac{2\pi \sin \theta d\theta}{4\pi} = \frac{1}{2} \sin \theta d\theta.$$

Using this value of the function $f(\theta)$, we get the expression for average relative velocity as

$$\langle v_{\text{rel}} \rangle = \left\langle 2\langle v \rangle \sin \frac{\theta}{2} \right\rangle = \int_0^\pi 2\langle v \rangle \sin \frac{\theta}{2} \left(\frac{1}{2} \sin \theta \right) d\theta = \frac{4\langle v \rangle}{3}.$$

5. Calculate the mean time between collisions for a nitrogen molecule in air at 27°C and a pressure of 1 atmosphere. Take $K_B = 1.38 \times 10^{-23}$ JK^{-1} and the diameter of nitrogen molecule $d_{N_2} = 10^{-10}$ m. Assume that nitrogen gas obeys Maxwell's distribution law of molecular speeds.

Answer: We know that the mean free path of the molecules of a gas obeying Maxwell's speed distribution law is given by

$$\lambda = \frac{1}{\sqrt{2}\pi d^2 n} = \frac{1}{\sqrt{2}\pi d^2} \times \frac{K_B T}{P}. \quad \text{[We have used the relation } P = nK_B T.]$$

Substituting the values of the parameters, we get

$$\lambda = \frac{1}{\sqrt{2}\pi(10^{-20} \text{ m}^2)} \times \frac{(1.38 \times 10^{-23} \text{ JK}^{-1}) \times (300 \text{ K})}{(1.01325 \times 10^5 \text{ Nm}^{-2})}$$
$$= 0.92 \times 10^{-6} \text{ m} = 92 \times 10^{-8} \text{ m}.$$

The average speed of the molecules of a Maxwellian gas is given by

$$\langle v \rangle = \sqrt{\frac{8K_B T}{\pi m}} = \sqrt{\frac{2.55 K_B T}{m}}.$$

The mass of a nitrogen molecule is given by

$$m = \frac{28}{6.023 \times 10^{26} \text{ kg}}; \qquad \Rightarrow \quad m = 4.65 \times 10^{-26} \text{ kg}.$$

So, the average velocity comes out to be

$$\langle v \rangle = \sqrt{\frac{2.55 \times (1.38 \times 10^{-23} \text{ JK}^{-1}) \times (300 \text{ K})}{4.65 \times 10^{-26} \text{ kg}}}$$
$$= \sqrt{2.27 \times 10^5} \text{ ms}^{-1} = 476.45 \text{ ms}^{-1}.$$

Hence, the mean time between collisions is given by

$$t = \frac{\lambda}{\bar{v}} = \frac{0.92 \times 10^{-6} \text{ m}}{476} \text{ ms}^{-1} = 1.93 \times 10^{-9} \text{ s}.$$

6. Calculate the mean free path of the molecules of a gas of diameter 0.2 nm in a closed chamber maintained at a pressure of 10^{-6} mm Hg and at a temperature of 273 K. One gram molecule of the gas occupies a volume of 22.4 l at N.T.P.

   Answer: We know that at a pressure of 0.76 m Hg and at a temperature of 273 K, Avogadro's number of molecules (6.023×10^{23}) of a gas occupy a volume of 22.4 l.

   The number of molecules per unit volume can be calculated using Clapeyron's relation: $P = n\, K_B T$. When temperature remains constant, we get from this relation: $\frac{P_1}{n_1} = \frac{P_2}{n_2}$.

   Let n_1 be the number of molecules per m^3 at a pressure of 0.76 m Hg and is given by

   $$n_1 = \frac{6.023 \times 10^{23}}{22.4 \times 10^{-3} \text{ m}^3} = 0.27 \times 10^{26} \text{ m}^{-3}.$$

   Using the above relation, the number of molecules n_2 per m^3 at a pressure of 10^{-6} mm Hg could be obtained as

   $$n_2 = \frac{P_2}{P_1} \times n_1 = \left[\frac{10^{-9} \text{ m}}{0.76 \text{ m}}\right] \times 0.27 \times 10^{26} \text{ m}^{-3} = 0.355 \times 10^{17} \text{ m}^{-3}.$$

   Further, the diameter of the molecule is given by $d = 0.2$ nm $= 2 \times 10^{-10}$ m. If the gas obeys Maxwell's speed distribution law, the expression for mean free path λ is given by

   $$\lambda = \frac{1}{\sqrt{2}\pi d^2 n} = \frac{1}{\sqrt{2}\pi (4 \times 10^{-20} \text{ m}^2) \times (0.355 \times 10^{17} \text{ m}^{-3})}$$

   $$= \frac{1}{6.305 \times 10^{-3}} \text{ m} = 1.58604 \times 10^2 \text{ m} = 158.60 \text{ m}.$$

   Note that the mean free path has a very large value that may be greater than the size of a typical chamber. This indicates the possibility of obtaining unhindered movement of molecules in a chamber.

7. At 27°C, the coefficient of viscosity of the helium molecule is 2×10^{-4} poise. Find out the diameter of the helium molecule with the following data:

   $N_A = 6 \times 10^{23}$ mol^{-1}, $K_B = 1.38 \times 10^{-23}$ JK^{-1}, and the atomic weight of helium is 4.

Answer: The expression for the coefficient of viscosity of a gas is given by $\eta = \frac{1}{3} \lambda \rho \langle v \rangle$. Using $\lambda = \frac{1}{\sqrt{2}\pi d^2 n}$ and $\rho = m\,n$, we get the coefficient of viscosity η as

$$\eta = \frac{m\langle v \rangle}{3\sqrt{2}\pi d^2}. \text{ This leads to } d^2 = \frac{m\langle v \rangle}{3\sqrt{2}\pi\eta}.$$

As per problem, we have $\eta = 2 \times 10^{-4}$ poise, and $T = (273 + 27)$ K $= 300$ K, the mass of the helium molecule is

$$m = \frac{At.wt}{N_A} = \frac{4 \text{ g mol}^{-1}}{6 \times 10^{23} \text{ mol}^{-1}} = 6.67 \times 10^{-24} \text{ g} = 6.67 \times 10^{-27} \text{ kg}$$

and the average velocity $\langle v \rangle$ is given by

$$\langle v \rangle = \left(\frac{2.55 \times K_B T}{m} \right)^{\frac{1}{2}} = \left(\frac{2.55 \times (1.38 \times 10^{-23} \text{ JK}^{-1}) \times 300 \text{ K}}{6.67 \times 10^{-27}} \right)^{\frac{1}{2}}$$

$$= 1.26 \times 10^3 \text{ ms}^{-1}.$$

Hence, the diameter d of helium molecule is

$$d = \sqrt{\frac{6.67 \times 10^{-27} \text{ kg} \times 1.26 \times 10^3 \text{ ms}^{-1}}{3\sqrt{2}\pi \times 2 \times 10^{-4} \text{ poise}}} = 0.56 \times 10^{-10} \text{ m} = 0.56 \text{ Å}.$$

8. The coefficient of viscosity of a gas is 16.6×10^{-6} N s m^{-1}. Calculate the mean free path, frequency of collision, and the diameter of the gas molecule. Given $v = 450$ ms^{-1}, $\rho = 1.25$ kg m^{-3}, and the number density is 2.7×10^{25} molecules per m^{-3}.

Answer: The expression for the coefficient of viscosity of a gas is given by $\eta = \frac{1}{3}\lambda\rho\langle v \rangle$. This leads to the expression for mean free path as

$$\lambda = \frac{3\,\eta}{\rho\,\langle v \rangle}.$$

Using the given data, the mean free path λ is given by

$$\lambda = \frac{3 \times (16.6 \times 10^{-6} \text{N sm}^{-2})}{(1.25 \text{ kg m}^{-3})(450 \text{ m s}^{-1})} = 8.85 \times 10^{-8} \text{ m}.$$

The frequency of collision p_c is given by

$$p_c = \frac{\langle v \rangle}{\lambda} = \frac{450 \text{ ms}^{-1}}{8.85 \times 10^{-8} \text{ m}} = 5.08 \times 10^9 \text{ s}^{-1}.$$

According to Maxwell's calculation, the mean free path is given by $\lambda = \frac{1}{\sqrt{2}\pi d^2 n}$.

This leads to the expression for diameter d as

$$d = \left(\frac{1}{\sqrt{2}\pi n\lambda}\right)^{\frac{1}{2}} = \left(\frac{1}{1.414 \times 3.14 \times 2.7 \times 10^{25} \text{ m}^{-3} \times 8.85 \times 10^{-8} \text{ m}}\right)^{\frac{1}{2}}$$

$$= 3.07 \times 10^{-10} \text{ m} = 3.07 \text{ Å}.$$

9. The viscosity of a gas at STP was measured to be 1.66×10^{-5} Nm^{-2} per unit velocity gradient. The average speed of the molecules is 450 ms^{-1} and the density of the gas is 1.25 kg m^{-3}. Calculate (A) the mean free path of the gas, (B) the collision frequency, and (C) the molecular diameter of the gas.

Answer: A. The coefficient of viscosity is given by

$$\eta = \frac{1}{3}\rho\langle v\rangle\lambda; \qquad \Rightarrow \qquad \lambda = \frac{3\eta}{\rho\langle v\rangle}.$$

Here, $\rho = 1.25$ kg m^{-3}, $\langle v\rangle = 450$ ms^{-1}, and $\eta = 1.66 \times 10^{-5}$ N m^{-2} s.

Hence, the mean free path λ is given by

$$\lambda = \frac{3 \times (1.66 \times 10^{-5} \text{ N m}^{-2} \text{ s})}{(1.25 \text{ kg m}^{-3}) \times (450 \text{ ms}^{-1})} = \frac{4.98 \times 10^{-5} \text{ kg m}^{-1}\text{s}^{-1}}{562.5 \text{ kg m}^{-2}\text{s}^{-1}} = 8.85 \times 10^{-8} \text{ m}.$$

B. Frequency of collisions p_c is given by

$$p_c = \frac{\langle v\rangle}{\lambda} = \frac{450 \text{ ms}^{-1}}{8.85 \times 10^{-8} \text{ m}} = 5.08 \times 10^9 \text{ s}^{-1}.$$

C. We know that for a gas obeying Maxwell's speed distribution law, the expression for mean free path λ is given by

$$\lambda = \frac{1}{\sqrt{2}\pi d^2 n}; \qquad \Rightarrow \qquad d^2 = \frac{1}{\sqrt{2}\pi n\lambda}.$$

In order to find the diameter d, we have to calculate the number of molecules n per unit volume. This can be calculated in the following way:

We recall that 22.4 l of every gas contains Avogadro's number (6.023×10^{23}) of molecules. Hence, the number density per cm^3 is given by

$$n = \frac{6.023 \times 10^{23}}{22400} = 0.269 \times 10^{20} \text{ cm}^{-3}.$$

This number density per m^3 is given by $n = 2.69 \times 10^{25}$ m^{-3}. On substituting these values, we get

$$d^2 = \frac{1}{\sqrt{2} \times 3.1417 \times (2.69 \times 10^{25} \text{ m}^{-3}) \times (8.85 \times 10^{-8} \text{ m})}$$
$$= 9.456 \times 10^{-20} \text{ m}^2;$$
$$\Rightarrow \quad d = 3.08 \times 10^{-10} \text{ m}.$$

10. The molecular cross-section is defined as $\sigma = \pi d^2$. For slow neutrons in hydrogen, $\sigma = 80 \times 10^{-28}$ m^2. Assume that neutrons obey Maxwell's speed distribution law. Calculate their mean free paths at $T = 273$ K and $P = 1$ atm. Take $K_B = 1.38 \times 10^{-23}$ JK^{-1}.

    Answer: We know that the mean free path λ is given by

$$\lambda = \frac{1}{\sqrt{2}\pi d^2 n} = \frac{1}{\sqrt{2}\sigma n} = \frac{K_B T}{\sqrt{2}\sigma P}.$$

On substituting the given values, we get

$$\lambda = \frac{1}{1.414} \times \frac{(1.38 \times 10^{-23} \text{ JK}^{-1}) \times 300 \text{ K}}{(1.013 \times 10^5 \text{ Nm}^{-2}) \times (80 \times 10^{-28} \text{ m}^2)}$$
$$= \frac{4.14 \times 10^{-21}}{1.414 \times 1.013 \times 80 \times 10^{-23}} \text{ m} = 3.613 \text{ m}.$$

11. The viscosity of oxygen gas at a temperature of 16° C is 169×10^{-6} poise. Calculate the diameter of the molecule of the oxygen gas. Take Avogadro's number $N_A = 6.023 \times 10^{23}$ mol^{-1}, molecular weight of oxygen is 32 g mol^{-1}, and $K_B = 1.38 \times 10^{-23}$ JK^{-1}.

    Answer: The coefficient of viscosity η is given by

$$\eta = \frac{1}{3} \rho \langle v \rangle \lambda = \frac{1}{3} (mn) \langle v \rangle \frac{1}{\sqrt{2}\pi d^2 n} = \frac{1}{3\sqrt{2}} \frac{m \langle v \rangle}{\pi d^2}.$$

By rearranging the terms, we can write $d^2 = \frac{1}{3\sqrt{2}} \frac{m \langle v \rangle}{\pi \eta}$.

According to the problem, we have

$\eta = 169 \times 10^{-6}$ poise $= 169 \times 10^{-7}$ kg m^{-1}s^{-1}, and $T = (273+16)$ K $= 289$ K.

and the mass m of one oxygen molecule is given by

$$m = \frac{32 \text{ g mol}^{-1}}{6.023 \times 10^{23} \text{ mol}^{-1}} = 5.313 \times 10^{-26} \text{ kg}.$$

The average velocity $\langle v \rangle$ is given by

$$\langle v \rangle = \sqrt{\frac{2.55\, K_B T}{m}} = \sqrt{\frac{2.55 \times (1.38 \times 10^{-23}\ \text{JK}^{-1}) \times (289\ \text{K})}{5.313 \times 10^{-26}\ \text{kg}}}$$

$$= \sqrt{191.4 \times 10^3}\ \text{ms}^{-1} = 437 \times 10^2\ \text{ms}^{-1}.$$

Hence, the square of the diameter of an oxygen molecule

$$d^2 = \frac{(5.313 \times 10^{-26}\ \text{kg}) \times (437\ \text{ms}^{-1})}{3 \times 1.414 \times 3.1417 \times (169 \times 10^{-7}\ \text{kg m}^{-1}\text{s}^{-1})} = 1.031 \times 10^{-19}\ \text{m}^2$$

This gives rise to the value of diameter as $d = 3.21 \times 10^{-10}$ m.

12. Calculate the coefficient of viscosity of hydrogen at STP. Given $\rho = 8.90 \times 10^{-2}$ kg m^{-3}, $\lambda = 2 \times 10^{-7}$ m, and $K_B = 1.38 \times 10^{-23}$ JK^{-1}.

Answer: The expression for coefficient of viscosity η is given by

$$\eta = \frac{1}{3}\, \rho\, \langle v \rangle\, \lambda.$$

Here, $\rho = 8.90 \times 10^{-2}$ kg m^{-3}, $\lambda = 2 \times 10^{-7}$ m, and $\langle v \rangle = \sqrt{\frac{2.55\, K_B T}{m}}$.

The mass of a hydrogen molecule is

$$m = \frac{2\ \text{g mol}^{-1}}{6.023 \times 10^{23}} = 3.32 \times 10^{-24}\ \text{g} = 3.32 \times 10^{-27}\ \text{kg}.$$

The average velocity $\langle v \rangle$ is given by

$$\langle v \rangle = \sqrt{\frac{2.55 \times (1.38 \times 10^{-23}\ \text{JK}^{-1}) \times (273\ \text{K})}{3.32 \times 10^{-27}\ \text{kg}}}$$

$$= \sqrt{289 \times 10^4}\ \text{ms}^{-1} = 1700\ \text{ms}^{-1}.$$

Hence, the coefficient of viscosity η is given by

$$\eta = \frac{1}{3} \times (8.9 \times 10^{-2}\ \text{kg m}^{-3}) \times (17 \times 10^2\ \text{ms}^{-1}) \times (2 \times 10^{-7}\ \text{m})$$

$$= 1.01 \times 10^{-5}\ \text{kg m}^{-1}\text{s}^{-1}.$$

13. The mean free path of the molecules of a gas is 2×10^{-7} m at a pressure P and at a temperature 200 K. Calculate its value at (i) P, 400 K, (ii) $2P$, 200 K, and (iii) $\frac{1}{2}P$, 400 K.

Answer: The mean free path λ, in terms of the number of gas molecules per unit volume (n) and molecular diameter (d), is given by

$$\lambda = \frac{1}{\sqrt{2}\pi d^2 n}.$$

From Clapeyron's equation, we have $P = nK_BT$, where K_B is the Boltzmann constant. In terms of pressure and temperature, we get λ as

$$\lambda = \frac{K_BT}{\sqrt{2}\pi d^2 P}.$$

At a pressure P and temperature $T = 200$ K, we have

$$\lambda = \frac{K_B\, 200\text{ K}}{\sqrt{2}\pi d^2 P} = 2 \times 10^{-7}\text{ m}.$$

(a) At a pressure P and temperature $T = 400$ K, we have

$$\lambda = \frac{K_B\, 400\text{ K}}{\sqrt{2}\pi d^2 P} = 4 \times 10^{-7}\text{ m}.$$

(b) At pressure $2P$ and temperature $T = 200$ K, we have

$$\lambda = \frac{K_B\, 200\text{ K}}{\sqrt{2}\pi d^2 2P} = 1 \times 10^{-7}\text{ m}.$$

(c) At pressure $\frac{P}{2}$ and temperature $T = 400$ K, we have

$$\lambda = \frac{K_B\, 400\text{ K}}{\sqrt{2}\pi d^2 \frac{P}{2}} = 8 \times 10^{-7}\text{ m}.$$

14. The molecular diameter of a gas is 3×10^{-10} m. Calculate the mean free path of the molecules of the gas at a temperature of $27°C$ and at a pressure of one atmosphere.

    Answer: The mean free path λ at temperature T and pressure P is given by

    $$\lambda = \frac{K_BT}{\sqrt{2}\pi d^2 P} = \frac{K_BT}{1.414\,\pi d^2 P}.$$

    Given, temperature $T = 27°C = 300$ K; pressure $P = 1$ atm. $= 1.013 \times 10^5$ N/m^2 and diameter of the molecule $d = 3 \times 10^{-10}$ m and Boltzmann constant $K_B = 1.38 \times 10^{-23}$ J/K.

Using these values, the mean free path λ comes out to be

$$\lambda = \frac{1.38 \times 10^{-23} \times 300}{1.414 \times 3.14 \times (3 \times 10^{-10})^2 \times (1.013 \times 10^5)} \text{ m} = 1.02 \times 10^{-7} \text{ m}.$$

15. At a given temperature, the average speed of the molecules of a gas is 1.0×10^5 cm s^{-1}. The effective molecular diameter and the number of molecules per cm^3 of the gas are, respectively, given by 2.0×10^{-8} cm and 3×10^{19} cm^{-3}. Calculate the mean free path and the corresponding collision frequency of the molecules of the gas.

Answer: The mean free path λ, in terms of molecular density n and molecular diameter d, is given by

$$\lambda = \frac{1}{\sqrt{2}\pi d^2 n}.$$

According to the problem, we have $n = 3 \times 10^{19}$ cm^{-3} and $d = 2.0 \times 10^{-8}$ cm. So, the mean free path λ is given by

$$\lambda = \frac{1}{1.414 \times 3.14 \times (2.0 \times 10^{-8})^2 \text{ cm}^2 \times (3 \times 10^{19} \text{ cm}^{-3})}$$

$$= 1.88 \times 10^{-5} \text{ cm} = 1.88 \times 10^{-7} \text{ m}.$$

The mean free path is the mean distance traveled by the molecule between two successive collisions. Therefore, the mean time-interval between two collisions is the mean free path divided by the mean molecular speed. Hence, the number of collisions per second, that is, the collision frequency p_c is given by

$$p_c = \frac{\langle v \rangle}{\lambda} = \frac{1.0 \times 10^5 \text{ cm s}^{-1}}{1.88 \times 10^{-5} \text{ cm}} = 5.3 \times 10^9 \text{ s}^{-1}.$$

16. The diameter of an oxygen molecule is roughly 3Å. Estimate the mean free path and the mean time between collisions for oxygen gas at N.T.P. The number of molecules per cm^3 at N.T.P. is 3×10^{19}. Given: Avogadro's number $N_A = 6.023 \times 10^{26}$ per kg mole, Boltzmann constant $K_B = 1.38 \times 10^{-23}$ JK^{-1}, and the molecular weight of oxygen $M = 32$ g mole^{-1}.

Answer: The mean free path λ, in terms of molecular density n and molecular diameter d, is given by

$$\lambda = \frac{1}{\sqrt{2}\pi d^2 n}.$$

Here, $n = 3 \times 10^{19}$ cm$^{-3} = 3 \times 10^{25}$ m^{-3} and $d = 3.0 \times 10^{-10}$ m. So, the mean free path λ is given by

$$\lambda = \frac{1}{1.414 \times 3.14 \times (3 \times 10^{-10})^2 \text{ m}^2 \times (3 \times 10^{25} \text{ m}^{-3})} = 8.36 \times 10^{-8} \text{ m}.$$

The mean free path is the distance traveled by the molecule between two successive collisions. Therefore, the mean time-interval between two collisions is $\lambda \langle v \rangle$, where $\langle v \rangle$ is mean molecular speed.

Now, the mean molecular speed is given by

$$\langle v \rangle = \sqrt{\frac{8K_B T}{m\pi}},$$

where m is the mass of the molecule and is given by

$$m = \frac{M}{N} = \frac{32}{6.023 \times 10^{26}} \text{ kg} = 5.31 \times 10^{-26} \text{ kg}.$$

The normal temperature is $T = 273$ K. Thus, we have

$$\langle v \rangle = \sqrt{\frac{8 \times (1.38 \times 10^{-23} \text{ JK}^{-1}) \times 273 \text{ K}}{3.14 \times (5.31 \times 10^{-26} \text{ kg})}} = 4.25 \times 10^2 \text{ ms}^{-1}.$$

Therefore, mean time t between collisions is

$$t = \frac{\lambda}{\langle v \rangle} = \frac{8.36 \times 10^{-8} \text{ m}}{4.25 \times 10^2 \text{ ms}^{-1}} = 1.97 \times 10^{-10} \text{ s}.$$

17. Calculate the mean free path of argon molecules at 25°C and 1 atmosphere pressure. Given by: $d = 2.56 \times 10^{-10}$ m and $K_B = 1.38 \times 10^{-23}$ JK^{-1}.

    Answer: Let us first calculate the number of molecules per unit volume of argon gas for the conditions stated. For 1 mole of gas, the equation of state is

    $$PV = RT.$$

    But $R = N_A K_B$, where N_A is the number of molecules in 1 mole of gas (Avogadro's number). Thus,

    $$PV = N_A K_B T.$$

The number of molecules per unit volume n is therefore

$$n = \frac{N_A}{V} = \frac{P}{K_B T}.$$

Here, $P = 1$ atm. $= 1.013 \times 10^5$ N m^{-2} and $T = 25°C = 298$ K.

$$n = \frac{1.013 \times 10^5 \text{ N.m}^{-2}}{1.38 \times 10^{-23} \text{ JK}^{-1} \times 298 \text{ K}} = 2.46 \times 10^{25} \text{ m}^{-3}.$$

Now, the mean free path λ is

$$\lambda = \frac{1}{\sqrt{2} \pi d^2 n}.$$

Substituting the above values, we get the mean free λ as

$$\lambda = \frac{1}{1.41 \times 3.14 \times (2.46 \times 10^{25}) \times (2.56 \times 10^{-10})^2} = 1.4 \times 10^{-7} \text{ m}.$$

18. The molecular diameter of nitrogen is $3.5\overset{\circ}{A}$. Calculate

   (a) the collision cross-section of nitrogen molecules
   (b) the mean free path of the nitrogen molecules and
   (c) the collision frequency for nitrogen at N.T.P.

   Given that: Boltzmann constant $K_B = 1.38 \times 10^{-23}$ JK^{-1}, Avogadro's number $N_A = 6.023 \times 10^{23}$ per gm mol and mass of hydrogen atom $= 1.67 \times 10^{-24}$ gm.

   Answer:

   (a) If a molecule of diameter d is moving with average speed $\langle v \rangle$, its sphere of influence sweeps out in one second a cylinder of base area πd^2 and length $\langle v \rangle$. The quantity πd^2 is called the collision cross-section σ, which is given by

   $$\sigma = 3.14 \times (3.5 \times 10^{-10})^2 \text{ m}^2 = 38.465 \times 10^{-20} \text{ m}^2.$$

   (b) According to Clapeyron's equation, the number of molecules per unit volume n in a gas is given by

   $$P = n K_B T; \quad \Rightarrow \quad n = \frac{P}{K_B T}.$$

At N.T.P., the pressure is $P = 0.76$ m Hg $= 0.76 \times 13.6 \times 10^3 \times 9.8$ N m^{-2} $= 1.013 \times 10^5$ N m^{-2}, and the temperature is $T = 273$ K. Hence, the number of molecules n per unit volume is given by

$$n = \frac{1.013 \times 10^5 \text{ N m}^{-2}}{(1.38 \times 10^{-23} \text{ JK}^{-1}) \times 273 \text{ K}} = 2.69 \times 10^{25} \text{ m}^{-3}.$$

Therefore, the mean free path λ is given by

$$\lambda = \frac{1}{\sqrt{2}\pi d^2 n} = \frac{1}{1.414 \times 3.14 \times (3.5 \times 10^{-10} \text{ m})^2 \times 2.69 \times 10^{25} \text{ m}^{-3}}$$
$$= 6.834 \times 10^{-8} \text{ m}.$$

(c) The mean speed of a molecule of mass m in a gas at temperature T Kelvin is given by

$$\langle v \rangle = \sqrt{\frac{8 K_B T}{\pi m}}.$$

Here, $m = (28 \times 1.67 \times 10^{-24})$ g $= 28 \times 1.67 \times 10^{-27}$ kg, where 28 is the molecular weight of nitrogen, and the temperature is $T = 273$ K. So, the average speed $\langle v \rangle$ of the molecules is given by

$$\langle v \rangle = \sqrt{\frac{8 \times (1.38 \times 10^{-23}) \times 273}{3.14 \times (28 \times 1.67 \times 10^{-27})}} \text{ m s}^{-1} = 4.53 \times 10^2 \text{ m s}^{-1}.$$

The frequency of collisions p_c is given by

$$p_c = \frac{\langle v \rangle}{\lambda} = \frac{4.53 \times 10^2 \text{ m s}^{-1}}{6.85 \times 10^{-8} \text{ m}} = 6.613 \times 10^9 \text{ s}^{-1}.$$

19. The coefficient of viscosity of oxygen at N.T.P. is 1.96×10^{-4} gm/cm/s and its molecular weight is $M = 32$. Calculate the diameter and collision cross-section of O_2 molecule. Given: Boltzmann constant $K_B = 1.38 \times 10^{-16}$ erg/degree and Avogadro's number $N_A = 6.023 \times 10^{23}$ per gram mole.

Answer: The mass of an O_2 molecule is

$$m = \frac{M}{N_A} = \frac{32}{6.025 \times 10^{23}} \text{ g} = 5.31 \times 10^{-23} \text{ g}.$$

The number of O_2 molecules per unit volume at N.T.P. (273 K and 1 atmpressure) is

$$n = \frac{P}{K_B T} = \frac{76 \times 13.6 \times 980 \text{ dyne/cm}^2}{1.38 \times 10^{-16} \text{ erg/degree} \times 273 \text{ K}} = 2.69 \times 10^{18} \text{ cm}^{-3}.$$

The average speed of O_2 molecules at 273 K is

$$\langle v \rangle = \sqrt{\frac{8 \times (1.38 \times 10^{-16}) \times 273}{3.14 \times (5.31 \times 10^{-23})}} \text{ cm/s} = 4.25 \times 10^4 \text{ cm/s}.$$

Now, according to kinetic theory, the viscosity of a gas in terms of mean free path is given by $\eta = \frac{1}{3} mn\langle v \rangle \lambda$. This provides the value of λ as

$$\lambda = \frac{3\eta}{mn\langle v \rangle} = \frac{3 \times 1.96 \times 10^{-4}}{5.31 \times 10^{-23} \times (2.69 \times 10^{18}) \times (4.25 \times 10^4)} \text{ cm}$$
$$= 9.68 \times 10^{-5} \text{ cm}.$$

The mean free path λ is given by

$$\lambda = \frac{1}{\sqrt{2}\pi d^2 n} = \frac{1}{1.414 \pi d^2 n}.$$

Therefore, the diameter of O_2 molecule is

$$d = \sqrt{\frac{1}{1.414 \, \pi \, n \, \lambda}} = \sqrt{\frac{1}{1.414 \times 3.14 \times (2.69 \times 10^{18}) \times (9.68 \times 10^{-5})}}$$
$$= 3. \times 10^{-8} \text{ cm} = 2.94 \text{Å}.$$

The collision cross-section σ of O_2 molecules is

$$\sigma = \pi d^2 = 3.14 \times (2.94 \times 10^{-8})^2 \text{ cm}^2 = 27.14 \times 10^{-16} \text{ cm}^2.$$

20. Calculate the radius of an oxygen molecule if its coefficient of thermal conductivity $K = 24 \times 10^{-3}$ (J/Km/s/deg) at 0°C and specific heat at constant volume $C_v = 20.9 \times 10^3$ J/kilo mol/deg. Boltzmann constant $K_B = 1.38 \times 10^{-23}$ J/deg and mass of an oxygen molecule 5.31×10^{-26} kg.

Answer: The average velocity of oxygen molecules at 0°C = (273 K) is given by

$$\langle v \rangle = \sqrt{\frac{8 K_B T}{\pi m}} = 425 \text{ m/s}.$$

If d is the molecular diameter, then the coefficient of thermal conductivity K is given

$$K = \frac{1}{3\sqrt{2}} \frac{m\langle v \rangle C_v}{\pi d^2}.$$

Hence, $C_V = \frac{C_v}{M} = \frac{20.9 \times 10^3}{32} = 0.653 \times 10^3$ J/kg/deg. From the given expression for thermal conductivity K, we have

$$d^2 = \frac{m\langle v \rangle C_v}{3\sqrt{2\pi}K} = \frac{5.31 \times 10^{-26} \times 425 \times 0.653 \times 10^3}{3 \times 1.41 \times 3.14 \times 24 \times 10^{-3}} = 4.71 \times 10^{-21}.$$

This gives rise to

$$d = 6.86 \times 10^{-11} \text{ m} = 0.686\text{Å}. \text{ So, radius will be} = 0.343\text{Å}.$$

21. Estimate the mean free path of an air molecule at 273 K and 1 atm assuming it to be a sphere of diameter 4.0×10^{-10} m. Estimate the mean time between collisions for an oxygen molecule under these conditions, using $v = v_{r.m.s.} = 517$ ms^{-1}.

Answer: The expression for mean free path λ is given by

$$\lambda = \frac{RT}{\sqrt{2}\pi d^2 N_A P} = \frac{8.31 \text{ JK}^{-1} \times 273 \text{ K}}{\sqrt{2}\pi(4.0 \times 10^{-10})^2 \times (6.02 \times 10^{23}) \times (1.01 \times 10^5 \text{ Pa})}$$
$$= 5.2324 \times 10^{-8} \text{ m}.$$

The mean time between collisions for an oxygen molecule under these conditions is given by

$$t = \frac{5.2324 \times 10^{-8} \text{ m}}{517 \text{ ms}^{-1}} \approx 1.0 \times 10^{-10} \text{ s}.$$

22. The coefficient of self-diffusion of oxygen at 273 K and 1 atmosphere pressure is 1.8×10^{-5} m^2/s. Estimate mean free path. The gas constant is 8.3×10^3 J/kg-mole-K and molecular weight of oxygen is 32.

Answer: The average velocity of an oxygen molecule at 273 K is given by

$$\langle v \rangle = \sqrt{\frac{8K_B T}{\pi m}}.$$

Mass of the molecule, $m = \frac{M}{N_A}$, where M is molecular weight, and N_A is Avogadro's number, and $K_B = \frac{R}{N_A}$. The average velocity is given by

$$\langle v \rangle = \sqrt{\frac{8\frac{R}{N_A}T}{\pi \frac{M}{N_A}}} = = \sqrt{\frac{8RT}{\pi M}} = \sqrt{\frac{8 \times 8.3 \times 10^3 \times 273}{3.14 \times 32}} = 4.25 \times 10^2 \text{ m/s}.$$

Now, the coefficient of self-diffusion D is related to the mean free path λ by

$$D = \frac{1}{3}\lambda\langle v\rangle \quad \Rightarrow \quad \lambda = \frac{3D}{\langle v\rangle} = \frac{3 \times 1.8 \times 10^{-5} \text{ m}^2/\text{s}}{4.25 \times 10^2 \text{ m/s}} == 1.27 \times 10^{-7} \text{ m}.$$

23. Find the probability that a one-dimensional random walker starting from the origin is back to the origin after taking ten steps. [Burdwan University 2017]

Answer: We know that, the probability that the walker is at the origin after n steps depends on the nature of n. If n is odd then, the probability will be zero. If the walker takes an even number of steps (say $2n$), then the probability can be expressed as,

$$P(S_{2n} = 0) = \frac{(2n)!}{n!n!}2^{-2n}.$$

Here, $2n = 10$, so we have $P = \frac{(10)!}{5!5!}2^{-10} = \frac{63}{256} = 0.25$.

24. Consider a random walk in one dimension. In a single step, the probability of displacement between x and $x + dx$ is given by

$$P(x) = \frac{1}{\sqrt{2\pi\sigma^2}}\exp\left(-\frac{(x-a)^2}{2\sigma^2}\right)dx. \tag{4.115}$$

After N steps, the displacement of the walker is $S = X_1 + X_3 + \cdots + X_N$, where X_i is the displacement after i-th step. After N steps,
(a) what is the probability density for the displacement, S, of the walker?
(b) what is his standard deviation?

Answer: (a) By definition,

$$P_S(s) = \int dx_1 P(x_1) \int dx_2 P(x_2) \cdots \int dx_N P(x_N)\delta(s - x_1 - x_2 \cdots - x_N).$$

To calculate this integral, we expand δ-function into Fourier integral

$$\delta(x) = \int_{-\infty}^{\infty} \frac{dk}{2\pi}e^{ikx}.$$

In this way, we get

$$P_S(s) = \int_{-\infty}^{\infty} \frac{dk}{2\pi}e^{iks} \int dx_1 e^{-ik_1 x}P(x_1) \int dx_2 e^{-ik_2 x}P(x_2) \cdots \int dx_N e^{-ik x_N}P(x_N)$$

$$= \int_{-\infty}^{\infty} \frac{dk}{2\pi}e^{iks}\left(\int_{\infty}^{\infty} dx e^{-ikx}P(x)\right)^N = \int_{-\infty}^{\infty} \frac{dk}{2\pi}e^{iks}[\mathbf{P}(k)]^N. \tag{4.116}$$

Here,

$$\mathbf{P}(k) = \int_{-\infty}^{\infty} dx \, e^{-ikx} P(x)$$

is the Fourier component of the single-step probability. From the previous equation, we obtain

$$\mathbf{P}(k) = \exp(-ika - k^2\sigma^2/2) \quad \rightarrow \quad [\mathbf{P}(k)]N = \exp(-iNka - k^2N\sigma^2/2).$$

Now,

$$P_S(s) = \int_{\infty}^{\infty} \frac{dk}{2\pi} \exp[ik(s - Na) - k^2N\sigma^2/2] = \frac{1}{\sqrt{2\pi N\sigma^2}} \exp\left(-\frac{(s - Na)^2}{2N\sigma^2}\right).$$

(b) We have

$$\langle S \rangle = \int ds \, s P_S(s) = Na, \quad \langle S^2 \rangle - \langle S \rangle^2 = N\sigma^2.$$

25. Consider a random walk in one dimension for which the walker at each step is equally likely to take a step with displacement anywhere in the interval $d - a \leq x \leq d + a$, where $a < d$. Each step is independent of the others. After N steps, the displacement of the walker is $S = X_1 + X_3 + \cdots + X_N$ where X_i is the displacement after the i-th step. After N steps,
(a) what is the average displacement, $\langle S \rangle$, of the walker, and
(b) what is his standard deviation?

Answer: Starting from the definition, we have

$$\langle S \rangle = \int s P_S(s) ds = \int s ds \int dx_1 P(x_1) \cdots \int dx_N P(x_N) \delta(s - x_1 - \cdots - x_N),$$

$$= \int dx_1 P(x_1) \cdots \int dx_N P(x_N)(x_1 + x_2 + \cdots + x_N) = N\langle X \rangle = Nd.$$

In a similar way,

$$\langle S^2 \rangle = \int dx_1 P(x_1) \cdots \int dx_N P(x_N)(x_1 + x_2 + \cdots + x_N)^2.$$

We know that

$$\left(\sum_i x_i\right)^2 = \sum_i x_i^2 + \sum_{i \neq j} x_i x_j.$$

The number of pairs with non-equal i and j is $N(N_1)$. After averaging, the expression for $\langle S^2 \rangle$ comes out to be

$$\langle S^2 \rangle = N \langle X^2 \rangle + N(N_1)\langle X \rangle^2.$$

To calculate $\langle X^2 \rangle$ let us specify the normalized probability as

$$P(x) = \frac{1}{2a} \qquad d - a \leq x \leq d + a$$
$$= 0 \qquad \text{otherwise.}$$

Then

$$\langle X^2 \rangle = \frac{1}{2a} \int\limits_{d-a}^{d+a} x^2 dx = d^2 + \frac{a^2}{3}.$$

Summarizing we get

$$\langle X^2 \rangle - \langle S \rangle^2 = N \left(d^2 + \frac{a^2}{3} \right) + N(N-1)d^1 - (Nd)^2 = \frac{Na^2}{3}.$$

4.7   Multiple choice questions with answers

1. One mole of an ideal gas with an average molecular speed v_0 is kept in a container of fixed volume. If the temperature of the gas is increased such that the average speed gets doubled, then   [JAM 2016]

   (a) the mean free path of the gas molecules will increase.

   (b) the mean free path of gas molecule will not change.

   (c) the mean free path of the gas molecule will decrease.

   (d) the collision frequency of the gas molecule with wall of the container remains unchanged.

   Answer: The correct choice is (b).

   Solution: The expression for mean free path is given by $\lambda = \frac{K_B T}{\sqrt{2}\pi d^2 P}$. For a fixed volume, if the temperature is increased, the pressure is also increased by the same amount such that the ratio $\frac{T}{P}$ does not change. Therefore, the expression for mean free path will also not be changed.

2. At a constant temperature, the collision frequency of a given gas is

   (a) proportional to the pressure of the gas.

   (b) inversely proportional to the pressure of the gas.

   (c) proportional to the square root of pressure of the gas.

   (d) independent of the pressure of the gas.

   Answer: The correct choice is (a).

3. At a constant pressure, the collision frequency of a given gas is

(a) proportional to the temperature of the gas.

(b) inversely proportional to the temperature of the gas.

(c) inversely proportional to the square root of the temperature of the gas.

(d) independent of the temperature of the gas.

Answer: The correct choice is (c).

4. The survival equation is given by

(a) $N = N_0 e^{-\frac{x^2}{\lambda}}$

(b) $N = N_0 e^{-\frac{x}{\lambda}}$

(c) $N = N_0 e^{-\frac{x}{2\lambda}}$

(d) $N = 2N_0 e^{-\frac{x}{\lambda}}$

where symbols have their usual meaning.

Answer: The correct choice is (b).

5. In a dilute gas, the number of molecules with free path length $\geq x$ is given by $N(x) = N_0\, e^{-\frac{x}{\lambda}}$, where N_0 is the total number of molecules, and λ is the mean free path. The fraction of molecules with free path lengths between λ and 2λ is [IIT-JAM 2022]

(a) $\frac{1}{e}$

(b) $\frac{e}{e-1}$

(c) $\frac{e^2}{e-1}$

(d) $\frac{e-1}{e^2}$

Answer: The correct choice is (d).

Solution: The number of molecules with free path λ is given by $N_1 = N_0\, e^{-1}$. Similarly, the number of molecules with free path 2λ is given by $N_2 = N_0\, e^{-2}$. Therefore, the fraction of molecules with free paths between λ and 2λ is given by

$$\frac{N_0(e^{-1} - e^{-2})}{N_0} = \frac{1}{e} - \frac{1}{e^2} = \frac{e-1}{e^2}.$$

6. Consider N_1 number of ideal gas particles enclosed in a volume V_1. If the volume is changed to V_2 and the number of particles is reduced to half, the mean free path becomes four times of its initial value. The ration $\frac{V_1}{V_2}$ is ... (round off to one decimal place)

(a) $\frac{V_1}{V_2} = \frac{3}{4}$

(b) $\frac{V_1}{V_2} = \frac{1}{4}$

(c) $\frac{V_1}{V_2} = \frac{1}{2}$

(d) $\frac{V_1}{V_2} = \frac{3}{5}$

Answer: The correct choice is (c).

Solution: According to the problem, we have $\lambda_2 = 4\lambda_1$ and $N_2 = \frac{N_1}{2}$. This shows that

$$\frac{V_2}{\sqrt{2}\pi d^2 N_2} = \frac{4V_1}{\sqrt{2}\pi d^2 N_1}; \quad \Rightarrow \quad \frac{V_1}{V_2} = \frac{1}{2}.$$

7. Two boxes A and B contain an equal number of molecules of the same gas. If the volumes are V_A and V_B and λ_A and λ_B denote respective mean free paths, then   [IIT-JAM 2018]

(a) $\lambda_A = \lambda_B$

(b) $\frac{\lambda_A}{V_A} = \frac{\lambda_B}{V_B}$

(c) $\frac{\lambda_A}{V_A^{\frac{1}{2}}} = \frac{\lambda_B}{V_B^{\frac{1}{2}}}$

(d) $\lambda_A V_A = \lambda_B V_B$

Answer: The correct choice is (b).

Solution: $\lambda = \frac{K_B T}{\sqrt{2}\pi d^2 P} = \frac{1}{\sqrt{2}\pi d^2 n} = \frac{V}{\sqrt{2}\pi d^2 N} \Rightarrow \lambda \propto V$, where $n = \frac{N}{V}$

$\lambda_A = K\, V_B, \lambda_B = K\, V_B$. This gives rise to $\frac{\lambda_A}{V_A} = \frac{\lambda_B}{V_B}$.

8. Two gases having molecular diameters d_1 and d_2 and mean free paths λ_1 and λ_2, respectively, are trapped separately in identical container. If $d_1 = 2d_2$, then $\frac{\lambda_1}{\lambda_2} = \dots$ (Assume that there are no changes in other thermodynamic parameters.)   [JAM 2019]

(a) 4

(b) 8

(c) 32

(d) 64

Answer: The correct choice is (a).

Solution: We know that the mean free path λ is inversely proportional to the square of diameter of the molecules of a gas, that is, $\lambda \propto \frac{1}{d^2}$.

This gives rise to

$$\frac{\lambda_1}{\lambda_2} = \left(\frac{d_2}{d_1}\right)^2 = 4.$$

9. The coefficient of viscosity of a gas depends on temperature as

(a) T

(b) $T^{\frac{1}{2}}$

(c) $T^{-\frac{1}{2}}$

(d) T^0

Answer: The correct choice is (b).

10. According to the transport phenomena, the thermal conductivity of a gas is found to be

(a) proportional to the pressure of the gas.

(b) inversely proportional to the pressure of the gas.

(c) independent of the pressure of the gas.

(d) proportional to the square of the pressure of the gas.

Answer: The correct choice is (c).

11. The diffusion coefficient of a gas D is related to the coefficient of viscosity η and the density of the gas ρ as

(a) $D = \frac{\eta}{\rho}$

(b) $D = \frac{\eta^2}{\rho}$

(c) $D = \frac{\eta}{\rho^2}$

(d) $D = \eta \times \rho$

Answer: The correct choice is (a).

12. In the case of Brownian motion, the mean square displacement of a particle $\langle s^2 \rangle$ depends on the time of diffusion τ as

(a) $\langle s^2 \rangle \propto \sqrt{\tau}$

(b) $\langle s^2 \rangle \propto \tau$

(c) $\langle s^2 \rangle \propto \tau^2$

(d) $\langle s^2 \rangle \propto \tau^3$

Answer: The correct choice is (b).

13. In the case of Brownian motion, the mean square displacement of a particle $\langle s^2 \rangle$ depends on the radius of the particle r as

(a) $\langle s^2 \rangle \propto \sqrt{r}$ (b) $\langle s^2 \rangle \propto r$ (c) $\langle s^2 \rangle \propto \frac{1}{r}$ (d) $\langle s^2 \rangle \propto r^3$

Answer: The correct choice is (c).

14. The number of particles per unit volume varies with height Z as

(a) $n = n_0\, e^{-\left(\frac{2\, m\, g}{K_B T}\right) Z}$

(b) $n = n_0\, e^{-\left(\frac{m\, g}{K_B T}\right) Z}$

(c) $n = n_0\, e^{-\left(\frac{m\, g}{2\, K_B T}\right) Z}$

(d) $n = n_0\, e^{-\left(\frac{m\, g}{K_B T}\right) Z^2}$

where symbols have their usual meaning.

Answer: The correct choice is (b).

15. A rigid triangular molecule consists of three noncollinear atoms joined by rigid rods. The constant pressure molar specific heat (C_p) of an ideal gas consisting of such molecules is   [JAM 2015]

(a) $6R$ (b) $5R$ (c) $4R$ (d) $3R$

Answer: The correct choice is (c).

Solution: The degrees of freedom of the rigid triangular molecule is $= 6$.

The internal energy is given by $U = \frac{6RT}{2}$. So, the specific heat at constant volume is $\Rightarrow C_V = (\frac{\partial U}{\partial T})_V = 3R \Rightarrow C_P = C_V + R = 4R$.

4.8 Exercise

4.8.1 Short answer type questions

1. What do you mean by "mean free path"?

2. What is collision frequency?

3. How does the collision frequency of the ideal gas molecules at absolute temperature T for an isochoric process depend upon the mean free path λ of it? [Burdwan University 2020]

4. Define diffusion, coefficient of viscosity, and thermal conductivity.

5. Write down the expression for the number of collisions per unit area per second.

6. What do you mean by the sphere of influence and collision cross-section?

7. Write down the expression for the survival equation.

8. Write down the expression for the variation of the number of molecules per unit volume as a function of height.

9. What are the expressions for mean free path in elementary, Clausius, and Maxwell's approach?

10. What are the effects of pressure and temperature on the mean free path?

11. How are the coefficient of viscosity and thermal conductivity related to each other?

12. Name the physical quantities being transported by the molecules of a gaseous system.

13. What is Brownian motion? State the characteristics of Brownian motion.

14. Why is Brownian motion observed only below a definite size of the particles? [Calcutta University 2021]

15. Cite some examples of Brownian motion in natural systems.

16. What is a random walk?

17. What is the relation between the mean square displacement of a particle and time in a random walk?

4.8.2 Long answer type questions

1. Calculate the number of molecules of a gas that strike or cross a surface per unit time.

2. Derive the survival equation of state.

3. What do you mean by "mean free path λ" of the molecules in a gas? Show that the probability of a gas molecule traversing a distance x without suffering a collision is $\exp\left(-\frac{x}{\lambda}\right)$, where λ is the mean free path of the molecules of the gas. [Vidyasagar University 2019]

4. Show that the mean free path λ of molecules of a gas can be expressed as

$$\lambda = \frac{K_B T}{\sqrt{2}\pi d^2 P},$$

where K_B is the Boltzmann constant, T is the temperature, P is the pressure, and d is the diameter of the molecule.

5. Show that the mean free path of an ideal gas is directly proportional to the absolute temperature of the gas and inversely proportional to its pressure. [Burdwan University 2021]

6. State the assumptions used in the derivation for the mean free path due to Clausius, and hence, derive the expression for mean free path.

7. Prove that the collision probability is the reciprocal of the mean free path λ.

8. Show that $f(x) = e^{-\frac{x}{\lambda}}$, where $f(x)$ is the probability of a free path x for molecules moving with velocity v with corresponding free path λ.

9. State the assumptions used in the derivation for the mean free path due to Maxwell, and hence, derive the expression for mean free path.

10. Calculate the expression for pressure from the concept of the mean free path in three dimensions.

11. What are transport phenomena? Deduce an expression for the thermal conductivity of a gas on the basis of kinetic theory. [Delhi University 2021]

12. Explain the transport phenomena. Derive an expression for the coefficient of viscosity of a gas on the basis of the KTG. How is the viscosity of a gas dependent on temperature and pressure? [Christ (Deemed to be University), Bengaluru 2018]

13. What is the quantity of transport responsible for the phenomenon of thermal conductivity? At any given temperature, a hydrogen gas has larger thermal conductivity than any other gas—explain. [Burdwan University 2021]

14. Establish the relation between the coefficient of viscosity, the coefficient of thermal conductivity, and the mean free path of the molecules of a gas.

15. Define the coefficient of diffusion in a gaseous system. Discuss the physical basis of the diffusion process according to KTG and obtain an expression for the diffusion coefficient.

16. Establish the relation between the coefficient of diffusion and the mean free path of the molecules of a gas.

17. What do you mean by Brownian motion? Give two examples of natural phenomena where Brownian motion is observed.   [Burdwan University 2020]

18. Give Einstein's theory of translational Brownian motion in gases.
[Delhi University 2021]

19. Derive the expression for the probability density function for a random walker.

20. Derive the expression for the mean square displacement of a particle executing a random walk.

4.8.3 Numerical problems

1. Prove that the number of particles striking per unit area per unit time on the surface of the wall of the container is equal to $\frac{n\langle v\rangle}{4}$, where n is the number of molecules per unit volume, and $\langle v\rangle$ is the mean speed.
[Burdwan University 2020]

2. A gas has an average speed of 10 ms^{-1} and a collision frequency of 10 s^{-1}. What is its mean free path?   [1 m]

3. Find the mean free path of a gas molecule whose diameter is 2Å and number of molecules per cc is 3×10^{19}.   [18.75×10^{-6} cm] [Delhi University 2021]

4. A gas has an average speed of 10 ms^{-1} and an average time of 0.1 s between collisions. What is its mean free path?   [1 m]

5. Calculate the fraction of molecules that will be traveling undeflected after traversing 0.693 times the mean free path. [0.50] [Vidyasagar University 2019]

6. A gas has a density of 10 particles m^{-3} and a molecular diameter of 0.1 m. What is its mean free path?   [2.25 m]

7. The average speed of oxygen molecules at room temperature is 450 ms^{-1}. The radius of the hydrogen molecule is 1.8×10^{-10} m. Calculate (i) collision cross-section σ, (ii) collision frequency P_c, (iii) mean free path λ, and (iv) mean time between two successive collisions τ. The volume density of the number of molecules is $n = 3 \times 10^{25}$ m^{-3}.
[40.7×10^{-20} m^2, 5.49×10^9 s^{-1}, 82 nm, and 1.8×10^{-10} s]

8. Show that the number of molecules striking the unit area of a wall per second is $\dfrac{P}{\sqrt{2\pi m K_B T}}$, where P is the pressure, T the temperature, m the mass of a molecule, and K_B the Boltzmann constant.

9. The mean free path of the molecules of a gas at 25°C is 2.63×10^{-5} m. If the radius of the molecule is 2.56×10^{-10} m, find the pressure of the gas. $K_B = 1.38 \times 10^{-23}$ JK^{-1}.   [1.0 mm of mercury]

10. The pressure within a cathode ray tube is one atmospheric pressure. The mean free path of air molecules at atmospheric pressure and at room temperature (27°C) is 10^{-5} cm. Estimate the mean free path of air molecules within the cathode ray tube at room temperature.   [3.03×10^{-10} m]

11. A shower of 5000 molecules, each moving with the same velocity originally, traverses a gas. Compute the number that would travel undeflected even after traversing a distance equal 0.5 and 1 time the mean free path λ.
                    [3032 and 1839]

12. The coefficients of viscosity of argon and helium are 22×10^{-6} Pa and 19×10^{-6} Pa, respectively. Calculate the mean free paths for these gases, if they are kept under identical conditions.   [$\frac{\lambda_{Ar}}{\lambda_{He}} = 0.367$]

13. A Brownian particle of radius 2.10×10^{-7} m moves in a liquid at 20°C. If the value of root mean square displacement is 6.5×10^{-6} m in 32 s, calculate the value of Boltzmann constant K_B. The coefficient of viscosity of the liquid is 1.2×10^{-3} N sm^{-2}.   [$K_B = 1.07 \times 10^{-23}$ JK^{-1}]

CHAPTER 5

Physical Properties of Real Gases

We suppose ... that the constituent molecules of any simple gas whatever (i.e., the molecules which are at such a distance from each other that they cannot exercise their mutual action) are not formed of a solitary elementary molecule, but are made up of a certain number of these molecules united by attraction to form a single one.

—Count of Quaregna Amedeo Avogadro

Learning Outcomes

After reading this chapter, the reader will be able to:

- List the differences between ideal and real gas

- List the experiments that depicted the behavior of real gases over a large range of pressures and temperatures

- Demonstrate the meaning of liquid–gas interface, critical volume, critical pressure, and critical temperature

- Derive the equation of state of a real gas considering the effect of pressure and volume

- Obtain the reduced equation of state, the law of corresponding state, and the compressibility factor

- Compare and contrast the Van der Waals equation of state with experimental results on CO_2 due to Andrews

- Solve numerical problems and multiple choice questions on the Van der Waals equation of state, reduced equation state, and critical constants of a gas

5.1 Introduction

The foundation of kinetic theory of gases (KTG) is based on two important assumptions: (i) the volume occupied by the molecules of the gas is negligible compared to the total volume of the container, and (ii) no appreciable intermolecular attractive or repulsive forces are present among the molecules. A gas is said to be an ideal one when it conforms exactly to these tenets of the KTG. According to the KTG, such an ideal gas of n mole obeys the equation of state: $PV = nRT$. It is the task of the experimental physicists to test the validity of this equation of state over the whole range of physical parameters such as pressure and temperature. There are a large number of direct and indirect experimental pieces of evidences which clearly indicate that in reality, gases do not behave ideally, that is, the equation $PV = nRT$ is not satisfied by the real gases over the entire range of the above-mentioned physical parameters. Real gases deviate from ideal behavior, especially at high pressures and low temperatures. The extent of deviation is measured with the help of a parameter known as the compressibility factor Z_c given by $Z_c = \left(\frac{PV}{RT} \right)$ for one mole of the given substance. Under ideal conditions, this ratio $\left(\frac{PV}{RT} \right)$ should be exactly equal to 1, that is, $Z_c = 1$.

With the increase in pressure, molecules of a gaseous system come closer to each other in the physical space. As a result of this crowding, these molecules experience larger attractive intermolecular forces that hold them together more. This lessens the force and frequency of collisions with the wall of the container, and as a result of this, the pressure becomes lower than that of the ideal value. Further, at higher pressures, molecules of the gas occupy a larger proportion of the volume of the container. Hence, other gas molecules can take up a larger proportion of the volume of the container, and the unoccupied volume of the container available to any one molecule will be smaller than that of in ideal conditions. This reduction in available volume will cause an increase in pressure beyond ideal conditions.

Temperature also influences deviations from the ideal gas behavior. With the decrease in temperature, the average kinetic energy of the molecules of the gas decreases. As a result, a larger proportion of the gas molecules will have insufficient kinetic energy to overcome attractive intermolecular forces from neighboring atoms. Under this situation, the gas molecules become "stickier" to each other and collide with the walls of the container with less frequency and force. Therefore, the pressure decreases below to that of the ideal value. Further, the molecules of real gases have velocity, mass, and volume. They liquefy when cooled to their boiling point. The space filled by the gas molecules is not small when compared to the total volume of the gas. It is expected that such characteristic features could be explained from the derived equation of state for the gas.

In this chapter, the experimental results on several real gases over a wide range of temperatures and pressures are presented. Comparisons between the behavior of ideal and real gases are also highlighted. Various attempts to explain such experimental results theoretically led to the development of several equations of state for the real gases. Such experimental results and the associated theoretical models are described in detail.

5.2 Real gases

The molecules of a real gas occupy space and have interactions among themselves. Thus, a real gas behaves like a nonideal one. Consequently, the ideal gas laws are not applicable to them. There are certain deviations from the ideal behavior of gases. These deviations originate from two basic factors:

1. The ideal gas laws assume that as pressure increases, the volume of a gas becomes very small and approaches zero. In reality, the volume does approach a small number, but it will not be zero because molecules have volume and do occupy a certain space. With the increase in pressure, volume cannot be compressed below a certain value.

2. There exist intermolecular forces in gases. These forces become increasingly important at low temperatures, and the translational motion of the molecules slows down, almost to a halt. However, at high temperatures, the intermolecular forces are very small and tend to be considered negligible.

These two factors: finite volume of the gas molecules and intermolecular interactions lead to various equations of state applicable to real gases under different conditions of pressures and temperatures.

5.2.1 Behavior of real gases: deviations from the ideal gas equation

It has been experimentally observed that only under certain special conditions of low pressures and high temperatures, gaseous systems obey the ideal gas laws. Deviations are observed primarily at high pressures and low temperatures. Such deviations of the real gases from the ideal gas behavior are traced out mainly due to wrong or incorrect assumptions in the postulates. Such postulates are the following:

1. It is assumed in the KTG that the molecules are point particles with no volume assigned to them. Then, by increasing pressure to a large extent, it is possible to compress the gases to zero volume. But this does not happen in reality. Gases cannot be compressed to zero volume whatever be the value of pressure, which indicates that molecules do have some volume, though small and cannot be neglected in the case of real gases.

2. Further, it is assumed in the KTG that molecules are independent and do not interact with each other. But experimentally, it is found that molecules do interact with each other depending upon their nature. These interactions affect

the pressure of the gas and differ from gas to gas. By incorporating correction factors in the pressure and volume of the gases, a number of gas laws have been developed for the real gases.

3. It is assumed in the KTG that collisions among the molecules are elastic. But actually, they are not elastic, and they exchange energy. Hence, the particles do not have the same energy and have a distribution of energy.

Deviations of the behavior of real gases from that of the ideal gas laws can be graphically presented in plots. From the ideal gas equation $PV = RT$ for one mole, we know that at constant temperature T, the PV versus P plot would be a straight line parallel to the pressure axis. In fact, this plot is an indication of the nature of the gas. But experimentally, it is observed that for real gases, the PV versus P plot at a constant temperature is not a straight line parallel to the pressure axis, indicating a significant amount of deviation from the ideal gas behavior. There are, in general, two types of PV versus P curves observed experimentally:

1. The value of the product PV increases with the increase in pressure P. This is the first type of the curves and observed for the gases, such as hydrogen and helium.

2. In the second type of plot, there is a decrease in the product PV first with the increase in pressure P, that is, there is a negative deviation from ideal behavior. This is observed in the case of gases, such as carbon monoxide and methane. With further increase in pressure P, the product PV reaches to a minimum value characteristic of the gas at a certain pressure. With further increase in P, the product PV starts increasing. The curve then crosses the line for ideal gas and after that shows positive deviation continuously.

Thus, it is experimentally observed that under all conditions, the real gases do not follow ideal gas equation perfectly. It is found that under all conditions, real gases do not follow Boyle's law, Charles' law, and Avogadro's law perfectly. Naturally, two questions arise:

1. Why do gases deviate from the ideal behavior?

2. What are the conditions under which gases deviate from ideal gas behavior?

To get the answer of the first question, we have to look into the postulates of the kinetic theory once again. Two assumptions of the KTG do not hold good. We know that the assumption – "there is no force of attraction between the molecules of a gas" – is not correct. If it is correct, the gas will never liquefy. However, it is known that gases do liquefy when cooled and compressed. Also, liquids so formed are very difficult to compress.

5.2.2   Experimental results on the real gases

The model of KTG described in Chapters 2 and 3 seems to be quite simple but still is widely applicable to explain the physical behaviors of ideal gases. This model does not hold good universally. For example, the concept of ideal gas breaks down at high pressures and low temperatures for common gases. Another major drawback of the ideal gas model is its inability to predict the liquefaction of gases. Such phase changes occur frequently in physical phenomena and play a vital role in technological applications. In this section, experimental results on various gaseous substances leading to deviations from the prediction of the ideal gas model are described in detail.

Regnault's experimental investigation on real gases

Regnault performed a series of experiments on real gases such as hydrogen, oxygen, nitrogen, and carbon dioxide by applying pressure up to 30 atmosphere and varying the temperature from 273 K to 373 K. The experimental results obtained by him for such real gases H_2, O_2, N_2, and CO_2 over the pressure range from 0 to 10 atmosphere and at temperature $T = 273$ K are shown in Figure 5.1. In this figure, the product of pressure P and volume V, that is, PV

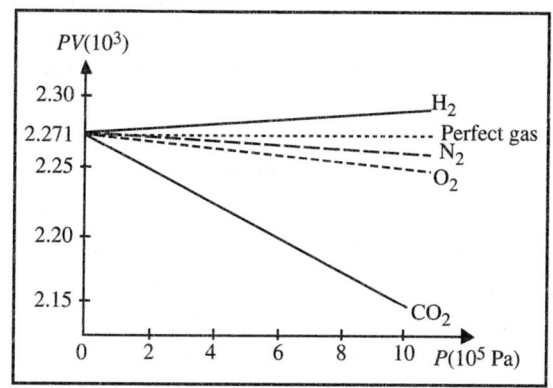

Figure 5.1 Variation of PV versus P for H_2, O_2, N_2, and CO_2 in the range 0–10 atm. The dotted horizontal line indicates values for an ideal gas.

is plotted as a function of the corresponding pressure P. It should be pointed out here that for one mole of an ideal gas (the equation of state is $PV = RT$), the corresponding plot will be a straight line parallel to the P-axis, and this is shown by a dotted line in the figure, indicated by perfect gas.

The following characteristic features of the behavior of real gas are observed from the curves:

1. In the given range of pressure, the curves are straight lines inclined to the pressure-axis at the given temperature of $T = 273$ K.

2. It is observed further from the figure that the product PV increases with P for hydrogen, whereas that decreases for nitrogen, oxygen, and carbon dioxide in the given range of pressure.

3. All the curves converge to the same point $(2.271 \times 10^3 \text{ J mol}^{-1})$ on the PV–axis at a temperature of 273 K. This value matches to 8.31 J mol^{-1} K^{-1} for $T = 273$ K. This is the universally accepted value of the gas constant R. Thus, it can be concluded that the real gases deviate from the perfect gas behavior, except for $P \to 0$.

These experimental results on several common gases indicate that there exist clear deviations between the behavior of real and ideal gases.

Andrews' experimental investigation on real gases

Andrews' experimental investigation of real gases, mainly on CO_2, provides a major contribution to the understanding of the behavior of real gases. The experiment was carried out over a wide range of temperatures and pressures as well. The isotherms obtained in this experiment clearly indicate the deviations between the behavior of real and ideal gases. The experimental techniques and the corresponding observation are described in detail below.

Principle of Andrews' experiment

The basic principle involved in Andrews' experiment is very simple. **At a given temperature, the volume of a fixed mass of a gas is measured as a function of pressure, and the variation of pressure with volume is plotted in a graph at that temperature.** Such curves at various temperatures are called isotherms. From such a given isotherm, it is possible to investigate the deviation of a real gas from the behavior predicted by the ideal gas equation.

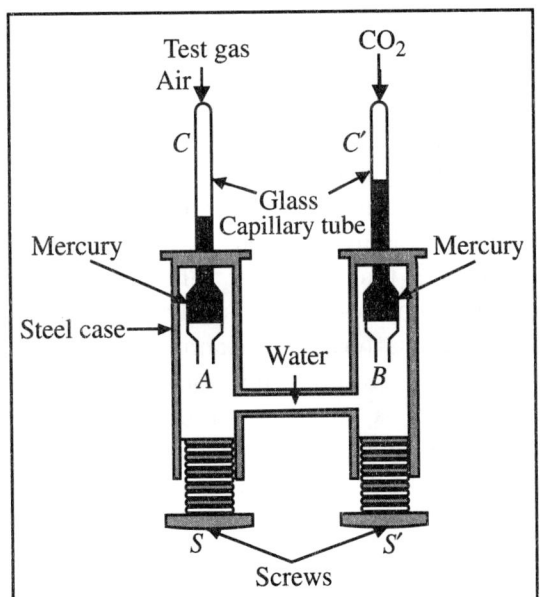

Figure 5.2 Apparatus of Andrews' experiment on CO_2 at different temperatures.

Andrews' experimental setup

In 1863, Thomas Andrews investigated thoroughly the liquefaction of gases. For this purpose, he designed an experimental setup capable of applying pressures up to 200 times the atmospheric pressure to a gas. A cross-section of Andrews' experimental setup is shown in Figure 5.2.

Andrews' apparatus consists of two identical glass tubes, A and B, initially open at both ends, with thick-walled capillary tubes C and C' at the upper side. The capillary tubes are graduated to measure their volume and length correctly. In tube A, pure dry air is passed for a long time, and both ends of the tube are sealed. In tube B, experimental gas CO_2 is passed for a long time, and both ends of the tube are sealed. The lower ends of both tubes are opened and immersed in mercury. When the tube is heated up by an ambient air gas burner (Bunsen burner) and allowed to cool, a small column of mercury (C') enters the tube and encloses the gas in the capillary portion of the tube.

In this way, by alternate heating and cooling the tubes, small pallets of mercury are drawn in both tubes. Both tubes A and B are fixed in H-shaped copper vessel having stoppers S and S', respectively. The vessel is filled with water. By screwing in a plunger, water is compressed and pressure up to 400 atmosphere can be applied.

Since the pressure on tubes A and B are the same, from the volume of air in tube A, pressure of CO_2 in tube B can be calculated. The volume of CO_2 in tube B can be recorded directly. The temperature of CO_2 can be maintained at any desired value from 0 to 100°C. Hence, using this experimental set up, isotherms can be recorded over a wide range of pressures and temperatures.

Andrews' experimental results

Andrews performed detailed experiments on the compressibility of a number of gases during their liquefaction, but his experimental findings were mainly focused on CO_2 gas. Such experimental results on CO_2 gas are shown in Figure 5.3 on a P–V diagram at selected temperatures. These isotherms exhibit structures and pass through various states of matter. For example, with the increase in pressure, the isotherm at 21.5° passes from a gaseous state at a higher volume to condensation (volume decreases) and finally to a liquid state at high pressure (volume almost remains constant). Further, the following conclusions can be drawn from Andews' experimental observations:

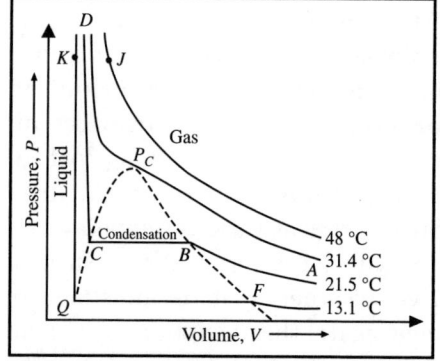

Figure 5.3 Variation of pressure with volume for CO_2 at selected temperatures.

1. The isotherms at 48°C (= 321 K) and above this temperature, are similar to those of the ideal gas, that is, the behavior of CO_2 resembles that of a perfect gas at a temperature 48°C (= 321 K) and above this temperature.

2. A kink appears in the isotherm corresponding to the temperature 31.4°C (= 304.4 K). It signifies that at this temperature, CO_2 gas has started to condense. The point P_c is called the point of inflexion, that is, at this point, there is a change in curvature from convex upward to concave downward.

3. The kink spreads into a horizontal line at 21.5°C (= 294.5 K). The region represented by this horizontal line exhibits the coexistence of gaseous and liquid phases. For a specified range of values of temperature and pressure, this region implies physically a discontinuous change in the density of the material.

4. This trend continues for temperatures below 21.5°C (= 294.5 K).

The set of values of temperature T_c, volume V_c, and pressure P_c at the point of inflexion are called critical values. One may ask the question: Are the gaseous and liquid phases identical at this point? It is to be noted that the dotted curve in Figure 5.3 passing through the extremities of horizontal portions of different isotherms indicates a vapor state on the right and a liquid state on the left. The region within the dotted curve marks the coexistence of vapor–liquid phase in equilibrium.

Each gas has its own characteristic critical temperature and pressure. It should be noted further that the pressure required to liquefy a gas is less, having a lower value of critical temperature. The following characteristic features are observed from the curves shown in Figure 5.3:

1. Liquefaction of a gas is possible only when the gas is cooled up to or below its characteristic critical temperature T_c. It is, therefore, clear that the increase in PV with P for hydrogen in Regnault's experiment (see Figure 5.1) arises because its critical temperature T_c is much lower than the room temperature.

2. There exists a continuity of liquid and gaseous states, which clearly indicates that these two distinct states of a substance belong to the class of a continuous physical phenomenon. This demonstrates the fact that it is possible to move from the gaseous state to the liquid state by compressing the gas to a high pressure and then gradually cooling it at the required temperature.

Amagat's experimental investigation on the real gases

Amagat and other experimental physicists investigated the compressive behavior of several gases at various temperatures and pressures up to 3000 atmosphere. Experimental results of their phenomenal works also lent support to the observations of Regnault and Andrews. According to Boyle's law, the volume V of a gas is inversely proportional to the pressure P exerted on it, provided its mass M and temperature T remain constant. This, in turn, implies that the product of pressure P, and volume V of a fixed amount of a gas is constant at a given temperature T. Therefore, the product PV for an ideal gas should be parallel to the pressure axis.

Amagat determined the volume of different real gases as a function of pressure P, keeping the temperature T of the gas constant. Then, he plotted PV against P in a graph, where he got several lines known as Amagat's curve. At a constant temperature and for a fixed quantity of an ideal gas, the product PV should remain constant in spite of the variation of P, and consequently, the line obtained by plotting PV against P will be a straight line parallel to the P-axis. But for real

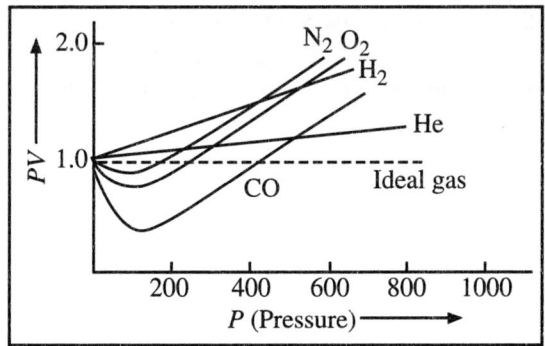

Figure 5.4 Variation of PV versus P of different volume of gases at 0°C in the range of pressure from 0 to 800 atm. The dotted horizontal line indicates the same for an ideal gas.

gases, it has been observed that the plots of PV against P are not parallel to the P-axis but exhibit structure depending upon the nature of the gas. Such characteristic features of PV against V for a number of real gases at 0°C are shown in Figure 5.4.

The following characteristic features are observed in Figure 5.4:

1. In the case of gases containing small molecules, such as H_2 and He, the value PV increases steadily with the increase of pressure P from the very beginning.

2. For O_2, N_2, CO_2, and other gases containing big molecules, product PV at first decreases, becomes minimum at a certain value of P, and then increases steadily with a further increase in pressure P.

The following conclusions can be drawn from the experiment due to Amagat:

1. No gas obeys Boyle's law at higher pressure P.

2. All gases show the same pattern of deviation from ideal gas behavior.

3. Greater deviation from ideal gas behavior is observed near critical temperature T_c.

4. The gas, which liquefied easily, shows more deviation from Boyle's law.

Thus, it is undoubtedly proved from the experimental observation that the real gases do not obey the ideal gas laws, that is, they deviate from the behavior of ideal gases. Hence, we can simply state that a gas that follows the gas laws **exactly**, is known as an ideal gas and one that does not is called a real gas. However, at high temperatures and low pressures, real gases behave like ideal gases.

Difference between ideal and real gases

Observation no.	Ideal gas	Real gas
1	The ideal gas obeys Boyle's law, Charles' law, and perfect gas equation.	A real gas does not obey Boyle's law, Charles' law, and perfect gas equation.
2	Equation of state of the gas is $PV = RT$.	The equation of state of the real (Van der Waals) gas is $\left(P + \frac{a}{V^2}\right)(V - b) = RT$.
3	The specific heat of an ideal gas is independent of temperature.	The specific heat of a real gas is dependent on temperature.
4	The internal energy of an ideal gas depends only on temperature.	The internal energy of a real gas depends on temperature as well as volume.
5	Real gases at high temperatures and low pressures behave as an ideal gas.	At low temperatures and high pressures, real gas behavior is perfect.

5.2.3 Critical constants

Thomas Andrews first presented complete data on pressure–volume–temperature relationships of carbon dioxide in both gaseous and liquid state. He plotted isotherms of carbon dioxide at various temperatures (such isotherms are shown in Figure 5.3). It was found later on that other real gases also behave in the same manner as carbon dioxide. Andrews observed that isotherms for real gases at high temperatures look like those of an ideal gas, and the gas cannot be liquefied even at very high pressure. As the temperature is lowered, the shape of the isotherm changes and considerable deviation from the ideal gas behavior is observed. For example, carbon dioxide remains gas up to 73 atmospheric pressure for the isotherm at 31.4°C (see the point P in Figure 5.3). Liquid carbon dioxide appears for the first time at 73 atmospheric pressure. This temperature of 31.4°C is called a critical temperature T_c for carbon dioxide and is the highest temperature at which liquid carbon dioxide is observed. Above this temperature, carbon dioxide is in gaseous state. The volume of one mole of the gas corresponding to the critical temperature is called the critical volume V_c, and the corresponding pressure is called the critical pressure P_c. The critical temperature, pressure, and volume are called critical constants of a given substance. The liquid carbon dioxide is simply compressed with a further increase in pressure, and the steep line represents the isotherm of the liquid. Even a slight compression produces a steep rise in pressure, indicating very low compressibility of the liquid.

On compression, the behavior of the gas is quite different below its critical temperature of 31.4°C. For example, carbon dioxide for the isotherm at 21.5°C remains as a gas-only up to point B. At this point B, liquid of a particular volume appears. Further, compression beyond point B does not change the pressure, that is, pressure remains constant though volume decreases following the straight line BC. In this region, liquid and gaseous carbon dioxide coexist, and further, an increase in pressure results in the condensation of more gas until point C is reached. At point C, condensation of all the gaseous phases of CO_2 becomes complete, and further, an increase in pressure merely compresses the liquid as shown by the steep line CD shown in Figure 5.3.

Below the critical temperature of 31.4°C, each isotherm shows similar characteristic features as that of isotherm at 21.5°C, except the length of the horizontal line increases with the decrease in temperature. It should be pointed out that at the critical point, this horizontal portion of the isotherm merges into one point (the point P), this point has been termed as the critical point. Thus, it is observed that a point like "A" in Figure 5.3 represents the gaseous state. A point like "D" represents the liquid state, and a point under the dome-shaped area represents the coexistence of liquid and gaseous phases of carbon dioxide in equilibrium. Almost all gases, upon compression at constant temperature (isothermal compression), exhibit similar characteristic features as those shown by carbon dioxide.

The above-mentioned results indicate that gases should be cooled below their critical temperature for liquefaction. The critical temperature of a gas is the highest temperature at which liquefaction of the gas first occurs. Liquefaction of so-called permanent gases (that is, gases that show continuous positive deviation in Z_c value) requires cooling as well as considerable compression. Compression brings the molecules in close vicinity, and cooling slows down the movement of the molecules. Therefore, intermolecular interactions may hold the closely spaced and slowly moving molecules together and the gas liquefies.

Another important characteristic feature of the isotherms below the critical point is that there is a continuity between the gaseous and liquid states. To recognize this continuity, the term fluid is used for either a liquid or a gas. Thus, a liquid can be viewed as a very dense gas. Liquid and gas can be distinguished from one another only when the fluid is below its critical temperature T_c and its pressure and volume lie under the dome. In this region, liquid and gas exist in a state of equilibrium, and a surface separating the two phases is visible. In the absence of this surface, there is no fundamental way of distinguishing between these two states. At critical temperature T_c, the liquid enters into the gaseous state imperceptibly and continuously, and the surface separating these two phases disappears. Hence, a gas below the critical temperature T_c can be liquefied only by applying pressure, and is called **vapor** of the substance. For example, carbon dioxide (CO_2) gas below its critical temperature T_c is called carbon dioxide vapor. Critical constants P_c, V_c, and T_c for some common substances are shown in Table 5.1.

Table 5.1 Critical constants for some gaseous substances

Substance	$T_c(K)$	$P_c(bar)$	$V_c(dm^3\ mol^{-1})$
He	5.3	2.29	0.0577
H_2	33.2	12.97	0.0650
O_2	154.3	50.4	0.0744
N_2	126	33.9	0.0900
H_2O	647.1	220.6	0.0450
CO_2	304.10	73.9	0.0956
NH_3	405.5	113.0	0.0723

5.2.4 Continuity of liquid and gaseous state

It is surprising to know the fact that there is no fundamental difference between a gas and a liquid. In general, a gas occupies a volume about 1600 times greater than that of a liquid with an equal amount of weight. Naturally, it is interesting to know the behavior of a gas that has been compressed to 1/1600 of its initial volume by applying a sufficiently high pressure. No change of phase occurs if the gas is compressed above a specific temperature called the critical temperature T_c, and the resulting substance remains as a gas that is just as dense as a liquid. This critical temperature T_c is different for different gases. Table 5.1 shows T_c for some gaseous substances. However, a liquid is suddenly formed at a particular pressure if the gas is compressed at a fixed temperature below the critical temperature T_c. If the gas is compressed further, simply the amount of forming liquid increases with a subsequent decrease in the amount of the gas. In this process, pressure remains constant to a specific value till all the gas is converted to the corresponding liquid. The applied pressure has to be increased by a huge amount to reduce the volume further, as liquids behave nearly as incompressible fluids.

Thus, it is observed that the condensation of a gas to a liquid occurs abruptly, and also it takes place at temperatures below a critical temperature. From the viewpoint of the KTG, these observations are something of a mystery. The questions – why does condensation of a gas to a liquid occur so abruptly and only at temperatures below a critical temperature – cannot be answered from the KTG. Considering the laws of ordinary mechanics and forces between molecules, people have written down equations to describe this condensation, but a satisfactory explanation is still lacking in the sense that no one has been able to show that it must occur. Condensation is an example of a first-order phase transition and remains one of the outstanding unsolved problems in statistical mechanics.

The critical temperature T_c traces the partition between a continuous change and an abrupt change. Near this critical temperature T_c, various interesting phenomena occur. As T_c is approached from below, the densities of the coexisting liquid and gas (which is usually called a vapor in this case) become closer, and they become identical at T_c. It should be mentioned that there is a unique point for every fluid at which liquid and vapor become identical. This point is called the critical point and is described by a critical temperature T_c, a critical volume V_c, and a critical pressure P_c. Above the critical temperature T_c, gas and liquid cannot be distinguished from each other; only a single fluid exists there. Moreover, without the occurrence of any abrupt condensation, it is possible to pass continuously from an apparently definite gas or vapor to an apparently definite liquid. This can be achieved by heating the vapor above the critical temperature T_c while keeping the volume V constant and then compressing the gas to a high density characteristic of the liquid. Finally, the gas is cooled at constant volume to its original temperature, and the gas clearly behaves as a liquid at this point.

5.2.5 Vapor and gas

A substance in its gaseous phase below the critical temperature is referred to as **vapor in physics**. The vapor can be frozen to a liquid by increasing the pressure without raising the temperature. It should be mentioned that a vapor is different from an aerosol. An aerosol is a substance that contains tiny particles of solid, liquid, or both within a gas. Amusingly, a large component of atmosphere of the earth is made of water vapor compared to other greenhouse gases. Some of the characteristic features of vapor are mentioned below:

1. A vapor is the gas phase below the critical temperature where either a liquid or a solid can exist at the same time.

2. The vapor phase is in a state of equilibrium with either the solid or the liquid phase in close proximity.

3. Clouds are produced when water vapor is condensed.

4. The molecules of vapor move in three directions with rotational, translational, and vibrational degrees of freedom.

5. The molecules of water vapor emit and absorb radiations in the infrared region with many more wavelengths than other greenhouse gases. This accounts for the largest greenhouse effect.

The gas is classified as one of the four states of matter; they have the distinct features of occupying the available space regardless of the shape and volume. This feature is because of the presence of very little intermolecular attraction between the molecules. The other three states of matter are solid, liquid, and plasma. The substances that exist in a gaseous form have neither a specific shape nor a specific volume. When they are encased in a container, they take up the entire space. They then exert a little pressure on the container walls. The following are the characteristic features of a gas:

1. Gases have a lower density and are highly compressible when compared to solids and liquids.

2. All of the gaseous particles exert the same amount of pressure on the wall's surfaces.

3. Gases have high kinetic energy. And the distance between each gas particle is considerable.

4. The intermolecular forces that exist between gas molecules are considered negligible.

5. Gases take up the entire volume of any container in which they are placed. The particles of gas move in all directions and collide with one another.

5.3   Theoretical results on the real gases

It has been experimentally found that there exist a number of discrepancies between the physical properties of the real gases and the predictions of ideal gas equation. To remove such discrepancies, a number of theoretical attempts have been made by several scientists, such as Van der Waals, Dieterici, Berthelot Saha, and Bose, to develop suitable equations of state applicable to such real gases over a wide range of temperatures and pressures. These theoretical models primarily focused on the explanation of experimental isotherms obtained on various real gases and attempted to provide a direction to understand the critical phenomena (phase transition) in real gases. In all these theoretical models, the key factor playing an important role is the nature of intermolecular force. Hence, a detail discussion on the nature of intermolecular interaction is presented below.

5.3.1   Intermolecular forces in real gases

Intermolecular forces are the forces of attraction and repulsion between molecules and atoms. It should be mentioned here that the electrostatic forces that exist between two oppositely charged ions are not included in intermolecular force. The electrostatic force holds atoms of a molecule together and is responsible for the formation of covalent bonds. Intermolecular forces which are **attractive in nature** are known as **Van der Waals forces**. This Van der Waals force varies considerably in magnitude and includes the following types of forces:

1. Dispersion forces or London forces

2. Dipole–dipole forces, and

3. Dipole-induced dipole forces.

A brief description of such forces is presented below.

Instantaneous dipole-induced dipole interaction or London interaction

Atoms and nonpolar molecules are electrically symmetrical and have no dipole moment. But, it may happen momentarily that the electronic charge distribution in one of the atoms is unsymmetrical, resulting in the creation of dipoles instantaneously on the atom for a very short time. Due to the development of such a transient dipole, the electron density of the neighboring atom is distorted and as a result, a dipole is induced in the neighboring atom. These temporary dipoles attract each other. A similar situation also occurs in the case of molecules. This force of attraction is known as the **London force**. It is always attractive and interaction energy due to this force is inversely proportional to the sixth power of the distance between two interacting atoms or molecules, that is, $E \propto \frac{1}{r^6}$, where r is the distance between two neighboring atoms or molecules. This force is dominant only at short distances ($r \sim 500$ pm), and their magnitude depends on the polarizability of the atoms or molecules.

Dipole–dipole forces or Keesom forces

Some molecules possess permanent dipole moments that exert force on each other. This force acting between the molecules with permanent dipoles is known as the **dipole–dipole force**. The interaction due to this force is stronger than the London force but is weaker than the ion–ion interaction because, in this case, only partial charges are involved. This dipole–dipole attractive force decreases with the increase in distance between the dipoles. Further, this interaction increases with the increase in the dipole moment of the molecules. In this case as well, the interaction energy is inversely proportional to the distance between polar molecules. When the polar molecules are stationary, dipole–dipole interaction energy is proportional to $\frac{1}{r^3}$, and that between rotating polar molecules is proportional to $\frac{1}{r^6}$. It is to be noted that in addition to this interaction, these polar molecules may also have London interaction.

Dipole-induced dipole forces or Debye forces

This type of attractive force acts between the polar molecules with permanent dipoles and the neutral molecules without permanent dipoles. This interaction is known as **Debye-type interaction**. A dipole is induced on the electrically neutral molecule when its electronic cloud is deformed by the permanent dipole of the polar molecule. Thus, an induced dipole is developed in the neutral molecule. In this case, the interaction energy is also proportional to $\frac{1}{r^6}$. The induced dipole moment on the neutral molecule depends on two factors: (i) the dipole moment of the permanent dipole and (ii) the polarizability of the electrically neutral molecule.

Atoms and molecules are attracted together by various types of Van der Waals forces. The importance of these forces follows from two unique properties:

1. Firstly, these forces are universal. All atoms and molecules attract one another through these types of forces. These forces are responsible for accounting physical phenomena such as the cohesion of the inert gases in the solid and the liquid states, and physical adsorption of molecules to solid surfaces. In the latter case, no normal chemical bonds are formed.

2. Secondly, Van der Waals force is additive for large numbers of molecules and is still significant when the molecules are comparatively far apart. Various physical properties of gases are also affected by the Van der Waals forces. This attractive force is also operative between two solid objects separated by a small gap and plays an important role in adhesion and in the stability of colloids.

Further, it should be noted that the theoretical expressions for the intermolecular forces become particularly simple when the molecules are some distance apart. This fact has been experimentally verified for two isolated molecules as well as for two solid objects separated by a small gap.

Characteristic features of Van der Waals interactions

Van der Waals interactions have several interesting characteristic features. Some of these features are listed below:

1. Forces due to Van der Waals interactions are much weaker than the normal covalent bond, ionic or even hydrogen bonding.

2. These interactions are additive in nature and cannot be saturated.

3. These interactions have no directional characteristics.

4. These forces are short-range. Only nearest-neighbor interactions are to be considered.

5. Van der Waals forces are independent of temperature except for dipole–dipole interactions or Keesom interactions.

6. These interactions depend on the molecular parameters, such as dipole moment, polarizability.

5.3.2 Van der Waals equation of state for real gases

According to the assumptions of KTG, the molecules of a gaseous system are treated as point particles that experience perfectly elastic collisions with each other and also with the walls of the container. These point particles have no interactions among themselves. In many experimental circumstances, these approximations work better for dilute gases at high temperatures and low pressures. But in reality, molecules of a gaseous system are not point masses; they possess finite volume, and they also exert intermolecular interactions among themselves. There are circumstances where these two factors have experimentally measurable effects on the properties of the gas. A modification of the ideal gas law was proposed by Johannes D. Van der Waals[1] in 1873 to take into account the molecular size and molecular interaction forces. This modification is usually referred to as the **Van der Waals equation of state**. For one mole of a real gas, this equation of state is given by

$$\left(P + \frac{a}{V^2}\right)(V - b) = RT, \tag{5.1}$$

where "a" and "b" are two constants, referred to as **Van der Waals constants**, and other terms have usual meaning. The Van der Waals equation of state is a thermodynamic equation of state based on the theory that fluids are composed of particles with nonzero volumes, and subject to an (not necessarily pairwise) interparticle attractive force.

The constants "a" and "b" have positive values and are characteristics of the individual gas. The constant "a" is related to the correction for the intermolecular forces. The constant "b" is related to the correction for finite molecular size, and its value is equal to four times the total volume of the atoms or molecules. It is interesting to note that the volume correction for each molecule is not the volume of the molecule but four times of it. The Van der Waals equation of state approaches the ideal gas law $PV = nRT$ as the values of these constants approach zero, that is, no volume is assigned to the molecules, and no interaction between the molecules is present. For "n" moles of the real gas, this equation takes the form

---

[1] J. D. Van der Waals (1873), Over de Constituiteit van den gas-en Vloeistoftoestand, doctoral dissertation, Leiden, Holland.

$$\left(P + \frac{an^2}{V^2}\right)(V - nb) = nRT. \tag{5.2}$$

In deriving equation (5.1), Van der Waals made the following assumptions:

1. Molecules of a real gas cannot be regarded as point masses; they must be assigned with finite size.

2. Molecules of the gas attract one another with a weak force that depends only on the distance between them. (This implies that gas molecules have both kinetic and potential energies and only nearest-neighbor interactions are important.)

3. The molecular density $n = \frac{N}{V}$ is small, where N is the total number of molecules of the gas enclosed in a volume V.

4. The number of collisions with the walls of the container is exactly the same for point and finite size molecules.

The equation of state is derived considering the following two assumptions: correction for finite size (volume) and correction for intermolecular attraction.

Johannes Diderik Van der Waals

Johannes Diderik Van der Waals (November 23, 1837–March 8, 1923) was a Dutch theoretical physicist and is famous for the pioneering work on the equation of state of real gases. This equation of state describes the behavior of gases and their condensation to the liquid phase. Van der Waals started his career as a school teacher and became the first physics professor at the University of Amsterdam in 1877. He is also famous for Van der Waals force, Van der Waals molecules, and Van der Waals radii. He first introduced the concept of nonideality of real gases in his 1873 thesis and attributing it to the existence of intermolecular interactions; he derived the equation of state of real gases. By comparing his equation of state with experimental data, Van der Waals was able to estimate the actual size of the molecules of a gas and the strength of mutual attraction between the molecules of a gas. The critical constants of a gaseous system are correctly predicted from the Van der Waals equation of state. The values of these critical constants match well to those obtained from thermodynamic measurements made at much higher temperatures. Using the knowledge of Van der Waals equation of state, it was subsequently possible to liquefy gases, such as nitrogen, oxygen, hydrogen, and helium. The pioneering work of Van der Waals greatly influenced Heike Kamerlingh. Van der Waals received the 1910 Nobel Prize in Physics for his work on the equation of state for gases and liquids.

Correction for the finite size of the molecules

We consider one mole of a real gas enclosed in a container of volume V. If this gas were composed of point masses, all the space would be available to them for free motion. When molecular size is taken into account, the volume available to a single molecule for free movement will be somewhat less than V. For one mole of the real gas, let us denote this reduction by "b". Therefore, the factor V occurring in the perfect gas equation should be replaced by $(V - b)$, and we will have

$$P(V - b) = RT. \tag{5.3}$$

The magnitude of "b" is equal to four times the total molecular volume for one mole of a Van der Waals gas. The expression for "b" is derived below in a simple manner. Let "r" be the radius of each molecule. At the instant of collision, the center-to-center distance of the two colliding molecules will be $d = 2r$. This is shown at the right of Figure 5.5. This implies that around any molecule, a spherical volume $V_s = \frac{4}{3}\pi d^3$ will be denied to every other molecules. This

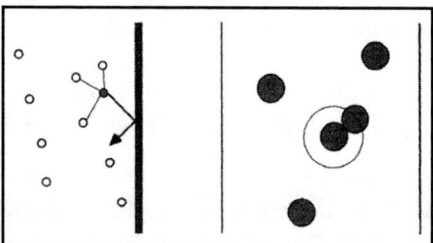

Figure 5.5 The pressure of a real gas is less than that of ideal gas because intermolecular attraction decreases the speed of the molecules approaching the wall (left), and the volume available to molecules is less than the volume of the container due to the finite size of the molecules (right).

volume is called the **volume of exclusion** and is eight times the volume of a molecule, that is, $V_s = 8V_m$, where $V_m = \frac{4}{3}\pi r^3$ is the volume of a single molecule.

Let us fill the container with $N(= N_A)$ molecules one-by-one. The volume available to the first molecule will be V. The volume available to the second molecule will be $V - V_s$, and by the method of induction, the volume available to the Nth molecule will be $V - (N - 1)V_s$. Hence, the average volume available to each molecule can be obtained from the arithmetic average and is given by

$$\langle V \rangle = \frac{1}{N}[V + (V - V_s) + \cdots + (V - (N-1)V_s)] = \frac{1}{N}\sum_{i=1}^{N}[V - (i-1)V_s]$$

$$= V - \frac{(N-1)}{2}V_s.$$

Here, we have used the result: $\sum_{i=1}^{N} (i-1) = \frac{N}{2}(N-1)$. For large value of N, the factor 1 on the right-hand side of the above expression can be neglected, that is, $(N-1) \approx N$, and we can write

$$\langle V \rangle = V - \frac{N}{2}V_s = V - 4\,N\,V_m = V - b, \tag{5.4}$$

where $b = 4\,N\,V_m$. This equation (5.4) shows that the volume V in the ideal gas equation $PV = RT$ should be replaced by $(V - b)$ for one mole of a Van der Waals gas.

Correction for intermolecular attraction

To account for the intermolecular attractive force due to Van der Waals interaction, we proceed in the following way: A molecule in the interior of the gas, on an average, is attracted equally in all directions so that there is no resultant force acting on it. Therefore, it will move in the interior as if there are no intermolecular interactions. This is shown at the left of Figure 5.5. But this is not true for a molecule in the outermost layer closer to the surface. Since the molecular distribution is only on one side, there will be a net inward force. So, whenever a molecule strikes the walls of the container, the momentum transferred will be less than that for an ideal gas. Thus, the intermolecular (cohesive) forces result in a decrease in pressure P. This drop in pressure ΔP is known as **cohesive pressure** and is found to be proportional to the number of molecules per unit volume $\frac{N}{V}$ in the surface layer (on which the inward forces act) and the number of molecules per unit volume $\frac{N}{V}$ in the layer just below the surface layer (which are pulling the striking molecules due to attraction).

Combining these two factors, we get

$$\Delta P \propto \left(\frac{N}{V}\right)^2, \quad \text{or} \quad \Rightarrow \Delta P = k\,\left(\frac{N^2}{V^2}\right).$$

Here, k is the constant of proportionality. Taking $kN^2 = a$, we get $\Rightarrow \Delta P = \left(\frac{a}{V^2}\right)$. Hence, in the ideal gas equation, we must replace pressure P by the sum of the observed pressure for a real gas and the cohesive pressure ΔP caused by intermolecular attractions. Hence, for one mole of the real gas, the pressure P will be given by

$$\left(P + \frac{a}{V^2}\right).$$

Using the results for the correction in volume and pressure, the equation of state for one mole of a real gas becomes

$$\left(P + \frac{a}{V^2}\right)(V - b) = RT. \tag{5.5}$$

This is known as Van der Waals equation of state for one mole of a real gas. Here, a and b are known as Van der Waals constants.

Comments on the constants a and b

Physical significance, units and dimensions of the two constants a and b appearing in Van der Waals equation of state for real gases are described in detail below.

Physical significance of the constants a and b

Van der Waals parameter a is a measure of the magnitude of intermolecular attraction among the gas molecules. The larger the value of a, the stronger will be the intermolecular force of attraction. It is assumed to be independent of pressure and temperature. But experimental observations show that the parameter a actually depends on temperature and pressure.

Van der Waals parameter b provides a measure of the excluded volume per mole. It represents the effective size of gas molecules and the intermolecular repulsion. With the increase in the value of b, both the size of the molecule and the intermolecular repulsion increase. It also slightly depends on temperature and pressure.

Units and dimensions of the constants a and b

The correction for pressure in the Van der Waals equation of state is given by $\frac{a\,n^2}{V^2}$. From the principle of homogeneity of dimension, unit and dimension of $\frac{a\,n^2}{V^2}$ will be equal to the unit and dimension of pressure. Hence, the unit of a is the unit of $\frac{P\,V^2}{n^2}$ and the dimension of a is $\left[\frac{P\,V^2}{n^2}\right]$. Therefore, the unit and dimension of a is atm L^2 mol^{-2} or bar L^2 mol^{-2} or Pa m^6 mol^{-2} (SI unit) and $[ML^5T^{-2}\text{ mol}^{-2}]$, respectively.

The correction for volume in the Van der Waals equation of state is given by $n\,b$. Therefore, the unit of b will be the unit of $\frac{V}{n}$, and $[b]$ is $\left[\frac{V}{n}\right]$. Hence, the unit and dimension of b are L mol^{-1} or m^3 mol^{-1} (SI unit) and $[L^3\text{ mol}^{-1}]$, respectively.

Temperature dependence of the constants a and b

The Van der Waals constant a is a measure of the intermolecular attraction. For non-polar molecules, a is more or less temperature independent. But for polar molecules, a decreases with increase in temperature. This is mainly due to dipole–dipole interaction or Keesom interaction that decreases with the increase in temperature.

The Van der Waals constant b represents the correction for finite volume and the intermolecular repulsion. In the hard sphere model, it is more or less temperature-independent. But, the molecules are not hard sphere. With an increase in temperature, an average speed of the molecules increases, and they may approach to a closer distance by penetration. This decreases the collisional diameter and the excluded volume. Hence, b slightly decreases with the increase in temperature. For example, the value of b decreases from 51 (L mol^{-1}) $\times$ 10^3 at 151.2 K to 41 (L mol^{-1}) $\times$ 10^3 at 183.2 K for nitrogen.

Values of the constants a and b

Values of the Van der Waals constants a and b for some gases are shown in Table 5.2.

Table 5.2 Values of the Van der Waals constants a and b for some gases. Values are calculated for 1 c.c. of the gas at N.T.P.

Gas	$a \times 10^5$ atoms/cm^6	$b \times 10^5$ (c.c.)
Helium	6.8	106
Argon	268.0	143
Oxygen	273.0	143
Nitrogen	272.0	173
Hydrogen	48.7	118
Carbon dioxide	717.0	191

5.3.3   Discussion on the Van der Waals equation of state

In this section, various aspects of Van der Waals equation of state are presented in detail.

For one mole of a real gas, Van der Waals equation of state can be written as

$$P + \frac{a}{V^2} = \frac{RT}{V - b}, \quad \Rightarrow PV^3 + aV - PbV^2 - ab = RTV^2.$$

This expression can be simplified to

$$V^3 - \left(b + \frac{RT}{P}\right)V^2 + \frac{a}{P}V - \frac{ab}{P} = 0. \tag{5.6}$$

Equation (5.6) is cubic in volume V. So, in each isotherm ($T = $ constant), there will be three values of V corresponding to each value of P. We know from the theory of equations that for such a cubic equation, either all three roots will be *real* (different or equal) or one real and two *imaginary*.

The Van der Waals equation of state can be expressed in the following form

$$P = \frac{RT}{V - b} - \frac{a}{V^2}. \tag{5.7}$$

From this equation (5.7), the following characteristic features about the behavior of a real gas may be noted:

1. For very large volume V, pressure P is very small, that is, in the limit, $P \to 0$ when $V \to \infty$.

2. When the volume V is very small, approaching b, the pressure $P \to \infty$. Hence, the curve must have a concavity upward, $P = 0$ line, that is, the volume axis is an asymptote to the curve. The other asymptote is $V = b$ line, that is, a line close and parallel to the pressure axis. Volume V cannot be less than b, for then pressure P becomes negative which is an absurdity.

3. For curve 1 (this is at a lower temperature) shown in Figure 5.6, there are three real roots for volume V which are different. Let these roots be V_1, V_2, and V_3.

4. For curve 2 (this is at a slightly higher temperature than curve 1) shown in Figure 5.6, there are still three different roots for volume V, but they are very close to each other.

5. For curve 3 (this is at a temperature T_c) shown in Figure 5.6, three real roots become equal, that is, they coincide with each other. The temperature at which this coincidence occurs is known as the critical temperature T_c. Pressure and molar volume corresponding to this temperature are, respectively, called critical pressure P_c and critical volume V_c. This particular point corresponding to $P = P_c$, $T = T_c$, and $V = V_c$ is called the critical point of the gas.

6. When the temperature is greater than the critical temperature T_c corresponding to curve 4, the P–V isotherm looks more or less similar to that of an ideal gas. In this region, there is only one real root for volume V for a particular value of P. The other two roots are complex and conjugate to each other.

7. James Clerk Maxwell replaced the isotherm 2 in Figure 5.6 between A and D with a horizontal line (pressure constant) positioned so that the areas of the two hatched regions are equal (means an area of AbB and BdD are equal). The flat line portion of the isotherm now corresponds to liquid–vapor equilibrium. The portions Ab and dD are interpreted as metastable states of **super heated liquid and super cooled vapor**, respectively. The equal area rule can be expressed as

$$P_V(V_G - V_L) = \int_{V_L}^{V_G} PdV,$$

where P_V is the vapor pressure (flat portion of the curve), V_L is the volume of the pure liquid phase at point "a" on the diagram, and V_G is the volume of the pure gas phase at point "C" on the diagram. The sum of these two volumes will be equal to the total volume V.

Experimental results and the Van der Waals equation of state

The **Van der Waals equation** is a thermodynamic equation describing the physical behavior of gases and liquids under a given set of pressure P, volume V, and temperature T. It is a thermodynamic equation of state used for a modification to and improvement of the ideal gas law, taking into account the nonzero size of atoms and molecules and the attraction between them. Van der Waals equation of state, when supplemented by the Maxwell construction (equal-area rule) provides, in principle, a complete description of the gas and its transition to the liquid, including the shape of the coexistence boundary curve.

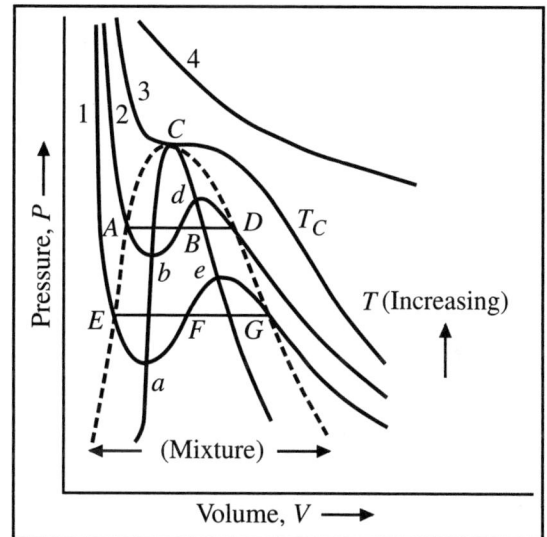

Figure 5.6 Theoretical curves for CO_2 according to the Van der Waals equation.

To verify various characteristic features of the Van der Waals equation of state with experimental results, we plot Van der Waals isotherms for CO_2 gas considering its known values of the parameters a and b. Once a and b are known, Van der Waals isotherms at different temperatures can be drawn theoretically from the equation: $\left(P + \frac{a}{V^2}\right)(V - b) = RT$. Four such isotherms are shown in Figure 5.6. For isotherms 1 and 2, the points (E, F, G) and (A, B, D) represent the three real but different roots. As one moves to isotherms at higher temperatures, the length of EG or AD decreases and the three roots approach each other. At the critical on isotherm 3, the three roots become equal (coincident) at the critical point C. For the isotherm above T_c, that is, the isotherm 4, there is only one real value of V for each value of P.

The portion Dd in curve 2 represents the *supersaturated* or *undercooled* state of vapor, and the portion bA is a *superheated* or *unsaturated* state of liquid. Each point of Dd normally belongs to some isotherm at a higher temperature, and when it forms a part of Van der Waals isotherm, it is in a supercooled state. This corresponds to the liquid state in stable equilibrium at a higher temperature, as is evident by a reference to Andrews' experimental curves. Though under certain conditions, an undercooled vapor can be obtained, its equilibrium state can be easily disturbed by mechanical shocks, introduction of dust particles, etc. Hence, the portion Dd that represents undercooled vapor in metastable equilibrium does not occur in Andrews' curves, which represent only states of stable equilibrium.

Similarly, bA normally corresponds to lower isotherm, and so in a Van der Waals curve, it represents a superheated liquid state in metastable equilibrium. Under certain experimental conditions, the portions Dd and bA may be realized, but the equilibrium of the vapor in such conditions is *metastable*. Hence, they do not occur in Andrews' curves that represent only states of stable equilibrium.

The most unphysical region is from $b \to B \to d$ along the isotherm 2 in Figure 5.6. In this region, both volume V and pressure P increase at a constant temperature. Hence, the isothermal coefficient of compression $\beta_T = -\frac{1}{V}\left(\frac{\partial V}{\partial P}\right)_T < 0$. This state is thermodynamically unstable. This is often called a **collapsible state**.

Van der Waals theory, however, does not tell us when the condensation begins, that is, the straight portion of Andrews curve commences. Although the metastable states $D \to d$ and $b \to A$ can be realized under specific experimental condition, but they are not at all stable. To resolve discrepancy between theoretical and experimental isotherms, particularly in the liquid–vapor interphase region, James Clerk Maxwell replaced the theoretical isotherm between D and A by a horizontal line positioned so that the areas enclosed above and below the constant pressure line are equal, that is, the area of DdB = area of BbA (on curve 2). This is known as **Maxwell's equal area construction**.[2] This construction is consistent with the prediction of the second law of thermodynamics.

Comments on the maxima and minima points of the Van der Waals curve

The maxima and minima points on the Van der Waals curves are obtained from the condition that $\left(\frac{\delta P}{\delta V}\right)_T = 0$. Differentiating equation (5.5) w.r.t volume V at constant temperature T, we get

$$\left(\frac{\delta P}{\delta V}\right)_T = -\frac{RT}{(V-b)^2} + \frac{2a}{V^3} = 0; \quad \Rightarrow \quad T = \frac{2a(V-b)^2}{RV^3}. \tag{5.8}$$

[2] J. C. Maxwell, On the Dynamical Evidence of the Molecular Constituent of Bodies, *Nature* 11, no. 279 (1875): 357–359.

This equation (5.8) is cubic in volume V. Each isotherm has three real, or one real, and two imaginary points of maxima or minima. The curves below the critical point C (see isotherm 3 in Figure 5.6) have one maximum and one minimum point. The isotherms above the critical point have none at all. The other point of minimum ties (as can be shown by a slight mathematical transformation) in the region $V < b$ and, thus, bears no physical meaning.

Using the value of T from equation (5.8) into the Van der Waals equation given by equation (5.5), we get the value of pressure P as

$$P = \frac{a(V - 2b)}{V^3}, \tag{5.9}$$

which is the equation to the curve abCde (see Figure 5.6) passing through the points of maxima and minima.

All isotherms lower than C cut the curve abCde at two points, and so the liquefaction can be seen by changing the pressure along them. For the critical isotherm, the two points coalesce into one. The isotherm passing through C, where C is **a point of inflexion** for the family of curves (or a maximum point for the curve abCde), is the critical isotherm, as every isotherm below C has maxima and minima points while there is none above C. Hence, C must be identified with the critical point. The pressure, volume, and temperature corresponding to this critical point are defined as the critical pressure P_c, critical volume V_c, and critical temperature T_c, respectively.

5.3.4 Comparison with experimental curves

The theoretical P–V curves (isotherms) for CO_2 obtained from Van der Waals' equation and the experimental P–V curves (isotherms) due to Andrews for CO_2 have the following similarities and dissimilarities:

1. The flat part of the experimental isotherm, below the critical point C, reveals an important fact. Since the pressure remains constant, while more and more of the gas condenses into liquid, the pressure of the gas in contact with the liquid must be always the same, quite independent of whether a small or a large fraction of the volume is occupied by liquid. It is also apparent from Andrews' curves that the value of this equilibrium pressure increases with the increase in temperature, that is, as one moves toward higher isotherms. Moreover, we also notice that the flat part becomes shorter until a singularly important isotherm is reached which has no true flat portion at all but just one point (the so-called critical point C) at which the direction of the curve changes its sign. The higher isotherms are now all ascending smoothly over the whole range of pressure and volume.

2. Below the critical temperature, the curves due to Van der Waals equation are different from Andrews' experimental curves. If the isotherms of Andrews' experimental curves are made to overlap with the theoretical curves for the same temperatures, it is found that above the critical temperature the two sets of curves coincide whereas below the critical temperature the two sets of curves are coincident only at the purely gaseous and liquid region. There is a prominent difference in the central region.

3. These two sets of P–V curves are similar only at and above the critical temperature T_c. If one goes to higher temperatures, that is, above the critical temperature T_c, the isotherms attain more and more the shape of a true rectangular hyperbola. This is the region, in which Boyle's law is valid.

4. The remarkable divergence between the two sets is that in the region where Andrews' curves show a straight line, the theoretical curves give maxima and minima. Experimentally, this is the region where condensation or vaporization sets in, and the pressure remains constant as long as the process continues.

5. The theoretical curves thus resemble, in general, the experimental curves of Andrews. But, the two sets of curves do not exactly overlap; the agreement is thus not at all quantitative. In fact, the agreement is only approximate and qualitative. This is due in part to errors in the assumed values of a and b and partly because the equation holds only approximately.

Explanation of the discrepancy between theoretical and experimental isotherms

An explanation for the discrepancy lies in the proper interpretation of the theoretical curve. The part dBb for the isotherm 2 in Figure 5.6 indicates that the volume decreases with a decrease of pressure. Apparently, it seems to be unphysical. This would correspond to a **collapsible state**—a decrease of volume with a decrease of pressure which would further decrease the volume. The state dBb, thus, represents an equilibrium state which is highly unstable to small changes of pressure (or density). It cannot, therefore, be realized with all the substance in one phase and, hence, was not observed by Andrews in the experiment.

5.3.5   Other equations of states for a real gas

There are certain deficiencies in Van der Waals equation of state. To overcome these deficiencies and to explain the physical behaviors of real gases, a number of equations of states have been proposed by some renowned physicists. Some of these are based on rigorous theoretical considerations, while others consist of adding more empirical constants chosen suitably. Some of these equations are briefly described below.

Clausius equation of state

One criticism of the Van der Waals equation is that it does not take into account the possibility of temperature dependence of the parameters a and b. Clausius[3] formulated an equation of state, in which the intermolecular attraction between the molecules of a real gas is described as inversely proportional to temperature T. He also considered three constants: a, b, and c in the equation of state. These constants are characteristic of the gas. This equation of state due to Clausius is given by

$$\left(P + \frac{a}{T\left(V_j + c\right)^2}\right)\left(V_j - b\right) = RT. \tag{5.10}$$

Though this equation of state shows several advantages over Van der Waals equation of state, its prediction of the temperature-dependent attraction is outweighed by problems associated with using this equation.

Berthelot equation of state

The term $(V_j + c)$ in the Clausius equation of state given by equation (5.10) is replaced by V_j in the equation of state proposed by Berthelot. Thus, the Berthelot[4] equation of state is given by

$$\left(P + \frac{a}{T\left(V_j\right)^2}\right)\left(V_j - b\right) = RT. \tag{5.11}$$

The equations of Van der Waals, Clausius, and Berthelot are the forerunners of a large family of cubic equations of states. In fact, these equations of states are cubic polynomials in molar volume.

Dieterici equation of state

In 1899, C. Dieterici[5] suggested a modification of the equation of state for a real gas. In this modification, he took into account the fact that the molecules at or near the walls of the containing vessel have higher potential energy than the molecules in the bulk gas, that is, the pressure gradient at the boundary of the gas was taken into account. Considering this physical argument, Dieterici proposed the following equation of state

$$P\left(V_j - b\right) = RTe^{-\frac{a}{RTV_j}}. \tag{5.12}$$

This Dieterici equation is a modification of Van der Waals equation of state. At low pressures, the Dieterici equation becomes identical to Van der Waals equation.

[3]R. Clausius, Ueber das Verhalten der Kohlensaure in Bezug auf Druck, Volumen, and Temperatur, *Annalen der Physik* 169 (1880): 337–357.

[4]D. J. Berthelot, Sur Une Méthode Purement Physique Pour La Détermination des Poids Moléculaires des Gaz et des Poids Atomiques de Leurs Éléments, *J. Phys.* 8 (1899): 263–274.

[5]C. Dieterici, Die spectrale Empfindlichkeit des Auges, *Ann. Phys. Chem. Wiedemanns* 69 (1899): 685.

Saha and Bose equation of state

Using the theory of probability, Saha and Bose derived an equation of state from thermodynamic considerations. The equation of state is given by

$$P = -\frac{RT}{2b} e^{-\left(\frac{a}{K_B TV}\right)} \ln\left(\frac{V - 2b}{V}\right). \tag{5.13}$$

This equation predicts the critical coefficient $\frac{RT_c}{P_c V_c} = 3.53$, which gives a better agreement with observed values for simpler gases.

Virial expansion of the equation of state

The perfect gas law is an imperfect description of a real gas. Therefore, the perfect gas law and the compressibility factors of real gases are combined to develop an equation of state to describe the isotherms of a real gas. This equation is expressed as a power series in the density $\rho = \frac{M}{V}$ and measures the deviation of a gas from the ideal behavior. This equation may be represented in terms of the compressibility factor, Z, as

$$Z = \frac{P}{\rho RT} = A + B\,\rho + C\,\rho^2 + \cdots \tag{5.14}$$

This equation was first proposed by Kamerlingh Onnes. The terms A, B, and C represent the virial coefficients. The leading coefficient A is defined as the constant value of 1, which enforces that the equation of state reduces to the ideal gas expression as the gas density approaches zero. B and C are known as the second and the third virial coefficients, respectively.

The second virial coefficient B represents the initial departure from the ideal-gas behavior and describes the contribution of the pair-wise potential to the pressure of the gas. The third virial coefficient C depends on the interactions between three molecules. The jth virial coefficient can be calculated in terms of the interaction of j molecules in a volume V. The second and third virial coefficients give most of the deviation from ideal $\frac{P}{\rho RT}$ up to 100 atm. It is important to note that the value of the virial coefficients is temperature-dependent. In terms of volume per molecule, $\langle V \rangle = \frac{V}{N}$, the equation of state in terms of virial coefficients can be rearranged to

$$P\langle V \rangle = RT\left(1 + \frac{B}{\langle V \rangle} + \frac{C}{\langle V \rangle^2} + \cdots\right). \tag{5.15}$$

The compressibility factor Z_c is defined as

$$Z_c = \frac{PV}{NRT} = \frac{P\langle V \rangle}{RT}. \tag{5.16}$$

Using equation (5.16) into equation (5.15), we get

$$Z_c = 1 + \frac{B}{\langle V \rangle} + \frac{C}{\langle V \rangle^2} + \cdots \tag{5.17}$$

The virial equation of state is a model that attempts to describe the properties of a real gas. If it were a perfect model, the virial equation would give results identical to those of the perfect gas law as the pressure of a gas sample approached zero. For the virial equation to collapse to the perfect gas law, all of the virial coefficients would need to have a value of zero at the same temperature. This is an unlikely occurrence, but because the second term in the virial equation, $\frac{B}{\langle V \rangle}$, is the largest term in the equation (5.15) as $\frac{1}{\langle V \rangle} \gg \frac{1}{\langle V \rangle^2} \gg \frac{1}{\langle V \rangle^3} \gg \cdots$, we can focus on the temperature at which B is zero. This temperature is known as the Boyle temperature, T_B, and **it is the temperature at which the repulsive forces between the gas molecules exactly balance the attractive forces between the gas molecules.**

5.3.6   Calculation of the critical constants P_c, V_c, and T_c for various equations of states

The critical constants of a gas are defined in the following way:

Critical temperature T_c

Gases can be converted to liquids by compressing the gas, that is, by increasing pressure at a suitable temperature. But, it becomes more difficult to liquefy as temperature increases because the kinetic energy of the particles also increases. If the temperature is above a critical value, then the gas cannot be liquefied by applying pressure only, even if the pressure is extremely high. **The critical temperature T_c of a substance is defined as the temperature at and above which the vapor of the substance cannot be liquefied by simply applying pressure, no matter how much is the applied pressure.**

Critical pressure P_c

A gas can be liquefied by applying pressure and keeping the temperature as constant. More and more pressure is required to liquefy the gas if it is at higher and higher temperatures. But if the temperature of the substance is above the critical temperature then it cannot be liquefied by applying pressure only. **The critical pressure P_c of a substance is the minimum pressure that must be required to liquefy a gas when it is at the critical temperature T_c of that particular substance.**

Critical volume V_c

The critical volume V_c of a substance is defined as the molar volume of the substance when it is at the critical pressure and critical temperature of that particular substance.

Critical constants P_c, V_c, and T_c for a Van der Waals gas

The Van der Waals equation for one mole of a real gas is given by

$$\left(P + \frac{a}{V^2}\right)(V - b) = RT, \tag{5.18}$$

where symbols have their usual meaning. At the critical point, the fluid does not exist in a particular state, either gas or liquid, but has characteristics of both. Hence, it is called a supercritical fluid. To see at what temperature, pressure, and volume, this supercritical behavior is observed, we use the fact that at the critical point, the isotherm is both horizontal (zero slope) and has no curvature. These two conditions are mathematically interpreted as

$$\left(\frac{\partial P}{\partial V}\right)_{T=T_c} = 0 = \left(\frac{\partial^2 P}{\partial V^2}\right)_{T=T_c}.$$

The Van der Waals equation (5.18) can be expressed in the following form to get the pressure P

$$P = \frac{RT}{V - b} - \frac{a}{V^2}. \tag{5.19}$$

Differentiating equation (5.19) with respect to volume V at the critical temperature $T = T_c$ and equating the result to zero at $V = V_c$, we get

$$\left(\frac{\partial P}{\partial V}\right)_{T=T_c} = \left[-\frac{RT}{(V - b)^2} + \frac{2a}{V^3}\right]_{T=T_c} = -\frac{RT_c}{(V_c - b)^2} + \frac{2a}{V_c^3} = 0. \tag{5.20}$$

This leads to the expression for T_c as

$$T_c = \frac{2a(V_c - b)^2}{RV_c^3}. \tag{5.21}$$

Again, differentiating equation (5.20) with respect to volume V at $T = T_c$ and equating the result to zero at $V = V_c$, we get

$$\left(\frac{\partial^2 P}{\partial V^2}\right)_{T=T_c} = \left[\frac{2RT}{(V - b)^3} - \frac{6a}{V^4}\right]_{T=T_c} = \frac{2RT_c}{(V_c - b)^3} - \frac{6a}{V_c^4} = 0. \tag{5.22}$$

Substituting the value of T_c from equation (5.21) into equation (5.22), we get

$$\frac{2RT_c}{(V_c - b)^3} = \frac{6a}{V_c^4}; \qquad \frac{2R}{(V_c - b)^3}\frac{2a(V_c - b)^2}{RV_c^3} = \frac{6a}{V_c^4}.$$

$$\frac{4}{(V_c - b)} = \frac{6}{V_c}; \qquad \Rightarrow \quad 4V_c = 6V_c - 6b; \qquad \Rightarrow \quad V_c = 3b. \tag{5.23}$$

Substituting this value of V_c into equation (5.21), we get the expression for critical temperature T_c as

$$T_c = \frac{2a(V_c - b)^2}{RV_c^3}; \quad \Rightarrow \quad T_c = \frac{2a(3b - b)^2}{R(3b)^3}; \quad \Rightarrow \quad T_c = \frac{8a}{27Rb}. \tag{5.24}$$

Again, substituting the values of V_c and T_c, respectively, from equations (5.23) and (5.24) into equation (5.19), we get the expression for critical pressure P_c as

$$P_c = \frac{RT_c}{V_c - b} - \frac{a}{V_c^2}; \quad \Rightarrow \quad P_c = \frac{8aR}{27Rb}\frac{1}{(3b - b)} - \frac{a}{(3b)^2} = \frac{4a}{27b^2} - \frac{a}{9b^2}; \quad \Rightarrow \quad P_c = \frac{a}{27b^2}. \tag{5.25}$$

Therefore, the critical constants of a Van der Waals gas are:

$$P_c = \frac{a}{27b^2}; \qquad V_c = 3b, \quad \text{and} \quad T_c = \frac{8a}{27Rb}. \tag{5.26}$$

Equation (5.26) shows that the values of the critical constants P_c, V_c, and T_c depend on the Van der Waals constants a and b and the universal gas constant R.

Critical constants for Dieterici's equation of state

The Dieterici's equation for one mole of a real gas is given by

$$P = \frac{RT}{(V - b)}e^{-\frac{a}{RTV}}, \tag{5.27}$$

where a and b are two constants. Following the similar arguments as stated in the Van der Waals equation, we have at the critical point,

$$\left(\frac{\partial P}{\partial V}\right)_{T=T_c} = 0 = \left(\frac{\partial^2 P}{\partial V^2}\right)_{T=T_c}. \tag{5.28}$$

Now, differentiating equation (5.27) with respect to volume V at the critical temperature $T = T_c$ and equating the result to zero at $V = V_c$, we get

$$\left(\frac{\partial P}{\partial V}\right)_{T=T_c} = \frac{RT}{V - b}e^{-a/RTV}\left[\frac{a}{RTV^2} - \frac{1}{(V - b)}\right] = 0.$$

This expression leads to

$$\frac{a}{RT_c V_c^2} - \frac{1}{(V_c - b)} = 0 \quad \Rightarrow \quad \frac{a}{RT_c V_c^2} = \frac{1}{(V_c - b)} \quad \Rightarrow \quad aV_c - ab = RT_c V_c^2;$$

$$\Rightarrow \quad RT_c V_c^2 - aV_c + ab = 0.$$

This leads to the expression for critical volume V_c as

$$V_c = \frac{a \pm \sqrt{a^2 - 4RT_c ab}}{2RT_c}. \tag{5.29}$$

Since only one permissible value of the critical volume is possible, the discriminant of equation (5.29) is taken to be zero at the critical point. Thus, we have

$$a^2 - 4RT_c ab = 0; \quad \Rightarrow \quad a = 4RT_c b.$$

Using this value of a, the expression for critical volume V_c becomes

$$V_c = \frac{a}{2RT_c} = \frac{4RT_c b}{2RT_c} = 2b; \quad \Rightarrow \quad V_c = 2b. \tag{5.30}$$

Now, substituting the value of V_c from equation (5.30) in the expression $\frac{a}{RT_c V_c^2} - \frac{1}{(V_c - b)} = 0$, we get the critical temperature T_c as

$$T_c = \frac{a(V_c - b)}{RV_c^2} = \frac{ab}{4Rb^2}; \quad \Rightarrow \quad T_c = \frac{a}{4Rb}. \tag{5.31}$$

Again, substituting the values of V_c and T_c from equations (5.30) and (5.31), respectively, into equations (5.27), we get

$$P = \frac{RT}{(V - b)} e^{-\left(\frac{a}{RTV}\right)}; \quad \Rightarrow \quad P_c = \frac{aR}{4Rb^2} e^{-\left(\frac{4abR}{2abR}\right)}; \quad \Rightarrow \quad P_c = \frac{a}{4b^2 e^2}. \tag{5.32}$$

Therefore, the critical constants of Dieterici's equation of state are

$$P_c = \frac{a}{4b^2 e^2}, \quad V_c = 2b, \quad \text{and} \quad T_c = \frac{a}{4Rb}. \tag{5.33}$$

Thus, in this case also, the values of the critical constants P_c, V_c, and T_c depend on the constants a and b and the universal gas constant R.

Critical constants for Berthelot's equation of state

The Berthelot's equation of state of one mole of a real gas is given by

$$\left(P + \frac{a}{TV^2}\right)(V - b) = RT, \tag{5.34}$$

where a and b are two constants. Following similar arguments as stated in Van der Waals equation, at the critical point, we have

$$\left(\frac{\partial P}{\partial V}\right)_{T=T_c} = 0 = \left(\frac{\partial^2 P}{\partial V^2}\right)_{T=T_c}. \tag{5.35}$$

Berthelot's equation can be expressed in the following form to get pressure p as

$$P = \frac{RT}{V-b} - \frac{a}{TV^2}. \tag{5.36}$$

Now, differentiating equation (5.36) with respect to volume V at the critical temperature $T = T_c$ and equating the result to zero at $V = V_c$, we get

$$\left(\frac{\partial P}{\partial V}\right)_{T=T_c} = -\frac{RT}{(V-b)^2} + \frac{2a}{TV^3} = 0. \tag{5.37}$$

At a critical point, this leads to

$$T_c^2 = \frac{2a(V_c - b)^2}{RV_c^3}. \tag{5.38}$$

Again, differentiating equation (5.37) with respect to volume V at $T = T_c$ and equating the result to zero at $V = V_c$, we get

$$\left(\frac{\partial^2 P}{\partial V^2}\right)_{T=T_c} = \frac{2RT_c}{(V_c - b)^3} - \frac{6a}{TV_c^4} = 0. \tag{5.39}$$

Solving equations (5.38) and (5.39), we get the values of the critical volume V_c and the critical temperature T_c as

$$V_c = 3b; \quad \text{and} \quad T_c = \sqrt{\frac{8a}{27Rb}}. \tag{5.40}$$

Substituting the values of V_c and T_c from equation (5.40) into equation (5.36), we get

$$P = \frac{RT}{V-b} - \frac{a}{TV^2}; \Rightarrow P_c = \frac{R}{(3b-b)}\sqrt{\frac{8a}{27Rb}} - \frac{a}{(3b)^2}\sqrt{\frac{27Rb}{8a}} = \frac{\sqrt{2aR}}{3\sqrt{3}b^{3/2}} - \frac{3\sqrt{3aR}}{6\sqrt{2}b^{3/2}}.$$

Multiplying the numerator and denominator of the second term by $\sqrt{2}$,

$$P_c = \frac{\sqrt{2aR}}{3\sqrt{3}b^{3/2}} - \frac{3\sqrt{6aR}}{12b^{3/2}} = \frac{4\sqrt{2aR} - 3\sqrt{2aR}}{12\sqrt{3}b^{3/2}} = \frac{1}{12b}\sqrt{\frac{2aR}{3b}}; \Rightarrow P_c = \frac{1}{b}\sqrt{\frac{aR}{216b}}.$$

Hence, we get the critical constants of Berthelot's equation as

$$P_c = \frac{1}{b}\sqrt{\frac{aR}{216b}}; \quad V_c = 3b \quad \text{and} \quad T_c = \sqrt{\frac{8a}{27Rb}}. \tag{5.41}$$

Thus, in this case as well, the values of the critical constants P_c, V_c, and T_c depend on the constants a and b and the universal gas constant R.

5.3.7 Van der Waals reduced equation of state

The Van der Waals constants a and b depend on temperature T and the nature of the gas. It is desirable to eliminate them to make the Van der Waals equation applicable to all gases. This is done in the reduced equation of state, the form to which Van der Waals equation of state reduces on expressing the pressure P, volume V, and temperature T as fractions of their corresponding critical values. The pressure, volume, and temperature thus expressed are called **reduced pressure, reduced volume**, and **reduced temperature**, and are symbolized by P_r, V_r, and T_r, respectively.

By definition, the reduced pressure is $P_r = \frac{P}{P_c}$, the reduced temperature is $T_r = \frac{T}{T_c}$, and the reduced volume is $V_r = \frac{V}{V_c}$. Note that P_r, V_r, and T_r, are pure numbers with no units. When an equation of state is expressed in terms of the reduced parameters, that is, in terms of P_r, V_r, and T_r, then it is called a reduced equation of state.[6]

Van der Waals equation of state can be expressed in the form of the reduced equation of state in the following way: Substituting $P = P_r P_c$, $T = T_r T_c$, and $V = V_r V_c$ in Van der Waals equation, we get

$$\left(P_r P_c + \frac{a}{V_r^2 V_c^2}\right)(V_r V_c - b) = RT_r T_c; \quad \Rightarrow \quad \left(P_r \frac{a}{27b^2} + \frac{a}{V_r^2\, 9b^2}\right) = RT_r\, \frac{8a}{27Rb}. \tag{5.42}$$

Substituting the values of P_c, V_c, and T_c from equation (5.26) into equation (5.42), we get

$$\frac{a}{27b_2}\left(P_r + \frac{3}{V_r^2}\right)3b\left(V_r - \frac{1}{3}\right) = T_r\, \frac{8a}{27b}; \quad \Rightarrow \quad \left(P_r + \frac{3}{V_r^2}\right)\left(V_r - \frac{1}{3}\right) = \frac{8}{3}T_r, \tag{5.43}$$

which is the Van der Waals *reduced equation of state*.

---

[6]C. A. Stevenson, Note on Van der Waals Reduced Equation of State, *Journal of Chemical Education* 23, no. 3 (1946): 148.

The following comments can be made from equation (5.43):

1. This equation (5.43) is more versatile than the original Van der Waals equation of state since it contains no characteristics of the parameters (apparently) of an individual substance; hence, it is equally valid for all gases. The isotherms of various substances built in accordance with the above equation will coincide at the same values of the reduced parameters P_r, V_r, and T_r. The states of different substances, which are described by the same reduced parameters, are called the corresponding states. The thermodynamic properties of the substances are the same in their corresponding states. In fact, the existence of such an equation implies that if two reduced variables are the same for a set of systems, then the third reduced variable will also be the same throughout the set. This universal property is called the law of corresponding states.

2. The law of corresponding states is an empirical law that encapsulates the finding that the equations of state for many real gases are remarkably similar when they are expressed in terms of reduced temperatures $(T_r = \frac{T}{T_c})$, pressures, $(P_r = \frac{P}{P_c})$, and volumes $(V_r = \frac{V}{V_c})$, where the subscript c represents the value of the property at the critical point. This law was first described by Johannes Diderik Van der Waals in his thesis in 1873.

3. The quantities a and b that are characteristics of a particular substance are not completely removed, but they are just hidden. It is, however, valid to the extent its parent – the Van der Waals equation – is obeyed.

4. The compressibility factor Z_c can be expressed in terms of the reduced parameters in the following way:

$$Z_c = \frac{PV}{RT} = \frac{P_r P_c \times V_r V_c}{R\, T_r T_c} = \frac{P_r V_r}{T_r} \times \frac{\frac{a}{27b^2} \times 3b}{R \times \frac{8a}{27bR}} = \frac{3}{8}\frac{P_r V_r}{T_r}.$$

5. So long as the equation of state contains only quantities characteristic of the substance, for example, Van der Waals a and b, it is always reducible to yield an equation of universal validity. This type of equation helps to find out the inherent similarity present in a system.

5.3.8 Variation of the compressibility factor Z_c with the reduced pressure P_r

It is observed that almost all the gases, when expressed in terms of the same reduced temperature and reduced pressure, have approximately the same compressibility factor, and all deviate from the ideal gas behavior to about the same degree. The compressibility factor (Z_c), also known as the compression factor or the gas deviation factor, is a correction factor that describes the deviation of a real gas from the ideal gas behavior. It is simply defined as the ratio of the molar volume of a real gas to the molar volume of an ideal gas at the same temperature and pressure. It is a useful thermodynamic property for modifying the ideal gas law to account for the real gas behavior. In general, deviation from the ideal gas behavior becomes more significant when a real gas is closer to a phase change, the temperature is lower and the pressure is larger. The value of the compressibility factor is usually obtained from the equation of state, such as the virial equation that takes compound-specific empirical constants as input. For a gas that is a mixture of two or more pure gases (air or natural gas, for example), the gas composition must be known before compressibility can be calculated.

The compressibility factor Z_c is defined by

$$Z_c = \frac{PV}{N K_B T} = \frac{Pv}{K_B T} = \frac{P_c P_r v_c v_r}{K_B T_c T_r} = \frac{P_c v_c}{K_B T_c} \times \frac{p_r v_r}{T_r} = \frac{3}{8} \times \frac{p_r v_r}{T_r}, \qquad (5.44)$$

where the result $\frac{P_c v_c}{K_B T_c} = \frac{3}{8}$ has been used. The following characteristic features are noticeable regarding the behavior of Z_c:

1. As per definition, the compressibility factor for an ideal gas is $Z_c = 1$. In many real-world applications, value of Z_c is considered as a measure of deviation of the real gas from the ideal gas behavior.

2. Z_c should be equal to $\frac{3}{8}$ at the critical point ($P_r = 1$, $V_r = 1, T_r = 1$) for the Van der Waals systems.

3. Figure 5.7 shows the collapsed data of the compressibility factor Z_c for a number of fluids (mentioned inside the graph) corresponding to the various values of reduced temperature. It is observed from the figure that at a particular value of reduced temperature T_r, say, $\tilde{T} = T_r = 1.20$, Z_c initially decreases with the increase in reduced pressure $\tilde{P} = P_r$, becomes minimum at a particular value of reduced pressure, say, $P_r = P_{rm}$ and increases steadily with further increase in P_r.

4. The value of P_{rm} increases with the increase in reduced temperature T_r.

5. Figure 5.7 shows that Z_c tends to 1 as the reduced temperature T_r is much larger than 1. This is expected from Boyle's law for the ideal gas (in the noninteracting limit).

6. At high pressures, the increase in Z_c could be understood in the following way: The collisions among the molecules increase at high pressures. This allows repulsive forces between molecules to have a noticeable effect, making the molar volume of the real gas (V_m) greater than the molar volume of the corresponding ideal gas ($(V_m)_{ideal} = \frac{RT}{P}$) that causes Z_c to exceed one. When pressures are lower, the molecules are free to move. In this case, attractive forces dominate, making $Z_c < 1$. The closer the gas is to its critical point or its boiling point, the more Z_c deviates from the ideal case.

5.3.9   The critical coefficient

The ratio $\frac{RT_c}{P_cV_c}$ for a gas is called the **critical coefficient** or sometimes the Kamerling Onnes ratio, symbolized by η_c. Substituting the values of the critical constants P_c, V_c, and T_c in terms of a and b for a Van der Waals gas, we get the critical coefficient η_c as

$$\eta_c = \frac{RT_c}{P_cV_c} = R\frac{8a}{27Rb}\frac{27b^2}{a}\frac{1}{3b}$$

$$= \frac{8}{3} = 2.67. \qquad (5.45)$$

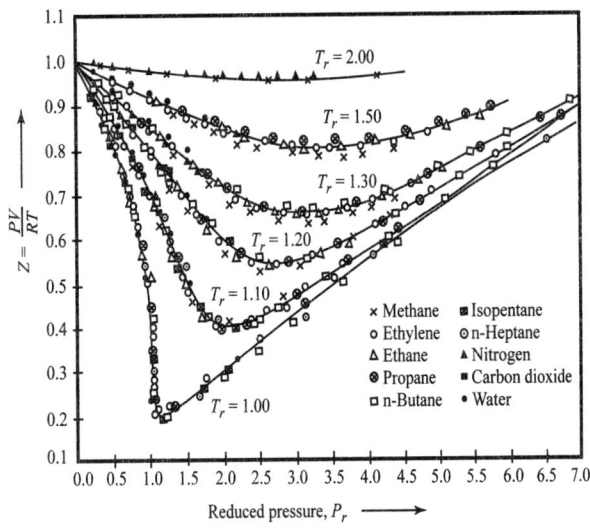

Figure 5.7 Compressibility factor $Z_c = Z$ as a function of reduced pressure $\tilde{P} = P_r$ for various values of reduced temperature $\tilde{T} = T_r$. The fact that the data for a wide variety of fluids fall on identical curves supports the law of corresponding states (H. E. Stanley, *Introduction to Phase Transition and Critical Phenomena* [Oxford University Press], p. 73).

It thus appears that η_c is a pure number, independent of the characteristics of the constants a and b of the gas. Experimentally, however, the values of η_c for real gases are found to vary from 3.27 to 4.99. This fact implies that the Van der Waals equation is only approximately true. For an ideal gas, $\eta_c = \frac{RT_c}{P_cV_c} = 1$, since $PV = RT$. The value of η_c is a measure of the amount of deviation of a real gas from the behavior of an ideal one.

These two dimensionless parameters Z_c and η_c have important physical significance. If the theoretical values of these two parameters, derived from an equation of state, are closer to the experimental values, then the corresponding equation of state is a better one to explain the behavior of the real gas. If we measure the value of Z_c or η_c for two real gases and compare them with the values obtained from the Van der Waals equation of state, the closer the value of Z_c to 0.375 or the closer the value of η_c to 2.66, the more accurately the gas obeys the Van der Waals equation of state.

5.3.10   The law of corresponding states

Equation (5.43) leads to a very useful generalization known as the *law of corresponding states*. It may be stated as follows: *If for two substances, any two of the reduced parameters P_r, V_r, and T_r be the same, then the* *third reduced parameter must be identical for them.* These two substances are then said to be in corresponding states. Thus, if the temperature and pressure of two substances bear the same ratio to the critical temperature and critical pressure, respectively, their volumes will also bear an identical ratio with the respective critical volumes, that is, by suitably adjusting (by expansion or contraction) the scales for pressure P and volume V, the isotherms of all substances can be made to coincide.

According to Van der Waals, the theorem of corresponding states (or principle/law of corresponding states) indicates that all fluids, when compared at the same reduced temperature and reduced pressure, have approximately the same compressibility factor Z_c, and all deviate from ideal gas behavior to about the same degree. The law of the corresponding state is true only approximately and not rigorously. According to this law, values of $\frac{RT_c}{P_c V_c}$ for various substances should remain constant. But, experimental results on these substances show that it varies from 3.2 to 4.9.

5.3.11 Values of the constants a and b in terms of the critical constants P_c, V_c, and T_c

The critical constants of a Van der Waals gas are given by

$$P_c = \frac{a}{27b^2}, \quad T_c = \frac{8a}{27Rb}, \quad \text{and} \quad V_c = 3b.$$

Using these expressions, we get

$$\frac{P_c}{T_c} = \frac{a}{27b^2} \times \frac{27bR}{8a} = \frac{R}{8b}.$$

This provides the value of b to be

$$b = \frac{RT_c}{8P_c}. \tag{5.46}$$

By squaring T_c and dividing by P_c, we get

$$\frac{T_c^2}{P_c} = \frac{64a^2}{27 \times 27b^2 R^2} \times \frac{27b^2}{a} = \frac{64a}{27R^2}.$$

This provides the value of a to be

$$a = \frac{27}{64} \frac{R^2 T_c^{\,2}}{P_c}. \tag{5.47}$$

Equations (5.46) and (5.47) can be utilized to determine the values of the Van der Waals constants a and b. Since the measurements of T_c and P_c are comparatively easier than estimating V_c, both a and b have been expressed in equations (5.46) and (5.47) in terms of T_c and P_c only.

5.3.12 Boyle temperature T_B for a Van der Waals gas

For one mole of a real gas, the Van der Waals equation of state is given by

$$\left(P + \frac{a}{V^2}\right)(V - b) = RT; \quad \Rightarrow \quad P = \frac{RT}{V - b} - \frac{a}{V^2}; \quad \Rightarrow \quad PV = \frac{RTV}{V - b} - \frac{a}{V}. \tag{5.48}$$

At the Boyle temperature T_B, we have $PV = RT_B$. Putting this value of PV in equation (5.48), we get

$$RT_B = \frac{RT_B V}{V - b} - \frac{a}{V}; \quad \Rightarrow \quad RT_B \left(\frac{V}{V - b} - 1\right) = \frac{a}{V},$$

$$\Rightarrow \quad RT_B \frac{b}{V - b} = \frac{a}{V}; \quad \Rightarrow \quad T_B = \frac{a}{V} \cdot \frac{V - b}{Rb} \approx \frac{a}{Rb}. \tag{5.49}$$

Equation (5.49) gives the expression for the Boyle temperature T_B of a Van der Waals gas.

Boyle temperature T_B for a Van der Waals gas from virial coefficient

Comparing with virial coefficient, the Boyle temperature T_B of a Van der Waals gas can be obtained in the following way: For one mole of a real gas, the Van der Waals equation of state can be expressed as

$$PV = RT \left(1 - \frac{b}{V}\right)^{-1} - \frac{a}{V}. \tag{5.50}$$

Using the binomial theorem, we get

$$\left(1 - \tfrac{b}{V}\right)^{-1} = 1 + \tfrac{b}{V} + \tfrac{b^2}{V^2} + \cdots.$$

Using this result in equation (5.50), we get

$$PV = RT + \frac{RTb - a}{V} + \frac{RTb^2}{V^2} + \cdots. \tag{5.51}$$

Comparing this equation (5.51) with equation (5.15), we get

$$B = RTb - a, \quad \text{and} \quad C = RTb^2,$$

where B and C are virial coefficients. At the Boyle temperature, the second virial coefficient B is zero. Hence, we have

$$B = RT_B\, b - a = 0, \quad \Rightarrow \quad T_B = \frac{a}{R\,b}.$$

Thus, we get the same expression for Boyle temperature T_B as given by equation (5.49).

Boyle temperature T_B for a Van der Waals gas from differentiation method

For one mole of a real gas, the Van der Waals equation of state is given by

$$\left(P + \frac{a}{V^2}\right)(V - b) = RT.$$

From this equation, we have

$$P = \frac{RT}{V - b} - \frac{a}{V^2}; \quad \Rightarrow \quad PV = \frac{RTV}{V - b} - \frac{a}{V}.$$

The definition of Boyle temperature indicates that the slope of the PV versus P curve is zero at the Boyle temperature T_B. Hence, we get

$$\left[\frac{\partial(PV)}{\partial P}\right]_{T=T_B} = 0; \quad \Rightarrow \quad \left[\frac{\partial}{\partial P}\left(\frac{RTV}{V - b} - \frac{a}{V}\right)\right]_{T=T_B} = 0.$$

Carrying out the derivative and putting $T = T_B$, we get

$$\left[\frac{RT_B}{V - b} - \frac{RT_B V}{(V - b)^2} + \frac{a}{V^2}\right]\left(\frac{\partial V}{\partial P}\right)_{T=T_B} = 0;$$

$$\Rightarrow \left[\frac{-RT_B b}{(V - b)^2} + \frac{a}{V^2}\right]\left(\frac{\partial V}{\partial P}\right)_{T=T_B} = 0.$$

As $\left(\frac{\partial V}{\partial P}\right)_{T=T_B} \neq 0$ at Boyle temperature T_B, we must have

$$\left[\frac{-RT_B b}{(V - b)^2} + \frac{a}{V^2}\right] = 0; \quad \Rightarrow \quad T_B = \frac{a}{Rb}.$$

It has been assumed that $(V - b) \approx V$. This is the expression for Boyle temperature T_B of a real gas obeying Van der Waals equation of state.

5.3.13   Relation between the Boyle temperature T_B and the critical temperature T_c

Again, the critical temperature T_c is given by $T_c = \frac{8a}{27Rb}$. Therefore,

$$T_B = \frac{a}{Rb} = \frac{a}{Rb}\frac{8}{27}\frac{27}{8} = T_c \times \frac{27}{8} = 3.375 T_c. \tag{5.52}$$

Equation (5.52) gives the relationship between the Boyle temperature T_B and the critical temperature T_c of a Van der Waals gas.

5.3.14   Important notes on the Boyle temperature T_B

The following points are to be noted about the Boyle temperature T_B:

1. At the Boyle temperature T_B, the second virial coefficient $B = b - \frac{a}{RT_B}$ becomes zero.

2. The Boyle temperature T_B and the inversion temperature T_i are related by $T_i = 2T_B$.

3. The Boyle temperature T_B and the critical temperature T_c are related by $T_B = 3.375\ T_c$.

4. For an ideal gas, there is no concept of the Boyle temperature T_B.

5.4   Methods of finding the values of the constants a and b

There are a large number of experimental methods for finding out the values of the Van der Waals constants a and b. Some of these methods are:

1. Isothermal method,

2. The constant volume method,

3. From critical data, and

4. From Joule–Thomson effect.

We shall now describe these methods one after another.

5.4.1   The isothermal method

From the Van der Waals equation for one mole, we have

$$P = \frac{RT}{V-b} - \frac{a}{V^2}. \tag{5.53}$$

Therefore,

$$\left(\frac{\delta P}{\delta V}\right)_T = -\frac{RT}{(V-b)^2} + \frac{2a}{V^3}. \tag{5.54}$$

Solving the simultaneous equations (5.53) and (5.54), the constants a and b could be evaluated if $\left(\frac{\delta P}{\delta V}\right)_T$ is known. It can, however, be found from an isotherm at a known temperature.

The constant volume method

This method is very simple as well as more accurate. From the Van der Waals equation for one mole of a real gas, we have

$$P = \frac{RT}{V-b} - \frac{a}{V^2}, \quad \Rightarrow \quad \left(\frac{\delta P}{\delta T}\right)_V = \frac{R}{V-b}. \tag{5.55}$$

Using this, the expression for pressure P becomes

$$P = T\left(\frac{\delta P}{\delta T}\right)_V - \frac{a}{V^2}. \tag{5.56}$$

Solving this equation (5.56) for a, we get

$$\Rightarrow \quad a = V^2\left\{T\left(\frac{\delta P}{\delta T}\right)_V - p\right\}. \tag{5.57}$$

Also, by solving equation (5.55) for b, we get

$$b = V - \frac{R}{\left(\frac{\delta P}{\delta T}\right)_V}. \tag{5.58}$$

By observing the rate of increase of pressure with temperature in a constant volume gas thermometer, $\left(\frac{\delta P}{\delta T}\right)_V$ can be measured, and hence, values of a and b can be evaluated using equations (5.57) and (5.58), respectively. It should be pointed out that the values of a and b are observed to vary with temperature T.

Values of the constants a and b from the critical data

For a gas obeying Van der Waals equation, we have the following relations among the critical constants:

$$V_c = 3b; \qquad P_c = \frac{a}{27b^2}; \qquad T_c = \frac{8a}{27Rb}.$$

From these relations, expressions for a and b come out to be

$$a = \frac{27R^2}{64} \times \frac{T_c^2}{P_c} \quad \text{and} \quad b = \frac{RT_c}{8P_c}.$$

Thus, if P_c and T_c are experimentally determined, values of a and b can be calculated in this method. This method, however, is not very reliable as the Van der Waals equation does not properly hold at the critical region where the density is high.

Values of the constants a and b from Joule–Thomson effect

From Joule–Thomson effect, the expression for the temperature of inversion T_i is given by

$$T_i = \frac{2a}{Rb}.$$

If the constant "a" is measured by some other method and the inversion temperature T_i is determined from Joule–Thomson effect, the constant b can be calculated from the above expression.

5.5 Derivation of Van der Waals equation of state from various approaches (optional)

Van der Waals equation of state for a real gas can be derived from various approaches in physics. Clausius deduced the Van der Waals equation rigorously by applying his famous *Virial theorem*. In this approach, the gas particles are assumed to obey the equations of classical dynamics. Further, using the concept of the partition function of a system under the framework of canonical ensemble in statistical mechanics, Van der Waals equation can also be derived. These two approaches are discussed in brief below.

5.5.1 Virial theorem

The virial theorem is an important theorem for a system of moving particles both in classical and quantum physics. This theorem states that the total kinetic energy T_K of the system and the potential energy V are related by the expression

$$2T_K + V = 0.$$

This theorem can be applied to central force problems, gravitationally bound galaxies, and clusters of galaxies. The virial theorem can be used to get the mass of an object.

This theorem can be mathematically derived in the following way: Consider a gaseous ensemble with the particles defined by the position vectors $\tilde{r}_1, \tilde{r}_2, ..., \tilde{r}_i$, etc. The equation of motion for the ith particle can be written as

$$m_i \frac{d^2 \mathbf{r_i}}{dt^2} = \mathbf{F_i} \quad \Rightarrow \quad \mathbf{F_i} = \frac{d}{dt}(m_i v_i), \tag{5.59}$$

where m_i is the mass of the ith particle, and $\mathbf{F_i}$ is the force acting upon it.

Let a function ϕ be given by: $\quad \phi = \sum \mathbf{F_i}.\mathbf{r_i}$, where the summation extends over all the particles of the ensemble.

$$\frac{d\phi}{dt} = \sum \frac{d\mathbf{F_i}}{dt}.\mathbf{r_i} + \sum \frac{d\mathbf{r_i}}{dt}.\mathbf{F_i} = \sum \mathbf{F_i}.\mathbf{r_i} + \sum m_i \mathbf{v_i}.\mathbf{v_i},$$
$$= \sum \mathbf{F_i}.\mathbf{r_i} + \sum m_i \mathbf{v_i^2} = \sum \mathbf{F_i}.\mathbf{r_i} + 2T_K. \tag{5.60}$$

Here, equation (5.59) has been used. Using the value of the total kinetic energy as $T_K = \frac{1}{2} \sum m_i \mathbf{v_i^2}$, and taking average value of equation (5.60) over a finite interval of time τ, we get

$$\frac{1}{\tau} \int_0^\tau \frac{d\phi}{dt} dt = \frac{1}{\tau} \int_0^\tau \sum \mathbf{F_i}.\mathbf{r_i} dt + \frac{1}{\tau} \int_0^\tau 2T_K dt,$$

$$\Rightarrow \quad \frac{1}{\tau}[\phi(\tau) - \phi(0)] = \sum \langle \mathbf{F_i}.\mathbf{r_i} \rangle + 2\langle T_K \rangle. \tag{5.61}$$

The time average is taken from 0 to τ. In periodic motions, all the coordinates are repeated after a regular time interval equal to the time period of the motion. If τ is the time period, we will have $\phi(\tau) = \phi(0)$. Using this condition in equation (5.61), we get

$$-\frac{1}{2} \sum \langle \mathbf{F_i}.\mathbf{r_i} \rangle = \langle T_K \rangle. \tag{5.62}$$

In nonperiodic motions, there are no repetitions of the coordinates. But, since a particle fluctuates regularly with time, being as often positive as negative, the left-hand side of equation (5.61) can be made effectively zero, by making τ large. So, equation (5.61) would then reduce to

$$-\frac{1}{2} \sum \langle \mathbf{F_i}.\mathbf{r_i} \rangle = \langle T_K \rangle, \tag{5.63}$$

which is the same as equation (5.62).

The expression $\frac{1}{2}\sum\langle \mathbf{F_i}.\mathbf{r_i}\rangle$ was termed by *Clausius* **the virial of the system.** The above equation then states that the average kinetic energy of translation of the particles of a gas is equal to its virial in the steady state. This is the celebrated **virial theorem of Clausius.**

The force $\mathbf{F_i}$ in equation (5.63) includes all the forces acting on the particles, and their contributions to the virial are to be evaluated. These forces may be of three different types:

1. The forces of impact on the molecules due to collision with the walls of the container,

2. The forces due to mutual collisions between the molecules, and

3. The forces of attraction (or repulsion) between the molecules of the gaseous system separated by some distance.

The forces of type (2) do not contribute anything to the virial, since the action and the reaction nullify each other. Again, for an ideal gas, the forces of type (3) do not exist. For a real gas, however, forces of both type (1) and type (3) exist and contribute to the virial.

Ideal gas equation from virial theorem

In the case of an ideal gas, the contribution of the forces of type (1) is only relevant to the virial. We consider the gas to be enclosed in a vessel of volume V and let P be the pressure exerted by the gas molecules due to the impact of the molecules on the surface of the walls. The pressure P is the force acting normally on the unit area of the surface. Hence, the force acting on the molecules is given by

$$\mathbf{F} = -\hat{n}\,Pds; \quad \Rightarrow \quad \mathbf{F}.\mathbf{r} = -\hat{n}\,Pds.\mathbf{r} = -P\,d\mathbf{s}.\mathbf{r},$$

where $\hat{n}$ is the unit vector drawn outward, $\mathbf{ds}$ is the vector surface area, and the force $\mathbf{F}$ acts inward. From equation (5.63), we have

$$\langle T_K\rangle = -\frac{1}{2}\sum\langle \mathbf{F_i}.\mathbf{r_i}\rangle = \frac{1}{2}P\int_S \mathbf{r_i}.\mathbf{ds_i} = \frac{1}{2}P\int_V \nabla.\mathbf{r}dV, \quad \text{[using divergence theorem]}$$

$$= \frac{1}{2}P\int_V \left(\mathbf{i}\frac{\delta}{\delta x} + \mathbf{j}\frac{\delta}{\delta y} + \mathbf{k}\frac{\delta}{\delta z}\right).(\mathbf{i}x + \mathbf{j}y + \mathbf{k}z)dV = \frac{1}{2}P\int_V 3dV = \frac{3}{2}PV.$$

Hence, for an ideal gas, we have from *virial theorem of Clausius*

$$\frac{3}{2}PV = N \times \frac{1}{2}m\langle v^2\rangle = \frac{1}{2}mN\langle v^2\rangle = \frac{1}{2}M_0 \times \frac{3RT}{M_0},$$

where $N(= N_A)$ is the number of molecules in one mole of the ideal gas of volume V. This expression leads to

$$PV = RT,$$

which is the equation of state for one mole of an ideal gas. This shows that the virial theorem can be successfully used to get the equation of state for an ideal gas.

Van der Waals equation from virial theorem

For a Van der Waals gas, both type (1) force, that is, impact on the particles due to collision with the walls of the container, and type (3) force, that is, the forces of attraction (or repulsion) between the particles separated by some distance contribute to the virial.

The contribution of the type (1) forces to the virial can be obtained from the similar approach considered for the ideal gas. Hence, we have

$$-\frac{1}{2}\sum\langle \mathbf{F_i}.\mathbf{r_i}\rangle = \frac{3}{2}\,PV. \tag{5.64}$$

The contribution of the type of (3) force to the virial can be obtained in the following way: The nature of the type (3) force is not definitely known. However, it is radial in nature, and depends upon the intermolecular distance. Let this dependency is expressed by some function $\phi(r)$, where r is the separation between two molecules. We consider two such molecules, i and j, and their contribution to the virial can be written as

$$-\frac{1}{2}\mathbf{F_{ij}}.\mathbf{r_i} = \frac{1}{2}\mathbf{F_{ji}}.\mathbf{r_j},$$

where $\mathbf{F_{ij}}$ is the force on the jth molecule due to ith one, $\mathbf{F_{ji}}$ is the force on jth molecule due jth one, $\mathbf{r_i}$ and $\mathbf{r_j}$, are the position vectors of the respective molecules with reference to origin O as shown in Figure 5.8.

But we have the relation

$$\mathbf{F_{ji}} = -\mathbf{F_{ij}}$$

$$\Rightarrow \quad -\frac{1}{2}\sum\langle \mathbf{F_i}.\mathbf{r_i}\rangle$$

$$= -\frac{1}{2}[\mathbf{F_{ij}}.(\mathbf{r_i} - \mathbf{r_j})] = -\frac{1}{2}\mathbf{F_{ij}}.\mathbf{r_{ij}}$$

$$= -\frac{1}{2}F_{ij}\,r_{ij}. \quad [\mathbf{F_{ij}} \text{ is along } \mathbf{r_{ij}}].$$

When all the pairs of molecules are taken into account, the contribution to the virial is equal to

$$-\frac{1}{2}\sum F_{ij}\,r_{ij} = -\frac{1}{2}r\phi(r). \tag{5.65}$$

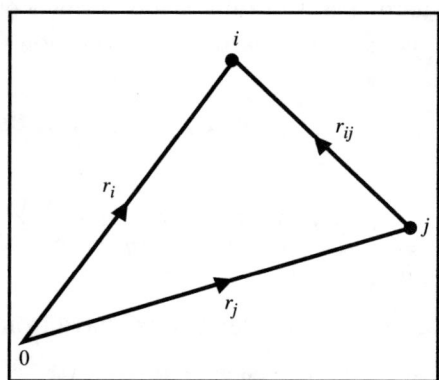

Figure 5.8 Position vectors of two molecules.

Here, the summation extends over all the pairs of the gas molecules.

Applying equations (5.64) and (5.65) to the virial theorem, we get

$$\frac{3}{2}PV - \frac{1}{2}\sum r\phi(r) = \frac{1}{2}mN\langle v^2 \rangle. \tag{5.66}$$

Using $\langle v^2 \rangle = \frac{3RT}{M_0}$ and rearranging equation (5.66), we get

$$\boxed{PV = RT + \frac{1}{3}\sum r\phi(r)}. \tag{5.67}$$

Equation (5.67) is the **general form of the equation of state** of a fluid as obtained from the virial theorem. This equation holds good in various systems.

Th factor $\frac{1}{3}\sum r\phi(r)$ in equation (5.67) can be evaluated in the following way. It should be pointed out that in order to get an exact equation of state, the factor $\frac{1}{3}\sum r\phi(r)$ must be evaluated correctly. Unfortunately, this is not possible as $\phi(r)$ is not precisely known. However, an approximate value of $\frac{1}{3}\sum r\phi(r)$ can be determined.

According to the Maxwell–Boltzmann distribution law, the number of molecules $n(v)dv$ with speeds between v and $v + dv$ per unit volume is given by

$$n(v)dv = 4\pi n \left(\frac{m}{2\pi K_B T} \right)^{\frac{3}{2}} e^{-\frac{\left(\frac{1}{2}mv^2 + \phi_P\right)}{K_B T}} v^2 dv,$$

where ϕ_P is the potential energy, and n is the number of molecules per unit volume in regions of potential energy zero. Integrating the above expression for all values of v from 0 to ∞, we get

$$n(r) = ne^{-\frac{\phi_P(r)}{K_B T}},$$

where $n(r)$ is the molecular density in the region of potential energy $\phi_P(r)$.

If N be the number of molecules in 1 mole of the real gas enclosed in volume V, the possible number of pairs of molecules will be given by

$$\frac{N(N-1)}{2} \simeq \frac{N^2}{2} \qquad (as \; N \text{ is very large}).$$

Now we address the question: Out of these pairs, how many of them are separated by a distance r and $r + dr$? Let us initially consider this for two particular molecules. The probability that the pair would lie between r and $r + dr$ is given by

$$P_r = \frac{4\pi r^2 dr}{V}.$$

We know that the molecular distribution will be changed due to the presence of inter-molecular forces. Taking this into consideration, the probability that the pair would lie at the specified distance will be given by

$$P_r e^{-\frac{\phi_p(r)}{K_B T}} = \frac{4\pi r^2 dr}{V} e^{-\frac{\phi_p(r)}{K_B T}}.$$

Then, the number of possible pairs with the above configuration will be given by

$$\frac{N^2}{2} \frac{4\pi r^2 dr}{V} e^{-\frac{\phi_p(r)}{KT}} = \frac{2\pi N^2 r^2}{V} e^{-\frac{\phi_p(r)}{KT}} dr; \quad \Rightarrow \quad \sum r\phi(r) = \frac{2\pi N^2}{V} \int r^3 e^{-\frac{\phi_p(r)}{K_B T}} \phi(r) dr.$$

The intermolecular force $\phi(r)$ is a short-range force; it is repulsive between $\sigma - \delta$ and σ, where σ is molecular diameter, and δ is a very small quantity. Between σ to ∞, however, $\phi(r)$ is attractive, falling off rapidly with distance. Denoting $\phi(r)$ by $f(r)$ when it is repulsive and by $F(r)$ when it is attractive, we have

$$\sum r\phi(r) = \frac{2\pi N^2}{V} \left[\int_{\sigma-\delta}^{\sigma} r^3 f(r) e^{-\frac{\phi_p(r)}{K_B T}} dr + \int_{\sigma}^{\infty} r^3 F(r) e^{-\frac{\phi_p(r)}{K_B T}} dr \right] = \frac{2\pi N^2}{V} [I_1 + I_2].$$

Now the integrand $I_1 = \int_{\sigma-\delta}^{\sigma} r^3 f(r) e^{-\frac{\phi_p(r)}{K_B T}} dr \simeq \sigma^3 \int_{\sigma-\delta}^{\sigma} f(r) e^{-\frac{\phi_p(r)}{K_B T}} dr.$

Since $\sigma - \delta \approx \sigma$, and r does not occur in the functional term, it may be replaced without much error by σ.

But $f(r) = -\frac{\delta\phi_p(r)}{\delta r}$ and when $r = \sigma - \delta$, $\phi_p(r) = \infty$ and when $r = \sigma, \phi_p(r) = 0$. So,

$$I_1 = -\sigma^3 \int_0^{\infty} e^{-\frac{\phi_p(r)}{K_B T}} d\phi_p(r) = K_B T \sigma^3 \left[e^{-\frac{\phi_p(r)}{K_B T}} \right]_{\infty}^{0} = K_B T \sigma^3,$$

$$I_2 = \int_{\sigma}^{\infty} r^3 F(r) e^{-\frac{\phi_p(r)}{K_B T}} dr = \int_{\sigma}^{\infty} r^3 F(r) e^{\left(-\frac{1}{K_B T} \int_r^{\infty} F(r) dr \right)} dr = \psi(T)$$

$$[\text{say } \phi_p(r) = \int_r^{\infty} F(r) dr]$$

$\psi(T)$ is some function of temperature, as is evident from the exponential term.

$$\sum r\phi(r) = \frac{2\pi N^2}{V} \left[K_B T \sigma^3 + \psi(T) \right]. \tag{5.68}$$

From equations (5.67) and (5.68), we get

$$\frac{3}{2}PV = \frac{1}{2}mN\langle v^2\rangle + \frac{\pi N^2}{V}K_BT\sigma^3 + \frac{\pi N^2}{V}\psi(T); \quad \Rightarrow$$

$$PV = \frac{1}{3}mN\langle v^2\rangle + \frac{2\pi N^2}{3V}K_BT\sigma^3 + \frac{2\pi N^2}{3V}\psi(T);$$

$$\Rightarrow \quad P = \frac{RT}{V} + \frac{bRT}{V^2} + \frac{2\pi N^2}{3V^2}\psi(T) = \frac{RT}{V}\left(1 + \frac{b}{V}\right) - \frac{a}{V^2}, \text{ where } b = \frac{2\pi\sigma^3 N}{3} \text{ and}$$

$$a = -\frac{2\pi N^2 \psi(T)}{3}.$$

This leads to

$$\left(P + \frac{a}{V^2}\right)\left(1 + \frac{b}{V}\right)^{-1}V = RT.$$

$$\Rightarrow \quad \boxed{\left(p + \frac{a}{V^2}\right)(V - b) = RT} \quad \text{as} \quad \frac{b^2}{V^2} << 1. \tag{5.69}$$

This equation (5.69) is the well-known **Van der Waals equation of state**. The Van der Waals constants a and b are given by

$$a = -\frac{2\pi N^2 \psi(T)}{3} \text{ and } b = \frac{2\pi\sigma^3 N}{3}.$$

Here, a is a positive quantity since $\phi_p(r)$ is negative. It should be mentioned that the Van der Waals equation is theoretically correct up to terms in $\frac{1}{V}$ and no further.

5.5.2 Van der Waals equation from the concept of partition function

In order to derive the Van der Waals equation of state, we consider first the partition function of an ideal gas and calculate Helmholtz free energy, and finally a term is added to the free energy due to Van der Waals interaction among the molecules of a real gas.

We consider an ideal gas in the canonical ensemble for which the Hamiltonian function is given by

$$H = \sum_{i=1}^{N} \frac{\vec{p}_i^{\,2}}{2m}. \tag{5.70}$$

As there is no potential energy, Hamiltonian H contains just the kinetic energy. The corresponding partition function $Z_N(V, T)$ is then given by

$$Z_N(T, V) = \int \frac{d^{3N}q\, d^{3N}p}{h^{3N}N!} e^{-\left(\beta\sum_{i=1}^{N}\frac{\vec{p}_i^{\,2}}{2m}\right)}. \tag{5.71}$$

Factorization: The integral leading to Z_N factorizes (equation 5.71) to

$$Z_N(T,V) = \frac{V^N}{N!} \left(\int_{-\infty}^{+\infty} \frac{dp}{h} e^{-\beta \frac{p^2}{2m}} \right)^{3N} = \frac{V^N}{N!} \left(\int_{-\infty}^{+\infty} \frac{\sqrt{2K_BTm}}{h} e^{-x^2} dx \right)^{3N}, \quad (5.72)$$

where we have used the variable substitution

$$x^2 = \frac{p^2}{2K_BTm}, \quad dx = \frac{dp}{\sqrt{2K_BTm}}, \quad \int_{-\infty}^{+\infty} e^{-x^2} dx = \sqrt{\pi}. \quad (5.73)$$

Evaluating the integral in equation (5.72) explicitly with the help of equation (5.73), we get

$$Z_N(T,V) = \frac{V^N}{N!} \left(\frac{2\pi m K_BT}{h^2} \right)^{\frac{3N}{2}} = \frac{Z_1^N}{N!}, \quad \text{where} \quad Z_1 = \left(\frac{2\pi m K_BT}{h^2} \right)^{\frac{3}{2}} V. \quad (5.74)$$

So, the free energy F is calculated as

$$F = -K_BT \ln\left(\frac{Z_1^N}{N!} \right) = -K_BT(N \ln Z_1 - \ln N!)$$
$$= -K_BT(N \ln Z_1 - N \ln N + N),$$
$$\Rightarrow F = -K_BTN \left[\ln\left(\frac{Z_1}{N} \right) + 1 \right] = -K_BTN \left[\ln\left(\frac{n_Q}{n} \right) + 1 \right], \quad (5.75)$$

where $n = \frac{N}{V}$ and n_Q is the quantum concentration;

$$n_Q = \left(\frac{2\pi m K_BT}{h^2} \right)^{\frac{3}{2}} = \frac{(2\pi m K_BT)^{3/2}}{h^3},$$

where m is the mass of a molecule. For the real gas, the factor $\frac{1}{n}$ can be written as

$$\frac{1}{n} = \frac{V - Nb}{N} = v - b,$$

where $v = \frac{V}{N}$ is the specific volume. Further, the energy per molecule due to the Van der Waals interaction is given by

$$= \left(-\frac{aN^2}{V^2} \times V \times \frac{1}{N} \right) = -\frac{a}{v}.$$

Considering these two factors and equation (5.75), the free energy per molecule can be written as

$$f = -K_BT\{\ln\left[n_Q(v-b)\right] + 1\} - \frac{a}{v} = -K_BT \ln(v-b) - \frac{a}{v} - K_BT[\ln(n_Q) + 1].$$

$$f = -K_BT \ln(v-b) - \frac{a}{v} + \Phi(T), \qquad (5.76)$$

where

$$\Phi(T) = -K_BT[\ln(n_Q) + 1] = -K_BT\left\{\ln\left[\frac{(2\pi mK_BT)^{3/2}}{h^3}\right] + 1\right\}$$

$$= -K_BT\left\{\ln T^{3/2} + \ln\frac{(2\pi mK_B)^{3/2}}{h^3} + 1\right\} = -K_BT\left(\ln T^{3/2} + \ln\alpha + 1\right)$$

$$= -K_BT\left(\frac{3}{2}\ln T + \ln\alpha + 1\right), \qquad (5.77)$$

with the constant α

$$\alpha = \frac{(2\pi mK_B)^{3/2}}{h^3}.$$

Using equation (5.77) into equation (5.76), the Helmholtz free energy f per molecule is obtained as

$$f = -\frac{a}{v} - K_BT\left[\ln(v-b) + \frac{3}{2}\ln T + \ln\alpha + 1\right].$$

The pressure P is obtained as

$$P = -\left(\frac{\partial F}{\partial V}\right)_{T,N} = -\left(\frac{\partial f}{\partial v}\right)_{T,N} = \frac{K_BT}{v-b} - \frac{a}{v^2}$$

or more simply we can write

$$\left(P + \frac{a}{v^2}\right)(v-b) = K_BT, \qquad (5.78)$$

which is the well-known Van der Waals equation.

Since $v = \frac{V}{N}$, the above Van der Waals equation can be rewritten as

$$\left(P + \frac{N^2a}{V^2}\right)(V - bN) = NK_BT. \qquad (5.79)$$

5.5.3   Important features of the Van der Waals equation of state

1. The most interesting feature of the Van der Waals equation is not its ability to fit small deviations from the ideal gas law, but the fact that it gives a remarkably good qualitative description of the behavior of the system near the gas–liquid critical point. In 1822, Cagniard de la Tour and others showed that for every substance, there is a particular pressure, volume, and temperature, called the critical point, at which the distinction between liquid and gas vanishes. (For water, $P = 218$ atm; $V = 3.2$ cc/g; $T = 374°$C.) It was initially assumed that the liquid simply changes to a gas above the critical point, but in 1863, the Irish physical chemist Thomas Andrews demonstrated that the supercritical substance can be changed continuously into either gas or liquid by appropriate variations of temperature and pressure.

2. As the first successful explanation of phase transitions, Van der Waals' theory demonstrated the fertility of the atomistic approach and stimulated much research on liquid–gas critical phenomena. It was a major breakthrough to show that the same model could be used to explain two different states of matter, for some scientists had previously attributed different properties to "gas molecules" and "liquid molecules". Even though the Van der Waals theory was eventually replaced by more sophisticated theories of the critical point in the 20th century, it played an important role by demonstrating that qualitative changes on the macroscopic level, such as changes from the liquid to the gaseous state, might be explained by quantitative changes on the microscopic level.

5.5.4   Limitations of the Van der Waals equation of state

1. Theoretical isotherms obtained from Van der Waals equation of state do not match exactly with the isotherms obtained in Andrews' experiment. Below the critical temperature, the experimental isotherms due to Andrews do not show any maxima or minima, but such maxima or minima, are exhibited by Van der Waals theoretical curves. No explanations are provided from this discrepancy.

2. Van der Waals constants a and b are not really constants. They are found experimentally to vary with different substances as well as with temperature for a particular substance. This variation may be as high as 30%.

3. In Van der Waals theoretical calculation, b is found to be $\frac{V_c}{3}$. But, the actual value of it is very close to $\frac{V_c}{2}$. Further, the calculation shows that the constant b is four times the volume occupied by the molecules of the gas. In fact, it is $4\sqrt{2}$ rather than 4.

4. From Van der Waals equation of state, the critical coefficient $\eta_c = \frac{RT_c}{P_c V_c}$ is 2.67, constant for all gases. But experimentally, it is found that this value ranges from 3.27 for helium to 4.99 for acetic acid.

5. From Van der Waals equation of state, Boyle temperature T_B is found to be $T_B = \frac{27}{8} T_c = 3.375 T_c$. But, it is also found to vary with different gases. For example, it is $3.65 T_c$ for helium and less than $3.3 T_c$ for other gases.

6. Though it is claimed that the reduced equation of state is independent of a, b, and R and is universally applicable to all gases, it is not experimentally observed in real data for the gases. This defect, however, is in-built in the parent equation.

5.6 Solved numerical problems

1. Consider one mole of a real gas enclosed in a vessel of volume V and obeying Van der Waals equation of state. The gas is subject to a pressure of P_1 atmosphere at a temperature T_1 and P_2 atmosphere at a temperature T_2. Find out the expressions for Van der Waals parameters a and b for this gas.

   Answer: The number of mole, in this case, is $n = 1$. Writing the Van der Waals equation for these two conditions, we get, respectively

   $$\left(P_1 + \frac{a}{V^2} \right)(V - b) = RT_1 \quad \text{and} \quad \left(P_2 + \frac{a}{V^2} \right)(V - b) = RT_2.$$

   Solving these two expressions for a and b, we get

   $$a = \frac{V^2(T_1 P_2 - T_2 P_1)}{T_2 - T_1} \quad \text{and} \quad b = V - \frac{R(T_2 - T_1)}{P_2 - P_1}.$$

   These are the respective expressions for the Van der Waals parameters a and b.

2. Calculate the pressure at which carbon dioxide (CO_2) of molar mass M possesses density ρ at a temperature T. Assume that CO_2 obeys the Van der Waals equation of state.

   Answer: Let M be the molar mass of CO_2 and V be its volume. Then, its density ρ is given by $\rho = \frac{M}{V}$.

   The Van der Waals equation for one mole of a real gas is given by

   $$\left(P + \frac{a}{V^2} \right)(V - b) = RT.$$

Replacing volume V in terms of molar mass M and density ρ, we get

$$\left(P + \frac{a\rho^2}{M^2}\right)\left(\frac{M}{\rho} - b\right) = RT; \quad \Rightarrow \quad P = \frac{RT}{\left(\frac{M}{\rho} - b\right)} - \frac{a\rho^2}{M^2};$$

$$\Rightarrow \quad P = \frac{\rho RT}{M - \rho b} - \frac{a\rho^2}{M^2}.$$

This is the expression for required pressure P.

3. At a temperature of 373 K, one mole of carbon dioxide gas occupies a volume 536 mL at 50.0 atmospheric pressure. Calculate the value of pressure using

   (a) Ideal gas equation and
   (b) Van der Waals equation of state.

   Also, calculate the % deviation in pressure from the observed value in each case.

   [Given: Van der Waals constants for carbon dioxide: $a = 3.61$ L^2 atm mol^{-2}; and $b = 0.0428$ L mol^{-1}]

   Answer (a) According to the problem, we have

   $$V = 0.536 \text{ L}; \quad n = 1 \text{ mol}; \quad T = 373 \text{ K}.$$

   Using the ideal gas equation, we have

   $$PV = nRT; \quad \Rightarrow \quad P = \frac{nRT}{V} = \frac{1 \times 0.0821 \times 373}{0.536} = 57.1 \text{ atm.}$$

   The actual pressure is 50 atm. So, the % deviation in pressure is

   $$= \frac{(57.1 - 50)}{50} \times 100 = 14.2\%.$$

   (b) Using the Van der Waals equation, we have

   $$\left(P + \frac{an^2}{V^2}\right)(V - nb) = nRT;$$

   $$\Rightarrow \quad \left(P + \frac{3.61 \times (1.00)^2}{0.536^2}\right)(0.536 - 1 \times 0.0428) = 1 \times 0.082 \times 373$$

   $$\Rightarrow (P + 12.57) \times (0.493) = 30.62; \quad \Rightarrow \quad P = 49.6 \text{ atm.}$$

The actual pressure is 50 atm. So, the % deviation in pressure is

$$= \frac{(50 - 49.6)}{50} \times 100 = 0.8\%.$$

This problem shows that the Van der Waals equation of state predicts more correctly the value of pressure under the given conditions.

4. Find the Boyle temperature of the gas obeying the equation of state

$$\left(P + \frac{a}{TV^2}\right)(V - b) = RT.$$

Answer: This is Berthelot's equation of state for one mole of a real gas. From this equation, we have

$$P = \frac{RT}{V - b} - \frac{a}{TV^2}; \quad \Rightarrow \quad PV = \frac{RTV}{V - b} - \frac{a}{TV}.$$

The definition of Boyle temperature indicates that the slope of PV versus P curve is zero at the Boyle temperature T_B. Hence, we get

$$\left[\frac{\partial(PV)}{\partial P}\right]_{T=T_B} = 0; \quad \Rightarrow \quad \left[\frac{\partial}{\partial P}\left(\frac{RTV}{V - b} - \frac{a}{TV}\right)\right]_{T=T_B} = 0.$$

Carrying out the derivative and putting $T = T_B$, we get

$$\left[\frac{RT_B}{V - b} - \frac{RT_B V}{(V - b)^2} + \frac{a}{T_B V^2}\right]\left(\frac{\partial V}{\partial P}\right)_{T=T_B} = 0;$$

$$\Rightarrow \quad \left[\frac{-RT_B b}{(V - b)^2} + \frac{a}{T_B V^2}\right]\left(\frac{\partial V}{\partial P}\right)_{T=T_B} = 0.$$

As $\left(\frac{\partial V}{\partial P}\right)_{T=T_B} \neq 0$ at Boyle temperature T_B, we must have

$$\left[\frac{-RT_B b}{(V - b)^2} + \frac{a}{T_B V^2}\right] = 0; \quad \Rightarrow \quad T_B = \left(\frac{V - b}{V}\right)\sqrt{\frac{a}{Rb}}.$$

This is the expression for the Boyle temperature T_B of a real gas obeying Berthelot's equation of state.

5. Calculate the critical temperature of a gas, for which Van der Waals constants are $a = 0.00874$ atm $-$ cm^6 and $b = 0.0023$ cm^3 for 1 c.c. of the gas at N.T.P.

Answer: Here, the value of a is 0.00874 atm $-$ cm^6 for 1 c.c. of the gas.

We know that 1 mole of a gas has a volume of 22400 cm^3. So, the value of a for one mole of the gas is

$$a = 0.00874 \times (22400)^2 \text{ atm cm}^6 \text{ mol}^{-2}$$

$$= 0.00874 \times (22400)^2 \times 76 \times 13.6 \times 981 \text{ dyne cm}^4 \text{ mol}^{-2}.$$

The value of b for one mole of the gas is

$$b = 0.0023 \text{ cm}^3 = 0.0023 \times 22400 \text{ cm}^3 \text{ mol}^{-1}.$$

We know that the critical temperature T_c is given by

$$T_c = \frac{8a}{27bR} = \frac{8 \times 0.00874 \times (22400)^2 \times 76 \times 13.6 \times 981}{27 \times 0.0023 \times 22400 \times 8.31 \times 10^7} = 307.7 \text{ K}.$$

So, the critical temperature for the given gas is $T_c = 307.7$ K.

6. Calculate the critical temperature for He gas for which Van der Waals constants are $a = 6.15 \times 10^{-5}$ and $b = 9.95 \times 10^{-4}$. Consider the unit of pressure as 1 atm and the unit of volume as the volume of the gas at N.T.P.

   Answer: Here, the values of a and b are

   $$a = 6.15 \times 10^{-5} \quad \text{and} \quad b = 9.95 \times 10^{-4}.$$

   Further, according to the problem, the unit of pressure is 1 atm, that is, $P = 1$ and the unit of volume is the volume of the gas at N.T.P., that is, at N.T.P. $V = 1$.

   From Van der Waals equation of state, we know

   $$\left(P + \frac{a}{V^2}\right)(V - b) = RT; \quad \Rightarrow \quad R = \frac{\left(P + \frac{a}{V^2}\right)(V - b)}{T}$$

   $$= \frac{\left(1 + \frac{6.15 \times 10^{-5}}{1}\right)(1 - 9.95 \times 10^{-4})}{273} = 3.66 \times 10^{-3}.$$

   We know that the critical temperature T_c is given by

   $$T_c = \frac{8a}{27bR} = \frac{8 \times 6.15 \times 10^{-5}}{27 \times 9.95 \times 10^{-4} \times 3.66 \times 10^{-3}} = 5 \text{ K} = -268°\text{C}.$$

   So, the critical temperature for the given gas is $T_c = -268°$C.

7. Calculate the Van der Waals constants a and b for one gm molecule of He from the following data:

   $$T_c = 5.3 \text{ K}; \quad P_c = 2.25 \text{ atm, and } R = 8.3 \times 10^7 \text{ erg/(K mol)}.$$

Answer: The values of critical pressure P_c, critical temperature T_c, and molar gas constant R are, respectively, given by

$$P_c = 2.25 \text{ atm} = 2.25 \times 1.013 \times 10^6 \text{ dyne cm}^{-2},$$

$$T_c = 5.3 \text{ K, and } R = 8.3 \times 10^7 \text{ erg (K mol)}^{-1}.$$

In terms of the critical constants P_c and T_c and the universal gas constant R, the Van der Waals constant a is given by

$$a = \frac{27R^2T_c^2}{64P_c} = \frac{27 \times (8.3 \times 10^7)^2(5.3)^2}{64 \times 2.25 \times 1.013 \times 10^6} = 3.58 \times 10^{10} \text{ dyne cm}^4 \text{ mol}^{-2}.$$

In terms of the critical constants P_c and T_c and the universal gas constant R, the Van der Waals constant b is given by

$$b = \frac{RT_c}{8P_c} = \frac{8.3 \times 10^7 \times 5.3}{8 \times 2.25 \times 1.013 \times 10^6} = 24.44 \text{ c.c. mol}^{-1}.$$

Hence, the values of a and b are, respectively, 3.58×10^{10} dyne cm^4 mol^{-2} and 24.44 c.c. mol^{-1}.

8. Show that at the critical point, the deviation of the Van der Waals gas law from the perfect gas law is 62.55%.

Answer: At the critical point, we have $\frac{RT_c}{P_cV_c} = 1$ for an ideal gas, whereas for a Van der Waals gas, we have

$$\frac{RT_c}{P_cV_c} = \frac{R \times \frac{8a}{27bR}}{\frac{a}{27b^2} \times 3b} = \frac{8}{3} = 2.67.$$

So, deviation from ideal gas behavior is $= 2.67 - 1 = 1.67$. Hence, the percentage of deviation is

$$\frac{1.67}{2.67} \times 100 = 62.55\%.$$

9. The critical temperature and pressure of argon are $-122°$C and 48 atmos, respectively. Calculate the radius of an argon atom.

Answer: In terms of the critical constants P_c and T_c and the universal gas constant R, the Van der Waals, constant b is given by

$$b = \frac{RT_c}{8P_c} = \frac{8.31 \times 151}{8 \times 48 \times 1.013 \times 10^5} = 3.22 \times 10^{-5} \text{ m}^3 \text{ mol}^{-1}.$$

Further, it is known that b is 4 times the volume occupied by all the molecules of 1 mole of argon. Hence, the expression for b is

$$b = 4 \times \frac{4}{3}\pi r^3 \times N_A,$$

where r is the radius of an argon atom, and N_A is Avogadro's number. From this expression, we can write

$$r^3 = \frac{3b}{16\pi N_A} = \frac{3 \times 3.22 \times 10^{-5}}{16 \times 3.14 \times 6.02 \times 10^{23}} = 3.19 \times 10^{-30}\ \text{m}^3.$$

So, the radius of an argon atom is

$$r = \left(3.19 \times 10^{-30}\right)^{\frac{1}{3}}\ m = 1.472 \times 10^{-10}\ m = 1.472\mathring{A}.$$

10. Calculate the critical temperature T_c of helium if the critical pressure P_c is 2.26 atm and critical density ρ_c is 0.069 g cm^{-3}.

   Answer: We know that the critical coefficient Z_c is given by

   $$\frac{RT_c}{P_c V_c} = \frac{8}{3}.$$

   This leads to

   $$T_c = \frac{8}{3} \times \frac{P_c V_c}{R}.$$

   Here, we have the critical pressure $P_c = 2.26$ atm $= 2.26 \times 1.013 \times 10^5$ Nm^{-2} and the critical density $\rho_c = 69$ kg m^{-3} so that the critical volume of 1 kg of the gas is $\frac{1}{69}$ m^3. Hence, the critical volume per kmol of the gas is $\frac{4}{69}$ m^3, that is, $V_c = \frac{4}{69}$ m^3 kmol^{-1}.

   Using these values of the parameters in the above expression, we get

   $$T_c = \frac{8}{3} \times \frac{2.26 \times 1.013 \times 10^5 \times \frac{4}{69}}{8.31 \times 10^3} = 4.26\ \text{K}.$$

11. One mole of a gas occupies a volume of 0.55 l at 0°C. Calculate the pressure it will exert if it behaves as (a) an ideal gas, and (b) as a Van der Waals gas. Given $a = 0.37$ Nm^4 mol^{-2}, $b = 43 \times 10^{-6}$ m^3 mol^{-1}, and $R = 8.31$ J mol^{-1}K^{-1}.

   Answer: Here, $V = 0.55$ l mol$^{-1} = 550$ cm^3 mol$^{-1} = 550 \times 10^6$ m^3 mol^{-1}.

(a) For one mole of an ideal gas,

$$p = \frac{RT}{V}.$$

On substituting the values of various quantities, we get

$$p = \frac{8.31 \text{ J mol}^{-1}\text{K}^{-1} \times 273 \text{ K}}{550 \times 10^{-6} \text{ m}^3 \text{ mol}^{-1}} = 4.12 \times 10^6 \text{ Nm}^{-2}.$$

(b) For one mole of Van der Waals gas,

$$p = \frac{RT}{V - b} - \frac{a}{V^2} = \frac{(8.31 \text{ J mol}^{-1}\text{K}^{-1}) \times (273 \text{ K})}{(550 - 43) \times 10^{-6} \text{ m}^3 \text{ mol}^{-1}} - \frac{0.37 \text{ Nm}^4\text{mol}^{-2}}{(550 \times 10^{-6})^2 \text{ m}^6 \text{ mol}^{-2}}$$

$$= \left(\frac{8.31 \times 273}{507 \times 10^{-6}} - \frac{0.37}{(550 \times 10^{-6})^2} \right) \text{ Nm}^{-2} = (4.48 \times 10^6 - 1.22 \times 10^6) \text{ Nm}^{-2}$$

$$= 3.26 \times 10^6 \text{ Nm}^{-2}.$$

As expected, the pressure exerted by a Van der Waals gas is less than that exerted by an ideal gas.

12. For a gas obeying Van der Waals equation, the constants are $a = 1.32 \text{ l}^2\text{atm.}$ mol^{-2} and $b = 3.12 \times 10^{-2} \text{ l mol}^{-1}$. Calculate the temperature at which 5 moles of the gas at 5 atmospheric pressures will occupy a volume of 20 l. Given: $R = 8.31 \times 10^7 \text{ erg mol}^{-1}\text{K}^{-1}$.

Answer: The Van der Waals equation for n mole of a gas is

$$\left(P + \frac{an^2}{V^2} \right)(V - nb) = nRT.$$

By inverting it, we can write

$$T = \frac{\left(P + \frac{an^2}{V^2} \right)(V - nb)}{nR}.$$

Here, $a = 1.32 \text{ l}^2 \text{ atm. mol}^{-2} = 1.34 \times 10^{12} \text{ dyne cm}^4 \text{ mol}^{-2}$
$b = 3.12 \times 10^{-2} \text{ l mol}^{-1} = 31.2 \text{ cm}^3 \text{ mol}^{-1}$;
$P = 5 \text{ atm} = 5.065 \times 10^6 \text{ dyne cm}^{-2}$;
$V = 20 \text{ l} = 20 \times 10^3 \text{ cm}^3 \quad n = 5, R = 8.31 \times 10^7 \text{ erg mol}^{-1} \text{ K}^{-1}$

$$T = \frac{\left[(5.065 \times 10^6) + \frac{25 \times (1.34 \times 10^{12})}{(20 \times 10^3)^2} \right] \times [(20 \times 10^3) - (5 \times 31.2)]}{5 \times 8.31 \times 10^7} \text{ K} = 245.9 \text{ K}.$$

5.7 Multiple choice questions with answers

1. The terms b and $\left(\frac{a}{V^2}\right)$ in Van der Waals equation of state are introduced to account for the

(a) total volume occupied by the gas and the intermolecular attraction, respectively.

(b) molecular size and the size of the containing vessel, respectively.

(c) intermolecular attraction and the total volume of the molecules, respectively.

(d) intermolecular attraction and the force exerted by the molecules on the walls of the container, respectively.

Answer: The correct choice is (a).

Solution: The equation for an ideal gas is given by $PV = RT$, whereas that for a for real gases is given by $\left(P + \frac{a}{V^2}\right)(V - b) = RT$. This equation is due to Van der Waals. In this equation, the factor b is due to the volume occupied by the molecules themselves, and the term $\left(\frac{a}{V^2}\right)$ is due to the interaction forces between the molecules.

2. If C_V is the specific heat of an ideal gas, then the change in internal energy dU of a Van der Waals gas having the same degree of freedom as that of the ideal gas is given by

(a) $dU = C_V dT$

(b) $dU = -\frac{a}{V^2} dV$

(c) $dU = C_V dT - \frac{a}{V^2} dV$

(d) $dU = C_V dT + \frac{a}{V^2} dV$

Answer: The correct choice is (d).

Solution: Considering internal energy U as a function of temperature T and volume V, we have

$$U = U(T, V) \Rightarrow dU = \left(\frac{\partial U}{\partial T}\right)_V dT + \left(\frac{\partial U}{\partial V}\right)_T dV \Rightarrow dU = C_V dT + \left(\frac{\partial U}{\partial V}\right)_T dV.$$

For one mole of a Van der Waals gas, we have $\left(\frac{\partial U}{\partial V}\right)_T = \frac{a}{V^2}$. Using this result, we get

$$dU = C_V dT + \frac{a}{V^2} dV.$$

3. "Critical temperature" is defined as the

   (a) lowest temperature at which the gas can be liquefied at constant pressure.

   (c) highest temperature at which the gas can be liquefied by the increase of pressure alone.

   (b) lowest temperature at which the gas can be liquefied by the increase of pressure alone.

   (d) highest temperature at which the gas can be liquefied at constant pressure.

   Answer: The correct choice is (c).

4. At the critical state of a Van der Waals gas, the compressibility factor Z_c is

   (a) 3.735

   (c) 3.375

   (b) 0.735

   (d) 0.375

   Answer: The correct choice is (d).

5. In the Van der Waals equation of state $\left(P + \frac{a}{V^2}\right)(V - b) = RT$, the constant a is introduced to compensate for

   (a) reduction in specific volume.

   (c) reduction in specific heat.

   (b) intermolecular forces.

   (d) force of adhesion.

   Answer: The correct choice is (b).

6. In terms of the variables T and V, the equation of the adiabatic process for a Van der Waals gas is given by

   (a) $T(V - b)^{\frac{R}{C_P}} = \text{constant}$

   (c) $T(V - b)^{-\frac{R}{C_P}} = \text{constant}$

   (b) $T(V - b)^{\frac{R}{C_V}} = \text{constant}$

   (d) $T(V - b)^{-\frac{R}{C_V}} = \text{constant}$

   Answer: The correct choice is (b).

   Solution: For one mole of a Van der Waals gas, we have $\left(P + \frac{a}{V^2}\right)(V - b) = RT$ and the change in internal energy is $dU = C_V dT + \frac{a}{V^2} dV$.

   For an adiabatic process, we have $\delta Q = 0 = dU + P dV$. Using the value of dU, we get

   $$dU + P dV = 0; \quad \Rightarrow \quad C_V dT + \frac{a}{V^2} dV + P dV = C_V dT + \left(P + \frac{a}{V^2}\right) dV = 0.$$

From Van der Waals equation for one mole, we get $\left(P + \frac{a}{V^2}\right) = \frac{RT}{V-b}$. This value leads to

$$C_V dT + \frac{RT}{V-b} dV = 0; \quad \Rightarrow \quad \frac{C_V}{R}\frac{dT}{T} + \frac{dV}{V-b} = 0.$$

Integrating this expression, we get

$$\int \frac{C_V}{R}\frac{dT}{T} + \int \frac{dV}{V-b} = 0$$

$$\Rightarrow \quad \frac{C_V}{R}\ln T + \ln(V-b) = k \text{ (constant)}; \quad \Rightarrow \quad \ln\left[(V-b)T^{\frac{C_V}{R}}\right] = k.$$

$$\Rightarrow \quad \left[(V-b)T^{\frac{C_V}{R}}\right] = e^k = k', \text{ (another constant)}; \quad \Rightarrow \quad T\,(V-b)^{\frac{R}{C_V}} = \text{constant}.$$

7. The unit of the constant "a" in Van der Waals equation of state for one mole is

(a) Nm^5

(b) Nm^4

(c) Nm^2

(d) Nm^3

Answer: The correct choice is (b).

8. Figure 5.9 shows the phase diagram for a pure substance in the regions I, II, III, and IV, respectively. These regions, respectively, represent

[JAM 2019]

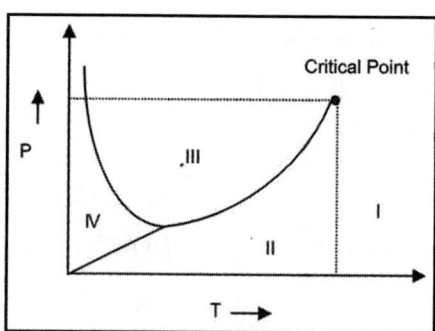

Figure 5.9 Phase diagram for the pure substance referred to problem 8.

(a) Vapor, gas, solid, liquid

(b) Gas, vapor, liquid, solid

(c) Gas, liquid, vapor, solid

(d) Vapor, gas, liquid, solid

Answer: The correct choice is (b).

Solution: IV— solid; III— liquid; II— vapor; I— gas (superheated dry vapor)

9. The Van der Waals equation for one mole of CO_2 gas at low pressure will be

(a) $\frac{\left(P+\frac{a}{V^2}\right)}{RT} = \frac{1}{V}$

(b) $P(V-b) = RT - -\frac{a}{V^2}$

(c) $P = \frac{RT}{V-b}$

(d) $P = \left(\frac{RT}{V-b} - \frac{a}{V^2}\right)$

Answer: The correct choice is (a).

Solution At low pressure, the volume correction for one mole of a real gas will be negligible. Hence, $b = 0$ and the Van der Waals equation of state for one mole will be given by

$$\left(P + \frac{a}{V^2}\right) V = RT; \quad \Rightarrow \quad \frac{\left(P + \frac{a}{V^2}\right)}{RT} = \frac{1}{V}.$$

10. If V is the volume of one molecule of gas under given conditions, then Van der Waals' constant b (also called excluded volume or effective volume) with N_0 number of molecules is

(a) $4V$

(b) $\frac{4V}{N_0}$

(c) $\frac{N_0}{4V}$

(d) $4VN_0$

Answer: The correct choice is (d).

11. In an experiment, a certain quantity of an ideal gas at temperature T_0, pressure P_0, and volume V_0 is heated by a current flowing through a wire for a duration of t seconds. The volume is kept constant, and the pressure changes to P_1. If the experiment is performed at constant pressure starting with the same initial conditions, the volume changes from V_0 to V_1. The ratio of the specific heats at constant pressure and constant volume is   [JEST 2018]

(a) $\frac{P_1-P_0}{V_1-V_0}\frac{V_0}{P_0}$

(b) $\frac{P_1-P_0}{V_1-V_0}\frac{V_1}{P_1}$

(c) $\frac{P_1V_1}{P_0V_0}$

(d) $\frac{P_0V_0}{P_1V_1}$

Answer: The correct choice is (a).

Solution: (I) Heating at constant volume leads to $\frac{P_0}{T_0} = \frac{P_1}{T_1} \Rightarrow T_1 = \frac{P_1}{P_0}T_0$. This provides

$$Q = C_V(T_1 - T_0) = C_V\left(\frac{P_1}{P_0} - 1\right)T_0.$$

(II) Heating at constant pressure leads to $\frac{V_0}{T_0} = \frac{V_1}{T_1} \Rightarrow T_1' = \frac{V_1}{V_0}T_0$. This provides

$$Q' = C_p(T_1' - T_0) = C_p T_0 \left(\frac{V_1}{V_0} - 1 \right).$$

Using the ideal gas equation, we get $PdV + VdP = RdT$. At constant pressure, this becomes $PdV = RdT$. Thus, we have

$$dT_p = \frac{P}{R}dV = \frac{P_0}{R} \times (V_1 - V_0) \text{ and } dT_v = \frac{V}{R}dP = \frac{V_0}{R}(V_1 - V_0)$$

$$C_v \times \frac{V_0}{R}(P_1 - P_0) = C_p \times \frac{P_0}{R}(V_1 - V_0)$$

$$\text{and } \frac{C_P}{C_V} = \frac{V_0(P_1 - P_0)}{P_0(V_1 - V_0)} = \left(\frac{P_1 - P_0}{V_1 - V_0} \right) \times \frac{V_0}{P_0}.$$

12. An ideal gas is the one for which   [JAM 2010]

   (a) C_V is a function of volume only.

   (b) C_V is a function of volume and temperature.

   (c) $\frac{C_P}{C_V}$ is not a function of temperature.

   (d) C_V and C_P depend on temperature only.

Answer: The correct choice is (c).

13. The temperature at which the second virial coefficient of a real gas is zero is called

   (a) Critical temperature

   (b) Eutectic point

   (c) Boiling point

   (d) Boyle temperature

Answer: The correct choice is (d).

14. A rigid triangular molecule consist of three noncollinear atoms joined by rigid rods. The constant pressure molar specific heat (C_P) of an ideal gas consisting of such molecules is   [JAM 2015]

   (a) $6R$

   (b) $5R$

   (c) $4R$

   (d) $3R$

Answer: The correct choice is (c).

Solution: The degrees of freedom of the rigid triangular molecule is $= 6$.

The internal energy is given by $U = \frac{6RT}{2}$. So, the specific heat at constant volume is $\Rightarrow C_V = (\frac{\partial U}{\partial T})_V = 3R \Rightarrow C_P = C_V + R = 4R$.

15. The temperature of the inversion of a gas is

(a) $\frac{a}{Rb}$

(c) $\frac{2a}{27Rb}$

(b) $\frac{2a}{Rb}$

(d) $\frac{27}{8}\frac{Rb}{a}$

Answer: The correct choice is (b).

16. The graph shown in Figure 5.10 represents the Van der Waals equation of state. Which portion of the graph cannot be explained?

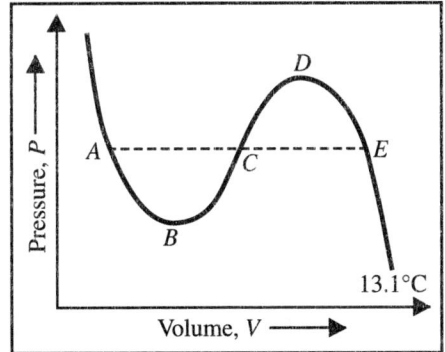

Figure 5.10 Typical $P-V$ diagram for problem 16.

(a) AB

(c) DE

(b) BC

(d) BCD

Answer: The correct choice is (d).

Solution: The Van der Waals equation of state is given by

$$\left(P + \frac{a}{V^2}\right)(V - b) = RT,$$

where a and b are Van der Waals constants.

The graph drawn is shown as the curve ABCDE. This does not agree with the experimental isothermal for CO_2 as obtained by Andrews. However, the portion DE has been explained as due to supercooling of vapor and the portion AB due to super heating of the liquid. But the portion BCD cannot be explained because it shows an increase in volume with an increase in pressure.

17. Critical temperature is defined as

   (a) the lowest temperature at which the gas can be liquefied at constant pressure.

   (b) the lowest temperature at which the gas can be liquefied at by an increase of pressure alone.

   (c) the highest temperature at which the gas can be liquefied at by an increase of pressure alone.

   (d) the highest temperature at which the gas can be liquefied at by an increase of constant pressure.

Answer: The correct choice is (c).

5.8   Exercise

5.8.1   Short answer type questions

1. What are the differences between an ideal and a real gas?

2. What do you mean by a critical point?

3. What are the differences between gas and vapor?

4. Write down the Van der Waals equation of state for the n-mole of a real gas.

5. In the Van der Waals equation, why is the term that corrects for volume negative and the term that corrects for pressure positive? Why is $\frac{n}{V}$ squared?

6. What are the units and dimensions of the Van der Waals constants a and b.

7. Comment on the validity of Van der Waals equation in explaining experimental results.

8. State the law of corresponding states for real gases.

                                                       [Vidyasagar University 2019]

9. What do you mean by compressibility factor? What is its value for an ideal gas?

10. How are the real gases classified in terms of compressibility factor?

11. What do you mean by "metastable" state?

12. State the relation between critical constants of a Van der Waals gas.

13. What is the reduced equation of state? What is its significance?

14. State the law of corresponding states. What is its significance?

15. Write down Onnes equation of state.

16. What are the virial coefficients?

17. What is the Boyle temperature T_B? Show that at the Boyle temperature, the second virial coefficient is zero.   [Burdwan University 2020]

18. How is the Boyle temperature T_B related to the inversion temperature T_i and the critical temperature T_c?

19. What is the relation between the Boyle temperature and the second virial coefficient?

5.8.2   Long answer type questions

1. What do you mean by equation of state? Derive the equation of state for an ideal gas.

2. What are the defects of an ideal gas equation?

3. What factors cause deviations from ideal gas behavior? Use a sketch to explain your answer based on interactions at the molecular level.

4. Explain the effect of nonzero atomic volume on the ideal gas law at high pressure. Draw a typical graph of volume V versus $\frac{1}{P}$ for an ideal and a real gas.

5. Describe Regnault's experiment for the determination of isotherms of a real gas. What are the implications of the results of this experiment?

6. Describe Andrews' experiment for the determination of isotherm of carbon dioxide clearly mentioning the range of pressure. What are the main conclusions of this experiment? How is the concept of critical parameters obtained from this experiment?

7. Discuss the results obtained by Andrews in his experiment on CO_2. Explain the term "critical temperature" of a gas.   [Delhi University 2021]

8. Explain the critical phenomenon on CO_2 in Andrews experiment.

9. Describe Amagat's experiment for the determination of isotherm of carbon dioxide clearly mentioning the range of pressure. What are the main conclusions of this experiment?

10. For an ideal gas, the product of pressure and volume should be constant, regardless of the pressure. Experimental data for methane, however, show that the value of PV decreases significantly over the pressure range 0 to 120 atm at 0°C. The decrease in PV over the same pressure range is much smaller at 100°C. Explain why PV decreases with increasing temperature. Why is the decrease less significant at higher temperatures?

11. What are the characteristic features of Van der Waals interactions?

12. Deduce Van der Waals equation of state for one mole of a real gas. Discuss the results obtained from Van der Waals equation and compare them with the experimental results obtained for a real gas.

13. What are the limitations of Van der Waals equation of state?

14. What is the effect of intermolecular forces on the liquefaction of a gas? At constant pressure and volume, does it become easier or harder to liquefy a gas as its temperature increases? Explain your reasoning. What is the effect of increasing the pressure on the liquefaction temperature?

15. What are critical constants? Find out the expressions for the critical constants in terms of Van der Waals constants "a" and "b" and universal gas constant "R". Also prove that $\frac{RT_c}{P_C V_c} = \frac{8}{3}$, where P_c, V_c, and T_c are critical pressure, critical volume, and critical temperature, respectively.

    [St. Joseph's College (Autonomous), Bengaluru 2020]

16. Calculate the critical constants of Van der Waals gas in terms of constants "a" and "b". Hence, derive the reduced equation of state.   [Delhi University 2021, WBSU 2021]

17. Establish Van der Waals reduced equation of state for one mole of a real gas. What is the importance of reduced equation of state?

18. What are the different ways by which the values of the Van der Waals constants "a" and "b" could be determined?

19. Find the values of Van der Waals constants "a" and "b" in terms of critical constants.

20. Give an explanation of the behavior of a real gas on the basis of Van der Waals equation.

21. Derive the relations between critical constants and constants of Van der Waals equation.

22. Using Van der Waals equation, explain the deviations of gases from Boyle's law.

23. What is Boyle temperature? Find an expression for it in terms of critical temperature T_c.

24. What do you mean by Boyle temperature? The equation of state of a real gas is

$$\left(P + \frac{a}{TV^2}\right)(V - b) = RT,$$

where a and b are constants. Find an expression for the Boyle temperature of the gas.

25. "The critical temperature must always be less than the Boyle temperature" – Explain.

26. Explain the significance of the critical constants of a gas.

27. Liquefaction of a gas depends strongly on two factors. What are they? As temperature is decreased, which gas will liquefy first – ammonia, methane, or carbon monoxide? Why?

28. What is a cryogenic liquid? Describe three uses of cryogenic liquids.

29. Air consists primarily of O_2, N_2, Ar, Ne, Kr, and Xe. Use the concepts discussed in this chapter to propose two methods by which air can be separated into its components. Which component of air will be isolated first?

30. How can gas liquefaction facilitate the storage and transport of fossil fuels? What are the potential drawbacks to these methods?

31. What do you mean by an ideal gas? In what situation does a real gas behave like an ideal gas?

32. Write down Van der Waals equation of state and plot its isotherms. Identify the critical isotherms. Show those regions of the isotherms that represent the meta-stable state of the substance.

33. Derive the reduced equation of state from Van der Waals equation and state its significance.   [Burdwan University 2021]

34. Express the Van der Waals equation in the following virial form:

$$PV = RT \left[1 + \frac{B}{V} + \frac{C}{V^2} + \cdots \right],$$

where $B, C, \ldots$ are virial coefficients. From the above expansion, find out the expression of the second virial coefficient "B" and the Boyle temperature T_B. What is the significance of the Boyle temperature?

5.8.3   Numerical problems

1. Consider the Van der Waals equation of state for n mole of a real gas: $\left(P + \frac{an^2}{V^2} \right) (V - nb) = nRT$, where symbols have their usual meaning. Calculate the values of the critical constants P_c, V_c, and T_c for such a gas.

$$[V_c = 3nb, T_c = (8a)/(27bR), P_c = a/(27b^2)]$$

2. One mole of a gas, stored at 300 K, occupies a volume of 1.2×10^{-3} m^3 mol^{-1}. Compare the pressures calculated on the basis of Van der Waals equation of state and ideal equation of state. Given $a = 1.32 \times 10^{-6}$ atm m^3 mol^{-2} and $b = 3.12 \times 10^{-5}$ m^3 mol^{-1}.   [1:1.018]

3. Under what pressure will carbon dioxide of molar mass M have the density ρ at the temperature T. If the given gas is obeying for a Van der Waals gas.

$$[P = \frac{\rho RT}{M - \rho b} - \frac{\rho^2 a}{M^2}]$$

4. Carbon dioxide gas (1.00 mol) at 373 K occupies 536 mL at 50.0 atmosphere pressure. What is the calculated value of the pressure using (i) the ideal gas equation, and (ii) Van der Waals equation?

   [Data – Van der Waals constants for carbon dioxide: $a = 3.61$ L^2 atm mol^{-2} and $b = 0.0428$ L mol^{-1}] $[P = 57.1$ atm, $P = 49.6$ atm]

5. The Van der Waals constants for hydrogen are $a = 0.0247$ N m^4 mol^{-2}, $b = 2.65 \times 10^{-5}$ m^3 mol^{-1}, and $R = 8.31$ J mol^{-1} K^{-1}. Find the inversion temperature of hydrogen.   [224.33 K] [BMS College Bengaluru 2017]

6. Using Van der Waals equation, calculate the temperature of 20.0 mol of helium in a 10.0 l cylinder at 120 atmosphere pressure.

   Compare this value with the temperature calculated from the ideal gas equation.

   [Data – Van der Waals constants for helium: $a = 0.0341$ L^2 atm mol^{-2} and $b = 0.0237$ L mol^{-1}] $[T = 696$ K, $P = 731$ K]

7. The critical temperature T_c and the critical pressure P_c are, respectively, $-122°C$ and 48 atmos for Argon atoms. Calculate the radius of the Argon atom.   [0.368 m] [Burdwan University 2020]

8. Calculate the critical temperature of helium if the critical pressure is 2.26 atm, and critical density is 0.069 g cm^{-3}. $[T_c = 4.26$ K]

9. Calculate Van der Waals constants a and b of helium if the critical temperature T_c is 5.3 K, the critical pressure P_c is 2.25 atm, and $R = 8.31$ J mol^{-1} K^{-1}.

   $[a = 3.59 \times 10^{-3}$ N m^4 mol^{-2} and $b = 2.42 \times 10^{-5}$ m^3 mol^{-1}]

10. Argon has critical temperature $T_c = -122°C$ and critical pressure $P_c = 48$ atm. Find the radius of the argon atom. $[1.47 Å]$

11. Nitric oxide obeys Berthelot's equation of state with critical pressure $P_c = 65.5$ atm, and critical temperature $T_c = 179$ K when its density is 1.37 kg m^{-3}. Calculate the molecular weight of the gas.   [30.15]

12. A gas obeys the equation of state: $PV = RT\left(1 + \frac{b}{V}\right)$. Would it be possible to liquefy the gas? Would it have a critical temperature? Explain your answers.

     [No, No]

13. Calculate the Boyle temperature from Dieterici's equation of state. $[T_B = \frac{a}{Rb}]$

14. A gas obeys the equation of state: $P(V - nb) = nRT$. Show that the gas does not have the Boyle temperature T_B.

15. Determine the critical constants P_c, V_c, and T_c in terms of B and C, and the critical compressibility factor of a gas that obeys the equation of state: $P = \frac{RT}{V} - \frac{B}{V^2} + \frac{C}{V^3}$. $[V_c = \frac{3C}{B}, T_c = \frac{B^2}{RC}, P_c = \frac{7B^3}{27C^2}$, and $Z_c = 0.78$.]

16. Consider a tank of rigid walls with a fixed volume of 40.0 L. The tank is filled with 1.000 kg of oxygen gas at a temperature of 298 K. Calculate the pressure exerted by this gas considering, it as ideal and Van der Waals' one. Given: $a = 1.36$ L^2 atm mol^{-2}, and $b = 0.0319$ L mol^{-1}.

$[P_{ideal} = 19.1$ atm and $P_{vander} = 18.8$ atm.$]$

CHAPTER 6

Conduction of Heat

Heat, like gravity, penetrates every substance of the universe, its rays occupy all parts of space.

—Jean-Baptiste-Joseph Fourier

Learning Outcomes

After reading this chapter, the reader will be able to

- Understand the meaning of three processes of heat flow: conduction, convection, and radiation

- Know about thermal conductivity, diffusivity, and steady-state condition of a thermal conductor

- Derive Fourier's one-dimensional heat flow equation and solve it in the steady state

- Derive the mathematical expression for the temperature distribution in a lagged bar

- Derive the amount of heat flow in a cylindrical and a spherical thermal conductor

- Solve numerical problems and multiple choice questions on the process of conduction of heat

---

6.1 Introduction

Heat is the thermal energy transferred between different substances that are maintained at different temperatures. This energy is always transferred from the hotter object (which is maintained at a higher temperature) to the colder one (which is maintained at a lower temperature). Heat is the energy arising due to the movement of atoms and molecules that are continuously moving around, hitting each other and other objects. This motion is faster for the molecules with a larger amount of energy than the molecules with a smaller amount of energy that causes the former to have more heat. Transfer of heat continues until both objects attain the same temperature or the same speed. This transfer of heat depends upon the nature of the material property determined by a parameter known as **thermal conductivity or coefficient of thermal conduction.** This parameter helps us to understand the concept of transfer of thermal energy from a hotter to a colder body, to differentiate various objects in terms of the thermal property, and to determine the amount of heat conducted from the hotter to the colder region of an object. The transfer of thermal energy occurs in several situations:

1. When there exists a difference in temperature between an object and its surroundings,

2. When there exists a difference in temperature between two objects in contact with each other, and

3. When there exists a temperature gradient within the same object.

It has been stated above that the flow of heat always occurs from a body at a higher temperature to a body at a lower temperature. One may then ask the following questions:

1. Why does the reverse process not occur in nature?

2. Which factor determines the direction of the flow of heat?

Answers to these questions can be provided from the second law of thermodynamics. This law states that for an isolated system, not in thermal equilibrium, the entropy will tend to increase over time and will ultimately approach toward a maximum value at equilibrium. According to the consequence of this second law of thermodynamics, the transfer of heat will thus always occur from a body at a higher temperature to a body at a lower temperature, and such flow of heat will continue until the state of thermal equilibrium is achieved by the two bodies.

There are three natural modes of thermal energy transfer: **conduction, convection, and radiation.** Each mode operates through a distinct mechanism, influencing both the process and the rate of heat transfer. In any given situation, the rate of heat transfer is determined by the dominance of a particular mode.

Conduction is the process in which the transfer of thermal energy takes place through direct contact with the concerned substances. When two substances are into contact, their particles collide with each other. The energy from the faster-moving particles of one substance is transferred to the slower-moving particles of another substance until they attain the same speed. In this situation, the temperatures of both substances will be the same. In the conduction process, the transfer of thermal energy involves a combination of diffusion of electrons and phonon vibrations and is, in general, applicable to solid materials. In a medium, the transfer of heat energy takes place through the conduction process by virtue of a temperature gradient in the medium. It is a microscopic mechanism and results from the exchange of translational, rotational, and vibrational energy among the particles comprising the medium. For example, a spoon is warmed up in the process of conduction when it is placed into a cup of hot tea.

Convection is the process in which the transfer of heat takes place in the air or fluid through currents. Consider the case of a pot of water placed on a hot stove. As the water in the pot is warmed up, the water particles spread out and it becomes less dense. The hot water on the bottom of the pot rises above and displaces the cold water. During this process, the cold water sinks. This cold water becomes hot and rises above. This again displaces the cold water, which becomes hot and again rises above. Further, the transfer of thermal energy occurs in a moving medium through the process of convection. Due to the difference in density, the hot gas/liquid moves through the cooler medium. Thus, convection can be considered as a macroscopic form of energy transfer through a fluid that occurs by the combined processes of conduction in the fluid and the bulk motion (mass transfer) of the fluid.

Radiation is the process in which the transfer of thermal energy takes place by electromagnetic radiation. It does not need any direct contact between the objects or movement of the particles as in conduction and convection. Radiation occurs through empty space as well. Through the process of radiation, the earth gets heat from the Sun, or we feel warm in front of a fire. The Sun is a good example of energy transfer through a (near) vacuum.

In this chapter, thermal conduction, coefficient of thermal conduction, and diffusivity are described in detail. The rate of transfer of heat through a solid medium is calculated using Fourier equation in various dimensions. This equation is solved in various physical cases using proper boundary conditions. These examples include rectilinear flow of heat in one-dimension, and radial flow of heat in spherical and cylindrical polar coordinate systems. The number of solved numerical problems and multiple choice questions with answers are presented at the end of this chapter.

6.2 Conduction

Conduction is the process in which heat is transferred from the hotter parts of a substance to the colder parts along the substance or from a hotter body to a colder body "in contact" without any physical transfer of material particles from one object to the other. The transfer of heat in this process follows a microscopic mechanism originated due to the exchange of translational, rotational, and vibrational energy among the particles in the medium. For instance, when one end of a metallic rod is heated by placing it in a furnace, the heat travels to the other end through the process of conduction. Since metals are excellent conductors of heat, the opposite end of the rod heats up rapidly.

6.2.1 Thermal conductivity

Consider two parts of a material substance maintained at two different temperatures by making contact with two sources maintained at two different constant temperatures. The temperature of each small volume element of the intervening substance is measured with the help of a thermometer. It is observed from these simple experimental results that a continuous distribution of temperature exists across the length of the intervening substance. This indicates that heat is transferred from one end of the substance to the other. **This process of transport of energy between neighboring volume elements by virtue of the temperature difference between them, is known as the conduction of heat**.

The fundamental law of conduction of heat can be established in the following way: Consider a piece of material made in the form of a slab of thickness Δx and area A. Heat flows linearly through this slab perpendicular to the faces of the material. Further, consider that one face of the slab is maintained at a temperature θ and the other face at a temperature $(\theta + d\theta)$. The amount of heat Q that passes perpendicular to the faces is measured for a time t with the help of some arrangements. The same experiment is repeated with other slabs of the same material but with different values of Δx and A. The results of such experiments show that, for a given value of $\Delta\theta$, the amount of heat Q is found to be proportional to both the time t and the cross-sectional area A. Also, for a given time t and area A, Q is found to be proportional to the temperature gradient $\frac{\Delta\theta}{\Delta x}$, provided that both $\Delta\theta$ and Δx are small. Further, Q is found to be inversely proportional to the thickness of the slab Δx keeping other parameters constant. Combining these experimental results, the amount of heat Q flowing per second through the slab can be expressed as

$$\frac{Q}{t} \propto \frac{A \times \Delta\theta}{\Delta x}. \tag{6.1}$$

It should be noted that equation (6.1) is only approximately true when $\Delta\theta$ and Δx are finite but is rigorously true in the limit as $\Delta\theta$ and Δx approach zero. If we generalize this result for an infinitesimal slab of thickness dx, across which there is a finite temperature difference $d\theta$, and introduce the concept of proportionality constant K, the fundamental law of conduction of heat can be written as

$$\frac{\delta Q}{dt} = -K \times A \times \frac{d\theta}{dx}. \tag{6.2}$$

The derivative $\frac{d\theta}{dx}$ in equation (6.2) is known as the temperature gradient. The negative sign in equation (6.2) is introduced to ensure that the positive direction of heat conduction aligns with the positive x-axis. For heat to flow in the positive x-direction, the temperature θ must decrease in that direction. The proportionality constant K is referred to as the "thermal conductivity". A substance with a large value of thermal conductivity K is defined as the "good conductor of heat or thermal conductor", whereas a substance with a small value of thermal conductivity K is defined as the "bad conductor of heat or thermal insulator". This equation (6.2) plays a very important role in the theory of conduction of heat. It is observed from equation (6.2) that δQ becomes equal to K when the cross-sectional area $A = 1$, temperature gradient $\frac{d\theta}{dx} = 1$, and time $dt = 1$. Thus, equation (6.2) can be used to define the thermal conductivity K in the following way: **The thermal conductivity of a material is the amount of heat that passes normally in unit time by conduction through a plane of the unit cross-sectional area under a temperature gradient of unity.**

In order to solve the problems related to the conduction of heat, it is necessary to transform the general equation into a second-order differential equation. The solution of this differential equation subject to given boundary conditions involves, as a rule, the use of functions and series that are beyond the scope of collegiate mathematics. However, there are three simple cases of rectilinear and radial flow of heat that can be solved in a very elementary way. In all these problems, it is assumed that the thermal conductivity K remains constant throughout the conducting substance. Also, K remains constant over the range of temperatures studied. It should be mentioned that diamond, a highly ordered crystalline solid, has the highest known value of thermal conductivity $K = 2200$ Wm^{-1} K^{-1}. However, it is a bad conductor of electricity, even worse than the semiconductor silicon. Diamond is used as heat sinks in cooling electronic components.

Unit and dimension of K

Considering the magnitude, equation (6.2) can be rearranged as

$$K = \frac{\delta Q}{dt} \times \frac{1}{A \frac{d\theta}{dx}}. \tag{6.3}$$

The unit of K can be obtained as follows:

$$[K] = \left[\frac{Q \cdot x}{A.(\theta_1 - \theta_2).t} \right] = \left[\frac{\text{calorie.cm}}{\text{cm}^2 \times {}^\circ\text{C} \times \text{s}} \right].$$

The unit of K in the C.G.S. system is cal s^{-1} $^\circ$C^{-1} cm^{-1}. In the S.I. unit, K is expressed in watt m^{-1} K^{-1} (W m^{-1} deg^{-1}).

It can be easily shown that the dimension of K is

$$[K] = \left[\frac{Q \cdot x}{A.(\theta_1 - \theta_2).t} \right] = \text{MLT}^{-3}\text{K}^{-1}.$$

6.2.2   Thermal diffusivity

The capacity to store heat energy and the ability to transfer heat energy of a given material depends upon factors, such as thermal conductivity, thermal diffusivity, and specific heat capacity. Therefore, for any given process or substance that experiences a large or fast temperature gradient, these properties should be understood thoroughly. Proper values of these parameters are required for modeling and managing the transfer of heat in the insulating or conducting state of the substance. Further, the measurements of the properties related to the transfer of heat provide important information about material composition, purity, and structure, as well as secondary performance characteristics such as tolerance to thermal shock.

In order to introduce the concept of thermal diffusivity, we proceed in the following way: Let us consider a metal bar with one end "A" being heated by an external source of heat and the other end "B" is maintained at ambient temperature. As the temperature of the end "A" rises, the immediate adjacent layer to "A" receives heat through the process of conduction. This received heat is utilized by the following three different processes:

1. The adjacent layer absorbs a part of the received heat that leads to an increase of its own temperature.

2. The adjacent layer loses another part of the received heat by way of radiation from its external surface and convection of gases around, and

3. The adjacent layer transmits the remainder part of the received heat to the next layer.

These processes continue from layer to layer until each layer attains a **steady or stationary temperature**. In this steady state, no layer absorbs any more heat passed on to it by the preceding layer, but the heat simply passes down the metal bar depending only on the value of thermal conductivity of its material. But in the variable state, that is, the state previous to the stationary one, each layer absorbs some amount of heat from what is received by it and its temperature rises. So, in the variable state, both absorption and conduction of heat occur. While conduction depends on thermal conductivity only, absorption depends on the specific heat of the material of the substance. If the specific heat is low, the temperature of any part rises quickly until the steady state is attained, even if the value of thermal conductivity is not high. On the other hand, if the specific heat is high, the temperature of any part will rise slowly to the temperature corresponding to the steady state, even if the value of thermal conductivity is high. This indicates that in the variable state, the absorption of heat is directly related to the specific heat of the material.

Let c be the specific heat of the material, ρ be the density, that is, mass per unit volume, $\frac{d\theta}{dt}$ be the temperature-increase per second, and Q be the amount of heat reaching the volume. The amount of heat Q absorbed by a particular layer of the material per unit volume per second is then given by

$$Q = \rho c \, \frac{d\theta}{dt}; \quad \Rightarrow \quad \frac{d\theta}{dt} = \frac{Q}{\rho c}. \tag{6.4}$$

This equation (6.4) implies that the rate of increase of temperature in the variable state per unit volume of the material is directly proportional to the quantity of heat reaching the volume by the process of conduction, that is, to the thermal conductivity K and inversely proportional to the product of density ρ and the specific heat c, that is, to the thermal capacity per unit volume. Hence, the rate of increase of temperature of a material depends on a parameter related to the thermal conductivity and the thermal capacity. This parameter is known as the thermal diffusivity h.

Thus, it is observed that thermal diffusivity h is related to the dissipation of heat in a material. However, it should be kept in mind that a high value of h does not necessarily mean that heat is better dissipated since it is the ratio between the thermal conductivity and the volumetric heat capacity at constant pressure. It provides information about the competition between the conducted heat and storing of heat, and both can be useful to keep the temperature as low as possible. Hence, the thermal diffusivity h can be mathematically expressed by the following equation

$$h = \frac{K}{\rho c}. \tag{6.5}$$

This ratio $\frac{K}{\rho c}$ has been termed as **thermal diffusivity** by Lord Kelvin and **thermometric conductivity** by Maxwell and is symbolized by h. It is known that the temperature governs the speed of chemical reactions, and flash temperature contributes to the rate of tribo-chemical reactions too. This thermal diffusivity of a material is related to the speed of such chemical reactions to reach the thermal equilibrium under variable thermal conditions. For this reason, lower values of h are better to reduce the possibility that a high temperature can be reached in-depth and damage the binder deeply. The thermal diffusivity says nothing about the flow of thermal energy to the surroundings.

In thermodynamics, a parameter known as "thermal effusivity", also known as thermal responsivity, is used to measure the ability of a material to exchange thermal energy with its surroundings. It is defined as the square root of the product of the thermal conductivity K of the material and its volumetric heat capacity ρc. Hence, the thermal effusivity e is given by

$$e = \sqrt{K\rho c}. \tag{6.6}$$

This thermal effusivity e characterizes the capacity of exchanging thermal energy with the surroundings of the given substance. In summary, the thermal effusivity and the thermal diffusivity are representative of two phenomena in competition: the former is related to the ability of the material to absorb heat, while the latter to the speed to reach thermal equilibrium, that is, to adapt itself to the surroundings. Several natural thermal phenomena can be explained by the concept of thermal effusivity. For example, when we touch a metal, it seems colder than other objects at the same temperature. This is because the heat transfer from the skin to the metal is very fast. In other words, the effusivity of a material is related to the exchange of heat on the surface, and diffusivity is related to the penetration of heat into the bulk. A huge amount of heat is created at the interface due to friction, which is dissipated inside the bulk. Thus, the most effective friction material can absorb (dissipate) more heat without raising the temperature. Therefore, the thermal effusivity is a surface property and plays a more important role in physical phenomena than the thermal diffusivity, which is a bulk property.

6.2.3   Steady-state conduction

Temperature difference is the driving force for the conduction of heat through a thermal conductor. When this temperature difference across the thermal conductor is constant, the spatial distribution of temperatures does not change any further after an equilibrium time. This situation is defined as the **steady-state conduction**. Under this situation, all partial derivatives of temperature with respect to space variables would be either zero or nonzero, but all derivatives of temperature with respect to time would be uniformly zero at all points. Hence, in the **steady-state conduction**, the amount of heat entering any region of the thermal conductor will be equal to the amount of heat coming out of it. This would happen; otherwise, the temperature would be increasing or decreasing with, respectively, tapping or trapping of thermal energy in a particular region of the thermal conductor.

For example, let us consider the case of a metal bar heated (higher temperature) at one end and maintained at a lower temperature at the other end. After a certain interval of time, a steady-state conduction in the metal bar would be reached, and the spatial gradient of temperatures along the bar will not change with time any further. It can be mathematically shown that the temperature in the metal bar will remain constant at any given cross-section normal to the direction of transfer of heat, and this temperature will vary linearly with length of the bar if no source of generation of heat is attached to the metal bar or no removal of heat takes place from the bar.

It should be mentioned that in the steady-state condition, the basic laws of direct current electrical conduction can be applied to "heat currents" flowing through a thermal conductor. A number of similarities are observed in both cases. For example, the "thermal resistance" of a heat conductor plays a similar role to that of an **electrical resistance** in an electric conductor. Further, the temperature difference plays the role of driving force in a thermal conductor, whereas the potential difference does the same in an electric conductor. Similarly, the transfer of heat per unit time (heat power) is analogous to the electric current. These similarity features indicate that thermal systems in the steady-state condition can be modeled by networks of thermal resistances connected in series and in parallel or in mixed combination, in exact analogy to electrical networks of resistors.

6.2.4 Heat transfer from an extended surface

An **extended surface** is the surface of a solid that experiences the transfer of energy by the conduction process within its boundaries as well as the transfer of energy by the convection and/or the radiation process between its boundaries and the surroundings. A **strut** is an example of an extended surface. It is used to provide mechanical support to two walls maintained at different temperatures. Due to this difference in temperature, a temperature gradient exists, say, along the x-direction. As a result, the transfer of heat takes place by the conduction process internally, and at the same time there is transfer of energy by convection from the surface. The direction of transfer of heat from the boundaries in an extended surface is perpendicular to the principal direction of transfer of heat in the solid.

The **extended surface** is frequently used in many practical applications. One such use of **extended surface** is to enhance the heat transfer rate between a solid and an adjoining fluid. This is known as **fin**. The fins take heat from the hot object by the process of conduction and give away heat by the process of convection (dominant process) and radiation. At constant temperature, the rate of heat transfer from a hot object can be enhanced by increasing the velocity of the fluid flowing over the surface of the hot object and also by decreasing the temperature of the surroundings.

Fins are used in various fields of applications. Some of these practical fields are the following:

1. Fins are used in the cooling arrangement for the engine heads on motorcycles and lawn mowers.

2. Fins are used for cooling electric power transformers.

3. Fins are attached to the tubes of an air conditioner to promote the exchange of heat between air and the working fluid.

6.2.5 Fourier equation for conduction of heat

The equation for the conduction of heat is generally a partial differential equation that describes the distribution of heat (or the temperature field) in a given substance over time. The concept of this distribution of temperature field plays a very important role in the process of thermal conduction through various substances. Fourier equation is one such differential equation to compute the conduction heat flux at any point in the substance or on its surface once the distribution of the temperature field is known.

The law of conduction of heat is known as Fourier's law. This law states that **the time rate of transfer of heat through a material is proportional to the negative gradient in the temperature and to the area at right angles to that gradient through which the heat flows.** This law can also be formulated in two equivalent forms: (i) The integral form that takes into account the amount of energy flowing into or out of a body as a whole, and (ii) the differential form that takes into account the flow rates or fluxes of energy locally. It should be pointed out that the discrete analog of Fourier's law is Newton's law of cooling, while the electrical analog of Fourier's law is Ohm's law.

6.2.6 Differential form of Fourier heat flow equation

The heat flux density is defined as the amount of thermal energy that flows through a unit area per unit time. According to the differential form of Fourier's law of thermal conduction, this local heat flux density q is equal to the product of thermal conductivity K, and the negative local temperature gradient $-\nabla\theta$. Hence, this q is given by

$$q = -K\nabla\theta. \tag{6.7}$$

In these types of problems, the thermal conductivity K is often assumed to be a constant. But K is found to vary with the following factors:

1. The thermal conductivity K of a material varies with temperature. For some common materials, this variation can be small over a significant range of temperatures.

2. The thermal conductivity K varies typically with orientation in the anisotropic materials. In this case, K is represented by a second-order tensor.

3. Further, it is observed that K varies with the spatial location in nonuniform materials.

For many simple applications, Fourier's law is used in its one-dimensional form. In the X-direction,

$$Q = -KA\frac{d\theta}{dx}. \tag{6.8}$$

In an isotropic medium, Fourier's law leads to the heat equation:

$$\frac{\partial u}{\partial t} = \alpha\left(\frac{\partial^2 u}{\partial x^2} + \frac{\partial^2 u}{\partial y^2} + \frac{\partial^2 u}{\partial z^2}\right), \tag{6.9}$$

with a fundamental solution famously known as the heat kernel.

6.2.7 Integral form of Fourier's heat flow equation

By integrating the differential form over the material's total surface area A, the integral form of Fourier's law is obtained as

$$\frac{\partial Q}{\partial t} = -K\,A\int\nabla\theta\cdot dA, \tag{6.10}$$

where $\dfrac{\partial Q}{\partial t}$ is the amount of heat transferred per unit time (in W), and dA is an oriented surface area element (in m^2). The rate of heat flow for a homogeneous material of one-dimensional geometry between two endpoints at constant temperature can be obtained by integrating the above differential equation. This leads to

$$\frac{Q}{\Delta t} = -KA\frac{\Delta\theta}{\Delta x}, \tag{6.11}$$

where A is the cross-sectional surface area, $\Delta\theta$ is the temperature difference between the ends, and Δx is the distance between the ends. This law forms the basis for the derivation of the heat equation.

The following points are to be noted about the above equation:

1. The negative sign in Fourier's equation indicates that the flow of heat takes place in the direction of a negative gradient of temperature, and this serves to make the **flow of heat** positive.

2. The thermal conductivity K is one of the transport properties. For example, the viscosity is associated with the transport of momentum, and the diffusion coefficient is associated with the transport of mass.

3. The thermal conductivity K provides an indication of the rate at which heat energy is transferred through a medium by the conduction process.

6.2.8 Derivation of one-dimensional heat-flow equation: Fourier theory

Fourier derived an equation related to the flow of heat due to mainly the process of conduction. This law states that the time rate of transfer of heat through a substance is found to be proportional to the negative temperature gradient as well as to the area, which is at right angles to the temperature gradient through which the heat flows. It has two equivalent forms–differential and integral. This law is based on experimental observations and expresses an empirical relationship between the involved thermal quantities. It should be noted that the transfer of heat in various processes is expressed in the form of rate equations. The rate equation for conduction of heat is deduced based on Fourier's law of thermal conduction.

In order to derive Fourier's heat flow equation, the following assumptions are made:

1. The flow of heat through the process of conduction is considered in the steady state only.

2. One directional flow of heat is only considered.

3. The bounding surfaces of the material are maintained at uniform temperatures at the two end faces, that is, these surfaces are isothermal in character.

4. The material under consideration is isotropic and homogeneous, and the thermal conductivity K of the material is assumed to be constant throughout.

5. The temperature gradient across the length of the material is assumed to be constant, and a linear temperature profile is maintained throughout.

6. No heat is generated internally.

Let us consider a thin rod of length L and cross-sectional area A (perpendicular to the diagram). The rod is heated at one end ($x = 0$). This is shown in Figure 6.1. Further, we consider a very thin slice of the rod with thickness δx at a distance x. Let the amount of heat flowing in time δt across the face at a distance x from the left be $Q(x)$. This heat will be utilized in three ways: (i) some part of it will be transmitted through the face at $x + \delta x$, (ii) some part will be absorbed by the thin slice, and (iii) the rest of the heat is radiated to the surrounding from the outer surface of the rod. According to Fourier's law, $Q(x)$ is given by

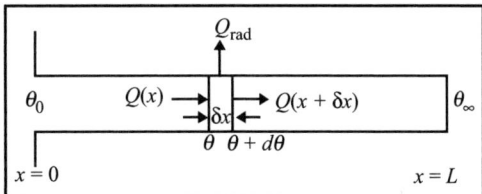

Figure 6.1 Schematic representation of heat flow in one dimension.

$$Q(x) = -KA\left(\frac{\partial \theta}{\partial x}\right)_x \delta t. \tag{6.12}$$

The amount of heat that flows across the face at $x + \delta x$ is given by

$$Q(x + \delta x) = -KA\left(\frac{\partial \theta}{\partial x}\right)_{x+\delta x} \delta t. \tag{6.13}$$

So, the amount of heat that remains within the slice in time δt is

$$Q(x) - Q(x + \delta x) = \left[-KA\left(\frac{\partial \theta}{\partial x}\right)_x \delta t\right] - \left[-KA\left(\frac{\partial \theta}{\partial x}\right)_{x+\delta x} \delta t\right],$$

$$\implies \quad Q(x) - Q(x + \delta x) = KA\delta t\left[-\left(\frac{\partial \theta}{\partial x}\right)_x + \left(\frac{\partial \theta}{\partial x}\right)_{x+\delta x}\right]. \tag{6.14}$$

Here, the operator "δ" is used because θ in the tiny volume will depend on both position x and time t.

Using Taylor's series expansion, we can expand $\left(\frac{\partial \theta}{\partial x}\right)_{x+\delta x}$ as

$$\left(\frac{\partial \theta}{\partial x}\right)_{x+\delta x} = \left(\frac{\partial \theta}{\partial x}\right)_x + \frac{\partial^2 \theta}{\partial x^2}\delta x + \frac{\partial^3 \theta}{\partial x^3}\frac{(\delta x)^2}{2!} + \cdots.$$

This simply means that any function can be approximated as linear when observed at a sufficiently small scale. Keeping only the first-order term in δx, we have

$$\left(\frac{\partial \theta}{\partial x}\right)_{x+\delta x} \approx \left(\frac{\partial \theta}{\partial x}\right)_{x} + \frac{\partial^2 \theta}{\partial x^2} \delta x. \tag{6.15}$$

Using equation (6.15) in equation (6.14), we get

$$Q(x) - Q(x+\delta x) = KA\delta t \left[-\left(\frac{\partial \theta}{\partial x}\right)_{x} + \left(\frac{\partial \theta}{\partial x}\right)_{x} + \frac{\partial^2 \theta}{\partial x^2} \delta x\right] = KA\delta t \; \delta x \; \frac{\partial^2 \theta}{\partial x^2}.$$

So, the amount of heat that remains within the thin slice per unit time is given by

$$\frac{Q(x) - Q(x+\delta x)}{\delta t} = KA \; \delta x \; \frac{d^2 \theta}{dx^2}. \quad \text{[Here, the total differential is used.]} \tag{6.16}$$

The amount of heat described by equation (6.16) is partly used to raise the temperature of the thin slice, while the remainder is radiated from its exposed surface.

If the rate of increase of temperature is $\frac{d\theta}{dt}$, then the heat required for heating the slice is

$$Q_{\text{absor}} = A \; \rho c \; \delta x \; \frac{d\theta}{dt}, \tag{6.17}$$

where ρ is the density of the material, and c is the specific heat of the material.

The amount of heat radiated Q_{rad} from the exposed surface of the slice per unit time is proportional to the difference in temperature between the body and the surroundings. So, if E be the emissive power of the surface, p the sectional perimeter, and θ_e the excess temperature of the surface over the surroundings, then the amount of heat lost by radiation is given by

$$Q_{\text{rad}} = Ep \; \theta_e \; \delta x. \tag{6.18}$$

Evidently, the sum of equations (6.17) and (6.18) will be equal to the result of equation (6.16). Hence, we have

$$KA \; \delta x \; \frac{d^2 \theta}{dx^2} = Q_{\text{absor}} + Q_{\text{rad}} = A \; \rho c \; \delta x \; \frac{d\theta}{dt} + Ep \; \theta_e \; \delta x. \tag{6.19}$$

But, we have $\theta_e = \theta - \theta_s;$ $\Rightarrow$ $\theta = \theta_e + \theta_s$, where θ_s is the temperature of the surroundings and is assumed to be constant. This leads to

$$\frac{d^2 \theta}{dx^2} = \frac{d^2 \theta_e}{dx^2} \quad \text{and} \quad \frac{d\theta}{dt} = \frac{d\theta_e}{dt}.$$

Using these results in equation (6.19), we get

$$KA \, \delta x \, \frac{d^2\theta_e}{dx^2} = A \, \rho c \, \delta x \, \frac{d\theta_e}{dt} + Ep \, \theta_e \, \delta x \qquad \Rightarrow \qquad \frac{d\theta_e}{dt} = \frac{K}{\rho c} \frac{d^2\theta_e}{dx^2} - \frac{Ep}{\rho Ac} \theta_e.$$

This leads to

$$\frac{d\theta_e}{dt} = h \frac{d^2\theta_e}{dx^2} - \mu \, \theta_e, \qquad (6.20)$$

where $h = \frac{K}{\rho c}$ = a constant, is known as the thermometric conductivity, and $\mu = \frac{EP}{\rho Ac}$. Equation (6.20) is the standard **Fourier equation** for one-dimensional flow of heat due to the process of conduction.

Characteristic features of Fourier's equation

Fourier's equation has the following characteristic features:

1. This equation is applicable to all forms of matter, such as solid, liquid, and gas.

2. The vector expression of Fourier's equation indicates that the rate of flow of heat is normal to a given isotherm and occurs in the direction of decreasing temperature.

3. This equation cannot be derived from any first principle; it is a generalization based on experimental evidence.

4. This equation can be used to define the thermal property of a material in terms of the thermal conductivity K.

Solution of Fourier's equation in the steady state

In the steady state, $\frac{d\theta_e}{dt} = 0$ and equation (6.20) reduces to

$$h \frac{d^2\theta_e}{dx^2} = \mu \, \theta_e, \qquad \Rightarrow \qquad \frac{d^2\theta_e}{dx^2} = m^2 \, \theta_e, \qquad (6.21)$$

where $m = \sqrt{\frac{\mu}{h}}$. The general solution of equation (6.21) is

$$\theta_e = Ae^{mx} + Be^{-mx}, \qquad (6.22)$$

where A and B are constants whose values depend on the boundary conditions of the specific problem under consideration.

An example

Let us examine the following boundary conditions for the one-dimensional heat flow through a conducting rod of length L:

1. At $x = 0$, $\theta_e = \theta_0$. This θ_0 is the temperature of the source in excess of the surroundings maintained at constant temperature θ_s.

2. At $x = \infty$, $\theta_e = 0$, that is, $\theta = \theta_s$.

3. At $x = L$, $\theta_e = \theta_L$, where θ_L is the excess temperature at the other end of the rod $(x = L)$.

Applying boundary conditions (1) and (2) to equation (6.22), we get

$$\theta_0 = A + B \quad \text{and} \quad 0 = Ae^\infty + Be^{-\infty}. \tag{6.23}$$

This leads to $A = 0$ and $B = \theta_0$. Equation (6.22) takes the form $\theta_e = \theta_0\, e^{-mx}$. Applying boundary condition (3) to this expression, we get

$$\theta_L = \theta_0\, e^{-L\,m} = \theta_0\, e^{-\sqrt{\frac{\mu}{h}}\,L}. \tag{6.24}$$

At any point x on the rod, the excess temperature is found to be

$$\theta_e = \theta_0\, e^{-\sqrt{\frac{\mu}{h}}\,x}. \tag{6.25}$$

Equation (6.25) shows that in such a case, the temperature decreases exponentially with distance x. This exponential decrease is controlled by the physical parameters, such as thermal conductivity, thermal diffusivity, and specific heat.

Solution of Fourier's equation for a lagged bar

A bar is said to be "lagged" when it is insulated in such a way that no heat is lost to the surroundings through its sides. In such a bar, the temperature distribution is uniform along the length of the conductor. Since the escape of heat through the sides of the lagged bar is negligible, the rate of flow of heat along the bar is safely assumed to be constant. Further, the temperature gradient along the bar is greatest when the rate of flow of heat is greatest.

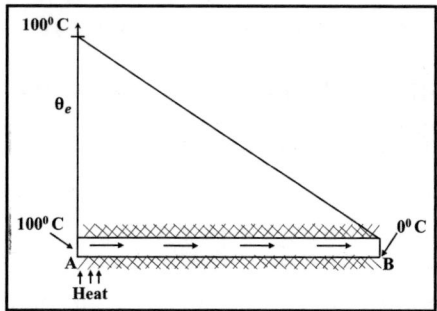

Figure 6.2 Temperature distribution in a lagged bar.

Fourier's equation can be used to get the temperature distribution in such a lagged bar in the following way: In a lagged bar, as there is no loss of heat due to radiation, the terms $\frac{d\theta_e}{dt}$ and μ in equation (6.20) would be zero. Hence, from this equation we have

$$h\,\frac{d^2\theta_e}{dx^2} = 0; \quad \Rightarrow \quad h\,\frac{d^2\theta}{dx^2} = 0. \tag{6.26}$$

Here, θ is the actual temperature of the bar at a distance x. Integrating this equation twice with respect to x, we get

$$\theta = B_1\,x + B_2,\tag{6.27}$$

where B_1 and B_2 are two constants, and their values are determined from the boundary conditions. In the lagged bar, the boundary conditions are

1. At $x = 0$, $\theta = \theta_0$. (Actual temperature of the bar at the hot end.)
2. At $x = L$, $\theta = \theta_L$. (Actual temperature of the bar at the other end.)

Using these boundary conditions, we get the values of the constants B_1 and B_2 as

$$B_1 = -\left(\frac{\theta_0 - \theta_L}{L}\right)\text{ and }B_2 = \theta_0.$$

Using these values of B_1 and B_2 in equation (6.27), we get the temperature θ at a distance x from the hot end of the lagged bar as

$$\theta = \theta_0 - \frac{\theta_0 - \theta_L}{L}\,x.\tag{6.28}$$

This equation (6.28) shows that the temperature in a lagged bar decreases linearly with distance x from the hot end. This is graphically shown in Figure 6.2. Further, it is to be noted that the temperature gradient, in this case, is constant (negative) and depends on the temperatures at both ends of the bar and its length L.

6.2.9   Cylindrical flow of heat

Cylindrical metal tubes are widely used as essential components in various technological fields, including power plants, oil refineries, and numerous process industries. These industries rely on a significant number of cylindrical metal tubes for their operations. For example, the boilers have

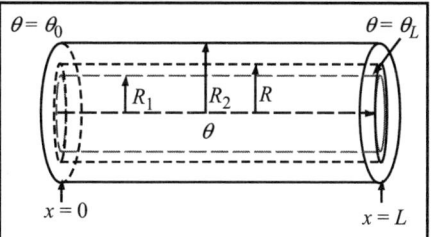

Figure 6.3 Cylindrical flow of heat.

cylindrical tubes in them, the condensers contain a bank of cylindrical tubes, the exchangers for heat are tubular, and all such units are connected by cylindrical tubes. Evidently, then the radial heat transfer rate through a cylindrical tube or any insulation that may surround it is quite important in practical fields. As basic information, the expression for thermal conductivity, and the amount of heat flow per second through such a cylindrical tube are derived below.

To derive the expression for the radial flow of heat through a cylindrical tube with a heat source along its axis, the following assumptions are made:

1. Conduction of heat is considered in the steady state.

2. Conduction of heat is considered only radially, and hence, one-dimensional heat flow equation can be applied.

3. The thermodynamic properties of the material of the cylindrical tube remain constant.

4. The source of volumetric generation of heat placed along the axis of the tube is assumed to be uniform.

5. There is no loss of heat from the outer surface of the tube, that is, the outer surface of the tube is adiabatic.

We consider the flow of heat through a cylindrical tube of inner radius R_1 and outer radius R_2. The length of the tube is L. The inside and outside surfaces of the tube are at constant temperatures θ_1 and θ_2, respectively. Both ends of the tube are kept perfectly insulated, and further it is assumed that the thermal conductivity K of the material of the tube is constant within the given temperature range. The temperature θ_1 at the inner surface is greater than the temperature θ_2 at the outer surface, and as a result, the heat flows radially outwards. Such a cylindrical tube with an indication of radii and the corresponding temperatures is shown in Figure 6.3.

We consider an infinitesimally thin cylindrical element at radius R, where $R_1 < R < R_2$. Let dR be the thickness of this elementary cylindrical element, and $d\theta$ be the change of temperature across it. As the cylinder is symmetrical about its axis, the isothermal surfaces are also cylindrical. Heat may be considered to flow across a surface represented by

$$R^2 = x^2 + y^2. \tag{6.29}$$

In this case, Fourier's equation in Cartesian coordinates is given by

$$\frac{\partial \theta}{\partial t} = h \left[\frac{\partial^2 \theta}{\partial x^2} + \frac{\partial^2 \theta}{\partial y^2} \right], \tag{6.30}$$

where θ stands for the temperature at a distance R from the axis of the tube, and h is the thermal diffusivity of the material.

It is advantageous here to work in a cylindrical coordinate system. So, equation (6.30) has to be converted to its cylindrical form. This is given by

$$\frac{\partial \theta}{\partial t} = h \left[\frac{2}{R} \frac{d\theta}{dR} - \frac{1}{R} \frac{d\theta}{dR} + \frac{d^2\theta}{dR^2} \right] = h \left[\frac{1}{R} \frac{d\theta}{dR} + \frac{d^2\theta}{dR^2} \right]. \tag{6.31}$$

In the steady state, $\frac{\partial \theta}{\partial t} = 0$. Hence, equation (6.31) in the steady state can be written as

$$\frac{1}{R}\frac{d\theta}{dR} + \frac{d^2\theta}{dR^2} = 0; \quad \Rightarrow \quad \frac{d\theta}{dR} + R\frac{d^2\theta}{dR^2} = 0; \quad \Rightarrow \quad \frac{d}{dR}\left(R\frac{d\theta}{dR}\right) = 0. \quad (6.32)$$

Integrating this equation (6.32) with respect to R, we get

$$\left(R\frac{d\theta}{dR}\right) = B_1; \quad \Rightarrow \quad \left(\frac{d\theta}{dR}\right) = \frac{B_1}{R}, \quad (6.33)$$

where B_1 is the constant of integration. Integrating this equation with respect to R again, we get

$$\theta = B_1 \ln R + B_2, \quad (6.34)$$

where B_2 is the constant of integration. The values of B_1 and B_2 are to be determined from the given boundary conditions. The boundary conditions are:

$$\theta = \theta_1 \text{ at } R = R_1 \text{ and } \theta = \theta_2 \text{ at } R = R_2.$$

Using these boundary conditions in equation (6.34), we get the values of B_1 and B_2 as

$$B_1 = \frac{(\theta_1 - \theta_2)}{\ln\left(\frac{R_1}{R_2}\right)} \text{ and } B_2 = \frac{(\theta_2 \ln R_1 - \theta_1 \ln R_2)}{\ln\left(\frac{R_1}{R_2}\right)}. \quad (6.35)$$

Using these values of B_1 and B_2 in equation (6.34), the temperature θ at a distance R from the axis of the cylinder comes out to be

$$\theta = \frac{1}{\ln\left(\frac{R_2}{R_1}\right)}\left[(\theta_1 \ln R_2 - \theta_2 \ln R_1) - (\theta_1 - \theta_2)\ln R\right]. \quad (6.36)$$

This equation (6.36) gives the temperature distribution along the radial direction in the case of a cylindrical tube.

Amount of heat flowing through a cylindrical conductor

The amount of heat flowing per second $\dot{Q}$, across the cylindrical isothermal surfaces of radius R and $R + dR$ is given by

$$\dot{Q} = -KA\frac{d\theta}{dR}, \quad (6.37)$$

where A is the conducting area, K is the thermal conductivity, θ is the temperature at R, and $\frac{d\theta}{dR}$ is the temperature gradient (which is negative).

Again, from equation (6.33), we have $\frac{d\theta}{dR} = \frac{B_1}{R}$ and the cross-sectional area is $A = 2\pi RL$. Using these values, the amount of heat flowing per second along the radial direction is given by

$$\dot{Q} = -K \ 2\pi RL \ \frac{B_1}{R} = -2\pi KLB_1.$$

Using the value of B_1 from equation (6.35), this expression comes out to be

$$\dot{Q} = -2\pi KL \ \frac{(\theta_1 - \theta_2)}{\ln\left(\frac{R_1}{R_2}\right)} = 2\pi KL \ \frac{(\theta_1 - \theta_2)}{\ln\left(\frac{R_2}{R_1}\right)}. \tag{6.38}$$

This equation (6.38) provides the rate of radial flow of heat in a cylindrical tube. From this equation, we get the value of thermal conductivity K to be

$$K = \frac{\dot{Q}\ln(R_2/R_1)}{2\pi L(\theta_1 - \theta_2)}. \tag{6.39}$$

This is an important equation frequently used in industry for the calculation of the outgoing amount of heat in various technological arrangements. This equation (6.39) can be used to determine the value of thermal conductivity K if the values of $\dot{Q}$, R_1, R_2, and L are known.

Thermal resistance R_{cyl} of a cylindrical conductor

Equation (6.38) can be rearranged as

$$\dot{Q} = \frac{(\theta_1 - \theta_2)}{\frac{\ln\left(\frac{R_2}{R_1}\right)}{2\pi KL}} = \frac{\theta_1 - \theta_2}{R_{\text{cyl}}}, \tag{6.40}$$

where R_{cyl} is the thermal resistance of the cylindrical conductor against heat conduction and is given by

$$R_{\text{cyl}} = \frac{\ln\left(\frac{R_2}{R_1}\right)}{2\pi KL}. \tag{6.41}$$

Thus, the thermal resistance R_{cyl} of a cylindrical conductor depends on the geometrical factors such as inner and outer radii R_1 and R_2, length L of the cylindrical conductor, and the thermal conductivity K of the material of the cylindrical conductor.

Conduction through a composite cylindrical wall

It should be mentioned that analysis of the steady-state conduction in a cylindrical tube can be straightforwardly extended to multiple layers in the cylindrical geometry.

For example, consider a cylindrical tube of length L carrying hot or cold fluid that needs to be insulated from the surroundings. We add insulation layers on both sides of the tube. Such a geometry is shown in Figure 6.4.

In the above figure, the region between r_1 and r_2 with thermal conductivity K_1 acts as inner insulation to the region between r_2 and r_3 with thermal conductivity K_2. Similarly, the region between r_3 and r_4 with thermal conductivity K_3 acts as outer insulation to the region between r_2 and r_3. Following the earlier procedure, we can write the rate of flow of heat Q in the steady state from the interior wall of the pipe with thermal conductivity K_2 to the outside surface of the insulation

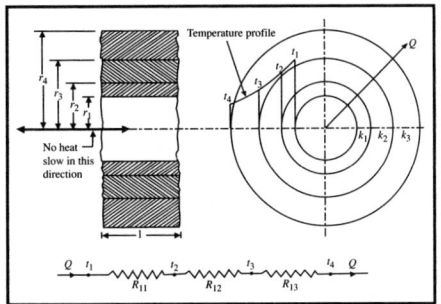

Figure 6.4 Conduction through a composite cylindrical wall.

with thermal conductivity K_3 in terms of the temperature difference for conduction in each of the two layers and the corresponding resistance of each layer.

Considering the steady-state conduction from the main material to the outer insulation, we get

$$\dot{Q} = \frac{\theta_2 - \theta_3}{R_{\text{tube}}} = \frac{\theta_3 - \theta_4}{R_{\text{ins}}},$$

where $R_{\text{tube}} = \dfrac{\ln\left(\frac{r_3}{r_2}\right)}{2\pi K_2 L}$ and $R_{\text{ins}} = \dfrac{\ln\left(\frac{r_4}{r_3}\right)}{2\pi K_3 L}$.

6.2.10   Flow of heat through a spherical conductor

Let us consider the flow of heat through a spherical conductor of inner radius r_1 and outer radius r_2. The inside and outside surfaces of the conductor are respectively at constant temperatures θ_1 and θ_2. We assume that the temperature θ_1 at the inner surface is greater than the temperature θ_2 at the outer surface, and the heat flows radially outwards. Further, it is assumed that the thermal conductivity K of the material of the spherical conductor is constant within the given temperature range. Such a

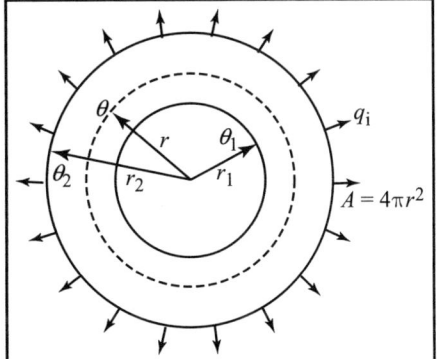

Figure 6.5 Spherical flow of heat.

spherical conductor with a mention of radii and the corresponding temperatures is schematically shown in Figure 6.5.

Consider an infinitesimally thin spherical element at radius r with temperature θ. Let dr be the thickness of this elementary ring, and $d\theta$ be the change in temperature across it. As the spherical conductor is symmetrical about its axis, the isothermal surfaces here are cylindrical in shape. Thus, in this case, heat may be considered to flow across a surface represented by

$$r^2 = x^2 + y^2 + z^2. \tag{6.42}$$

Fourier's equation in Cartesian coordinates can be written as

$$\frac{\partial \theta}{\partial t} = h \left[\frac{\partial^2 \theta}{\partial x^2} + \frac{\partial^2 \theta}{\partial y^2} + \frac{\partial^2 \theta}{\partial z^2} \right], \tag{6.43}$$

where θ stands for the temperature at a distance r from the center of the spherical conductor, and h is the diffusivity of the material.

In this problem, it is advantageous here to work in a spherical polar coordinate system. Hence, equation (6.43) has to be converted to the corresponding spherical form. This is given by

$$\frac{\partial \theta}{\partial t} = h \left[\frac{2}{r} \frac{d\theta}{dr} + \frac{d^2 \theta}{dr^2} \right]. \tag{6.44}$$

In the steady state, we have $\frac{\partial \theta}{\partial t} = 0$. So, equation (6.44) can be written as

$$\frac{2}{r} \frac{d\theta}{dr} + \frac{d^2 \theta}{dr^2} = 0; \quad \Rightarrow \quad 2r \frac{d\theta}{dr} + r^2 \frac{d^2 \theta}{dr^2} = 0; \quad \Rightarrow \quad \frac{d}{dr} \left(r^2 \frac{d\theta}{dr} \right) = 0. \tag{6.45}$$

Integrating this equation with respect to r, we get

$$r^2 \frac{d\theta}{dr} = B_1 \text{ (constant)}; \quad \Rightarrow \quad d\theta = \frac{B_1}{r^2} dr,$$

where B_1 is the constant of integration. Integrating this expression again with respect to r, we get

$$\theta = -\frac{B_1}{r} + B_2, \tag{6.46}$$

where B_2 is another constant of integration. The values of the constants of integration B_1 and B_2 are to be determined from the boundary conditions that are given as

$$\theta = \theta_1 \text{ when } r = r_1 \text{ and } \theta = \theta_2 \text{ when } r = r_2.$$

Using these boundary conditions in equation (6.46), we get the values of B_1 and B_2 as

$$B_1 = \frac{\theta_1 - \theta_2}{r_1 - r_2} r_1 r_2 \quad \text{and} \quad B_2 = \theta_1 + \frac{\theta_1 - \theta_2}{r_1 - r_2} r_2. \tag{6.47}$$

Using these values of the constants B_1 and B_2 in equation (6.46), we get the temperature θ at a distance r from the center of the spherical conductor as

$$\theta = \frac{1}{r_2 - r_1}\left[\left(\frac{(\theta_1 - \theta_2)r_1 r_2}{r}\right) + (\theta_2 r_2 - \theta_1 r_1)\right]. \tag{6.48}$$

This gives the temperature distribution along the radial direction of the spherical conductor.

Amount of heat flowing through a spherical conductor

The amount of heat flowing through a spherical conductor per second $\dot{Q}$, across the spherical isothermal surfaces of radius r and $r + dr$ is given by

$$\dot{Q} = -KA\frac{d\theta}{dr}, \tag{6.49}$$

where A is the conducting area $4\pi r^2$, K is the thermal conductivity, θ is the temperature at r, and $\frac{d\theta}{dr}$ is the temperature gradient (which is negative) $\frac{B_1}{r^2}$.

Using the value of B_1, the amount of heat flowing per second is given by

$$\dot{Q} = K\, 4\pi r^2\, \frac{B_1}{r^2} = 4\pi K B_1 = 4\pi K\, \frac{(\theta_1 - \theta_2)}{r_2 - r_1}r_1 r_2. \tag{6.50}$$

Simplifying this equation, we get the expression for thermal conductivity K as

$$K = \frac{\dot{Q}(r_2 - r_1)}{4\pi r_1 r_2(\theta_1 - \theta_2)}. \tag{6.51}$$

This is an important equation. It helps us to determine the value of K if the values of $\dot{Q}$, r_1, and r_2 are known.

Thermal resistance R_{sph} of a spherical conductor

Equation (6.50) can be rearranged as

$$\dot{Q} = \frac{\theta_1 - \theta_2}{\frac{r_2 - r_1}{4\pi K r_1 r_2}} = \frac{\theta_1 - \theta_2}{R_{sph}}, \tag{6.52}$$

where R_{sph} is the thermal resistance of the spherical conductor against heat conduction and is given by

$$R_{sph} = \frac{r_2 - r_1}{4\pi K r_1 r_2}. \tag{6.53}$$

Thus, the thermal resistance R_{sph} of a spherical conductor depends on the geometrical factors like inner and outer radii r_1 and r_2, and the thermal conductivity K of the material of the spherical conductor.

6.3   One-dimensional flow of heat

We consider a homogeneous bar of uniform cross-section a. Further, we assume that the stream lines of the heat flow are all parallel and perpendicular to the cross-section a. The origin of the coordinate axis is taken at one end of the bar, and the direction of flow of heat is considered to be along the positive x-axis. Let ρ be the density, c the specific heat, and K the thermal conductivity of the material used for the study of one-dimensional heat flow. Such a material is shown schematically in Figure 6.6.

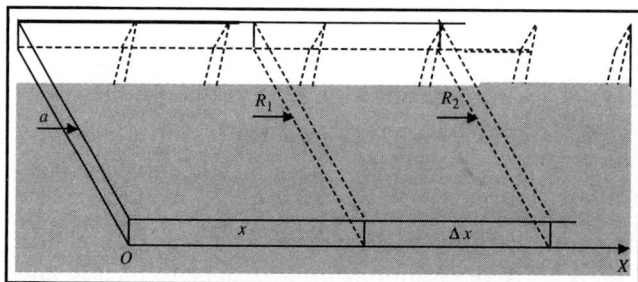

Figure 6.6 One-dimensional heat flow where the stream lines of the heat flow are all parallel and perpendicular to the area.

Let $u(x, t)$ be the temperature at a distance x from O. If Δu is the change in temperature in a slab of thickness Δx of the bar, then the quantity of heat in this slab is $c\rho a \Delta x \Delta u$. Hence, $c\rho a \Delta x(\partial u/\partial t) = R_1 - R_2$, where R_1 and R_2 represent the rate of inflow and outflow of heat, respectively. Now

$$R_1 = -Ka\left(\frac{\partial u}{\partial x}\right)_x, \qquad R_2 = -Ka\left(\frac{\partial u}{\partial x}\right)_{x+\Delta x}. \tag{6.54}$$

Thus,

$$c\rho a \Delta x \frac{\partial u}{\partial t} = -Ka\left(\frac{\partial u}{\partial x}\right)_x + Ka\left(\frac{\partial u}{\partial x}\right)_{x+\Delta x}$$

$$\frac{\partial u}{\partial t} = \frac{K}{c\rho}\left[\frac{(\partial u/\partial x)_{x+\Delta x} - (\partial u/\partial x)_x}{\Delta x}\right]. \tag{6.55}$$

Now, taking the limit as $\Delta x \to 0$, we get

$$\frac{\partial u}{\partial t} = h^2 \frac{\partial^2 u}{\partial x^2}. \tag{6.56}$$

Here, $h^2 = K/(c\rho)$ is called the *diffusivity* of the material. Equation (6.56) is called the *one-dimensional heat flow equation.*

To solve equation (6.56), we assume that the solution is of the form

$$u(x,t) = X(x)T(t). \tag{6.57}$$

Now, X is a function of x only, and T is a function of t only. Substituting this into equation (6.56), we get

$$\frac{d^2 X}{dx^2} - k_1 X = 0, \quad \frac{dT}{dt} - k_1 h^2 T = 0. \tag{6.58}$$

Here, we are dealing with the problem of heat condition; the solution of equation (6.58) must be a transient solution, that is, u must decrease with time. Solving ordinary differential equation (6.58) by taking $k_1 = -\omega^2$, we get

$$u = (c_1 \cos \omega x + c_2 \sin \omega x) e^{-h^2 \omega^2 t}, \tag{6.59}$$

which is the only possible solution of equation (6.56). Here, c_1 and c_2 are two constants.

Example-1: Consider an insulated rod of length L. Its ends A and B are maintained, respectively, at temperatures 0°C and 100°C until the steady-state conditions are achieved. If the temperature of the end B is suddenly reduced to 0°C and the end A is still maintained at 0°C, find the temperature at a distance x from the end A at an instant of time t.

Solution: We know that the solution of equation (6.56) of the heat conduction is given by equation (6.59).

Now, prior to the change in temperature at the end B, when $t = 0$, the heat flow is independent of time (steady-state condition). When u depends on x only, equation (6.56) reduces to

$$\frac{\partial^2 u}{\partial x^2} = 0, \tag{6.60}$$

whose general solution is

$$u = Ax + B. \tag{6.61}$$

As $u = 0$ and $u = 100$ for $x = 0$ and $x = L$, respectively, equation (6.61) gives $B = 0$ and $A = 100/L$. Thus, the initial conditions are

$$u(x,0) = \frac{100}{L}x. \tag{6.62}$$

The boundary conditions for the subsequent flow are

$$u(0,t) = 0, \text{ for all } t, \tag{6.63}$$

and

$$u(L, t) = 0, \text{ for all } t. \tag{6.64}$$

We now find the solution of equation (6.56) subject to the conditions (6.62)–(6.64). The solution of equation (6.56) is

$$u(x, t) = (c_1 \cos \omega x + c_2 \sin \omega x)e^{-h^2\omega^2 t}. \tag{6.65}$$

From equation (6.63), $u(0, t) = 0 = c_1 e^{-h^2\omega^2 t}$. Hence, $c_1 = 0$ and equation (6.65) reduces to

$$u(x, t) = c_2(\sin \omega x)e^{-h^2\omega^2 t}, \tag{6.66}$$

which, on applying equation (6.64), gives $u(L, t) = c_2 \sin \omega L e^{-h^2\omega^2 t} = 0$ and this requires $\sin \omega L = 0$, that is, $\omega L = n\pi$, as $c_2 \neq 0$. Thus, $\omega = \frac{n\pi}{L}$, where n is an integer. Hence, equation (6.66) becomes

$$u(x, t) = b_n \sin \frac{n\pi x}{L} e^{-n^2\pi^2 h^2 t/L^2}, \qquad b_n = c_2. \tag{6.67}$$

Adding all such solutions, the most general solution of equation (6.56) satisfying equations (6.63) and (6.64) is

$$u(x, t) = \sum_{n=1}^{\infty} b_n \sin \frac{n\pi x}{L} e^{-h^2 n^2 \pi^2 t/L^2}. \tag{6.68}$$

Put $t = 0$ to obtain

$$u(x, 0) = \sum_{n=1}^{\infty} b_n \sin \frac{n\pi x}{L}. \tag{6.69}$$

Now, in order that the condition (6.62) be satisfied, equations (6.62) and (6.69) must be the same. This needs the expansion of $100x/L$ as the Fourier series in $(0, L)$. Thus,

$$\frac{100x}{L} = \sum_{n=1}^{\infty} b_n \sin \frac{n\pi x}{L},$$

where

$$b_n = \frac{2}{L} \int_0^L \frac{100x}{L} \sin \frac{n\pi x}{L} dx = \frac{200}{L^2} \frac{-L^2}{n\pi} \cos n\pi = \frac{200(-1)^{n+1}}{n\pi}.$$

Thus, the most general solution (6.68) becomes

$$u(x, t) = \frac{200}{\pi} \sum_{n=1}^{\infty} \frac{(-1)^{n+1}}{n} \sin \frac{n\pi x}{L} e^{-h^2 n^2 \pi^2 t/L^2},$$

which is the desired result.

6.4   Two-dimensional flow of heat

Consider the flow of heat in a metal plate of uniform thickness a. Let the density, specific heat, and thermal conductivity be denoted by ρ, c, and K, respectively, and XOY represent one face of the plate. The flow of heat is said to be two-dimensional if the temperature at any point depends only on x and y coordinates and time t, independent of the z-coordinate, and in this case, the flow of heat is in the XY plane and zero along the normal to the plane.

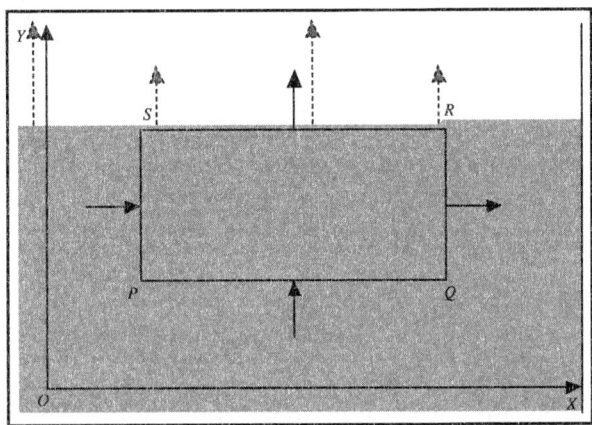

Figure 6.7 Flow of heat in a metal plate of uniform thickness—two-dimensional heat flow.

Consider a rectangular element $PQRS$ as shown in Figure 6.7 with the sides Δx and Δy. The amount of heat entering the element per second from the side $PQ = -Ka\Delta x \left(\frac{\partial u}{\partial y}\right)_y$, and the amount of heat entering the element per second from the side PS is $-Ka\Delta y \left(\frac{\partial u}{\partial x}\right)_x$. Also, the amount of heat coming out through side RS per second is $-Ka\Delta x \left(\frac{\partial u}{\partial y}\right)_{y+\Delta y}$, and the amount of heat coming out through side RQ is $-Ka\Delta y \left(\frac{\partial u}{\partial x}\right)_{x+\Delta x}$.

Thus, the total amount of heat gained by $PQRS$ per second is given by

$$-Ka\Delta x \left(\frac{\partial u}{\partial y}\right)_y - Ka\Delta y \left(\frac{\partial u}{\partial x}\right)_x + Ka\Delta x \left(\frac{\partial u}{\partial y}\right)_{y+\Delta y} + Ka\Delta y \left(\frac{\partial u}{\partial x}\right)_{x+\Delta x}$$

$$= Ka\Delta x\Delta y \left[\frac{\left(\frac{\partial u}{\partial x}\right)_{x+\Delta x} - \left(\frac{\partial u}{\partial x}\right)_x}{\Delta x} + \frac{\left(\frac{\partial u}{\partial y}\right)_{y+\Delta y} - \left(\frac{\partial u}{\partial y}\right)_y}{\Delta y} \right]. \quad (6.70)$$

Again the rate of gain of heat by $PQRS$ is

$$\rho \Delta x \Delta y \left(ac \, \frac{\partial u}{\partial t} \right). \tag{6.71}$$

Equations (6.70) and (6.71) yield

$$K a \Delta x \Delta y \left[\frac{\left(\frac{\partial u}{\partial x} \right)_{x+\Delta x} - \left(\frac{\partial u}{\partial x} \right)_{x}}{\Delta x} + \frac{\left(\frac{\partial u}{\partial y} \right)_{y+\Delta y} - \left(\frac{\partial u}{\partial y} \right)_{y}}{\Delta y} \right] = \rho \Delta x \Delta y \left(ac \, \frac{\partial u}{\partial t} \right),$$

which, on taking limit as $\Delta x \to 0$, $\Delta y \to 0$, becomes

$$K \left(\frac{\partial^2 u}{\partial x^2} + \frac{\partial^2 u}{\partial y^2} \right) = \rho c \, \frac{\partial u}{\partial t}$$

or,

$$\frac{\partial u}{\partial t} = h^2 \left(\frac{\partial^2 u}{\partial x^2} + \frac{\partial^2 u}{\partial y^2} \right), \tag{6.72}$$

where $h^2 = \frac{K}{\rho c}$ is the *diffusivity*. Equation (6.72) represents the distribution of temperature in the plate in the transient state.

6.4.1 Solution of Laplace's equation

In the steady state, u is independent of time t, and $\partial u / \partial t = 0$. Equation (6.72) thus reduces to the well-known *Laplace's equation*.

$$\frac{\partial^2 u}{\partial x^2} + \frac{\partial^2 u}{\partial y^2} = 0 \tag{6.73}$$

in two dimensions.

When the flow of heat is three-dimensional, we get the following equation [in a similar manner as that of equation (6.72)]:

$$\frac{\partial u}{\partial t} = h^2 \left(\frac{\partial^2 u}{\partial x^2} + \frac{\partial^2 u}{\partial y^2} + \frac{\partial^2 u}{\partial z^2} \right), \tag{6.74}$$

which in a steady-state condition yields Laplace's equation

$$\frac{\partial^2 u}{\partial x^2} + \frac{\partial^2 u}{\partial y^2} + \frac{\partial^2 u}{\partial z^2} = 0 \tag{6.75}$$

in three dimensions.

Let $u = X(x)Y(y)$ be a solution of equation (6.73). Substitute it in equation (6.73) to obtain

$$\frac{d^2 X}{dx^2} Y + X \frac{d^2 Y}{dy^2} = 0.$$

Separation of variables yields

$$\frac{1}{X} \frac{d^2 X}{dx^2} = -\frac{1}{Y} \frac{d^2 Y}{dy^2} = A \text{ (constant).} \tag{6.76}$$

The solution of the ordinary differential equation (6.76) depends upon the choice of the constant, and the possible solutions of equation (6.73) are

$$u = (c_1 e^{\omega x} + c_2 e^{-\omega x})(c_3 \cos \omega y + c_4 \sin \omega y), \tag{6.77}$$

[where in equation (6.76) A is positive and $= \omega^2$ (say)]

$$u = (c_5 \cos \omega x + c_6 \sin \omega x)(c_7 \, e^{\omega y} + c_8 \, e^{-\omega y}), \tag{6.78}$$

[where in equation (6.76) A is negative and $= -\omega^2$ (say)]. This gives the value of u as

$$u = (c_9 \, x + c_{10})(c_{11} \, y + c_{12}), \tag{6.79}$$

[where $A = 0$ in (6.76)]. Out of these solutions, we have to choose that solution which is constant with the physical nature of the problem. The solutions of Laplace equation are known as *harmonic functions*.

An example: An infinitely long plane uniform plate is bounded by two parallel edges and an end at right angles to them. The breadth is π: this end is maintained at temperature u_0 at all points, and other edges are at zero temperature (see Figure 6.8). Find the temperature at any point of the plate in the steady state.

Solution: In the steady state, the temperature $u(x, y)$ at any point $P(x, y)$ satisfies the Laplace equation (6.73). The boundary conditions are

$$u(0, x) = 0, \qquad \text{for all } y, \tag{6.80}$$
$$u(\pi, y) = 0, \qquad \text{for all } y, \tag{6.81}$$
$$u(x, \infty) = 0, \qquad \text{in } 0 < x < \pi, \tag{6.82}$$
$$u(x, 0) = u_0, \qquad \text{in } 0 < x < \pi. \tag{6.83}$$

The three possible solutions of equations (6.73) are (6.77)–(6.79). Solution (6.77) cannot verify equation (6.80) as $u \neq 0$ for $x = 0$, for all values of y. The solution (6.79) cannot verify (6.83). Thus, the only possible solution of equation (6.73) is of the form

$$u(x, y) = (A_1 \cos \omega x + A_2 \sin \omega x)(A_3 e^{\omega y} + A_4 e^{-\omega y}). \tag{6.84}$$

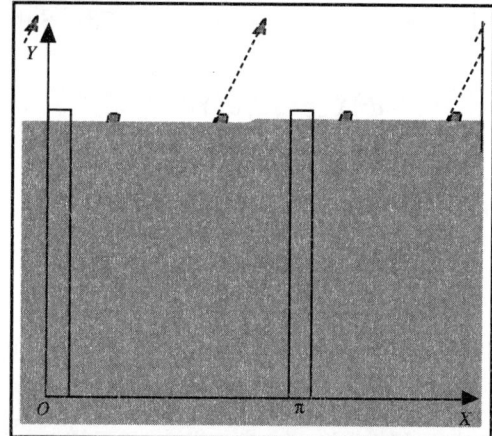

Figure 6.8 Schematic diagram of an infinitely long uniform plate bounded by two parallel edges and an end at right angles to them.

From equation (6.80), we have $u(0, y) = A_1(A_3 e^{\omega y} + A_4 e^{-\omega y}) = 0$. Hence, $A_1 = 0$, and equation (6.84) reduces to

$$u(x, y) = A_2 \sin \omega x (A_3 e^{\omega y} + A_4 e^{-\omega y}). \tag{6.85}$$

By equation (6.81), $u(\pi, y) = A_2 \sin \omega \pi (A_1 e^{\omega y} + A_2 e^{-\omega y}) = 0$, which requires $\sin \omega \pi = 0$, that is, $\omega \pi = n\pi$, as $A_2 = 0$. Thus, $\omega = n$, an integer. Also, in order to satisfy equation (6.82) $A_3 = 0$. Hence, equation (6.85) becomes

$$u(x, y) = B_n \sin nx e^{-ny}, \quad B_n = A_2 A_4.$$

Therefore, the most general solution satisfying (6.80)–(6.82) is of the form

$$u(x, y) = \sum_{n=1}^{\infty} B_n \sin nx e^{-ny}. \tag{6.86}$$

Put $y = 0$,

$$u(x, 0) = \sum_{n=1}^{\infty} B_n \sin nx. \tag{6.87}$$

Now, equations (6.83) and (6.87) must be the same in order that the condition (6.83) is to be satisfied, and this requires the expansion of $u_0 = u(x, 0)$ in a Fourier series in $(0, x)$. Thus,

$$u_0 = \sum_{n=1}^{\infty} B_n \sin nx,$$

where

$$B_n = \frac{2}{\pi} \int\limits_0^\pi u_0 \sin nx dx = \frac{2u_0}{n\pi} \left[1 - (-1)^n\right],$$

that is,

$$B_n = 0 \qquad \text{if } n \text{ is even}$$
$$= \frac{4u_0}{n\pi} \qquad \text{if } n \text{ is odd.}$$

Hence, equation (6.86) takes the form

$$u(x, y) = \frac{4u_0}{\pi} \left(e^{-y} \sin x + \frac{1}{3} e^{-3y} \sin 3x + \frac{1}{5} e^{-5y} \sin 5x + \cdots \right). \qquad (6.88)$$

An example: Consider a π units wide and infinitely long thin plate. Determine the temperature u at any point $P(x, y)$ on the plate assuming that Fourier equation $\partial^2 u/\partial x^2 + \partial^2 u/\partial y^2 = 0$ is valid in the steady state. Further, assume that the short edge of the plate is maintained constantly at temperature unity, and the long edge at temperature zero.

Answer: Take the Y-axis along an infinite edge and the X-axis along the short edge. The boundary conditions are [(a)]

1. $u = 0$, when $x = 0$.

2. $u = 0$, when $x = \pi$.

3. $u = 0$, when $y = \infty$.

4. $u = 1$, when $y = 0$.

These boundary conditions are the same as those of the above example except for the condition given by equation (6.83). Now take $u_0 = 1$ in equation (6.83); then from the above example we have

$$u(x, y) = \frac{4}{\pi} \left(\frac{1}{1} e^{-y} \sin x + \frac{1}{3} e^{-3y} \sin 3x + \cdots \right), \qquad (6.89)$$

which is the required temperature at any point $P(x, y)$.

6.5   Some simple questions with answers

1. Why do metals feel both colder in the winter and hotter in the summer?

   Answer: Materials like metals and stones have high thermal conductivity K and conduct heat well in both ways—into or out of the material. So, when our skin comes into contact with a metal that is colder than the temperature of our skin, heat energy will be transferred from our hand to the metal rapidly, making the metal feel particularly cold. Similarly, if the temperature of the metal is larger than the temperature of our skin, heat energy will be transferred from the metal into our hand rapidly, making the metal feel particularly hot.

For the same reason, concrete will feel especially cold to our bare feet in winter (heat energy is transferred from our feet to the concrete rapidly) and especially hot to our bare feet in summer (heat energy is transferred from the concrete into our feet rapidly).

2. A greenhouse has an enclosure with a high transmission at short wavelengths and a very low transmission (almost opaque) for high wavelengths. Why does a greenhouse get warmer than the surrounding air during clear days? Will it have a similar effect during clear nights?

Answer: On a clear day, the glass of the greenhouse allows a considerable part of the incident radiation to pass through it. The various surfaces (plants, for example) inside the greenhouse reflect the radiation, but the reflected radiation is spectroscopically different, with a higher proportion of long wavelengths. As a result, the reflected radiation does not pass through the glass wall and is reflected back into the greenhouse. Because of the "trapped" radiation, the interior of the greenhouse is heated up. Because there is no Sun radiation on a clear night, the same effect will not be visible.

6.6 Solved numerical problems

1. Calculate the time for the formation of ice 3-cm thick on the surface of a lake when the surrounding air temperature drops to $-20°C$. The density of ice $=$ 0.917 g/c.c., latent heat of ice $=$ 80 cal/g, and thermal conductivity of ice $=$ 0.005 cal cm^{-1}sec^{-1} °C^{-1}.

Solution: The required time τ is given by the relation

$$\tau = \frac{1\rho L}{2K\theta}(x_2^2 - x_1^2).$$

Here $\rho = 0.917$ g/c.c., $L = 80$ cal/g, $x_2 = 3$ cm, $x_1 = 0$; $\theta = 20°C$, and $K = 0.005$ c.g.s. units. So, the required time is

$$\tau = \frac{0.917 \times 80 \times 9}{2 \times 0.005 \times 20} \text{ sec} \approx 55 \text{ min.}$$

2. A pond is covered with ice 0.04 m thick. The temperature of the air above is 261 K. At what rate will the ice thicken? Given: K of ice = 2.184 W/m/K; density of ice = 920 kg/m^3, and latent heat = 333 kJ/kg.

Solution: For a thickness dx to grow, when the thickness is already x, in time dt, $L\rho dx$ calorie must be transmitted across a unit area of the layer x.

$$L\rho dx = \frac{K\theta}{x}dt, \text{ where } \frac{\theta}{x} \text{ is the temperature gradient.}$$

Rate of growth of thickness of ice is then give by

$$\frac{dx}{dt} = \frac{K\theta}{L\rho x} = \frac{2.184 \times (273 - 261)}{333 \times 10^3 \times 920 \times 0.04} \text{ m/s} = \frac{2.184 \times 12 \times 3600}{333 \times 10^3 \times 920 \times 0.04} \text{ m/h}$$

$$= 7.699 \times 10^{-3} \text{ m/h}.$$

3. Calculate the loss of heat per second per unit length of a rubber tube carrying steam at 100°C when the outer surface is at a temperature 20°C. (Thermal conductivity of rubber = 4.5 $\times 10^{-4}$ c.g.s. unit, inner diameter of tube = 1 cm, outer diameter = 2 cm.)

Answer: Here we have

$$r_2 = 1 \text{ cm}, \ r_1 = 0.5 \text{ cm}, \ \theta = 100°C, \ \theta_2 = 20°C,$$
$$K = 4.5 \times 10^{-4} \text{ c.g.s. unit, } l = 1 \text{ cm}.$$

Therefore, from the relation, we have

$$Q = \frac{2\pi l K(\theta_1 - \theta_2)}{\ln(r_2/r_1)}.$$

The rate of heat loss is

$$(Q)_{t=1} = \frac{2 \times 3.14 \times 1 \times 4.5 \times 10^{-4} \times (100 - 20)}{\ln(1/0.5)}$$

$$= \frac{2 \times 3.14 \times 4.5 \times 10^{-4} \times 80}{\ln 2},$$

$$= \frac{2 \times 3.14 \times 4.5 \times 10^{-4} \times 80}{0.6931} = 0.3262 \text{ cal/s/cm}.$$

4. Calculate the difference in temperature between the inner and outer surfaces of an annular cylinder of aluminum of length 5 cm when it is being heated internally by an axial coil delivering 10 W power. Given: Thermal conductivity

of aluminum $= 0.5$ c.g.s., inner diameter $= 3$ cm, outer diameter $= 6$ cm, $J = 4.2$ J/cal.

Answer: We have here

$$Q = 10 \text{ J/s}, = (10/4.2) \text{ cal/s}, r_2/r_1 = 3/1.5 = 2, l = 5 \text{ cm, and } K = 0.5.$$

$$\theta_1 - \theta_2 = \frac{\ln(r_2/r_1)}{2\pi l K} Q = \frac{10 \times \ln 2}{4.2 \times 2 \times 3.14 \times 3.14 \times 5 \times 0.5}$$

$$= \frac{10 \times 0.6923}{4.2 \times 3.14 \times 5} = 0.105°\text{C}.$$

5. A composite slab is made with two parallel layers of different materials A and B. Their conductivities are 70 $\text{Wm}^{-1}\text{K}^{-1}$ and 200 $\text{Wm}^{-1}\text{K}^{-1}$, respectively, and thicknesses of 0.045 m and 0.025 m, respectively. Find the temperature at the interface of A and B, when their other surfaces are maintained at 373 K and 273 K respectively.

Answer: Let θ be the temperature of the interface at the steady state. In this state, the quantity of beat passing per sec, Q, across a unit area of B will be the same as that across A.

$$Q = K_1 \frac{373 - \theta}{x_1} = K_2 \frac{\theta - 273}{x_2} \frac{70 \times (373 - \theta)}{0.045} = \frac{200(\theta - 273)}{0.025};$$

$$\Rightarrow \quad 7 \times (373 - \theta) \times 25 = 45 \times 20 \times (\theta - 273)$$

$$\Rightarrow \quad 7 \times 5 \times (373 - \theta) = 45 \times 20 \times (\theta - 273);$$

$$\Rightarrow \quad 215\theta = 62195; \quad \Rightarrow \quad \theta \simeq 289 \text{ K}.$$

6. Find the thermal resistance of an aluminum rod of length 20 cm and area of cross-section 1 cm^2. Heat flows along the length of the rod. Thermal conductivity of aluminum $= 200 \text{ Wm}^{-1} \text{ K}^{-1}$.

Answer: The thermal resistance is

$$R = \frac{l}{KA} = \frac{20 \times 10^{-2} \text{ m}}{(200 \text{ Wm}^{-1} \text{ K}^{-1})(1 \times 10^{-4} \text{ m}^2)} = 10 \text{ KW}^{-1}. \tag{6.90}$$

7. Three slabs of equal thickness having cross-sectional areas normal to the planes A_1, A_2, and A_3, respectively, are so arranged that heat flows from the upper to the lower face without loss, covering the entire cross-sectional area. Show that the combination will behave order steady state as a single slab of conductivity

$$K = \frac{K_1 A_1 + K_2 A_2 + K_3 A_3}{A_1 + A_2 + A_3},$$

where K_1, K_2, and K_3 are the conductivities of the slab 1, 2, and 3, respectively.

Answer: If θ_1 and θ_2 be the temperatures $(\theta_1 > \theta_2)$ of the upper and lower face, we have

$$\frac{Q_1}{t} = \frac{K_1 A_1 (\theta_1 - \theta_2)}{x}; \frac{Q_2}{t} = \frac{K_2 A_2 (\theta_1 - \theta_2)}{x}; \text{ and } \frac{Q_3}{t} = \frac{K_3 A_3 (\theta_1 - \theta_2)}{x},$$

where x is the thickness of the slab and Q_i's the quantities of heat flow,

$$\frac{Q}{t} = \frac{Q_1 + Q_2 + Q_3}{t} = \frac{(K_1 A_1 + K_2 A_2 + K_3 A_3)(\theta_1 - \theta_2)}{x}$$

$$= \frac{K(A_1 + A_2 + A_3)(\theta_1 - \theta_2)}{x}$$

$$K(A_1 + A_2 + A_3) = K_1 A_1 + K_2 A_2 + K_3 A_3; \quad \Rightarrow \quad K = \frac{K_1 A_1 + K_2 A_2 + K_3 A_3}{A_1 + A_2 + A_3}.$$

8. A spherical hot water tank, fitted with an electrical heater, has an internal radius of 0.2 m and has a wall of 0.05 m thick made of a poor thermal conductor $(K = 0.84 \text{ Wm}^{-1}\text{deg}^{-1})$. If the temperature of the outside of the wall is 15°C when the water is at 95°C, calculate the power that must be dissipated in the heater so as to maintain the water temperature at 95°C.

   Answer: To maintain the temperature of water constant, heat must be supplied at the same rate at which it is conducted away through the wall.

   The amount of heat conducted away per second is

   $$Q = 4\pi K \frac{\theta_1 - \theta_2}{r_2 - r_1} r_1 r_2 = \frac{4\pi \times 0.84 \times (95 - 15) \times 0.2 \times 0.25}{0.25 - 0.2}$$

   $$= 4\pi \times 0.84 \times 80 = 844.03 \text{ W}.$$

   This is the rate at which heat is conducted through the wall and must be equal to the power dissipated in the heater.

9. Two thin concentric spherical shells of radii 5 cm and 15 cm, respectively, have their annular cavity filled with charcoal. When energy is supplied at the steady rate of 10.8 W to a heater at the center, a temperature difference of 50°C is set up between the spheres. Find the thermal conductivity of charcoal.

   Answer: In the case of a spherical conductor, the expression for the amount of heat flowing per second is given by

   $$Q = 4\pi K \frac{\theta_1 - \theta_2}{r_2 - r_1} r_1 r_2 \quad \Rightarrow \quad K = \frac{Q(r_2 - r_1)}{4\pi (\theta_1 - \theta_2) r_1 r_2}.$$

Substituting the values for $r_1, r_2, \theta_1, \theta_2$, and Q, we obtain the value for thermal conductivity as

$$K = \frac{10.8 \times (15 - 5)}{4 \times 3.14 \times 50 \times 15 \times 5} = 0.0023 \text{ W cm}^{-1}\text{deg}^{-1}.$$

10. In the periodic flow method, a rod is heated at one end with a heating cycle of 4 min. The temperature maximum travels at the rate of 6 cm/min. Calculate the thermal conductivity of the metal. Given that the density of the metal is 7.8 g/c.c. and the specific heat is 0.11.

Answer: The velocity of propagation of the temperature maximum is given by $V = \frac{x}{t} = 2\sqrt{\frac{\pi h}{T}}$, where the diffusivity of the material h is given by

$$h = \frac{K}{\rho c} = \frac{x^2 T}{4\pi t^2}.$$

This leads to

$$K = \frac{x^2 \rho c T}{4\pi t^2}.$$

The parameters given here are $V = x/t = 6/60$ cm/s; $\rho = 7.8$ g/c.c.; $c = 0.11$; $T = 4 \times 60$ s. Hence, the value of K is given by

$$K = \frac{6 \times 6 \times 7.8 \times 0.11 \times 4 \times 60}{60 \times 60 \times 4 \times 3.14} = 0.164 \text{ c.g.s. unit.}$$

11. A cylindrical cement tube of radii 0.05 cm and 1.0 cm has a wire embedded into it along its axis. To maintain a steady temperature difference of 120°C between the inner and outer surfaces, a current of 5 ampere is made to flow in the wire. Make calculations for the amount of heat generated per meter length and the thermal conductivity of cement. Assume that the resistance of the wire is equal to 0.1 Ω per cm of length.

Answer: Resistance of wire is $R = 0.1$ Ω per cm of length $= 10$ Ω per m length.

Heat generated in the resistor R is given by

$$H = (I^2 R) \text{ Wm}^{-1} = (5^2 \times 10) \text{ Wm}^{-1} = 250 \text{ Wm}^{-1}.$$

Under steady-state conditions, the heat generated equals the heat transfer through the cylindrical element. So,

$$Q = \frac{2\pi K l (t_1 - t_2)}{\ln\left(\frac{r_2}{r_1}\right)}; \qquad \Rightarrow \qquad 250 = \frac{2\pi \times K \times 1 \times 120}{\ln\left(\frac{1.0}{0.05}\right)}.$$

This gives

$$K = \frac{250 \times \ln\left(\frac{1.0}{0.05}\right)}{2\pi \times 120} = 0.994 \text{ Wm}^{-1} \text{ deg}^{-1}.$$

12. If the temperature at one end of a bar being heated periodically with a period of 1 hour varies between 288 K and 318 K, to what distance could the variation of temperature be detected by a thermometer reading to 0.1°C? Given that the thermal conductivity K of the metal is 126 Wm^{-1}K^{-1}, density 7×10^3 kg/m^3, specific heat is given by 4.2×10^2 J/kg/K.

Answer: Given parameters of the problem are

$$\theta = 0.1°\text{C}; \theta_0 = 318 - 288 = 30 \text{ K or } 30°\text{C}; \rho = 7 \times 10^3 \text{ kg/m}^3,$$

$$c = 4.2 \times 10^2 \text{ J/kg/K}, K = 126 \text{ Wm}^{-1}\text{K}^{-1}, T = 5 \text{ hours}.$$

$$b = \sqrt{\frac{\pi}{hT}} = \sqrt{\frac{\pi \rho c}{KT}} = \sqrt{\frac{3.14 \times 7 \times 10^3 \times 4.2 \times 10^2}{126 \times 5 \times 3600}} = 2.017.$$

Let x be the required distance. From the relation, $\theta = \theta_0 e^{-bx}$, we have

$$b \times x = \ln(\theta_0/\theta) \quad \Rightarrow \quad 2.017 \times x = \ln 300$$

$$\Rightarrow \quad x = \frac{\ln 300}{2.017} = \frac{5.7037}{2.017} = 2.83 \text{ m}.$$

13. At depths of 2 m, 4 m, and 8 m in the earth's crust, the annual ranges of fluctuation of temperature are 5.6°C, 2.8°C, and 0.7°C, respectively. Find the velocity of propagation of the temperature-wave and the diffusivity.

Answer: The amplitude of fluctuation at a distance x is given by

$$\theta_0 e^{-\sqrt{\omega/2h}x} = \theta_0 e^{-ax}, \text{ where } a = \sqrt{\omega/2h}.$$

Now, by the problem, $5.6 = \theta_0 e^{-2a}$; $2.8 = \theta_0 e^{-4a}$; $0.7 = \theta_0 e^{-8a}$.

From any two of the above relations, we get

$$e^{2a} = 2 \quad \Rightarrow \quad a = \frac{1}{2} \ln 2 = 0.3465.$$

This shows that the temperature wave propagates in a regular manner and a is given by 0.3465. The velocity of propagation is given by

$$V = \sqrt{2h\omega} = \omega\sqrt{\frac{2h}{\omega}} = \frac{\omega}{a} = \frac{2\pi}{aT} = \frac{2 \times 3.14}{0.3465 \times 365} = 0.0496 \text{ m/day}$$

Again, we have

$$h = \frac{\omega}{2a^2} = \frac{2\pi}{T} \times \frac{1}{2a^2} = \frac{2 \times 3.14}{365} \times \frac{1}{2 \times (0.3465)^2} = 0.0716 \text{ sq.m/day.}$$

14. A lagged bar is made in two parts, each having the same cross-sectional area of $4 \times 10^{-2} \text{m}^2$. The first part is 0.15 m long and has a thermal conductivity of 385 Wm^{-1} deg^{-1}, while the second part is 0.05 m long and has a thermal conductivity of 100 Wm^{-1} deg^{-1}. Find the temperature of the join and the quantity of heat conducted per sec along the bar when one end is at 100°C and the other end at 0°C.

Answer: The quantity of heat that flows per sec down the first part is

$$\left(\frac{dQ}{dt} \right)_1 = \frac{KA(\theta_1 - \theta_2)}{l} = \frac{385 \times 4 \times 10^{-2} \times (100 - \theta)}{0.15},$$

where θ is the temperature of the join.

This must also be the same as the heat flowing down the second part each second.

$$\left(\frac{dQ}{dt} \right)_2 = \frac{100 \times 4 \times 10^{-2} \times (\theta - 0)}{0.05};$$

$$\Rightarrow \quad \frac{385 \times 4 \times 10^{-2} \times (100 - \theta)}{0.15} = \frac{100 \times 4 \times 10^{-2} \times \theta}{0.05};$$

$$\Rightarrow \quad \theta = 56.2°\text{C}.$$

Substituting this value of θ

$$\frac{dQ}{dt} = \frac{100 \times 4 \times 10^{-2} \times 56.2}{0.05} = 4496 \text{ W.}$$

15. One face of a copper cube of edge 10 cm is maintained at 100°C and the opposite face is maintained at 0°C. All other surfaces are covered with an insulating material. Find the amount of heat flowing per second through the cube. The thermal conductivity of copper is 385 Wm^{-1} °C^{-1}.

Answer: The heat flows from the hotter face toward the colder face. The area of the cross-section perpendicular to the heat flow is $A = (10 \text{ cm})^2$.

The amount of heat flowing per second is

$$\frac{\delta Q}{\delta t} = KA \frac{T_2 - T_1}{l} = 385 \text{ Wm}^{-1} °\text{C}^{-1} \times (0.10 \text{ m})^2 \times \frac{100°\text{C} - 0°\text{C}}{0.1 \text{ m}} = 3850 \text{ W.}$$

$$(6.91)$$

16. A body initially at 353 K cools down to 337 K in 5 min and to 325 K in 10 min. What will be its temperature after 15 min, and what is the temperature of the surroundings?

Answer: If a body cools from a temperature of $\theta_1°$C to $\theta_2°$C in time t, the temperature of the surroundings being $\theta_0°$C, then from Newton's law of cooling,

$$t = \frac{1}{K} \log_e \frac{\theta_1 - \theta_0}{\theta_2 - \theta_0}. \tag{6.92}$$

Now, 353 K = 80°C, 337 K = 64°C, and 325 K = 52°C.

In the first case,

$$5 \times 60 = \frac{1}{K} \log_e \frac{80 - \theta_0}{64 - \theta_0}. \tag{6.93}$$

In the second case, we can say that the body cools down from 337 K, that is, 64°C to 325 K, that is, 52°C in (10–5) or 5 min. So,

$$5 \times 60 = \frac{1}{K} \log_e \frac{64 - \theta_0}{52 - \theta_0}. \tag{6.94}$$

Equating equations (6.93) and (6.94), we get

$$\frac{80 - \theta_0}{64 - \theta_0} = \frac{64 - \theta_0}{52 - \theta_0}; \quad \Rightarrow \quad \theta_0 = 16°C = 289 \text{ K}.$$

If the temperature of the body after 15 min be θ, then proceeding as above, we get

$$5 \times 60 = \frac{1}{K} \log_e \frac{52 - \theta_0}{\theta - \theta_0}. \tag{6.95}$$

Equating (6.95) and (6.94), we get

$$\frac{64 - \theta_0}{52 - \theta_0} = \frac{52 - \theta_0}{\theta - \theta_0} \quad \Rightarrow \quad \frac{64 - 16}{52 - 16} = \frac{52 - 16}{\theta - 16},$$

$$\theta = 43°C = 316 \text{ K}.$$

17. A solid copper sphere (density ρ and specific heat capacity c) of radius r at an initial temperature of 200 K is suspended inside a chamber whose walls are at almost 0 K. Find the time required for the temperature of the sphere to drop to 100 K.

Answer: The net rate of loss of heat per unit area of the copper sphere,

$$E = \epsilon \sigma \left(T^4 - T_0^4 \right). \tag{6.96}$$

As the chamber is closed, the copper sphere will act as a black body. So, $\epsilon = 1$. Also, $T = 200$ K, $T_0 = 0$ K.

If the radius of the copper sphere be r, then its surface area $A = 4\pi r^2$ and its mass $m = \frac{4}{3}\pi r^2 \rho$. So, the net rate of heat loss by the copper sphere is

$$-\frac{dQ}{dt} = EA = \sigma A \left(T^4 - T_0^4\right) = \sigma A T^4 \qquad [as \quad T_0 = 0 \text{ K}]$$

$$\text{But } -\frac{dQ}{dt} = -mc\frac{dT}{dt}; \quad \Rightarrow \quad -mc\frac{dT}{dt} = \sigma A T^4 \qquad \text{or, } dt = \frac{mc\, dT}{\sigma A\, T^4}.$$

Integrating, we get

$$\int_0^t dt = -\frac{mc}{\sigma A} \int_{200}^{100} \frac{dT}{T^4}.$$

$$\Rightarrow \quad t = \frac{7mc}{24 \times 10^6 \sigma A} = \frac{7\left(\frac{4}{3}\pi r^3 \rho\right)c}{24 \times 10^6 \sigma \left(4\pi r^2\right)} = \frac{7r\rho c}{72 \times 10^6 \sigma}.$$

6.7 Multiple choice questions and answers

1. In a room containing air, heat can go from one place to another

(a) by conduction only.
(b) by convection only.
(c) by radiation only.
(d) by all the three modes.

Answer: The correct choice is (d).

2. Thermal conductivity of solid metals normally ... with rise in temperature.

(a) increases
(b) decreases
(c) remains constant
(d) may increase or decrease depending on temperature

Answer: The correct choice is (b).

3. Which material has the highest thermal conductivity?

(a) Silver
(b) Copper
(c) Diamond
(d) Bronze

Answer: The correct choice is (c).

Solution: Around room temperature, diamond has the highest thermal conductivity along with graphite and graphene. The thermal conductivity of diamond is 2000 Wm^{-1} K^{-1}, which is five times higher than copper.

4. Which is the best insulator?

(a) Potassium aerogels

(c) Manganese aerogels

(b) Lead aerogels

(d) Silica aerogels

Answer: The correct choice is (d).

Solution: The thermal conductivities of silica aerogel is 0.03 W/m/K. The liquid is removed from the gel without collapsing its structure, thus creating air pockets in the medium. This slows down the heat transfer and acts as the insulator.

5. The expression for thermal resistance is

(a) $R = \frac{2KA}{l}$

(c) $R = \frac{l^2}{KA}$

(b) $R = \frac{l}{KA}$

(d) $R = \frac{2l}{KA^2}$

Answer: The correct choice is (b).

6. A body cools down from 65°C to 60°C in 5 min. It will cool down from 60°C to 55°C in

(a) 5 min

(d) less than or more than 5 min depending on whether its mass is more than or less than 1 kg

(b) less than 5 min

(c) more than 5 min

Answer: The correct choice is (c).

7. One end of a metal rod is dipped into boiling water, and the other end is dipped in melting ice.

(a) All parts of the rod are in thermal equilibrium with each other.

(c) We can assign a temperature to the rod after a steady state is reached.

(b) We can assign a temperature to the rod.

(d) The state of the rod does not change after a steady state is reached.

Answer: The correct choice is (d).

8. An aluminum plate of mass 1 kg at 95°C is immersed in 0.5 l of water at 20°C kept inside an insulating container, and then removed. If the temperature of the water is found to be 23°C, then the temperature of the aluminum plate is ...° C, (the specific heat of water and aluminum are 4200 J/kg−K and 900 J/kg−K, respectively, the density of water is 1000 kg/m^3)   [JAM 2016]

(a) 94.36

(c) 104.36

(b) 84.36

(d) 100.16

Answer: The correct choice is (a).

Solution: $-M_a S_a(T_{af} - T_{ai}) = M_w S_w(T_{wf} - T_{wi}) \Rightarrow -0.1 \times 4200(T_{af} - 368) = 0.5 \times 900(296 - 293)$,

$\Rightarrow -2100(T_{af} - 368) = 450 \times 3 \Rightarrow (T_{af} - 368) = \frac{450}{700} = -0.64$.

$\Rightarrow T_{af} = 368 - 0.64 = 367.36 = 367.36 - 273 = 94.36$.

6.8   Exercise

6.8.1   Short answer type questions

1. What is thermal conduction?

2. State Fourier's law of conduction.

3. What do you mean by the thermal conductivity of a substance?

4. What are the unit and dimension of thermal conductivity?

5. What are the factors affecting the thermal conductivity of a substance?

6. What do you understand by stationary state?

7. What do you mean by thermometric conductivity?

8. What do you mean by a lagged bar?

9. What are the differences between a lagged and unlagged bar?

10. What is thermal resistance? What is its unit?

11. Does thermal conductivity depend upon temperature?

12. What are fins?

13. Mention some applications of fins.

6.8.2   Long answer type questions

1. What are the differences between thermal conductivity and thermometric conductivity?

2. Set up the differential equation for the flow of heat through a metal bar of uniform cross-section heated at one end. Solve the equation with appropriate boundary conditions.

3. Apply Fourier's heat flow equation to a lagged bar to find the temperature as a function of distance from the hot end of the lagged bar.

4. Set up the differential equation in spherical polar coordinates for the radial flow of heat through a spherical conductor of internal radius r_1 and external radius

r_2 with temperatures, respectively, θ_1 and θ_2. Solve the differential equation with appropriate boundary conditions to find the temperature at a distance r from the center of the spherical conductor.

5. Set up the differential equation in a cylindrical coordinate system for the radial flow of heat through a cylindrical conductor of internal radius R_1 and external radius R_2 with temperatures, respectively, θ_1 and θ_2. Solve the differential equation with appropriate boundary conditions to find the temperature θ at a distance r from the axis of the cylindrical conductor.

6. Consider a pipe of inner and outer radii r_1 and r_2, respectively. The thermal conductivity of the material of the pipe is K. The inner and outer surfaces of the pipe are maintained at temperatures T_1 and T_2, respectively. Obtain an expression for the temperature at a distance r along the radial distribution inside the pipe, that is, $r_1 < r < r_2$.

6.8.3   Numerical problems

1. A rod CD of thermal resistance $5.0~\text{KW}^{-1}$ is joined at the middle of an identical rod AB, that is, C is the mid-point of AB. The ends A, B, and D are maintained at 100°C, 0°C, and 25°C, respectively. Find the heat current in the CD.
   [4.0 W]

2. A glass tube of internal and external radii 0.8075 and 09775 cm and length 54.4 cm is enclosed in a steam jacket. The temperatures of inflowing and outflowing water through the tube are 18.8°C and 28.4°C. The thermal conductivity of glass is 0.00154 c.g.s. unit. Find the rate of flow of water through the tube.
   [23.92 c.c./s]

3. The hot combustion gases of a furnace are separated from the ambient air and surroundings, which are at 25°C, by a brick wall 0.15 m thick. The brick has a thermal conductivity of $1.2~\text{Wm}^{-1}~\text{K}^{-1}$ and a surface emissivity of 0.8. Under steady-state conditions, an outer surface temperature of 100°C is measured. Free convection heat transfer to the air adjoining the surface is characterized by a convection coefficient of $20~\text{Wm}^2~\text{K}^{-1}$. What is the brick's inner surface temperature?
   [625 K]

4. A 3 mm diameter and 6 m long electric wire is tightly wrapped with a 2 mm thick plastic cover whose thermal conductivity is $0.15~\text{Wm}^{-1}~\text{K}^{-1}$. Electrical measurements indicate that a current of 10 A passes through the wire and there is a voltage drop of 8 V along the wire. If the insulated wire is exposed to a medium at 27°C with a heat transfer coefficient of $12~\text{Wm}^{-2}~\text{K}^{-1}$, determine the temperature at the interface of the wire and the plastic cover in steady operation. Also, determine whether doubling the thickness of the plastic cover will increase or decrease this interface temperature.
   [89.5°C, 77.5°C]

5. A 25 mm diameter egg roll ($K = 1$ W/m degree) is roasted with the help of microwave heating. For good quality roasting, it is desired that the temperature at the center of the roll is maintained at 100°C when the surrounding temperature is 25°C. What should be the heating capacity in Wm^{-3} of the microwave if the heat transfer coefficient on the surface of the egg roll is 20 Wm^{-2} degree^{-1}?   [213.31 kW m^{-3}]

CHAPTER 7

Basic Formalism of Thermodynamics and the Zeroth Law

Thermodynamics is the only physical theory of universal content which, within the framework of the applicability of its basic concepts, I am convinced will never be overthrown.

—Albert Einstein

Learning Outcomes

After reading this chapter, the reader will be able to

- Know various types of thermodynamic systems such as open, closed, and isolated, and the surroundings

- Classify between intensive and extensive thermodynamic variables

- Understand various types of equilibrium conditions satisfied by a thermodynamic system

- State the zeroth law of thermodynamics and highlight its physical significance

- Comprehend the idea of temperature from the zeroth law of thermodynamics

- Solve numerical problems and multiple choice questions on thermodynamic equilibrium and the zeroth law of thermodynamics

7.1 Introduction

Heat is a form of energy. It can be transformed from one form to another as well as can be transferred between various objects maintained at suitable temperatures. For example, in an electric motor, heat is transformed into mechanical energy by the turbine to power the motor. This mechanical energy is then transformed into electrical energy by the engine to illuminate light bulbs. "Thermodynamics" **is a branch of physics that deals with heat and the transformation of heat from one form to another, work, temperature, and their relation to energy, entropy, and other physical properties of matter and radiation.** It establishes the relation between heat and various forms of energy and describes the transformations that occur in thermal energy from one energy state to another and how this transformation affects matter.

A thermodynamic system is described within a framework based on the **four laws of thermodynamics** that facilitate a quantitative description of the **average macroscopic properties** of the system in equilibrium. Macroscopic matter refers to large objects that consist of many atoms and molecules. The average properties of such macroscopic systems are determined by the physical quantities such as volume, pressure, and temperature that do not depend upon the detailed microscopic positions and velocities of the atoms and the molecules comprising the macroscopic system. In the equilibrium state of a thermodynamic system, these average properties also do not change with time. These physical quantities are called thermodynamic coordinates, variables, or parameters. If a subset of these properties are experimentally measured, the rest of them can be calculated using thermodynamic relations. Thermodynamics not only gives the exact description of the state of equilibrium but also provides an approximate description (to a very high degree of precision!) of relatively slow processes. This branch of physics can be successfully applied to a wide variety of topics in science, such as physics, physical chemistry, biochemistry, chemical engineering, and mechanical engineering, but also in other complex fields, such as meteorology.

According to Callen, "Thermodynamics is the study of the restrictions on the possible properties of matter that follow from the symmetry properties of the fundamental laws of physics". With the development of the subject, this statement seems to link thermodynamics with statistical physics and modernize the term. Thermodynamics is concerned with the relationship between thermal, mechanical, and chemical interactions and equilibrium (and small fluctuations about equilibrium) on a macroscopic scale. It describes the macroscopic behavior of systems but does not always describe the mechanisms behind that behavior. The related theory is restricted to closed thermodynamic systems. A closed thermodynamic system is a quantity of matter separated from its environment by a container. The system has a set of equilibrium states. These equilibrium states are the basic elements of the theory. A transition is a change from one such equilibrium state to another. The theory in thermodynamics is about what transitions are possible and how much energy exchanges occur between the system and its environment during these transitions. During a transition, a system may pass through a number of nonequilibrium states. In such cases, the theory deals only with the relation between the end states and the total effect of the transition. To deal with such nonequilibrium states between the end states and to derive physical properties, advanced theoretical concepts in physics have to be adopted.

Classical thermodynamics is the branch of physics that studies the general properties of a substance connected to thermal motion in equilibrium conditions. It deals with the study of energy, energy transformation, and its relation to matter. The analysis of a thermodynamic system is achieved through the application of the governing conservation equations such as conservation of mass, conservation of energy, and the laws of thermodynamics, and the relations associated with these laws. The laws of thermodynamic are explained in terms of microscopic constituents by statistical mechanics. Thermodynamic applies to a wide variety of topics in science and engineering, especially physical sciences, physical chemistry, chemical engineering, and mechanical engineering. It occupies a special position in physics for any form of energy on its transformation finally changes into energy of thermal motion. This branch of physics is based on empirical laws, that is, there is no way to prove it and is, therefore, phenomenological, and nevertheless, it is exact and powerful. Each of these empirical laws introduces a new concept, for example, temperature, internal energy, and entropy that provides a definite meaning to physically measurable quantities and gives useful correlations between them. The macroscopic parameters of a system are determined in thermodynamics in equilibrium.

The laws of thermodynamics are very general and used to describe a wide varieties of physical and chemical systems in equilibrium. It deals with the study of internal motions of many-body systems and is a very powerful tool for accounting the observed features of the physical world. We mention here some natural phenomena that could be answered easily from the knowledge of thermodynamics:

1. Why is the sky blue?

2. Why are raindrops spherical?

3. Why do we not fall through the floor?

4. Why does heat flow spontaneously from a hotter to a colder body?

5. Why is it impossible to measure a temperature below $-273.16°C$?

6. Why is there a maximum theoretical efficiency of a power generation unit that can never be exceeded, no matter what the design is?

7. Why does the atmosphere of the earth become thinner and colder at higher altitudes?

8. Why does the Sun appear yellow, whereas colder stars appear red, and hotter stars appear bluish-white?

9. Why are high mass stars ultimately collapse to form black holes?

The objective of this chapter is to introduce the fundamental concepts in thermodynamics, basic definitions of various quantities and processes, equilibrium conditions, and the zeroth law of thermodynamics. A number of solved numerical problems and multiple choice questions with answers are given at the end of this chapter.

7.2 Postulates of thermodynamics

The average values of the macroscopic thermodynamic parameters of a system in equilibrium are determined in this branch of physics. In order to calculate these average values, certain postulates are made that are briefly presented below:

1. *An isolated macroscopic system has a definite and precise total energy E that obeys the microscopic conservation laws.*

   This postulate is highly acknowledged in the case of microscopic laws of physics. The system, whether classical or quantum, as a whole evolves in time and satisfies the law of conservation of energy. This is true only for so-called conservative systems (the definition is circular), but this suffices as the associated Hamiltonian describes the fundamental laws of nature.

2. *The equilibrium states of a thermodynamic system are completely characterized by their total energy E, volume V, and the number of moles $n = n_1, n_2, \cdots, n_n$ of the chemical components of the system under consideration.*

   This postulate states that the macroscopic states in thermodynamics can be described or characterized uniquely by amazingly few variables. These variables are generally chosen to be extensive. This is but one possibility, and it leads to the so-called entropic representation of thermodynamics. Depending upon the position and momenta of each particle, one macroscopic state would have many microscopic realizations (classical view). Equilibrium states do not change with time (but there may be changes in the microscopic realizations). In practice, such states are often found to behave as metastable, meaning that changes in such systems occur on a very long timescale, and therefore, they can be viewed as equilibrium states. Equilibrium states satisfy the postulates of equilibrium thermodynamics. Further, it is to be noted that equilibrium thermodynamics only refers to the properties of equilibrium states and does not immediately apply to the nonequilibrium processes that might connect various equilibrium states. As the equilibrium states are defined in terms of the variables U, V, and n, it follows that it does not matter how these states are prepared. But, it should be mentioned that the process of preparation does matter for many systems. For example, there are systems that show hysteresis or glassy behavior. Such systems are not in thermodynamic equilibrium.

3. *It is possible to connect the equilibrium states $A(U_A, V, n) \leftrightarrow B(U_B, V, n)$, by transforming either A to B or B to A using exclusive mechanical work (which can be determined quantitatively), enclosing the states in an adiabatic environment: $W_{A \rightarrow B} = U_B - U_A$.*

Joule first made this observation. It should be noted that in thermodynamics, mechanical work, for example, PV work or electrical work can be quantified easily, and work can be performed on a system isolated from the rest of the world. This type of experiment is the quintessential process to measure the change in energy between two equilibrium states, say, A and B, and it is important to keep heat out of the equation. Most transformations required to change an equilibrium system A into another equilibrium system B would be associated with an increase in entropy and, hence, cannot be reversed under adiabatic conditions.

4. *It is possible to transform state A to B (or conversely) by other processes. The mechanical work $W_{A \rightarrow B}$ can be measured. In addition, the process involves heat in general, such that $q_{A \rightarrow B} = (U_B - U_A) - W_{A \rightarrow B}$, or $q_{A \rightarrow B} + W_{A \rightarrow B} = (U_B - U_A)$.*

This postulate is the essence of the first law of thermodynamics. The processes described in postulate (3) are needed to measure the change in energy associated with the transformation $A \rightarrow B$. The equilibrium states themselves are unique (from postulate 1), and any other process that affects the same transformation must preserve this difference in energy. The difference in energy that is not supplied or taken out by work is accounted for by heat. The mechanical work can always be measured, and we then talk about the quantity of heat. From the above definition, it is interesting to note that work and heat depend on the particulars of the process and can therefore not be functions of state.

5. *Using the suitable nature of walls, thermodynamic systems can be partitioned into subsystems, and various thermodynamic principles can be applied to these subsystems for extracting out the physical properties of the system. The nature of the walls can be of the following types:*

   (a) *Adiabatic versus diathermic (allowing the flow of heat among the subsystems of the system).*

   (b) *Fixed versus moveable (allowing the total volume V of the system to be partitioned among different subsystems), and*

   (c) *Impermeable versus permeable to specific chemical compounds (allowing redistribution of matter among various subsystems of the system).*

The use of walls is an effective way to impose constraints on the subsystems and is an essential ingredient of formal analysis of thermodynamics. The subsystems will reach to equilibrium to make their temperature equal when the separating wall is changed from adiabatic to diathermic that allows the flow of heat. When the flow of heat stops, entropy will reach a maximum value, but the volume and the number of particles for each subsystem would be unchanged. If the wall is changed from a fixed to a moveable one, the wall will continue its movement until the pressure on both sides becomes equal. Further, when one subsystem does P–V work on another subsystem, it is observed that both volumes and energies of the subsystems change. If the moveable wall is kept adiabatic, there is no unique solution for the temperatures of the subsystems, and the system would keep oscillating. The final constraint to be considered in this case is the flow of particles. If particles flow from one subsystem to another, the energy content of the subsystems also changes. Moreover, a permeable wall always allows the transfer of heat. In this case, the equilibrium condition of the system is such that the temperature and the chemical potential of each migrated chemical species are equal in every subsystem. If all constraints are lifted, the situation is as if there are no walls at all. But still, there are imaginary boundaries that allow one to calculate the number of particles for each subsystem; there is a finite volume, and one might assign certain amount of energy to each subsystem. The assignment of energy is problematic as it depends on interactions between particles, possibly from different subsystems. Thus, under this situation, the temperature, pressure, and chemical potentials of a thermodynamic system would be homogeneous throughout equilibrium.

6. *The nature of the thermodynamic variables can be extensive or intensive for a system of subsystems. These variables are:*

   (a) *Extensive (or additive) thermodynamic quantities for the total system are the sums of the quantities for the individual subsystems. For example, internal energy U, volume V, and the number of moles n are the extensive thermodynamic variables.*

   (b) *Intensive thermodynamic quantities can only be defined individually for the subsystems. Some examples of intensive thermodynamic quantities are temperature T, pressure P, and the chemical potential μ.*

It should be noted that some interesting results follow from the extensive property of entropy. It is known that entropy is a function of purely extensive quantities, and this entropy function reaches its maximum at equilibrium. Under this situation, one finds equality of the intensive variables temperature T, pressure P, and the chemical potential μ_i in the system. Extensive variables are always associated with their conjugate intensive variables, for example, P and V, n_i and μ_i, U and T, or S and T, and most thermodynamic quantities are expressed as functions of either the intensive or extensive partner of a pair. However, this is not necessarily true.

7.3   Thermodynamic system

The study of any special branch of physics starts with a separation of a restricted region of space or a finite position of matter from its surroundings. The position that is set aside (in imagination) and on which the attention is focused is called the system. A system may be simple or complex. It may be homogeneous or inhomogeneous. It will, however, always be finite. A system that may be described in terms of thermodynamic coordinates is called a thermodynamic system. Such a thermodynamic system, surroundings, and the corresponding boundary between these two are shown in Figure 7.1.

The system is a macroscopically identifiable collection of matter on which attention is paid to extracting physical information. The immediate surroundings of the system, in reality, also interact with it directly and, therefore, have a much stronger influence on the behavior and physical properties of the system. For example, a car engine and the burning gasoline inside the cylinder of an engine constitute a thermodynamic system, whereas the piston, exhaust system, radiator, and outside air form the

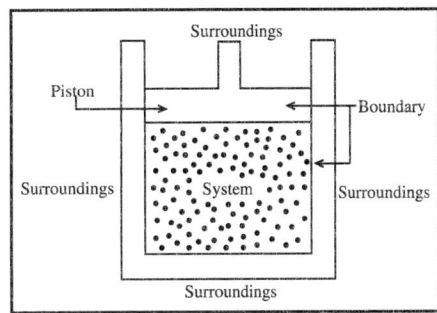

Figure 7.1 The burning of gasoline in the cylinder of a car engine is an example of a thermodynamic system.

surroundings of the system. There exists a fair amount of interaction between the thermodynamic system and the surroundings. The inner surfaces of the cylinder and the piston are the boundary between the system and the surroundings. This boundary plays a vital role in determining the interaction and the physical properties of the system.

Some examples of thermodynamic systems are:

1. A gas such as air in a cylinder (fitted with a frictionless, leakproof piston). It constitutes a simple homogeneous system.

2. A vapor such as steam.

3. A mixture, such as gasoline vapor and air or a phenol-water, constitutes a more complex heterogeneous system.

4. A vapor in contact with its liquid. For example, the combination of liquid and vaporized ammonia serves as a thermodynamic system.

5. Other thermodynamic systems are stretched wires, electric capacitors, thermocouples, etc. In these cases, the thermodynamic variables will be different.

For the discussion of the laws of thermodynamics, basic principles, physical properties, and practical applications of thermodynamic systems, we will be dealing with these types of equilibrium thermodynamic systems in the forthcoming chapters. This indicates that the thermodynamic systems under consideration are in a state of thermal, mechanical, and chemical equilibrium.

7.3.1   Classification of thermodynamic system

Interactions between the systems and the surroundings play an important role in thermodynamics and determine the physical properties of the thermodynamic systems. Depending upon the way the systems interact with the surroundings, they are classified into three main categories: open system, closed system, and isolated system. These types of systems with examples are briefly described below.

An open system

In thermodynamics, an open system is defined as a system in which the transfer of mass (matter) as well as energy can take place across the boundary existing between the system and the surroundings. In this case, the boundary plays an important role, and its nature would be such that it could be able to transfer both matter and energy. A schematic representation of an open thermodynamic system with its surroundings is shown in Figure 7.2. The arrows indicate the exchange of heat/work

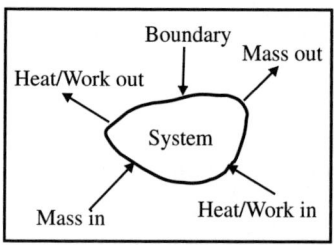

Figure 7.2 An open thermodynamic system with a provision for exchange of energy as well as mass.

as well as mass between the system and the surroundings. Some examples of open systems in thermodynamics are the following:

1. The boiling water in an open vessel is an example of an open system. In this case, the transfer of heat energy as well as mass takes place in the form of steam between the vessel and the surroundings.

2. A human body and an automobile engine are also examples of open systems. The human body exchanges both matter and energy with the surroundings.

3. In a bonfire, it is necessary to provide the fire with flammable material to keep the fire burning. For example, charcoal or dry branches have to be added to consume otherwise the fire will go out.

A closed system

A closed system is defined as a system in which the transfer of energy takes place across the boundary between the system and the surroundings, but there is no provision for the exchange of mass. Hence, a closed system is a fixed mass system. A schematic representation of a closed thermodynamic system is shown in Figure 7.3. Some examples of closed systems in thermodynamics are the following:

1. Compressed or expanded fluid such as air or gas in a piston and cylinder arrangement is an example of a closed system. In this case, the number of gas molecules remains constant but can be heated or cooled by a suitable arrangement.

2. Heating of water inside a closed vessel is an example of a closed system. For example, cooking something in a pressure cooker is an example of this type of closed system. In this case, water is heated, but its mass remains the same.

3. The content of a thermometer is an example of a closed system since the thermometer is hermetically sealed and its content never varies, but it does react according to the temperature it perceives.

An isolated system

In thermodynamics, an isolated system is defined as a system in which there is no transfer of mass or energy across the boundary between the system and the surroundings. This system is completely sealed, and both matter and energy are not allowed to be exchanged with the surroundings. The boundary separating the system and the surroundings plays a vital role in this case. Heat cannot be transferred from the system to the surroundings. It is of fixed mass and fixed energy. In this case, no interaction is allowed to take place between the system and the surroundings. Some examples of isolated systems in thermodynamics are the following:

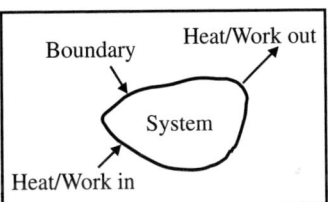

Figure 7.3 Schematic representation of a closed thermodynamic system. There is no provision for entry or exit of mass, but energy can be exchanged.

1. Compressed or expanded fluids like air or gas in a piston and cylinder arrangement serve as an example of an isolated system if the whole arrangement is insulated. Similarly, hot water, coffee, or tea kept in the thermos flask are examples of isolated systems.

2. Igloos of the Eskimos are examples of isolated systems. These are designed in such a way that no heat or matter enters or egresses.

3. The universe is an example of isolated system since neither matter nor energy enters it or leaves from it.

Table 7.1 A classification of the thermodynamic systems as a function of their interactions with their surroundings is shown in a tabular form.

System type	Exchange of mass	Exchange of heat	Exchange of work
Open	Yes	Yes	Yes
Closed	No	Yes	Yes
Thermally isolated	No	No	Yes
Isolated	No	No	No

7.3.2 Concept of surroundings in thermodynamics

The part of the universe under investigation is defined as the **system**, and the remainder of the universe that lies outside the boundaries of the system is referred to as the **surroundings**. Thus, everything external to the system is the surroundings. A thermodynamic system, surroundings, and the boundary separating these two from each other are shown in Figure 7.4. The surroundings are also known as the **environment**, and the **reservoir**. The surroundings interact with the system by

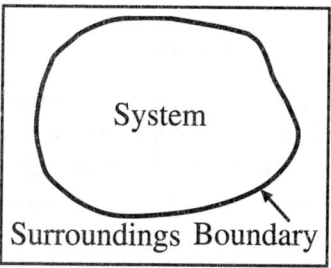

Figure 7.4 A thermodynamic system is separated from the surroundings by a boundary.

exchanging mass, energy (including heat and work), momentum, electric charge, or other conserved quantities. The physical quantity to be exchanged between the system and the surroundings depends upon the nature of the boundary. Except in regards to these interactions, the effect of the surroundings is ignored in the analysis of the thermodynamic system. The main focus in thermodynamic analysis is given to the system that forms the subject of primary investigation. A boundary, as shown in Figure 7.4, is a closed surface surrounding a system through which energy and mass may enter or leave the system. The boundary can greatly influence the behavior of the system; hence, it has a direct bearing on the physical properties of the system.

7.3.3 The boundary of a thermodynamic system

The three-dimensional envelope that encloses a system and separates it from the surroundings is known as **the boundary** of the system. Such a boundary between a system and the surroundings is shown in Figure 7.4. The boundary of a system can be either physical or imaginary. It can also be either fixed or movable. Subsequently, the volume of a system enclosed by the boundary can be fixed or varying. The boundary may or may not allow the system to interact with the surroundings. A boundary that prevents any exchange of matter or energy between the system and the surroundings is termed as an **isolating boundary**, and the associated system is known as **an isolated system**. The behavior of such systems is not affected by their surroundings. The mass and total energy of an isolated system remain constant over time as an isolated system does not allow any exchange of matter, heat, or work with the surroundings.

It is to be noted that the boundary of an open system permits the exchange of both matter and energy between the system and the surroundings, whereas the boundary of a closed system permits the exchange of energy only between the system and the surroundings but not the matter.

It is a well-established fact that heat is one form of energy (thermal energy). Sometimes, the boundary of a system allows the transfer of heat across it, while sometimes, it does not allow the flow of heat. Depending on the feasibility of the transfer of heat across the boundary of the system, the boundary can be classified into two distinct categories:

1. Diathermal boundary and

2. Adiabatic boundary

Diathermal boundary

Diathermal boundaries between thermodynamic systems are defined as the boundaries that permit the flow of heat but do not allow the transfer of matter across themselves. As a result, a system with diathermal wall can absorb/reject heat from/to the surroundings. As usual, a diathermal wall also allows energy transfer in other forms, such as work transfer. Any physical or imaginary boundary is theoretically a diathermal boundary. If the transfer of heat by the process of conduction dominates, a boundary made of thermally conductive material can be a perfect example of diathermal boundary.

The equilibrium state of a thermodynamic system is defined as the state in which the thermodynamic variables, such as pressure, volume, temperature, and entropy, have definite values that remain constant so long as the external conditions are unchanged. Experimental results indicate that the existence of an equilibrium state in one system depends on the nature of the wall separating the system from other systems and also on the proximity of other systems. Let x and y be two thermodynamic variables representing, respectively, the temperature and volume of a thermodynamic system. If two such systems are separated by a diathermic wall, values of x for the first system and x' for the second system will change spontaneously until an equilibrium state of the combined system is attained. These two systems are then said to be in thermal equilibrium with each other. Thus, thermal equilibrium is the state achieved by two or more systems characterized by restricted values of the coordinates of the systems after they have been in communication with each other through a diathermal boundary.

Adiabatic wall or adiabatic boundary

Adiabatic boundaries between thermodynamic systems are defined as the boundaries that do not permit the flow of heat across themselves. As a result, a system with an adiabatic wall can neither absorb heat from the surroundings nor reject heat to the surroundings. An adiabatic wall, however, allows the transfer of energy in other forms, such as transfer of work. It should be mentioned that a perfect adiabatic wall has no existence in this world as the transfer of heat by radiation cannot be made zero so long as there exists a temperature difference between the system and the surroundings. However, a boundary made of thermally insulating material with integrated radiation shields can be taken into consideration as an adiabatic wall for practical purposes. A closed system with an adiabatic boundary behaves as an isolated system as it is constrained to do no work and to have no work done on it under this situation.

The concept of adiabatic wall would be clear from the following simple example. Consider a state of a system represented by the thermodynamic variables (x, y), and another state of another system represented by (x', y'). Suppose the systems are separated from each other by an adiabatic wall. These systems may coexist as equilibrium states for any attainable values of the four thermodynamic quantities provided the wall is able to withstand the stress associated with the difference between the two sets of thermodynamic variables. In this case, neither heat energy nor matter will be transferred between the two systems. Some examples of adiabatic walls include thick layers of wood, concrete asbestos, felt, etc. These adiabatic walls are frequently used in experimental situations as good approximations of the same.

Further classification of boundary

Further, boundaries may be classified into four categories: **fixed, movable, real, and imaginary**. These boundaries are briefly described below:

1. **Fixed boundary:** When the piston in a piston-cylinder arrangement enclosing a gas is locked at one fixed position, it serves as a fixed boundary to the system. In this situation, a thermodynamic process occurs at constant volume.

2. **Movable boundary:** In Figure 7.1, if the piston is allowed to move from one position to another inside the cylinder, the boundary of the system acts as a movable one. In this case, the cylinder and the cylinder head boundaries are kept fixed.

3. **Real boundary:** In the case of closed systems, real boundaries are observed. This boundary does not allow the exchange of energy and matter both, between its different parts or between the system and the surroundings.

4. **Imaginary boundary:** The boundaries for open systems are often imaginary. Such boundaries are imagined in case of a jet engine. There are several such boundaries: a fixed imaginary boundary at the intake of the engine, fixed boundaries along the surface of the case, and a second fixed imaginary boundary across the exhaust nozzle.

7.4   Thermodynamic variables and the state of a thermodynamic system

Thermodynamic description of a system is accomplished with the help of thermodynamic variables. Classification of these variables and their nature are described below.

7.4.1   Intensive and extensive thermodynamic variables

A thermodynamic system is recognized by the property it exhibits. Such properties are characteristics or attributes of matter, and their values can be determined experimentally and evaluated quantitatively theoretically. From the thermodynamic point of view, the changes in the properties of the system help us to evaluate the amount of energy transferred in a given process, work done, energy stored, etc. Such a thermodynamic property depends only on the state of the system and is independent of the path followed by the system to arrive at the specified state. This signifies that all thermodynamic properties are state functions. It is to be mentioned that thermodynamic properties relevant to refrigeration and air conditioning systems are temperature, pressure, volume, density, specific heat, enthalpy, entropy, etc. Some of these thermodynamic properties are independent of the size/mass of the system, whereas some of which depend on them. For example, temperature, pressure, density, etc., do not depend on the size/mass of the system but mass, volume, etc., belong to the latter class.

A macroscopic point of view in terms of the physical quantities is adopted in thermodynamics to get information about the interior of a system. These macroscopic quantities are called thermodynamic coordinates and have a bearing on the internal state of a system. For example, such coordinates help one to determine the internal energy of a thermodynamic system. One of the objectives of thermodynamics is to find the general relations among these thermodynamic coordinates that are consistent with the laws of thermodynamics.

Intensive variable

Thermodynamic variables can be divided into two categories: extensive and intensive variables. Let y be a macroscopic parameter specifying the state of a homogeneous system. Let us further divide the system into two parts by introducing a partition. y_1 and y_2 are the values of this parameter for the two subsystems created by the partition. y is then said to be extensive if $y = y_1 + y_2$ and intensive if $y = y_1 = y_2$.

Thus, the thermodynamic variables of a system in a given state, which are independent of its mass or the number of particles, are called intensive variables. It is a characteristic of the substance present in the system. The values of intensive variables remain the same if the system is divided into several subsystems. Examples of such intensive variables are temperature, pressure, density, specific volume, viscosity, refractive index, magnetic induction, surface tension, electromagnetic force, etc. If x_i and y_i are two arbitrary intensive variables, then the mathematical combinations such as $(x_i + y_i)$, $(x_i \times y_i)$, $\frac{x_i}{y_i}$, and $\frac{\partial x_i}{\partial y_i}$ will also be intensive in nature.

In thermodynamics, specific variables are sometimes used to describe the physical properties of a system. These specific variables are defined as the required variable per unit mass of the system. The specific variables are thus independent of mass and always intensive in character. Examples of specific variables are specific volume v (the system volume per unit system mass) and density ρ (system mass per unit system volume). Other examples of specific variables are: molar density $\rho_m = \frac{N}{N_A V}$ and molar specific volume $v_m = \frac{N_A V}{N}$. Here, N and N_A are, respectively, the total number of molecules and Avogadro's number.

Extensive variable

An extensive variable of a thermodynamic system is defined as the macroscopic parameter that has a value equal to the sum of its values in each part of the system. This variable describes the thermodynamic system in equilibrium conditions. The extensive variables depend upon the size or the mass of the substance present in the system. Some examples of extensive variables are length, area, mass, volume, internal energy, entropy, heat capacity, magnetization, electric charge, etc.

If x_e and y_e be two arbitrary extensive variables, then the sum of these two variables $(x_e + y_e)$ will also be an extensive variable. But the mathematical combinations $\frac{x_e}{y_e}$ and $\frac{\partial x_e}{\partial y_e}$ will be intensive in character. Further, it is to be mentioned that if x_i is an intensive variable and x_e is an extensive variable, then the mathematical combinations such as $(x_i \times x_e)$, $\frac{x_i}{x_e}$, and $\frac{\partial x_i}{\partial x_e}$ will be extensive variables.

7.4.2 State of a thermodynamic system

In thermodynamics, the state of a thermodynamic system at any instant of time represents its condition at that time. The state of a system is fully described by the values of a suitable set of parameters known as state parameters, state variables, or thermodynamic variables. If such a set of values of thermodynamic variables is specified for a system, the values of all thermodynamic properties of the system can be uniquely determined. A thermodynamic state is usually found to be in a thermodynamic equilibrium. This means that the state of the system at a specific time continues to be the same over an indefinitely long period of time. Examples of such experimentally measurable thermodynamic variables are temperature, pressure, volume, surface area, electric field, etc.

Path and processes of a thermodynamic system

The succession of states through which a system passes during the change of state from one to another is called the **path of the system**. Each and every point on this path corresponds to an equilibrium state of the system. While going through a series of changes in state, a system follows a particular **process**. During a particular process following through certain paths, the following things may happen:

1. Some or all the properties of a system may undergo changes.

2. A process can be construed to be the locus of changes of state.

Processes in thermodynamics play the roles of streets in a city. For example, in a city, streets are spread from north to south, east to west, roundabouts, and crescents.

Quasi-static processes

A quasi-static process is defined as a process that happens at an infinitesimally slow rate so that the deviation from thermodynamic equilibrium is infinitesimal. In this process, all states of the system are in thermodynamic equilibrium with its surroundings at all times. An example of quasi-static process is the quasi-static expansion of a mixture of oxygen and hydrogen gas. Here, the volume of the system changes so slowly that the pressure of the system can be safely assumed to remain uniform throughout the system at each instant of time during the process. Such an idealized process is a succession of physical equilibrium states characterized by infinite slowness.

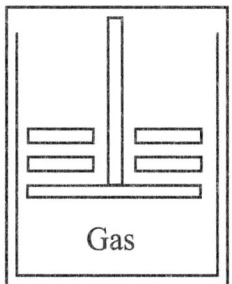

Figure 7.5 An example of a quasi-static process.

There are some weights attached over the piston shown in Figure 7.5. If these weights are removed slowly one by one, the pressure on the gas will decrease, and the piston will be displaced upward gradually due to the pressure of the gas. This is an example of **a quasi-static process**. On the other hand, if all the weights are removed at once, the piston will be kicked up by the gas pressure. This is an example of an unrestrained expansion. The work done in this case is not considered as it does not take place in a sustained manner. But, it should be mentioned that in both cases, the systems have undergone a change of state.

Climbing down a ladder from the roof to the ground by a man is the simplest example of a quasi-static process. On the other hand, if the man jumps from the roof to the ground, then it is not a quasi-static process. **It leads to severe damage to the man that cannot be explained by the concept of thermodynamics!!**

Indicator diagram

When the values of thermodynamic variables associated with a system change from one equilibrium state to another, the system is said to undergo a thermodynamic process. In order to analyze such a process, it is necessary to plot the variation of one thermodynamic variable with respect to another in a graph. Such a graphical representation of the cyclic variations of pressure and volume in a thermodynamic process is known as the **indicator diagram**. Such an indicator diagram is shown in Figure 7.6. The state of

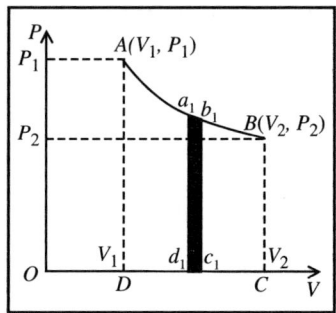

Figure 7.6 Indicator diagram.

a thermodynamic system is uniquely determined by a point on this diagram, and a process just means a line joining a series of such points. The shape of the indicator diagram shall depend on the nature of the thermodynamic process the system undergoes.

The state of a thermodynamic system is uniquely determined by a point on this diagram. A particular process in thermodynamics just means a line joining a series of such points in the indicator diagram. Let us consider one mole of an ideal gas enclosed in a cylinder fitted with a perfectly frictionless piston. Let $A(V_1, P_1, T)$ and $B(V_2, P_2, T)$ be the initial and final states of the gas. If dV is an infinitesimally small increase in the volume of the gas during which the pressure P is assumed to be constant, then the small amount of work done by the gas is $\delta W = P dV$. In the indicator diagram, such a small amount of work is shown by $\delta W = $ area $a_1 b_1 c_1 d_1$. The total work done by the gas during expansion from initial volume V_1 to final volume V_2 is

$$W = \int_{V_1}^{V_2} P \, dV = \text{area } ABCD, \text{ in the indicator diagram.}$$

Hence, in an indicator diagram, the area under the curve represents the work done. It may be emphasized here that the process of joining the initial and final points with a line has an important implication in that the indeterminate states are also equilibrium states.

State postulate

This postulate provides an idea about the number of independent intensive thermodynamic variables required to specify the state of a closed thermodynamic system. Subject to the conditions of local equilibrium, this number, according to this postulate, is $(n + m)$, where "n" is the number of different (nonchemical) work modes of energy transport, and "m" is the number of different pure substances present in the closed system.

For a pure substance $(m = 1)$ subjected to only one work mode $(n = 1)$, two independent intensive properties $(n + m = 2)$ are required to fix the state of the system completely. Such a system is called a **simple system**. A pure gas or vapor under compression or expansion is an example of a simple system. In this case, the work mode is moving the system boundary work.

A mathematical relation frequently used in thermodynamics

Thermodynamic variables are used to describe the infinitesimal changes in a thermodynamic process from an initial equilibrium state to another. This enables us to use an equation of equilibrium (equation of state) and to solve it for any coordinate in terms of the other two. The differentials dP, dV, and dT, therefore, are differentials of actual functions and are called exact differentials.

If dZ is an exact differential of a function of, say, X and Y, then dZ may be written as

$$dZ = \left(\frac{\partial Z}{\partial X}\right)_Y dX + \left(\frac{\partial Z}{\partial Y}\right)_X dY.$$

An infinitesimal that is not the differential of an actual function is called an **inexact differential** and cannot be expressed by an equation of the type shown above.

There are two simple theorems in partial differential calculus that are used frequently in thermodynamics. The proofs are as follows:

Let us assume that there exists a relation among the three coordinates X, Y, and Z; thus,

$$f(X, Y, Z) = 0. \tag{7.1}$$

Then X can be imagined to be a function of Y and Z, and the differential change dX in X can be written as

$$dX = \left(\frac{\partial X}{\partial Y}\right)_Z dY + \left(\frac{\partial X}{\partial Z}\right)_Y dZ. \tag{7.2}$$

Also, Y can be imagined to be a function of X and Z, and the differential change dY in Y can be written as

$$dY = \left(\frac{\partial Y}{\partial X}\right)_Z dX + \left(\frac{\partial Y}{\partial Z}\right)_X dZ. \tag{7.3}$$

Using equation (7.3) in equation (7.2), we get

$$dX = \left(\frac{\partial X}{\partial Y}\right)_Z \left[\left(\frac{\partial Y}{\partial X}\right)_Z dX + \left(\frac{\partial Y}{\partial Z}\right)_X dZ\right] + \left(\frac{\partial X}{\partial Z}\right)_Y dZ. \tag{7.4}$$

This leads to

$$dX = \left(\frac{\partial X}{\partial Y}\right)_Z \left(\frac{\partial Y}{\partial X}\right)_Z dX + \left[\left(\frac{\partial X}{\partial Y}\right)_Z \left(\frac{\partial Y}{\partial Z}\right)_X + \left(\frac{\partial X}{\partial Z}\right)_Y\right] dZ. \qquad (7.5)$$

Now, of the three coordinates, only two are independent. Choosing X and Z as the independent coordinates, the equation above must be true for all sets of values of dX and dZ. Thus, if $dZ = 0$ and $dX \neq 0$, it follows that

$$\left(\frac{\partial X}{\partial Y}\right)_Z \left(\frac{\partial Y}{\partial X}\right)_Z = 1; \quad \Rightarrow \quad \left(\frac{\partial X}{\partial Y}\right)_Z = \frac{1}{\left(\frac{\partial Y}{\partial X}\right)_Z}. \qquad (7.6)$$

This is a very useful relation in thermodynamics.

Thus, if $dX = 0$ and $dZ \neq 0$, it follows that

$$\left(\frac{\partial X}{\partial Y}\right)_Z \left(\frac{\partial Y}{\partial Z}\right)_X + \left(\frac{\partial X}{\partial Z}\right)_Y = 0; \quad \Rightarrow \quad \left(\frac{\partial X}{\partial Y}\right)_Z \left(\frac{\partial Y}{\partial Z}\right)_X = -\left(\frac{\partial X}{\partial Z}\right)_Y. \qquad (7.7)$$

Using equation (7.6) in equation (7.7), we get

$$\left(\frac{\partial X}{\partial Y}\right)_Z \left(\frac{\partial Y}{\partial Z}\right)_X \left(\frac{\partial Z}{\partial X}\right)_Y = -1. \qquad (7.8)$$

In terms of P, V, and T, this equation (7.8) becomes

$$\left(\frac{\partial P}{\partial V}\right)_T \left(\frac{\partial V}{\partial T}\right)_P \left(\frac{\partial T}{\partial P}\right)_V = -1. \qquad (7.9)$$

These relations, given by equations (7.8) and (7.9), are very frequently used in thermodynamics.

State function

A state function, function of state, or point function for a system in thermodynamic equilibrium is defined as a function relating several state variables that depends only on the present thermodynamic state of the system, not on the path followed by the system to reach the present state. Being able to describe the equilibrium state of a system, a state function could explain the behavior of various types of systems. For example, a state function could describe the following systems:

1. An atom or a molecule in a gaseous, liquid, or solid form,

2. A homogeneous or heterogeneous mixture,

3. The amounts of energy required to create such systems,

4. The amount of energy required to change such systems into a different equilibrium state.

There are certain thermodynamic quantities required to define the state of a thermodynamic system. Examples of such state quantities include internal energy, enthalpy, entropy, etc. These quantities quantitatively describe an equilibrium state of a thermodynamic system, regardless of how the system arrived in that state. Two such states (1) and (2) with thermodynamic coordinates, respectively, (V_1, P_1) and (V_2, P_2) are shown in Figure 7.7. It should be mentioned here that heat and mechanical work are path functions because their values depend on the specific "transition" (or "path") between two equilibrium states. However, heat (in certain discrete amounts) can describe a state function such as enthalpy, but in general, does not truly describe the system unless it is defined as the state function of a certain system. Enthalpy is such a state function of the amount of heat. A similar argument can also be applied to entropy that behaves as a state function when heat is compared to temperature.

Path function

A path function for a thermodynamic system is defined as a function whose value depends on the path followed by a thermodynamic process in going from an initial to a final state of the process. In other words, a path function depends on the path taken to reach a final state from an initial state. The path function is also called a process function. Examples of path functions are heat and work. These functions cannot be defined for a state.

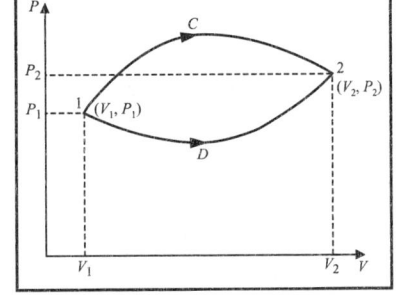

Figure 7.7 Two possible paths between states 1 and 2.

Since a path function depends on the path, it gives different values for different paths followed by the thermodynamic system. Therefore, multiple integrals and limits are required to express mathematically and to integrate the path function.

Figure 7.7 shows two possible paths in going from state 1 to state 2 of a thermodynamic system. These paths are identified, respectively, as $(1C2)$ and $(1D2)$. The work done by the system following these two paths can be written as

$$W_C = \int_C PdV \quad \text{and} \quad W_D = \int_D PdV.$$

It is easy to notice that the work done on these two different paths is different as the work done is given by the integral giving the area beneath the curve. These two shaded areas are shown in Figure 7.8. Path functions are not properties of the system, while point or state functions are properties of the system. Any change in the value of a point function can be obtained from the initial and final values of the function,

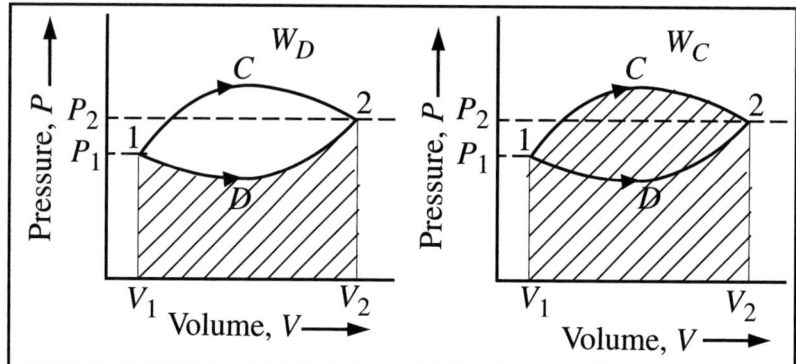

Figure 7.8 Work done along two different paths between states 1 and 2.

whereas the path followed by the thermodynamic process has to be defined in order to evaluate the value of the path functions. Some examples of path functions are arc length, mechanical work, and heat.

Let us consider, for example, the mathematical expression for the internal energy of a thermodynamic system. It is given by (basically, it is the mathematical form of the first law of thermodynamics)

$$\Delta U = \delta Q + \delta W.$$

Here, ΔU is the change in internal energy, δQ is the supplied heat, and δW is the mechanical work done. It should be mentioned here that the internal energy U is a state function, but heat Q and work W are both path functions.

Differences between state and path functions are summarized in the table below

No. of observations	State function	Path function
1.	State function is a thermodynamic term used to specify a property whose value is independent of the path taken to reach that specific value.	Path function is a thermodynamic term used to specify a property whose value does depend on the path taken to reach that specific value.
2.	These functions are also called point functions.	These functions are also called process functions.
3.	These functions do not depend on the path or process.	These functions depend on the path or process.

4.	This function can be integrated from the initial to the final value of the thermodynamic property of the system.	Path function requires multiple integrals and limits of integration to integrate the property.
5.	The value of the state function is independent of the number of steps involved.	The value of the path function of a single-step process is different from a multiple-step process.
6.	Some examples of state functions are entropy, enthalpy, mass, volume, temperature, etc.	Examples of path functions are heat and mechanical work.

7.4.3 Thermodynamic equilibrium

The equilibrium state of a thermodynamic system refers to that state, which does not undergo any change in itself with the passage of time without the influence of any external factors. This state of a system can be investigated by observing whether any change in the state of the system occurs or not. If there is no change with time or due to any external influence observed in the state of the system, the corresponding state of the system will then be in equilibrium.

The state of a thermodynamic system undergoes a change when a certain amount of work is involved in it, and there is thermal interaction with its surroundings. After a certain period of time, a state is reached when all the observable changes in its measurable properties cease to occur as long as external conditions remain unaltered. This state is known as the equilibrium state of the combined configuration, and the system is said to be in thermodynamic equilibrium.

The following mapping can be made regarding the existence of an equilibrium between the system and the surroundings. The type of equilibrium and the corresponding responsible parameters for a thermodynamic system are mentioned below:

1. Mechanical equilibrium $\longrightarrow$ pressure

2. Electrical equilibrium $\longrightarrow$ potential

3. Species equilibrium $\longrightarrow$ concentration of species

4. Thermal equilibrium $\longrightarrow$ temperature

Thus, the absence of any type of interactions between the system and the surroundings leads to the equilibrium state of a system. When all the abovementioned conditions of equilibrium exist together, a system will be in thermodynamic equilibrium. Since a thermodynamic system can be hydrostatic, physical, or chemical in general, thermodynamic equilibrium is defined as the state achieved by a system when it is in thermal, mechanical, and chemical equilibrium with its surroundings. It is important to note that a system in thermodynamic equilibrium does not deliver anything.

Mechanical equilibrium

In thermodynamics, a mechanical equilibrium is defined as the state of a thermodynamic system in which it experiences no pressure or elastic stress within it, and there is no unbalanced force or torque acting between the system and its surroundings. These factors should not change with time. There will not be any bulk movement of the fluids because there is no pressure gradient inside the system. In classical mechanics, equilibrium is defined by the condition that the sum of external forces acting on a system equals to zero. The link between the two is that no external forces perform work in an equilibrium situation. For example, a gas in a cylinder fitted with a piston is said to be in mechanical equilibrium if there is no unbalanced force acting on the piston. When a person presses a spring to a defined point, the compressive load and the spring reaction become equal at that point, and the spring remains stationary. In this state, the spring-system is in mechanical equilibrium.

Thermal equilibrium

Temperature is a property that distinguishes thermodynamics from other sciences. This property can distinguish between hot and cold objects. When two or more bodies at different temperatures are brought into contact with each other, then after some time, these bodies attain a common temperature, and they are said to exist in thermal equilibrium. Hence, thermal equilibrium is defined as the state attained by two or more systems placed in thermal contact with each other through a diathermic wall.

Thermodynamic systems are said to be in thermal equilibrium if there is no transfer of heat between the systems, even if they are in a position to transfer heat, based on other factors. For example, food kept in the refrigerator overnight is in thermal equilibrium with the air of that refrigerator. Inside the refrigerator, no heat flows from food to the air or from the air to the food. Thus, food and air inside the refrigerator are in the state of thermal equilibrium.

Chemical equilibrium

The chemical equilibrium in a thermodynamic system indicates a point at which the concentrations of reactants and products of a chemical reaction do not change with time. It appears from the outside that the reaction has stopped, but in fact, the rates of the forward and reverse reactions are equal such that reactants and products in the chemical chamber are being created at the same rate. Hence, **chemical equilibrium** is defined as the state attained by a system if the chemical composition of the system is the same throughout. No observable chemical changes are noticed when a system is in chemical equilibrium. Thus, if a system is such that it neither shows a tendency to undergo a spontaneous change in its internal structure (such as during a chemical reaction) nor allows the transfer of matter from one position of it to another (such as in diffusion), it is said to be in chemical equilibrium.

It should be mentioned that at constant temperature and pressure, the chemical potential of every species must be the same. For example, in the case of the vapor–liquid equilibrium for a system, the chemical potential for both liquid and vapor will have the same value. This can be generalized to any number of phases, for which the chemical potential of every species must be the same in all phases. It is known that this chemical potential is the driving force that moves a species from one phase to the other. If the chemical potential of a species in one phase is the same as that in the other, there is no net driving force in the system. As a result, there is no net transfer of species at equilibrium.

When a system satisfies all the abovementioned equilibrium criteria, that is, mechanical, thermal, and chemical equilibrium, the system will then be in **thermodynamic equilibrium**. It can be shown that at equilibrium, the total Gibbs free energy of the system G_{total} must have a minimum value. This indicates that **the differential of Gibbs free energy** dG_{total} **must be zero at equilibrium**. Actually, this is a restatement of the second law of thermodynamics that the Gibbs free energy will be minimum at equilibrium. This also signifies that considering all of the possible states for equilibrium, the entropy of a system must be at its maximum in equilibrium.

7.5   The zeroth law of thermodynamics

There are four laws in thermodynamics. The zeroth law is one of them, though it was invented later than the original three laws. The credit for formulating the zeroth law in thermodynamics goes to Ralph H. Fowler. There was little confusion regarding the nomenclature of this law – whether it should be named **the fourth law or be given some other name**. The confusion arose because a much clear definition of temperature was obtained from this new law. It basically replaced the definition of temperature what the other three laws had to state. Fowler finally came up with the name "zeroth law of thermodynamics" to end this confusion. This zeroth law provides an idea about temperature as an indicator of thermal equilibrium between two or more thermodynamic systems.

7.5.1   Statement of the zeroth law of thermodynamics

The zeroth law of thermodynamics states that "when a body 'A' is in thermal equilibrium with another body 'B', and also separately in thermal equilibrium with a third body 'C', then both bodies 'B' and 'C' will also be in thermal equilibrium with each other". This law is based on the measurement of temperature. It should be mentioned that there are a number of ways to state the zeroth law of thermodynamics. In simple terms, this law can be expressed as "systems that are in thermal equilibrium exist at the same temperature".

It should be mentioned that though it is termed as the zeroth law of thermodynamics, it is the last law of thermodynamics that we know of so far. The zeroth law of thermodynamics provides the idea of temperature and the thermal equilibrium. This thermal equilibrium indicates that when two objects are brought into contact with each other and separated by a barrier that is permeable to heat, no transfer of heat takes place from one to the other. This eventually indicates that the two objects are at the same temperature. James Clerk Maxwell paraphrased this in a more simple way by saying that "All heat is of the same kind". From the viewpoint of thermodynamics, it is important to mention that the zeroth law establishes that temperature is a fundamental and measurable property of matter.

The zeroth law of thermodynamics takes into account that temperature is something worth measuring because it predicts whether heat will be transferred between objects or not. This is true regardless of how the objects interact. Even if two objects are not in physical contact, heat still can flow between them, by means of radiation mode of heat transfer. Thus, the zeroth law of thermodynamics can be stated in the following way: **If the systems are in thermal equilibrium, no heat flow will take place between them.**

7.5.2   Concept of temperature from the zeroth law of thermodynamics

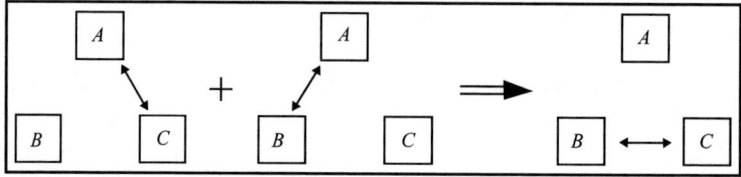

Figure 7.9 The double arrow indicates the thermal equilibrium between systems. The left figures show that systems A and C and A and B are in thermal equilibrium with each other. The figure on the right then shows that systems B and C are in thermal equilibrium.

The zeroth law of thermodynamics is one of the four laws of thermodynamics. It provides a much clearer definition of the thermodynamic parameter "temperature" and essentially underpins the other three laws. The zeroth law of thermodynamics frames the idea of temperature as an indicator of thermal equilibrium and is based on the measurement of temperature. The concept of the zeroth law of thermodynamics is schematically presented in Figure 7.9. The system "A" is in thermal equilibrium with "B" and "C" separately, then according to the zeroth law of thermodynamics, "B" and "C" will also be in thermal equilibrium with each other. Further, it should be mentioned that this law is valid for more than three thermodynamic systems. Thus, the temperature is defined as the quantity that remains the same for all fluids in thermal equilibrium and is a measure of heat or thermal energy contained within a body.

Temperature plays a fundamental role in any subjects in thermodynamics. It is, therefore, necessary to have an objective way of measuring temperature. In general, when two objects are brought into thermal contact, heat will flow between them until they come into equilibrium with each other. When the flow of heat stops, they are said to be at the same temperature. Having this definition of temperature, a temperature scale is established by assigning numerical values to certain easily reproducible fixed points. For example, in the Celsius (°C) scale of temperature, these two fixed points are chosen to be the freezing point of pure water and the boiling point of water. The former is arbitrarily assigned a temperature of 0°C while the latter is assigned the value of 100°C; the pressure in both cases is maintained at 1 standard atmosphere. Similarly, these same two fixed points in the Fahrenheit (°F) scale of temperature are assigned the values 32°F and 212°F, respectively. Besides these two temperature scales, there are absolute temperature scales related to the second law of thermodynamics. The absolute scale related to the Celsius scale is known as the Kelvin (K) scale, and that related to the Fahrenheit scale is known as the Rankine (°R) scale. These scales are related by the equations $K = °C + 273.15$, $°R = °F + 459.67$, and $°R = 1.8 K$. Zero in both the Kelvin and the Rankine scales is at absolute zero.

7.5.3   Applications of the zeroth law of thermodynamics

The significance of the zeroth law of thermodynamics is that it provides the mathematical definition of temperature. Hence, this law can be used to set up the mathematical formulation of the thermodynamic properties where temperature is involved. This law is mostly used to compare the temperatures of different objects. Accurate determination of the temperature of some unknown object requires a reference body and a certain characteristic of that body, known as a thermodynamic property that changes with temperature. When properly calibrated, this change in the thermometric characteristic is considered as an indication of a change in temperature. There are different kinds of thermometers that can be used depending on their thermometric property. They are shown in Table 7.2.

The most common application of the zeroth law of thermodynamics can be seen in thermometers. The zeroth law can be seen in action by taking a common thermometer having mercury in a tube. As the temperature is increased, this mercury expands since the area of the tube is constant. Due to this expansion, the height is increased, which, in turn, shows the changes in temperature and helps one to measure temperature.

Table 7.2 Thermometers made with various properties.

Thermometer	Thermometric property
Constant volume gas thermometer	Pressure
Constant pressure gas thermometer	Volume
Electrical resistance thermometer	Resistant
Thermocouple	Thermal electromotive force
Mercury in glass thermometer	Length

Similarly, the zeroth law of thermodynamics can be seen in action when two glasses of water maintained at two different thermodynamic conditions are kept on a table. Suppose one glass contains hot water and the other glass contains cold water. Now, if these two glasses are kept on the table for a few hours, they will attain thermal equilibrium with the temperature of the room. This is a consequence of the zeroth law of thermodynamics.

7.6   Equation of state

An equation of state in thermodynamics is a thermodynamic equation that depicts the relationship among the state variables used to describe the state of matter under a given set of physical conditions. These state variables are volume, pressure, temperature, internal energy, or density. Modern equations of state are formulated mainly in terms of the Helmholtz free energy.

Such equations of state find a wide range of applications in describing the properties of pure substances and mixtures in gases, liquids, and solid states. Equation of state is also used to describe the state of matter in the interior of stars. With the help of the equation of state for a substance, one can calculate further the amount of work done in taking the substance from one equilibrium state to another along some specified path. The Clapeyron–Mendeleev equation is a simple example of the equation of state for ideal gases. This equation is given by

$$PV = RT,$$

where P, V, and T are the state variables, and V is the volume of one mole of the ideal gas. For a fixed amount of gas, one can choose two independent variables, either volume and temperature or volume and pressure, and can formulate the equation of state. It is not necessary to consider all the variables independent at the same time. The above expression is known as the equation of state for ideal gas. Considering the interaction between particles, the virial equation of state for real gases is given by

$$PV = RT \left[1 + \frac{B(T)}{V} + \frac{C(T)}{V^2} + \cdots \right],$$

where $B(T)$, $C(T)$, $\cdots$, are the second, third, etc. virial coefficients. These coefficients are the functions of temperature alone and depend on the forces of two, three, or more particles interacting in the system. This equation of state is widely used and is the most theoretically substantiated one for real gas.

The Van der Waals equation of state provides a qualitatively accurate picture of the phase diagram of real gases. This equation of state for one mole of a real gas is given by

$$\left(P + \frac{a}{V^2} \right) (V - b) = RT,$$

where a and b are two constants known as the Van der Waals constants. Values of these two constants are determined from the experimental data taken on real gases. This equation of state takes into account both the existence of attractive forces between the molecules of the real gas, which is responsible for the decrease in pressure, and the presence of repulsive forces, which appears when the distances between the molecules are small. The latter prevents the gas from infinite compression. This equation of state has made it possible for the first time to obtain a thermodynamically consistent description of the liquid–gas phase transition in real gases.

The above discussion shows that the equation of state, in general, can be written in the form

$$f(P, V, T) = 0. \tag{7.10}$$

It is to be noted that the form of the equation of state depends on the special properties of the substance. Any one of the three variables mentioned in equation (7.10) can be expressed as a function of the other two. Hence, the thermodynamic state of a system can be specified by stating the values of two parameters. The value of the third parameter can be determined from the equation of state of the given system.

Further, it should be mentioned that the function f in equation (7.10) is a single-valued function of pressure P and volume V at a given temperature T. This is a parametric form of the equation of state of the substance under consideration. The equation of state is a fundamental characteristic of a substance and is very useful for the determination of the physical properties of particular physical objects by applying the general principles of thermodynamics and hydrodynamics.

7.6.1 Some deductions from the equation of state

We can use the equation of state in parametric form to study the behavior of any substance under different conditions. To show this, we consider the following cases.

Hydrostatic system

We consider a hydrostatic system whose equation of state in parametric form is given by equation (7.10). Let us solve this parametric equation for pressure P. Considering pressure P as a function of volume V and temperature T, we can write

$$P = P(V, T). \tag{7.11}$$

The differential change dP in pressure P can be written as

$$dP = \left(\frac{\partial P}{\partial V}\right)_T dV + \left(\frac{\partial P}{\partial T}\right)_V dT. \tag{7.12}$$

For an isobaric process, $dP = 0$, and equation (7.12) reduces to

$$\left(\frac{\partial P}{\partial T}\right)_V = -\left(\frac{\partial P}{\partial V}\right)_T \left(\frac{\partial V}{\partial T}\right)_P. \tag{7.13}$$

In order to understand equation (7.12) physically, we introduce the coefficient of isobaric volume expansion or expansivity, α_P, of a substance in an isobaric process defined as

$$\alpha_P = \frac{1}{V} \left(\frac{\partial V}{\partial T}\right)_P, \tag{7.14}$$

and the isothermal elasticity κ_T of a substance is defined as

$$\kappa_T = -V \left(\frac{\partial P}{\partial V}\right)_T. \tag{7.15}$$

Another parameter, isothermal compressibility β_T is the inverse of the isothermal elasticity κ_T and is given by

$$\beta_T = \frac{1}{\kappa_T} = -\frac{1}{V} \left(\frac{\partial V}{\partial P}\right)_T. \tag{7.16}$$

We know that at constant temperature, volume decreases as pressure increases. So, the term $\left(\frac{\partial P}{\partial V}\right)_T$ is always a negative quantity. Therefore, a negative sign has been included in the definition of κ_T so as to keep this quantity positive and physically meaningful.

On combining equations (7.13), (7.14), and (7.15), we get

$$\left(\frac{\partial P}{\partial T}\right)_V = \frac{\alpha_P \, V \, \kappa_T}{V} = \alpha_P \, \kappa_T. \tag{7.17}$$

On using the result in equation (7.12), we find that when a hydrostatic substance undergoes an infinitesimal change, the change in pressure is given by

$$dP = -\left(\frac{\kappa_T}{V}\right) dV + \alpha_P \, \kappa_T \, dT. \tag{7.18}$$

This result shows that if a system is allowed to expand, the change in pressure will be less than that for an isochoric process. For an isochoric process, equation (7.18) reduces to

$$dP = \alpha_P \, \kappa_T \, dT. \tag{7.19}$$

Thus, for a finite change in temperature, the change in pressure is given by

$$\int_{P_1}^{P_2} dP = \int_{T_1}^{T_2} \alpha_P \, \kappa_T \, dT. \tag{7.20}$$

If α_P and κ_T are independent of temperature in the small range of temperatures, we can write

$$P_2 - P_1 = \alpha_P \, \kappa_T \, (T_2 - T_1). \tag{7.21}$$

Equation (7.21) shows that if α_P and κ_T are known experimentally, then for an isochoric process, we can calculate the final pressure for a given rise in temperature and vice versa.

It would be shown that the equation of state can be used to calculate the change in pressure when the temperature of a system is raised during an isochoric process. In order to find this, let us solve the parametric equation (7.10) for volume V. Considering volume V as a function of pressure P and temperature T, we can write

$$V = V(P, T). \tag{7.22}$$

The differential change dV in volume V can be written as

$$dV = \left(\frac{\partial V}{\partial P}\right)_T dP + \left(\frac{\partial V}{\partial T}\right)_P dT. \tag{7.23}$$

In terms of $\alpha_P = \frac{1}{V}\left(\frac{\partial V}{\partial T}\right)_P$ and $\beta_T = -\frac{1}{V}\left(\frac{\partial V}{\partial P}\right)_T$, this equation (7.23) reduces to

$$dV = \alpha_P \, V \, dT - V \, \beta_T \, dP; \quad \Rightarrow \quad \frac{dV}{V} = \alpha_P \, dT - \beta_T \, dP. \tag{7.24}$$

For an isobaric process, $dP = 0$. Equation (7.24) then reduces to

$$dV = \alpha_P \, V \, dT. \tag{7.25}$$

For a finite change in temperature, we can write

$$V_2 - V_1 = \alpha_P \, V \, (T_2 - T_1). \tag{7.26}$$

Over the entire range of change in the thermodynamic variables considered here, α_P has been assumed to be constant.

A wire attached to a rigid support at its two ends

To illustrate the use of the equation of state, we consider a wire attached to a rigid support. The state of the wire will be defined by the tension, length, and the temperature since P as well as V can be assumed to be constant. Hence, we can write

$$F = F(L, T). \tag{7.27}$$

The differential change in tension dF can be written as

$$dF = \left(\frac{\partial F}{\partial L}\right)_T dL + \left(\frac{\partial F}{\partial T}\right)_L dT. \tag{7.28}$$

If the tension remains constant, $dF = 0$, and equation (7.28) reduces to

$$\left(\frac{\partial F}{\partial L}\right)_T \left(\frac{\partial L}{\partial T}\right)_F = -\left(\frac{\partial F}{\partial T}\right)_L. \tag{7.29}$$

We now introduce isothermal Young's modulus defined by

$$Y = \frac{L}{A}\left(\frac{\partial F}{\partial L}\right)_T, \tag{7.30}$$

where A is the cross-sectional area of the wire. Similarly, the coefficient of linear expansion α is defined as

$$\alpha = \frac{1}{L}\left(\frac{\partial L}{\partial T}\right)_F. \tag{7.31}$$

On combining equations (7.29), (7.30), and (7.31), we get

$$\left(\frac{\partial F}{\partial T}\right)_L = -AY\alpha. \tag{7.32}$$

On using the result in equation (7.28), we find that when a wire undergoes an infinitesimal change, the change in tension is given by

$$dF = \left(\frac{AY}{L}\right) dL - AY\alpha dT. \tag{7.33}$$

If we consider a wire vibrating between two fixed rigid supports, $dL = 0$, and equation (7.33) reduces to

$$dF = -AY\alpha dT. \tag{7.34}$$

It is to be noted that this does not depend on the length of the wire. For a small change in temperature, we get

$$\int_{F_1}^{F_2} dF = -A \int_{T_1}^{T_2} Y\alpha \, dT. \tag{7.35}$$

If Y and α are independent of temperature, we can write

$$F_2 - F_1 = AY\alpha(T_1 - T_2). \tag{7.36}$$

Equation (7.36) shows that the temperature of a wire vibrating between two fixed supports will drop if tension is increased and vice versa.

Table 7.3 Equations of state for some physical systems.

Systems	Parametric form of equation of state	Equation of state
Hydrostatic	$f(P,V,T) = 0$	Ideal gas $PV = RT$; real gas $\left(P + \frac{an^2}{V^2}\right)(V - nb) = nRT$.
Paramagnetic substance	$f(M,B,T) = 0$	$M = \frac{kB}{T}$.
Electric cell	$f(V_d, Z, T) = 0$	$V_d = c_0 + c_1 T + c_2 T^2 + c_3 T^3$.
Surface film	$f(\sigma, A, T) = 0$	$\sigma = \sigma_0 \left[\frac{T_C - T}{T_C - T_0}\right]$.
Stretched wire	$f(L, F, T) = 0$	$L = L_0 \left(1 + \frac{F}{YA} + \alpha(T - T_0)\right)$.
Dielectric solid	$f(P, E, T) = 0$	$P = \left(a + \frac{b}{T}\right) E$.

7.7 Solved numerical problems

1. What is the zeroth law of thermodynamics? On the basis of this law, introduce the concept of temperature. [Calcutta University 2013, 2018]

   Answer: See text for the statement of the zeroth law of thermodynamics.

This law defines thermal equilibrium between different systems and also between different parts of an isolated system. The essence of zeroth law is: When two objects at thermal equilibrium are in contact with each other, there is no net transfer of heat between the two objects; therefore, they are at the same temperature. Another way to state the zeroth law of thermodynamics is that if two objects are separately in thermal equilibrium with a third one, then they are all in thermal equilibrium with each other.

The zeroth law of thermodynamics forms the foundation of measurement of temperature of the material objects. For example, a thermometer is an easily accessible simple instrument used to measure the temperature, and its working principle is based on the zeroth law of thermodynamics. Using the thermometer, one can measure the temperature of a water bath following a simple process. By inserting the thermometer into the water bath, one has to wait for a sufficiently long time so that the reading of the thermometer remains constant to a value with time, and that value is found to be accurate (well calibrated). This signifies that the thermometer and the water bath reach thermal equilibrium. Under this condition of thermal equilibrium, the temperature of the thermometer bulb and the water bath will be the same, and there should not be any net transfer of heat from one object to the other (it is assumed that there are no other losses of heat to the surroundings).

2. The equation of state for the magnetization M of a paramagnetic substance is $M = \frac{kB}{T}$, where k is a constant, and B is the magnetic field. Show that

$$\left(\frac{\partial M}{\partial B}\right)_T \left(\frac{\partial B}{\partial T}\right)_M \left(\frac{\partial T}{\partial M}\right)_B = -1.$$

Answer: According to the problem, magnetization M is found to be a function of magnetic field B and temperature T, that is, we can write

$$M = M(B, T).$$

The differential change dM in magnetization M can be written as

$$dM = \left(\frac{\partial M}{\partial B}\right)_T dB + \left(\frac{\partial M}{\partial T}\right)_B dT.$$

Now, M increases with an increase in B, whereas M decreases with an increase in T. These two variables can be so changed that M remains constant. In this situation, we can write

$$0 = \left(\frac{\partial M}{\partial B}\right)_T dB + \left(\frac{\partial M}{\partial T}\right)_B dT; \quad \Rightarrow \quad \left(\frac{\partial M}{\partial B}\right)_T \left(\frac{\partial B}{\partial T}\right)_M = -\left(\frac{\partial M}{\partial T}\right)_B.$$

This leads to

$$\left(\frac{\partial M}{\partial B}\right)_T \left(\frac{\partial B}{\partial T}\right)_M \left(\frac{\partial T}{\partial M}\right)_B = -1.$$

3. Derive the equation of state of a substance if the coefficient of volume expansion, or expansivity α_P and the isothermal compressibility β_T of the substance are given.

Answer: To find the equation of state, we consider pressure P and temperature T as independent variables, and an attempt is made to determine V uniquely. So,

$$V = V(P, T).$$

On partial differentiation, we obtain

$$dV = \left(\frac{\partial V}{\partial P}\right)_T dP + \left(\frac{\partial V}{\partial T}\right)_P dT.$$

In terms of $\alpha_P = \frac{1}{V}\left(\frac{\partial V}{\partial T}\right)_P$ and $\beta_T = -\frac{1}{V}\left(\frac{\partial V}{\partial P}\right)_T$, the above expression takes the form

$$dV = -\beta_T V\, dP + \alpha_P V\, dT; \quad \Rightarrow \quad \frac{dV}{V} = -\beta_T\, dP + \alpha_P\, dT. \qquad (7.37)$$

From equation (7.37), we note that once α_P and β_T are known experimentally, one can calculate the change in volume for a given change in temperature and pressure. Alternatively, one may use this information to arrive at the equation of state for the system of interest.

4. Derive the equation of state if the volume expansion coefficient $\alpha_P = \frac{2bT}{V}$ and the isothermal compressibility $\beta_T = \frac{a}{V}$, where a and b are two constants.

Answer: From equation (7.37), we have the relation

$$\frac{dV}{V} = -\beta_T\, dP + \alpha_P\, dT.$$

Using the values of α_P and β_T in the above expression, we get

$$\frac{dV}{V} = -\frac{a}{V}\, dP + \frac{2bT}{V}\, dT; \quad \Rightarrow \quad dV = -a\,dP + 2bT\,dT.$$

Integrating the above expression, we get

$$V + aP - bT^2 = \text{constant}.$$

This is the required equation of state.

5. The isothermal compressibility and expansivity of a hypothetical substance are, respectively, given by $\beta_T = \frac{aT^3}{P^2}$ and $\alpha_P = \frac{bT^2}{P}$, where a and b are constants. Derive the equation of state for the system under consideration.

Answer: From equation (7.37), we have the relation

$$\frac{dV}{V} = -\beta_T \, dP + \alpha_P \, dT.$$

Using the values of α_P and β_T in the above expression, we get

$$\frac{dV}{V} = -\frac{aT^3}{P^2} \, dP + \frac{bT^2}{P} \, dT. \tag{7.38}$$

Here, $M = -\frac{aT^3}{P^2}$ and $N = \frac{bT^2}{P}$. Hence, we have

$$\frac{\partial M}{\partial T} = \frac{\partial}{\partial T}\left(-\frac{aT^3}{P^2}\right) = -\frac{3aT^2}{P^2} \text{ and } \frac{\partial N}{\partial P}$$

$$= \frac{\partial}{\partial P}\left(\frac{bT^2}{P}\right) = -\frac{bT^2}{P^2}.$$

As it is an exact differential, we have $\frac{\partial M}{\partial T} = \frac{\partial N}{\partial P}$. This leads to $b = 3a$. Using this value into equation (7.38), we get

$$\frac{dV}{V} = -\frac{aT^3}{P^2} \, dP + \frac{3aT^2}{P} \, dT; \quad \Rightarrow \quad \frac{dV}{V} = d\left(\frac{aT^3}{P}\right).$$

Integrating this expression, we get

$$\ln V = \frac{aT^3}{P} + \ln V_0; \quad \Rightarrow \quad V = V_0 e^{\left(\frac{aT^3}{P}\right)},$$

where V_0 is the constant of integration. This is the equation of state.

6. Calculate the minimum attainable pressure of an ideal gas during a process governed by the relation $T = T_0 + \alpha V^2$, where T_0 and α are positive constants, and V is the volume of one mole of the gas.

Answer: The equation of state for one mole of an ideal gas is given by

$$PV = RT.$$

Using the expression of T from the problem, we get

$$PV = R\left(T_0 + \alpha V^2\right) = RT_0 + \alpha RV^2. \quad \Rightarrow \quad P = \frac{RT_0}{V} + \alpha RV.$$

Differentiating this w.r.t temperature T, we get

$$\frac{\partial P}{\partial T} = -\frac{RT_0}{V^2}\frac{\partial V}{\partial T} + \alpha R\frac{\partial V}{\partial T}.$$

As P to be minimum, $\frac{\partial P}{\partial T} = 0$. Hence, from the above expression, we have

$$0 = -\frac{RT_0}{V^2}\frac{\partial V}{\partial T} + \alpha R\frac{\partial V}{\partial T}; \quad \Rightarrow \quad \frac{\partial V}{\partial T}\left[\alpha R - \frac{RT_0}{V^2}\right].$$

$$\alpha R - \frac{RT_0}{V^2} = 0; \quad \Rightarrow \quad T_0 = \alpha V^2; \quad \Rightarrow \quad V = \sqrt{\frac{T_0}{\alpha}}.$$

Using this value of V, temperature T becomes

$$T = T_0 + \alpha V^2 = T_0 + T_0 = 2T_0.$$

So, the minimum attainable pressure is given by

$$P_{\min} = \frac{RT}{V} = \frac{R\,2T_0}{\sqrt{\frac{T_0}{\alpha}}} = 2R\sqrt{\alpha T_0}.$$

7. Calculate the maximum attainable temperature of an ideal gas during a process governed by the relation $P = P_0 e^{-\beta V}$, where P_0 and β are positive constants, and V is the volume of one mole of the gas.

   Answer: The equation of state for one mole of an ideal gas is given by

   $$P = P_0\,e^{-\beta V}; \quad \Rightarrow \quad \frac{RT}{V} = P_0\,e^{-\beta V}.$$

This gives

$$T = \frac{P_0 V}{R}\,e^{-\beta V}; \quad \Rightarrow \quad \frac{dT}{dV} = \frac{P_0}{R}\left[e^{-\beta V} - \beta V e^{-\beta V}\right].$$

Temperature T to be maximum, $\frac{\partial T}{\partial V} = 0$. Hence, from the above expression, we have

$$0 = \frac{P_0}{R}\left[e^{-\beta V} - \beta V e^{-\beta V}\right]; \quad \Rightarrow \quad V = \frac{1}{\beta}.$$

Using this value of V, maximum temperature T becomes

$$T_{\max} = \frac{P_0}{e\beta R}.$$

8. Calculate the coefficient of volume expansion α_P for one mole of a Van der Waals gas.

   Answer: From definition, the coefficient of volume expansion α is given by

   $$\alpha_P = \frac{1}{V}\left(\frac{\partial V}{\partial T}\right)_P.$$

   The equation of state for one mole of a Van der Waals gas is given by

   $$\left(P + \frac{a}{V^2}\right)(V - b) = RT.$$

   Differentiating the above expression w.r.t temperature T at constant pressure P, we get

   $$\left[-\frac{2a}{V^3}\left(\frac{\partial V}{\partial T}\right)_P\right](V - b) + \left(P + \frac{a}{V^2}\right)\left(\frac{\partial V}{\partial T}\right)_P = R;$$

   $$\Rightarrow \left(\frac{\partial V}{\partial T}\right)_P\left[-\frac{2a(V - b)}{V^3} + \left(P + \frac{a}{V^2}\right)\right] = R;$$

   $$\Rightarrow \left(\frac{\partial V}{\partial T}\right)_P\left[-\frac{2a(V - b)}{V^3} + \frac{RT}{V - b}\right] = R;$$

   $$\Rightarrow \alpha_P V\left[-\frac{2a(V - b)^2 + RTV^3}{V^3(V - b)}\right] = R.$$

   This leads to

   $$\alpha_P = \left[\frac{RV^2(V - b)}{RTV^3 - 2a(V - b)^2}\right].$$

9. Calculate the isothermal compressibility β_T for one mole of a Van der Waals gas.

   Answer: From definition, the isothermal compressibility β_T is given by

   $$\beta_T = -\frac{1}{V}\left(\frac{\partial V}{\partial P}\right)_T; \quad \Rightarrow \quad \left(\frac{\partial V}{\partial P}\right)_T = -V\,\beta_T.$$

   The equation of state for one mole of a Van der Waals gas is given by

   $$\left(P + \frac{a}{V^2}\right)(V - b) = RT.$$

   Differentiating the above expression w.r.t pressure P at constant temperature T, we get

$$\left[1 - \frac{2a}{V^3}\left(\frac{\partial V}{\partial P}\right)_T\right](V-b) + \left(P + \frac{a}{V^2}\right)\left(\frac{\partial V}{\partial P}\right)_T = 0;$$

$$\Rightarrow (V-b) - \frac{2a(V-b)}{V^3}\left(\frac{\partial V}{\partial P}\right)_T + \left(\frac{RT}{V-b}\right)\left(\frac{\partial V}{\partial P}\right)_T = 0;$$

$$\Rightarrow \left(\frac{\partial V}{\partial P}\right)_T\left[\left(\frac{RT}{V-b}\right) - \left(\frac{2a(V-b)}{V^3}\right)\right] = -(V-b).$$

Thus, we have

$$\left(\frac{\partial V}{\partial P}\right)_T = -\frac{(V-b)}{\left(\frac{RT}{V-b}\right) - \left(\frac{2a(V-b)}{V^3}\right)} = -\frac{V^3(V-b)^2}{RTV^3 - 2a(V-b)^2};$$

$$\Rightarrow -V\beta_T = -\frac{V^3(V-b)^2}{RTV^3 - 2a(V-b)^2}.$$

This leads to

$$\beta_T = \frac{V^2(V-b)^2}{RTV^3 - 2a(V-b)^2}.$$

10. Show that for a hydrostatic system $\frac{dV}{V} = \alpha_P \, dT - \frac{dP}{\kappa_T}$, where α_P is the coefficient of volume expansion at constant pressure, and κ_T is the isothermal bulk modulus. [Calcutta University 2013]

Answer: From the equation of state, we have

$$V = V(T, P).$$

Any infinitesimal change from one state to another state of equilibrium involves a dV, a dT, and a dP, all of which we shall assume to satisfy the condition laid down in the definition of a quasi-static process. A fundamental theorem in partial differential calculus enables us to write

$$dV = \left(\frac{\partial V}{\partial T}\right)_P dT + \left(\frac{\partial V}{\partial P}\right)_T dP,$$

where each partial derivative is itself a function of T and P. Both partial derivatives have an important physical meaning. If the change in temperature is made infinitesimal, then the change in volume also becomes infinitesimal, and we have what is known as the infinitesimal volume expansivity, or just the coefficient of volume expansion, which is denoted by α. Thus,

$$\alpha_P = \frac{1}{V}\left(\frac{\partial V}{\partial T}\right)_P.$$

The effect of a change of pressure on the volume of a hydrostatic system when the temperature is kept constant is expressed by a quantity called isothermal compressibility and is represented by the symbol β_T. The isothermal bulk modulus κ_T is defined as the inverse of β_T. Thus,

$$\beta_T = -\frac{1}{V}\left(\frac{\partial V}{\partial P}\right)_T; \quad \Rightarrow \quad \kappa_T = -V\left(\frac{\partial P}{\partial V}\right)_T.$$

Using these two definitions, we have

$$dV = \alpha_P\, V\, dT + (-V\,\beta_T\, dP) = \alpha_P\, V\, dT - \frac{V}{\kappa_T}\, dP.$$

This leads to

$$\frac{dV}{V} = \alpha_P\, dT - \frac{1}{\kappa_T}\, dP.$$

11. Assuming the Van der Waals equation for one mole of a real gas, calculate $\left(\frac{\partial P}{\partial T}\right)_V$. [Calcutta University 2014]

Answer: For one mole of a Van der Waals gas, we have

$$\left(P + \frac{a}{V^2}\right)(V - b) = RT; \quad \Rightarrow \quad P + \frac{a}{V^2} = \frac{RT}{(V - b)}; \quad \Rightarrow \quad P = -\frac{a}{V^2} + \frac{RT}{(V - b)}.$$

Differentiating the above expression w.r.t temperature T at constant volume V, we get

$$\left(\frac{\partial P}{\partial T}\right)_V = \frac{R}{(V - b)} - \frac{2a}{V^3}\left(\frac{\partial V}{\partial T}\right)_P = -\frac{RT}{(V - b)^2}\left(\frac{\partial V}{\partial T}\right)_P + \frac{R}{V - b}.$$

7.8   Multiple choice questions with answers

1. Which law of thermodynamics suggests that there is a tendency for equalization of temperature throughout the system

   (a) Zeroth law   (c) Second law

   (b) First law   (d) Third law

   Answer: The correct choice is (a).

2. The working of a thermostat is based on the ... law of thermodynamics.

   (a) Zeroth   (c) Second

   (b) First   (d) Third

   Answer: The correct choice is (a).

3. Gas A is in thermal equilibrium with gases B and C. Which of the following is a valid conclusion?

(a) Thermal equilibrium of gas B is directly proportional to that of gas C.

(b) Gases A and B have equal amounts of entropy.

(c) Thermal equilibrium of gas B is indirectly proportional to that of gas C.

(d) Gases B and C are in thermal equilibrium with each other.

Answer: The correct choice is (d).

4. Most of the real processes are

(a) quasi-static   **(c)** adiabatic

(b) non-quasi-static   **(d)** isothermal

Answer: The correct choice is (b).

5. In a cyclic process, the change in internal energy

(a) is infinity   **(c)** is equal to area of cycle

(b) 0   **(d)** cannot be determined

Answer: The correct choice is (c).

6. An ideal gas is the one for which   [JAM 2010]

(a) C_V is a function of volume only.

(b) C_V is a function of volume and temperature.

(c) $\frac{C_P}{C_V}$ is not a function of temperature.

(d) C_V and C_P depend on temperature only.

Answer: The correct choice is (c).

7. Two gases separated by an impermeable but movable partition are allowed to exchange energy freely. At equilibrium, the two sides will have the same   [GATE 2013]

(a) pressure and temperature

(b) volume and temperature

(c) pressure and volume

(d) volume and energy

Answer: The correct choice is (a).

8. If the mean square fluctuations in energy of a system in equilibrium at temperature T is proportional to T^α, then the energy of the system is proportional to   [JEST 2017]

(a) $T^{\alpha-2}$ **(b)** $T^{\frac{\alpha}{2}}$ **(c)** $T^{\alpha-1}$ **(d)** T^{α}

Answer: The correct choice is (c).

Solution: $(\Delta E)^2 = kT^2 C_v \Rightarrow T^{\alpha-2} \propto C_v \Rightarrow T^{\alpha-2} \propto (\frac{\partial U}{\partial T})_V \Rightarrow U \propto T^{\alpha-1}$.

9. A monoatomic gas is described by the equation of state $P(V - nb) = nRT$, where b and R are constants and other quantities have their usual meanings. The maximum density (in moles per unit volume) to which this gas can be compressed is

(a) $\frac{1}{b}$ **(b)** b **(c)** $\frac{1}{n\,b}$ **(d)** ∞

Answer: The correct choice is (a).

10. If heat is supplied to an ideal gas in an isothermal process,

(a) the internal energy of the gas will increase.

(b) the gas will do positive work.

(c) the gas will do negative work.

(d) the said process is not possible.

Answer: The correct choice is (b).

11. A gas is contained in a metallic cylinder fitted with a piston. The piston is suddenly moved in to compress the gas and is maintained at this position. As time passes, the pressure of the gas in the cylinder

(a) increases.

(b) decreases.

(c) remains constant.

(d) increases or decreases depending on the nature of the gas.

Answer: The correct choice is (c).

After some time, the system will be in mechanical equilibrium. So, the pressure will remain the same throughout the system.

12. The pressure and density of a gas with $\gamma = \frac{7}{5}$ change adiabatically from (P_1, d_1) to (P_2, d_2). If $\frac{d_2}{d_1} = 32$, then the value of $\frac{P_2}{P_1}$ is

(a) 32 **(b)** 128 **(c)** $\frac{1}{32}$ **(d)** $\frac{1}{128}$

Answer: The correct choice is (b).

Solution: If there is no exchange of heat between a system and its surroundings, the process is known as adiabatic. The equation of state for an adiabatic change is given by $PV = $ constant.

Further, we know that volume $\propto \frac{1}{\text{density}}$, pressure $P \propto (\text{density})^\gamma$. This shows that

$$\Rightarrow P_1 = k \, d_1^\gamma; \quad P_2 = k \, d_2^\gamma; \quad \frac{P_2}{P_1} = \left(\frac{d_2}{d_1}\right)^\gamma.$$

Since we have $\frac{d_2}{d_1} = 32$, the required ratio in pressure is

$$\frac{P_2}{P_1} = (32)^{7/5} = (2^5)^{7/5} = 128.$$

13. Under an isothermal condition, the pressure P of a gas varies with volume V as $P \propto V^{-\left(\frac{5}{3}\right)}$. The bulk modulus B of the gas is proportional to

[GATE 2014]

(a) $V^{-\left(\frac{1}{2}\right)}$ (b) $V^{-\left(\frac{2}{3}\right)}$ (c) $V^{-\left(\frac{3}{5}\right)}$ (d) $V^{-\left(\frac{5}{3}\right)}$

Answer: The correct choice is (d).

Solution: According to the problem, we have $P = K V^{-\left(\frac{5}{3}\right)}$. The expression for bulk modulus B is given by $B = -V \left(\frac{dP}{dV}\right)_T$. Therefore, we have $B \propto V^{-\left(\frac{5}{3}\right)}$.

14. The molar specific heat of a gas as given from the kinetic theory is $\frac{5}{2}R$. If it is not specified whether it is C_V or C_P, one can conclude that the molecules of the gas

[JAM 2005]

(a) are definitely monoatomic.

(b) are definitely rigid diatomic.

(c) are definitely nonrigid diatomic.

(d) can be monoatomic or rigid diatomic.

Answer: The correct choice is (d).

Solution: If a molecule is monoatomic, then $C_P = \frac{5R}{2}$.

And if a molecule is a rigid diatomic, then $C_V = \frac{5R}{2}$.

15. If the number of degrees of freedom of a molecule in a gas is n, then the ratio of specific heats is given by

(a) $1 + \frac{1}{n}$ (b) $1 + \frac{1}{2n}$ (c) $1 + \frac{2}{n}$ (d) $\frac{2n}{2n-1}$

Answer: The correct choice is (c).

Solution: We know that the energy associated with each degree of freedom of one mole of gas is given by $\frac{1}{2}RT$. Hence, if there are n degrees of freedom, we have the internal energy as $U = \frac{n}{2}RT$, where R is the universal gas constant.

We know that the specific heat at constant volume is given as

$$C_V = \frac{dU}{dT}; \quad \Rightarrow \quad C_V = \frac{n}{2}R.$$

Hence, we have

$$C_P - C_V = R; \quad \Rightarrow \quad C_P = R + \frac{n}{2}R; \quad \Rightarrow \quad C_P = \left(1 + \frac{n}{2}\right)R.$$

By Mayor's formula, $\gamma = \frac{C_P}{C_V}$, we get $\gamma = 1 + \frac{2}{n}$.

16. One mole of a perfect gas expands adiabatically. As a result of this, its pressure, temperature, and volume change from P_1, T_1, V_1 to P_2, T_2, V_2, respectively. If molar specific heat at constant volume is C_V, then the work done by the gas is:

(a) $2.303 P_1 V_1 \log\frac{V_2}{V_1}$

(c) $\frac{P_1^2 V_1^2 - P_2^2 V_2^2}{R(T_2 - T_1)}$

(b) $RT_1 \log\frac{V_2}{V_1}$

(d) $C_V(T_1 - T_2)$

Answer: The correct choice is (d).

Solution: The specific heat at constant volume C_V is given by $C_V = \left(\frac{dU}{dT}\right)_V$. From the first law of thermodynamics, we get $\Delta Q = dU + \Delta W$. For an adiabatic process, $\Delta Q = 0$. Hence, $dU + \Delta W = 0$. Thus, we get

$$C_V dT + \Delta W = 0; \quad \Rightarrow \quad C_V(T_2 - T_1) + \Delta W = 0; \quad \Rightarrow \quad \Delta W = C_V(T_1 - T_2).$$

17. A thermally insulated vessel containing a gas whose molar mass is equal to M and the ratio of specific heat $\frac{C_P}{C_V} = \gamma$ moves with a velocity v. If the vessel is suddenly stopped, the increment of temperature is given by

(a) $\Delta T = \frac{(\gamma - 1)}{R} M v^2$

(b) $\Delta T = \frac{(\gamma - 1)}{2R} M v^2$

(c) $\Delta T = \frac{2(\gamma - 1)}{R} M v^2$

(d) $\Delta T = \frac{(2\gamma - 1)}{R} M v^2$

Answer: The correct choice is (b).

Solution: Suppose the number of moles of gas $= n$. The directional kinetic energy of gas is given by $\frac{1}{2}(nM)v^2$. When the vessel is suddenly stopped, after a long time, thermodynamic equilibrium is achieved, and this directional kinetic energy of a gas is converted into random kinetic energy and then $\frac{1}{2}(nM)v^2 = nC_V \Delta T \Rightarrow \Delta T = \frac{\gamma - 1}{2R} M v^2$.

7.9 Exercise

7.9.1 Short answer type questions

1. What do you mean by a thermodynamic system?

2. Explain clearly the difference between "surroundings" and "boundary" of a system.

3. What are "isolated" and "open" systems? Is a "closed system" an "isolated system"?

4. What do you mean by "equilibrium state"?

5. What do you mean by mechanical, thermal, and chemical equilibrium of a system? Explain with examples.

6. What do you mean by thermodynamic variables? Distinguish between "intensive" and "extensive" thermodynamic variables. [Burdwan University 2020, Calcutta University 2022]

7. What do you mean by "path" of a thermodynamic system?

8. What is a "quasi-static" process?

9. What is the reason for considering quasi-static process in the context of thermodynamics? [Calcutta University 2022]

10. What are reversible and irreversible processes? Explain with examples.

11. What is an indicator diagram? What is its utility in thermodynamics?

12. What do you mean by a "state function"?

13. State zeroth law of thermodynamics.

14. Explain the concept of temperature on the basis of the zeroth law of thermodynamics. [Calcutta University 2022]

15. What do you mean by "equation of state"?

7.9.2 Long answer type questions

1. Explain the concept of thermodynamic equilibrium.

2. Consider a system described by three thermodynamic variables X, Y, and Z. For such a system, prove that

$$\left(\frac{\partial X}{\partial Y}\right)_Z \left(\frac{\partial Y}{\partial Z}\right)_X \left(\frac{\partial Z}{\partial X}\right)_Y = -1.$$

3. State differences between "state" and "path" functions.

4. Using the indicator diagram shown in Figure 7.10, show that the work done is not a state function. [Calcutta University 2022]

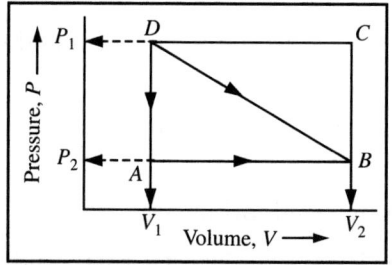

Figure 7.10 Indicator diagram of the thermodynamic process for problem 4.

5. State and explain the zeroth law of thermodynamics.

[St. Joseph's College (Autonomous), Bengaluru 2019]

6. How does the concept of "temperature" arise from the zeroth law of thermodynamics?

7. State some applications of the zeroth law of thermodynamics.

8. Show that the temperature of a wire vibrating between two fixed supports will decrease if the tension in the wire is increased and vice versa.

9. What is the importance of an "indicator diagram"?

10. State the "zeroth law of thermodynamics" and explain its importance.

[Burdwan University 2021]

11. Explain the concept of "temperature" and "thermal equilibrium."

12. Establish the condition of equilibrium of a closed composite system consisting of two simple systems separated by a movable diathermic wall that is impervious to the flow of matter. [Burdwan University 2021]

7.9.3 Numerical problems

1. Three substances are added to a mug to make coffee: the coffee, which is 65°C, the milk, which is 65°C, and the sugar, which is in thermal equilibrium with the coffee. Describe the thermal state of the sugar.

[Based on the zeroth law of thermodynamics, sugar is in equilibrium with the milk.]

2. Three bodies A, B, and C having temperature 30°C, 40°C, and 50°C, respectively, are kept in contact with each other. What will be the direction of the flow of heat in these three bodies?   [C to B to A]

3. If body A is at 50°C , body B is at 60°C, and body C is at 40°C, and they are thermally insulated from each other, then what will be the direction of heat flow?   [No heat flow]

4. An imaginary engine receives heat and performs work on a slowly moving piston at such a rate that the cycle of operation of 1 kg of fluid can be represented as a circle of 10 cm diameter on a P–V diagram. The scale is 1 cm = 300 kPa on Y-axis and 1 cm = 0.1 m³ on X-axis. Find the net work done. [2356.2 kJ]

5. A monoatomic ideal gas ($\gamma = 1.67$ and molecular weight = 40) is compressed adiabatically from 0.1 Mpa at 300 K to 0.2 Mpa. The universal gas constant is 8.314 kJ $(\text{kmol})^{-1}$ K^{-1}. What will be the work of compression of the gas in kJ/kg?   [-29.86 kJ/kg]

6. Calculate the coefficient of volume expansion α and isothermal compressibility β_T of a Van der Waals gas.

$$\left[\alpha = \frac{RV^2(V-b)}{RTV^3 - 2a(V-b)^2} \right] \text{ and } \left[\beta_T = \frac{V^2(V-b)^2}{RTV^3 - 2a(V-b)^2} \right]$$

7. An ideal gas undergoes the cyclic process $A \rightarrow B \rightarrow C \rightarrow D$. Indicator diagram P–T of this cyclic process is shown in Figure 7.11. Represent the same process in the P–V and V–T indicator diagrams.   [Burdwan University 2020]

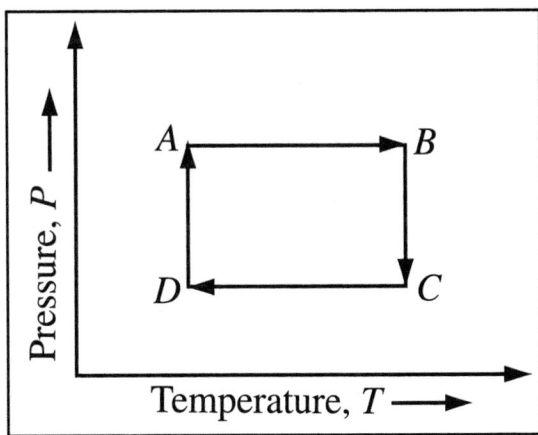

Figure 7.11 P–T indicator diagram of the process for problem 7.

8. Figure 7.12 shows an indicator diagram. During path $1 \rightarrow 2 \rightarrow 3$, 100 cal is given to the system, and 40 cal worth of work is done. During path $1 \rightarrow 4 \rightarrow 3$, the work done is 10 cal. If the system is brought from 3 to 1 along the straight line path $3 \rightarrow 1$, calculate the work done. $[W_{31} = -25 \text{ cal}]$

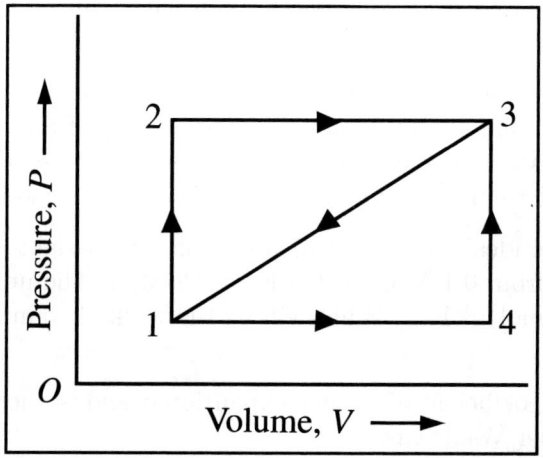

Figure 7.12 Indicator diagram for problem 8.

CHAPTER 8

First Law of Thermodynamics

A theory is the more impressive the greater the simplicity of its premises, the more different kinds of things it relates, and the more extended its area of applicability. Therefore the deep impression that classical thermodynamics made upon me. It is the only physical theory of universal content which I am convinced will never be overthrown, within the framework of applicability of its basic concepts.

—Albert Einstein

Learning Outcomes

After reading this chapter, the reader will be able to

- Gather knowledge about state function and its properties

- Understand the meaning of internal energy and its significance in formulating the first law of thermodynamics

- Formulate the first law of thermodynamics and apply it to various thermodynamic processes

- Grasp the idea of various thermodynamic processes and the related work done in these processes

- Find a relation between specific heats at constant volume and constant pressure for ideal and real gases

- Find expression for isothermal compressibility and volume expansion coefficient for ideal and real gases

- Understand how the temperature of air varies with height assuming an adiabatic process

- Solve numerical problems and multiple choice questions on the first law of thermodynamics

8.1 Introduction

Systems, surroundings, and interactions between them play a vital role in the development of the subject—thermodynamics. To extract out the physical properties of a thermodynamic system, it is essential to have knowledge about the fundamental laws and concepts of thermodynamics. For example, heat and work are two interrelated concepts. Heat is the transfer of thermal energy between two bodies that are at different temperatures and are not equal to thermal energy. Work is the external physical parameter used to transfer energy between a system and its surroundings. Further, work is needed to create heat and to transfer the thermal energy. Thus, work and heat together allow systems to exchange energy. The relationship between these two physical quantities, heat and work, can be analyzed through the laws and concepts of thermodynamics. The interaction between heat and other types of energy is primarily focused on the topic of thermodynamics.

To understand the relationship between heat and work, it is required to have an idea about a third linking factor, known as **the change in internal energy**. This internal energy includes all forms of energy within a given thermodynamic system, such as the kinetic energy of the atoms and molecules and the energy stored in all of the chemical bonds between molecules. When a certain change is made in a thermodynamic system, the transfers and/or conversions of energy take place with the interactions of heat, work, and internal energy. However, net energy is not created or lost during these transfers and/or conversions. Only conversion from one form of energy to the other takes place.

8.2 Concept of heat and work

Joule conducted a classical experiment on the equivalence of heat and mechanical work in 1840. In the experimental set up, he took water in a cylinder containing brass paddles. The heat was produced by churning this water by means of brass paddles. The mechanical energy of the paddles due to rotation was thus converted into heat. Joule opined that the heat must have been produced through a chaotic motion of the molecules of water. Through this simple experiment, Joule thus established that heat is associated with the molecular motion.

Very often, we say that the coffee in the cup is very hot or there is tremendous heat outside today. But in thermodynamics, we shall use the word "heat" only when it enters or leaves a system. We shall use the concept that heat is a form of energy in transit. Thus, "heat in a body" is not a correct statement. We essentially mean the internal energy of a system by the term "heat in a system". Since heat is a directional quantity in thermodynamics, a sign convention is adopted to represent it. We consider heat to be **positive** when it is transferred to a system, and heat is considered to be **negative** when it is taken out of a system. If no transfer of heat takes place in a process, it is said to be adiabatic.

Heat is a form of energy that is transferred from one part of the body to another or from one body to another by virtue of temperature difference. This transfer of heat can take place by one or more of the three processes: conduction, convection, and radiation. Also, the amount of heat (thermal energy) produced is always proportional to the amount of work done. Both are transient phenomena and have the same unit.

8.2.1   Heat is a path function

Heat is a path function. This implies that when a system changes its state from 1 to 2 (see Figure 8.1), the amount of heat δQ transferred between these two states will depend upon the intermediate stages through which the system passes, that is, it will depend on the path followed by the system. Hence, heat is an **inexact differential** and is written as δQ. On integrating this δQ between these two states, we get

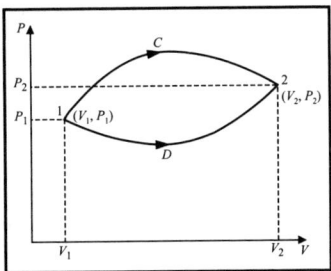

Figure 8.1 Two possible paths between states 1 and 2.

$$\int \delta Q = Q_1^2.$$

Here, Q_1^2 represents the heat transferred during the given process between states 1 and 2 along a particular path. Thus, the amount of heat Q_1^2 will be different if the system is transferred between states 1 and 2 via "path C" or "path D".

8.2.2   Work

It is an established fact that if the transfer of energy between the system and the surroundings occurs due to the difference in temperature, then that form of energy is heat. But if the transfer of energy is not due to the difference in temperature between the system and the surroundings, it is not in the form of heat, but it is work. An external work is said to be done by or on the system if a system as a whole exerts force on its surroundings and a displacement takes place. If a gas contained in a cylinder expands at constant pressure, it exerts force on the piston and does external work on its surroundings.

On the other hand, internal work is defined as the work done by a part of the system on another part. This internal work has no significance in thermodynamics. For example, when a storage battery is not in operation, internal diffusion of chemicals takes place inside the cell. But such changes are not accompanied by the performance of any work; hence, they are not significant for us.

The following sign convention about work is adopted: work done by a system is considered as "positive", whereas the work done on a system is considered as "negative". According to this sign convention, the work produced by the engine of a car or gas turbine is positive, whereas the work consumed by compressors or mixers is negative.

8.2.3   Comparison of heat and work

It is known that the transfer of energy from one system to another takes place in two different ways: heat and work. But in the field of thermodynamics, the distinction between heat and work is important and has to be considered with great care. Heat is the transfer of thermal energy between two objects or two systems, while work is the transfer of mechanical energy between two such systems. Further, the difference between heat (the microscopic motion) and work (macroscopic motion) plays an important role in how thermodynamic processes work. Heat can be transformed into work and vice versa, but they are not the same thing. It is demonstrated (the first law of thermodynamics) that heat and work both contribute to the total internal energy of a system, but there is a limit to the amount of heat that can be converted into work (the second law of thermodynamics). The second law of thermodynamics allows work to be transformed fully into heat, but it forbids heat to be totally converted into work. If heat could be transformed fully into work it would violate the laws of entropy. The maximum amount of work that can be obtained from heat is given by the efficiency of a Carnot engine.

Heat is the energy associated with the random motion of particles, while work is the energy of ordered motion in one direction. Therefore, heat is the "low-quality" energy, whereas work is the "high-quality" energy. This assertion supports the entropy statement of the second law of thermodynamics.

8.3   Idea about the state functions

The "state" of the specific molecule or compound or system plays a very important role in thermodynamics. A "state" of a thermodynamic system is characterized by thermodynamic variables such as temperature, pressure, and the amount and type of substance present. Once the state is established, state functions can be defined. State functions are the values of the thermodynamic variables that depend on the state of the substance and not on how that state has been reached by the system, that is, not on the path followed by the system.

A state function is a property of a thermodynamic system whose value at any given instant depends only on the state of the system at that instant. State functions can also be looked upon as integrals. The value of an integral generally depends only on three factors: the function, the lower, and the upper limit of the integral. Similarly, the value of a state function depends on three things: the property of the function, the initial, and the final value of the state variables. Thus, how state functions depend only on the final and initial value and not on the object's history or the path taken to get from the initial to the final value can be illustrated by the integrals.

Examples of state functions are pressure, entropy, volume, and others. The properties of a state function can be stated mathematically in the following way:

1. The differential dF of a state function F is an infinitesimal change of the corresponding state function F. Since, by definition, the value of a state function depends only on the state of the system, integrating the differential dF between an initial state "i" and a final state "f" results in the change of F, and this change is independent of the path followed by the system. Thus, we have

$$\int_{F_i}^{F_f} = F_f - F_i = dF.$$

   An **exact differential** satisfies this property. Therefore, the differential of a state function is always exact.

2. A state function F can be treated as a dependent variable when it is a function of a certain number of independent variables, say, x, y, and z, which are also state functions. The total differential dF of the state function F can be expressed in terms of the differentials of the independent variables and has the form

$$dF = \left(\frac{\partial F}{\partial x}\right)_{y,z} dx + \left(\frac{\partial F}{\partial y}\right)_{z,x} dy + \left(\frac{\partial F}{\partial z}\right)_{x,y} dz.$$

   Thus, it is observed that the number of terms in the expression on the right side equals the number of independent variables. In the expression, for each partial derivative, all independent variables except the variable shown in the denominator are held constant.

3. The second partial derivative $\left(\frac{\partial^2 F}{\partial y \partial x}\right)$ is defined as the partial derivative with respect to y of the partial derivative of F with respect to x. According to the theorem of differential calculus, a well-known result is that if a function F is single-valued and has continuous derivatives, the order of differentiation in a mixed derivative is immaterial. Therefore, the mixed derivatives $\left(\frac{\partial^2 F}{\partial y \partial x}\right)$ and $\left(\frac{\partial^2 F}{\partial x \partial y}\right)$ are equal in any given state for a thermodynamic system.

4. If the coefficients of the total differential of a dependent variable are known as the functions of independent variables, the expression for the total differential can be integrated to get an expression for the dependent variable as a function of the independent variables.

5. A **Legendre transform** of a state function is a linear change of one or more of the independent variables made by subtracting products of conjugate variables.

It is to be noted that "an infinitesimal amount of work cannot be represented by an exact differential". This can be understood in the following way: For an infinitesimal change, the first law of thermodynamics can be written as: $\delta Q = dU + \delta W$. Here, dU is the change in internal energy of the system, δQ is the heat absorbed by the system from its surroundings, and δW is the work done by the system on the surroundings. Here, Q is a function of path, that is, it depends on the path followed by the system. If the initial and final states of a system are fixed, but the path of the process is changed, different amounts of heat will be exchanged between the system and the surroundings. Therefore, it is not a function of the thermodynamic states and cannot give rise to an exact differential, as functions having the property of exact differentials are path-independent. So, an infinitesimal work cannot be represented by an exact differential.

If the state functions pressure P, volume V, and temperature T are related by an equation of state $f(P, V, T) = 0$, then they must satisfy the following important relation:

$$\left(\frac{\partial P}{\partial V}\right)_T \left(\frac{\partial V}{\partial T}\right)_P \left(\frac{\partial T}{\partial P}\right)_V = -1. \tag{8.1}$$

This is a very important relation and is frequently used in thermodynamics.

Equation (8.1) can be verified in the following way: Among these three variables, we consider any two of them as independent variables and the third one as dependent variable. Let us consider pressure and volume as independent variables and the temperature as dependent variable, so that the temperature is some function of pressure and volume, that is, $T = T(P, V)$.

Hence, the differential dT can be written as

$$dT = \left(\frac{\partial T}{\partial P}\right)_V dP + \left(\frac{\partial T}{\partial V}\right)_P dV. \tag{8.2}$$

Now we consider temperature and volume as the independent variables and pressure as the dependent variable, that is, $P = P(T, V)$. So, the differential dP for pressure P can be written in the similar way as

$$dP = \left(\frac{\partial P}{\partial T}\right)_V dT + \left(\frac{\partial P}{\partial V}\right)_T dV; \quad \Rightarrow \quad \left(\frac{\partial P}{\partial T}\right)_V dT = dP - \left(\frac{\partial P}{\partial V}\right)_T dV. \tag{8.3}$$

Comparing the coefficients of equations (8.2) and (8.3), we have

$$\left(\frac{\partial P}{\partial T}\right)_V = \frac{1}{\left(\frac{\partial T}{\partial P}\right)_V} = \frac{-\left(\frac{\partial P}{\partial V}\right)_T}{\left(\frac{\partial T}{\partial V}\right)_P};$$

$$\Rightarrow \quad -\left(\frac{\partial P}{\partial V}\right)_T \left(\frac{\partial T}{\partial P}\right)_V \frac{1}{\left(\frac{\partial T}{\partial V}\right)_P} = 1.$$

$$\text{Or,} \implies \left(\frac{\partial P}{\partial V}\right)_T \left(\frac{\partial T}{\partial P}\right)_V \left(\frac{\partial V}{\partial T}\right)_P = -1;$$

$$\text{Or,} \implies \left(\frac{\partial P}{\partial V}\right)_T \left(\frac{\partial V}{\partial T}\right)_P \left(\frac{\partial T}{\partial P}\right)_V = -1.$$

8.4 Internal energy and the first law of thermodynamics

In this section, internal energy and its characteristics features, statement, significance, and limitations of the first law of thermodynamics are presented in detail.

8.4.1 Internal energy

The internal energy of a thermodynamic system is the energy contained within the system. It is the energy required to create the system in a given thermodynamic state. Internal energy is a state function that defines the energy of a substance in the absence of external field factors, such as electric, magnetic, and other fields. The internal energy of a system is proportional to the temperature of the system. Therefore, we can control changes in the internal energy by observing what happens to the temperature of the system. Whenever there is an increase in the temperature of the system, there is also an increase in the internal energy. We know that temperature is a state function, that is, it depends only on the state of the system at any moment in time, not on the path used to get the system to that state. As the internal energy of the system is proportional to temperature, the internal energy is also a state function. Hence, its value depends upon the state of the substance and is independent of the processes by which it attains the given thermodynamic state.

In accordance with the first law of thermodynamics, when a system undergoes a change of state as a result of a process in which **only work** is involved, the work is equal to the change in internal energy. The first law also implies that if **both heat and work** are involved in the change of state of a system, then the change in internal energy is equal to the heat supplied to the system minus the work done by the system. The internal energy keeps an account of the gains and losses of energy of the system that are due to changes in its internal state. Thus, any change in internal energy ΔU of the system is equal to the difference between its initial and final values, that is,

$$\Delta U = U_f - U_i,$$

where U_i and U_f are the values of internal energy, respectively, in the initial and final states of a given thermodynamic process followed by the system.

Further, it should be mentioned that the internal energy of a system can be increased by several techniques, such as by introducing matter, by heat, or by doing thermodynamic work on the system. When the transfer of matter is prevented by impermeable containing walls, the system is said to be closed, and the change in internal energy can be defined from the first law of thermodynamics as the sum of the heat added to the system and the thermodynamic work done by the surroundings on the system. If the containing walls pass neither matter nor energy, the system is said to be isolated, and its internal energy remains constant. The change in internal energy ΔU is macroscopically given by the first law of thermodynamics as $\Delta U = \delta Q - \delta W$. In microscopic terms, the internal energy can be explained by the random kinetic energy of the microscopic motion of the system's particles from translations, rotations, and vibrations and by the potential energy associated with microscopic forces, including chemical bonds.

Characteristic features of internal energy

1. The internal energy is a state function of the thermodynamic state variables. Its value depends only on the present state of the system and is independent of the processes undergone to prepare it.

2. It is an exact differential dU. Its fundamental variables are V and T. So, dU can be written as $dU = \left(\frac{\partial U}{\partial V}\right)_T \, dV + \left(\frac{\partial U}{\partial T}\right)_V \, dT$.

3. It is an extensive quantity. It depends on the size of the system, or on the amount of substance it contains.

4. For a cyclic process $\implies \oint dU = 0$.

5. It is the one and only cardinal thermodynamic potential. All other thermodynamic potentials can be formulated from the internal energy of the system.

6. At any temperature greater than absolute zero, the potential energy and the kinetic energy are constantly converted into one another microscopically, but the sum remains constant in an isolated system. In classical thermodynamics, the kinetic energy of a thermodynamic system vanishes at zero temperature, and the internal energy becomes purely the potential energy. However, according to quantum mechanics, even at zero temperature, the molecules of the system possess a residual energy of motion, known as the zero point energy. Thus, at absolute zero, a system is in the quantum-mechanical ground state with its lowest available energy. Further, it should be stated that at absolute zero, a system of a given composition attains its minimum value of entropy.

7. The microscopic kinetic energy part of the internal energy gives rise to the temperature of the system. According to statistical mechanics, we know that the pseudorandom kinetic energy of individual particles is related to the mean kinetic energy of the entire ensemble of particles comprising the system. Furthermore, the mean microscopic kinetic energy is related to the macroscopically observed empirical property, the temperature of the system.

It should be mentioned that temperature is an intensive variable, whereas the energy that expresses the temperature is an extensive property of the system.

It is to be noted that the internal energy of a system can be thought of in two different but consistent ways: (i) It is the sum of the kinetic and potential energies of its atoms and molecules that is called the mechanical energy of the system. Further, it should be stated that one has to deal with averages and distributions as it is impossible to keep track of all individual atoms and molecules. (ii) A second way to view the internal energy of a system is in terms of its macroscopic characteristics, which are very similar to the average values of atomic and molecular motions.

It has been experimentally found that the internal energy U of a system depends only on the state of the system and not on the way it reached that state. The internal energy is also found to be a function of a few macroscopic variables, such as pressure, volume, and temperature. It is independent of past history such as whether there has been heat transfer or work done. This independence means that if we know the state of a system, we can calculate changes in internal energy from the knowledge of a few macroscopic variables.

8.4.2   The first law of thermodynamics

The first law of thermodynamics is basically related to the principle of conservation of energy. This conservation principle states that "energy can neither be created nor destroyed; it can be transformed from one form to another, the total amount of energy (in the universe) remaining constant". According to this principle, we cannot get energy out of nothing. If there is an increase in energy in a system, there must have been an equivalent loss of energy in the surroundings. This is basically the essence of the first law of thermodynamics.

Differential form of the first law of thermodynamics

The first law of thermodynamics can be stated in the following way: "when a system is constrained to undergo a change by mechanical, diffusive, or thermal interactions, its internal energy changes by an amount equal to the heat transferred to it, work done on it, and matter exchanged". Let dU be the change in the internal energy of a system when an amount of heat δQ is added to it, an amount of work δW is done by it, and dN is the change in a number of the particles of the system. The first law of thermodynamics can then be mathematically expressed as

$$\delta Q = dU + \delta W - \mu dN, \tag{8.4}$$

where μ is the chemical potential. The sign convention for heat transfer indicates the direction of energy flow. Usually in physics, it is assumed that heat absorbed by a system is positive ($Q > 0$), indicating that the internal energy of the system increases. Conversely, heat released by a system is considered to be negative $(Q < 0)$, signifying a decrease in the internal energy of the system. Similarly, work done on the system is considered to be negative, and work done by the system is positive.

If more than one kind of work is done by the system, δW is replaced by $\sum_j \delta W_j$. Similarly, if there are several different kinds of particles in the system, the last term in equation (8.4) will be $\sum_i \mu_i dN_i$. In the case of a diffusive equilibrium, there is no exchange of particles; the system may undergo only thermal and mechanical interactions. Under this situation, the first law of thermodynamics takes the form

$$dU = \delta Q - \delta W. \tag{8.5}$$

This is the differential form of the first law of thermodynamics for non-diffusive interacting systems.

Significance of the first law of thermodynamics

This first law of thermodynamics is significant in physical sciences for various reasons. Some of them are:

1. The concept of internal energy is introduced by this law.

2. It provides a means to determine the change in internal energy of a thermodynamic system.

3. It is applicable to any process by which a system undergoes a physical or chemical change.

4. According to this law, a fixed quantity of heat is needed to obtain a fixed amount of work or vice versa.

5. When one form of energy is lost, an exact equivalent amount of another form of energy is generated.

Limitations of the first law of thermodynamics

Though the first law of thermodynamics retells the law of conservation of energy, it lacks to explain some experimentally observed facts. Some of the major drawbacks of the first law of thermodynamics are the following:

1. The first law of thermodynamics does not indicate the direction in which a particular thermodynamic process would occur.

2. When brakes are applied to stop a car in winter, work is done against the frictional force, and this work is converted into heat. However, when it cools down, the heat stored is not utilized in starting the car. One may ask the question: Where does the produced heat go? This is one of the limitations of the first law of thermodynamics.

3. The potential energy of water is converted into the kinetic energy when water flows down the Himalayas. But the reverse of this natural process does not occur. There is no answer for this phenomenon from the first law of thermodynamics.

4. In a room heater, heat is produced by the flow of electric current through an electric resistor. If we take away heat from the room, the electric current will not be generated through the wire. Thus, our observations and experiences tell us that the heat is not converted to other forms of energy by itself; an external agency, that is, a heat engine, is required for this purpose.

5. The first law of thermodynamics places no restriction on the direction of the flow of heat. This law does not indicate whether heat can flow from a cold end to a hot end or not. For example, heat cannot be extracted from ice by cooling it to a low temperature. Some external work has to be done for this purpose.

6. The first law does not specify whether a process (chemical reaction) is feasible or not.

To overcome these limitations, another law was required to be formulated that is known as the **second law of thermodynamics**. This law introduces a very important physical property known as the entropy. It is a state function. This second law of thermodynamics helps us to predict whether a particular reaction is feasible or not and also helps us to predict the direction of the flow of heat. It also signifies that energy cannot be completely converted into equivalent work.

8.5 Various thermodynamic processes

In classical mechanics, we adopt the following scheme to find out the physical behavior of a system: The body under consideration is first isolated, the external forces acting on it are then analyzed, and finally, Newton's laws of motion are used to predict its physical behavior. A similar approach is also followed in thermodynamics. A part of the universe is identified as the system. Once the system is selected, the interaction between the system and the surroundings is determined. Finally, the thermodynamic behavior of the system is investigated with the help of the laws of thermodynamics. This thermodynamic behavior of a system is described in terms of thermodynamic variables. For a thermodynamic system, these variables are pressure, volume, temperature, and the number of molecules or moles of the gas. Different types of systems are generally characterized by different sets of variables. For example, the thermodynamic variables for a stretched rubber band are tension, length, temperature, and mass.

The state of a system can change as a result of its interaction with the surroundings. This change in the state of a system can be slow or fast and small or large. The manner in which such a change in the state of the system occurs from an initial state to a final state is called a thermodynamic process. In such a process, a particular thermodynamic variable is kept constant, and other variables are varied depending on the requirement of the process. There are various types of thermodynamic processes used to represent such changes in the states of a thermodynamic system. These processes are as follows:

1. Isothermal, where the temperature of the system is held constant;

2. Adiabatic, where no heat is exchanged between the system and the surroundings;

3. Isobaric, where the pressure of the system is held constant, and

4. Isochoric, where the volume of the system is held constant.

The equation of state, indicator diagram, and the expression for work done in these thermodynamic processes are described in detail below.

8.5.1   Isothermal process

In thermodynamics, an isothermal process is defined as the process in which the temperature of the system remains constant, that is, temperature T = constant and $\Delta T = 0$. This process is typically carried out by keeping the thermodynamic system in contact with an outside thermal reservoir, and the change occurs slowly enough to allow the system to continually adjust to the temperature of the reservoir through an exchange of heat. Such an arrangement depicting

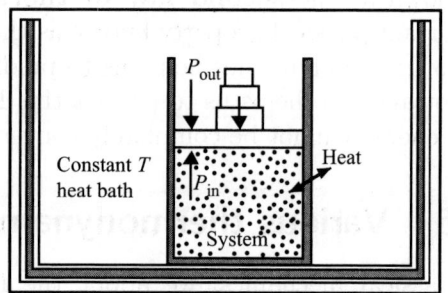

Figure 8.2 Expansion of a system at a constant temperature—an example of an isothermal process.

the essence of an isothermal process is schematically shown in Figure 8.2. It should be mentioned that a thermal reservoir is an idealized "infinitely" large system maintained at constant temperature for a sufficiently long time (much larger than the time of observation). In practice, the temperature of a **finite bath** is controlled by either adding or removing a finite amount of thermal energy to/from the system. If the weights on the piston bed shown in Figure 8.2 are removed in infinitesimal steps, the pressure on the system enclosed inside the container decreases infinitesimally slowly. In this way, an isothermal (expansion) process can be carried out quasistatically that cools the gaseous system to a lower temperature. At this stage, heat from the thermal reservoir enters the gaseous system until the temperature of the system is adjusted to that of the thermal heat bath. As a result, the system expands in an isothermal condition. On the other hand, if weights are added on the piston bed in infinitesimal steps, the gaseous system undergoes isothermal contraction quasistatically.

Work done by an ideal gas in an isothermal process

Consider an ideal gas of n moles enclosed in a container of volume V at a temperature T. The container is fitted with a piston on which different amounts of weights can be added or removed. When the ideal gas expands from an initial volume V_1 to the final volume V_2, the work done by the ideal gas on the piston is given by

$$W = \int_{V_1}^{V_2} P \, dV. \tag{8.6}$$

The process is carried out quasistatically so that the pressure remains almost constant to P. The equation for the n-mole of an ideal is

$$PV = nRT; \quad \Rightarrow \quad P = \frac{nRT}{V}.$$

Using this expression for P in equation (8.6), the expression for work done by the gas in the isothermal process comes out to be

$$W = \int_{V_1}^{V_2} \frac{nRT}{V} dV = nRT \, \ln\left(\frac{V_2}{V_1}\right) = nRT \ln\left(\frac{P_1}{P_2}\right). \tag{8.7}$$

This is the expression for work done by the ideal gas in an isothermal process in terms of volume or pressure. It should be mentioned that the change in internal energy in an isothermal process is zero as the temperature remains constant in this process. Thus, it readily follows from the first law of thermodynamics that in an isothermal process, the heat absorbed by the system is equal to the work done by it.

Work done by a real gas (Van der Waals type) in an isothermal process

Consider a real gas (Van der Waals gas) of n moles enclosed in a container of volume V at a temperature T. The container is fitted with a piston on which different amounts of weights can be added or removed. When the real gas expands from an initial volume V_1 to the final volume V_2, the work done by the real gas on the piston is given by

$$W = \int_{V_1}^{V_2} P \, dV. \tag{8.8}$$

Let us calculate the work done by n moles of a Van der Waals gas obeying the equation

$$\left(P + \frac{a\,n^2}{V^2}\right)(V - n\,b) = n\,R\,T; \quad \Rightarrow \quad P = \frac{nRT}{V - n\,b} - \frac{a\,n^2}{V^2}, \qquad (8.9)$$

and undergoing an isothermal change in volume from V_1 to V_2. Using equation (8.9) into equation (8.8), the expression for work done by n moles of a real gas in the isothermal expansion comes out to be

$$W = \int_{V_1}^{V_2} \left[\frac{nRT}{V - n\,b} - \frac{a\,n^2}{V^2}\right] dV = nRT \ln\left[\frac{V_2 - n\,b}{V_1 - n\,b}\right] + a\,n^2\left[\frac{1}{V_2} - \frac{1}{V_1}\right]. \qquad (8.10)$$

Equation (8.10) shows that in an isothermal process, the work done by n moles of a Van der Waals gas depends on the volume V, number of moles n, temperature T, and on the Van der Waals constants a and b.

Free expansion of a real gas (Van der Waals gas)

For a real gas (Van der Waals gas), the internal energy U is a function of both temperature T and volume V. Therefore, we can write

$$U = U(T, V).$$

An infinitesimal change in internal energy in terms of the changes in temperature T and volume V can be written as

$$dU = \left(\frac{\partial U}{\partial T}\right)_V dT + \left(\frac{\partial U}{\partial V}\right)_T dV.$$

If it is assumed that the internal energy remains constant during free expansion, we can write

$$\left(\frac{\partial U}{\partial T}\right)_V \left(\frac{\partial T}{\partial V}\right)_U + \left(\frac{\partial U}{\partial V}\right)_T = 0; \quad \Rightarrow \quad \left(\frac{\partial T}{\partial V}\right)_U = -\frac{\left(\frac{\partial U}{\partial V}\right)_T}{\left(\frac{\partial U}{\partial T}\right)_V} = -\frac{1}{C_V}\frac{a}{V^2}. \qquad (8.11)$$

In this equation (8.11), we have used $\left(\frac{\partial U}{\partial V}\right)_T = \frac{a}{V^2}$ for a Van der Waals gas. This term $\left(\frac{\partial U}{\partial V}\right)_T$ gives the change in internal energy with volume at constant temperature. Since all the quantities on the right-hand side of equation (8.11) are positive definite, the temperature will decrease when a Van der Waals gas is made to undergo **free expansion**.

8.5.2   Adiabatic process

In thermodynamic, an adiabatic process is one that occurs without the transfer of heat or matter between a thermodynamic system and its surroundings. An adiabatic (expansion) process is shown in Figure 8.3. In this process, energy is transferred to its surroundings only as work. An adiabatic process can be conducted either quasistatically or nonquasistatically. When a system expands adiabatically (as in the case of a system shown in Figure 8.3), it must do work against the outside world, and therefore, its energy goes down, which is reflected in the lowering of the temperature of the system. Thus, an

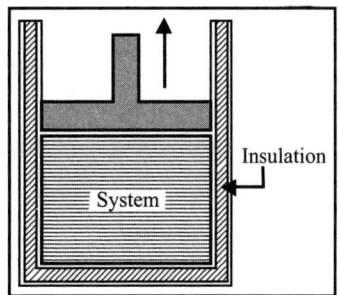

Figure 8.3 An insulated piston with hot, compressed gas in an insulating chamber is released—an example of an adiabatic (expansion) process.

adiabatic expansion leads to a decrease in temperature, and on the other hand, an adiabatic compression leads to an increase in temperature. This adiabatic process furnishes a rigorous conceptual basis for the theory used to put forward the first law of thermodynamics, and fundamentally, it is a key concept in thermodynamics.

Equation for an ideal gas in an adiabatic process

In an adiabatic process, the total heat content Q is constant, so that $\delta Q = 0$. Thus, for an adiabatic process, the first law of thermodynamics can be written as

$$dU + \delta W = 0. \tag{8.12}$$

This equation (8.12) shows that if a system enclosed within an adiabatic boundary is allowed to expand adiabatically, it does work at the cost of its internal energy. Hence, such an adiabatic expansion leads to decrease in internal energy which, in turn, results in a fall in temperature during an adiabatic expansion. Thus, it can be concluded that adiabatic expansion produces cooling and adiabatic compression, on the other hand, produces heating. For this reason, adiabatic demagnetization of a paramagnetic substance is used to produce temperatures below 1 K.

It should be mentioned that the first law of thermodynamics can be used to calculate the changes in any two of the three thermodynamic coordinates involved in an adiabatic process. Using the relations $dU = C_V\, dT$ and $\delta W = P\, dV$, equation (8.12) can be written as

$$C_V\, dT + P\, dV = 0.$$

During such an adiabatic expansion, the gas (thermodynamic system) passes through an infinite number of equilibrium states. Considering the gas to be ideal, we can write

$$C_V \, dT + \frac{RT}{V} \, dV = 0. \quad \text{[We have used } P = \frac{RT}{V}.\text{]}$$

Dividing throughout by $C_V \, T$ and integrating the resulting expression, we get

$$\ln T + \frac{R}{C_V} \ln V = K; \quad \Rightarrow \quad TV^{\frac{R}{C_V}} = K,$$

where K is the constant of integration. Further, we know that

$$C_P - C_V = R; \quad \Rightarrow \quad \frac{C_P}{C_V} - 1 = \frac{R}{C_V}. \quad \Rightarrow \quad \frac{R}{C_V} = \gamma - 1.$$

This provides $TV^{\gamma-1} = K$. Using the ideal gas equation for one mole, we can write

$$PV^{\gamma} = \text{constant} \quad \text{and} \quad T^{\gamma} P^{1-\gamma} = \text{constant.} \tag{8.13}$$

Regarding the adiabatic process, the following points should be noted:

1. The relation $T^{\gamma}P^{1-\gamma}$=constant indicates that if a gas is compressed adiabatically, the pressure increases that is accompanied by a corresponding increase in temperature.

2. The relation PV^{γ}=constant predicts that in adiabatic compression, volume decreases and since no heat leaves the system, its temperature rises.

3. During an adiabatic expansion, the work done by the gas occurs at the cost of its internal energy; hence, the temperature of the gas drops.

The results in equation (8.13) are highly useful for deriving various characteristic features of the adiabatic process in thermodynamics. These characteristic features are described in detail below.

An adiabatic is steeper than an isotherm

For an adiabatic process, we have PV^{γ}=constant. Differentiating this with respect to volume V, we get

$$\gamma P V^{\gamma-1} dV + V^{\gamma} dP = 0; \quad \Rightarrow \quad \left(\frac{\partial P}{\partial V}\right)_S = -\gamma \left(\frac{P}{V}\right), \tag{8.14}$$

where S is the entropy to be introduced in Chapter 9.

For an isothermal process, $PV = RT$.

$$PdV + VdP; \quad \Rightarrow \quad \left(\frac{\partial P}{\partial V}\right)_T = -\left(\frac{P}{V}\right). \tag{8.15}$$

Equations (8.14) and (8.15) show that an adiabatic is steeper than an isotherm since $\gamma > 1$. This is because the gas loses internal energy as it expands adiabatically. This also implies that the relative change in the volume in an adiabatic process is less than that in an isothermal process. An adiabatic and an isotherm of an ideal gas are shown in Figure 8.4. Further, it can be shown that the ratio of adiabatic elasticity κ_S and isothermal elasticity κ_T is equal to γ, that is,

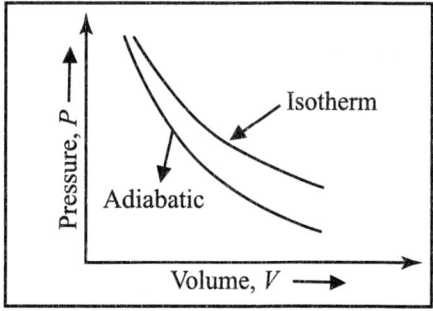

Figure 8.4 Adiabatic and isotherm of an ideal gas.

$$\frac{\kappa_S}{\kappa_T} = \frac{\beta_T}{\beta_S} = \gamma.$$

Here, β_S and β_T are adiabatic and isothermal compressibility, respectively. It should be mentioned that in an adiabatic process, entropy S remains constant (see Chapter 9).

Work done by an ideal gas in an adiabatic process

We consider one mole of an ideal gas undergoing an adiabatic change. During such an adiabatic process, the system is thermally insulated from the surroundings. In this process, the volume of the system changes from V_1 to V_2, and the corresponding change in pressure occurs from P_1 to P_2. When the ideal gas expands from volume V_1 to V_2, the work done by it is given by

$$W = \int_{V_1}^{V_2} P \, dV. \tag{8.16}$$

From the relation $PV^\gamma = K(=\text{constant})$ in the adiabatic process, we get $\quad \Rightarrow \quad P = \frac{K}{V^\gamma}$. Using this relation, the expression for work done by the gas comes out to be

$$W = \int_{V_1}^{V_2} \frac{K}{V^\gamma} dV = \frac{K}{1-\gamma} \left[\frac{1}{V_2^{\gamma-1}} - \frac{1}{V_1^{\gamma-1}}\right] = \frac{1}{1-\gamma}[P_2 V_2 - P_1 V_1] = \frac{R}{1-\gamma}[T_2 - T_1].$$

$$\tag{8.17}$$

Equation (8.17) shows that the work done by an ideal gas in an adiabatic process depends only upon the initial and final temperatures T_1 and T_2 and the nature of the gas through the factor γ. Further, it should be mentioned that the work done by an ideal gas along any adiabatic between two isotherms is independent of the particular adiabatic.

Equation for a Van der Waals gas in an adiabatic process

The equation for one mole of an ideal gas in the adiabatic process is given by $PV^\gamma = $ constant, where symbols have their usual meaning. Let us calculate such an equation for one mole of a Van der Waals gas in the adiabatic process. The equation of state of one mole of a Van der Waals gas is given by

$$\left(P + \frac{a}{V^2}\right)(V - b) = R\,T; \quad \Rightarrow \quad P = \frac{RT}{V - b} - \frac{a}{V^2}. \tag{8.18}$$

The first law of thermodynamics for an adiabatic process can be written as

$$C_V dT + P dV = 0. \tag{8.19}$$

Using this equation (8.18) in equation (8.19), we get

$$C_V dT + \left(\frac{RT}{V - b} - \frac{a}{V^2}\right) dV = 0. \tag{8.20}$$

On dividing throughout by C_V, we get

$$dT + \left[\frac{RT}{C_V(V - b)} - \frac{a}{C_V\,V^2}\right] dV = 0. \tag{8.21}$$

The second term within the third bracket is very small compared to the first term within the third bracket. Hence, equation (8.21) can be written as

$$\left[\frac{R}{C_V(V - b)}\right] dV = -\frac{dT}{T}. \tag{8.22}$$

Integrating this equation (8.22), we obtain

$$\frac{R}{C_V} \ln(V - b) = -\ln T + C, \tag{8.23}$$

where C is the constant of integration. Taking the antilog of equation (8.23), we get the following result

$$T(V - b)^{\gamma - 1} = \text{constant}. \tag{8.24}$$

This is the equation for one mole of a real gas obeying the Van der Waals equation of state in an adiabatic process.

Adiabatic lapse rate

The ground absorbs heat coming from the Sun, and the air is heated up by this absorbed heat as it is in immediate contact with the ground. When the heated air rises upward, a density gradient is established in the vertical direction. This results in convection currents. As a result, cooler air is transported downwards, and hot air is transported upwards. As hot air rises above, it expands. Now, one may ask the question: Will the hot air exchange heat with its surroundings? It may not do so because air is not a good conductor of heat. In fact, air is a very poor conductor of heat. This means that in the intermixing of air, the process involved is an adiabatic expansion.

We know that in an adiabatic expansion, the temperature of the system decreases. In order to calculate this decrease in temperature with height in air, we assume that air behaves as a perfect gas, and the existence of water vapor in the atmosphere is completely ignored. To calculate the drop in temperature with height for such an adiabatic expansion, we use equation (8.13), which is $T^\gamma P^{1-\gamma} = K$(constant). For one mole of air, equation (8.13) in logarithmic form can be written as

$$\gamma \ln T + (1-\gamma) \ln P = \ln K; \quad \Rightarrow \quad \gamma \ln T - (\gamma-1) \ln P = \ln K.$$

Differentiating this expression, we can write

$$\frac{dT}{T} - \frac{\gamma-1}{\gamma}\frac{dP}{P} = 0,$$

which can be rearranged as

$$\frac{dP}{P} = \frac{\gamma}{\gamma-1}\frac{dT}{T}. \tag{8.25}$$

To calculate the variation of temperature with height, that is, $\frac{dT}{dh}$, the pressure must be related to height. We know that as we go up, pressure decreases according to the following relation

$$dP = -\rho g dh,$$

where ρ is the average density of air, and g is the acceleration due to gravity. The negative sign indicates that pressure decreases with the increase in height.

Since we have considered air to behave as a perfect gas, the equation of state $PV = RT$ for one mole of air can be used in the above expression. This gives rise to

$$\frac{dP}{P} = -\frac{\rho g dh}{\frac{RT}{V}} = -\frac{\rho V g}{RT}dh = -\frac{Mg}{RT}dh, \tag{8.26}$$

where $M = \rho V$ is the mass of one mole of air. ($M = mN_A$, where m is the average mass of one air molecule.) On combining equations (8.25) and (8.26), we get

$$\frac{\gamma}{\gamma-1}\frac{dT}{T} = -\frac{Mg}{RT}dh \quad \Rightarrow \quad \frac{dT}{dh} = -\left(\frac{\gamma-1}{\gamma}\right)\frac{Mg}{R}. \tag{8.27}$$

Equation (8.27) gives the expression for adiabatic lapse rate. The negative sign on the right-hand side of equation (8.27) indicates that temperature decreases with height. Assuming air to be diatomic, the typical value of this drop in temperature comes out to be $\frac{dT}{dh} = 9.8 \times 10^{-3}$ K m^{-1}. This means that the temperature falls by about 9.8 K over one kilometer.

Polytropic process

A thermodynamic process in which the relation $PV^n = K$ is satisfied is known as the **polytropic process**. Here, P is the pressure, V is volume, n is the polytropic index, and K is a constant. This equation of state for the polytropic process can be used to describe multiple expansion and various compression processes. For example, it can also be used in the transfer of heat. In case of an ideal gas, a process is polytropic if and only if the ratio (K_P) of energy transfer as heat to energy transfer as work at each infinitesimal step of the process is kept constant, that is,

$$K_P = \frac{\delta Q}{\delta W} = \text{constant.} \tag{8.28}$$

This leads to the heat capacity $C = \frac{\delta Q}{dT}$=constant. For different processes, n will have different values. For example,

1. $n = 0$; $\Rightarrow$ isobaric process

2. $n = 1$; $\Rightarrow$ isothermal process

3. $n = \gamma$; $\Rightarrow$ adiabatic process

4. $n \to \infty$; $\Rightarrow$ isochoric process

8.5.3   Other thermodynamic processes

Other quasistatic processes of interest for gases are isobaric and isochoric processes.

Isobaric process: An isobaric process is a process in which the pressure of the system is kept constant. In case of a gas enclosed in a piston–cylinder arrangement, this process may be accomplished by allowing the piston to move freely so that it is always in equilibrium between the net force from the gas pushing upward, and the weight of the piston plus the force due to atmospheric pressure pushing downward. In this case, values of heat and work are generally both nonzero. The work done is $W = -P(V_f - V_i)$, where P is the constant pressure.

Isovolumetric process: A process in which the volume of the system is held constant is defined as the isovolumetric process. In the case of a gas enclosed in a piston–cylinder arrangement, this may be accomplished by fixing the piston at a particular position. Since the volume does not change, the work done is zero. So, from the first law of thermodynamics, we have $\Delta U = \delta Q$. If energy is added in the form of heat to a system kept at constant volume, all of the transferred energy remains in the system as an increase in the internal energy of the system.

Cyclic process: If the state of a system at the end is the same as the state at the beginning, it is said that a system goes through a cyclic process. We know that temperature, pressure, volume, and internal energy are state properties. In a cyclic process, the internal energy of the system does not change over a complete cycle: $\Delta U = 0$. When the first law of thermodynamics is applied to a cyclic process, a simple relation between heat into the system and the work done by the system over the cycle is then given by $Q = W$ (cyclic process).

Reversible and irreversible processes: Thermodynamic processes can also be reversible or irreversible. A reversible process is defined as the process in which the system and the environment can be made to retrace their paths to the same initial states that they were in before the process occurred. The necessary condition for a reversible process is that it must occur quasistatically. This is to be noted further that a system can be easily restored to its original state, but it is quite tough to restore its environment to its original state at the same time. For example, when an ideal gas expands into a vacuum to twice its original volume, the gaseous system can be easily push back with a piston and restore its temperature and pressure by removing some heat from the gas. But the problem is that this cannot be done without changing something in its surroundings, such as dumping some amount of heat there. It must be mentioned further that a quasistatic process is not necessarily reversible since there may be dissipative forces present. For example, if there is friction present between the piston and the walls of the cylinder containing a gaseous system, the energy lost due to friction would prevent us from reproducing the original states of the system. It should be noted that **a reversible process is truly an ideal process that rarely happens**.

On the other hand, we encounter an irreversible process in reality almost all the time. Systems do not go back to their original states during such irreversible processes. It is really difficult to restore the original states of the system and its environment at the same time. Almost all the natural processes are irreversible in nature. The direction of evolution of an irreversible process comes from the finite gradient between the states occurring in the actual process. For example, a finite temperature difference (gradient) between two objects is responsible for the flow of heat from one object to another. At any given moment of the process, the system is most likely not at equilibrium or in a well-defined state. This unidirectional phenomenon is known as **irreversibility** in thermodynamics and is the essence of the second law of thermodynamics. This formulation of the second law of thermodynamics is credited to German physicist Rudolf Clausius (1822–1888) and is referred to as the **Clausius statement of the second law of thermodynamics**.

8.6   Applications of the first law of thermodynamics

The most common practical application of the first law of thermodynamics is the heat engine. Most of the heat engines are categorized as open systems. In such heat engines, thermal energy is converted into mechanical energy and vice versa. The first law of thermodynamics is also successfully used in refrigerators. In sweating, the heat of the body is transferred to sweat. Hence, sweating is a great example of the first law of thermodynamics.

The first law of thermodynamics is used in heating and cooling systems in homes and in the design of buildings and vehicles. This law is also applied to ideal as well as real gases to extract physical information regarding the specific heats, the compressibility factor, and the volume expansion coefficient. Some of the applications of the first law of thermodynamics are discussed in detail below.

8.6.1   Gaseous system: General relation between C_P and C_V

For a gaseous system, the expression for work done is $\delta W = P \, dV$. Equation (8.5) can be expressed as

$$\delta Q = dU + P \, dV. \tag{8.29}$$

It should be noted that the state of a gas can be described in terms of only two coordinates out of pressure P, volume V, and temperature T. In order to find the expression for the specific heat of a gaseous system, we consider volume V and temperature T as independent variables. Since the internal energy U is a function of state, we can express it in terms of these two variables, that is, $U = U(T, V)$. A small change in U can then be expressed as

$$dU = \left(\frac{\partial U}{\partial T}\right)_V dT + \left(\frac{\partial U}{\partial V}\right)_T dV. \tag{8.30}$$

Using equation (8.30) in equation (8.29) and rearranging, we get

$$\delta Q = \left(\frac{\partial U}{\partial T}\right)_V dT + \left[P + \left(\frac{\partial U}{\partial V}\right)_T\right] dV. \tag{8.31}$$

At constant volume (V = constant), the second term in equation (8.31) is zero as $dV = 0$. Hence, the specific heat at constant volume C_V is found to be

$$C_V = \left(\frac{\delta Q}{\partial T}\right)_V = \left(\frac{\partial U}{\partial T}\right)_V. \tag{8.32}$$

From equation (8.31), the specific heat at constant pressure C_P is given by

$$C_P = \left(\frac{\delta Q}{\partial T}\right)_P = \left(\frac{\partial U}{\partial T}\right)_V + \left[\left(\frac{\partial U}{\partial V}\right)_T + P\right] \left(\frac{\partial V}{\partial T}\right)_P. \tag{8.33}$$

Using equation (8.32) in equation (8.33), we get

$$C_P = C_V + \left[\left(\frac{\partial U}{\partial V} \right)_T + P \right] \left(\frac{\partial V}{\partial T} \right)_P.$$

This leads to

$$C_P - C_V = \left[\left(\frac{\partial U}{\partial V} \right)_T + P \right] \left(\frac{\partial V}{\partial T} \right)_P. \tag{8.34}$$

Let us interpret the result given by equation (8.34). The first term in the third bracket on the right-hand side of this equation indicates how the internal energy U changes with volume V at constant temperature T. This is associated with the work done against intermolecular forces acting in the gaseous system. On the other hand, the second term in the third bracket tells how much work is done in pushing back the surroundings at constant pressure. These two factors clearly indicate that a knowledge of the difference in heat capacities at constant pressure and at constant volume will provide us information about the variation of internal energy U of a substance with volume V.

Relation between C_P and C_V in case of an ideal gas

One of the ways for determination of the relation between the specific heats C_P and C_V is to allow free expansion of the gas into a larger volume, that is, perform Joule expansion. Since there is no inflow of heat and no work is done in free expansion, we do not expect any change in the internal energy of the gas. This physical fact helps us to define a perfect gas as being characterized by the following properties:

1. There is no change in the internal energy of the perfect gas in Joule expansion. That is, there are no intermolecular attractions, and the internal energy is wholly kinetic.

2. The equation of state for n moles of the perfect gas is: $PV = nRT$.

Property (1) implies that $\left(\frac{\partial U}{\partial V} \right)_T = 0$, so that equation (8.34) reduces to

$$C_P - C_V = P \left(\frac{\partial V}{\partial T} \right)_P. \tag{8.35}$$

This equation (8.35) shows that the difference in the two heat capacities $(C_P - C_V)$ depends on how the volume of a system changes as its temperature increases at constant pressure. We expect this difference to be large because a small change in temperature gives rise to a large change in the volume of a gaseous system at a constant pressure.

From property (2), we can write

$$P\left(\frac{\partial V}{\partial T}\right)_P = nR, \quad \text{equation (8.35) then gives rise to} \quad C_P - C_V = nR. \quad (8.36)$$

The result given by equation (8.36) is known as **Mayer's formula**. It demonstrates that the internal energy of a perfect gas is a function of temperature T only. It should be mentioned that we started by expressing the internal energy U as a function of temperature T and volume V since it is a function of state. But the Joule expansion clearly demonstrates that the internal energy U of an ideal gas is independent of volume V. Further, it can be stated that the internal energy must also be independent of pressure because the pressure decreases in Joule expansion.

Using **Mayer's formula**, the first law of thermodynamics can be written in a different form. To illustrate this idea, we note that for a perfect gas, the internal energy U is a function of temperature T only. Hence, from equation (8.32), we can say that

$$C_V = \left(\frac{dU}{dT}\right)_V, \quad \text{so that} \quad dU = C_V \, dT.$$

Using this result in equation (8.29), we get

$$\delta Q = C_V \, dT + P \, dV. \quad (8.37)$$

Again, using the equation of state for one mole of a perfect gas, we can correlate infinitesimal changes in thermodynamic variables as

$$P \, dV + V \, dP = R \, dT, \quad \text{so that} \quad P \, dV = R \, dT - V \, dP.$$

Using this result in equation (8.37), we can rewrite the first law of thermodynamics as

$$\delta Q = C_V \, dT + R \, dT - V \, dP = (C_V + R) \, dT - V \, dP = C_P \, dT - V \, dP. \quad (8.38)$$

Eliminating dT from equations (8.37) and (8.38), we get

$$\delta Q = \frac{1}{R} \left(C_P \, P \, dV + C_V \, V \, dP\right). \quad (8.39)$$

Equations (8.37), (8.38), and (8.39) are the equivalent mathematical forms of the first law of thermodynamics. These equations are very helpful in solving mathematical problems in thermodynamics.

Relation between C_P and C_V in case of a real gas (Van der Waals type)

There exist intermolecular forces (e.g., Van der Waals force) in a real gas and work has to be done against these internal forces. The internal energy of such a gas depends on both temperature T and volume V of the gas and the intermolecular forces. A repulsive force also exists due to hard-core repulsion between the molecules at high pressures. Hence, one may expect a small change in temperature in the expansion against these intermolecular forces.

We consider one mole of a real gas obeying Van der Waals equation of state given by

$$\left(P + \frac{a}{V^2}\right)(V - b) = RT. \tag{8.40}$$

For one mole of such a real gas, we have

$$\left(\frac{\partial U}{\partial T}\right)_T = \frac{a}{V^2}. \tag{8.41}$$

Using the result of equation (8.41) in equation (8.34) and combining equation (8.40), we get

$$C_P - C_V = \frac{RT}{V - b}\left(\frac{\partial V}{\partial T}\right)_P. \tag{8.42}$$

We differentiate the Van der Waals equation to get the value of $\left(\frac{\partial V}{\partial T}\right)_P$. This provides

$$\left(P + \frac{a}{V^2} - (V - b)\frac{2a}{V^3}\right)\left(\frac{\partial V}{\partial T}\right)_P = R,$$

so that

$$\left(\frac{\partial V}{\partial T}\right)_P = \frac{R}{\left(P + \frac{a}{V^2} - (V - b)\frac{2a}{V^3}\right)}.$$

Multiplying the numerator and denominator of the right-hand side by $(V - b)$, we get

$$\left(\frac{\partial V}{\partial T}\right)_P = \frac{R(V - b)}{\left(RT - \frac{2a}{V^3}(V - b)^2\right)},$$

so that

$$\frac{1}{V - b}\left(\frac{\partial V}{\partial T}\right)_P = \frac{1}{T\left(1 - \frac{2a}{RTV^3}(V - b)^2\right)}.$$

Using binomial expansion, we can write for very small "a"

$$\frac{1}{V - b}\left(\frac{\partial V}{\partial T}\right)_P = \frac{1}{T}\left[1 + \frac{2a}{RTV^3}(V - b)^2\right], \tag{8.43}$$

where terms only up to the first order in a are retained. Substituting the value of $\left(\frac{\partial V}{\partial T}\right)_P$ from equation (8.43) into equation (8.42), we get

$$C_P - C_V = R \left[1 + \frac{2a}{RTV^3} (V - b)^2 \right].$$ (8.44)

It is observed from equations (8.36) and (8.44) that the difference between heat capacities at constant pressure and at constant volume is larger for a real gas than that for a perfect gas. However, if the gas is not very densely packed, that is, for large molecular separations, the second term within the bracket in equation (8.44) can be neglected. Under this situation, we get back the result valid for an ideal gas.

For a real gas, a variation of $\frac{C_P - C_V}{R}$ with volume V is shown in Figure 8.5. It

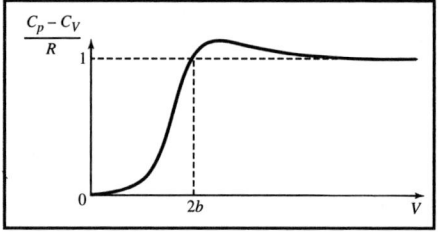

Figure 8.5 $\frac{C_P - C_V}{R}$ as a function of volume V for one mole of a Van der Waals gas.

is observed from the figure that in the limit of large V, the behavior of a real gas exactly matches to that of the ideal one.

8.6.2 Magnetic system: General relation between C_B and C_M

Thermodynamics is not just about the ideal gases described in terms of thermodynamic variables, such as pressure and volume. It is also capable of describing the thermodynamic properties of magnetic materials analogous to what is observed in an ideal gas.

In the presence of an external magnetic field, we get three types of possible magnetic responses of a magnetic material. These responses mainly depend on the magnetic dipole moments of the material and they differ according to whether there are atoms with permanent magnetic dipole moments or not in the material. For example, in the case of diamagnetic materials, atoms do not have any permanent dipole moments. When a magnetic field is applied to a diamagnetic material, an induced dipole moment is created that opposes the applied external magnetic field. **This phenomenon is temperature-independent and universally observed, however small, in almost all materials.** From the thermodynamic point of view, it is not so interesting. Another class of magnetic material is paramagnetic that possesses atoms with permanent dipole moments. These dipoles do not interact with each other. When an external magnetic field is applied to such systems, these permanent dipole moments tend to line up themselves along the direction of the applied magnetic field to some extent (swamping other diamagnetic effects and overcoming thermal agitation), leading to a magnetic moment that reinforces with the field. This effect is proportional to the applied magnetic field (until the system saturates). With the increase in temperature, this effect decreases. On the other hand, thermal agitation tends to randomize these dipoles. At sufficiently low temperatures, the thermal agitation is small enough such that the various atomic dipole moments can interact to establish and maintain a long-range order. These types of materials are known as **ferromagnetic**. The temperature at which the transition from paramagnetic to ferromagnetic occurs is known as the **Curie temperature**. For temperatures well above the Curie temperature, the interactions among the dipole moments in a paramagnetic material can be ignored. This is analogous to ignoring the molecular interactions in an ideal gas.

First law of thermodynamics for a magnetic system

We consider a paramagnetic system. Let us calculate the energy of such a paramagnetic system using the first law of thermodynamics and analyze how work and heat can change that energy. When a magnetic field is applied to a paramagnetic system, energy from the applied magnetic field is used up to align the magnetic moments of the atoms. Thus, a certain amount of work is performed on the system. Let the volume of the region occupied by the paramagnetic system be denoted by V. Let the magnetic field be represented by $\vec{B}$. This magnetic field is created due to a given current configuration in the volume V if the paramagnetic material was not present. The energy in this magnetic field is given by

$$E_B = \frac{1}{2\mu_0} \int_V B^2 \, dV. \tag{8.45}$$

Thus, in the absence of the paramagnetic material, to create the magnetic field $\vec{B}$, one has to do the work $W_B = E_B$.

Now, we consider the response of a paramagnetic material to the applied external magnetic field $\vec{B}$. In the presence of this field, let the total induced magnetic moment in the paramagnetic substance be $\vec{M}$. According to the electromagnetic theory, the additional work needed to establish this magnetic moment $\vec{M}$ is

$$W_M = \int_{M_1}^{M_2} \vec{B}.d\vec{M}. \tag{8.46}$$

The quantity E_B does not pertain to the "system" – the paramagnetic material – and it is convenient to define the internal energy U of the system relative to E_B. Thus, we have

$$U = Energy - E_B. \tag{8.47}$$

The mathematical form of the first law of thermodynamics for such a magnetic system can now be written as

$$dU = \delta Q - PdV + \vec{B}.d\vec{M}. \tag{8.48}$$

The volume of the material is held constant in many applications, and compressional work can be completely ignored. Hence, equation (8.48) can be written as

$$dU = \delta Q + \vec{B}.d\vec{M}; \quad \Rightarrow \quad \delta Q = dU - B \, dM. \tag{8.49}$$

It is assumed that the directions of B and M are the same. For a paramagnetic crystal in a uniform magnetic field $\vec{B}$, with total magnetic dipole moment $\vec{M}$, there are two forms for the work done when $\vec{B}$ and $\vec{M}$ change:

$$\delta W_{\mathrm{ms}} = B \, dM \quad and \quad \delta W_s = -M \, dB. \tag{8.50}$$

When the mutual field energy is included in the system, the expression for δW_{ms} is applied, and in the absence of the mutual field energy, the expression for δW_s is used. The latter one thus applies to the system whose internal energy is just the potential energy of the spins in the field of the lattice energy as well. Considering the potential energy of the spins in a paramagnetic system, the first law of thermodynamics can be written as

$$dU = \delta Q - B \, dM. \tag{8.51}$$

We thus see that there is a close similarity between the magnetic system and the gaseous system. As far as magnetic work is concerned, $\vec{B}$ is analogous to P, and $\vec{M}$ is analogous to V. It is to be noted further that both P and $\vec{B}$ are intensive variables, while V and $\vec{M}$ are extensive variables. According to the microscopic model of a paramagnetic substance (due to Langevin), under the assumption of weak magnetic field $\vec{B}$ and small intrinsic magnetic moments $\vec{\mu}$ of the atoms/molecules, the induced magnetization $\vec{M}$ is given by

$$\vec{M} = \frac{N\mu^2}{3K_B T}\vec{B}. \tag{8.52}$$

This is the equation of state for a magnetic system analogous to the ideal gas law.

General relation between C_B and C_M

The specific heat of a paramagnetic system can be defined under two different situations: at a constant magnetic field and at constant magnetization. Hence, the specific heats are

$$C_B = \left(\frac{\delta Q}{dT}\right)_B \quad and \quad C_M = \left(\frac{\delta Q}{dT}\right)_M. \tag{8.53}$$

Using equation (8.49), we can write

$$C_B = \left(\frac{dU}{dT}\right)_B - B\left(\frac{dM}{dT}\right)_B \tag{8.54}$$

and

$$C_M = \left(\frac{dU}{dT}\right)_M. \tag{8.55}$$

Hence, we can write

$$C_B - C_M = \left(\frac{dU}{dT}\right)_B - B\left(\frac{dM}{dT}\right)_B - \left(\frac{dU}{dT}\right)_M = -B\left(\frac{dM}{dT}\right)_B.$$

The property that the internal energy is a state function has been utilized above. Using equation (8.52), the above expression takes the form

$$C_B - C_M = -B\left(\frac{dM}{dT}\right)_B = -B\left[\frac{d}{dT}\left\{\frac{N\mu^2}{3K_B T}B\right\}\right]_B = \frac{N\mu^2}{3K_B}\frac{B^2}{T^2}. \tag{8.56}$$

This equation (8.56) plays a very important role in magnetism.

8.6.3   Adiabatic and isothermal elasticity

A gas is a highly compressible substance. Therefore, it possesses volume elasticity. But, the magnitude of the volume elasticity of a gas depends on the conditions under which it is compressed. If the gas is compressed under adiabatic conditions, that is, no heat is allowed to enter or leave the system, the corresponding elasticity is then known as the **adiabatic elasticity** and is represented by κ_S. On the other hand, if the gas is compressed under isothermal conditions, that is, the temperature of the system remains constant, then the corresponding volume elasticity is known as the **isothermal elasticity** and is represented by κ_T. For such a gaseous system, the first law of thermodynamics can be used to show that the ratio of the adiabatic elasticity κ_S to the isothermal elasticity κ_T is equal to γ, that is, $\frac{\kappa_S}{\kappa_T} = \gamma$, where γ is the ratio of two specific heats C_P and C_V.

These elasticity constants, that is, the adiabatic and isothermal elasticity are, respectively, given by

$$\kappa_S = -V\left(\frac{\partial P}{\partial V}\right)_S \quad \text{and} \quad \kappa_T = -V\left(\frac{\partial P}{\partial V}\right)_T.$$

Dividing these two expressions, we get

$$\frac{\kappa_S}{\kappa_T} = -\frac{\left(\frac{\partial P}{\partial V}\right)_S}{\left(\frac{\partial P}{\partial V}\right)_T}. \tag{8.57}$$

In order to find this ratio of two elasticity, we assume that temperature is a function of pressure and volume, that is, $T = T(V, P)$. Then an infinitesimal change dT in temperature can be written as

$$dT = \left(\frac{\partial T}{\partial V}\right)_P dV + \left(\frac{\partial T}{\partial P}\right)_V dP. \tag{8.58}$$

In the case of isothermal process, $dT = 0$. This leads to

$$\left(\frac{\partial P}{\partial V}\right)_T = -\frac{\left(\frac{\partial T}{\partial V}\right)_P}{\left(\frac{\partial T}{\partial P}\right)_V}. \tag{8.59}$$

Using equation (8.31) and the definition of specific heat at constant volume $C_V = \left(\frac{\partial U}{\partial T}\right)_V$, we get

$$\delta Q = C_V dT + \left[P + \left(\frac{\partial U}{\partial V}\right)_T\right] dV. \tag{8.60}$$

Again from equation (8.34), we have

$$C_P - C_V = \left[\left(\frac{\partial U}{\partial V}\right)_T + P\right]\left(\frac{\partial V}{\partial T}\right)_P; \quad \Rightarrow \quad P + \left(\frac{\partial U}{\partial V}\right)_T = \frac{C_P - C_V}{\left(\frac{\partial V}{\partial T}\right)_P}.$$

Using this result in equation (8.60), we get

$$\delta Q = C_V dT + \frac{C_P - C_V}{\left(\frac{\partial V}{\partial T}\right)_P} dV = C_V dT + (C_P - C_V)\left(\frac{\partial T}{\partial V}\right)_P dV;$$

$$\Rightarrow \quad \frac{\delta Q}{C_V} = dT + (\gamma - 1)\left(\frac{\partial T}{\partial V}\right)_P dV.$$

Using the expression for dT from equation (8.58) and simplifying, we get

$$\frac{\delta Q}{C_V} = \left(\frac{\partial T}{\partial P}\right)_V dP + \gamma\left(\frac{\partial T}{\partial V}\right)_P dV.$$

For an adiabatic process, we have $\delta Q = 0$, and this expression leads to

$$\left(\frac{\partial P}{\partial V}\right)_S = -\gamma\frac{\left(\frac{\partial T}{\partial V}\right)_P}{\left(\frac{\partial T}{\partial P}\right)_V}; \quad \Rightarrow \quad \gamma = -\left(\frac{\partial P}{\partial V}\right)_S\left(\frac{\partial T}{\partial P}\right)_V\left[\left(\frac{\partial T}{\partial V}\right)_P\right]^{-1}$$

$$\text{Or}, \gamma = -\left(\frac{\partial P}{\partial V}\right)_S\left(\frac{\partial V}{\partial P}\right)_T = -\frac{\left(\frac{\partial P}{\partial V}\right)_S}{\left(\frac{\partial P}{\partial V}\right)_T}; \quad \Rightarrow \quad \frac{\kappa_S}{\kappa_T} = -\frac{\left(\frac{\partial P}{\partial V}\right)_S}{\left(\frac{\partial P}{\partial V}\right)_T} = \gamma.$$

8.6.4   Compressibility and expansion coefficient

In thermodynamics, it is observed that every macroscopic property of a system can be expressed by a set of a few measurable quantities in equilibrium. For example, every monoatomic or monomolecular nonmagnetic system is completely described by a set of three quantities and their temperature dependence:

1. Coefficient of volume expansion

2. Isothermal compressibility and

3. Heat capacity.

Coefficient of thermal expansion α_P of an ideal gas

The coefficient of volume expansion of a gas is defined as the increase in volume per unit volume at 0°C for each Celsius degree rise in temperature, the pressure being constant.

Mathematically, this is given by

$$\alpha_P = \frac{1}{V} \left(\frac{\partial V}{\partial T} \right)_P.$$

For an ideal gas, we have $PV = nRT$. Hence, α_P for this gas is given by

$$\alpha_P = \frac{1}{V} \left(\frac{\partial V}{\partial T} \right)_P = \frac{1}{V} \frac{\partial}{\partial T} \left(\frac{nRT}{P} \right)_P = \frac{1}{T}.$$

Thus, for an ideal gas, α_P is inversely proportional to temperature, and its value is $\alpha_P = 3.7 \times 10^{-3} \text{ K}^{-1}$ at 0°C, which is greater than solid and liquid.

For one mole of a real gas (Van der Waals), we have $\left(P + \frac{a}{V^2} \right) (V - b) = RT$. In order to calculate α_P for this gas, let us calculate the factor $\left(\frac{\partial V}{\partial T} \right)_P$ for one mole of a Van der Waals gas. Using this equation for one mole, we get

$$\left[0 - \frac{2a}{V^3} \left(\frac{\partial V}{\partial T} \right)_P \right] (V - b) + \left(P + \frac{a}{V^2} \right) \left(\frac{\partial V}{\partial T} \right)_P = R;$$

$$\Rightarrow \quad \left(\frac{\partial V}{\partial T} \right)_P = \frac{RV^3(V - b)}{RTV^3 - 2a(V - b)^2}.$$

This leads to the value of α_P as

$$\alpha_P = \frac{1}{V} \left(\frac{\partial V}{\partial T} \right)_P = \frac{1}{V} \times \left[\frac{RV^3(V - b)}{RTV^3 - 2a(V - b)^2} \right] = \frac{RV^2(V - b)}{RTV^3 - 2a(V - b)^2}.$$

Thus, for a real gas, α_P depends on volume and temperature both.

Isothermal compressibility β_T of a gas

The isothermal compressibility β_T of a system is the inverse of isothermal elasticity κ_T and is given by the expression

$$\beta_T = -\frac{1}{V} \left(\frac{\partial V}{\partial P} \right)_T.$$

The negative sign is important in order to keep the value of the isothermal compressibility β_T positive since an increase in pressure P will lead to a decrease in volume V at constant temperature T. The term $\frac{1}{V}$ is needed to make the property intensive so that it can be tabulated in a useful manner. The factor β_T is also the reciprocal of the bulk modulus of elasticity κ_T. Gas usually is the most compressible medium in the reservoir. However, care should be taken so that it is not confused with the gas deviation factor, Z_c, which is sometimes called the compressibility factor.

β_T **for an ideal gas:** For an ideal gas, we have $PV = nRT$. Hence, β_T for this gas is given by

$$\beta_T = -\frac{1}{V}\left(\frac{\partial V}{\partial P}\right)_T = -\frac{1}{V}\frac{\partial}{\partial P}\left(\frac{nRT}{P}\right)_P = -\frac{1}{V} \times nRT \times \left(-\frac{1}{P^2}\right) = \frac{1}{P}.$$

Thus, for an ideal gas, β_T is inversely proportional to pressure.

β_T **for a Van der Waals gas:** The equation for one mole of a Van der Waals gas is given by

$$\left(P + \frac{a}{V^2}\right)(V - b) = RT.$$

The coefficient β_T is defined under isothermal conditions so that $T = $ constant. In order to find the factor β_T for the real gas, we have to calculate the partial derivative $\left(\frac{\partial V}{\partial P}\right)_T$. Differentiating the above expression w.r.t P at constant temperature T, we get

$$\left[1 - \frac{2a}{V^3}\left(\frac{\partial V}{\partial P}\right)_T\right](V - b) + \left(P + \frac{a}{V^2}\right)\left(\frac{\partial V}{\partial P}\right)_T = 0;$$

$$\Rightarrow \left[1 - \frac{2a}{V^3}\left(\frac{\partial V}{\partial P}\right)_T\right](V - b) + \left(\frac{RT}{V - b}\right)\left(\frac{\partial V}{\partial P}\right)_T = 0.$$

$$\Rightarrow \left(\frac{\partial V}{\partial P}\right)_T = -\frac{V - b}{\frac{RT}{V-b} - \frac{2a(V-b)}{V^3}} = -\frac{V^3(V - b)^2}{RTV^3 - 2a(V - b)^2}.$$

Using this value in the expression for β_T, we get

$$\beta_T = -\frac{1}{V}\left(\frac{\partial V}{\partial P}\right)_T = -\frac{1}{V} \times \left[-\frac{V^3(V - b)^2}{RTV^3 - 2a(V - b)^2}\right] = \frac{V^2(V - b)^2}{RTV^3 - 2a(V - b)^2}.$$

This is the expression for isothermal compressibility β_T for one mole of a Van der Waals gas.

The heat capacity has already been discussed in earlier section of this chapter.

8.7   Solved Problems

1. A system absorbs 1.5×10^3 J of energy as heat and produces 500 J of work. Calculate the change in the internal energy of the system.

   Answer: From the first law of thermodynamics, we have

   $$\delta Q = dU + \delta W.$$

   Here, we have $\delta Q = 1.5 \times 10^3$ J $= 1500$ J and $\delta W = 500$ J. So, the change in internal energy of the system is given by

   $$dU = \delta Q - \delta W = 1000 \text{ J}.$$

2. Show that during a certain thermodynamic process, the work done by a system depends on the path followed by the system. Also, calculate the work done in an isothermal process.

   Answer:

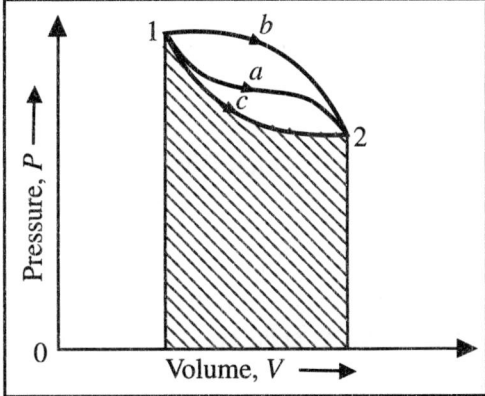

Figure 8.6 Indicator diagram for different paths for problem 2.

For a hydrostatic system, the work done during a thermodynamic process occurring between states 1 and 2 is given by

$$W = \int_{1}^{2} P \, dV.$$

Geometrically, the work done W is equal to the area under the curve along which a thermodynamic process occurs from the initial to the final state. As shown in Figure 8.6, one can go from state 1 to state 2 following different paths. It means that the value of W will be different for different paths. This implies that work is a path function, and in a cyclic thermodynamic process, the work done is nonzero.

In an isobaric process, pressure is constant, and the work done is given by

$$W = \int_{V_1}^{V_2} P \, dV = P(V_2 - V_1).$$

But for an isochoric process, V=constant. Thus, $dV = 0$. Therefore, the work done in an isochoric process is zero.

For n mole of an ideal gas, we have the expression for pressure P as $P = \frac{nRT}{V}$. Therefore, the work done by an ideal gas in the isothermal expansion is given by

$$W = \int_{V_1}^{V_2} P \, dV = nRT \int_{V_1}^{V_2} \frac{dV}{V} = nRT \ln\left[\frac{V_2}{V_1}\right].$$

Using Boyle's law, we can write the expression for work done by an ideal gas during an isothermal process as

$$W = nRT \ln\left[\frac{P_1}{P_2}\right].$$

It is to be noted that the factor $\ln\left[\frac{V_2}{V_1}\right]$ is positive in an expansion, so that the work done is positive. On the other hand, for compression, the ratio $\left[\frac{V_2}{V_1}\right]$ is less than one, and work done is negative, that is, work done is done on the gas.

3. A gaseous system is maintained at constant pressure. The surroundings around the system lose an amount of heat 63 J and do an amount of work 475 J onto the system. Calculate the change in internal energy of the system.

Answer: To find the change in internal energy, ΔU, we must consider the relationship between the system and the surroundings. Since the first law of thermodynamics states that energy is neither created nor destroyed, we know that anything lost by the surroundings is gained by the system. The surrounding area loses heat and does work on the system. Therefore, Q and W are positive in the equation $\Delta U = Q + W$ because the system gains heat and gets work done on itself. So, the change in internal energy is

$$\Delta U = 63 \text{ J} + 475 \text{ J} = 538 \text{ J}.$$

4. A gaseous system has constant volume and the heat around the system increases by 45 J.

   (a) What is the sign for heat (Q) for the system?

   (b) What is ΔU equal to?

   (c) What is the value of the internal energy of the system in Joules?

   Answer: Since the system has constant volume, that is, $\Delta V = 0$, the term $-P\Delta V = 0$. So, the work done is equal to zero. Thus, using equation $\Delta U = Q + W$, we get $\Delta U = Q$ as $W = 0$. The change in internal energy is equal to the heat supplied to the system. The surrounding heat increases, so the heat of the system decreases. Therefore, heat is taken away from the system making it exothermic and negative. The value of internal energy will be the negative value of the heat absorbed by the surroundings. Hence, we have

   (a) The sign for heat Q is negative.

   (b) $\Delta U = Q + (-P\Delta V) = Q + 0 = Q.$

   (c) $\Delta U = -45$ J.

5. n moles of a certain ideal gas at temperature T_0 are cooled isochorically so that the pressure of the gas is reduced n times. Then as a result of the isobaric process, the gas is expanded till its temperature gets back to its initial value. Find the total amount of heat absorbed by the gas in the process.

   Answer: Let, at the initial state A, pressure, volume, and temperature of the gas be P_0, V_0, and T_0. The number of moles of the gas is n.

   According to the question, after isochoric cooling at state B, the values of the parameters become $\frac{P_0}{n}$, V_0, and T_B (say). Further, we assume that the values of the parameters become $\frac{P_0}{n}$, V_C, and T_0 after the isobaric process at the state C.

   Applying the ideal gas equation between the states A and B, we get

   $$\frac{P_0 V_0}{T_0} = \frac{P_0 V_0}{n T_B}; \quad \Rightarrow \quad T_B = \frac{T_0}{n}.$$

   Again, applying the ideal gas equation between the states B and C, we get

   $$\frac{P_0 V_0}{n T_B} = \frac{P_0 V_C}{n T_0}; \quad \Rightarrow \quad \frac{P_0 V_0 n}{n T_0} = \frac{P_0 V_C}{n T_0}; \quad \Rightarrow \quad V_C = n V_0.$$

   In the process A to B, $\delta Q = nC_V dT = nC_V \left(\frac{T_0}{n} - T_0\right).$

   In the process B to C, $\delta Q = nC_V dT + P dV = nC_V \left(T_0 - \frac{T_0}{n}\right) + nR \left(T_0 - \frac{T_0}{n}\right).$

So, the total heat absorbed by the gas is

$$\delta Q_{\text{total}} = \delta Q(A \to B) + \delta Q(B \to C)$$
$$= nC_V \left(\frac{T_0}{n} - T_0 \right) + nC_V \left(T_0 - \frac{T_0}{n} \right) + nR \left(T_0 - \frac{T_0}{n} \right)$$
$$= nRT_0 \left(1 - \frac{1}{n} \right).$$

6. A sample of an ideal gas has pressure P_0, volume V_0, and temperature T_0. The gas is isothermally expanded to twice its original volume. It is then compressed at constant pressure to have the original volume V_0. Finally, the gas is heated at constant volume to get the original temperature.

(a) Show the process in a V–T indicator diagram.

(b) Calculate the heat absorbed in the process.

Answer: (a) The process is represented in Figure 8.7.

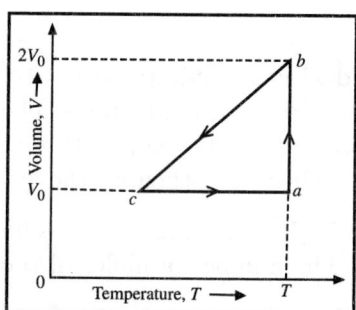

Figure 8.7 V–T indicator diagram for the process in problem 6a.

(b) As the process is cyclic, the change in internal energy in the cycle is zero. So, the heat supplied in the cycle is equal to the work done.

In the isothermal process from $a \to b$, work done is

$$W_{ab} = nRT_0 \ln \frac{2V_0}{V_0} = nRT_0 \ln 2.$$

In the isobaric process from $b \to c$, work done is PdV.

Applying the ideal gas equation between the states a and b, we get

$$P_a V_a = P_b V_b; \quad \Rightarrow \quad P_0 V_0 = P_b 2V_0; \quad \Rightarrow \quad P_b = \frac{P_0}{2}.$$

So, the work done in the process from $b \to c$ is

$$W_{b \to c} = P_b\,(V_0 - 2V_0) = \frac{P_0}{2}\,(V_0 - 2V_0) = -\frac{P_0\,V_0}{2} = -\frac{R\,T_0}{2}.$$

In the isochoric process from $c \to a$, $dV = 0$, and hence, $W_{c \to a} = 0$. So, the total work done is given by

$$W_{\text{total}} = nRT_0 \, \ln 2 + \left(-\frac{R\,T_0}{2}\right) = RT_0 \left(n \ln 2 - \frac{1}{2}\right).$$

7. The internal energy of a simple compressible system is a function of temperature T and pressure P. Show that the change in internal energy as a function of temperature at constant pressure is given by

$$\left(\frac{\partial U}{\partial T}\right)_P = C_P - P\,V\,\alpha_P,$$

where α_p is the isobaric volume expansion coefficient given by $\alpha_p = \frac{1}{V}\left(\frac{\partial V}{\partial T}\right)_P$, and C_P is the specific heat at constant pressure.

[Calcutta University 2012, 2014]

Answer: We consider internal energy U as a function of temperature T and pressure P, that is, $U = U(T, P)$. Then, the differential dU can be written as

$$dU = \left(\frac{\partial U}{\partial P}\right)_T dP + \left(\frac{\partial U}{\partial T}\right)_P dT.$$

At constant pressure, $dP = 0$. From the first law of thermodynamics, we have $\delta Q = dU + P\,dV$. Using the value of dU in this expression, we get

$$\delta Q = \left(\frac{\partial U}{\partial P}\right)_T dP + \left(\frac{\partial U}{\partial T}\right)_P dT + PdV; \quad \Rightarrow \quad \left(\frac{\delta Q}{\partial T}\right)_P = \left(\frac{\partial U}{\partial T}\right)_P + P\left(\frac{\partial V}{\partial T}\right)_P.$$

Therefore, the specific heat at constant pressure C_P is given by

$$C_P = \left(\frac{\partial U}{\partial T}\right)_P + PV\alpha_P; \quad \Rightarrow \quad \left(\frac{\partial U}{\partial T}\right)_P = C_P - PV\,\alpha_P.$$

The definition of $\alpha_P = \frac{1}{V}\left(\frac{\partial V}{\partial T}\right)_P$ is used in the above expression.

8. Using the fact that dS is an exact differential, derive the following relation: $\left(\frac{\partial U}{\partial V}\right)_T = T\left(\frac{\partial P}{\partial T}\right)_V - P$. Hence, for one mole of a Van der Waals gas, show that the internal energy is not a function of temperature alone.

[Calcutta University 2013]

Answer: Considering the internal energy U as a function of volume V and temperature T, we get

$$U = U(V, T); \quad \Rightarrow \quad dU = \left(\frac{\partial U}{\partial V}\right)_T dV + \left(\frac{\partial U}{\partial T}\right)_V dT.$$

Combining the first and the second laws of thermodynamics and the above relation, we get

$$dS = \frac{\delta Q}{T} = \frac{dU + PdV}{T} = \frac{1}{T}\left[\left(\left(\frac{\partial U}{\partial V}\right)_T + P\right)dV + \left(\frac{\partial U}{\partial T}\right)_V dT\right].$$

As entropy S is a state function, dS is an exact differential. Therefore, we have

$$\frac{\partial}{\partial V}\left[\left(\frac{1}{T}\frac{\partial U}{\partial T}\right)_V\right] = \frac{\partial}{\partial T}\left[\left(\frac{1}{T}\left(\frac{\partial U}{\partial V}\right)_T + P\right)\right];$$

$$\Rightarrow \frac{1}{T}\frac{\partial^2 U}{\partial U \partial T} = \frac{1}{T}\frac{\partial^2 U}{\partial T \partial V} - \left[\left(\frac{1}{T^2}\left(\frac{\partial U}{\partial V}\right)_T + P\right) + \left(\frac{\partial P}{\partial T}\right)_V \frac{1}{T}\right].$$

As the internal energy U is a state function, we have $\frac{\partial^2 U}{\partial T \partial V} = \frac{\partial^2 U}{\partial V \partial T}$. Using this condition, the above expression reduces to

$$P + \left(\frac{\partial U}{\partial V}\right)_T = T\left(\frac{\partial P}{\partial T}\right)_V.$$

For one mole of a Van der Waals gas, we have the equation of state as

$$\left(P + \frac{a}{V^2}\right)(V - b) = RT; \quad \Rightarrow \quad P + \frac{a}{V^2} = \frac{RT}{(V-b)}; \quad \Rightarrow \quad P = -\frac{a}{V^2} + \frac{RT}{(V-b)}.$$

We know that

$$\left(\frac{\partial U}{\partial V}\right)_T = T\left(\frac{\partial P}{\partial T}\right)_V - P; \quad \Rightarrow \quad \left(\frac{\partial U}{\partial V}\right)_T = T\frac{R}{(V-b)} - \frac{RT}{(V-b)} + \frac{a}{V^2} = \frac{a}{V^2}.$$

Thus, for a Van der Waals gas, the internal energy U depends on the volume V and the Van der Waals constant a. It is to be noted that for an ideal gas, U does not depend on the volume V.

9. Two identical bodies of constant heat capacity C_P at temperatures T_1 and T_2 are, respectively, used as reservoirs for a heat engine. If the bodies are at constant pressure without any change of phase, show that the amount of work obtainable is $W = C_P(T_1 + T_2 - 2T_f)$, where T_f is the final temperature

attained by both bodies. Further, show that the amount of work obtained will be maximum when the final temperature T_f is $T_f = \sqrt{T_1 T_2}$.

[Calcutta University 2012]

Answer: Let T_f be the final temperature of both the source and the sink. Heat lost by the source is Q_1, and the heat gained by the sink is Q_2. Then, we have the expressions for these two quantities as

$$Q_1 = C_P(T_1 - T_f) \quad \text{and} \quad Q_2 = C_P(T_f - T_2).$$

Hence, the work done W is given by

$$W = Q_1 - Q_2 = C_P(T_1 + T_2 - 2T_f).$$

Work W can be expressed as

$$W = C_P[(\sqrt{T_1} - \sqrt{T_2})^2 - (2T_f - 2\sqrt{T_1 T_2})].$$

Therefore, the amount of work W will be maximum when the second term within the third bracket is zero. This provides the final temperature T_f as $T_f = \sqrt{(T_1 T_2)}$.

10. In what respect does an ideal gas differ from a real gas? When does this difference become negligible?   [Calcutta University 2015, 2016]

   Answer: In the case of an ideal gas, there is no force of attraction or repulsion between the molecules. Therefore, an ideal gas cannot be compressed or liquefied. But this is not true for the molecules of a real gas. There is a force of attraction between the molecules of a real gas, and they can be liquefied.

   Further, the volume occupied by the molecules of an ideal gas is negligible compared to the volume of the gas container. This approximation is not true for the molecules of a real gas.

   The difference between an ideal and a real gas becomes negligible at low pressure and at high temperature.

11. Express the work done by a gas in terms of isothermal compressibility β_T and the isobaric volume expansion coefficient α_P.

   Answer: The work done by a gas is given by $\delta W = PdV$. If we take V as a function of T and P, we can write $V = V(T, P)$. Therefore, an infinitesimal change in volume can be written as

$$dV = \left(\frac{\partial V}{\partial T}\right)_P dT + \left(\frac{\partial V}{\partial P}\right)_T dP.$$

Using $\beta_T = -\frac{1}{V}\left(\frac{\partial V}{\partial P}\right)_T$ and $\alpha_P = \frac{1}{V}\left(\frac{\partial V}{\partial T}\right)_P$, the above expression can be written as

$$dV = \alpha_P \, V \, dT - V \, \beta_T \, dP.$$

Using this result in the expression $\delta W = PdV$, we get the expression for work done δW in terms of β_T and α_P as

$$\delta W = P \, \alpha_P \, V \, dT - PV \, \beta_T \, dP.$$

12. The equation of state of an ideal elastic substance is

$$T = K\theta \left(\frac{L}{L_0} - \frac{L_0^2}{L^2}\right),$$

where k is a constant, L_0 is the length at zero tension and is a function of temperature θ only. Derive an expression for the work done required to change the length from L_0 to $\frac{L_0}{3}$ quasistatically and isothermally.

Answer: The required work W is given by

$$W = -\int_{L_0}^{\frac{L_0}{3}} TdL = -\int_{L_0}^{\frac{L_0}{3}} K\theta \left(\frac{L}{L_0} - \frac{L_0^2}{L^2}\right) dL = -\frac{K\theta}{L_0}\int_{L_0}^{\frac{L_0}{3}} LdL + K\theta L_0^2 \int_{L_0}^{\frac{L_0}{3}} \frac{dL}{L^2},$$

$$= -\frac{K\theta}{L_0}\left[\frac{L_0^2}{18} - \frac{L_0^2}{2}\right] - K\theta L_0^2 \left[\frac{3}{L_0} - \frac{1}{L_0}\right]$$

$$= \frac{4}{9}K\theta L_0 - 2K\theta L_0 = -\frac{14}{9}K\theta L_0.$$

The negative sign indicates that the work is done on the system.

13. The thermodynamic variables pressure P, volume V, and temperature T for a certain material are related by

$$P = \frac{AT - BT^2}{V},$$

where A and B are two constants. Calculate the amount of work done by the material if the temperature changes from T to $2T$ at constant pressure P.

Answer: From the given expression, we have

$$P = \frac{AT - BT^2}{V}; \quad \Rightarrow \quad PV = AT - BT^2; \quad \Rightarrow \quad PdV = AdT - 2BTdT.$$

So, the work done by the material when the temperature changes from T to $2T$ at constant pressure P is given by

$$W = \int PdV = \int_{T}^{2T} (AdT - 2BTdT) = [AT - BT^2]_{T}^{2T} = AT - 3BT^2.$$

14. A horizontal cylinder closed from one end is rotated with a constant angular velocity ω about a vertical axis passing through the open end of the cylinder as shown in Figure 8.8. The outside air pressure is equal to P_0, the temperature to T, and the molar mass of air to M. Find the air pressure as a function of the distance r from the axis of rotation. The molar mass is assumed to be independent of r.

Answer:

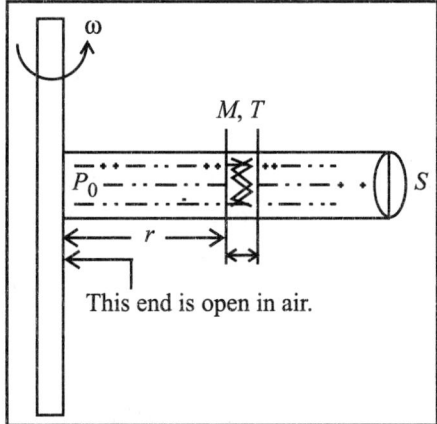

Figure 8.8 Geometry of problem 14.

Consider an element dr at a distance r from the open end of the cylinder, as shown in Figure 8.8.

Force equation of the dr element is: $dF = (dm)r\omega^2$. So, the differential change in pressure is

$$dP = \frac{dF}{A} = \frac{dm}{A}r\omega^2; \quad \Rightarrow \quad dm = \left(\frac{A}{r\omega^2}\right)dP,$$

where A is the cross-section. The work done is given by

$$P(Adr) = \frac{dm}{M}RT; \quad \Rightarrow \quad P(Adr) = \frac{RT}{M}\frac{A}{r\omega^2}dP.$$

Integrating both sides within proper limits, we get

$$Mw^2 \int_0^r r\, dr = RT \int_{P_0}^P \frac{dP}{P}; \quad \Rightarrow \quad \frac{Mw^2 r^2}{2} = RT \ln \frac{P}{P_0}.$$

This gives $P = P_0 e^{\frac{Mw^2 r^2}{2RT}}$.

15. A diatomic gas at N.T.P. is expanded to thirty-two times its volume under adiabatic conditions. Find the resulting temperature up to two decimal points.

Answer: For a diatomic gas, $\gamma = \frac{7}{5}$. The equation of an adiabatic process is

$$TV^{\gamma-1} = \text{constant}; \quad \Rightarrow \quad T_2 = T_1 \left(\frac{V_1}{V_2}\right)^{\gamma-1}.$$

Here, $V_1 = V$, $V_2 = 32V$, and $T_1 = 273$ K. Hence, the final temperature T_2 is given by

$$T_2 = \frac{273}{(32)^{\gamma-1}} = \frac{273}{(32)^{7/5-1}} = \frac{273}{4} = 68.25 \text{ K}.$$

16. A certain gas has an equation of state, $P = \frac{\alpha n^2 T}{V^2}$, where P is the pressure, n is the number of moles, V is the volume, and T is the temperature, and α is a constant. One mole of the gas undergoes expansion from volume V to $2V$ at constant temperature T. If the change in energy in the isothermal expansion is $\beta \frac{\alpha T}{V}$, find the value of β. [Calcutta University 2022]

Answer: We know the relation

$$\left(\frac{\partial U}{\partial V}\right)_T = T \left(\frac{\partial P}{\partial T}\right)_V - P.$$

Using $P = \frac{\alpha\, n^2\, T}{V^2}$, we get $\left(\frac{\partial P}{\partial T}\right)_V = \frac{\alpha\, n^2}{V^2}$. Hence, the above relation reduces to

$$\left(\frac{\partial U}{\partial V}\right)_T = T \left(\frac{\partial P}{\partial T}\right)_V - P = T\frac{\alpha N^2}{V^2} - P = P - P = 0.$$

Thus, according to the problem, we have

$$\frac{\beta\, \alpha\, T}{V} = 0. \quad \text{This leads to } \beta = 0.$$

17. A thermally insulated vessel containing a gas whose molar mass is equal to M, and the ratio of specific heats $\frac{C_P}{C_V} = \gamma$ moves with a velocity v. Calculate the increment in temperature if the vessel is suddenly stopped.

Answer: Let n be the number of moles of the gas. The translational kinetic energy of the gas is $\frac{1}{2}(n\,M)v^2$.

When the vessel is suddenly stopped, the translational kinetic energy of the gas $\frac{1}{2}(n\,M)v^2$ is converted into random kinetic energy $nC_V\Delta T$. After a long time, the system attains a state of thermodynamic equilibrium. Hence, we can write

$$\frac{1}{2}(nM)v^2 = nC_V\Delta T; \quad \Rightarrow \quad \Delta T = \frac{\gamma - 1}{2R}Mv^2.$$

This is the increment in temperature if the vessel is suddenly stopped.

8.8 Multiple choice questions and answers

1. The internal energy of an ideal gas depends on

   (a) pressure   **(c)** temperature

   (b) volume   **(d)** temperature and volume both

   Answer: The correct choice is (c).

2. The internal energy of a real gas depends on

   (a) pressure   **(c)** temperature

   (b) volume   **(d)** temperature and volume both

   Answer: The correct choice is (d).

3. An ideal gas is allowed to expand freely against a vacuum in a rigid insulated container (adiabatic condition). The gas undergoes

   (a) increase in its internal energy.

   (b) decrease in its internal energy.

   (c) neither increase nor decrease in its temperature and internal energy.

   (d) decrease in temperature.

   Answer: The correct choice is (c).

4. An ideal gas has initial volume V, and pressure P. In doubling its volume, the minimum work will be done in which of the following process?

   (a) Isobaric process

   (b) Isothermal process

   (c) Adiabatic process

   (d) Same in all given processes.

   Answer: The correct choice is (c).

5. If heat is supplied to an ideal gas in an isothermal process, then which of the following is correct?

   (a) The internal energy of the gas will increase.

   (b) The gas will do positive work.

   (c) The gas will do negative work.

   (d) The process is not possible.

   Answer: The correct choice is (b).

   For an isothermal process, $\delta Q = \delta W$. Since δQ is positive, δW is also positive, and the gas will do positive.

6. A monoatomic ideal gas at $170°C$ is adiabatically compressed to $1/8$ of its original volume. The temperature after compression is   [JEST 2012]

   (a) $2.1°C$ (b) $17°C$ (c) $-200.5°C$ (d) $887°C$

   Answer: The correct choice is (d).

   Solution: The equations of states for an adiabatic and an isothermal process are, respectively, $PV^\gamma =$constant and $PV = RT$. From these two relations, we get

   $$\frac{TV^\gamma}{V} = \text{constant}; \quad \Rightarrow \quad TV^{\gamma-1} = \text{constant};$$

   $$\Rightarrow \quad T_1 V_1^{\gamma-1} = T_2 V_2^{\gamma-1}; \quad \Rightarrow \quad T_2 = T_1 \left(\frac{V_1}{V_2}\right)^{\gamma-1};$$

   $$\Rightarrow \quad 443(8)^{\frac{5}{3}-1} = 443 \times (8)^{\frac{2}{3}} = 443 \times 4.$$

   Hence, the temperature in $°C$ is $(1772 - 273)°C = 1499°C$.

   The most appropriate answer is an option (d).

7. During the melting of a slab of ice at 273 K at atmospheric pressure -

   (a) Positive work is done by the ice-water system

   (b) Internal energy of ice-water system remains constant

   (c) Internal energy of ice-water system increases

   (d) Internal energy of ice-water system decreases

   Answer: The correct choice is (c).

   [Hint: During the melting of ice, the volume of ice decreases; therefore, work is being done on the system by atmosphere. The system also absorbs heat from surroundings, so internal energy $\Delta U = Q - W = Q + P\Delta V$ is positive, that is, U increases.]

8. At a temperature T, let α_P and β_T denote the volume expansivity and isothermal compressibility of a gas, respectively. Then the ratio $\frac{\alpha_P}{\beta_T}$ is equal to
   [IIT-JAM 2022]

   (a) $\left(\frac{\partial P}{\partial T}\right)_V$

   (c) $\left(\frac{\partial T}{\partial P}\right)_V$

   (b) $\left(\frac{\partial P}{\partial V}\right)_T$

   (d) $\left(\frac{\partial T}{\partial V}\right)_P$

   Answer: The correct choice is (a).

   Solution: We know that the volume expansivity α_P is given by $\alpha_P = \frac{1}{V}\left(\frac{\partial V}{\partial T}\right)_P$, and the isothermal compressibility β_T is given by $\beta_T = -\frac{1}{V}\left(\frac{\partial V}{\partial P}\right)_T$. As a result, we get

   $$\frac{\alpha_P}{\beta_T} = -\left(\frac{\partial P}{\partial V}\right)_T \left(\frac{\partial V}{\partial T}\right)_P.$$

   Now from chain rule, we have

   $$\left(\frac{\partial P}{\partial V}\right)_T \left(\frac{\partial V}{\partial T}\right)_P \left(\frac{\partial T}{\partial P}\right)_V = -1.$$

   Hence, we get

   $$\frac{\alpha_P}{\beta_T} = \left(\frac{\partial P}{\partial T}\right)_V.$$

9. An isolated ideal gas is kept at a pressure P_1 and volume V_1. The gas undergoes free expansion and attains a pressure P_2 and volume V_2. Identify the correct statement(s).
   [IIT-JAM 2021]

   (a) This is an adiabatic process

   (c) $P_1 V_1^\gamma = P_2 V_2^\gamma$

   (b) $P_1 V_1 = P_2 V_2$

   (d) This is an isobaric process

   Answer: The correct choices are (a) and (b).

   Solution: In the case of free expansion, the change in internal energy is zero, and the work done is also zero. As a result, the change in heat is also zero. $\delta Q = 0$ is possible for adiabatic process. The work done in an adiabatic process is given by $W = \frac{R}{\gamma-1}\left(P_1 V_1 - P_2 V_2\right) = 0$. Hence, $P_1 V_1 = P_2 V_2$.

10. A polyatomic gas $\left(\gamma = \frac{4}{3}\right)$ at pressure P is compressed to one-eighth of its initial volume adiabatically, then the pressure will change to

    (a) 4P   (b) 8P   (c) 16P   (d) 32P

    Answer: The correct choice is (c).

11. The molar specific heat of a gas, as given from the kinetic theory, is $\frac{5}{2}R$. If it is not specified whether it is C_V or C_P, one can conclude that the molecules of the gas   [IIT-JAM 2005]

   (a) are definitely monoatomic.
   (b) are definitely rigid diatomic.
   (c) are definitely nonrigid diatomic.
   (d) can be monatomic or rigid diatomic.

   Answer: The correct choice is (d).

   Solution: If a molecule is monoatomic, then $C_P = \frac{5R}{2}$.

   And if a molecule is a rigid diatomic, then $C_V = \frac{5R}{2}$.

12. In a thermodynamic process, the volume of one mole of an ideal gas is varied as $V = aT^{-1}$, where a is a constant. The adiabatic exponent of the gas is γ. What is the amount of heat received by the gas if the temperature of the gas increases by ΔT in the process?   [JEST 2018]

   (a) $R\Delta T$ (b) $\frac{R\Delta T}{1-\gamma}$ (c) $\frac{R\Delta T}{2-\gamma}$ (d) $R\Delta T \frac{2-\gamma}{\gamma-1}$

   Answer: The correct choice is (d).

   Solution: From the given problem, we have $V = \frac{a}{T}$ $\Rightarrow dV = -(a/T^2)dT$. Again, we have $PV = RT$; $\Rightarrow$ $P = \frac{RT}{V}$.

   So, the work done is given by

   $$W = \int P dV = \int \frac{RT}{V}dV \Rightarrow W = \int \frac{RT^2}{a} \times (-\frac{a}{T^2})dT$$

   $$\Rightarrow W = -\int R dT = -R\Delta T.$$

   This leads to

   $$\Delta U = C_V \Delta T = \frac{R}{\gamma - 1}\Delta T; \quad \Rightarrow \quad Q = W + \Delta U = \frac{R\Delta T}{\gamma - 1} - R\Delta T$$

   $$= R\Delta T \left(\frac{1}{\gamma - 1} - 1\right) = R\Delta T \left(\frac{2-\gamma}{\gamma - 1}\right).$$

13. After the detonation of an atom bomb, the spherical ball of gas is found to be of 15 m radius at a temperature of 3×10^5 K. Given the adiabatic expansion coefficient $\gamma = \frac{5}{3}$, what will be the radius of the ball when its temperature reduces to 3×10^3 K?   [JEST 2017]

   (a) 156 m   (b) 50 m   (c) 150 m   (d) 100 m

   Answer: The correct choice is (c).

Solution: Applying the equation of state for an adiabatic process, we get

$$T_1 V_1{}^{\gamma-1} = T_2 V_2{}^{\gamma-1} \quad \Rightarrow \quad V_2 = \left(\frac{T_1}{T_2}\right)^{\frac{1}{\gamma-1}} V_1 \quad \Rightarrow \quad V_2 = \left(\frac{T_1}{T_2}\right)^{\frac{3}{2}} V_1$$

$$\Rightarrow \quad R_2 = \left(\frac{T_1}{T_2}\right)^{\frac{1}{2}} R_1 \Rightarrow R_2 = \left(\frac{3 \times 10^5}{3 \times 10^3}\right)^{\frac{1}{2}} \times 15 \text{ m} = 150 \text{ m}.$$

14. An ideal gas consists of three-dimensional polyatomic molecules. The temperature is such that one vibrational mode is excited. If R denotes the gas constant, then the specific heat at a constant volume of one mole of the gas at this temperature is   [IIT-JAM 2018]

   (a) $3R$ (b) $\frac{7}{2}R$ (c) $4R$ (d) $\frac{9}{2}R$

   Answer: The correct choice is (c).

   Solution: For a polyatomic gas, $C_P = (4 + f)R$, $C_V = (3 + f)R$, as $f = 1$, $C_V = 4R$.

15. In the thermodynamic cycle shown in Figure 8.9, one mole of a monoatomic ideal gas is taken through a cycle. AB is a reversible isothermal expansion at temperature of 800 K in which the volume of the gas is doubled. BC is an isobaric contraction to the original volume in which the temperature is reduced to 300 K. CA is a constant volume process in which the pressure and temperature return to their initial values. The net amount of heat (in Joules) absorbed by the gas in one complete cycle is ...   [IIT-JAM 2015]

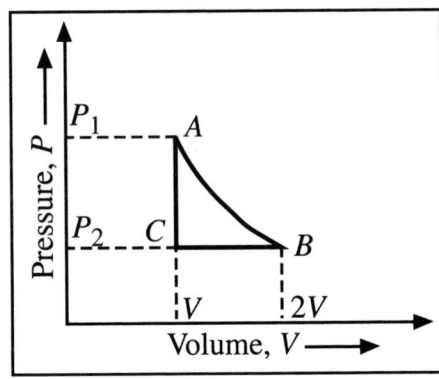

Figure 8.9 Indicator diagram for the thermodynamic cycles in problem 15.

   (a) 904   (b) 226   (c) 352   (d) 452

   Answer: The correct choice is (d).

   Solution: Process $A \to B$ is an isothermal expansion.

$T_A = 800$ K, V_A, P_A and $T_B = 800$ K, $V_B = 2V_A$, $P_B = \frac{P_A}{2}$, $R = 8.314$ J/K

Process $B \rightarrow C$ is isobaric. Hence, $P_C = P_B = \frac{P_A}{2}$, $V_C = V_A$, $T_C = 300$ K, $n = 1$, $\gamma = \frac{5}{2}$; $C \rightarrow A$ is isochoric. For process $A \rightarrow B$, $\Delta Q_1 = nRT_A \ln \frac{V_B}{V_A} = 4610$ J

$\Delta Q_2 = nC_P \Delta T = \frac{n\gamma R \Delta T}{(\gamma - 1)} = \left(\frac{\gamma}{\gamma - 1}\right) R(300 - 800) = -10392$

$\Delta Q_3 = \frac{R}{\gamma - 1}(800 - 300) = \frac{R}{\gamma - 1} \times 500 = 6235.5$ J. The total heat exchange is $\Delta Q_1 + \Delta Q_2 + \Delta Q_3 = 452$.

16. A trapped air of volume V_0 is released from a certain depth h measured from the water surface in a large water tank. The volume of the bubble grows to $2V_0$ as it reaches just below the surface. The temperature of water and the pressure above the surface of water (10^5 N/m^2) remains constant throughout the process. If the density of water is 1000 kg/m^3 and the acceleration due to gravity is 10 m/s^2, then the depth h is given by   [IIT-JAM 2010]

(a) 1 m   (b) 10 m   (c) 50 m   (d) 100 m

Answer: The correct choice is (b).

Solution: At a depth h, the pressure is given by $P_1 = P_0 + \rho g h$, and the volume is $V_1 = V_0$.

At the surface, pressure and volume are given by, respectively, $P_2 = P_0$ and $V_2 = 2V_0$.

Then, using the relation $P_1 V_1 = P_2 V_2$, and $P_0 = 10^5$ N/m^2, $g = 10$ m/s^2, $\rho = 1000$ kg/m^3, we get

$$(P_0 + \rho g h)V_0 = P_0 \, 2V_0 \Rightarrow h = \frac{P_0 \, V_0}{\rho g \, V_0} = \frac{P_0}{\rho g} = \frac{1 \times 10^5}{10 \times 10^3} = 10 \text{ m}.$$

17. A box containing 2 mol of a diatomic ideal gas at temperature T_0 is connected to another identical box containing 2 mol of a monoatomic ideal gas at temperature $5T_0$. There are no thermal loss, and the heat capacity of the boxes is negligible. Find the final temperature of the mixture of the gases. (Ignore the vibrational degrees of freedom for the diatomic molecules.)

[GATE 2013]

(a) T_0 (b) $1.5T_0$ (c) $2.5T_0$ (d) $3T_0$

Answer: The correct choice is (c).

Solution: Internal energy of the system remains conserved, that is, $U_{\text{monoatomic}} + U_{\text{diatomic}} = U_{\text{mixture}}$.

$U_{\text{monoatomic}} = n_1 C_{V_1} T_1$, $U_{\text{diatomic}} = n_2 C_{V_2} T_2$. $C_{V_1} = \frac{5R}{2}$, $C_{V_2} = \frac{3R}{2}$, $n_1 = n_2 = 2$, $T_1 = T_0$, $T_2 = 5T_0$.

Let the common temperature of the mixture be T. The specific heat is $C_V = \frac{n_1 C_{V_1} + n_2 C_{V_2}}{n_1 + n_2}$ and the number of moles of the mixture is $n = n_1 + n_2$, then

$$n_1 C_{V_1} T_1 + n_2 C_{V_2} T_2 = n C_V T; \quad \Rightarrow \quad T = 2.5 T_0.$$

18. An ideal gas is compressed adiabatically from an initial volume V to a final volume αV, and a work W is done on the system in doing so. The final pressure of the gas will be $\left(\gamma = \frac{C_P}{C_V} \right)$ [JEST 2015]

    (a) $\frac{W}{V^\gamma} \frac{1-\gamma}{\alpha - \alpha^\gamma}$

    (b) $\frac{W}{V^\gamma} \frac{\gamma-1}{\alpha - \alpha^\gamma}$

    (c) $\frac{W}{V} \frac{1-\gamma}{\alpha - \alpha^\gamma}$

    (d) $\frac{W}{V} \frac{\gamma-1}{\alpha - \alpha^\gamma}$

    Answer: The correct choice is (c).

    Solution: Work done in an adiabatic process is $W = \frac{P_2 V_2 - P_1 V_1}{1 - \gamma}$.

    Since $P_2 V_2^\gamma = P_1 V_1^\gamma \Rightarrow P_1 = P_2 \left(\frac{V_2}{V_1} \right)^\gamma \Rightarrow P_1 = P_2(\alpha)^\gamma$; $W = \frac{P_2 \alpha V - P_2 \alpha^\gamma V}{(1-\gamma)} \Rightarrow P_2 = \frac{W}{V} \left(\frac{1-\gamma}{\alpha - \alpha^\gamma} \right)$.

19. The isothermal compressibility β_T of an ideal gas at a temperature T_0 and volume V_0 is given by [GATE 2012]

    (a) $-\frac{1}{V_0} \left(\frac{\partial V}{\partial P} \right)_{T_0}$

    (b) $\frac{1}{V_0} \left(\frac{\partial V}{\partial P} \right)_{T_0}$

    (c) $-V_0 \left(\frac{\partial P}{\partial V} \right)_{T_0}$

    (d) $V_0 \left(\frac{\partial P}{\partial V} \right)_{T_0}$

    Answer: The correct choice is (a).

20. Th isothermal compressibility β_T of a substance is defined as $\beta_T = -\frac{1}{V} \left(\frac{\partial V}{\partial P} \right)_T$. Its value for n moles of an ideal gas will be [IIT-JAM 2009]

    (a) $\frac{1}{P}$

    (b) $\frac{n}{P}$

    (c) $-\frac{1}{P}$

    (d) $-\frac{n}{P}$

    Answer: The correct choice is (a).

    Solution: For n mole of an ideal gas, we have $PV = nRT \Rightarrow V = \frac{nRT}{P}$.

    According to the definition of isothermal compressibility β_T, we have

    $$\beta_T = -\frac{1}{V} \left(\frac{\partial V}{\partial P} \right)_T = -\frac{1}{V} nRT \left(-\frac{1}{P^2} \right) = -\frac{1}{V} \frac{nRT}{P} \left(-\frac{1}{P} \right)$$

    $$= -\frac{1}{V} \times V \times \left(-\frac{1}{P} \right) = \frac{1}{P}.$$

21. A diatomic gas at room temperature is expanded at constant pressure P_0. If the heat absorbed by the gas is $Q = 14$ J, what is the maximum work in Joules that can be extracted from the system?   [JEST 2019]

(a) 8 J   (b) 4 J   (c) 2 J   (d) 32 J

Answer: The correct choice is (b).

Solution: For a diatomic gas, we have $C_V = \frac{5}{2}R$ and $C_P = \frac{7}{2}R$. So, at constant pressure, we have the relation: $\delta Q = C_P \Delta T \Rightarrow 14 = \frac{7}{2}R\Delta T$. This provides $\Delta T = \frac{14 \times 2}{7 \times 8.314} = 0.481°C$. Further, the change in internal energy is given by

$$\Delta U = C_V \Delta T = \frac{5}{2}R \times \Delta T = \frac{5}{2} \times 8.314 \times 0.481 = 9.99 \text{ J}.$$

Therefore, the maximum work in Joules that can be extracted from the system is given by

$$W_{\text{max}} = Q - \Delta U = 14 - 9.99 = 4 \text{ J}.$$

22. For a diatomic ideal gas near room temperature, what fraction of the heat supplied is available for external work if the gas is expanded at constant pressure?   [JEST 2013]

(a) $\frac{1}{7}$ (b) $\frac{5}{7}$ (c) $\frac{3}{4}$ (d) $\frac{2}{7}$

Answer: The correct choice is (d).

Solution: It is an isobaric process (constant pressure). Then the supplied heat is $\delta Q = nC_P \Delta T$, and the amount of work done is $\Delta W = nRT$.

In this process, δQ is the amount of heat exchange during the process. The function of heat supplied

$$= \frac{\delta W}{\Delta Q} = \frac{n\,R\,\Delta T}{n\,C_p\,\Delta T} = \frac{R}{R\frac{\gamma}{\gamma-1}} = \frac{\gamma-1}{\gamma} = 1 - \frac{1}{\gamma}$$

$$= 1 - \frac{1}{1+\frac{2}{f}} = 1 - \frac{f}{f+2} = 1 - \frac{5}{5+2} = \frac{2}{7}.$$

[Since $\gamma = \frac{C_p}{C_V} \Rightarrow C_p = \frac{\gamma R}{\gamma-1}$, and $f =$ is the degree of freedom, for a diatomic molecule $f = 5$].

23. During an experiment, an ideal gas is found to obey a condition $VP^2 = $ constant. The gas is initially at a temperature T, pressure P, and volume V. If the gas expands to volume $4V$,

(a) the pressure of the gas changes to $\frac{P}{3}$.

(b) the graph of the above process on the $P - T$ diagram is a hyperbola.

(c) the temperature of the gas changes to $4T$.

(d) the graph of the above process on the $P - T$ diagram is a parabola.

Answer: The correct choice is (b).

24. An ideal fluid is subject to a thermodynamic process described by $\rho = CV^{-\alpha}$ and $P = n\rho^{\Gamma}$, where ρ is the energy density and P is pressure. For what values of n and Γ is the process adiabatic if the volume is changed slowly?

[JEST 2018]

(a) $\Gamma = \alpha - 1$, $n = 1$

(c) $\Gamma = 1$, $n = \alpha - 1$

(b) $\Gamma = 1 - \alpha$, $n = \alpha$

(d) $\Gamma = \alpha$, $n = 1 - \alpha$

Answer: The correct choice is (c).

Solution: As $\rho = \frac{U}{V} \Rightarrow U = \rho V = CV^{1-\alpha}$. $\rho = n\rho^{\Gamma} \Rightarrow \rho = n(CV^{-\alpha})^{\Gamma} = ne^{\Gamma}(V^{-\alpha})^{\Gamma}$.

We know that $TdS = dU + PdV$. As $TdS = 0$, we have $dU + PdV = 0$.

$$dU = C(1 - \alpha)V^{-\alpha}dV \quad PdV = ne^{\Gamma}V^{-\alpha\Gamma}dV$$
$$\Rightarrow C(1 - \alpha)V^{-\alpha}dV + ne^{\Gamma}V^{-\alpha\Gamma}dV = 0;$$
$$\Rightarrow CV^{-\alpha}\left[1 - \alpha + V(1 - \Gamma)ne^{\Gamma-1}\right]dV = 0.$$

This is true only if $\Gamma = 1$ and for $\Gamma = 1$, we have $1 - \alpha + n = 0$; $\Rightarrow n = \alpha - 1$.

25. The pressure P of a fluid is related to its density ρ by the equation of the state $P = a\rho + b\rho^2$, where a and b are constants. If the initial volume of the fluid is V_0, the work done on the system when it is compressed so as to increase the number density from an initial value of ρ_0 to $2\rho_0$ is

(a) $(a\ln 2 + b\rho_0)\rho_0 \, V_0$

(b) $2\rho_0 \, V_0$

(c) $(a + b\rho_0) \, \rho_0 \, V_0$

(d) $\left(\frac{3a}{2} + \frac{7\rho_0 b}{3}\right)\rho_0 \, V_0$

Answer: The correct choice is (a).

Solution: Let M be the mass of the fluid under consideration. Hence, $M = \rho_0 V_0 = 2\rho_0 \times \frac{V_0}{2}$. We have $dV = \frac{M}{\rho^2}d\rho$ (considering the positive value). Hence, the work done on the fluid is

$$W = M \int_{\rho_0}^{2\rho_0} (a\rho + b\rho^2) \frac{d\rho}{\rho^2} = M\left[a\ln 2 + b\rho_0\right] = \left[a\ln 2 + b\rho_0\right]\rho_0 \, V_0.$$

26. Pressure P, volume V, and temperature T for a certain material are related by $PV = AT^3$, where A is a constant. The expression for the work done by the material if its temperature changes from T_1 to T_2 at constant pressure is

    (a) $\frac{A}{3}(T_2^3 - T_1^3)$

    (c) $A(T_2 - T_1)^3$

    (b) $\frac{A}{3}(T_2 - T_1)^3$

    (d) $A(T_2^3 - T_1^3)$

    Answer: The correct choice is (d).

27. A gas at a pressure P_A and volume V_A is compressed adiabatically to a volume V_B at pressure P_B. If the pressure—volume relation for this gas during adiabatic compression is $PV^{5/3} = K$ (where K is a constant), the work done during the compression is   [JNU 2010]

    (a) $K\left(\frac{2}{V_B^{2/3} - V_A^{4/3}}\right)$

    (b) $\frac{5}{2}K\left(\frac{2}{V_B^{2/3} - V_A^{2/3}}\right)$

    (c) $\frac{5}{3}K\left(P_B V_B - P_A V_A\right)$

    (d) $\frac{3}{2}K\left(P_B V_B - P_A V_A\right)$

    Answer: The correct choice is (d).

28. A diatomic gas undergoes adiabatic expansion against the piston of a cylinder. As a result, the temperature of the gas drops from 1150 K to 400 K. The number of moles of the gas required to obtain 2300 J of work from the expansion is ... (the gas constant $R = 8.314$ J mol^{-1}K^{-1}). (Round off to 2 decimal places.)   [IIT-JAM 2019]

    Answer: The correct choice is 0.14.

    Solution: For a diatomic molecule, we have $\gamma = \frac{7}{5}$. The expression of work done in adiabatic expansion is $W = \frac{nR(T_2 - T_1)}{1 - \gamma}$.

29. A cylinder contains 16 g of O_2. The work done when the gas is compressed to 75% of the original volume at a constant temperature of 27°C is ... J. [Universal gas constant $R = 8.31$ J/(mol K).]   [IIT-JAM 2016]

    (a) 258   (b) 358   (c) 458   (d) 385

    Answer: The choice is (b).

    Solution: The expression for work done is given by $\delta W = P\,dV$. Integrating this expression, we get $W = nRT \int\limits_{V}^{.75V} \frac{dV}{V} = nRT \ln \frac{75}{100} = \frac{1}{2} \times 8.31 \times 300 \times \ln 0.75 = -358$ J.

30. Two gases separated by an impermeable but movable partition are allowed to exchange energy freely. At equilibrium, the two sides will have the same

[GATE 2013]

(a) pressure and temperature.

(c) pressure and volume.

(b) volume and temperature.

(d) volume and energy.

Answer: The correct choice is (a).

31. An ideal gas has a specific heat ratio $\frac{C_P}{C_V} = 2$. Starting at temperature T_1, the gas undergoes an isothermal compression to increase its density by a factor two. After this, adiabatic compression increases its pressure by a factor of two. The temperature of the gas at the end of the second process would be

[JEST 2016]

(a) $\frac{T_1}{2}$ 　　　　**(b)** $\sqrt{2}T_1$ 　　　　**(c)** $2T_1$ 　　　　**(d)** $\frac{T_1}{\sqrt{2}}$

Answer: The correct choice is (b).

Solution: During the isothermal process, $T = T_1$ is constant. Let us assume that the adiabatic process started at point $A(P_1, T_1)$, and at point B the coordinate is (P_2, T_2). They are given by

$$P_1^{1-\gamma}T_1^{\gamma} = P_2^{1-\gamma}T_2^{\gamma} \Rightarrow T_2 = \left(\frac{P_1}{P_2}\right)^{\frac{1-\gamma}{\gamma}} T_1 \Rightarrow T_2$$

$$= \left(\frac{P_1}{2P_1}\right)^{\frac{1-2}{2}} T_1. \quad \Rightarrow T_2 = \sqrt{2}T_1.$$

32. A real gas has a specific volume V at a temperature T. Its volume expansion coefficient and isothermal compressibility are, respectively, given by α and β_T. Its molar specific heat at constant pressure C_P and molar specific heat at constant volume C_V are related by 　　　　[JAM 2014]

(a) $C_P = C_V + R$

(c) $C_P = C_V + \frac{T V \alpha^2}{\beta_T}$

(b) $C_P = C_V + \frac{T V \alpha}{\beta_T}$

(d) $C_P = C_V$

Answer: The correct choice is (c).

33. The total energy E of an ideal nonrelativistic Fermi gas in three dimensions is given by $E \propto \frac{N^{\frac{5}{3}}}{V^{\frac{2}{3}}}$, where N is the number of particles, and V is the volume of the gas. Identify the correct equation of state (P being the pressure).

[GATE 2012]

(a) $PV = \frac{1}{3}E$ 　　　**(b)** $PV = \frac{2}{3}E$ 　　　**(c)** $PV = \frac{5}{3}E$ 　　　**(d)** $PV = E$

Answer: The correct choice is (b).

Solution: $P = -\left(\frac{\partial E}{\partial V}\right)_N = \frac{2}{3}\left(\frac{N}{V}\right)^{\frac{5}{3}}; \Rightarrow PV = \frac{2}{3}\frac{N^{\frac{5}{3}}}{V^{\frac{2}{3}}} = \frac{2}{3}E.$

34. A metal bullet comes to rest after hitting its target with a velocity of 80 m/s. If 50% of the heat generated remains in the bullet, what is the increase in its temperature? (The specific heat of the bullet = 160J/Kg/°C)   [JEST 2013]

(a) 14°C   (b) 12.5°C   (c) 10°C   (d) 8.2°C

Answer: The correct choice is (c).

Solution: Conservation of momentum 50% of $\frac{1}{2}mv^2 = mc\Delta T \Rightarrow$
$(1/2)\frac{80 \times 80}{2} = 160\Delta T$

$\Rightarrow \Delta T = \frac{80 \times 80}{4} \times \frac{1}{160} = 10°C.$

35. If the mean square fluctuations in energy of a system in equilibrium at temperature T is proportional to T^α, then the energy of the system is proportional to   [JEST 2017]

(a) $T^{\alpha-2}$ (b) $T^{\frac{\alpha}{2}}$ (c) $T^{\alpha-1}$ (d) T^α

Answer: The correct choice is (c).

Solution: $(\Delta E)^2 = K_B T^2 C_V \Rightarrow T^{\alpha-2} \propto C_V \Rightarrow T^{\alpha-2} \propto \left(\frac{\partial U}{\partial T}\right)_V \Rightarrow U \propto T^{\alpha-1}.$

36. For which gas, the ratio of specific heats $\left(\frac{C_P}{C_V}\right)$ will be the largest?
[JEST 2014]

(a) monoatomic   (c) triatomic

(b) diatomic   (d) hexatomic

Answer: The correct choice is (a).

Solution: $C_P/C_V = \gamma = \left(1 + \frac{2}{f}\right)$, where f is a degree of freedom. For monoatomic gas: $f = 3$, for diatomic gas: $f = 6$, for triatomic gas: $f = 9$, and for hexagonal: $f = 18$.

37. The isothermal compressibility $\beta_T = -\frac{1}{V}\frac{\partial V}{\partial P}$ of one mole of Van der Waals gas as a function of temperature and volume is (a and b are constants in the Van der Waals equation of state)

(a) $\frac{V^2(V-b)}{(K_B TV^2 - a(V-b))}$

(c) $\frac{V^2(V-b)^2}{(K_B TV^3 - 2a(V-b)^2)}$

(b) $\frac{V^2(V-b)^2}{(K_B TV^2 - 2a(V-b))}$

(d) $\frac{(V-b)}{K_B T}$

Answer: The correct choice is (c).

38. A monoatomic gas is described by the equation of state $P(V - nb) = nRT$, where b and R are constants and other quantities have their usual meanings. The maximum density (in moles per unit volume) to which this gas can be compressed is

(a) $\frac{1}{b}$ (b) b (c) $\frac{1}{b\,n}$ (d) ∞

Answer: The correct choice is (a).

39. The equation of the state of a gas is given by $V = \frac{RT}{P} - \frac{b}{T}$, where R is the gas constant, and b is another constant parameter. The specific heat at constant pressure C_P and the specific heat at constant volume C_V for this gas is related by $C_P - C_V =$ [TIFR 2015]

(a) R

(b) $R\left(1 + \frac{RT^2}{bP}\right)^2$

(c) $R\left(1 + \frac{bP}{RT^2}\right)^2$

(d) $R\left(1 - \frac{bP}{RT^2}\right)^2$

Answer: The correct choice is (c).

40. A gas is contained in a metallic cylinder fitted with a piston. The piston is suddenly moved in to compress the gas and is maintained at this position. As time passes, the pressure of the gas in the cylinder

(a) increases.

(b) decreases.

(c) remains constant.

(d) increases or decreases depending on the nature of the gas.

Answer: The correct choice is (c).

Solution: After some time the system will be in mechanical equilibrium, so the pressure will remain the same.

41. The pressure and density of gas $\left(\gamma = \frac{7}{5}\right)$ change adiabatically from (P_1, d_1) to (P_2, d_2). If $\frac{d_2}{d_1} = 32$, then the value of $\frac{P_2}{P_1}$ is

(a) 32   (b) 128   (c) $\frac{1}{32}$ (d) $\frac{1}{128}$

Answer: The correct choice is (b).

Solution: If there is no exchange of heat between a system and its surroundings, the process is known as adiabatic. The gas equation for adiabatic change is given as $PV = $ constant.

Since, volume $\propto \frac{1}{\text{density}}$, $P \propto (\text{density})^\gamma$.

$\Rightarrow P_1 = k d_1^\gamma$ $P_2 = k d_2^\gamma$ $\frac{P_2}{P_1} = \left(\frac{d_2}{d_1}\right)^\gamma$

Since, $\frac{d_2}{d_1} = 32$ $\frac{P_2}{P_1} = (32)^{7/5} = (2^5)^{7/5} = 128.$

42. The pressure P, volume V, and temperature T for a certain material are related by $P = \frac{(AT - BT^2)}{V}$, where A and B are constants. The work done by the material when the temperature changes from T to $2T$ while the pressure P remains constant is:

(a) $AT - BT^2$ (b) $AT - 2BT^2$ (c) $AT - 3BT^2$ (d) $2AT - 2BT^2$

Answer: The correct choice is (c).

43. Figure 8.10 shows the P–V plot of an ideal gas taken through a cycle ABCDA. Part ABC is a semicircle, and CDA is half of an ellipse. Then,

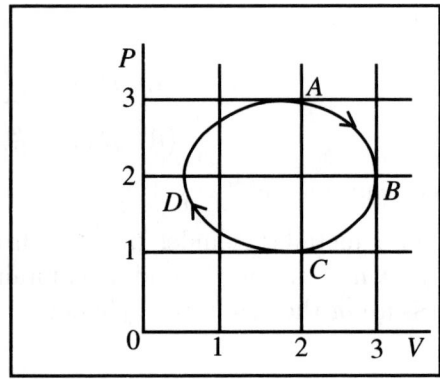

Figure 8.10 Indicator $(P$–$V)$ diagram of an ideal gas for problem 43.

(a) the process during the path $A \rightarrow B$ is isothermal.
(b) work done during the path $A \rightarrow B \rightarrow C$.
(c) negative work is done by the gas in the cycle ABCDA.
(d) heat flows out of the gas during the path $B \rightarrow C \rightarrow D$.

Answer: The correct choice is (d).

44. The specific heat C of a substance is found to vary with temperature T as $C = \alpha + \beta T^2$, where T is the temperature in the Celsius scale. At what temperature does the specific heat of the substance become equal to the mean specific heat in a temperature range between 0 and T?

(a) $\frac{T}{\sqrt{2}}$ (b) $\frac{T}{2}$ (c) $\frac{T}{\sqrt{3}}$ (d) $\frac{T}{3}$

Answer: The correct choice is (c).

Solution: The amount of heat required to increase the temperature of the unit mass of a substance is defined as the specific heat. The expression for heat is $Q = mc\Delta T$, where m is the mass of the substance, ΔT is the increment in temperature, and c is the specific heat given by $c = \alpha + \beta T^2$.

Hence, the mean specific heat is given by

$$\langle c \rangle = \frac{1}{T} \int_0^T c \, dT = \frac{1}{T} \int_0^T (\alpha + \beta T^2) dT = \frac{1}{T} \left[\alpha T + \frac{\beta T^3}{3} \right] = \alpha + \frac{\beta T^2}{3}.$$

Let T_1 be the temperature at which the specific heat of the substance becomes equal to the mean specific heat of the substance. Then we have

$$\langle c \rangle = c; \quad \Rightarrow \quad \alpha + \frac{\beta T^2}{3} = \alpha + \beta T_1^2 \quad \Rightarrow \quad \frac{\beta T^2}{3} = \beta T_1^2 \quad \Rightarrow \quad T_1 = \frac{T}{\sqrt{3}}.$$

45. The equation of state of a real gas is given as $P(V - b) = nRT$, where b is a constant, n is the number of moles, and R is the universal gas constant. When 2 mol of this gas undergo reversible isothermal expansion from volume V to volume $2V$, what is the work done by the gas?

(a) $2RT \ln \left[\frac{(V-b)}{2V-b} \right]$

(b) $2RT \ln \left[\frac{(2V-b)}{V-b} \right]$

(c) $2RT \ln \left[\frac{(V-b)}{2V} \right]$

(d) $2RT \ln \left[\frac{(2V)}{V-b} \right]$

Answer: The correct choice is (b).

Solution: The equation of state of the gas is given by $P(V - b) = nRT$. The general expression for work done is $\delta W = P dV$. Using the value of $P = \frac{nRT}{V-b}$, the total work done is given by

$$W = \int P \, dV = n \, RT \int_V^{2V} \frac{dV}{V - b} = n \, R \, T \, [\ln(2V - b) - \ln(V - b)].$$

Since $n = 2$, we have $W = 2RT \ln \left(\frac{2V-b}{V-b} \right)$.

46. If the temperature of a gas is increased in the process $PV^2 =$constant,

(a) the change in internal energy of the gas is negative.

(b) work done by the gas is positive.

(c) heat is given to the gas.

(d) heat is taken out from the gas.

Answer: The correct choice is (c).

47. What is the minimum attainable pressure of an ideal gas in the process given by $T = a + bV^2$, where a and b are constants, R is the universal gas constant, and V is the volume of one mole of ideal gas?

(a) $\sqrt{ab}$

(c) $2R\sqrt{ab}$

(b) $R\sqrt{ab}$

(d) $\sqrt{\frac{a}{b}}$

Answer: The correct choice is (c).

We have $T = a + bV^2$; $\Rightarrow \frac{PV}{R} = a + bV^2$.   This leads to $\Rightarrow P = \frac{aR}{V} + \frac{V^2 Rb}{V}$, $P = \frac{aR}{V} + VbR$.

To find the minimum pressure, we put $\frac{dP}{dV} = 0$. This provides

$$\frac{d}{dV}\left[\frac{a\,R}{V} + V\,b\,R\right] = 0 \quad \Rightarrow -\frac{a\,R}{V^2} + b\,R = 0 \quad \Rightarrow V^2 = \frac{a}{b}.$$

Again by differentiation of P, we get $\frac{d^2 P}{dV^2}$ is +ve for $V = \sqrt{\frac{a}{b}}$. Hence, P is minimum at $V^2 = \frac{a}{b}$. So, the minimum pressure is given by

$$P_{\min} = \frac{aR}{\sqrt{\frac{a}{b}}} + bR\sqrt{\frac{a}{b}} = 2R\sqrt{ab}.$$

48. A gas with 6000 cm^3 volume and 100 Pa pressure is compressed as per the law $PV^2 = C$. The volume of gas becomes 2000 cm^3. The work done on the gas is

(a) −120 KJ

(c) 0 KJ

(b) 200 KJ

(d) 300 KJ

Answer: The correct choice is (a).

$$P_1 V_1^2 = P_2 V_2^2 = C \implies P_2 = 100 \times \left(\frac{6000}{2000}\right)^2 = 900 \; kPa$$

$$W = \int_{V_1}^{V_2} PdV = \int_{V_1}^{V_2} \frac{CdV}{V^2} = -C\left(\frac{1}{V_2} - \frac{1}{V_1}\right)$$

As $C = (6000 \times 10^{-6})^2 \times 100 \times 10^5 = 360$

$$W = -360\left(\frac{1}{2000 \times 10^{-6}} - \frac{1}{6000 \times 10^{-6}}\right) = -120 \; \text{KJ}.$$

49. Two cylinders, A and B, fitted with frictionless pistons, contain an equal amount of ideal gas (diatomic) at 300 K. The piston in A is free to move, and that in B is fixed. The same amount of heat is given to both cylinders. If the temperature of gas in A rises by 30 K, then the rise in temperature of B is

(a) 18 K

(c) 42 K

(b) 30 K

(d) 50 K

Answer: The correct choice is (c).

$n = 2$ mol for each. For A, pressure is constant, and for B, volume is constant. For A, $Q = nC_P\delta T$.

For diatomic gas, $C_P = \frac{7}{2}R \implies Q = 2 \times \frac{7}{2} \times R \times 30 \implies Q = 210\,R$.

For B, $C_V = \frac{5}{2}R$ and $Q = nC_V\delta T \implies Q = 2 \times \frac{5}{2}R \times \delta T. \implies 210\,R = 2 \times \frac{5}{2}R \times \delta T. \implies \delta T = 42$ K.

50. The molar specific heat of a gas as given from the kinetic theory is $\frac{5}{2}R$. If it is not specified whether it is C_V or C_P, one can conclude that the molecules of the gas [JAM 2005]

(a) are definitely monoatomic.

(b) are definitely rigid diatomic.

(c) are definitely nonrigid diatomic.

(d) can be monoatomic or rigid diatomic.

Answer: The correct choice is (d).

Solution: If the molecule is monoatomic, then $C_P = \frac{5R}{2}$.

And if the molecule is rigid diatomic, then $C_V = \frac{5R}{2}$.

51. Consider two identical, finite, isolated systems of constant heat capacity C at temperatures T_1 and T_2 with $(T_1 > T_2)$. An engine works between them until their temperature becomes equal. Taking into account that the work performed by the engine will be maximum $(= W_{\max})$ if the process is reversible (equivalently, the entropy change of the entire system is zero), the value of $W_{\max}$ is given by [JAM 2017]

(a) $C(T_1 - T_2)$.

(b) $\frac{C(T_1 - T_1)}{2}$.

(c) $C(T_1 + T_2 - \sqrt{T_1 T_2})$.

(d) $C(\sqrt{T_1} - \sqrt{T_2})^2$.

Answer: The correct choice is (d).

Solution: $dS = C \left(\ln \frac{T_f^2}{T_1 T_2} \right) = 0 \Rightarrow T_f = \sqrt{T_1 T_2}$. So, the maximum work done is given by

$$dW_{\text{max}} = C(T_1 - T_f) + C(T_2 - T_f) = C(T_1 + T_2 - 2T_f)$$
$$= C(T_1 + T_2 - 2\sqrt{T_1 T_2}) = C(\sqrt{T_1} - \sqrt{T_2})^2.$$

52. The ratio $\frac{\text{Slope of isotherm curve}}{\text{Slope of adiabatic curve}}$ is equal to

(a) 1   (b) γ (c) $\frac{1}{\gamma}$ (d) 2

Answer: The correct choice is (c).

53. One mole of a perfect gas expands adiabatically. As a result of this, its pressure, temperature, and volume change from P_1, T_1, and V_1 to P_2, T_2, and V_2, respectively. If molar-specific heat at constant volume is C_V, then the work done by the gas is

(a) $2.303 P_1 V_1 \log \frac{V_2}{V_1}$ (c) $\frac{P_1^2 V_1^2 - P_2^2 V_2^2}{R(T_2 - T_1)}$

(b) $RT_1 \log \frac{V_2}{V_1}$ (d) $C_V(T_1 - T_2)$

Answer: The correct choice is (d).

We know $C_V = (\frac{dV}{dT})_V$. From the first law of thermodynamics, we have $\Delta Q = dU + \Delta W$. Again, for an adiabatic process $\Delta Q = 0$. So, we get $dU + \Delta W = 0$; $\Rightarrow C_V dT + \Delta W = 0$;

$$\Rightarrow C_V(T_2 - T_1) + \Delta W = 0; \Rightarrow \Delta W = C_V(T_1 - T_2).$$

54. What is the contribution of the conduction electrons in the molar entropy of a metal with an electronic coefficient of specific heat   [JEST 2014]

(a) γT (c) γT^3

(b) γT^2 (d) γT^4

Answer: The correct choice is (a).

We know that the expression for specific heat is $C_V = \gamma T + BT^3$, where γT is the contribution to the specific heat due to electrons.

55. A system of N noninteracting classical point particle is constrained to move on the two-dimensional surface of a sphere. The internal energy of the system is [GATE 2010]

(a) $\frac{3}{2} N K_B T$ (c) $N K_B T$

(b) $\frac{1}{2} N K_B T$ (d) $\frac{5}{2} N K_B T$

Answer: The correct choice is (c).

In this case, there are 2N degrees of freedom. So, the internal energy of the system is given by

$$\frac{NK_BT}{2} + \frac{NK_BT}{2} = NK_BT.$$

8.9   Exercise

8.9.1   Short answer type questions

1. Define heat and work.

2. What are the important characteristic features of heat and work?

3. What is an indicator diagram?

4. What do you mean by thermodynamic equilibrium?

5. Explain the mechanical, thermal, and chemical equilibrium of a thermodynamic system.

6. What is meant by a "thermodynamic process"? How is it represented in an indicator diagram?

7. What do you mean by "internal energy"?

8. What do you mean by "state function" and "path function"?

9. State the first law of thermodynamics.

10. What is "free expansion"? Is it an adiabatic process? Is it quasistatic?

11. Prove that in an adiabatic process, PV^γ is constant for an ideal gas, where symbols have their usual meaning. Is this relation valid for an irreversible process?   [Calcutta University 2020]

8.9.2   Long answer type questions

1. Sketch a general account of the background for the formulation of the first law of thermodynamics.

2. Write down the differences between state and path functions.

3. Show that the work done by a thermodynamic system is a path function and not a state function.

4. Internal energy is a state function—explain.

5. Show that for an ideal gas, the internal energy depends on temperature only and is independent of pressure or volume.

6. What is the change in internal energy due to the adiabatic expansion of an ideal gas from (P_1, V_1) to (P_2, V_2)? [Burdwan University 2021]

7. Starting from the equation of state, prove that the temperature of a wire vibrating between two rigid supports will drop on increase of tension and conversely.

8. Give the mathematical (differential) formulation of the first law of thermodynamics. What is the importance of this law?

9. Show that near the absolute zero temperature, an isothermal and an adiabatic process are identical. [Burdwan University 2020]

10. Show that the ratio of the adiabatic and isothermal elasticity of an ideal gas is equal to the ratio of the two specific heats. [Burdwan University 2020]

11. Derive expressions for the work done in an isothermal and adiabatic process of an ideal gas. [St. Joseph's College (Autonomous), Bengaluru 2019]

12. What is meant by the "adiabatic lapse rate"? Obtain an expression for the adiabatic lapse rate of the earth's atmosphere. [Delhi University 2021, Calcutta University 2022]

13. What is the "adiabatic lapse rate"? Graphically represent how the temperature of the earth's atmosphere varies with height above sea level, considering the atmosphere as an ideal gas system. [Burdwan University 2021]

8.9.3 Numerical problems

1. The equation of state of an ideal gas is $PV = nRT$, where symbols have their usual meaning. Show that (i) the isobaric volume expansivity α_P is equal to $\frac{1}{T}$, and (ii) the isothermal compressibility β_T is equal to $\frac{1}{P}$. [Vidyasagar University 2019]

2. A cylinder contains 1 mol of oxygen gas at a temperature of 27°C and 1 atmospheric pressure. If the gas expands isothermally from 1 L to 3 L, calculate

   (a) work done by the gas,
   (b) change in internal energy, and
   (c) heat transfer to the gas.

   [(a) 2740.65 J, (b) 0, and (c) 2740.65 J]

3. Using the first law of thermodynamics, derive the relation $C_P - C_V = R$. [Delhi University 2021]

4. A substance has an isothermal compressibility $\beta_T = \frac{a}{V}$ and the isobaric volume expansion coefficient $\alpha_P = \frac{2bT}{V}$, where a and b are constants. Find out the equation of state of the substance.

$$[V + aP - bT^2 = \text{ constant}] \text{ [Burdwan University 2020]}$$

5. The speed of longitudinal waves of small amplitude in an ideal gas is $v = \sqrt{\frac{P}{\rho}}$.

Show that for an adiabatic process, $v = \sqrt{\frac{\gamma RT}{M}}$, where ρ is the density, and M is the mass of the gas molecules. [Calcutta University 2020]

6. During an experiment, an ideal gas is found to obey a condition: $\frac{P^2}{\rho} = \text{constant}$ (ρ is the density of the gas). The gas is initially at temperature T, pressure P, and density ρ. The gas, then, expands such that density ρ changes to $\frac{\rho}{2}$. Draw a $P - T$ graph of the above process.

7. A cylinder contains 16 g of O_2. Calculate (up to the nearest integer) the work done when the gas is compressed to 75% of the original volume at a constant temperature of 27°C. Given that universal gas constant $R = 8.31$ J mol^{-1} K^{-1}.

[−359 J] [JAM 2016]

8. A mixture contains the same number of moles of two ideal gases A and B, with adiabatic constants γ_A and γ_B, respectively. Find the adiabatic constant γ of the mixture. $[\frac{1}{\gamma-1} = \frac{1}{2}\left(\frac{1}{\gamma_A-1} - \frac{1}{\gamma_B-1}\right)]$ [IISC 2016]

9. n moles of an ideal gas with constant volume heat capacity C_V undergo an isobaric expansion by a certain volume. Find the ratio of the work done in the process to the heat supplied. $[\frac{nR}{C_V+nR}]$

10. A box containing 2 mol of diatomic ideal gas at temperature T_0 is connected to another identical box containing 2 mol of a monoatomic ideal gas at temperature $5T_0$. There are no thermal losses and the heat capacity of the boxes is negligible. Find the final temperature of the mixture of the gases. (Ignore the vibrational degrees of freedom for the diatomic molecules.)

[$2.5T_0$] [JAM 2009]

11. In a thermodynamic process, the volume of one mole of an ideal varied as $V = aT^{-1}$, where a is a constant. The adiabatic exponent of the gas is γ. What is the amount of heat received by the gas if the temperature of the gas increases by ΔT in the process? $[R\Delta T \frac{2-\gamma}{\gamma-1}]$ [JEST 2018]

12. In an experiment, a certain quantity of an ideal gas at temperature T_0, pressure P_0, and volume V_0 is heated by a current flowing through a wire for a duration of t seconds. The volume is kept constant and the pressure changes to P_1. Under the same initial condition, the experiment is performed at constant pressure, and the volume changes from V_0 to V_1. Find the ratio of the specific heats at constant pressure and at constant volume. $[\frac{P_1-P_0V_0}{V_1-V_0P_0}]$ [JEST 2018]

CHAPTER 9

Second Law of Thermodynamics

It is a remarkable fact that the second law of thermodynamics has played in the history of science a fundamental role far beyond its original scope. Suffice it to mention Boltzmann's work on kinetic theory, Planck's discovery of quantum theory or Einstein's theory of spontaneous emission, which were all based on the second law of thermodynamics.

—Ilya Prigogine

Learning Outcomes

After reading this chapter, the reader will be able to

- Demonstrate the meaning of reversible, irreversible, and quasi-static processes used in thermodynamics
- Explain heat engines, and their efficiency and indicator diagram
- Formulate the second law of thermodynamics and apply it to various thermodynamic processes
- Demonstrate an idea about entropy and its variation in various thermodynamic processes
- State and compare various statements of the second law of thermodynamics
- Elucidate the thermodynamic scale of temperature and its equivalence to the perfect gas scale
- Explain the principle of increase of entropy
- Understand the third law of thermodynamics and explain the significance of unattainability of absolute zero
- Solve numerical problems and multiple choice questions on the second law of thermodynamics

9.1   Introduction

The first law of thermodynamics states that only those processes can occur in nature in which the law of conservation of energy holds good. But our daily experience shows that this cannot be the only restriction imposed by nature, because there are many possible thermodynamic processes that conserve energy but do not occur in nature. For example, when two objects are in thermal contact with each other, the heat never flows from the colder object to the warmer one, even though this is not forbidden by the first law of thermodynamics. This simple example indicates that there are some other basic principles in thermodynamics that must be responsible for controlling the behavior of natural processes.

One such basic principle is contained in the formulation of the second law of thermodynamics. This principle limits the use of energy within a source and elucidates that energy cannot be arbitrarily passed from one object to another, just as heat cannot be transferred from a colder object to a hotter one without doing any external work. Similarly, cream cannot be separated from coffee without a chemical process that changes the physical characteristics of the system or its surroundings. Further, the internal energy stored in the air cannot be used to propel a car, or the energy of the ocean cannot be used to run a ship without disturbing something (surroundings) around that object.

The second law of thermodynamics deals with the direction followed by spontaneous processes. Under a given set of conditions, many processes in nature occur spontaneously in one direction only, that is, they are irreversible. Irreversibility is a naturally associated characteristic of systems observed in day-to-day life. For example, a broken glass does not resume its original state. It should be mentioned that complete irreversibility is a statistical statement that cannot be seen during the lifetime of the universe. More precisely, an irreversible process is one that depends on the path. If the process can proceed in only one direction, then the reverse path differs fundamentally, and the process cannot be reversible. For example, mechanical energy, such as kinetic energy, can be completely converted to thermal energy by friction, but the reverse is impossible. A hot stationary object never spontaneously cools off and starts moving. Another example is the expansion of a puff of gas introduced into one corner of a vacuum chamber. The gas expands to fill the chamber, but it never regroups in the corner. The random motion of the gas molecules could take them all back to the corner, but this never happens.

All these above-mentioned processes occur spontaneously. This means that they will proceed to the end of the process if there is no external intervention to them. The reverse of these processes never happens in nature. In other words, it would be inconceivable that these processes could be reversed without tampering with the external conditions. Naturally the question arises: What determines the direction of evolution of a particular process under a given set of conditions? The answer can be obtained from a very important thermodynamic parameter known as "entropy". This parameter "entropy" signifies "whether or not a process or a reaction is going to be spontaneous" and is at the heart of the second law of thermodynamics.

In this chapter, various thermodynamic processes, heat engines and their efficiency, entropy and its value in different cases, and various formulations of the second law of thermodynamics are addressed in detail. A number of numerical problems are solved for the benefit of the students.

9.2 Thermodynamic processes

In solving problems in mechanics, we generally isolate the body under consideration, analyze the external forces acting on it, and then use Newton's laws of motion to predict its behavior. We adopt a similar approach in thermodynamics. We start by identifying the part of the universe we wish to study; it is known as **our system**. Once our system is selected, we determine how the system interacts with the environments or surroundings. Once the nature of interaction is understood, the thermal behavior of the system is investigated with the help of the laws of thermodynamics.

Such thermal behavior of a physical system is described in terms of **thermodynamic variables**. For an ideal gas, these variables are pressure, volume, temperature, and the number of molecules or moles of the gas. Different types of systems are generally characterized by different sets of thermodynamic variables. For example, the thermodynamic variables for a stretched rubber band are tension, length, temperature, and mass, whereas those for a gaseous system are pressure, volume, temperature, and the number of molecules or moles. These variables are used to define the state of the system.

The state of a system can change as a result of its interaction with the environment. This change in a system can be fast or slow and large or small. The manner in which a state of a system can change from an initial state to a final one is called a **thermodynamic process**. For example, the expansion of a gas within a cylinder at constant pressure due to heating is a thermodynamic process. A graphical representation of such a thermodynamic process is called a path. For analytical purposes in thermodynamics, it is helpful to divide up processes as either quasi-static or non-quasi-static, reversible, or irreversible. Such quasi-static, reversible and irreversible processes are described in detail in the next section.

9.2.1 Quasi-static process

When a system is in thermodynamic equilibrium, and the surroundings are kept unchanged, there will be no motion of the system, and no work will be done by the system. Under the influence of a finite unbalanced force, the system may pass through non-equilibrium states. During the occurrence of a particular process, if it is required to describe every state of the system by means of thermodynamic coordinates, the external forces should be varied only infinitesimally so that the system always remains near its thermodynamic equilibrium. A thermodynamic process carried out in this way is said to be **quasi-static**.

A quasi-static process thus refers to an idealized process where the change in state is made infinitesimally slowly so that at each instant of time, the system can be assumed to be at thermodynamic equilibrium with itself and also with the environment. For example, let us consider a case of heating of 1 kg water from a temperature 30°C to 31°C at a constant pressure of 1 atmosphere. To heat this amount of water very slowly, we

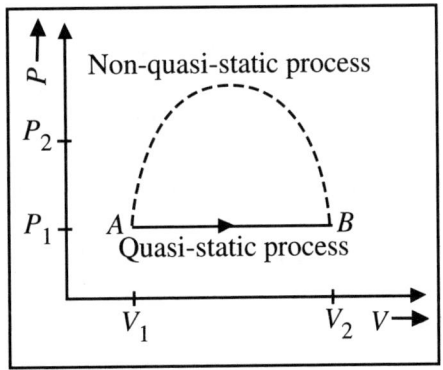

Figure 9.1 Shows schematically the quasi-static and non-quasi-static processes.

put the water-container in a large heat bath that can be heated slowly such that the temperature of the bath rises infinitesimally slowly from 30°C to 31°C. This is an example of a quasi-static process. On the other hand, if we put 1 kg water at 30°C directly into a bath at 31°C, the temperature of the water will rise rapidly to 31°C in a non-quasi-static way.

During a quasi-static process, a thermodynamic system is infinitesimally near the state of thermodynamic equilibrium. This is true for all instant of time the systems passes through the states. Therefore, we can have an equation of state for all these states. A quasi-static process can be applied to all thermodynamic systems including electric and magnetic systems. It should be mentioned that the conditions for such a quasi-static process can never be rigorously satisfied in the laboratory, but they can be approached with almost any degree of accuracy. The thermodynamic equilibrium of the system is necessary to have well-defined values of macroscopic properties such as temperature and pressure of the system at each instant of the process. Therefore, quasi-static processes can be represented by well-defined paths in state space of the system. Figure 9.1 shows quasi-static and non-quasi-static processes between states A and B of a gaseous system. In the quasi-static process, the path of the process between A and B can be drawn in a state diagram since all the states through which the system passes are known, whereas in the non-quasi-static process, the states between A and B are not known, and hence, no path can be drawn. It may follow the path represented by the dashed line, as shown in Figure 9.1, or take a different path.

During the evolution of a system following a quasi-static process, the system undergoes a small change of state, from an initial state of equilibrium to another state of equilibrium very close to the initial one. Under this situation, all three coordinates, such as temperature, pressure, and volume, required to describe a state undergo very small changes. This small change of any thermodynamic quantity, say volume V, is very small in comparison with V itself and very large in comparison to the space occupied by a few molecules and is denoted by a differential dV. Similarly, the change of pressure P is very small in comparison with pressure P itself and very large in comparison with molecular fluctuations, so the change of pressure may also be represented by the differential dP. It should be noted that every infinitesimal in thermodynamics must satisfy the requirement that it represents a change in a quantity that is small with respect to the quantity itself and large in comparison with the effect produced by a few molecules. This is the reason for which the thermodynamic coordinates such as volume, pressure, and temperature have no meaning when applied to a few molecules. These thermodynamic coordinates are applied to find out the physical properties of a system with a very large number of molecules and are termed as **macroscopic coordinates**.

9.2.2 Reversible processes with examples

There are certain thermodynamic processes that can be made to retrace its original path from the final to the initial state through some intermediate equilibrium states. Such thermodynamic processes are executed very slowly such that every possible state of the forward process (without affecting the system and the surroundings) can be retraced. These types of thermodynamic processes are said to be **reversible**.

In Figure 9.2, we consider an ideal gas that is held in half of a thermally insulated container by a **wall** in the middle of the container. The other half of the container is under vacuum with no molecules inside. If the wall in the middle is removed quickly, the gas expands and fills up the entire container immediately. From the microscopic point of view, a particle described by Newton's second law of motion can go backward if we flip the direction of time. But in practical terms, this is not the case in a macroscopic system with more than 10^{23} particles or molecules in motion. A huge number of collisions between these molecules tend to erase any trace of memory of

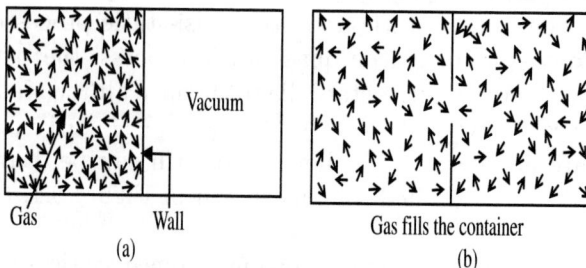

Figure 9.2 Expansion of a gas from half of a container to the entire volume: (a) before expansion and (b) after removing the wall in the middle.

the initial trajectory of each of the particles. This makes it quite impossible to track back all molecules in the first half. For example, shown in Figure 9.2, we can actually estimate the chance for all the particles in the expanded gas to go back to the original half of the container, but the current age of the universe is not long enough for it to happen even once.

The infinitesimally slow isothermal expansion and compression of a gas is a reversible process. In fact, all isothermal and adiabatic operations are reversible when carried out very slowly. A reversible process is, however, an ideal abstraction, though useful in theoretical calculations. We can, however, make a process approximate to a reversible one in the laboratory. If a gas is made to expand in a cylinder fitted with a piston that moves without friction (well lubricated), and the expansion be made very slowly against the opposing force supplied by an elastic spring, the process becomes approximately a reversible one. In a reversible process, there are certain conditions of reversibility. These conditions are as follows:

1. At any stage of the cycle of operation, the pressure and temperature of the working substance must not differ appreciably from those of the surroundings.

2. In the cycle of operation, all the processes must take place infinitely slowly.

3. In a cycle of operation, all the working parts of the engine must be completely free from dissipative forces such as friction.

4. During the cycle of operation, there should not be any loss of energy due to conduction or radiation in the whole system.

5. A process must be quasi-static (quasi-equilibrium) to be a reversible one.

Some points about the reversible process

A reversible process has some important characteristic features. These features are mentioned below:

1. There is an infinitesimal difference between the driving and opposing forces in this process.

2. A reversible process occurs in an infinite number of steps.

3. A reversible process can be reversed at any point during its progress by making infinitesimal changes in conditions.

4. The system achieves mechanical equilibrium at the end of every step of the process.

5. The maximum amount of work is obtained in this process.

6. A reversible process takes place so slowly that the system is always in temperature–pressure equilibrium with its surroundings.

9.2.3 Irreversible processes with examples

All natural processes are carried out in such a way that the conditions for thermodynamic equilibrium are not satisfied because a natural process does not take place quasi-statically. Dissipative forces, such as friction, viscosity, elastic resistance, and eddy formulations, are always present. Such processes are known as irreversible processes.

The phenomenon associated with a system undergoing an irreversible process is called irreversibility. This phenomenon of irreversibility results from the fact that if a thermodynamic system of interacting molecules is brought from one thermodynamic state to another, the configuration of the atoms and molecules in the system will change in a way that is not easily predictable. Some "transformation energy" will be used as the molecules of the "working body" do work on each other when they change from one state to another. During this transformation, there will be some heat energy loss or dissipation due to intermolecular friction and collisions. This energy will not be recoverable if the process is reversed. This is a very good demonstration of the example of an irreversible process.

The irreversible processes occur frequently in nature. Therefore, it is sometimes called a **natural process**. The sign of an irreversible process comes from the finite gradient between the states occurring in the actual process. For example, when heat flows from one object to another, there is a finite temperature difference (gradient) between the two objects. At any given moment of the process, the system most likely is not at equilibrium or in a well-defined state. This phenomenon is called irreversibility. The irreversibility of any natural process results from the second law of thermodynamics.

Some examples of irreversible process

Some examples of irreversible processes frequently used in thermodynamics are described below.

1. Spontaneous expansion of a gas into an evacuated chamber.

2. The conduction of heat spontaneously along a metal bar that is hot at one end and cold at the other.

3. The process of transfer of heat by radiation from a hot body to a cold one.

4. Production of heat by the passing of an electric current through an electrical resistor.

5. An explosion.

6. Inelastic deformation of an object.

7. Magnetization of a magnetic material or polarization of a dielectric medium with a hysteresis.

8. Chemical reactions that occur spontaneously.

9. Mixing of matter of varying states spontaneously.

Some points about the irreversible process

We know that in an irreversible process, a substance cannot return to its initial state, and the chemical properties of the substance change. There are also certain important points to be noted about an irreversible process. These are mentioned below:

1. The initial stage of the system and surroundings cannot be restored from the final stage in an irreversible process.

2. During an irreversible process, various states of the system are not in thermodynamic equilibrium with each other as the system changes its path from the initial to the final state.

3. The entropy of a system increases decisively during an irreversible process, and it cannot be reduced back to its initial value.

4. All complex natural processes are irreversible.

9.2.4   Difference between reversible and irreversible processes

The important differences between reversible and irreversible processes are presented below in a tabular form.

No. of observations	Reversible process	Irreversible process
1.	This process is carried out very slowly and goes through various smaller stages, which maintain equilibrium between the system and the surroundings.	This process is carried out relatively fast. In this process, no equilibrium is maintained between the system and the surroundings.
2.	Reversible processes can take place either in the forward direction or in the backward direction.	Irreversible processes can take place only in one direction.
3.	As the driving force is small, the reversible process proceeds in smaller steps.	As there is a definite driving force, the irreversible process proceeds in larger steps than the reversible one.

4.	Work done in a reversible process is greater than that of an irreversible process.	Work done in an irreversible process is always lower than that of a reversible process.
5.	A reversible process can be retraced to the initial state without making any change in its surroundings.	An irreversible process cannot be retraced to its initial state without making a change in its surroundings.

9.3 Conversion of work into heat and vice versa

Heat and work both describe the transfer of energy from one system to another. Heat is defined as the transfer of energy by virtue of a difference in temperature between two systems. Heat flows spontaneously. It carries entropy. In fact, we define heat in terms of the associated entropy transfer (for a reversible process) as follows: $\delta Q = TdS$, where the δ differential denotes the fact that heat is not a state function of a system and, therefore, is not a true differential quantity in the mathematical sense. It makes no sense to talk about the "heat content" of an object. Heat and entropy flow are inseparable.

Work is the transfer of energy between two systems by all means other than a difference in temperature. Reversible work carries no entropy. Work does not occur spontaneously but requires an "agent" to apply a force through a distance to generate the work. Again, a system does not have a work state function; there is no function W that is a unique function of the variables of a system (temperature, pressure, volume, etc.) that represents the "work content" of an object. Hence, energy transfer by means of work δW is another example of a pseudo-differential.

We expect the conservation of energy to hold in an energy transfer between two systems: $dU = \delta Q + \delta W$. Here is a true differential since the energy of a system is a unique and well-defined function of the state variables, for example, $U(N, T) = \frac{3}{2}NT$ for the ideal gas. This equation says that energy can be transferred into a system through either heat or work, or both.

Work can be completely converted into heat. Joule did such an experiment when he measured the mechanical equivalent of heat. However, heat cannot be completely converted into work. Heat carries entropy; work carries none. Hence, the complete conversion of heat to work would require the destruction of entropy. In all processes, entropy either stays the same or increases.

A work-around can be achieved by transferring heat from a hot reservoir to a cold reservoir, stripping off some of the heat, and converting it to work in the process. This works because the entropy delivered to the cold reservoir in a reversible process $(S_{\text{cold}} = \frac{Q_{\text{cold}}}{T_{\text{cold}}})$ requires less heat than the entropy acquired from the hot reservoir in a reversible process $(S_{\text{hot}} = \frac{Q_{\text{hot}}}{T_{\text{hot}}})$ since $T_{\text{cold}} < T_{\text{hot}}$. (We can't allow the entropy to accumulate without bound; otherwise, the system will quickly become unwieldy and useless.) The difference in heat can, in principle, be converted into work to do useful things. When this process is iterated, it is called a thermodynamic heat engine.

How good can a heat engine be? We know it cannot completely convert heat into work, so just how close can it get? Carnot made the following simple argument. Assuming that the heat engine expels all of the entropy that it acquires from the high-temperature reservoir and does not generate any additional entropy in the cyclic process, we can equate the incoming and outgoing entropy values:

$$S_{\text{cold}} = \frac{Q_{\text{cold}}}{T_{\text{cold}}} = S_{\text{hot}} = \frac{Q_{\text{hot}}}{T_{\text{hot}}}.$$

Also, we expect energy to be conserved in the heat engine, so that $W = Q_{\text{hot}} - Q_{\text{cold}}$. Using the entropy equality, we can write the work as

$$W = Q_{\text{hot}} - \frac{T_{\text{cold}}}{T_{\text{hot}}} Q_{\text{hot}}.$$

The Carnot efficiency of the heat engine is the ratio of the work performed to the heat acquired from the high-temperature reservoir:

$$\eta_C = \frac{W}{Q_{\text{hot}}} = \frac{T_{\text{hot}} - T_{\text{cold}}}{T_{\text{hot}}}.$$

The efficiency should be less than 1 because one cannot convert all the heat to work. Hence, the Carnot efficiency of the heat engine depends on the temperature difference of the hot and cold reservoirs.

Let us consider the case of a refrigerator that is a heat engine run in reverse. It "lifts" heat from a cold reservoir and delivers it to a hot reservoir by doing some work δW. From the working principle, one can define the coefficient of refrigerator performance as the ratio of the heat lifted to the work done as

$$\gamma_C = \frac{Q_{\text{cold}}}{W} = \frac{T_{\text{cold}}}{T_{\text{hot}} - T_{\text{cold}}}.$$

It should be noted that the coefficient of performance of a refrigerator may be greater than one.

9.4 The second law of thermodynamics

It is well known that there are certain processes in nature that never occur. This suggests that there is a law providing conditions for the occurrence of processes. The first law of thermodynamics would allow those processes to occur for which there is no violation of conservation of energy. The law that forbids some of these processes to occur is known as **the second law of thermodynamics**. It will be shown later that this law can be stated in many different forms that may seem different but which, in fact, are equivalent. Like other natural laws, the second law of thermodynamics provides insights into nature. Its several but equivalent statements imply that it is broadly applicable to fundamental but apparently different processes in nature.

9.4.1   Various statements of the second law of thermodynamics

It is interesting to note that the second law of thermodynamics can be formulated in various forms. In this section, we will introduce major statements of the second law of thermodynamics and show that all these statements are equivalent. Also, various statements of the second law of thermodynamics lead to the irreversibility of the spontaneous flow of heat between macroscopic objects consisting of a very large number of molecules or particles.

Clausius statement of the second law of thermodynamics

We know from common experience that heat flows from a hotter object to a colder one. Based on the results of such experiments that have been carried out on the spontaneous transfer of heat, Clausius made the following statement regarding the second law of thermodynamics: **Heat never flows spontaneously from a colder object to a hotter object.**

This statement is one of several different ways of stating the second law of thermodynamics. The form of this statement is credited to German physicist Rudolf Clausius (1822–1888) and is referred to as the **Clausius statement of the second law of thermodynamics**. The word "spontaneously" here means that no other effort has been made by a third party or one that is neither the hotter nor the colder object.

Kelvin statement of the second law of thermodynamics

In the last section, we mentioned the Clausius statement of the second law of thermodynamics, which is based on the irreversibility of spontaneous heat flow. In terms of the operational mechanism of heat engines, the second law of thermodynamics may be stated as follows: **It is impossible to convert the heat from a single source into work in a cycle without any other effect.**

This is known as the **Kelvin statement of the second law of thermodynamics**. This statement refers to the fact that a "perfect engine" cannot be attainable in nature. The phrase "without any other effect" is a very strong restriction between the system and the surroundings. For example, an engine can absorb heat and turn it all into work in a cycle. Without completing a cycle, the working substance in the engine will not come back to its original state, and therefore, a "other effect" has to occur. Another example is a gas enclosed in a chamber that can absorb heat from a heat reservoir and does isothermal work against the piston in the chamber as it expands isothermally. However, in order to make a complete cycle, the gas has to return to its initial state, that is, it has to be compressed, and heat would have to be extracted from it.

The Kelvin statement is thus a manifestation of a well-known engineering problem. It clearly demonstrates that despite the advancement of technology, we are not able to build a 100% efficient heat engine. It should be mentioned that the first law of thermodynamics does not exclude the possibility of constructing such a 100% efficient heat engine, that is, a perfect engine. However, the second law of thermodynamics clearly indicates that such a 100% efficient heat engine cannot be realized in practice.

Statement of the second law of thermodynamics based on Carnot principle

Sadi Carnot investigated the Carnot engine and Carnot cycle in detail and summarized the results into what is now known as the **Carnot principle**. This principle can be stated as: **No engine working between two reservoirs at constant temperatures can have a greater efficiency than a reversible engine**.

This principle can be viewed as one of the statements of the second law of thermodynamics and can be shown to be equivalent to the Kelvin statement and the Clausius statement.

Entropy statement of the second law of thermodynamics

The second law of thermodynamics is best expressed in terms of a change in the thermodynamic variable known as entropy, which is universally expressed by the symbol "S". Like internal energy, entropy is also a state function. This means that when a thermodynamic system makes a transition from one state into another, the change in entropy ΔS is independent of the path followed by the system and depends only on the thermodynamic variables of the two states. Hence, in terms of entropy, the second law of thermodynamics can also be stated as—**The entropy of a system and the entire universe never decreases.**

It can be shown that this entropy statement of the second law of thermodynamics is consistent with the Kelvin statement, the Clausius statement, and the Carnot principle. After presenting a detailed description of the heat engines, equivalence between the statements of the second law of thermodynamics due to Clausius and Kelvin would be established.

9.5   Heat engines

In the early 19th century, steam engines came to play an increasingly important role in industry and transportation. However, a systematic set of theories of the conversion of thermal energy to motive power by steam engines had not yet been developed. Nicolas Sadi Carnot (1796–1832), a French military engineer, published a book *Reflections on the Motive Power of Fire* in 1824. The book proposed a generalized theory of heat engines, as well as an idealized model of a thermodynamic system for a heat engine that is now known as the **Carnot cycle**. Carnot developed the foundation of the second law of thermodynamics and is often described as the "father of thermodynamics".

We know from the second law of thermodynamics that a heat engine cannot be 100% efficient since there must always be some heat transfer Q_L to the environment, which is often called waste heat. Naturally, the question arises: How efficient can a heat engine be? This question was answered by a young French engineer Sadi Carnot at a theoretical level. He performed this study of the then-emerging heat engine technology crucial to the industrial revolution in 1824. He devised a theoretical cycle,

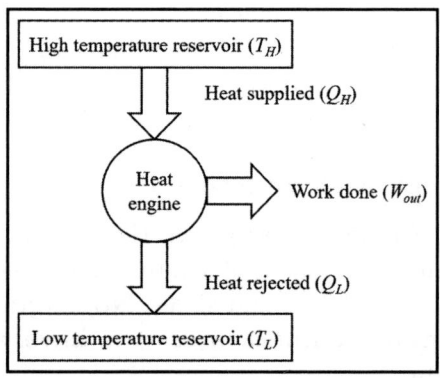

Figure 9.3 Shows schematically different parts of a heat engine.

now called the **Carnot cycle**, which is the most efficient cyclical process possible. The second law of thermodynamics was then restated in terms of the Carnot cycle, which signifies that Carnot actually discovered this fundamental law. Any heat engine employing the Carnot cycle is called a **Carnot engine**. A schematic diagram of a heat engine is shown in Figure 9.3.

The requirement of a reversible process is crucial to the Carnot cycle. It is known that irreversible processes involve dissipative factors, such as friction and turbulence. Therefore, a larger amount of heat Q_c is transferred to the surroundings in this process, and the efficiency of a heat engine is reduced. Reversible processes are obviously then considered superior to the operation of a heat engine.

9.5.1 Essential components of a heat engine

A heat engine is a system that converts the chemical energy of a fuel into thermal energy. This thermal energy is used to do other types of works. A heat engine converts the inherent energy in the fuel into force and motion. Various types of fuels used in heat engines are coal, gasoline, natural gas, wood, and peat. When these fuels are burnt in a heat engine, they release the stored energy in the fuel, and this energy is utilized to power the locomotives and the machinery in the factory.

Working substance

In a heat engine, the steam and gas are usually used as working substances. Steam is used in steam engines, steam turbines, and nuclear power plants, whereas gas or air is used in internal combustion (IC) engines, gas turbine power plants, jet engines, etc.

Heat reservoir

A heat reservoir is defined as the source of infinite heat energy. When a finite amount of heat is absorbed from the heat reservoir or rejected to the heat reservoir, the temperature of the heat reservoir is not affected, that is, the temperature of the heat reservoir remains constant. Typical examples of heat reservoirs are the ocean and the atmosphere.

A heat reservoir that supplies heat to a system is called a **source**, and a heat reservoir that absorbs heat from the system is called a **sink**. The sink must have finite thermal capacity and maintain a constant high temperature so that withdrawing or adding any amount of heat does not affect its temperature. The high-temperature source and the low-temperature sink are shown in Figure 9.3.

Cylinder

A cylinder in a heat engine is made up of nonconducting walls and a conducting bottom. A perfect gas is used as a working substance. The cylinder is fitted with a perfectly nonconducting (adiabatic) and frictionless piston that can move out and deliver work. The cylinder can receive work by moving the piston inward.

Insulating stand

The insulating stand in a heat engine is made up of nonconducting material so that it can be successfully used to perform adiabatic operations.

9.5.2   Carnot engine

The Carnot engine is a theoretical thermodynamic cycle proposed by Leonard Carnot in 1824. The Carnot engine model was mathematically explored by Clausius in 1857 and introduced the fundamental thermodynamic concept of entropy. This engine model estimates the maximum possible efficiency of a heat engine. This heat engine works between two reservoirs maintained at two different temperatures and converts heat into work and vice versa. The working fluid in a Carnot cycle is an ideal gas, and it operates in a reversible process. This engine has the maximum efficiency than any other engines working in the same temperature range.

Operation of the Carnot cycle

In a given thermodynamic cycle, a system is taken through a series of different states and finally returned to its initial state. In the process of going through this cycle, the system performs a certain amount of work on the surroundings, thereby acting as a heat engine. There are four stages in a Carnot cycle: isothermal expansion, adiabatic expansion, isothermal compression, and adiabatic compression. These four steps are

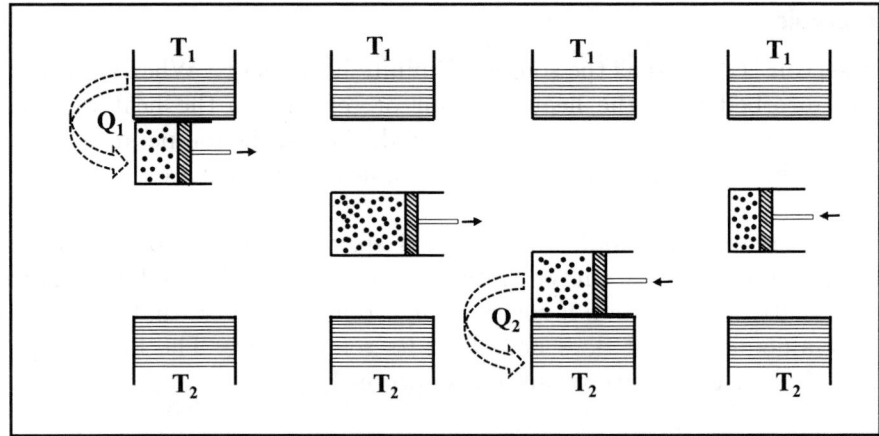

Figure 9.4 Four stages of operation in a Carnot cycle.

shown in a schematic diagram in Figure 9.4. Let T_1 and T_2 be the temperatures of the hot and cold reservoirs, respectively. Further, let a gas be the working substance confined in a cylinder fitted with a frictionless piston. The working of the real engine can be simulated in the following four reversible sequences:

1. The cylinder is placed in thermal contact with the hot reservoir source maintained at a temperature T_1. The working substance absorbs a certain amount of heat, say Q_1, from the source and undergoes a reversible isothermal expansion. The necessary arrangement is shown at the extreme left of Figure 9.4.

2. In the second stage, the working substance is thermally isolated, and is allowed to undergo a reversible adiabatic expansion. Its temperature falls from T_1 to T_2, the temperature of the sink. This is shown on the second left of Figure 9.4.

3. Due to the reversible adiabatic expansion in the second stage, the working substance is in a state of high volume and low pressure. To use it in a cyclic process, it has to be restored to its initial state. Hence, the working substance has to be compressed in two stages: first isothermally, and then adiabatically. In the third stage, the cylinder is placed in thermal contact with the sink at T_2, and the working substance is compressed isothermally and reversibly. This is shown on the second right of Figure 9.4.

4. Finally, in the fourth stage, the working substance is again thermally isolated and compressed under adiabatic conditions till the original state is restored. This is shown at the extreme right of Figure 9.4.

Efficiency of a Carnot cycle using a *P–V* indicator diagram

A Carnot cycle is shown in a *P–V* diagram in Figure 9.5. It consists of four processes: isothermal expansion, reversible adiabatic (isentropic) expansion,

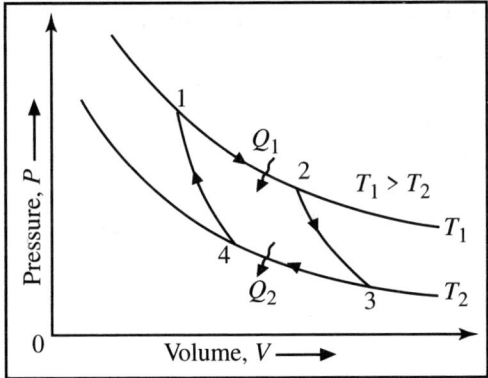

Figure 9.5 Carnot cycle in a P–V diagram.

isothermal compression, and reversible adiabatic compression. These are described in detail below:

1. **A reversible isothermal expansion process:** In this process (from 1 to 2 in Figure 9.5), the ideal gas in the system absorbs Q_1 amount of heat from a heat source at a high temperature T_1, expands, and does work on the surroundings. The temperature of the gas does not change during the process, and thus the expansion is isothermal. The expansion of the gas is propelled by the absorption of heat energy Q_1 given by

$$Q_1 = W_{12} = nRT_1 \ln \left(\frac{V_2}{V_1} \right) \quad \text{as } dT = 0, \text{ and hence, } dU = 0.$$

2. **A reversible adiabatic expansion process:** In this process (from steps 2 to 3 in Figure 9.5), the system is thermally insulated. The gas continues to expand adiabatically and does work on surroundings, which causes the system to cool to a lower temperature, T_2. The required work done W_{23} is

$$W_{23} = \frac{nR}{\gamma - 1}(T_1 - T_2),$$

where γ is the ratio of specific heats.

3. **A reversible isothermal compression process:** In this process (from steps 3 to 4 in Figure 9.5), the gas is exposed to the cold temperature T_2 of the reservoir while the surroundings do work on the gas by compressing it (such as through the return compression of a piston) while causing an amount of heat energy Q_2 to flow out of the gas to the low-temperature reservoir. This heat Q_2 is given by

$$Q_2 = W_{34} = nRT_2 \ln \left(\frac{V_4}{V_3} \right) \quad \text{and } dT = 0, \text{ hence, } dU = 0.$$

4. **A reversible adiabatic compression process:** In this process (from steps 4 to 1 in Figure 9.5), the system is thermally insulated, and the cold temperature reservoir is removed. During this step, the surroundings continue to do work to the gas, which causes the temperature to rise back to T_1. So, the work done by the surroundings is

$$W_{41} = \frac{nR}{\gamma - 1}(T_2 - T_1) \text{ and } \delta Q = 0.$$

The total work done W during the cyclic process is given by

$$W = W_{12} + W_{23} + W_{34} + W_{41} = W_{12} + W_{34} = Q_1 + Q_2.$$

Hence, the efficiency of the Carnot cycle is given by

$$\eta = \frac{W}{Q_1} = \frac{Q_1 + Q_2}{Q_1} = 1 + \frac{Q_2}{Q_1}.$$

Using the values of Q_1 and Q_2, the efficiency η comes out to be

$$\eta = 1 + \frac{nRT_2 \ln \frac{V_4}{V_3}}{nRT_1 \ln \frac{V_2}{V_1}}.$$

Now for the two adiabatic $2 \to 3$ and $4 \to 1$, we have

$$T_1 V_2^{\gamma-1} = T_2 V_3^{\gamma-1}, \text{ and } T_1 V_1^{\gamma-1} = T_2 V_4^{\gamma-1}; \quad \Rightarrow \quad \frac{V_1}{V_2} = \frac{V_4}{V_3}.$$

Using this relation, we get the expression for efficiency η as

$$\eta = 1 + \frac{nRT_2 \ln \frac{V_4}{V_3}}{nRT_1 \ln \frac{V_2}{V_1}} = 1 - \frac{nRT_2 \ln \frac{V_1}{V_2}}{nRT_1 \ln \frac{V_1}{V_2}} = 1 - \frac{T_2}{T_1}. \tag{9.1}$$

It can be concluded from equation (9.1) that we need two sources at different temperatures T_1 and T_2 for the operation of a heat engine; otherwise, the ratio $\frac{T_2}{T_1}$ will be equal to 1, and the efficiency of the heat engine will be zero. The Carnot cycle has the greatest possible efficiency that a heat engine can possess. This is based on the assumption of the absence of incidental wasteful processes, such as friction, and the assumption that there is no conduction of heat between different parts of the engine at different temperatures.

The Carnot cycle is used in thermal devices or machines. Some application fields of the Carnot cycle include heat pumps for generating heating, refrigerators for cooling, steam turbines used in ships, combustion engines in vehicles, and reaction turbines in airplanes.

Carnot cycle in terms of $T-S$ diagram

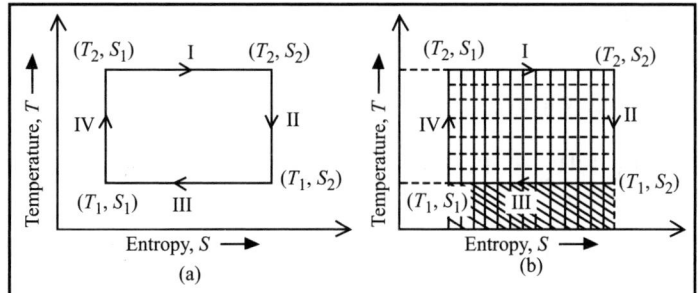

Figure 9.6 (a) Carnot cycle in $T-S$ diagram and (b) Area enclosed in $T-S$ diagram representing the net work done.

The first law of thermodynamics is mathematically expressed as

$$\delta Q = dU + \delta W = TdS + \delta W. \qquad \text{(For the reversible process.)}$$

Now, for an isothermal process of an ideal gas, $dU = 0$. Hence, we have $\delta W = -TdS$. For an adiabatic reversible process of an ideal gas, we have $\delta Q = 0$. Hence, we get

$$\delta W = dU = n\langle C_V \rangle dT.$$

The Carnot cycle in $T-S$ diagram is shown in Figure 9.6(a). The isothermal and adiabatic expansion and compression processes are represented by respective steps I, II, III, and IV. The efficiency of the Carnot cycle can be calculated in the following way:

Step 1: Here, isothermal reversible expansion occurs from (T_2, S_1) to (T_2, S_2). Hence, we have

$$\Delta U_1 = 0, \ W_1 = -T_2(S_2 - S_1) \text{ and } Q_2 = T_2(S_2 - S_1).$$

Step 2: Here, adiabatic reversible expansion occurs from (T_2, S_2) to (T_1, S_2). Adiabatic reversible process is an isentropic process, and hence, we have

$$\Delta S_2 = \ S_3 - S_2 = 0, \text{ or } S_3 = S_2; \ Q_2 = 0, \text{ and } W_2 = \Delta U_2 = nC_V(T_1 - T_2).$$

Step 3: Here, isothermal reversible compression occurs from (T_1, S_2) to (T_1, S_4). From step 4, it will be shown that $S_4 = S_1$. Hence, we have

$$\Delta U_3 = 0, \ W_3 = -T_1(S_4 - S_3) = -T_1(S_1 - S_2), \text{ and } Q_2 = T_1(S_4 - S_3) = T_1(S_1 - S_2).$$

Step 4: Here, adiabatic reversible compression occurs from (T_1, S_4) to (T_2, S_1). As it is an isentropic process, $S_4 = S_1$. For this step, we have

$$Q_1 = 0 \text{ and } W_4 = \Delta U_4 = nC_V(T_2 - T_1).$$

Therefore, the total change in internal energy,

$$\oint dU = U_1 + U_2 + U_3 + U_4 = 0 + nC_V(T_1 - T_2) + 0 + nC_V(T_2 - T_1) \text{ or, } \oint dU = 0.$$

The total work done during this cyclic process is given by

$$\begin{aligned} W_{\text{net}} &= W_1 + W_2 + W_3 + W_4 = -T_2(S_2 - S_1) + nC_V(T_1 - T_2) \\ &\quad - T_1(S_1 - S_2) + nC_V(T_2 - T_1) \\ &= -(T_2 - T_1)(S_2 - S_1); \quad \text{or, } |W_{\text{net}}| = (T_2 - T_1)(S_2 - S_1). \end{aligned}$$

This net work done is shown in Figure 9.6(b). Hence, the efficiency of the Carnot engine from the $T-S$ diagram is given by

$$\eta = \frac{|W_{\text{net}}|}{Q_2} = \frac{(T_2 - T_1)(S_2 - S_1)}{T_2(S_2 - S_1)} = 1 - \frac{T_1}{T_2}.$$

It is clear from the $T-S$ diagram that

$$|W_1| = |Q_2|, \quad |W_3| = |Q_2|, \quad \text{and } |W_2| = |W_4|.$$

They are opposite in sign and must be canceled out. Therefore, the net work done is given by

$$|W_{\text{net}}| = |W_1| - |W_3|.$$

That is the area enclosed in the $T-S$ diagram as shown in Figure 9.6(b). It is also clear that the efficiency η is given by

$$\eta = \frac{|W_{\text{net}}|}{Q_2},$$

that is, the efficiency in the $T-S$ diagram can be expressed as a ratio of two area elements.

Carnot cycle in terms of $U-S$ diagram

We get the same expression for efficiency η for the Carnot cycle in the $U-S$ diagram, as shown in Figure 9.7(a). The isothermal and adiabatic expansion and compression processes are represented by respective steps I, II, III, and IV in the diagram. The efficiency η of the Carnot cycle from this $U-S$ diagram can be calculated in the following way:

Step 1: In this step, isothermal reversible expansion occurs from S_1 to S_2 at an internal energy U_1 and a temperature T_2. Hence, for an ideal gas, we have

$$\Delta U_1 = 0 \text{ or, } U_2 - U_1 = 0 \text{ or, } U_2 = U_1.$$

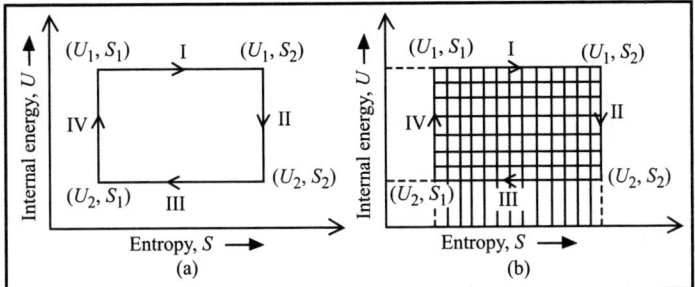

Figure 9.7 (a) Carnot cycle in $U-S$ diagram and (b) area enclosed in $U-S$ diagram representing the net work done.

From the first law of thermodynamics, we have

$$\Delta U_1 = Q + W \text{ or, } W_1 = -Q_2 = -T_2(S_2 - S_1) \text{ or, } Q_2 = T_2(S_2 - S_1).$$

Step 2: In this step, adiabatic reversible expansion occurs, temperature decreases from T_2 to T_1, and internal energy decreases from U_1 to U_2. Therefore,

$$Q = 0, \ \Delta S = 0 \text{ or, } S_3 = S_2, \text{ and } \Delta U_2 = U_2 - U_1 = W_2.$$

Step 3: In this step, isothermal reversible compression occurs at thermodynamic temperature T_1 and thermal energy U_2. Hence, for an ideal gas $\Delta U_3 = 0$. Again, from the first law of thermodynamics, we have

$$\Delta_3 = Q_1 + W_3 \text{ or, } W_3 = -Q_1 = -T_1(S_4 - S_3) \text{ or,}$$
$$W_3 = -T_1(S_1 - S_2); \text{ or, } Q_1 = T_1(S_4 - S_3) = T_1(S_1 - S_2).$$

Step 4: In this step, adiabatic reversible compression occurs, temperature increases from T_1 to T_2, and internal energy changes from U_2 to U_1. Therefore, we have

$$Q = 0, \Delta S = 0 \text{ or, } S_4 = S_1, \text{ and } \Delta U_4 = U_1 - U_4 = U_2 - U_2 = W_4.$$

The total work done during this cyclic process is given by

$$W_{\text{net}} = -T_2(S_2 - S_1) + (U_2 - U_1) - T_1(S_1 - S_2) + (U_1 - U_2),$$
$$= -T_2(S_2 - S_1) + T_1(S_2 - S_1) = -(T_2 - T_1)(S_2 - S_1); \text{ or,}$$
$$|W_{\text{net}}| = (T_2 - T_1)(S_2 - S_1).$$

This net work done is shown in Figure 9.7(b). Again, in step 2, we can write

$$\Delta U = U_2 - U_1 = n \, C_V(T_1 - T_2) \text{ or, } T_2 - T_1 = \frac{U_1 - U_2}{n \, C_V}$$

$$\text{or, } n \, C_V|W_{\text{net}}| = (U_1 - U_2)(S_2 - S_1) = \text{area enclosed.}$$

Now,

$$dU = n\, C_V dT \text{ or, } \int_0^{U_1} dU = n\, C_V \int_0^{T_2} dT \text{ (For an ideal gas, } U = 0 \text{ when } T = 0 \text{ K)}$$

or, $U_1 = n\, C_V T_2$. Hence, we have, $T_2 = \dfrac{U_1}{n\, C_V}$ or, $Q_2 = \dfrac{U_1}{n\, C_V}(S_2 - S_1)$ or,

$n\, C_V Q_2 = U_1(S_2 - S_1) = $ area under the curve.

Therefore, the efficiency of Carnot engine is given by

$$\eta = \frac{|W_{\text{net}}|}{Q_2} = \frac{n\, C_V |W_{\text{net}}|}{n\, C_V |Q_2|} = \frac{(U_1 - U_2)(S_2 - S_1)}{U_1(S_2 - S_1)}; \text{ or,}$$

$$\eta = 1 - \frac{U_2}{U_1} = 1 - \left(\frac{n\, C_V T_1}{n\, C_V T_2}\right) = 1 - \frac{T_1}{T_2}.$$

It is also clear that in the $U-S$ diagram, efficiency can be expressed as a ratio of two area elements. Here, a point is to be noted that the $U-S$ diagram is very similar to the $T-S$ diagram. In the $T-S$ diagram, area enclosed is equal to the net work done per cycle. But in the $U-S$ diagram, area enclosed is equal to $n\, C_V$ times the net work done per cycle.

Carnot cycle in terms of $H-S$ diagram

Here also, we get the same expression for efficiency η for the Carnot cycle in the $H-S$ diagram, as shown in Figure 9.8(a). The isothermal and adiabatic expansion and compression processes are represented by respective steps I, II, III, and IV in the diagram. The efficiency η of the Carnot cycle from this $H-S$ diagram can be calculated in the following way:

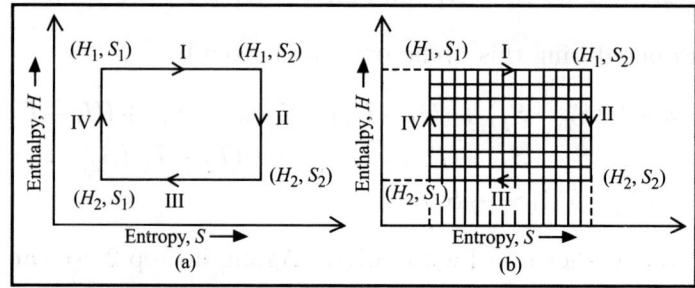

Figure 9.8 (a) Carnot cycle in $H-S$ diagram and (b) area enclosed in $H-S$ diagram representing the net work done.

Step 1: In this step, isothermal reversible expansion occurs from S_1 to S_2 at enthalpy H_1 and temperature T_2. Hence, for an ideal gas,

$$\Delta H_1 = 0 \text{ or, } H_2 - H_1 = 0 \text{ or, } H_2 = H_1.$$

From the first law of thermodynamics, we have

$$\Delta U_1 = Q + W \text{ or, } W_1 = -Q_2 = -T_2(S_2 - S_1) \text{ or, } Q_2 = T_2(S_2 - S_1).$$

Step 2: In this step, adiabatic reversible expansion occurs, temperature decreases from T_2 to T_1, and enthalpy decreases from H_1 to H_2. Therefore,

$$Q = 0, \Delta S = 0, \text{ or, } S_3 = S_2, \ \Delta H_2 = H_2 - H_1, \text{ and } W_2 = U_2 - U_1.$$

Step 3: In this step, isothermal reversible compression occurs at temperature T_1 and enthalpy H_2. Hence, for an ideal gas $\Delta H_3 = 0$. Using the first law of thermodynamics, we get

$$\Delta U_3 = Q_1 + W_3 \text{ or, } W_3 = -Q_1 = -T_1(S_4 - S_3) \text{ or, } W_3 = -T_1(S_1 - S_2) \text{ or,}$$
$$Q_1 = T_1(S_4 - S_3) = T_1(S_1 - S_2).$$

Step 4: In this step, adiabatic reversible compression occurs, temperature increases from T_1 to T_2, and enthalpy changes from H_2 to H_1. Therefore, it leads to

$$Q = 0, \Delta S = 0 \text{ or, } S_4 = S_1, \text{ and } \Delta U_4 = U_1 - U_4 = U_{1-2} = W_4.$$

The net total work done during the cyclic process is given by

$$W_{\text{net}} = -T_2(S_2 - S_1) + (U_2 - U_1) - T_1(S_1 - S_2) + (U_1 - U_2)$$
$$= -T_2(S_2 - S_1) + -T_1(S_2 - S_1) = -(T_2 - T_1)(S_2 - S_1); \text{ or,}$$
$$|W_{\text{net}}| = (T_2 - T_1)(S_2 - S_1).$$

This work done is represented in Figure 9.8(b).

Again in step 2, we get

$$\Delta H = H_2 - H_1 = n \, C_P(T_1 - T_2); \quad \Rightarrow \quad T_2 - T_1 = \frac{H_1 - H_2}{n \, C_P}.$$

Or, $n \, C_P|W_{\text{net}}| = (H_1 - H_2)(S - 2 - S_1) = $ area enclosed in Figure 9.8(b) and

$$|Q_2| = T_2(S_2 - S_1).$$

Now,

$$dH = n \, C_P dT \text{ or, } \int_0^{H_1} dH = n \, C_P \int_0^{T_2} dT.$$

Here, we have used the fact that for an ideal gas, $H = 0$ when $T = 0$ K. This gives

$$H_1 = n\, C_P T_2 \text{ or, } T_2 = \frac{H_1}{n\, C_P}$$

or, $|Q_2| = \dfrac{H_1}{nC_P}(S_2 - S_1)$ or, $nC_P|Q_2| = H_1(S_2 - S_1) = $ area under curve shown in

Figure 9.8(b).

Therefore, the efficiency η of the Carnot engine is given by

$$\eta = \frac{|W_{\text{net}}|}{Q_2} = \frac{nC_P|W_{\text{net}}|}{nC_P|Q_2|} = \frac{(H_1 - H_2)(S_2 - S_1)}{H_1(S_2 - S_1)}; \text{ or,}$$

$$\eta = 1 - \frac{H_2}{H_1} = 1 - \left(\frac{nC_P T_1}{nC_P T_2}\right) = 1 - \frac{T_1}{T_2}.$$

It is also clear that in the $H-S$ diagram, efficiency can be expressed as a ratio of two area elements. In this case, the area enclosed represents nC_P times the work involved in the cycle.

Carnot cycle in terms of U versus T and H versus T diagram

For an ideal gas, internal energy U and enthalpy H are only functions of temperature T. It is known that both U and H are proportional to temperature T, and the proportionality constants are, respectively, C_V and C_P, that is, $\left(\frac{\partial U}{\partial T}\right)_V = C_V$ and $\left(\frac{\partial H}{\partial T}\right)_P = C_P$.

In steps 1 and 3 of the Carnot cycle, we have isothermal processes, and hence, both internal energy U and enthalpy H will remain constant.

In step 2 of the Carnot cycle (adiabatic expansion), temperature decreases, and hence, both internal energy U and enthalpy H decrease. Again in step 4 (adiabatic compression), temperature increases, and hence, both U and H are increased. Variations of U versus T and H versus T in the Carnot cycle are, respectively,

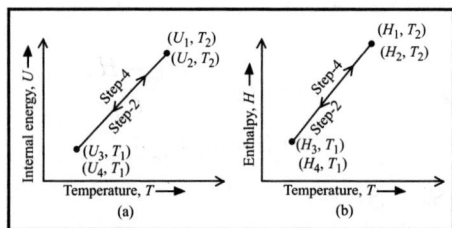

Figure 9.9 (a) Carnot cycle in $U-T$ diagram and (b) Carnot cycle in $H-T$ diagram.

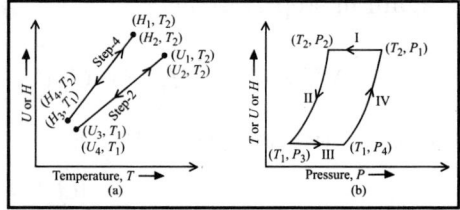

Figure 9.10 (a) Carnot cycle in $U-T$ or $H-T$ diagram and (b) Carnot cycle in $U-P$ or $H-P$ or $T-P$ diagram.

shown in Figures 9.9(a) and (b). If these working cycles are plotted in the same figure, then H versus T plot will be stepper than U versus T plot. This is shown in Figure 9.10(a). The reason of this is $C_P > C_V$. H versus P, T versus P, and U versus P plots are shown in Figure 9.10(b).

Carnot cycle in terms of $\ln P$ versus $\ln V$ diagram

The Carnot cycle consists of four reversible steps: isothermal expansion, adiabatic expansion, isothermal compression, and adiabatic compression. As the working substance of the Carnot cycle, the n-mole of an ideal gas is considered.

In step 1, isothermal reversible expansion of n-mole of the ideal gas occurs from (P_1, V_1) to (P_2, V_2) at temperature T_2. Since this is an isothermal step, $P_1 V_1 = P_2 V_2$. Hence, we have

$$\ln P_1 + \ln V_1 = \ln P_2 + \ln V_2. \tag{9.2}$$

In step 2, adiabatic reversible expansion of the working substance occurs from (P_2, V_2) to (P_3, V_3) when the temperature changes from T_2 to T_1. The equation for the adiabatic process for this step leads to $P_2 V_2^\gamma = P_3 V_3^\gamma$. Hence, we have

$$\ln P_2 + \gamma \ln V_2 = \ln P_3 + \gamma \ln V_3. \tag{9.3}$$

In step 3, isothermal reversible compression occurs from (P_3, V_3) to (P_4, V_4) at temperature T_1. For this isothermal step, we have $P_3 V_3 = P_4 V_4$. Hence, this leads to

$$\ln P_3 + \ln V_3 = \ln P_4 + \ln V_4. \tag{9.4}$$

In step 4, adiabatic reversible compression occurs from (P_4, V_4) to (P_1, V_1) when temperature changes from T_1 to T_2. For this step, we have $P_4 V_4^\gamma = P_1 V_1^\gamma$. This leads to

$$\ln P_4 + \gamma \ln V_4 = \ln P_1 + \gamma \ln V_1. \tag{9.5}$$

From equations (9.3) and (9.5), we get

$$\ln \left(\frac{P_2}{P_1} \right) + \gamma \ln \left(\frac{V_2}{V_1} \right) = \ln \left(\frac{P_3}{P_4} \right) + \gamma \ln \left(\frac{V_3}{V_4} \right). \tag{9.6}$$

From equations (9.2) and (9.4), we get $\ln \left(\frac{P_2}{P_1} \right) = -\ln \left(\frac{V_2}{V_1} \right)$ and $\ln \left(\frac{P_3}{P_4} \right) = -\ln \left(\frac{V_3}{V_4} \right)$. Using these values in equation (9.6), we get

$$(\gamma - 1) \ln \left(\frac{V_2}{V_1} \right) = (\gamma - 1) \ln \left(\frac{V_3}{V_4} \right); \quad \Rightarrow \quad \ln \left(\frac{V_2}{V_1} \right) = \ln \left(\frac{V_3}{V_4} \right). \tag{9.7}$$

The Carnot cycle in logarithm scale $\ln P$ versus $\ln V$ is represented by a parallelogram consisting of four line segments, as shown in Figure 9.11. The parallel lines AB and CD correspond to two isothermal processes with slope $= -1$, whereas as other two parallel lines, BC and DA, correspond to two adiabatic processes with slope $= -\gamma$. For an ideal gas $\gamma > 1$ and hence, adiabatic are steeper than isotherms even in logarithmic scale. From Figure 9.11, it is clear that

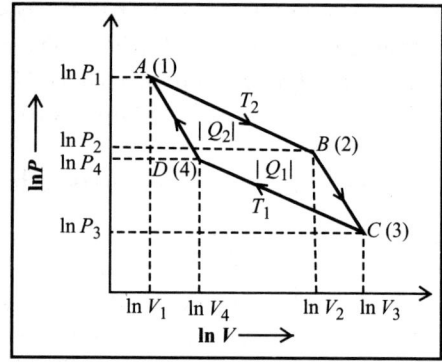

Figure 9.11 Carnot cycle plotted using logarithm scales. Corresponding steps are shown in the figure.

$$\ln V_2 - \ln V_1 = \ln V_3 - \ln V_4, \quad \text{or,} \quad \ln\left(\frac{V_2}{V_1}\right) = \ln\left(\frac{V_3}{V_4}\right). \tag{9.8}$$

Thus, equation (9.8) simply describes the geometric properties of the Carnot cycle as shown in Figure 9.11. The efficiency of the cycle can be easily calculated with heat absorbed in step 1(AB) and heat released in step 3 (CD). Hence, the efficiency of the Carnot engine is given by

$$\eta = 1 - \frac{|Q_1|}{|Q_2|} = 1 - \frac{nRT_1 \ln\left(\frac{V_3}{V_4}\right)}{nRT_2 \ln\left(\frac{V_2}{V_1}\right)} = 1 - \frac{T_1}{T_2}. \tag{9.9}$$

There are two essential geometric properties of the logarithmic plot of the Carnot cycle that are used in determining the efficiency. The first one is the equal lengths of opposite sides of a parallelogram, which correspond to the equality of the ratio between volumes or pressures of the ideal gas at the corner of the cycle. The second one is that the slope of the line segment associated with an adiabatic process is $-\gamma$, corresponding to the well-known power law relation between the pressure and volume of an ideal gas during an adiabatic reversible process.

Some important points regarding the Carnot cycle

The following points are to be noted about the Carnot cycle:

1. It is a hypothetical device.

2. The thermal efficiency of Carnot cycle depends on the temperature of source and sink and not on the working substance.

3. A practical engine based on Carnot cycle cannot be built because of the following reasons:

   (a) Practically it is not possible to bring in contact and remove alternately the heat reservoirs and adiabatic cover in order to complete the cycle.

   (b) The isothermal process could only be achieved by very slow motion of the piston, while for achieving reverse adiabatic process, we must do the piston movement very quickly. It is really tough to achieve such a large variation of speed of the piston in practice.

   (c) In order to get sufficient work, a large range of pressure and volume is required since the slope difference between isothermal and adiabatic is small.

4. Carnot engine has the maximum efficiency compared to other engines while operating within the given temperature range.

9.5.3   Refrigerator and the coefficient of performance

A **refrigerator** is a thermodynamic device designed to remove heat from a low-temperature space and to deliver the same at a high-temperature space. This device can also be used to heat a given volume that is at a higher temperature than the surroundings. In this case, the device is called a **heat pump**.

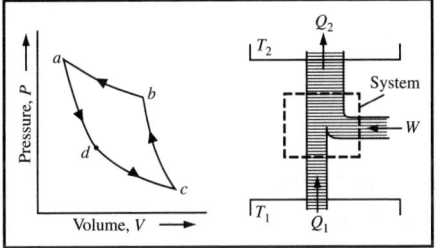

Figure 9.12 An ideal refrigerator—a reverse heat engine cycle.

   The Clausius statement of the second law of thermodynamics asserts that **it is impossible to construct a device that, operating in a cycle, has no effect other than the transfer of heat from a colder to a hotter body**. This signifies that thermal energy will not flow from a colder region to a hotter one without the assistance of external sources. The refrigerator and the heat pump both satisfy the **Clausius requirement of external action** through the application of mechanical power or equivalent natural transfers of heat.

   Several techniques are used to achieve continuous refrigeration. It should be mentioned that when the cycle of operation of any heat engine is effectively reversed, it becomes a refrigeration cycle. The most commonly used cycle of operation in refrigeration and air condition applications is the vapor compression cycle. The vapor absorption cycle provides an alternative system, particularly in applications where heat is economically available. Steam-jet systems are also being successfully used in many cooling applications, while air-cycle refrigeration is often used for aircraft cooling.

A Carnot cycle operating in the reverse direction acts as an ideal refrigerator. This is shown in Figure 9.12. In this cycle, an amount of heat Q_1 is extracted from the reservoir at a low-temperature T_1, and $Q_2(=Q_1+W)$ amount of heat is transferred to the reservoir maintained at a higher temperature T_2 by doing some external work W on the working substance. The indicator diagram of the refrigeration cycle is shown at the left of Figure 9.12.

The **coefficient of performance** β of a refrigerator is defined as

$$\beta = \frac{\text{heat extracted from cold reservoir}}{\text{work done on the working substance}} = \frac{Q_1}{W} = \frac{Q_1}{Q_2 - Q_1}. \tag{9.10}$$

Again, we have

$$\frac{Q_2}{Q_1} = \frac{T_2}{T_1}. \tag{9.11}$$

Using equation (9.11) into equation (9.10), we get

$$\beta = \frac{T_1}{T_2 - T_1}; \quad \Rightarrow \quad \beta\% = \frac{T_1}{T_2 - T_1} \times 100. \tag{9.12}$$

This is the expression for the coefficient of performance of a refrigerator. Equation (9.12) shows that β only depends on the temperatures of the reservoir and external high temperature surroundings.

Relation between the efficiency of a Carnot engine η and the coefficient of performance of a refrigerator β

The efficiency of a Carnot heat engine is

$$\eta = 1 - \frac{T_1}{T_2}; \quad \Rightarrow \quad \frac{T_1}{T_2} = 1 - \eta. \tag{9.13}$$

Using this equation (9.13) into equation (9.12), we get

$$\beta = \frac{T_1}{T_2 - T_1} = \frac{\frac{T_1}{T_2}}{\left[1 - \frac{T_1}{T_2}\right]} = \frac{1 - \eta}{\eta} = \frac{1}{\eta} - 1. \tag{9.14}$$

This leads to

$$\eta = \frac{1}{1 + \beta}. \tag{9.15}$$

This is the relation between η and β. As the temperature of the cold reservoir T_1 decreases, the difference in temperature $\Delta T = T_2 - T_1$ increases, and this leads to a decrease in β. For a given amount of extracted heat Q_1, β decreases with the increase in W as $\beta = \frac{Q_1}{W}$. Hence, to extract a given amount of heat Q_1 from a cold reservoir, the work done W increases with a decrease in temperature T_1 of the cold reservoir.

Refrigerants

Refrigerants are the working fluids in refrigeration systems. They must have the following characteristics:

1. The performance of refrigerants should be good.

2. Refrigerants should have low flammability and toxicity.

3. Refrigerants should be compatible with compressor lubricating oils and metals, and

4. Refrigerants should have good heat transfer properties.

For the identification of refrigerants, generally a number relating to their molecular composition is assigned. The ASHRAE Handbook of Fundamentals (1993) lists a large number of refrigerants with their suitable properties. In recent years, it has been observed that the use of chlorofluorocarbons (CFCs) as the working fluids in refrigeration and air-conditioning plants has given rise to grave concern due to pollution. To get rid of this severe pollution problem in the atmosphere, people are trying to develop alternative fluids as refrigerants. The majority of the used refrigerants fall into two categories: (i) hydrofluorocarbons (HDCs), which contain no chlorine and have zero ozone depletion potential, and (ii) hydrochlorofluorocarbons (HCFCs), which do contain chlorine. But by adding hydrogen to the CFC structure, all the chlorine in HCFCs is allowed to disperse virtually in the lower atmosphere before it can reach the ozone layer. Therefore, HCFCs have much lower ozone depletion potentials ranging from 2% to 10% that of CFCs.

9.6 Equivalence of Kelvin–Planck and Clausius statements

Statements of the second law of thermodynamics due to Kelvin–Planck and Clausius and their equivalence are discussed in detail below.

9.6.1 Kelvin–Planck statement

The Kelvin–Planck Statement: It is impossible to construct an engine which, working on a cycle, shall produce no other effect than the extraction of heat from a reservoir and the performance of an equal amount of work.

This statement clearly signifies that we cannot devise a machine that just absorbs a certain amount of heat and converts it fully into work, that is, it is 100% efficient. Let us consider the isothermal expansion of an ideal gas. When the gas expands isothermally, it does work on the piston. The first law of thermodynamics tells us that during an isothermal expansion of an ideal gas, the internal energy remains constant. That is,

$$\Delta U = \delta Q - \delta W \quad \text{so that} \quad \delta W = \delta Q.$$

This shows that we have a complete conversion of heat into work! Now we ask the question: Does this violate Kelvin–Planck statement? The answer is: No, it does not violate Kelvin–Planck statement because the state of the gas at the end of the process is different from that at the beginning (due to a change in volume). Hence, the conversion of heat into work is not the sole effect of the process.

There are other processes in which energy is conserved, but they do not occur. For example, heat does not flow on its own from a body at a lower temperature to a body at a higher temperature. This indicates that the spontaneous flow of heat is unidirectional. This fact of natural occurrence is contained in the Clausius statement of the second law of thermodynamics.

9.6.2 Clausius statement

Clausius statement: No process is possible whose sole result is the transfer of heat from a body at a lower temperature to a body at a higher temperature.

It is to be noted that this is a fact of experience and needs no proof. However, we may ask the following question: Does heat flow in a refrigerator from cold items kept inside (a colder body) to the surroundings (a hotter body) in violation of Clausius statement? Certainly, this does not happen because some work has to be done by an external source on the working substance to transfer a certain amount of heat from a colder body to a hotter body. Therefore, an important implication of this law is that it is not possible to transfer heat from a cold body to a hot body without any change in somewhere, including the working substance/surroundings of the system.

9.6.3 Equivalence of Kelvin–Planck and Clausius statements

It is to be noted that the two statements of the second law of thermodynamics apparently seem different or unconnected, but in fact, this is not true; these statements are equivalent to each other. It can be shown that if one statement is true, then the other statement must necessarily be true. On the other hand, if one statement is found to be untrue, then the other will also be untrue. Hence, the truth of either form is both a necessary and sufficient condition for the truth of the other.

Violation of Clausius statement is equivalent to violation of Kelvin–Plank statement

Let us suppose that the Clausius statement of the second law is violated by a hypothetical refrigerator RI that transfers Q units of heat in each cycle from a cold

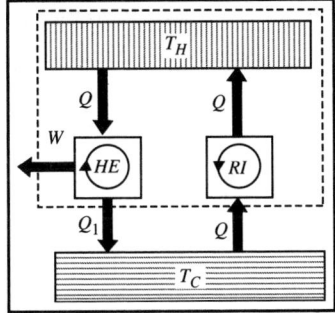

Figure 9.13 Equivalence of Clausius and Kelvin–Planck statements of the second law of thermodynamics.

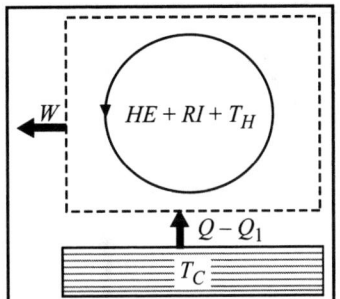

Figure 9.14 Equivalence of Clausius and Kelvin–Planck statements of the second law of thermodynamics.

reservoir at temperature T_C to a hot reservoir at temperature T_H without expenditure of any work. Such a refrigerator is shown in Figure 9.13.

Further, we consider a heat engine HE working between the temperature range T_H and T_C. It takes Q heat from the high-temperature source and converts some part of it to work W and rejects the rest amount of heat $Q_1 = (Q - W)$ to the low-temperature sink. If these two engines (the refrigerator and the heat engine) are combined together (see Figure 9.14), the net result would be to transfer $Q - Q_1$ amount of heat from the cold reservoir and convert it completely into work. Such an engine obviously violates Kelvin–Planck statement.

Violation of Kelvin–Planck statement is equivalent to violation of Clausius statement

To prove that if Kelvin–Planck statement is violated, the Clausius statement is also violated. We consider a hypothetical engine HE, which extracts heat Q from the hot reservoir at a temperature T_H, and converts this heat completely into work W. In this process, no heat is rejected to the cold reservoir maintained at a low-temperature T_C. This is shown in Figure 9.15. It is further assumed that the work performed by the heat engine HE is used to drive a refrigerator working between the same two reservoirs maintained at temperatures T_H and T_C, respectively. Further, we consider that the refrigerator absorbs Q_1 amount of heat from the low-temperature reservoir at T_C and rejects $Q_2 = (W + Q_1)$ amount of heat to the hot reservoir at T_H per cycle.

It is further assumed that the refrigerator completes one cycle in the same period as the heat engine. According to the principle of operation of a refrigerator, the refrigerator by itself does not violate any law. But when the refrigerator is made to form a composite engine $(HE + R)$, as shown in Figure 9.16, it is observed that the net result of the operation of the composite engine $(HE + R)$ working as the refrigerator is the transfer of heat Q_1 from the low-temperature reservoir at T_C to the high-temperature reservoir at T_H without the aid of any external work. This definitely

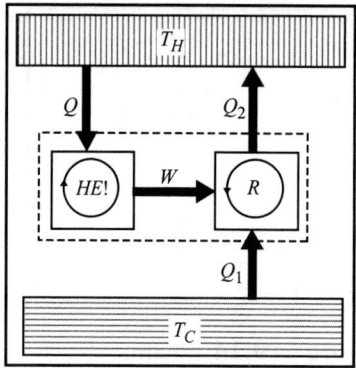

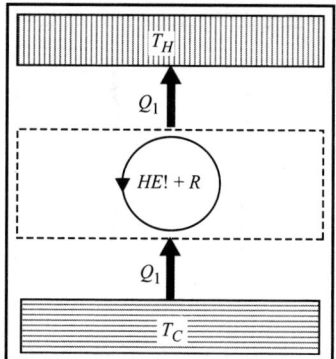

Figure 9.15 Equivalence of Kelvin—Planck and Clausius statements of the second law of thermodynamics.

Figure 9.16 Equivalence of Kelvin—Planck and Clausius statements of the second law of thermodynamics.

violates the Clausius statement of the second law of thermodynamics. Hence, it can be concluded that both the statements of the second law of thermodynamics are equivalent to each other.

9.7 Carnot theorem

From the second law of thermodynamics, two conclusions are obtained that are used to constitute Carnot theorem:

1. **Of all heat engines working between the same (constant) temperatures, the reversible Carnot engine has the maximum efficiency.**

2. **The efficiency of all reversible engines working between the same two temperatures (the same source and the sink) is the same whatever be the working substance.**

Proof: **First part:** In order to prove the first part of the theorem, we proceed in the following way: We consider two heat engines, R (reversible engine) and I (irreversible engine), working between the same source T_H and the same sink T_C, where $T_H > T_C$.

Suppose reversible engine R takes heat Q_1 from the source at a temperature T_H, performs work W, and rejects heat $Q_2 = Q_1 - W$ to the sink at a temperature T_C. Hence, the efficiency of the reversible engine R is given by

$$\eta_R = \frac{W}{Q_1}. \tag{9.16}$$

Similarly, suppose the irreversible engine I takes heat Q_1' from the source (T_H), performs work W, and rejects heat $Q_2' = Q_1' - W$ to the sink (T_C). Hence, the efficiency of the irreversible engine I is given by

$$\eta_I = \frac{W}{Q_1'}. \tag{9.17}$$

Suppose the irreversible engine I is more efficient than the reversible engine R, i.e., $\eta_I > \eta_R$. Using the expressions for the efficiency of these two engines, we have

$$\eta_I > \eta_R; \quad \Rightarrow \quad \frac{W}{Q_1'} > \frac{W}{Q_1}; \quad \Rightarrow \quad Q_1 > Q_1'. \tag{9.18}$$

This indicates that $Q_1 - Q_1'$ is a positive quantity. Suppose the two engines are coupled together so that the engine R works in opposite directions, that is, the engine R works as a refrigerator. Engine I absorbs heat Q_1' from the source (T_H), performs work W, and rejects heat $Q_2' = Q_1' - W$ to the sink T_C. The engine R takes heat Q_2 from the sink T_C, W work is done on the engine R, and $Q_1 = Q_2 + W$ heat is rejected to the source T_H. Here, we can say that work done on the gas by the engine I. Thus, both the engines R and I work as a self-acting machine.

Heat lost by the sink is

$$Q_2 - Q_2' = (Q_1 - W) - (Q_1' - W); \quad \Rightarrow \quad (Q_1 - Q_1' = \text{a positive quantity}), \tag{9.19}$$

that is, external work done on the system $= 0$. Thus, the coupled engines forming a self-acting machine unaided by any external agency transfer heat continuously from a body at a lower temperature to the body at a higher temperature. This conclusion is contrary to the second law of thermodynamics, that is, our assumption $\eta_I > \eta_R$ is wrong. Therefore, no engine will be more efficient than a perfectly reversible engine working between the same two temperatures (the same source and the same sink).

Second part: We consider two reversible engines R_1 and R_2 working between the same source and the same sink. Suppose R_1 drives R_2 backward, then R_1 cannot be more efficient that R_2, and suppose if R_2 drives R_1 backward, then R_2 cannot be more efficient than R_1. Hence, the two reversible engines, R_1 and R_2, are equally efficient. Thus, all reversible engines working between the same two temperatures will have the same efficiency.

9.7.1 Applications of second law of thermodynamics

The second law of thermodynamics finds a wide range of practical applications in everyday life as well as in industry. Some of these applications are mentioned below:

1. According to the second law of thermodynamics, heat flows spontaneously from a body at a higher temperature to a body at a lower temperature. This law is

applicable to all types of heat engines, such as Otto and Diesel and also for all types of working fluids used in the engines.

2. This law is also applied to refrigerators and heat pumps based on the principle of reversed Carnot cycle. In these cycles, heat is forced to flow from a body at a lower temperature to a body at a higher temperature by doing some external work on the substance. Thus, in the original Carnot cycle, work is produced at the expense of heat, while in the reversed Carnot cycle, some external work is performed on the working substance to transfer heat from a low-temperature reservoir to a higher temperature reservoir.

It should be mentioned that in the refrigerator used in daily life, heat is removed from the food items and thrown away to the high-temperature atmosphere. This is achieved at the expense of an external supply of work via the compressor.

3. Air conditioners and heat pumps follow a similar law of thermodynamics. Normally, in summer, the air conditioner removes heat from the room and maintains it at a lower temperature, and the absorbed heat is thrown into the atmosphere. Similarly, in winter, the heat pump absorbs heat from the atmosphere and supplies it to the room making it warmer. In both cases, external work/energy in the form of electricity is supplied. When the temperature difference is larger, larger external work is required.

9.7.2   Thermodynamic scale of temperature and its equivalence with perfect gas scale

The thermodynamic temperature and its consequences are contained within the framework of the second law of thermodynamics. This thermodynamic temperature is found to exhibit some special properties, and in particular, it can be uniquely defined (up to some constant multiplicative factor) by considering the Carnot principle. According to this principle, the efficiency of idealized heat engines does not depend on the working fluid. It depends only on the temperatures of the reservoirs between which it operates. As the efficiency of the idealized heat engine remains the same in all temperature scales, the ratio $\frac{T_1}{T_2}$ of two temperatures, T_1 and T_2, is the same in all absolute scales. This provides a clue to define a thermodynamic scale of temperature. It should be mentioned further that this temperature scale does not depend on the thermometric property of the substance.

The temperature of a system is specifically defined when the system is at thermal equilibrium. From a microscopic point of view, a material is at thermal equilibrium if the quantities of heat between its individual particles cancel out. There are many possible scales of temperature, derived from a variety of observations of physical phenomena.

In order to introduce the thermodynamic scale of temperature, we consider the operation of three reversible engines, 1, 2, and 3, as shown in Figure 9.17. The reversible engine 1 absorbs an amount of heat Q_1 from the reservoir maintained at a constant temperature T_1, does W_1 amount of work, and rejects a Q_2 amount of heat to the reservoir maintained at a constant temperature T_2. Then, the efficiency η_1 of this reversible engine is given by

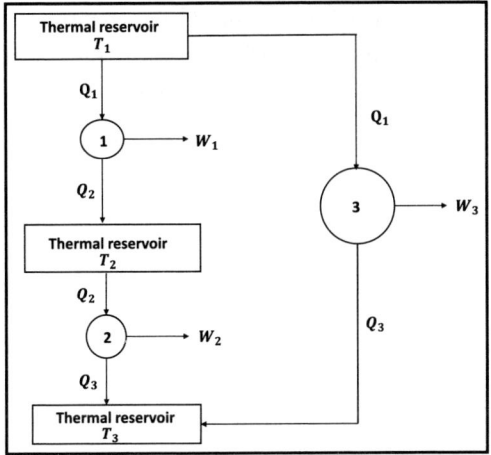

Figure 9.17 Schematic diagram of reversible engines to introduce thermodynamic temperature scale.

$$\eta_1 = \frac{W_1}{Q_1} = \frac{Q_1 - Q_2}{Q_1} = 1 - \frac{Q_2}{Q_1} = f(T_1, T_2), \tag{9.20}$$

where $f(T_1, T_2)$ is some function depending only on the temperatures T_1 and T_2. This equation (9.20) shows that the efficiency of a reversible heat engine depends only on the ratio $\frac{Q_2}{Q_1}$ and, hence T_1 and T_2. Further, equation (9.20) can be expressed as

$$\frac{Q_1}{Q_2} = F(T_1, T_2), \tag{9.21}$$

where $F(T_1, T_2)$ is some function depending only on the temperatures T_1 and T_2.

Let engine 2 absorb a Q_2 amount of heat from the reservoir at a constant temperature T_2, do a W_2 amount of work, and rejects a Q_3 amount of heat to the reservoir maintained at a constant temperature T_3. Then, the efficiency η_2 of this reversible engine is given by

$$\eta_2 = \frac{W_2}{Q_2} = \frac{Q_2 - Q_3}{Q_2} = 1 - \frac{Q_3}{Q_2} = f(T_2, T_3), \tag{9.22}$$

where $f(T_2, T_3)$ is some function depending only on the temperatures T_2 and T_3. Further, equation (9.22) can be expressed as

$$\frac{Q_2}{Q_3} = F(T_2, T_3), \tag{9.23}$$

where $F(T_2, T_3)$ is some function depending only on the temperatures T_2 and T_3.

The third reversible engine 3 absorbs an amount of heat Q_1 from the reservoir at T_1, does an amount of work W_3, and rejects an amount of heat Q_3 to the reservoir at T_3. Then, the efficiency η_3 of this reversible engine is given by

$$\eta_3 = \frac{W_3}{Q_1} = \frac{Q_1 - Q_3}{Q_1} = 1 - \frac{Q_3}{Q_1} = f(T_1, T_3), \tag{9.24}$$

where $f(T_1, T_3)$ is some function depending only on the temperatures T_1 and T_3. Further, equation (9.24) can be expressed as

$$\frac{Q_1}{Q_3} = F(T_1, T_3), \tag{9.25}$$

where $F(T_1, T_3)$ is some function depending only on the temperatures T_1 and T_3.
We can write

$$\frac{Q_1}{Q_2} = \frac{Q_1}{Q_3} \times \frac{Q_3}{Q_2} = \frac{\frac{Q_1}{Q_3}}{\frac{Q_2}{Q_3}}; \quad \Rightarrow \quad F(T_1, T_2) = \frac{F(T_1, T_3)}{F(T_2, T_3)}.$$

Since T_3 does not appear on the left side of the above expression; therefore, T_3 should cancel out on the right-hand side. This is possible if the function F can be written in the form

$$F(T_1, T_2) = \phi(T_1)\psi(T_2).$$

Hence, we can write $\phi(T_1)\psi(T_2) = \frac{\phi(T_1)\psi(T_3)}{\phi(T_2)\psi(T_3)} = \frac{\phi(T_1)}{\phi(T_2)}$.
This gives rise to $\psi(T_2) = \frac{1}{\phi(T_2)}$.
Hence, from equation (9.21), we have

$$\frac{Q_1}{Q_2} = F(T_1, T_2) = \phi(T_1)\psi(T_2) = \frac{\phi(T_1)}{\phi(T_2)}.$$

Now, there are several functional relations that will satisfy this equation. For the thermodynamic scale of temperature, Kelvin selected the relation

$$\frac{Q_1}{Q_2} = \frac{T_1}{T_2}. \tag{9.26}$$

Equation (9.26) shows that the ratio of thermal energy absorbed to the thermal energy rejected by a reversible heat engine is equal to the ratio of the temperatures of the source and the sink. This equation (9.26) can be used to determine the temperature of any reservoir by operating a reversible engine between that reservoir and another easily reproducible reservoir and by measuring the efficiency of the reversible heat engine. The temperature of the easily reproducible thermal reservoir can be arbitrarily assigned a numerical value (the reproducible reservoir can be at triple point of water and the temperature value assigned is 273.16 K).

The efficiency of a Carnot engine operating between two thermal reservoirs, the temperatures of which are measured on the thermodynamic temperature scale, is given by

$$\eta_1 = 1 - \frac{Q_2}{Q_1} = 1 - \frac{T_2}{T_1}. \tag{9.27}$$

It is to be noted that for $T_2 = 0$, the efficiency is 100%, and the efficiency becomes greater than 100% for $T_2 < 0$, but such a situation is unrealistic. The efficiency of a Carnot engine, using an ideal gas as the working medium and the temperature measured on the ideal gas temperature scale, is also given by a similar expression.

It is to be noted that such a definition of the thermodynamic temperature enables us to represent the Carnot efficiency in terms of T_1 and T_2 and hence, to derive that the (complete) Carnot cycle is isentropic:

$$\frac{Q_2}{Q_1} = f(T_1, T_2) = \frac{T_2}{T_1}. \tag{9.28}$$

This equation (9.28) can be expressed as

$$\frac{Q_1}{T_1} - \frac{Q_2}{T_2} = 0, \tag{9.29}$$

where the negative sign indicates that heat is ejected from the system in a reversible process. The generalization of this equation is **Clausius theorem**, which suggests the existence of a state function S, that is, a function that depends only on the state of the system, not on how it reached that state. This function S is defined (up to an additive constant) by

$$dS = \frac{\delta Q_{\text{rev}}}{T}, \tag{9.30}$$

where the subscript indicates heat transfer in a reversible process. The function S corresponds to the entropy of the system, and the change of S around any cycle is zero (as is necessary for any state function). Equation (9.30) can be rearranged to get an alternative definition for temperature in terms of entropy and heat (to avoid logic loop, we should first define entropy through Statistical mechanics):

$$T = \frac{\delta Q_{\text{rev}}}{dS}. \tag{9.31}$$

For a system in which the entropy S is a function of its energy E, the thermodynamic temperature T is therefore given by

$$\frac{1}{T} = \frac{dS}{dE}, \tag{9.32}$$

so that the reciprocal of the thermodynamic temperature is the rate of increase of entropy with energy.

9.8   Entropy in thermodynamics

The concept of entropy was first introduced by Clausius in 1854 while working on the formulation of the second law of thermodynamics. Its literal meaning is "transformation".

The quantity **entropy** can be defined in the following way: It is a measure of disorder of the molecular motion. Greater is the disorder of molecular motion of the system, greater is the entropy. With the supply of heat to the system, the randomness of molecular motion increases, and therefore, the entropy of the system increases. On the contrary, if heat is extracted from the system, disorder or randomness of molecular motion decreases, and thereby entropy decreases.

Thus, entropy is a measure of the degree of randomness of the molecules comprising the system. The higher the disorder, the greater is the increase in entropy. Entropy is a function of the quantity of heat, which shows the possibility of conversion of that heat into work. The increase in entropy is small when heat is added at a higher temperature and is greater when heat is added at lower temperature. Entropy is an extensive property, that is, it depends on the mass of the system, not on the path by which the process is carried out.

9.8.1   Properties of entropy

The essential properties of entropy are briefly mentioned below:

1. Entropy is an extensive quantity. The entropy is proportional to the number of particles in a system and, hence, proportional to the size of the system. Thus, entropy S is an extensive quantity. It should be mentioned here that temperature T and the average energy per spin ϵ are intensive quantities as these are independent of the mass or number of particles of a system, whereas the total energy E is an extensive quantity as it depends on the mass and number of particles.

2. According to Boltzmann prescription, entropy is given by $S(E, \Delta E) = K_B \ln$ (number of states with energy between E and ΔE). We thus see that entropy also depends on the energy window ΔE. However, in the thermodynamic limit $N \to \infty$, such a dependence can be dropped out, and we can regard S as a function of energy E only.

3. The entropy has additive property. Due to its additive property, entropy is a homogeneous function of the extensive coordinates of the system which implies that

$$S(\lambda E, \lambda V, \lambda N_1, \lambda N_2, \lambda N_3, \cdots, \lambda N_m) = \lambda S(E, V, N_1, N_2, N_3, \cdots, N_m).$$

This additive property shows that entropy can be written as a function of the total number of particles and of intensive coordinates like mole fractions and molar volume.

4. In all adiabatic processes, entropy remains constant.

5. Entropy is a state function, meaning its value depends only on the state, and not on the path.

6. In a reversible process, the total entropy of the system and surroundings remains constant.

9.8.2 Physical significance of entropy

In this section, some of the physical and philosophical aspects of the concept of entropy are briefly narrated.

1. We know that entropy increases during an irreversible process that evolves only in one direction. This means that entropy serves as an effective vehicle for communicating our observations with respect to the evolution of irreversible processes, such as heat conduction along a rod or the cooling of a cup of coffee.

2. The entropy of a system increases when heat is added to it. However, the continuous withdrawal of heat makes the molecular motion of the system increasingly ordered, and entropy decreases. This decrease in entropy implies a transition from a chaotic or more random state to an ordered state. This indicates that the more disordered the state is, the larger the entropy.

3. Entropy is related to the probability of occurrence of a state. The increase in entropy during an irreversible process can be associated with a change of state from a less probable to a more probable state. When a cup of coffee cools, the most probable state is when it reaches the same temperature as its surroundings. Hence, the coffee cools, entropy increases, and the system changes from less probable to more probable state.

4. A system left to itself shows a tendency to change in the direction of increasing entropy, that is, to maximize entropy. On the other hand, from a statistical point of view, a system approaches the state of larger probability from a lesser one. Again, the larger probability state has a larger thermodynamic probability. Hence, a system proceeds in the direction of maximum entropy and maximum thermodynamic probability. This led Boltzmann to relate entropy with thermodynamic probability.

5. The second law of thermodynamics is intimately related to the concept of entropy. In fact, the second law of thermodynamics can be stated in terms of entropy in the following way: Every physical or chemical process in nature takes place in such a way as to increase the entropy of the system.

6. In 1870, Ludwig Boltzmann formulated the statistical definition of entropy by analyzing the microscopic behavior of the components of a statistical system.

Boltzmann further showed that this definition of entropy is equivalent to the thermodynamic entropy within a constant number that has since been known as Boltzmann constant. It can be concluded that the thermodynamic definition of entropy provides the experimental definition of entropy, while the statistical definition of entropy extends the concept of entropy, providing an explanation and a deeper understanding of its very nature.

9.8.3   Boltzmann hypothesis: Entropy and thermodynamic probability

In this section, concepts of entropy, thermodynamic probability, and the relation between them are introduced. The thermodynamic probability Ω denotes the total number of micro-states consistent with the given constraint and available to a given macro-state of a thermodynamic system. It is also known as the statistical weight of the macro-state. The entropy and the thermodynamic probability are related by the Boltzmann hypothesis as

$$S = K_B \ln \Omega. \tag{9.33}$$

Following Carnot Ω is also treated as the degree of disorder or simply disorder of the system. Hence, we can restate the Boltzmann hypothesis as **Entropy** $= K_B \ln$ **(Disorder)**. The entropy S being a monotonically increasing function of disorder, can itself be considered as a measure of disorder of the system. Thus, the entropy introduced by Boltzmann provides a measure of the disorder of the distribution of the states over permissible microstates.

9.8.4   Entropy of an ideal gas

Ideal gas is an important model in statistical mechanics and thermodynamics. It refers to N molecules in a container of volume V and at a temperature T. The interaction between the particles is sufficiently weak that it could be ignored in many calculations. But conceptually, the interaction cannot be exactly zero; otherwise, the system would no longer be **ergodic**—a particle would never be able to transfer energy to another particle and reach equilibrium when there were no interactions at all.

We consider such an ensemble of gas containers containing ideal gas particles (monoatomic molecules) that can be described by the micro-canonical ensemble. In this section, we will calculate the macroscopic thermodynamic properties of the ideal gas using the formalism of a micro-canonical ensemble.

Number of accessible micro-states to the ideal gas

For an ideal gas with N number of particles, each of mass m, confined in volume V at a temperature T, the number of micro-states accessible to the gaseous system is given by

$$\Omega = \frac{V^N V_{\text{mom}}}{h^{3N}} = \frac{V^N (2\pi m E)^{\frac{3N}{2}}}{h^{3N} \left(\frac{3N}{2}\right)!} = \frac{V^N}{\left(\frac{3N}{2}\right)!} \left[\frac{2\pi m E}{h^2}\right]^{\frac{3N}{2}}. \tag{9.34}$$

This equation (9.34) has been derived in Appendix I and can be used to obtain an expression for the entropy of an ideal gas as a function of temperature T.

Expression for entropy of a classical ideal gas

According to Boltzmann hypothesis, the entropy and the thermodynamic probability are related by $S = K_B \ln \Omega$. Using equation (9.34), the expression for entropy of a classical ideal gas can be written as

$$S = K_B \ln \Omega = K_B \ln \left[\frac{V^N}{\left(\frac{3N}{2}\right)!} \left(\frac{2\pi m E}{h^2}\right)^{\frac{3N}{2}} \right]$$

$$= K_B \ln \left[V^N \left(\frac{2\pi m E}{h^2}\right)^{\frac{3N}{2}} \right] - K_B \ln \left[\left(\frac{3N}{2}\right)! \right].$$

The factorial of a large number can be simplified using the Stirling formula: $N! = N \ln N - N$. Applying this formula to the above expression, we get

$$S = K_B \ln \left[V^N \left(\frac{2\pi m E}{h^2}\right)^{\frac{3N}{2}} \right] - K_B \left[\frac{3N}{2} \ln \left(\frac{3N}{2}\right) - \frac{3N}{2} \right],$$

$$= N K_B \ln \left[V \left(\frac{2\pi m E}{h^2}\right)^{\frac{3}{2}} \right] - \frac{3N K_B}{2} \ln \left(\frac{3N}{2}\right) + \frac{3N K_B}{2},$$

$$= N K_B \ln \left[V \left(\frac{2\pi m E}{h^2}\right)^{\frac{3}{2}} \right] - N K_B \ln \left(\frac{3N}{2}\right)^{\frac{3}{2}} + \frac{3N K_B}{2}$$

$$= N K_B \ln \left[V \left(\frac{4\pi m E}{3N h^2}\right)^{\frac{3}{2}} \right] + \frac{3N K_B}{2}.$$

This leads to the expression for entropy S as

$$S = N K_B \ln \left[V \left(\frac{E}{N}\right)^{\frac{3}{2}} \left(\frac{4\pi m}{3 h^2}\right)^{\frac{3}{2}} \right] + \frac{3N K_B}{2}. \tag{9.35}$$

This equation (9.35) provides the expression for the entropy of an ideal gas in terms of energy E, volume V, and number of molecules N.

This equation (9.35) can be expressed in terms of temperature T in the following way: The total translational energy of an ideal gas consisting of N number of molecules at a temperature T is given by $E = \frac{3}{2} N K_B T$. Using this expression, equation (9.35) can be written as

$$S = NK_B \ln \left[V (E)^{\frac{3}{2}} \left(\frac{4\pi m}{3Nh^2} \right)^{\frac{3}{2}} \right] + \frac{3}{2} NK_B$$

$$= NK_B \ln \left[V \left(\frac{3}{2} NK_B T \right)^{\frac{3}{2}} \left(\frac{4\pi m}{3Nh^2} \right)^{\frac{3}{2}} \right] + \frac{3}{2} NK_B,$$

$$= NK_B \left[\ln \left\{ V \left(\frac{3}{2} NK_B T \right)^{\frac{3}{2}} \left(\frac{4\pi m}{3Nh^2} \right)^{\frac{3}{2}} \right\} + \frac{3}{2} \right],$$

$$= NK_B \left[\ln V + \frac{3}{2} \ln T + \ln \left\{ \frac{2\pi m K_B}{h^2} \right\}^{\frac{3}{2}} + \frac{3}{2} \right].$$

This leads to the expression for entropy S as

$$S = n_m R \left[\ln V + \frac{3}{2} \ln T + S_\sigma \right], \tag{9.36}$$

where $S_\sigma = \ln \left(\frac{2\pi m \, K_B}{h^2} \right)^{\frac{3}{2}} + \frac{3}{2}$ and n_m is the number of moles. This is the expression for entropy S of an ideal gas as a function of temperature T at constant volume V.

Inconsistency of equation (9.35)

It should be mentioned that equation (9.35) provides certainly a new result for the entropy S of an ideal gas. But this equation (9.35) has some problems. It suffers from the following inconsistencies:

1. Equation (9.35) does not satisfy the additive property of entropy and leads to the Gibbs paradox.

2. Further, it is observed from equation (9.36) (derived from equation 9.35) that entropy $S \to -\infty$ as $T \to 0$, which contradicts the essence of the third law of thermodynamics. This contradiction, however, can be removed from the following physical argument. Equation (9.35) has been derived using the principle of classical physics, which breaks down at low temperatures. Thus, it is not expected that this equation (9.36) would provide a sensible expression for the entropy of an ideal gas at a temperature very close to absolute zero.

Extensive property of entropy is not satisfied by equation (9.35)

Thermodynamic quantities are divided into two groups: extensive and intensive. If extensive quantities increase by a factor c_m, the size of the system under consideration is also increased by the same factor but the intensive quantities remain the same. Energy and volume are the typical examples of extensive quantities, whereas pressure and temperature are typical examples of intensive quantities. Entropy is very definitely an extensive quantity, and the entropy of two weakly interacting systems is additive. Thus, if we double the size of a system, we expect the entropy to double as well.

In order to check the additive nature of entropy given by equation (9.35), we consider an ideal gaseous system consisting of n_m moles in volume V and at temperature T. Doubling the size of the system is like joining two identical systems together to form a new system of volume $2V$ containing $2n_m$ moles of the gas at the same temperature T. Further, we consider that the number of molecules in the system is N with m as the mass of each particle. According to equation (9.35), the entropy of the system is given by

$$S = NK_B \ln\left[V\left(\frac{E}{N}\right)^{\frac{3}{2}}\left(\frac{4\pi m}{3h^2}\right)^{\frac{3}{2}}\right] + \frac{3NK_B}{2}. \qquad (9.37)$$

In order to check the additive property of entropy given by this equation, we increase volume V, number of particles N, and energy E of the gas by a common factor, say c_m. Then, the entropy S_{new} of the new system (using equation 9.35) will be given by

$$S_{\text{new}} = c_m NK_B \ln\left[c_m V\left(\frac{c_m E}{c_m N}\right)^{\frac{3}{2}}\left(\frac{4\pi m}{3h^2}\right)^{\frac{3}{2}}\right] + \frac{3c_m NK_B}{2},$$

$$= c_m NK_B \ln\left[c_m V\left(\frac{4\pi m E}{3N h^2}\right)^{\frac{3}{2}}\right] + \frac{3c_m NK_B}{2}.$$

This leads to

$$S_{\text{new}} = c_m\left[NK_B \ln\left\{V\left(\frac{4\pi m E}{3N h^2}\right)^{\frac{3}{2}}\right\} + \frac{3NK_B}{2}\right] + c_m NK_B \ln c_m. \qquad (9.38)$$

Using the value of S from equation (9.37) to equation (9.38), we get

$$S_{\text{new}} = c_m S + c_m NK_B \ln c_m. \quad \Rightarrow \quad \Delta S = S_{\text{new}} - c_m S = c_m NK_B \ln c_m. \qquad (9.39)$$

In the new system, particle density $\frac{c_m N}{c_m V} = \frac{N}{V}$ and the energy per particle $\frac{c_m E}{c_m N} = \frac{E}{N}$ remain the same as in the case of the old system. In order to satisfy the additive property of entropy, the entropy of the new system S_{new} would have been $c_m S$, but equation (9.39) shows that the entropy S_{new} of the c_m-times the original system is more than c_m-times the entropy of the original system by a factor of $c_m NK_B \ln c_m$. Hence, this calculation shows that the entropy given by equation (9.35) does not satisfy the additive property. This extra entropy $\Delta S = c_m NK_B \ln c_m$ is termed as the **entropy of mixing**. If the original system is doubled, that is, $c_m = 2$, the change in entropy ΔS of the system will be given by

$$\Delta S = 2n_m R \ln 2. \tag{9.40}$$

This change in entropy ΔS given by equation (9.40) represents the **extra entropy** of the system. The origin of this extra entropy and its subsequent effects on the mixing of two same ideal gases or two different ideal gases are discussed elaborately in the next section. Gibbs paradox and its resolution by introducing the counting factor are discussed. The Sackur–Tetrode equation is also derived.

Entropy of mixing of two different ideal gases using equation (9.35)

Let us now calculate the change in entropy due to the mixing of two different, but similar ideal gases using equation (9.35). We consider a box of volume V, which is partitioned as shown in Figure 9.18. The left side of the partition contains the first ideal gas with volume V_1, number of particles N_1, and energy E_1, whereas the right side of the partition contains the second ideal gas with volume V_2, number of particles N_2, and energy E_2 at the same temperature T and

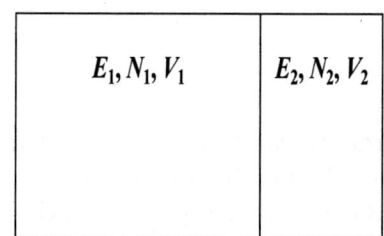

Figure 9.18 Two gases at the same temperature and pressure are mixed.

pressure P. The entropy of the two gases before mixing (when the partition is present) is just the sum of the entropy of the two gases and is given by

$$S_{\text{initial}} = N_1 K_B \ln\left[V_1 \left(\frac{E_1}{N_1}\right)^{\frac{3}{2}} \left(\frac{4\pi m}{3h^2}\right)^{\frac{3}{2}}\right] + \frac{3}{2} N_1 K_B$$

$$+ N_2 K_B \ln\left[V_2 \left(\frac{E_2}{N_2}\right)^{\frac{3}{2}} \left(\frac{4\pi m}{3h^2}\right)^{\frac{3}{2}}\right] + \frac{3}{2} N_2 K_B. \tag{9.41}$$

If the partition is removed, two gases will diffuse into each other. Under this situation, the energy, pressure, and temperature of the gases will not change, as they were already at the same temperature and pressure. The only difference is that the total volume V will now be available to both gases. So, the final entropy, after mixing, will be given by

$$S_{\text{final}} = N_1 K_B \ln\left[V \left(\frac{E_1}{N_1}\right)^{\frac{3}{2}} \left(\frac{4\pi m}{3h^2}\right)^{\frac{3}{2}}\right] + \frac{3}{2} N_1 K_B$$

$$+ N_2 K_B \ln\left[V \left(\frac{E_2}{N_2}\right)^{\frac{3}{2}} \left(\frac{4\pi m}{3h^2}\right)^{\frac{3}{2}}\right] + \frac{3}{2} N_2 K_B. \tag{9.42}$$

The change in entropy ΔS is given by

$$\Delta S = S_{\text{final}} - S_{\text{initial}} = N_1 K_B \ln\left(\frac{V}{V_1}\right) + N_2 K_B \ln\left(\frac{V}{V_2}\right). \qquad (9.43)$$

Equation (9.43) gives the increase in entropy ΔS when two different gases of identical nature are mixed with each other at the same temperature T and pressure P. This change in entropy ΔS is called the **entropy of mixing**.

Entropy of mixing of two same ideal gases using equation (9.35): Gibbs paradox

Let us now calculate the change in entropy due to the mixing of two same, but similar ideal gases using equation (9.35). We consider a box of volume V, which is partitioned as shown in Figure 9.18. The left side of the partition contains some part of the ideal gas with volume V_1, number of particles N_1, and energy E_1, whereas the right side of the partition contains the rest part of the ideal gas with volume V_2, number of particles N_2, and energy E_2 at the same temperature T and pressure P. The entropy of the two parts of the gas before partition is removed is just the sum of the entropy of the two parts and is given by

$$S_{\text{initial}} = N_1 K_B \ln\left[V_1 \left(\frac{E_1}{N_1}\right)^{\frac{3}{2}} \left(\frac{4\pi m}{3h^2}\right)^{\frac{3}{2}}\right] + \frac{3}{2} N_1 K_B$$

$$+ N_2 K_B \ln\left[V_2 \left(\frac{E_2}{N_2}\right)^{\frac{3}{2}} \left(\frac{4\pi m}{3h^2}\right)^{\frac{3}{2}}\right] + \frac{3}{2} N_2 K_B. \qquad (9.44)$$

If the partition is removed, two parts of the gas will diffuse into each other. Under this situation, volume per molecule $\frac{V_1}{N_1}$, $\frac{V_2}{N_2}$ and energy per molecule $\frac{E_1}{N_1}$ and $\frac{E_2}{N_2}$ will remain the same as they are at the same temperature T and pressure P. The only difference is that the total volume V is now available to both parts of the same gas. So, the final entropy, after removing the partition, will be given by

$$S_{\text{final}} = N_1 K_B \ln\left[V \left(\frac{E_1}{N_1}\right)^{\frac{3}{2}} \left(\frac{4\pi m}{3h^2}\right)^{\frac{3}{2}}\right] + \frac{3}{2} N_1 K_B$$

$$+ N_2 K_B \ln\left[V \left(\frac{E_2}{N_2}\right)^{\frac{3}{2}} \left(\frac{4\pi m}{3h^2}\right)^{\frac{3}{2}}\right] + \frac{3}{2} N_2 K_B. \qquad (9.45)$$

The change in entropy ΔS is given by

$$\Delta S = S_{\text{final}} - S_{\text{initial}} = N_1 K_B \ln\left(\frac{V}{V_1}\right) + N_2 K_B \ln\left(\frac{V}{V_2}\right). \qquad (9.46)$$

Equation (9.46) gives the increase in entropy ΔS when two parts of the same gas are allowed to mix with each other at the same temperature T and pressure P.

If the partition is placed in the middle of the box, then $V_1 = \frac{V}{2}$, $V_2 = \frac{V}{2}$, and $N_1 = N_2 = \frac{N}{2}$. According to equation (9.46), the change in entropy ΔS will be given by

$$\Delta S = N_1 K_B \ln \left(\frac{V}{\frac{V}{2}} \right) + N_2 K_B \ln \left(\frac{V}{\frac{V}{2}} \right) = N K_B \ln 2. \qquad (9.47)$$

This is in agreement with equation (9.40) for the prediction of entropy of mixing for the same ideal gases.

But we know that for the case of the same gas in the two parts, removing the partition is a reversible process because one can reinsert the partition later, and one will not be able to make out if the partition was removed before. Hence, the change in entropy on removing the partition should be zero, that is, $\Delta S = 0$. But, equation (9.46) yields a nonzero entropy of mixing for identical gases, that is, $\Delta S = finite$. This implies that the prediction of equation (9.35) regarding the mixing of two same ideal gases in not tenable to the physical situation. This is known as **Gibbs paradox**.

The origin of entropy of mixing

The entropy of mixing in thermodynamics is the increase in total entropy when several initially separated systems of different compositions are mixed without chemical reaction by the thermodynamic operation of removal of impermeable partition(s) between them. Before mixing, each of these systems is in a thermodynamic state of internal equilibrium and possesses some entropy. After mixing, the systems attain a new thermodynamic state of internal equilibrium and possess new entropy. The difference between these two entropy is defined as **the entropy of mixing**.

In general, the mixing of different systems may be constrained to occur under various prescribed conditions. The systems are each initially at a common temperature and pressure. The new system may change its volume, but the temperature, pressure, and masses of the chemical components should be maintained at the same constant level. In the process of mixing, the volume available for each system is increased from that of its initially separate compartment to the total common final volume. It is to be noted further that the final volume need not be the sum of the initially separate volumes, as work may be done on or by the new closed system during the process of mixing. At the same time, heat can be transferred to or from the surroundings, because the variables pressure and temperature remain constant during the process of mixing. The internal energy of the new closed system is equal to the sum of the internal energies of the initially separate systems. The reference values for the internal energies should be specified in a way that is constrained to make this so, maintaining also that the internal energies are, respectively, proportional to the masses of the systems.

In the general case of mixing nonideal systems, however, the total final common volume may be different from the sum of the separate initial volumes, and there may be a transfer of work or heat to or from the surroundings. There may be a departure of the entropy of mixing from that of the corresponding ideal case. That departure is the main reason for interest in the entropy of mixing. These energy and entropy variables and their temperature dependence provide valuable information about the properties of the systems. On a molecular level, the entropy of mixing is of interest because it is a macroscopic variable that provides information about constitutive molecular properties. In ideal systems, intermolecular forces are the same between every pair of molecules, so that a molecule feels no difference between other molecules of its own kind and of those of the other kind. In nonideal systems, there may be differences in intermolecular forces or specific molecular effects between different species, even though they are chemically non-reacting. The entropy of mixing provides information about constitutive differences of intermolecular forces or specific molecular effects in the systems. The statistical concept of randomness is used for the statistical mechanical explanation of the entropy of mixing. The mixing of ideal systems is regarded as random at a molecular level, and, correspondingly, the mixing of nonideal systems may be nonrandom.

Entropy of mixing for distinguishable particles

Consider that in the mixing problem, the second system is identical to the first system in all respects except that its molecules are in some way slightly different from the molecules in the first system so that the two sets of molecules are distinguishable. In this case, we would certainly expect an overall increase in entropy when the partition is removed. Before the partition is removed, it separates type 1 molecules from type 2 molecules. After the partition is removed, molecules of both types become jumbled together. This is clearly an irreversible process. We cannot imagine the molecules spontaneously sorting themselves out again. Thus, disorder in the system increases, leading to an increase in entropy. This increase in entropy associated with the jumbling of molecules is called the **entropy of mixing**. This entropy of mixing can be easily calculated in the following way:

We know that the number of accessible states of an ideal gas varies with volume as $\Omega \propto V^N$, that is, $\Omega = AV^N$, where A is a constant. The volume accessible to type 1 molecules clearly doubles after the partition is removed, as does the volume accessible to type 2 molecules. Before removing the partition, the entropy of two systems were, respectively,

$$S_1 = K_B \ln \left\{ A\left(V_1\right)^{N_1} \right\} \qquad \text{and} \qquad S_2 = K_B \ln \left\{ A\left(V_2\right)^{N_2} \right\}. \qquad (9.48)$$

The fundamental relation $S = K_B \ln \Omega$ is used in this case. As the two systems are identical, we have $N_1 = N_2 = N$ and $V_1 = V_2 = V$. This leads to the expression for the total entropy of these two systems before removing the partition as

$$S_{\text{initial}} = S_1 + S_2 = K_B \ln \left\{ A \left(V_1 \right)^{N_1} \right\} + K_B \ln \left\{ A \left(V_2 \right)^{N_2} \right\} = 2K_B \ln \left(AV^N \right). \tag{9.49}$$

Let us now suddenly remove the partition. Each system thereby will occupy $2V$ volume with N number of particles. So, the expression for final entropy S_{final} will be given by

$$S_{\text{final}} = K_B \ln \left\{ A \left(2V \right)^N \right\} + K_B \ln \left\{ A \left(2V \right)^N \right\} = 2K_B \ln \left[A(2V)^N \right]. \tag{9.50}$$

Hence, the increase in entropy due to mixing is

$$\Delta S = S_{\text{final}} - S_{\text{initial}} = 2K_B \ln \left[A(2V)^N \right] - 2K_B \ln \left[AV^N \right].$$

This leads to

$$\Delta S = 2K_B \left[\ln \frac{A(2V)^N}{AV^N} \right] = 2K_B \ln 2^N = 2NK_B \ln 2 = 2n_m R \ln 2. \tag{9.51}$$

The result given by equation (9.51) is identical to that given by equation (9.40). It is clear that this additional entropy $2n_m R \ln 2$ appears when we double the size of an ideal gas system by joining together two identical systems. This is the entropy of the **mixing of the molecules contained in the original systems.**

Removal of Gibbs paradox in a micro-canonical ensemble using Boltzmann counting factor

Gibbs empirically realized that while counting the number of microstates Ω if one divides the number of microstates Ω by $N!$, the paradox disappears. This suggests that the earlier method of counting of microstates must be thus wrong. The way the microstates have been counted, interchanging two particles gives a new microstate. However, from quantum mechanics, we know that elementary particles and atoms should be treated as identical particles. So, in the earlier method of counting the microstates, an overcounting was made by assuming the particles to distinguishable. This argument led to divide the number of microstates Ω by $N!$ to get the correct the number of microstates Ω_{corr}. Hence, we get

$$\Omega_{\text{corr}} = \frac{\Omega}{N!}. \tag{9.52}$$

Using the value of Ω from equation (9.34) into equation (9.52), we get

$$\Omega_{\text{corr}} = \frac{1}{N!} \frac{V^N}{\left(\frac{3N}{2} \right)!} \left[\frac{2\pi m E}{h^2} \right]^{\frac{3N}{2}}. \tag{9.53}$$

Using this equation (9.53) and Boltzmann entropy relation: $S = K_B \ln \Omega_{\text{corr}}$, the entropy of the system can be written as

$$S = K_B \ln \Omega_{\text{corr}} = K_B \ln \left[\frac{1}{N! \left(\frac{3N}{2}\right)!} \frac{V^N}{} \left(\frac{2\pi m E}{h^2}\right)^{\frac{3N}{2}} \right],$$

$$= K_B \ln \left[V^N \left(\frac{2\pi m E}{h^2}\right)^{\frac{3N}{2}} \right] - K_B \ln \left[\left(\frac{3N}{2}\right)! \right] - - K_B \ln N!.$$

Using Stirling formula, we get

$$S = K_B \ln \left[V^N \left(\frac{2\pi m E}{h^2}\right)^{\frac{3N}{2}} \right] - K_B \left[\frac{3N}{2} \ln \left(\frac{3N}{2}\right) - \frac{3N}{2} \right] - K_B \left[N \ln N - N \right],$$

$$= N K_B \ln \left[V \left(\frac{2\pi m E}{h^2}\right)^{\frac{3}{2}} \right] - \frac{3N K_B}{2} \ln \left(\frac{3N}{2}\right) + \frac{3N K_B}{2} - N K_B \ln N + N K_B,$$

$$= N K_B \ln \left[\frac{V}{N} \left(\frac{2\pi m E}{h^2}\right)^{\frac{3}{2}} \right] - N K_B \ln \left(\frac{3N}{2}\right)^{\frac{3}{2}} + \frac{5}{2} N K_B$$

$$= N K_B \ln \left[\frac{V}{N} \left(\frac{4\pi m E}{3N h^2}\right)^{\frac{3}{2}} \right] + \frac{5}{2} N K_B.$$

This leads to

$$S = N K_B \ln \left[\frac{V}{N} \left(\frac{E}{N}\right)^{\frac{3}{2}} \left(\frac{4\pi m}{3h^2}\right)^{\frac{3}{2}} \right] + \frac{5}{2} N K_B. \tag{9.54}$$

This equation (9.54) is called the Sackur—Tetrode equation and describes the entropy of an (monoatomic) ideal gas.[1] It is named after Hugo Martin Tetrode[2] and Otto Sackur,[3] who developed it independently as a solution of Boltzmann gas statistics and entropy equations at about the same time in 1912.

The entropy of mixing of two different ideal gases using Boltzmann counting factor

Let us now calculate the change in entropy due to the mixing of two different but similar ideal gases using equation (9.54). We consider a box of volume V, which is partitioned as shown in the figure. The left side of the partition contains the first ideal gas with volume V_1, number of particles N_1, and energy E_1, whereas the right side of the partition contains the second ideal gas with volume V_2, number of particles

---

[1]O. Sackur, The Application of the Kinetic Theory of Gases to Chemical Problems, *Annalen der Physik* 36 (1911): 958–980.

[2]H. M. Tetrode, The Chemical Constant of Gases and the Elementary Quantum of Action, *Annalen der Physik* 39 (1912): 255–256.

[3]O. Sackur, The Significance of the Elementary Quantum of Action to Gas Theory and the Calculation of the Chemical Constant, *Halle an der Saale*, (1912): 405–423.

N_2, and energy E_2 at the same temperature T and pressure P. The entropy of two gases before mixing is given by

$$S_{\text{initial}} = N_1 K_B \ln \left[\frac{V_1}{N_1} \left(\frac{E_1}{N_1} \right)^{\frac{3}{2}} \left(\frac{4\pi m}{3h^2} \right)^{\frac{3}{2}} \right] + \frac{5}{2} N_1 K_B$$

$$+ N_2 K_B \ln \left[\frac{V_2}{N_2} \left(\frac{E_2}{N_2} \right)^{\frac{3}{2}} \left(\frac{4\pi m}{3h^2} \right)^{\frac{3}{2}} \right] + \frac{5}{2} N_2 K_B. \qquad (9.55)$$

If the partition is removed, two gases will diffuse into each other. Under this situation, the energy, pressure, and temperature of the gases will not change, as they were already at the same temperature and pressure. The only difference is that the total volume V will now be available to both gases. So, the final entropy, after mixing, will be given by

$$S_{\text{final}} = N_1 K_B \ln \left[\frac{V}{N_1} \left(\frac{E_1}{N_1} \right)^{\frac{3}{2}} \left(\frac{4\pi m}{3h^2} \right)^{\frac{3}{2}} \right] + \frac{5}{2} N_1 K_B$$

$$+ N_2 K_B \ln \left[\frac{V}{N_2} \left(\frac{E_2}{N_2} \right)^{\frac{3}{2}} \left(\frac{4\pi m}{3h^2} \right)^{\frac{3}{2}} \right] + \frac{5}{2} N_2 K_B. \qquad (9.56)$$

The change in entropy ΔS is given by

$$\Delta S = S_{\text{final}} - S_{\text{initial}} = N_1 K_B \ln \left(\frac{V}{V_1} \right) + N_2 K_B \ln \left(\frac{V}{V_2} \right). \qquad (9.57)$$

Clearly, after mixing the two gases, the increase in entropy ΔS given by this equation (9.57) is the same as that given by equation (9.46). **This ΔS is called the entropy of mixing**.

Removal of Gibbs paradox in mixing of two same ideal gases using Boltzmann counting factor

If the two ideal gases are the same, we have $\frac{V_1}{N_1}$, $\frac{V_2}{N_2}$ and energy per particle $\frac{E_1}{N_1}$ and $\frac{E_2}{N_2}$ are also the same as they are at the same temperature T and pressure P. So, the above equation (9.55) reduces to

$$S_{\text{initial}} = N_1 K_B \ln \left[\frac{V}{N} \left(\frac{E}{N} \right)^{\frac{3}{2}} \left(\frac{4\pi m}{3h^2} \right)^{\frac{3}{2}} \right] + \frac{5}{2} N_1 K_B$$

$$+ N_2 K_B \ln \left[\frac{V}{N} \left(\frac{E}{N} \right)^{\frac{3}{2}} \left(\frac{4\pi m}{3h^2} \right)^{\frac{3}{2}} \right] + \frac{5}{2} N_2 K_B.$$

$$= (N_1 + N_2) K_B \ln \left[\frac{V}{N} \left(\frac{E}{N} \right)^{\frac{3}{2}} \left(\frac{4\pi m}{3h^2} \right)^{\frac{3}{2}} \right] + \frac{5}{2} (N_1 + N_2) K_B.$$

This leads to

$$S_{\text{initial}} = NK_B \ln \left[\frac{V}{N} \left(\frac{E}{N} \right)^{\frac{3}{2}} \left(\frac{4\pi m}{3h^2} \right)^{\frac{3}{2}} \right] + \frac{5}{2} NK_B, \qquad (9.58)$$

where $N = N_1 + N_2$. This shows that equation (9.58) is identical to equation (9.54), which is the equation for the entropy of the gas of N particles in a volume V at a temperature T.

If the partition is removed in the case of mixing of two same ideal gases, we will have a system of N particles in a volume V at a temperature T, and the final entropy will be given by equation (9.58). Thus, there will be no change in entropy in this case, that is, $\Delta S = 0$. The Gibbs paradox for the mixing of two same ideal gases is resolved.

9.9   Change in entropy in terms of various thermodynamic parameters

But for the sake of the introduction of the second law of thermodynamics, a few comments about entropy are presented here. Working substances play a major role in thermodynamics. In order to have the complete description of the condition of a working substance, Clausius introduced the idea of entropy in addition to volume, pressure, temperature, and internal energy. This quantity entropy remains constant in an adiabatic process and can be defined as the thermal property of a working substance that remains constant during an adiabatic process.

The change in entropy between two adiabatic is mathematically expressed as $dS = \frac{\delta Q}{T}$. The expression $\int_A^B \frac{\delta Q}{T} = \int_{S_1}^{S_2} dS$ is a function of the thermodynamic coordinates of a system. This function is called entropy. It is an exact differential. Entropy is an extensive property since it depends on the mass of the working substance. We see that heat has the same dimension as the product of entropy and absolute temperature. We can have the following analogy: since the gravitational potential energy of a body is proportional to the product of its mass and height above some zero level, likewise, we may take temperature analogous to height and entropy as analogous to mass or inertia. Thus, entropy can be taken as *thermal inertia* to the heat motion, a relation similar to that in which mass presents inertia to linear motion or moment of inertia presents to rotational motion. The following important features are to be noted regarding entropy:

1. Entropy is a measure of disorder in a thermodynamic system.

2. In an adiabatic process, there is no change in entropy.

3. In a cycle of reversible process, the total change in entropy is always zero.

4. In an irreversible process, there is always a net increase in entropy for the whole system.

5. As all natural processes in the universe are irreversible, the entropy of the universe increases.

6. The unavailable energy in the universe increases due to an increase in entropy.

7. Various diagrams such as P–V, T–S, U–S, and H–S can be used to calculate the efficiency of a heat engine.

9.9.1 T–dS Equations

The three T–dS equations seem to be the "tedious equations" to the generations of students. But these equations are not at all tedious to a true lover of thermodynamics. Among other things, these equations are frequently used to calculate the change in entropy during various reversible processes in terms of either dV and dT, or dP and dT, or dV and dP, and sometimes even in terms of directly measurable quantities such as the isobaric volume expansion coefficient α_P and the isothermal bulk modulus κ_T. We can express the change in entropy in terms of any two of the thermodynamic variables pressure P, volume V, and temperature T, and these subsequently lead to three T–dS equations. These equations are derived below.

First T–dS equation

The entropy S of a pure substance can be considered as a function of temperature T and volume V so that we can write

$$dS = \left(\frac{\partial S}{\partial T}\right)_V dT + \left(\frac{\partial S}{\partial V}\right)_T dV.$$

Multiplying both sides by T, we get

$$TdS = T\left(\frac{\partial S}{\partial T}\right)_V dT + T\left(\frac{\partial S}{\partial V}\right)_T dV.$$

Using the definition of specific heat at constant volume $C_V = T\left(\frac{\partial S}{\partial T}\right)_V$ and one of Maxwell's thermodynamic relation $\left(\frac{\partial S}{\partial V}\right)_T = \left(\frac{\partial P}{\partial T}\right)_V$, we get

$$TdS = C_V dT + T\left(\frac{\partial P}{\partial T}\right)_V dV.$$

This is the first T–dS equation. The change in entropy can be calculated if the changes in temperature and volume are available in a thermodynamic process.

Second $T-dS$ equation

The entropy S of a pure substance can be taken as a function of temperature T and pressure P so that we can write

$$dS = \left(\frac{\partial S}{\partial T}\right)_P dT + \left(\frac{\partial S}{\partial P}\right)_T dP.$$

Multiplying both sides by T, we get

$$TdS = T\left(\frac{\partial S}{\partial T}\right)_P dT + T\left(\frac{\partial S}{\partial P}\right)_T dP.$$

Using the definition of specific heat at constant pressure $C_P = T\left(\frac{\partial S}{\partial T}\right)_P$ and one of Maxwell's thermodynamic relation $\left(\frac{\partial S}{\partial P}\right)_T = -\left(\frac{\partial V}{\partial T}\right)_P$, we get

$$TdS = C_P dT - T\left(\frac{\partial V}{\partial T}\right)_P dP.$$

This is the second $T-dS$ equation. The change in entropy can be calculated if the changes in temperature and pressure are available in a thermodynamic process.

Third $T-dS$ equation

In order to derive the third $T-dS$ equation, we consider the entropy S as a function of pressure P and volume V. Thus, considering P and V as independent variables, we get the third $T-dS$ equation as

$$TdS = C_P \left(\frac{\partial T}{\partial V}\right)_P dV + C_V \left(\frac{\partial T}{\partial P}\right)_V dP.$$

This equation can be derived in the following way: The energy difference between two neighboring equilibrium states in which the pressure and volume differ by dP and dV is given by

$$dU = \left(\frac{\partial U}{\partial P}\right)_V dP + \left(\frac{\partial P}{\partial V}\right)_P dV. \tag{9.59}$$

In equation (9.59), the partial derivatives do not involve any properties other than those introduced in considering T and V as independent variables, that is,

$$dU = \left(\frac{\partial U}{\partial T}\right)_V dT + \left(\frac{\partial U}{\partial V}\right)_T dV. \tag{9.60}$$

Again, the differential change in temperature dT can be written as

$$dT = \left(\frac{\partial T}{\partial P}\right)_V dP + \left(\frac{\partial T}{\partial V}\right)_P dV. \tag{9.61}$$

Using equation (9.61) into equation (9.60), we get

$$dU = \left(\frac{\partial U}{\partial T}\right)_V \left[\left(\frac{\partial T}{\partial P}\right)_V dP + \left(\frac{\partial T}{\partial V}\right)_P dV\right] + \left(\frac{\partial U}{\partial V}\right)_T dV, \tag{9.62}$$

$$= \left(\frac{\partial U}{\partial T}\right)_V \left(\frac{\partial T}{\partial P}\right)_V dP + \left(\frac{\partial U}{\partial T}\right)_V \left(\frac{\partial T}{\partial V}\right)_P dV + \left(\frac{\partial U}{\partial V}\right)_T dV, \tag{9.63}$$

$$= \left(\frac{\partial U}{\partial T}\right)_V \left(\frac{\partial T}{\partial P}\right)_V dP + \left[\left(\frac{\partial U}{\partial T}\right)_V \left(\frac{\partial T}{\partial V}\right)_P + \left(\frac{\partial U}{\partial V}\right)_T\right] dV. \tag{9.64}$$

Comparing equation (9.59) with equation (9.64), we get

$$\left(\frac{\partial U}{\partial P}\right)_V dP = \left(\frac{\partial U}{\partial T}\right)_V \left(\frac{\partial T}{\partial P}\right)_V dP \Rightarrow \left(\frac{\partial U}{\partial P}\right)_V = \left(\frac{\partial U}{\partial T}\right)_V \left(\frac{\partial T}{\partial P}\right)_V. \tag{9.65}$$

Further, we have

$$\left(\frac{\partial U}{\partial V}\right)_P = \left(\frac{\partial U}{\partial T}\right)_V \left(\frac{\partial T}{\partial V}\right)_P + \left(\frac{\partial U}{\partial V}\right)_T. \tag{9.66}$$

Recall that

$$C_V = \left(\frac{\partial U}{\partial T}\right). \tag{9.67}$$

Putting equation (9.67) into equation (9.65), we have

$$\left(\frac{\partial U}{\partial P}\right)_V = C_V \left(\frac{\partial T}{\partial P}\right)_V. \tag{9.68}$$

Further, we know the relation that

$$H = U + PV, \tag{9.69}$$

which provides

$$dH = dU + PdV + VdP. \tag{9.70}$$

At constant volume and constant pressure, we have $dV = 0$ and $dP = 0$. Hence, equation (9.70) can be written as

$$dH = dU. \tag{9.71}$$

Putting equation (9.71) into equation (9.66), we have

$$\left(\frac{\partial H}{\partial V}\right)_P = \left(\frac{\partial H}{\partial T}\right)_V \left(\frac{\partial T}{\partial V}\right)_P + \left(\frac{\partial H}{\partial V}\right)_T. \tag{9.72}$$

At constant temperature $\left(\frac{\partial H}{\partial V}\right)_T = 0$. Hence, equation (9.72) becomes

$$\left(\frac{\partial H}{\partial V}\right)_P = \left(\frac{\partial H}{\partial T}\right)_V \left(\frac{\partial T}{\partial V}\right)_P. \tag{9.73}$$

Let us recall that

$$C_P = \left(\frac{\partial H}{\partial T}\right)_V. \tag{9.74}$$

Using equation (9.74) into equation (9.73), we have

$$\left(\frac{\partial H}{\partial V}\right)_P = C_P \left(\frac{\partial T}{\partial V}\right)_V. \tag{9.75}$$

From equation (9.71), substituting dH for dU in equation (9.75) we have

$$\left(\frac{\partial U}{\partial V}\right)_P = C_P \left(\frac{\partial T}{\partial V}\right)_P. \tag{9.76}$$

Hence, putting equations (9.68) and (9.76) into equation (9.59), we have

$$dU = C_V \left(\frac{\partial T}{\partial P}\right)_V dP + C_P \left(\frac{\partial T}{\partial V}\right)_P dV. \tag{9.77}$$

From the combined first and second law of thermodynamics, we have

$$dS = \frac{1}{T} \left(dU + PdV\right). \tag{9.78}$$

At constant volume process $dV = 0$. Hence, equation (9.78) becomes

$$dS = \frac{1}{T} dU. \tag{9.79}$$

Putting equation (9.77) into equation (9.79), we have

$$dS = \frac{1}{T} \left[C_V \left(\frac{\partial T}{\partial P}\right)_V dP + C_P \left(\frac{\partial T}{\partial V}\right)_P dV \right], \tag{9.80}$$

$$dS = \frac{C_V}{T} \left(\frac{\partial T}{\partial P}\right)_V dP + \frac{C_P}{T} \left(\frac{\partial T}{\partial V}\right)_P dV. \tag{9.81}$$

But we can also write

$$dS = \left(\frac{\partial S}{\partial P}\right)_V dP + \left(\frac{\partial S}{\partial V}\right)_P dV. \tag{9.82}$$

Comparing equation (9.81) with equation (9.82), we get

$$\left(\frac{\partial S}{\partial P}\right)_V = \frac{C_V}{T}\left(\frac{\partial T}{\partial P}\right)_V, \tag{9.83}$$

$$\left(\frac{\partial S}{\partial V}\right)_P = \frac{C_P}{T}\left(\frac{\partial T}{\partial V}\right)_P. \tag{9.84}$$

Differentiating equation (9.83) partially w.r.t V, we have

$$\frac{\partial^2 S}{\partial V \partial P} = \frac{C_V}{T}\frac{\partial}{\partial V}\left(\frac{\partial T}{\partial P}\right)_V = \frac{C_V}{T}\frac{\partial^2 T}{\partial V \partial P}. \tag{9.85}$$

Differentiating equation (9.84) w.r.t P we have

$$\frac{\partial}{\partial P}\left(\frac{\partial S}{\partial V}\right)_P = \frac{\partial^2 S}{\partial P \partial V} = \frac{\partial}{\partial P}\left(\frac{C_P}{T}\left(\frac{\partial T}{\partial V}\right)_P\right) = \frac{C_P}{T}\frac{\partial^2 T}{\partial P \partial V}. \tag{9.86}$$

Equating the R.H.S. of equations (9.85) and (9.86), we have

$$\frac{C_V}{T}\frac{\partial^2 T}{\partial V \partial P} = \frac{C_P}{T}\frac{\partial^2 T}{\partial P \partial V}. \tag{9.87}$$

From equation (9.87), we get

$$\frac{C_V}{T} = \frac{C_P}{T}; \quad \Rightarrow \quad C_V = C_P. \tag{9.88}$$

From equation (9.81), we have

$$dS = \frac{C_V}{T}\left(\frac{\partial T}{\partial P}\right)_V dP + \frac{C_P}{T}\left(\frac{\partial T}{\partial V}\right)_P dV. \tag{9.89}$$

Multiplying equation (9.89) by T, we have

$$TdS = C_V\left(\frac{\partial T}{\partial P}\right)_V dP + C_P\left(\frac{\partial T}{\partial V}\right)_P dV, \tag{9.90}$$

$$TdS = C_P\left(\frac{\partial T}{\partial V}\right)_P dV + C_V\left(\frac{\partial T}{\partial P}\right)_V dP. \tag{9.91}$$

This equation (9.91) is known as the third T–dS equation in thermodynamics.

9.9.2   Clausius theorem

Clausius established an important theorem in thermodynamics in 1855. This is regarding the change in entropy between a thermodynamic system (heat engine or heat pump) and the surroundings in a thermodynamic process (reversible or irreversible). This theorem states that **for a system exchanging heat with the** surroundings (external reservoirs) and undergoing a thermodynamic cyclic process, the following inequality holds.

$$-\oint dS_{\text{sur}} = \oint \frac{\delta Q}{T} \leq 0, \tag{9.92}$$

where $\oint dS_{\text{sur}}$ is the total change in entropy of the surroundings, δQ is the infinitesimal amount of heat absorbed by the system from the surroundings ($\delta Q < 0$ if heat is rejected from the system to the surroundings and $\delta Q > 0$ if heat is absorbed by the system from the surroundings) and T is the temperature of the surroundings at a particular instant in time. The equality sign holds for a reversible thermodynamic process.

The integral in equation (9.92) represents the change in entropy of a heat engine in a complete cycle. It would be confusing to consider this integral as entropy integral because we know that for any irreversible engine cycle, the total entropy always increases. But the change in entropy given by equation (9.92) describes what happens to the engine itself and provides that the entropy on the output side, given to the surroundings, by the engine in an irreversible cycle is always greater than the entropy added to the engine. Thus, equation (9.92) is consistent with the general understanding of an increase in entropy for an irreversible process (also consistent with the statement of the second law of thermodynamics).

The equality sign in equation (9.92) applies only to the ideal thermodynamic cycle (Carnot cycle). Since the integral represents the net change in entropy in one complete cycle, it gives a zero entropy change to the most efficient engine cycle and clearly supports the statement that **entropy does not decrease at all in any thermodynamic process whether it is reversible or irreversible.**

Concept of entropy from Clausius theorem

The concept of entropy can be demonstrated from the Clausius theorem in the following way: Let us consider a reversible thermodynamic process shown in Figure 9.19. In this process, a thermodynamic system proceeds from state 1 to state 2 along path A and completes the cycle along another reversible path B. The following relation follows from the knowledge of the Carnot cycle:

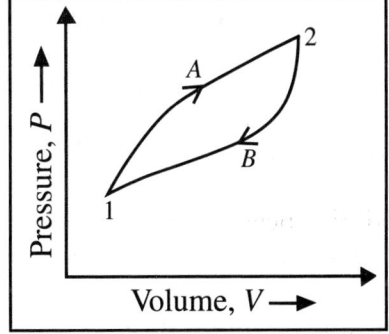

Figure 9.19 A reversible thermodynamic process to verify the statement of Clausius theorem.

$$\frac{Q_1}{T_1} - \frac{Q_2}{T_2} = 0. \tag{9.93}$$

For the cyclic path $1A2B1$, the above equation (9.93) can be written as

$$\int_{1A2B1} \frac{\delta Q_R}{T} = 0 = \int_{1}^{2} \frac{\delta Q_R}{T} + \int_{2}^{1} \frac{\delta Q_R}{T}. \tag{9.94}$$

Thus, it is observed from equation (9.94) that the value of the integral $\int_{1}^{2} \frac{\delta Q_R}{T}$ is the same between state 1 and state 2, that is, it is independent of the path followed by the system. It should be noted that for any coordinate system, we can always move between two points (states) by an infinite number of infinitesimally small Carnot cycles, and according to equation (9.94), the same value of the integral, independent of these paths, will be achieved. Mathematically, the factor $\frac{\delta Q_R}{T}$ represents a state function. We denote this function as dS, and it is called the **entropy**. It is a function of state.

Thus, for a reversible process, the relation $\int_{1}^{2} \frac{\delta Q_R}{T}$ given by equation (9.94) indicates that the ratio of heat to temperature has characteristics of a property that does not change in a cycle, but it is associated with heat transfer (a path function). In a paper published in 1865, *On various forms of the laws of thermodynamics that are convenient for applications*, Clausius named the ratio $\frac{\text{heat}}{\text{temperature}}$ as entropy. This is a Greek word used to mean "transformation" and is represented by the letter **"S"**. Clausius, in another paper published in 1865, made a bold statement regarding the broadest possible application of the laws of thermodynamics:

1. The energy of the universe is constant (first law).

2. The entropy of the universe tends toward a maximum (second law).

The statement of the second law of thermodynamics and the discovery of entropy as the ratio of two macroscopic components, heat and absolute temperature, are truly two remarkable contributions of Clausius in thermodynamics. It should be pointed out that the true nature of entropy was only discovered much later when physicists understood the behavior of individual atoms and molecules, that is, when the knowledge of the microscopic world started unfolding itself to the physicists.

9.9.3   Clausius inequality

We consider an irreversible cyclic engine working between the temperatures T_1 and T_2 (assume $T_1 > T_2$). If a reversible engine operates between the same temperature range, we know from Carnot theorem that the efficiency of the reversible engine (η_{rev}) will always be greater than the efficiency of the irreversible engine (η_{irr}).

Mathematically,

$$\eta_{rev} > \eta_{irr} \quad \Rightarrow \quad \frac{Q_1^{rev} - Q_2^{rev}}{Q_1^{rev}} > \frac{Q_1^{irr} - Q_2^{irr}}{Q_1^{irr}};$$

$$\Rightarrow \quad 1 - \frac{Q_2^{rev}}{Q_1^{rev}} > 1 - \frac{Q_2^{irr}}{Q_1^{irr}}; \quad \Rightarrow \quad 1 - \frac{T_2}{T_1} > 1 - \frac{Q_2^{irr}}{Q_1^{irr}};$$

$$\Rightarrow \quad \frac{Q_2^{irr}}{T_2} > \frac{Q_1^{irr}}{T_1}.$$

Thus, for an irreversible cyclic engine operating between the temperatures T_1 and T_2, we get

$$\Rightarrow \quad \frac{Q_1^{irr}}{T_1} - \frac{Q_2^{irr}}{T_2} < 0 \quad \Rightarrow \quad \oint \frac{\delta Q}{T} < 0.$$

This relation is known as **Clausius inequality**.

9.9.4   Second law of thermodynamics in terms of entropy

For an infinitesimal quasi-static process, the second law of thermodynamics can be expressed mathematically as

$$\delta Q = T \, dS, \tag{9.95}$$

where δQ is the change in associated heat in the process, T is the temperature, and dS is the change in entropy of the system. Furthermore, the work done by a system on the environment is given by

$$\delta W = P \, dV, \tag{9.96}$$

where P is the pressure, and dV is the change in the external parameter volume of the system. Combining the first and the second laws of thermodynamics, we have the following fundamental thermodynamic relation

$$TdS = dU + PdV. \tag{9.97}$$

This equation (9.97) shows a connection between the mathematical formulation of the second law of thermodynamics and the change in entropy in a thermodynamic process.

Using this idea of change in entropy, the second law of thermodynamics can be stated in two different ways:

Kelvin Statement: *It is impossible to construct an engine which, working on a cycle, shall produce no other effect than the extraction of heat from a reservoir and the performance of an equal amount of work.*

Clausius Statement: *It is impossible to construct an engine which, working on a cycle, shall produce no other effect than a transfer of heat from a hotter to a colder body.*

These two statements of the second law of thermodynamics are actually equivalent and state the same thing.

9.9.5 Principle of increase in entropy

Consider the process in which a quantity of heat δQ flows from the surroundings at temperature T_{surr} to the system at temperature T_{sys}. Let δW be the work done by the system during the process.

From the inequality of Clausius, we can write, $\Delta S_{\text{sys}} \geq \frac{\delta Q}{T_{\text{sys}}}$. For the surroundings, δQ is negative, and we can write, $\Delta S_{\text{surr}} = -\frac{\delta Q}{T_{\text{surr}}}$.

The net change in entropy ΔS_H of the universe is, therefore, given by

$$\Delta S_H = \Delta S_{\text{sys}} + \Delta S_{\text{surr}} \geq \frac{\delta Q}{T_{\text{sys}}} - \frac{\delta Q}{T_{\text{surr}}} = \delta Q \left(\frac{1}{T_{\text{sys}}} - \frac{1}{T_{\text{surr}}} \right).$$

Since $T_{\text{surr}} \geq T_{\text{sys}}$, the quantity within the brackets is positive, and we conclude that

$$\Delta S_H \geq 0.$$

If $T_{\text{sys}} \geq T_{\text{surr}}$, the transfer of heat occurs from the system to the surroundings, and both δQ and the quantity $(\frac{1}{T_{\text{sys}}} - \frac{1}{T_{\text{surr}}})$ will be negative, and the result will be the same.

Therefore, we may conclude that for all processes that a system in a given surroundings can undergo, the increase in entropy is

$$\Delta S_{\text{sys}} + \Delta S_{\text{surr}} = \Delta S_H \geq 0.$$

The equality and greater than signs hold, respectively, for reversible and irreversible processes. Since the natural processes are irreversible in character, the above expression indicates that the entropy of the universe is increasing. This result is known as **the principle of increase of entropy**. The significance of this expression is that it guides the unidirectional evolution of a thermodynamic process. This also helps us to make a quantitative general statement of the second law of thermodynamics in the following way: **The entropy of the universe can never decrease.** This principle of increase of entropy can be successfully applied to many natural phenomena, such as, the cooling of a cup of coffee, warming up the brakes of a car during stop, and all other unidirectional processes. This principle is valid for open, closed, and isolated system.

9.9.6   Entropy change in a reversible process with examples

When a thermodynamic system undergoes a thermodynamic process, the entropy of the system changes. This change in entropy helps to understand the physical meaning of entropy and its significance. In a particular thermodynamic process, one has to consider the change in entropy of the system, the change in entropy of the local surroundings, and the change in entropy of the universe. It should be noted that the change in entropy of the universe in a process is equal to the sum of the changes in entropy of the system and the local surroundings in that process.

In order to calculate the change of entropy in a reversible process, we consider a complete reversible Carnot cycle, as shown in Figure 9.20. In this cycle, an ideal gas is used as a working substance. The cycle consists of two isothermal, that is, AB at a temperature T_1 in which some amount of heat is absorbed by the working substance from the high-temperature source, and CD at temperature T_2, in which some amount of heat is rejected

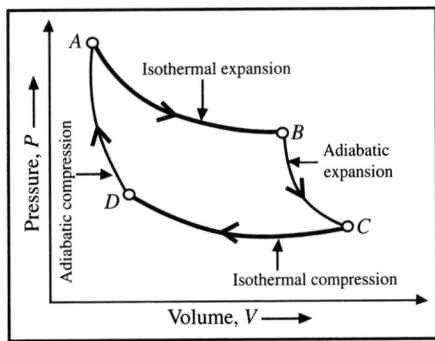

Figure 9.20 Reversible Carnot cycle.

by the working substance to the low-temperature reservoir and two adiabatic BC and DA in between these two isothermal. The changes in entropy in these four reversible processes can be calculated in the following way:

1. **Isothermal expansion** AB: Let δQ_1 be the amount of heat absorbed by the working substance from the reservoir at a constant temperature T_1 in going from state A to state B during isothermal expansion. Hence, the increase in entropy of the working substance is given by

$$\int_A^B dS = +\frac{\delta Q_1}{T_1},$$

and the decrease in entropy of the reservoir (local surroundings) is given by

$$\int_A^B dS = -\frac{\delta Q_1}{T_1}.$$

So, the entropy change of the universe in this isothermal process is

$$+\frac{\delta Q_1}{T_1} - \frac{\delta Q_1}{T_1} = 0.$$

2. **Adiabatic expansion BC:** In going from state B to state C along the adiabatic BC, there is no change in entropy of the working substance as well as of the local surroundings, but the temperature falls from T_1 to T_2 due to adiabatic expansion. So, the entropy change of the universe is again zero in this process.

3. **Isothermal compression CD:** In going from state C to state D along the isothermal CD, the working substance rejects δQ_2 amount of heat to the sink at temperature T_2. In this process, the entropy of the working substance decreases and is given by

$$\int_C^D dS = -\frac{\delta Q_2}{T_2},$$

and the increase in entropy of the sink (local surroundings) is given by

$$\int_C^D dS = +\frac{\delta Q_2}{T_2}.$$

So, the entropy change of the universe in this isothermal process along CD is given by

$$-\frac{\delta Q_2}{T_2} + \frac{\delta Q_2}{T_2} = 0.$$

4. **Adiabatic compression DA:** In going from state D to state A along the adiabatic DA, there is no change in entropy of the working substance as well as of the local surroundings, but the temperature rises from T_2 to T_1 due to adiabatic compression. So, the entropy change of the universe is again zero in this process.

Considering the contributions of all these thermodynamic reversible processes, we can arrive at the following conclusions:

1. The total change in entropy of the universe in a reversible process is zero.

2. The total change in entropy of the system, that is, the working substance, is given by

$$\int_A^B dS + \int_B^C dS + \int_C^D dS + \int_D^A dS = +\frac{\delta Q_1}{T_1} + 0 - \frac{\delta Q_2}{T_2} + 0 = \frac{\delta Q_1}{T_1} - \frac{\delta Q_2}{T_2}.$$

But for a reversible Carnot's cycle, $\frac{\delta Q_1}{T_1} = \frac{\delta Q_2}{T_2}$. Hence, in a cycle of a reversible process, the change in entropy of a system is zero, that is, its entropy remains constant.

3. However, it should be noted that all natural processes are irreversible.

9.9.7 Entropy change in an irreversible process with examples

So far, we have considered the change of entropy only in reversible processes. We now turn our attention to the calculation of change of entropy in irreversible processes. We focus primarily on the irreversible processes proceeding in an isolated system, as such types of systems are of great interest in thermodynamics. We know that an isolated system has a rigid boundary as well as an ideal insulation of heat. The rigid boundary protects the system against any exchange of expansion work with the surroundings, that is, $(P\,dV)_{sys} = 0$, and the ideal heat insulation prevents any exchange of heat between the system and the surroundings, that is, $\delta Q_{sys} = 0$. Since $(P\,dV)_{sys} = 0$ and $\delta Q_{sys} = 0$, it is clear from the first law of thermodynamics that the change in internal energy $dU_{sys} = 0$, that is, internal energy $U_{sys} = $ constant. This analysis of the processes proceeding in an isolated system is of interest mostly because, in the limit, any nonisolated system and its surroundings can be treated as an integral isolated system. Inasmuch as for an isolated system $\delta Q_{sys} = 0$, any process proceeding in such a system is an adiabatic process for the system as a whole.

Now, it will be shown that in an irreversible process, the entropy will increase in the case of an isolated system. Let us consider an isolated system comprising of two bodies at different temperatures, T_1 and T_2, with $T_1 > T_2$. The transfer of heat will take place between these two bodies of the system. Heat will flow from the body at a higher temperature T_1 to the body at a lower temperature T_2. If an amount of heat δQ flows from the first body to the second, the entropy of the first body will decrease by an amount of

$$dS_1 = -\frac{\delta Q}{T_1} \tag{9.98}$$

and the entropy of the second body will increase by an amount of

$$dS_2 = \frac{\delta Q}{T_2}. \tag{9.99}$$

So, the total change in entropy of the entire system can be written as

$$dS_{sys} = \delta Q \left[\frac{1}{T_2} - \frac{1}{T_1} \right]. \tag{9.100}$$

Since $T_2 < T_1$, we have $dS_{sys} > 0$. This shows that an irreversible process results in an increase in entropy in the case of an isolated system.

9.9.8 Entropy of the universe

In everyday science, we notice some very interesting observations. These observations include cooling down of coffee mugs, shining of stars, explosion of supernovae, formation of black holes, and blowing of dust around by the winds on the planetary surfaces, etc. It is known that since 1929, the universe is expanding. This expansion is isentropic, indicating that the entropy of the relativistic particles such as photons, gravitons, and neutrinos does not increase or decrease with this expansion. This is because the entropy of a gas of relativistic particles is proportional to the number of particles N, which remains constant as the universe expands. If we follow the forward or backward motion in time of a co-moving volume of the universe, the number of photons in that volume does not change, and hence, entropy also does not change.

In all these processes, the entropy of the universe S_{uni} continues to increase and has been increasing since the occurrence of the hot big bang nearly 13.8 billion years ago. The universe obeys the second law of thermodynamics with

$$dS_{uni} \geq 0. \tag{9.101}$$

The equality sign holds in the case of a reversible process, whereas the inequality sign prevails in the case of a spontaneous process. As the entropy is a state function, a change in entropy of the system ΔS and that of the surroundings $\Delta S'$ change sign when the direction of a process is reversed. Therefore, one can say that a process for which $\Delta S + \Delta S' < 0$ will not occur in nature, that is, it is an impossible process. It should be mentioned that these classifications—reversible, spontaneous, and impossible—are exhaustive and mutually exclusive. Hence, it can be concluded that $\Delta S_{uni} = \Delta S + \Delta S' = 0$ is necessary and sufficient for a process to be reversible, and $\Delta S_{uni} = \Delta S + \Delta S' > 0$ is necessary and sufficient for a process to be spontaneous.

9.9.9 Temperature–entropy diagrams for cycle

A Carnot cycle is an idealized reversible cycle. The efficiency of such a cycle is calculated based on certain common assumptions. Some of these assumptions are mentioned below.

1. The cycles do not have friction in any steps. Thus, there are no pressure drops in the working fluid due to friction.

2. All expansions and compressions involved in the process are considered in the quasi-equilibrium states.

3. There are no heat losses in the pipes and other components of the cycles of the engine.

4. It is assumed that changes in kinetic and potential energies are negligible. But in the case of nozzles and diffusers, there are changes in kinetic and potential energies as these are specifically designed to change the velocity of passing fluid.

Various steps of an idealized reversible Carnot cycle are shown in Figure 9.21 in an "entropy–temperature $(T{-}S)$" diagram.

The first step is the addition of heat at a constant temperature T_H. Thus, this heat addition process proceeds in the direction of increasing entropy (from S_I to S_h) at the constant temperature and vice versa. This is shown by the process from $1 \rightarrow 2$ in the $T{-}S$ diagram in Figure 9.21. The second step from $2 \rightarrow 3$ is an isentropic expansion of the working substance. As no change of heat occurs in this process, entropy remains constant. This is shown by the vertical line $2 \rightarrow 3$ in the diagram. Here, the temperature drops from T_H to T_c.

The third step from $3 \rightarrow 4$ is the isothermal heat transfer (condenser or heat rejection) from the working substance. As heat is rejected by the substance at constant temperature T_c, entropy decreases from S_h to S_I in this process. This is shown by the horizontal line $3 \rightarrow 4$ in the diagram. Here, temperature remains constant at T_c. The fourth

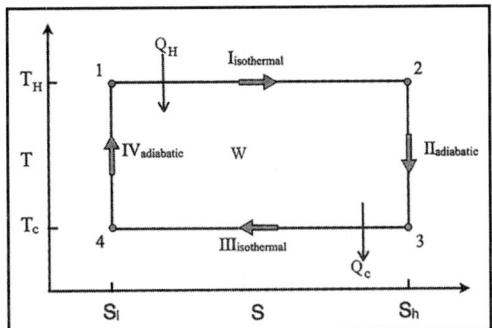

Figure 9.21 Entropy–temperature diagram for a reversible Carnot cycle.

step from $4 \rightarrow 1$ is an isentropic compression of the working substance to high pressure and temperature. As no change of heat occurs in this process, entropy remains constant at S_I. This is shown by the vertical line $4 \rightarrow 1$ in the diagram. Here, temperature increases from T_c to T_H.

The area enclosed by the process curve on the $T{-}S$ diagram represents the net work done in the cycle. The ratio of the area enclosed by the cyclic curve to the area under the heat addition process curve represents the thermal efficiency of the cycle. This thermal efficiency can be calculated in the following way:

Here, the amount of heat taken Q_H is $Q_H = T_H \Delta S$, and the amount of heat rejected Q_c is $Q_c = T_c \Delta S$. Hence, the efficiency η of the Carnot cycle in the T_S diagram can be expressed as

$$\eta = \frac{Q_H - Q_c}{Q_H} = \frac{T_H \Delta S - T_c \Delta S}{T_H \Delta S} = \frac{T_H - T_c}{T_H} = 1 - \frac{T_c}{T_H}.$$

This is the same expression as obtained from the P–V diagram of the Carnot cycle.

9.9.10 Change in entropy for various systems

Changes in entropy for systems like an ideal gas, a Van der Waals gas, a gas mixture, and due to diffusion are calculated in this section.

Change in entropy for an ideal gas

For a hydrostatic system, the combination of first and second law of thermodynamics leads to

$$T dS = C_V dT + P dV. \tag{9.102}$$

Here, we have used $dU = C_V dT$ for an ideal gas. This is one of the many forms of the first law of thermodynamics for a PVT system. We shall now deduce other equivalent forms.

Since for an ideal gas, $PV = RT$, we have $P dV + V dP = R dT$, and also for one mole of an ideal gas, $C_P - C_V = R$. Eliminating dV and dT from equation (9.102), we get

$$T dS = C_P dT - V dP \quad \text{and} \quad T dS = \frac{1}{R}(C_P \, P dV + C_V \, V dP). \tag{9.103}$$

These expressions given by equation (9.103) can be used to calculate the change in entropy for an ideal gaseous system.

Equations (9.102) and (9.103) can also be expressed in the following forms:

$$dS = C_V \frac{dT}{T} + R \frac{dV}{V}, \tag{9.104}$$

$$dS = C_P \frac{dT}{T} - R \frac{dP}{P}, \tag{9.105}$$

and

$$dS = C_P \frac{dV}{V} + C_V \frac{dP}{P}. \tag{9.106}$$

Entropy change between two states can thus be expressed by either of the following.

$$S_2 - S_1 = \int_1^2 C_V \frac{dT}{T} + R \int_1^2 \frac{dV}{V} = C_V \ln \frac{T_2}{T_1} + R \ln \frac{V_2}{V_1}, \qquad (9.107)$$

$$S_2 - S_1 = \int_1^2 C_P \frac{dT}{T} - R \int_1^2 \frac{dP}{P} = C_P \ln \frac{T_2}{T_1} - R \ln \frac{P_2}{P_1}, \qquad (9.108)$$

and

$$S_2 - S_1 = \int_1^2 C_P \frac{dV}{V} + \int_1^2 C_V \frac{dP}{P} = C_P \ln \frac{V_2}{V_1} + C_V \ln \frac{P_2}{P_1}. \qquad (9.109)$$

We have assumed here C_P and C_V remain constant over the temperature-interval of interest.

From these equations, the entropy change for an *isothermal expansion* can be, respectively, expressed as

$$\Delta S_T = R \ln \frac{V_2}{V_1}; \quad \Delta S_T = R \ln \frac{P_1}{P_2}, \quad \text{and} \quad \Delta S_T > 0,$$

which is in agreement with the principle of increase of entropy.

Similarly, from these equations, the entropy change for *isobaric* processes can be expressed as

$$\Delta S_P = C_P \ln \frac{T_2}{T_1} = C_P \ln \frac{V_2}{V_1},$$

and for *isochoric* processes, the entropy change can be expressed as

$$\Delta S_V = C_V \ln \frac{T_2}{T_1} = C_V \ln \frac{P_2}{P_1}.$$

These results are very useful in thermodynamics.

Change in entropy of a Van der Waals gas

We consider one mole of a real gas obeying Van der Waals equation of state. Two states of the system are, respectively, described by the parameters: pressure P_1, volume V_1, and temperature T_1 for state 1, and pressure P_2, volume V_2, and temperature T_2 for state 2. The change is entropy is given by

$$S_2 - S_1 = \int \frac{\delta Q}{T}. \qquad (9.110)$$

But, from the first law of thermodynamics, we have

$$\frac{\delta Q}{T} = \frac{dU}{T} + \frac{P}{T}dV = C_V\frac{dT}{T} + \frac{R}{V-b}dV. \tag{9.111}$$

Using $P = \frac{RT}{V-b}$ for a Van der Waals gas as the first approximation. Assuming C_V remains constant in the temperature-interval, the change in entropy for such a real gas can be obtained from equations (9.110) and (9.111) as

$$S_2 - S_1 = C_V\int_{T_1}^{T_2}\frac{dT}{T} + R\int_{V_1}^{V_2}\frac{dV}{V-b} = C_V\ln\frac{T_2}{T_1} + R\ln\frac{V_2-b}{V_1-b}. \tag{9.112}$$

This is the change in entropy for a real gas.

Change in entropy of a gas mixture

From the first law of thermodynamics, $\delta Q = dU + PdV$. For one mole of an ideal gas, we can then write

$$\frac{\delta Q}{T} = \frac{dU}{T} + P\frac{dV}{T} = C_V\frac{dT}{T} + R\frac{dV}{V}; \quad \Rightarrow \quad dS = C_V d(\ln T) + Rd(\ln V).$$

Integrating this expression, we get

$$S = C_V\ln T + R\ln V + K, \tag{9.113}$$

where K is a constant whose value depends on mass. Here, we assume that the C_V is constant. For 1 mol of an ideal gas, using the relation $PV = RT$, we get

$$S = C_V\ln T + R\ln\left(\frac{RT}{P}\right) + K = C_V\ln T + R\ln\left(\frac{T}{P}\right) + R\ln R + K$$

$$= C_V\ln T + R\ln\left(\frac{T}{P}\right) + C,$$

where $C = R\ln R + K = a$ constant.

Consider a mixture of gases where the first gas has n_1 moles, the second n_2 moles, etc. Entropy of the gas mixture S is given by

$$S = n_1\left[C_{V_1}\ln T + R\ln\left(\frac{T}{P_1}\right) + C_1\right] + n_2\left[C_{V_2}\ln T + R\ln\left(\frac{T}{P_2}\right) + C_2\right] + \cdots + \text{etc.},$$

where P_1 is the partial pressure of gas 1, P_2 that of gas 2, $\cdots$, etc. This partial pressure P_1 is given by

$$P_1 = \frac{n_1}{n_1 + n_2 + \cdots}P = \frac{n_1}{\sum n_i}P = K_1P \text{ and } P_2 = \frac{n_2}{\sum n_i}P = K_2P, \cdots \text{ and so on.}$$

Here, P is the total pressure, and $K_i = \frac{n_i}{\sum n_i}$. So, the entropy of the gas mixture can be written as

$$S = \sum_i n_i \left[C_{V_i} \ln T + R \ln \left(\frac{T}{K_i P} \right) + C_i \right]. \tag{9.114}$$

Change in entropy in mixing two gases

We consider two vessels of volumes V_1 and V_2 containing molecules of the same gas, respectively, at temperatures T_1 and T_2. The pressures of the gases are also the same. The two vessels are then connected with each other, and the gases mix with each other and subsequently attain a state of equilibrium. The change in entropy during this process can be calculated in the following way:

When the two gases are allowed to mix with each other, the final temperature and volume will be, respectively, $\frac{T_1+T_2}{2}$ and $V_1 + V_2$. According to equation (9.107), the change in entropy ΔS_1 for the gas initially enclosed in volume V_1 is given by

$$\Delta S_1 = C_V \ln \left(\frac{T_1 + T_2}{2\,T_1} \right) + R \ln \left(\frac{V_1 + V_2}{V_1} \right).$$

Similarly, for the gas initially enclosed in volume V_2, the change in entropy ΔS_2 is given by

$$\Delta S_2 = C_V \ln \left(\frac{T_1 + T_2}{2\,T_2} \right) + R \ln \left(\frac{V_1 + V_2}{V_2} \right).$$

So, the total change in entropy ΔS in the process is given by

$$\Delta S = \Delta S_1 + \Delta S_2 = C_V \ln \frac{(T_1 + T_2)^2}{4T_1 T_2} + R \ln \frac{(V_1 + V_2)^2}{V_1 V_2}$$

$$= C_V \ln \frac{(T_1 + T_2)^2}{4T_1 T_2} + R \ln \frac{\frac{R^2 (T_1+T_2)^2}{4p^2}}{\frac{R^2 T_1 + T_2}{4p^2}},$$

$$= C_V \ln \frac{(T_1 + T_2)^2}{4T_1 T_2} + R \ln \frac{(T_1 + T_2)^2}{4T_1 T_2} = C_P \ln \frac{(T_1 + T_2)^2}{4T_1 T_2}. \quad (\text{as } R = C_P - C_V)$$

Change in entropy of steam

Let 1 g of ice be dropped into a system of steam at temperature T_s K. We are to compute the change in entropy due to the process. The transformation takes place in the following three successive steps:

1. Ice at T_i K changes into water at T_i K,

2. Water at T_i K changes into water at T_s K, and finally

3. Water at T_s K changes into steam at T_s K.

The change of entropy in step (1) is given by

$$\Delta S_1 = R\frac{\delta Q}{T} = \frac{L_i}{T_i}, \text{ where } L_i = \text{latent heat of ice.}$$

The change of entropy in step (2) is given by

$$\Delta S_2 = \int_{T_i}^{T_s} \frac{\delta Q}{T} = \int_{T_i}^{T_s} \frac{1.c.dT}{T} = c \ln\frac{T_s}{T_i}, \text{ } c \text{ is the specific heat of water.}$$

The change of entropy in step (3) is given by

$$\Delta S_3 = \int \frac{\delta Q}{T} = \frac{L_s}{T_s}, \text{ where } L_s \text{ is the latent heat of steam.}$$

So, the net change in entropy ΔS in this process is given by

$$\Delta S = \Delta S_1 + \Delta S_2 + \Delta S_3 = \frac{L_i}{T_i} + c \ln\frac{T_s}{T_i} + \frac{L_s}{T_s}.$$

Change in entropy due to diffusion

Entropy of a mixed system increases due to diffusion. We consider a number of gases at pressure P and temperature T, and these gaseous systems are allowed to mix up *spontaneously* by diffusion. The increase in entropy due to this process can be calculated in the following way:

Using the expression for entropy $S = C_V \ln T + R \ln\left(\frac{T}{P}\right) + C$, the total entropy S_i **before** diffusion is given by

$$S_i = n_1 C_{V_1} \ln T + n_1 R \ln\left(\frac{T}{P}\right) + n_1 C_1 + n_2 C_{V_2} \ln T + n_2 R \ln\left(\frac{T}{P}\right) + n_2 C_2 + \cdots$$

$$= \sum_i n_i \left[C_{V_i} \ln T + R \ln\left(\frac{T}{P}\right) + C_i \right],$$

where n_i is the number of moles of the ith gas.

After the mixing of all the gases due to diffusion, the total entropy S_f of the mixture can be written as

$$S_f = \sum_i n_i \left[C_{V_i} \ln T + R \ln\left(\frac{T}{K_i P}\right) + C_i \right].$$

Hence, the increase in entropy can be written as

$$\Delta S = S_f - S_i = -\sum_i n_i R \ln K_i = \sum_i n_i R \ln\frac{\sum n_i}{n_i} \text{ as } \left(K_i = \frac{n_i}{\sum n_i} \right),$$

which is obviously positive. For two gases, we have the increase in entropy as

$$\Delta S = n_1 R \ln \frac{n_1 + n_2}{n_1} + n_2 R \ln \frac{n_1 + n_2}{n_2}. \tag{9.115}$$

It may be remarked that equation (9.115) holds good for mixing two different gases but not for two identical gases. If equation (9.115) is applied to two portions of the same gas (at the same pressure and temperature), obviously there can be no increase in entropy. This is called a **paradox**. The reason for this failure can be resolved only from quantum statistical mechanics.

9.10   Third Law of Thermodynamics

The third law of thermodynamics states something regarding the properties of systems in equilibrium at absolute zero temperature. This law states that "the equilibrium entropy of all systems and the change in entropy in all reversible isothermal processes tend to zero as temperature approaches absolute zero". From the knowledge of quantum mechanics, we know that a system must be in a state with the minimum possible energy at absolute zero temperature. Thus, for a system consisting of many particles, there is only **one unique state (called the ground state)** with this minimum energy at absolute zero. Hence, the number of accessible states Ω by the system at absolute zero temperature is only **one**. Further, it is known that entropy is related to the number of accessible microstates Ω by the famous Boltzmann relation $S = K_B \ln \Omega$. From this relation, we have $S = K_B \times \ln \Omega = K_B \times \ln(1) = K_B \times 0 = 0$. This result indicates that as absolute zero is approached, the entropy of a system also approaches zero. This result is in perfect agreement with systems having ordered crystal structures and is the essence of the third law of thermodynamics.

The situation is somewhat different for the systems that do not have a well-defined order. A good example of such type of system is amorphous glass. When this type of system is brought to very low temperatures, there will be some finite entropy, not zero. The origin of such finite entropy at very low temperatures can be understood in the following way: The glassy system becomes locked into a configuration with non-minimal energy at such very low temperatures, resulting in a number of microstates greater than one. Then, according to Boltzmann relation, the entropy of the system will be finite (constant depending upon the configuration). This constant value of entropy is called the residual entropy of the system. It should be mentioned here that although there are formally four laws of thermodynamics, that is, the zeroth to the third, the zeroth law is really a consequence of the second law, and the third law is only important at temperatures close to absolute zero.

9.10.1 Unattainability of Absolute Zero

In literature, the third law of thermodynamics has been expressed in various forms:

1. **As temperature approaches absolute zero, the entropy of any system vanishes.**

2. **The entropy of a system approaches a constant value as its temperature approaches absolute zero.** However, there are many systems like glass materials and random alloys that possess residual entropy. With the progress of material sciences, many systems are also being found to exhibit this property.

3. **It is impossible to reach absolute zero temperature.** This expression is called the unattainability of absolute zero temperature. In this case, there is no exception.

One more statement of the third law of thermodynamics is that the lowering of the temperature of an object to absolute zero is impossible in a finite number of steps. To understand the meaning of this statement, we take the concept of the Carnot refrigerator. This refrigerator is basically just the reverse of the Carnot heat engine, that is, the Carnot heat engine working in a reverse cycle. Since a Carnot heat engine provides work through reversible isothermal-adiabatic compressions and expansions, a net amount of work must be done in the Carnot refrigerator, making it an electricity consumer.

Let Q_1 be the amount of heat absorbed by a Carnot refrigerator from a body at a lower temperature T_1 and Q_2 as the amount of heat rejected by the same refrigerator to a body at a higher temperature T_2. The coefficient of performance β of the reversible Carnot refrigerator can be given by the following relation:

$$\beta = \frac{1}{Q_2/Q_1 - 1}. \tag{9.116}$$

For a Carnot cycle, we know that

$$\frac{Q_1}{T_1} = \frac{Q_2}{T_2}; \quad \Rightarrow \quad \frac{Q_2}{Q_1} = \frac{T_2}{T_1}. \tag{9.117}$$

Using this equation (9.117) in equation (9.116), we get

$$\beta = \frac{1}{T_2/T_1 - 1}. \tag{9.118}$$

Again, the coefficient of performance β of an ideal Carnot refrigerator is simply the ratio of the cooling effect to the work done, that is,

$$\beta = \frac{Q_1}{W}. \tag{9.119}$$

This indicates the amount of heat removed from the body at a lower temperature per unit of work done. Now, from equations (9.118) and (9.119), we have

$$\frac{Q_1}{W} = \frac{1}{T_2/T_1 - 1} = \frac{T_1}{T_2 - T_1}; \quad \Rightarrow \quad \frac{W}{Q_1} = \frac{T_2 - T_1}{T_1}. \tag{9.120}$$

Thus, it is obvious from this equation that as the lower temperature T_1 approaches zero, more and more work will be needed to remove the same amount of heat, which is the **unattainability of the absolute zero**.

9.11   Solved Problems

1. "A reversible process in quasistatic but the reverse statement is not true"— Justify the statement.   [Calcutta University 2013]

   Answer: A reversible process is always quasi-static because in this process the concerned parameters vary infinitesimally, and equilibrium is maintained at every instant of time. Hence, this process can be represented in a *P–V* diagram. But a process which is quasi-static must not necessarily be reversible like in the case of a cyclic process that goes from one state to another via one process and returns to its original state via a different process, and here also equilibrium is maintained all the time.

2. What do you mean by quasi-static process? All the reversible processes are quasi-static but all quasi-static processes are not reversible—explain.
   [Calcutta University 2015]

   Answer: When a system is in thermodynamic equilibrium, and the surroundings are kept unchanged, no motion will take place, and no work will be done. If a finite unbalanced force acts on the system, the system may pass through nonequilibrium states. If it is desired during a process to describe every state of a system by means of thermodynamic coordinates referring to the system, the external forces acting on the system should be varied only slightly so that the unbalanced force is infinitesimal. A small process performed in this way is said to be quasi-static.

   During a quasi-static process, the system is at all times infinitesimally near a state of thermodynamic coordinates, referring to the system as a whole. Therefore, for all these states, an equation of state is valid. A quasi-static process is an idealization that is applicable to all thermodynamic systems including electric and magnetic systems. The conditions for such a process can never be rigorously satisfied in the laboratory, but they can be approached with almost any degree of accuracy.

3. Explain the meaning of the quasi-static process. Is "free expansion" quasi-static?   [Calcutta University 2014]

Answer: When a system is in thermodynamic equilibrium, and the surroundings are kept unchanged, no motion will take place and no work will be done. If a finite unbalanced force acts on the system, the system may pass through nonequilibrium states. If it is desired during a process to describe every state of a system by means of thermodynamic coordinates referring to the system, the external forces acting on the system should be varied only slightly so that the unbalanced force is infinitesimal. A small process performed in this way is said to be quasi-static.

Free expansion is not a quasi-static process as it occurs rapidly, and only the initial and final states are in equilibrium.

4. A 100 Ω resistor carrying a current 5A for 1 s is kept at a constant temperature of 27°C by a stream of cooling water. Find the change in entropy of (i) the resistor and (ii) the universe.   [Calcutta University 2012]

Answer: Now, since the temperature of the resistor is constant, and the current through the resistor is constant, then there is no net change in heat in the resistor. Therefore, $\delta Q = 0$ and $\Delta S_{\text{resistor}} = 0$ J/K.

We know that

$$\delta Q Pt = i^2 \, R \, t.$$

The change in entropy of the surroundings is given by

$$\Delta S_{\text{surrounding}} = \frac{\Delta Q}{T} = \frac{i^2 Rt}{T} = \frac{5 \times 5 \times 100 \times 1}{27 + 273} = 8.33 \text{ J/K.}$$

So, the change in entropy of the universe is

$$\Delta S_{\text{universe}} = \Delta S_{\text{resistor}} + \Delta S_{\text{surrounding}} = 0 + 8.33 \text{ J/K} = 8.33 \text{ J/K.}$$

5. Each of the two isolated vessels, A and B, of the fixed volume contains N molecules of a perfect monoatomic gas at a pressure P. The temperatures of A and B are, respectively, T_1 and T_2. The two vessels are brought into thermal contact with each other. Find the change in entropy in this process at equilibrium.

Answer: At equilibrium, the final temperature of each vessel is $T = \frac{T_1 + T_2}{2}$.

So, the change in entropy in the process is given by

$$\Delta S = \int\limits_{T_1}^{T} \frac{C_V dT}{T} + \int\limits_{T_2}^{T} \frac{C_V dT}{T} = \frac{3}{2} N K_B \ln \left[\frac{(T_1 + T_2)^2}{4 T_1 T_2} \right],$$

where $C_V = \frac{3K_B}{2}$ for monoatomic gas.

6. Using the inequality $\oint \frac{dQ}{T} \leq 0$ shows that the entropy of an isolated system undergoing an irreversible process will always increase.

[Calcutta University 2012, 2015]

Answer: Let R be a reversible path, and I be an irreversible path, and the system undergoes a change from state A to B via I and back to A via R such that $AIBRA$ is an irreversible cycle.

For irreversible process, we know

$$\oint \frac{\delta Q}{T} < 0; \quad \Rightarrow \quad \int_{IA}^{B} \frac{\delta Q}{T} + \int_{RB}^{A} \frac{\delta Q}{T} < 0; \quad \Rightarrow \quad \int_{IA}^{B} \frac{\delta Q}{T} < - \int_{RB}^{A} \frac{\delta Q}{T};$$

$$\Rightarrow \int_{IA}^{B} \frac{\delta Q}{T} < \int_{RA}^{B} \frac{\delta Q}{T}; \quad \Rightarrow \quad \int_{IA}^{B} \frac{\delta Q}{T} < S(B) - S(A).$$

For an irreversible isolated system, $\oint \frac{\delta Q}{T}$ is less than zero. Hence,

$$\Rightarrow 0 < S(B) - S(A); \quad \Rightarrow \quad S(B) > S(A).$$

Hence, the entropy of the final state is greater than the entropy of the initial state.

7. Is the isothermal expansion of an ideal gas violates the second law of thermodynamics?   [Calcutta University 2017]

Answer: It is possible to convert heat 100% into work provided that something else happens to your heat machine (i.e., the heat-to-work conversion system). In your case, the gas has expanded, so the state of the heat machine is not the same as in the beginning. The second law of thermodynamics states that it is not possible to convert heat 100% into work and have the heat machine in a state that is identical to its initial state. Further, the second law of thermodynamics states that complete conversion of heat into work is not possible in a cycle. Of course, the first step in the Carnot cycle is the isothermal expansion where heat absorbed is converted into work, but the final state after isothermal expansion is not as the initial one, or it is not cyclic. However, when the Carnot cycle is completed, you will find that the part of heat absorbed is rejected, and remaining part is converted into work, which is clearly what the second law states.

8. The internal energy $U(T)$ of a system at a fixed volume is found to depend on the temperature as $U(T) = \frac{aT^2}{2} + \frac{bT^4}{4}$. Find the entropy $S(T)$ of the system as a function of temperature.

   Answer: From the combination of the first and second laws of thermodynamics, we have

   $$TdS = dU + PdV; \quad \Rightarrow \quad dU = TdS - PdV.$$ At a fixed volume, $dV = 0$ and

   $$dS = \frac{1}{T}dU.$$

   As

   $$U(T) = \frac{aT^2}{2} + \frac{bT^4}{4}; \quad \Rightarrow \quad dU = aTdT + bT^3dT.$$

   Using this value of dU, the differential change in entropy dS becomes

   $$dS = \frac{1}{T}\left(aTdT + bT^3dT\right) = adT + bT^2dT.$$

   Integrating this expression, we get the entropy $S(T)$ of the system as

   $$S = aT + \frac{bT^3}{3}.$$

9. Show that the entropy of n moles of an ideal gas of n moles heat capacity C_V at constant volume at temperature T and volume V is given by $S = nC_V \ln T + nR \ln V + S_0$. Assuming that C_V is independent of temperature.
   [Calcutta University 2013]

   Answer: Combining the first and second laws of thermodynamics, we get $\delta Q = TdS = dU + PdV$.

   For an ideal gas, $dU = nC_V dT$. This leads to $\Rightarrow \quad TdS = nC_V dT + PdV$.

   Using the equation for n mole of an ideal gas, $PV = nRT$, we get

   $$\Rightarrow \quad TdS = nC_V dT + \frac{nRT}{V}dV; \quad \Rightarrow \quad dS = nC_V \frac{dT}{T} + \frac{nR}{V}dV.$$

   Integrating both sides, we get $S = nC_V \ln T + nR \ln V + S_0$.

10. In the free expansion of an ideal gas with initial volume V_1 and final volume V_2, compute the change in entropy. Is this process reversible?
    [Calcutta University 2013]

    Answer: Adiabatic free expansion of an ideal gas is irreversible. Therefore, $\delta Q \neq TdS$.

Volume increases V_1 to V_2. Since the free expansion is adiabatic, both ΔQ and ΔW are zero. Hence, from the first law of thermodynamics, we have $\Delta U = 0$ for such a free expansion. As the internal energy of an ideal gas depends only on the temperature T, we get $\Delta T = 0$. This means the initial and final states have the same temperature and can be connected by a reversible isothermal path.

$$dS = \frac{\delta Q}{T} = C_V \frac{dT}{T} + P \frac{dV}{T} = R \frac{dV}{V}.$$

Since $dT = 0$, integrating the above equation, we have

$$\Delta S = \int_{V_1}^{V_2} R \frac{dV}{V} = R \ln \frac{V_2}{V_1}.$$

11. A Carnot engine with a sink at 10°C has an efficiency of 30%. By how much must the temperature of the source be change to increase its efficiency to 50%?
   [Calcutta University 2012]

   Answer: The expression for efficiency is given by $\eta = 1 - \frac{T_2}{T_1}$. Here, η is $= 30\%$.

   Solving for T_1, we get

   $$T_1 = \frac{T_2}{1 - \eta} = \frac{10 + 273}{1 - 0.3} = 404.2 \text{ K} = 131.2°\text{C}.$$

   Therefore, T_1 has to be increased by 161.8 K.

12. The $T-S$ diagram of a reversible heat engine is given in Figure 9.22. Calculate the efficiency of the engine.
   [Calcutta University 2013]

   Answer:

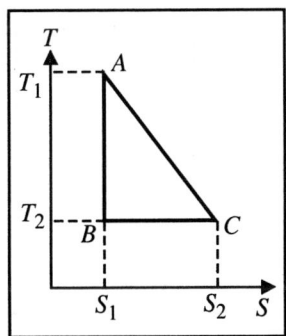

Figure 9.22 $T-S$ diagram for the problem 12.

Work done = Area of $\triangle ACB$. Heat absorbed = Area under $ACS_2 S_1 B$.

$$\eta = \frac{W}{Q_1} = \frac{\frac{1}{2}(T_2 - T_1)(S_2 - S_1)}{\frac{1}{2}(T_2 - T_1)(S_2 - S_1) + T_1(S_2 - S_1)} = \frac{T_2 - T_1}{T_2 + T_1}.$$

13. Calculate the maximum efficiency of a Carnot engine working between two temperatures: $T_1 = 400°C$ and $T_2 = 120°C$.   [Calcutta University 2015]

Answer: Efficiency $\eta = 1 - \frac{T_2}{T_1}$, where T_1 and T_2 are the temperatures of source and sink, respectively. So, we have the efficiency as

$$\eta = 1 - \frac{T_2}{T_1} = 1 - \frac{(120 + 273)}{(400 + 273)} = 1 - \frac{393}{673} = 1 - 0.584 = 0.416;$$

$$\Rightarrow \quad \eta = 0.416 \times 100 = 41.6\%.$$

14. An ideal gas engine operates in a cycle, which when represented on a $P-V$ diagram is a rectangle. This is shown in Figure 9.23. Here, P_1 and P_2 are the lower and higher pressures, respectively, and V_1 and V_2 are the lower and higher volumes, respectively.

(a) Calculate the work done in one complete cycle.

(b) Indicate in which parts of the cycle heat is absorbed and in which heat is liberated. Calculate the quantity of heat flowing into the gas in one cycle.

(c) Show that the efficiency of the engine is

$$\eta = \frac{\gamma - 1}{\frac{\gamma P_2}{P_2 - P_1} + \frac{V_1}{V_2 - V_1}}.$$

Answer:

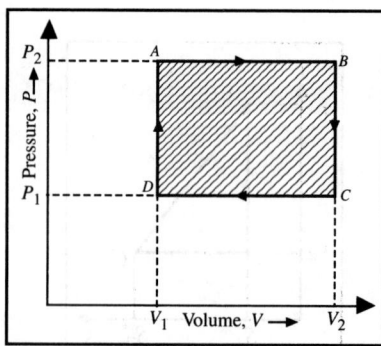

Figure 9.23 $P-V$ diagram for the thermodynamic process in problem 14.

(a) Work done in the complete cycle is equal to the area of the cycle $=$ $(P_2 - P_1)(V_2 - V_1)$ units.

(b) Heat is absorbed in parts DA and AB of the full cycle, and heat is rejected in parts BC and CD of the full cycle.

(c) The work done in the process is $(P_2 - P_1)(V_2 - V_1)$, and the amount of heat received is

$$Q_1 = C_V(T_A - T_D) + C_P(T_B - T_A).$$

So, the efficiency of the engine is given by

$$\begin{aligned}
\eta &= \frac{W}{Q_1} = \frac{(P_2 - P_1)(V_2 - V_1)}{C_V(T_A - T_D) + C_P(T_B - T_A)} \\
&= \frac{(P_2 - P_1)(V_2 - V_1)}{C_V\left[\frac{P_2 V_1}{R} - \frac{P_1 V_1}{R}\right] + C_P\left[\frac{P_2 V_2}{R} - \frac{P_2 V_1}{R}\right]} \\
&= \frac{R(P_2 - P_1)(V_2 - V_1)}{C_V(P_2 V_1 - P_1 V_1) + C_P(P_2 V_2 - P_2 V_1)} \\
&= \frac{(C_P - C_V)(P_2 - P_1)(V_2 - V_1)}{C_V V_1(P_2 - P_1) + C_P P_2(V_2 - V_1)} \\
&= \frac{\gamma - 1}{\frac{V_1}{V_2 - V_1} + \frac{\gamma P_2}{P_2 - P_1}} = \frac{\gamma - 1}{\frac{\gamma P_2}{P_2 - P_1} + \frac{V_1}{V_2 - V_1}}.
\end{aligned}$$

15. An ideal gas goes through a cycle consisting of

(a) isochoric, adiabatic, and isothermal lines (see Figure 9.24), and

(b) isobaric, adiabatic, and isothermal lines (see Figure 9.26).

When the isothermal occurs at the minimum temperature, calculate the efficiency of each cycle if the absolute temperature varies n-fold within the cycle.

Answer:

(a) Let the coordinates of the points are $a(P_1, V_1, T_1)$, $b(P_2, V_1, T_2)$, and $c(P_3, V_2, T_1)$.

We have $T_2 = nT_1$; $V_2 = nV_1$; $Q_1 = C_V(T_2 - T_1)$, and $Q_2 = RT_1 \ln\left(\frac{V_2}{V_1}\right)$.

The efficiency η is given by

$$\eta = 1 - \frac{Q_2}{Q_1} = 1 - \frac{RT_1 \ln\left(\frac{V_2}{V_1}\right)}{C_V(T_2 - T_1)}.$$

Again, $V_1^{\gamma-1} T_2 = V_2^{\gamma-1} T_1$. This provides $\frac{V_2}{V_1} = \left(\frac{T_2}{T_1}\right)^{\frac{1}{\gamma-1}}$.

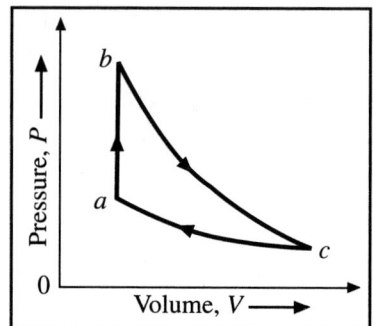

Figure 9.24 P–V diagram for the thermodynamic process in problem 15a.

Hence, the efficiency η is given by

$$\eta = 1 - \frac{RT_1 \frac{1}{\gamma-1} \ln\left(\frac{T_2}{T_1}\right)}{C_V T_1 \left(\frac{T_2}{T_1} - 1\right)} = 1 - \frac{R}{C_V(\gamma-1)} \frac{\ln n}{n-1}$$

$$= 1 - \frac{R}{C_V\left(\frac{C_P}{C_V} - 1\right)} \frac{\ln n}{n-1} = 1 - \frac{\ln n}{n-1}.$$

(b) Let the coordinates of the points are $a(P_1, V_1, T_1)$, $b(P_1, V_2, T_2)$, and $c(P_2, V_3, T_1)$.

We have $T_2 = nT_1$; $V_2 = nV_1$; $Q_1 = C_P(T_2 - T_1)$, and $Q_2 = RT_1 \ln\left(\frac{V_3}{V_1}\right)$.

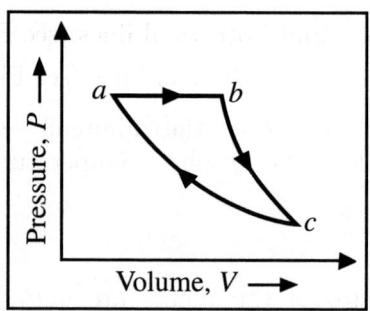

Figure 9.25 P–V diagram for the thermodynamic process in problem 15b.

The efficiency η is given by

$$\eta = 1 - \frac{Q_2}{Q_1} = 1 - \frac{RT_1 \ln\left(\frac{V_3}{V_1}\right)}{C_P(T_2 - T_1)}.$$

For the adiabatic bc, we have $V_2^{\gamma-1} T_2 = V_3^{\gamma-1} T_1$.

From the ideal gas equation:

$$\frac{P_1 V_1}{T_1} = \frac{P_1 V_2}{T_2}; \quad \Rightarrow \quad V_2 = \frac{T_2 V_1}{T_1}.$$

This provides

$$\left(\frac{T_2 V_1}{T_1}\right)^{\gamma-1} T_2 = V_3^{\gamma-1} T_1; \quad \Rightarrow \quad \frac{V_3}{V_1} = \left(\frac{T_2}{T_1}\right)^{\frac{\gamma}{\gamma-1}}.$$

Hence, the efficiency η is given by

$$\eta = 1 - \frac{RT_1 \ln \left(\frac{T_2}{T_1}\right)^{\frac{\gamma}{\gamma-1}}}{C_P T_1 \left(\frac{T_2}{T_1} - 1\right)} = 1 - \frac{R}{C_P} \frac{\gamma}{\gamma-1} \frac{\ln n}{n-1}$$

$$= 1 - \frac{R}{C_P} \frac{C_P/C_V}{(C_P - C_V)/C_V} \frac{\ln n}{n-1} = 1 - \frac{\ln n}{n-1}.$$

Thus, the efficiencies are the same in both cases.

16. Compare the efficiencies of the cycles $ACBA$ shown in Figures 9.26(a) and 9.26(b). [Calcutta University 2017]

Answer:

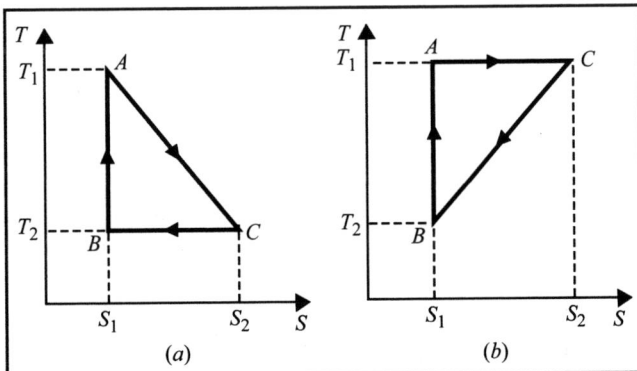

(a)   (b)

Figure 9.26 T–S diagram for the thermodynamic processes in problems 16(a) and (b).

For the first figure, work done = area of $\triangle ABC = \frac{1}{2}(S_2 - S_1)(T_1 - T_2)$.

Heat input = area under $AB = \triangle ABC + (S_2 - S_1)T_2$.

Hence, the efficiency of the first $ACBA$ cycle, shown in Figure 9.26(a), is given by efficiency =

$$\frac{\text{work done}}{\text{heat input}} = \frac{\Delta ABC}{\Delta ABC + (S_2 - S_1)T_2} = \frac{\frac{1}{2}(S_2 - S_1)(T_1 - T_2)}{\frac{1}{2}(S_2 - S_1)(T_1 - T_2) + (S_2 - S_1)T_2}$$
$$= \frac{T_1 - T_2}{T_1 + T_2}.$$

Similarly, the efficiency of the second $ACBA$ cycle, shown in Figure 9.26(b), is given by efficiency =

$$= \frac{\Delta ABC}{\Delta ABC + \frac{1}{2}(S_2 - S_1)(T_2 + T_1)}$$
$$= \frac{\frac{1}{2}(S_2 - S_1)(T_1 - T_2)}{\frac{1}{2}(S_2 - S_1)(T_1 - T_2) + \frac{1}{2}(S_2 - S_1)(T_1 + T_2)} = \frac{T_1 - T_2}{T_1 - T_2 + T_1 + T_2}$$
$$= \frac{1}{2}\left(1 - \frac{T_2}{T_1}\right).$$

So, the ratio of the efficiencies is $\frac{\frac{T_1 - T_2}{T_1 + T_2}}{\frac{1}{2}(1 - \frac{T_2}{T_1})} = \frac{2T_1}{T_1 + T_2}.$

17. Figure 9.27 shows the $P-V$ diagram of an ideal gas scale. Assume that all processes are quasistatic and the specific heat at constant pressure C_P is constant. Prove that the efficiency of the engine is given by

$$\eta = 1 - \left(\frac{P_{DA}}{P_{BC}}\right)^{\frac{\gamma-1}{\gamma}}.$$

Answer:

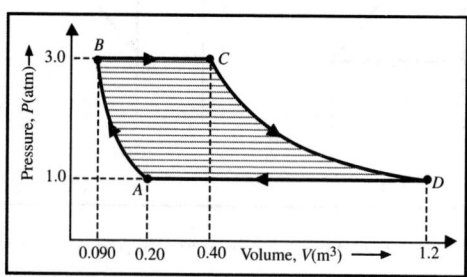

Figure 9.27 $P-V$ diagram for the thermodynamic process in problem 17.

CD and AB are two adiabatic. So, the work done in these two adiabatic are equal and opposite.

The heat received along the isobaric BC is $Q_1 = P_{BC}(V_C - V_B)$, and the heat rejected along the isobaric DA is given by $Q_2 = P_{DA}(V_D - V_A)$.

So, the efficiency in this case is given by

$$\eta = \frac{Q_1 - Q_2}{Q_1} = 1 - \frac{Q_2}{Q_1} = 1 - \frac{P_{DA}(V_D - V_A)}{P_{BC}(V_C - V_B)} = 1 - \frac{P_{DA}}{P_{BC}} \frac{V_D \left(1 - \frac{V_A}{V_D}\right)}{V_C \left(1 - \frac{V_B}{V_C}\right)}.$$

Now, along the adiabatic CD, we have $P_{BC}V_C^\gamma = P_{DA}V_D^\gamma$ and along the adiabatic AB, we have $P_{BC}V_B^\gamma = P_{DA}V_A^\gamma$. This leads to

$$\left(\frac{V_C}{V_B}\right)^\gamma = \left(\frac{V_D}{V_A}\right)^\gamma; \quad \Rightarrow \quad \left(\frac{V_B}{V_C}\right) = \left(\frac{V_A}{V_D}\right).$$

Using this relation, we get the expression for efficiency as

$$\eta = 1 - \frac{P_{DA}}{P_{BC}} \frac{V_D}{V_C}.$$

Again, $\left(\frac{V_D}{V_C}\right)^\gamma = \frac{P_{BC}}{P_{DA}} \quad \Rightarrow \quad \frac{V_D}{V_C} = \frac{1}{\left(\frac{P_{DA}}{P_{BC}}\right)^{\frac{1}{\gamma}}}$. Using this result, the efficiency becomes

$$\eta = 1 - \frac{P_{DA}}{P_{BC}} \frac{1}{\left(\frac{P_{DA}}{P_{BC}}\right)^{\frac{1}{\gamma}}} = 1 - \left(\frac{P_{DA}}{P_{BC}}\right)^{\frac{\gamma-1}{\gamma}}.$$

18. Find the thermal efficiency of a reversible heat engine shown in Figure 9.28 operating in a cycle consisting of two isotherms and two isobars, using ideal gas as the working substance. Assume the specific heats C_P and C_V are constants throughout the process.

Answer:

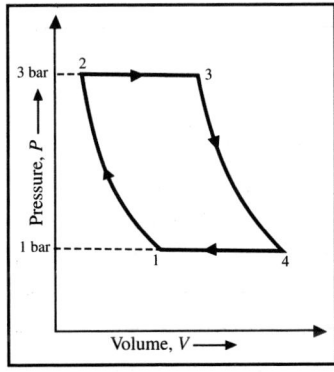

Figure 9.28 $P-V$ diagram for the thermodynamic process in problem 18.

In the above figure, $3 \to 4$ and $1 \to 2$ are two isotherms, and $2 \to 3$ and $4 \to 1$ are two isobars.

The total amount of heat received along the isobar $2 \to 3$ and the isotherm $3 \to 4$ is given by

$$Q_1 = C_P(T_1 - T_2) + RT_1 \ln \left(\frac{V_4}{V_3} \right),$$

where T_1 and T_2 are, respectively, the temperatures of the isotherms $3 \to 4$ and $1 \to 2$. The above expression can be written as

$$Q_1 = C_P(T_1 - T_2) + RT_1 \ln \left(\frac{P_{23}}{P_{41}} \right).$$

The total amount of heat rejected along the isobar $4 \to 1$ and the isotherm $1 \to 2$ is given by

$$Q_2 = C_P(T_1 - T_2) + RT_2 \ln \left(\frac{V_1}{V_2} \right) = C_P(T_1 - T_2) + RT_2 \ln \left(\frac{P_{23}}{P_{41}} \right).$$

So, the efficiency is given by

$$\eta = \frac{Q_1 - Q_2}{Q_1} = 1 - \frac{Q_2}{Q_1} = \frac{R(T_1 - T_2) \ln \left(\frac{P_{23}}{P_{41}} \right)}{C_P(T_1 - T_2) + RT_1 \ln \left(\frac{P_{23}}{P_{41}} \right)}.$$

19. Find the thermal efficiency of a reversible heat engine shown in Figure 9.29 operating in a cycle consisting of two isotherms T_{23} and T_{41} with $(T_{23} > T_{41})$ and two isochores V_{41} and V_{23} with $V_{41} > V_{23}$, using ideal gas as working substance. Assume the specific heats C_P and C_V are constant throughout the process.

Answer:

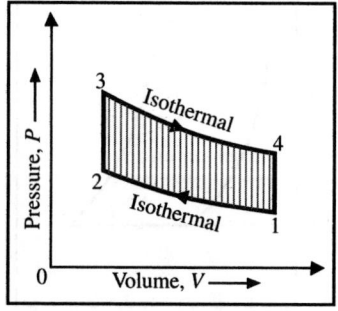

Figure 9.29 P–V diagram for the thermodynamic process in problem 19.

In the above figure, $3 \rightarrow 4$ and $1 \rightarrow 2$ are two isotherms, and $4 \rightarrow 1$ and $2 \rightarrow 3$ are two isochores.

The total amount of heat received along the isotherm $3 \rightarrow 4$ and the isochoric $4 \rightarrow 1$ is given by

$$Q_1 = RT_{34} \ln \left(\frac{V_{41}}{V_{23}} \right) + C_V(T_{34} - T_{12}).$$

The total amount of heat rejected along the isotherm $1 \rightarrow 2$ and the isochore $2 \rightarrow 3$ is given by

$$Q_2 = RT_{12} \ln \left(\frac{V_{41}}{V_{23}} \right) + C_V(T_{34} - T_{12}).$$

So, the efficiency is given by

$$\eta = \frac{Q_1 - Q_2}{Q_1} = 1 - \frac{Q_2}{Q_1} = \frac{R(T_{34} - T_{12}) \ln \left(\frac{V_{41}}{V_{23}} \right)}{C_V(T_{34} - T_{12}) + RT_{34} \ln \left(\frac{V_{41}}{V_{23}} \right)}.$$

20. Find the thermal efficiency of a reversible heat engine shown in Figure 9.30 operating in cycle $ABCA$.

    Answer:

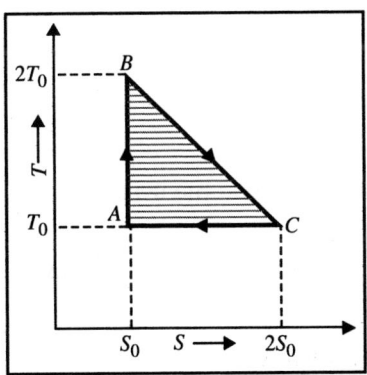

Figure 9.30 $T-S$ diagram for the thermodynamic process in problem 20.

The efficiency is given by $\eta = \frac{W}{Q_1} = \frac{\text{area } ABC}{\text{area under } BC}.$

As AB is adiabatic ($\delta Q = 0$), heat is added in BC and rejected in AC.

So, the efficiency of the heat engine operating in the cycle $ABCA$ is given by

$$\eta = \frac{\frac{1}{2} \times (2S_0 - S_0) \times (2T_0 - T_0)}{\frac{1}{2} \times (2S_0 - S_0) \times (2T_0 - T_0) + T_0 \times (2S_0 - S_0)}; \quad \Rightarrow \quad \eta = \frac{\frac{S_0 T_0}{2}}{\frac{S_0 T_0}{2} + T_0 S_0};$$

$$\Rightarrow \quad \eta = \frac{1}{3}.$$

21. Find the efficiency of the cycle $ABCDA$ shown in Figure 9.31 if the helium gas is used as a working fluid. Assume that helium is an ideal gas. Given that for helium, $C_V = \frac{3R}{2}$ and $C_P = \frac{5R}{2}$.

Answer:

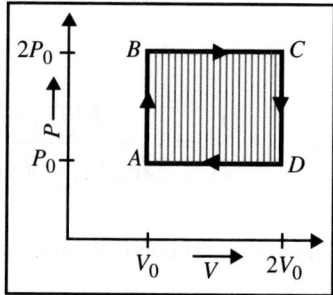

Figure 9.31 P–V diagram for the thermodynamic process in problem 21.

From the figure, we have the expression for work done as

$$W = \text{area of cycle } ABCDA = P_0 V_0.$$

Heat is added to the thermodynamic system.

$$Q_1 = C_V dT + C_P dT = \frac{3R}{2} \Delta T + \frac{5R}{2} \Delta T.$$

Assuming helium as an ideal gas,

$$PV = RT; \quad \Rightarrow \quad PdV + VdP = RdT; \quad \Rightarrow \quad PdV = RdT. \text{ [for constant } P,$$
$$dP = 0] \text{ and } \quad \Rightarrow \quad VdP = RdT \text{ [for constant } V, dV = 0].$$

Hence, the amount of heat added to the system is

$$Q_1 = \frac{3R}{2} \times \frac{V_0}{R} \times \Delta P + \frac{5R}{2} \times \frac{2P_0}{R} \times \Delta V = \frac{3V_0}{2} \times P_0 + 5P_0 \times V_0 = \frac{13}{2} P_0 V_0.$$

Hence, the efficiency η of the cycle is given by

$$\eta = \frac{W}{Q_1} = \frac{P_0V_0}{\frac{13}{2}P_0V_0} = \frac{2}{13}.$$

22. In the cycle $ABCA$ shown in Figure 9.32, the amount of heat added to a thermodynamic system in the process AB and BC are, respectively, 400 J and 100 J. Heat rejected during the process CA is 460 J. Find the efficiency of the cycle.

Answer:

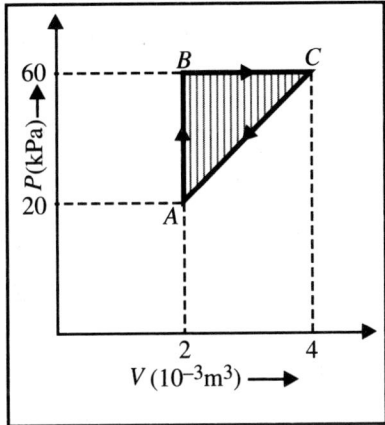

Figure 9.32 P–V diagram for the thermodynamic process in problem 22.

From the figure, we have the expression for work done as

$$W = \text{area of cycle} = \frac{1}{2} \times 2 \times 10^{-3} \times 40 \times 10^3 = 40 \text{ J}.$$

Heat added to the thermodynamic system $Q_1 = 400$ J $+ 100$ J $= 500$ J.

Hence, the efficiency η of the cycle is given by

$$\eta = \frac{W}{Q_1} = \frac{40}{500} = 8\%.$$

23. Compare the thermal efficiencies of the cycles $ABCA$ and $EFGE$ shown in Figures 9.33 (a) and 9.33 (b), respectively. If $T_1 = 1000$ K, and $T_2 = 500$ K, find the ratio $\frac{\eta_1}{\eta_2}$, where η_1 and η_2 are the efficiencies of the cycles ABC and EFG, respectively.

Answer:

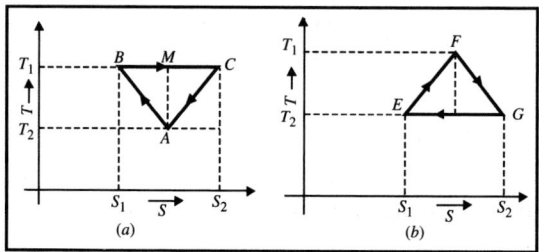

Figure 9.33 T–S diagram for the thermodynamic processes in problems 23.

From the figure at the left, we have the efficiency given by $\eta_1 = \frac{W}{Q_1} = \frac{\text{area } ABC}{\text{area under } BC}$.

Area of $ABC = \frac{1}{2}.BC.AM = \frac{1}{2} \times (S_2 - S_1) \times (T_1 - T_2)$.

Hence, the efficiency η_1 is given by

$$\eta_1 = \frac{\frac{1}{2} \times (S_2 - S_1) \times (T_1 - T_2)}{T_1 \times (S_2 - S_1)} = \frac{1}{2}\left(1 - \frac{T_2}{T_1}\right).$$

From the figure at the right, we have the efficiency given by

$$\eta_2 = \frac{W}{Q_2} = \frac{\frac{1}{2} \times (S_2 - S_1) \times (T_1 - T_2)}{\frac{1}{2} \times (S_2 - S_1) \times (T_1 - T_2) + T_2(S_2 - S_1)} = \frac{T_1 - T_2}{(T_1 - T_2) + 2T_2}$$
$$= \frac{(T_1 - T_2)}{(T_1 + T_2)}.$$

Hence, the ratio of the efficiencies given by

$$\frac{\eta_1}{\eta_2} = \frac{\frac{1}{2}\left(1 - \frac{T_2}{T_1}\right)}{\frac{(T_1 - T_2)}{(T_1 + T_2)}} = \frac{T_1 + T_2}{2T_1} = \frac{1}{2}\left(1 + \frac{T_2}{T_1}\right).$$

Using the values of T_1 and T_2, we get

$$\frac{\eta_1}{\eta_2} = \frac{1}{2}\left(1 + \frac{500}{1000}\right) = \frac{3}{4}.$$

24. A box of gas is joined by an identical box of the same gas. What is the change in total entropy?

Answer: This is a reversible process (as nothing changes when removing the partition and putting it back). Total entropy change is therefore zero, that is, $\Delta S_{total} = 0$, where $S_{total} = S_1 + S_2 = 2S_1$.

25. An insulated container is originally divided in half by a partition, and each half is occupied by n moles of a different ideal gas at temperature T_0. What is the total change in entropy when the partition is removed without doing any work?

Answer: Since both gases are ideal, they don't interact with each other, each species is oblivious of the other, and ΔS for each gas is exactly $nR\ln 2$. Thus, the total change in entropy is $\Delta S_{total} = 2nR\ln 2$.

Note: The total mixing—which we know will happen eventually—is precisely the change that now maximizes the entropy, that is, a clear hint that entropy means disorder. It is obvious that the process is an irreversible one as the situation is quite different from the initial state when putting back the partition after mixing.

26. A lump of steel of mass 30 kg at 427°C is dropped in 100 kg oil at 27°C. The specific heats of steel and oil are 0.5 $Jkg^{-1}K^{-1}$ and 3.0 $Jkg^{-1}K^{-1}$, respectively. Calculate the change in entropy of steel, oil, and the universe.

Answer: Let T be the final temperature. We have then

$$(mC_P\Delta T)_{steel} = (mC_P\Delta T)_{oil}; \quad \Rightarrow \quad 300 \times 0.5(700 - T)$$
$$= 100 \times 3 \times (T - 300) \quad \Rightarrow \quad T = 319 \text{ K}.$$

So, the change in entropy of steel is

$$(\Delta S)_{steel} = \int_1^2 \frac{\delta Q}{T} = \int_1^2 \frac{mC_P}{T}dT = \left[mC_P\ln\frac{T_2}{T_1}\right]_{steel} = 30 \times 0.5 \times \ln\left(\frac{319}{700}\right)$$
$$= -11.7883 \text{ kJ K}^{-1}.$$

The change in entropy of oil is

$$(\Delta S)_{oil} = \int_1^2 \frac{\delta Q}{T} = \int_1^2 \frac{mC_P}{T}dT = \left[mC_P\ln\frac{T_2}{T_1}\right]_{oil} = 100 \times 3 \times \ln\left(\frac{319}{300}\right)$$
$$= 18.4226 \text{ kJ K}^{-1}.$$

Hence, the change in entropy of the universe is given by

$$(\Delta S)_{\text{universe}} = (-11.7883 + 18.4226) \text{ kJ K}^{-1} = 6.6343 \text{ kJ K}^{-1}.$$

27. Two thermally isolated identical systems have heat capacity, which varies as $C_V = \beta T^3$ (where $\beta > 0$). Initially, one system is at 300 K and the other at 400 K. The systems are then brought into thermal contact, and the combined system is allowed to reach thermal equilibrium. The final temperature of the combined system is $\cdots$. [JAM 2013]

Answer: The final temperature is within $357 \text{ K} \leq T \leq 360 \text{ K}$.

There is not a unique value of temperature; instead, there is a range of temperature. The maximum temperature will occur when the work done is zero. So, $\delta Q_1 + \delta Q_2 = 0$.

$$m\beta \int_{300}^{T_{\text{max}}} T^3 dT + m\beta \int_{400}^{T_{\text{max}}} T^3 dT = 0 \Rightarrow 2\frac{T_{\text{max}}^4}{4} - \frac{300^4}{4} - \frac{400^4}{4} = 0;$$

$$\Rightarrow T_{\text{max}}^4 = \frac{(300)^4 + (400)^4}{2}; \quad \Rightarrow T_{\text{max}} = 360 \text{ K}.$$

The minimum temperature of the system will occur when the process is reversible. So, the change in entropy of the system is zero, that is, $\Rightarrow \Delta S_1 + \Delta S_0 = 0$. So, we have

$$m\beta \int_{300}^{T_{\text{min}}} \frac{T^3 dT}{T} + m\beta \int_{400}^{T_{\text{min}}} \frac{T^3 dT}{T} = 0 \Rightarrow T_{\text{min}}^3 = \frac{(300)^2 + (400)^2}{2} \Rightarrow T_{\text{min}} = 357 \text{ K}.$$

So, the temperature range is within $357 \text{ K} \leq T \leq 360 \text{ K}$.

28. Show that entropy increases when two gases diffuse into each other at the same temperature and pressure. [Calcutta University 2013, 2014]

Answer: Let a number of gases at the same pressure P and the same temperature T get spontaneously mixed up by diffusion.

The total entropy before diffusion is

$$S_A = n_1 C_{V_1} \ln T + n_1 R \ln \left(\frac{T}{P}\right) + n_1 C_1 + n_2 C_{V_2} \ln T + n_2 R \ln \left(\frac{T}{P}\right)$$

$$+ n_2 C_2 + \cdots.$$

$$= \sum_i n \left(n_i C_{V_i} \ln T + n_i R \ln \left(\frac{T}{P} + n_i C_i\right)\right),$$

where n_i is the number of moles of the ith gas.

Total entropy, after diffusion, is given by

$$S_B = \sum_i \left(n_i C_{V_i} \ln T + n_i R \ln \left(\frac{T}{K_i P} \right) + n_i C_i \right).$$

Hence, the increase in entropy is given by $\Delta S = S_B - S_A = -\sum_i (n_i R \ln(K_i))$.

As $K_i = \frac{n_i}{\sum_i n_i}$, we have the increase in entropy as

$$\Delta S = S_A - S_B = -\sum_i n_i R \ln \left(\frac{n_i}{\sum_i n_i} \right),$$

which is obviously positive. Therefore, the entropy increases when two gases at the same temperature and pressure are diffused into each other.

29. Calculate ΔS due to diffusion of 1 mol of Helium and 1 mol of Nitrogen at the same pressure and temperature.   [Calcutta University 2017]

    Answer: The mole fraction of both Helium and Nitrogen is 1/2. So, the change in entropy is given by

    $$\Delta S_{\text{diffusion}} = -R(1 \times \ln \frac{1}{2} + 1 \times \ln \frac{1}{2}) = -2R \ln \frac{1}{2} = 2R \ln 2 = 16.63 \text{ J/K},$$

    where R is the universal gas constant, $R = 8.314 \text{ JK}^{-1}$.

    Generally, the change in entropy is given by

    $$\Delta S_{\text{diffusion}} = -R(n_1 \ln X_1 + n_2 \ln X_2),$$

    where n_1, n_2 are the number of moles participating in diffusion and x_1, x_2 are the mole fractions of the gases.

30. For a thermodynamic system, show that $C_P - C_V = \alpha^2 \kappa_T T V$, where α is the coefficient of volume expansion, and κ_T is the isothermal bulk modulus.
                                                                [Calcutta University 2018]

    Answer: From Maxwell relation, we have $(\frac{\partial S}{\partial P})_T = (\frac{\partial V}{\partial T})_P$. Multiplying both sides by T, we have

    $$T \left(\frac{\partial S}{\partial P} \right)_T = T \left(\frac{\partial V}{\partial T} \right)_P.$$

Considering entropy S as a function of temperature T and volume V, that is, $S = S(T, V)$, we can write

$$dS = \left(\frac{\partial S}{\partial T}\right)_V dT + \left(\frac{\partial S}{\partial V}\right)_T dV; \Rightarrow \left(\frac{\partial S}{\partial T}\right)_P = \left(\frac{\partial S}{\partial T}\right)_V + \left(\frac{\partial S}{\partial V}\right)_T \left(\frac{\partial V}{\partial T}\right)_P.$$

Multiplying both sides by T, $\quad T\left(\frac{\partial S}{\partial T}\right)_P = T\left(\frac{\partial S}{\partial T}\right)_V + T\left(\frac{\partial S}{\partial V}\right)_T \left(\frac{\partial V}{\partial T}\right)_P.$

Using Maxwell's relation, we have $\Rightarrow (\frac{\partial S}{\partial V})_T = (\frac{\partial P}{\partial T})_V.$

This leads to

$$C_P - C_V = T\left(\frac{\partial P}{\partial T}\right)_V \left(\frac{\partial V}{\partial T}\right)_P.$$

From the definitions, we have $\alpha = \frac{1}{V}(\frac{\partial V}{\partial T})_P$ and $\kappa_T = -V(\frac{\partial P}{\partial V})_T.$

Hence, we get

$$\alpha^2 \kappa_T = -\left(\frac{\partial P}{\partial V}\right)_T \left(\frac{\partial V}{\partial T}\right)_P \left(\frac{\partial V}{\partial T}\right)_P \frac{1}{V} = \left(\frac{\partial P}{\partial T}\right)_V \left(\frac{\partial V}{\partial T}\right)_P \frac{1}{V};$$

$$\Rightarrow \quad \alpha^2 \kappa_T TV = \left(\frac{\partial P}{\partial T}\right)_V \left(\frac{\partial V}{\partial T}\right)_P T; \quad \Rightarrow \quad C_P - C_V = \alpha^2 \kappa_T TV.$$

9.12   Multiple choice questions and answers

1. The efficiency of a perfectly reversible (Carnot) heat engine operating between absolute temperature T and zero is equal to   [JEST 2012]

   (a) 0

   (b) 0.5

   (c) 0.75

   (d) 1

   Answer: The correct choice is (d).

   Solution: $\eta = 1 - \frac{T_2}{T_1} = 1 - \frac{0}{T} = 1.$

2. The efficiency of a heat engine can never be

   (a) 10%

   (b) 40%

   (c) 60%

   (d) 100%

   Answer: The correct choice is (d).

3. Consider a Carnot engine operating between temperatures of 600 K and 400 K. The engine does 1000 J of work per cycle. The heat (in Joules) extracted per cycle from the high-temperature reservoir is $\cdots$. (Specify your answer to two digits after the decimal point)   [JAM 2017]

(a) 300 J

(c) 3000 J

(b) 150 J

(d) 30 J

Answer: The correct choice is (c).

Solution: The efficiency is $\eta = 1 - \frac{T_2}{T_1} = 1 - \frac{400}{600} = \frac{1000}{Q_1} \Rightarrow \frac{1000}{Q_1} = \frac{2}{6} \Rightarrow Q_1 = 3000$ J.

4. Consider an ideal gas of mass m at temperature T_1, which is mixed isobarically (i.e., at constant pressure) with an equal mass of the same gas at temperature T_2 in a thermally insulated container. What is the change in entropy of the universe?   [JEST 2012]

(a) $2mC_p \ln \left(\frac{T_1 + T_2}{2\sqrt{T_1 T_2}} \right)$

(c) $2mC_p \ln \left(\frac{T_1 + T_2}{2T_1 T_2} \right)$

(b) $2mC_p \ln \left(\frac{T_1 - T_2}{2\sqrt{T_1 T_2}} \right)$

(d) $2mC_p \ln \left(\frac{T_1 - T_2}{2\sqrt{T_1 T_2}} \right)$

Answer: The correct choice is (a).

Solution: Let the final temperature be T. This can be calculated in the following way:

$$mC(T_1 - T_2) = mC(T - T_2) \Rightarrow T = (T_1 + T_2)/2.$$

Using the expression $\Delta S_1 = mC_p \frac{\Delta T}{T}$, the change in entropy is given by

$$\Delta S = \Delta S_1 + \Delta S_2 \Rightarrow \Delta S = mC_p \int_{T_1}^{T} \frac{\delta T}{T} + mC_p \int_{T_2}^{T} \frac{\delta T}{T}$$

$$= mC_p \ln \left(\frac{T}{T_1} \right) + mC_p \ln \left(\frac{T}{T_2} \right); \quad \Rightarrow \Delta S = 2mC_p \ln \frac{T}{\sqrt{T_1 T_2}}$$

$$= mC_p \ln \left(\frac{T_1 + T_2}{2\sqrt{T_1 T_2}} \right)^2 = 2mC_p \ln \left(\frac{T_1 + T_2}{2\sqrt{T_1 T_2}} \right).$$

5. The entropy–temperature diagram of two Carnot engines, A and B, is shown in Figure 9.34. The efficiencies of the engines are η_A and η_B, respectively. Which one of the following equalities is correct?   [JEST 2015]

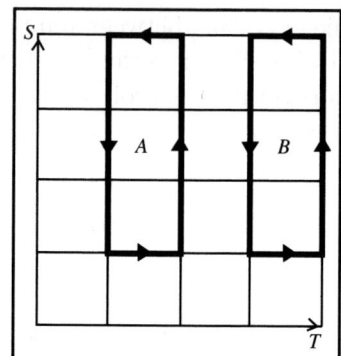

Figure 9.34 S–T diagram for the thermodynamic processes in problem 5.

(a) $\eta_A = \frac{\eta_B}{2}$ **(b)** $\eta_A = \eta_B$ **(c)** $\eta_A = 3\eta_B$ **(d)** $\eta_A = 2\eta_B$

Answer: The correct choice is (d).

Solution: We know the efficiency is $\eta = \frac{\Delta W}{Q_1}$, where $\Delta W=$ area under the curve, $Q_1 =$ area under high temperature. The efficiencies of the two Carnot engines are

$$\eta_A = \frac{(2T - T)(3S - 0)}{2T(3S - 0)} = \frac{T}{2T} = \frac{1}{2} \text{ and } \eta_B = \frac{(4T - 3T)}{4T}\frac{(3S - 0)}{(3S - 0)} = \frac{T}{4T} = \frac{1}{4}.$$

Hence, the ratio of the efficiency is

$$\frac{\eta_A}{\eta_B} = \frac{1/2}{1/4} = 2 \quad \Rightarrow \quad \eta_A = 2\eta_B.$$

6. Experimental measurements of heat capacity per mole of aluminum at low-temperatures show that the data can be fitted to the formula $C_V = aT + bT^2$, where $a = 0.00135 \text{ JK}^{-2}\text{mol}^{-1}$, $b = 2.48 \times 10^{-5} \text{ JK}^{-4}\text{mol}^{-1}$, and T is the temperature in Kelvin. The entropy of a mole of aluminum at such temperatures is given by the formula   [JAM 2007]

   (a) $aT + \frac{b}{3}T^3 + c$, where $c > 0$ is a constant

   (b) $\frac{aT}{2} + \frac{b}{4}T^3 + c$, where $c > 0$ is a constant

   (c) $aT + \frac{b}{3}T^3$

   (d) $\frac{aT}{2} + \frac{b}{4}T^3$

   Answer: The correct choice is (a).

   Solution: $dS = \int \frac{C_v dT}{T} \Rightarrow S = \int \frac{aT + bT^3}{T} dT = aT + \frac{b}{3}T^3 + C.$

7. The entropy of a gas containing N particles enclosed in a volume V is given by $S = NK_B \ln \left(\frac{aVE^{\frac{3}{2}}}{N^{\frac{5}{2}}} \right)$, where E is the total energy, a is a constant, and K_B is the Boltzmann constant. The chemical potential μ of the system at a temperature T is given by [GATE 2015]

(a) $\mu = -K_B T \left[\ln \left(\frac{aVE^{\frac{3}{2}}}{N^{\frac{5}{2}}} \right) - \frac{5}{2} \right]$

(c) $\mu = -K_B T \left[\ln \left(\frac{aVE^{\frac{3}{2}}}{N^{\frac{3}{2}}} \right) - \frac{5}{2} \right]$

(b) $\mu = -K_B T \left[\ln \left(\frac{aVE^{\frac{3}{2}}}{N^{\frac{5}{2}}} \right) - \frac{3}{2} \right]$

(d) $\mu = -K_B T \left[\ln \left(\frac{aVE^{\frac{3}{2}}}{N^{\frac{3}{2}}} \right) - \frac{3}{2} \right]$

Answer: The correct choice is (a).

Solution: The entropy S is related to Gibbs free energy G by $\left(\frac{\partial G}{\partial T} \right)_P = -S$. According to the problem, entropy is given by

$$S = N\, K_B\, \ln \left(\frac{aVE^{\frac{3}{2}}}{N^{\frac{5}{2}}} \right).$$

Therefore, we have

$$\left(\frac{\partial G}{\partial T} \right)_P = -NK_B \ln \left(\frac{aVE^{\frac{3}{2}}}{N^{\frac{5}{2}}} \right); \quad \Rightarrow \quad G = -NK_B T \ln \left(\frac{aVE^{\frac{3}{2}}}{N^{\frac{5}{2}}} \right) + \ln A.$$

The chemical potential μ now becomes

$$\mu = \left(\frac{\partial G}{\partial N} \right) = - \left[K_B T \ln \left(\frac{aVE^{\frac{3}{2}}}{N^{\frac{5}{2}}} \right) + NK_B T \frac{N^{\frac{5}{2}}}{aVE^{\frac{3}{2}}} \frac{\left(-\frac{5}{2} \right)}{N^{\frac{7}{2}}} aVE^{\frac{3}{2}} \right]$$

$$= -K_B T \left[\ln \left(\frac{aVE^{\frac{3}{2}}}{N^{\frac{5}{2}}} \right) - \frac{5}{2} \right].$$

8. A reversible Carnot engine is operated between the temperatures T_1 and T_2 $(T_2 > T_1)$ with a phonon gas as a working substance. The efficiency of the engine is [GATE 2017]

(a) $1 - \frac{3T_1}{4T_2}$

(c) $1 - \left(\frac{T_1}{T_2} \right)^{\frac{3}{4}}$

(b) $1 - \frac{T_1}{T_2}$

(d) $1 - \left(\frac{T_1}{T_2} \right)^{\frac{4}{3}}$

Answer: The correct choice is (b).

Solution: The efficiency of a Carnot engine does not depend on the nature of the working substance but rather depends on the temperatures of the source and the sink. Hence, we have $\eta = 1 - \frac{T_1}{T_2}$.

9. Two gases separated by an impermeable but movable partition are allowed to exchange energy freely. At equilibrium, the two sides will have the same

   [GATE 2013]

   (a) Pressure and temperature
   (b) Volume and temperature
   (c) Pressure and volume
   (d) Volume and energy

   Answer: The correct choice is (a).

10. For a certain thermodynamic system, the internal energy $U = PV$ and P is proportional to T^2. The entropy of the system is proportional to

    [IIT-JAM 2022]

    (a) UV

    (b) $\sqrt{\frac{U}{V}}$

    (c) $\sqrt{\frac{V}{U}}$

    (d) $\sqrt{UV}$

    Answer: The correct choice is (d).

    Solution: Here, $U = PV$ and $dU = PdV + VdP$. As $P = \beta T^2$, $dP = 2\beta TdT$. Then, from the first law of thermodynamics, $\delta Q = dU + PdV$,

    $$\delta Q = dU + PdV = PdV + VdP + PdV = 2PdV + VdP$$
    $$= 2\beta T^2 dV + 2V\beta TdT; \quad \Rightarrow \quad TdS = 2\beta T^2 dV + 2V\beta TdT.$$

    This leads to $S = 2\beta TdT + 2V\beta dT = d(2\beta VT)$.

    Thus, $S \propto 2\beta VT$. Again, $U = \beta T^2 V$. This gives $T = \sqrt{\frac{U}{\beta V}}$.

    Hence, we get $S \propto 2\beta V \sqrt{\frac{U}{\beta V}} \propto \sqrt{UV}$.

11. If the mean square fluctuations in energy of a system in equilibrium at temperature T is proportional to T^α, then the energy of the system is proportional to

    [JEST 2017]

    (a) $T^{\alpha-2}$

    (b) $T^{\frac{\alpha}{2}}$

    (c) $T^{\alpha-1}$

    (d) T^α

    Answer: The correct choice is (c).

    Solution: $(\Delta E)^2 = kT^2 C_v \Rightarrow T^{\alpha-2} \propto C_v \Rightarrow T^{\alpha-2} \propto (\frac{\partial U}{\partial T})_V \Rightarrow U \propto T^{\alpha-1}$.

12. Consider the following statements regarding the characteristics of entropy. Which of the following options is/are correct?

    (a) Entropy is a measure of disorder.
    (b) Entropy changes during a reversible adiabatic process.
    (c) Entropy of a system decreases in all irreversible processes.

(d) Change in entropy for a complete reversible thermodynamic cycle is zero.

Answer: The correct choices are (a) and (d).

13. A Carnot engine works between a hot reservoir at temperature T_1 and a cold reservoir at temperature T_2. To increase its efficiency

(a) T_1 and T_2 both should be increased.
(b) T_1 and T_2 both should be decreased.
(c) T_1 should be increased and T_2 decreased.
(d) T_1 should be decreased and T_2 increased.

Answer: The correct choice is (c).

14. Which is the most effective to increase the efficiency of a Carnot engine?

(a) Increase of temperature of source by $50°C$.
(b) Decrease of temperature of source by $50°C$.
(c) Increase of temperature of source by $25°C$ and decrease of temperature of sink by $25°C$.
(d) All of the above are equally effective.

Answer: The correct choice is (b).

15. The internal energy $E(T)$ of a system at a fixed volume is found to depend on the temperature T as $E(T) = \frac{aT^2}{2} + \frac{bT^4}{4}$. Then the entropy $S(T)$, as a function of temperature, is

(a) $\frac{1}{2} aT^2 + \frac{1}{4} bT^4$ (c) $2aT + \frac{4}{3} bT^3$
(b) $2aT^2 + 4bT^4$ (d) $aT + \frac{bT^3}{3}$

Answer: The correct choice is (d).

Solution: From the first and second law of thermodynamics, we have

$$TdS = dE + PdV; \quad \Rightarrow \quad dE = TdS - PdV.$$

As the volume remains constant, $dV = 0$. Hence, we have

$$dS = \frac{1}{T}dE = \frac{1}{T}d\left[\frac{aT^2}{2} + \frac{bT^4}{4}\right] = a\ dT + bT^2 dT.$$

Integrating we get

$$S = a\ T + \frac{bT^3}{3}.$$

16. It is not possible in a cyclic process to extract heat from a reservoir and convert completely into work. This statement is due to

    (a) Maxwell and Boltzmann **(c)** Clausius

    (b) Kelvin and Planck **(d)** Wien

    Answer: The correct choice is (b).

17. The entropy of a system in an irreversible process

    (a) Increases

    (b) Decreases

    (c) Remains constant

    (d) May increase or decrease depending on the nature of process

    Answer: The correct choice is (a).

18. In regard to entropy, which of the following statements is false? [JAM 2009]

    (a) In a reversible process, the entropy change of the universe is zero.

    (b) For any process, the entropy of the universe never decreases.

    (c) In an irreversible process, the entropy of the universe increases.

    (d) When a system changes state, the resulting entropy change depends upon the process by which the change of state occurs.

    Answer: The correct choice is (d).

19. Which relation represents Clausius theorem

    (a) $\oint \frac{\delta Q}{T} = 0$ **(c)** $\oint \frac{\delta Q}{T} > 0$

    (b) $\oint \frac{\delta Q}{T^2} \neq 0$ **(d)** $\oint \frac{\delta Q}{T} < 0$

    Answer: The correct choice is (a).

20. An ideal gas with an adiabatic exponent γ undergoes a process in which the pressure P is related to its volume V by $P = P_0 - \alpha V$, where P_0 and α are positive constants. The volume starts from being very close to zero and increases monotonically to $\frac{P_0}{\alpha}$. At what value of the volume during the process does the gas have maximum entropy? [JEST 2016]

    (a) $\frac{P_0}{\alpha(1+\gamma)}$ **(c)** $\frac{\gamma P_0}{\alpha(1+\gamma)}$

    (b) $\frac{\gamma P_0}{\alpha(1-\gamma)}$ **(d)** $\frac{P_0}{\alpha(1-\gamma)}$

    Answer: The correct choice is (c).

    Solution: We know that $TdS = nC_v dT + PdV \Rightarrow TdS = \frac{nRdT}{(\gamma-1)} + PdV$.

For maximum entropy, we have $dS = 0$. For n mole of an ideal gas,

$$PV = nRT; \quad \Rightarrow \quad PdV + VdP = nRdT,$$

$$\Rightarrow TdS = \frac{PdV + VdP}{(\gamma - 1)} + PdV \Rightarrow \frac{PV}{nR}dS = \frac{\gamma}{(\gamma - 1)}PdV + \frac{VdP}{(\gamma - 1)}.$$

Since $P = P_0 - \alpha V \quad \Rightarrow \quad dP = -\alpha dV$. From the above expression, we get

$$\frac{PV}{nR}dS = \frac{\gamma}{(\gamma - 1)}PdV - \frac{\alpha V dV}{(\gamma - 1)} \Rightarrow \frac{dS}{dV} = \frac{\gamma nRP}{(\gamma - 1)PV} - \frac{nR}{(\gamma - 1)PV}\alpha V.$$

For maximum entropy,

$$\frac{dS}{dV} = 0 \Rightarrow \gamma P - \alpha V = 0 \Rightarrow \gamma(P_0 - \alpha V) = \alpha V; \quad \Rightarrow \quad V = \frac{\gamma P_0}{\alpha(1 + \gamma)}.$$

21. A solid metallic cube of heat capacity S is at temperature 300 K. It is brought in contact with a reservoir at 600 K. If the heat transfer takes place only between the reservoir and the cube, the entropy change of the universe after reaching the thermal equilibrium is   [JAM 2014]

   (a) 0.69 S   (b) 0.54 S   (c) 0.27 S   (d) 0.19 S

   Answer: The correct choice is (d).

   Solution: Heat taken by the cube is given by $= S(600 - 300)\text{J} = 300 \text{ SJ}$.

   The change in entropy of the reservoir $= -\frac{dQ}{T} = -\frac{300}{600}S = -0.5 \text{ S}$

   Hence, the change in entropy of the cube $= 5\ln\frac{600}{300} = 0.69 \text{ S}$.

   So, the total change of entropy of the universe is: $\Delta_{(\text{univ})} = -0.5 \text{ S} + 0.69 \text{ S} = 0.19 \text{ S}$.

22. A heat engine converts a given quantity of heat into work with maximum efficiency during which one of the following processes?

   (a) Isobaric process
   (b) Isochoric process
   (c) Isoenthalpic process
   (d) Isothermal process

   Answer: The correct choice is (c).

   Solution: The first law of thermodynamics $dQ = dU + PdV$. The maximum efficiency can be obtained, if the process is isoenthalpic.

23. The entropy function of a system is given by $S(E) = a\,E(E_0 - E)$, where a and E_0 are positive constants. The temperature of the system is [GATE 2013]

(a) Negative for some energy.

(b) Increase monotonically with energy.

(c) Decreases monotonically with energy.

(d) Zero.

Answer: The correct choice is (a).

From the first and second laws of thermodynamics, we have

$TdS = dU + PdV \Rightarrow dS = \frac{1}{T}(dU + PdV) \Rightarrow (\frac{\partial S}{\partial E})_V = \frac{1}{T}$ as here $E = U$.

$S(E) = aE(E_0 - E) \Rightarrow (\frac{\partial S}{\partial E})_V = a(E_0 - E) - aE = a(E_0 - 2E) \Rightarrow T = \frac{1}{a(E_0-E)}$.

24. Each of two isolated vessels, A and B of fixed volumes, contains N molecules of a perfect monoatomic gas at a pressure P. The temperatures of A and B are T_1 and T_2, respectively. The two vessels are brought into thermal contact. At equilibrium, the change in entropy is

(a) $\frac{3}{2}NK_B \ln\left[\frac{(T_1^2+T_2^2)}{4T_1T_2}\right]$

(c) $\frac{3}{2}NK_B \ln\left[\frac{(T_1+T_2)^2}{4T_1T_2}\right]$

(b) $\frac{3}{2}NK_B \ln\left[\frac{T_2}{T_1}\right]$

(d) $2NK_B$

Answer: The correct choice is (c).

Solution: At equilibrium, the final temperature of each vessel is $T = \frac{T_1+T_2}{2}$. So, the change in entropy ΔS will be given by

$$\Delta S = N \times \int_{T_1}^{T} \frac{C_V dT}{T} + N \times \int_{T_2}^{T} \frac{C_V dT}{T} = \frac{3}{2}NK_B \ln\left[\frac{(T_1+T_2)^2}{4T_1T_2}\right],$$

where the specific heat at constant volume C_V for monoatomic gas is $C_V = \frac{3K_B}{2}$.

25. The temperature of water (mass, m) increases from T_1 to T_2. If c is the specific heat capacity of water, then the total increase in entropy of water is given by:

(a) $mc(T_2 - T_1)$

(c) $mc(T_1 - T_2)$

(b) $mc \ln\left(\frac{T_1}{T_2}\right)$

(d) $mc \ln\left(\frac{T_2}{T_1}\right)$

Answer: The correct choice is (d).

The change of entropy of a system is given as $dS = \int_{T_1}^{T_2} \frac{\delta Q}{T}$.

We know that $\delta Q = mcdT$. Hence, the change in entropy is given by

$$dS = \int_{T_1}^{T_2} \frac{mcdT}{T} = mc \ \ln\left(\frac{T_2}{T_1}\right).$$

26. One mole of an ideal gas is carried from temperature T_1 and molar volume V_1 to T_2, V_2. Then, the change in entropy is given by

(a) $\Delta S = R \ln \frac{V_2}{V_1}$

(c) $\Delta S = C_V \ln \frac{T_2}{T_1} - R \ln \frac{V_2}{V_1}$

(b) $\Delta S = C_V \ln \frac{T_2}{T_1}$

(d) $\Delta S = C_V \ln \frac{T_2}{T_1} + R \ln \frac{V_2}{V_1}$

Answer: The correct choice is (d).

Solution: The equation for ideal gas is $PV = RT$ and the change in entropy is given by

$$dS = \frac{1}{T}(dU + PdV) = \frac{1}{T}(C_V dT + PdV).$$

Integrating the above expression within proper limits, the expression for the change in entropy comes out to be

$$\Delta S = C_V \ln \frac{T_2}{T_1} + R \ln \frac{V_2}{V_1}.$$

27. A Carnot cycle operates on a working substance between two reservoirs of temperature T_1 and T_2 with $T_1 > T_2$. During each cycle, an amount of heat Q_1 is extracted from the reservoir at T_1, and an amount of Q_2 is delivered in the reservoir at T_2. Which of the following statements is incorrect?

[GATE 2011]

(a) Work done in one cycle is $Q_1 - Q_2$

(b) $\frac{Q_1}{T_1} = \frac{Q_2}{T_2}$

(c) Entropy of the hotter reservoir decreases

(d) Entropy of the universe (consisting of the working substance and the two reservoirs) increases

Answer: The correct choice is (c).

Solution: Entropy of the hotter reservoir decreases.

28. Consider an engine working in a reversible cycle and using an ideal gas with constant heat capacity C_P as the working substance. The cycle consists of two processes at constant pressure, joined by two adiabatic processes. Then, the efficiency of this engine in terms of P_1, and P_2 is given by

(a) $\eta = 1 - (\frac{P_2}{P_1})^{\frac{\gamma-1}{\gamma}}$

(c) $\eta = 1 - (\frac{P_1}{P_2})^{\frac{\gamma-1}{\gamma}}$

(b) $\eta = 1 - (\frac{P_2}{P_1})^{\frac{1}{\gamma}}$

(d) $\eta = 1 - (\frac{P_2}{P_1})^{\frac{\gamma}{\gamma-1}}$

Answer: The correct choice is (a).

Solution: In the cycle, the working substance absorbs energy from the source at higher temperature T_b. This heat energy is given by $Q_{ab} = C_P(T_b - T_a)$. The energy it gives to the source of lower temperature is $Q_{reject} = C_P(T_c - T_d)$. Thus, we have

$$\eta = 1 - (\frac{Q_{reject}}{Q_{ab}}) = 1 - (\frac{T_c - T_d}{T_b - T_a}).$$

From the equation of state $PV = nRT$ and the adiabatic equations, we have

$$P_2 V_d^\gamma = P_1 V_a^\gamma, P_2 V_c^\gamma = P_1 V_b^\gamma, \text{we have } \eta = 1 - \left(\frac{P_2}{P_1}\right)^{\frac{\gamma-1}{\gamma}}.$$

29. An isolated box is divided into two equal compartments by a partition (see Figure 9.35). One compartment contains a Van der Waals gas, while the other compartment is empty. The partition between the two compartments is now removed. After the gas has filled the entire box and equilibrium has been achieved, which of the following statement(s) is (are) correct?   [JAM 2017]

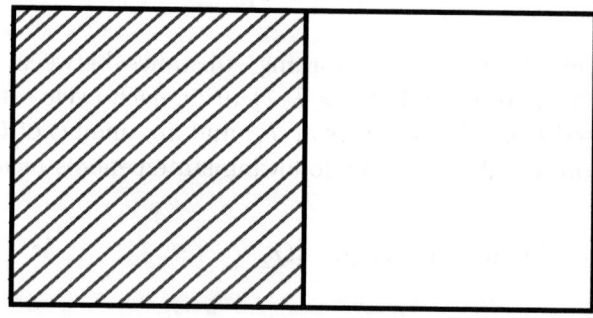

Figure 9.35 Schematic representation of an isothermal magnetization in problem 29.

(a) Internal energy of the gas has not changed

(b) Internal energy of the gas has decreased

(c) Temperature of the gas has increased

(d) Temperature of the gas has decreased

Answer: The correct choices are (a) and (d).

Solution: It is the example of free expansion, so, internal energy of the gas has not changed. $dU = C_V T - \frac{a}{V} dV$.

For Van der Waals gas, $dU = C_V dT + \frac{a}{V^2} dV$. For keeping internal energy constant, if dV increases, then dT must decrease.

30. Experimental measurements of heat capacity per mole of aluminum at low temperatures show that the data can be fitted to the formula $C_V = aT + bT^2$, where $a = 0.00135 \text{JK}^{-2}\text{mol}^{-1}$, $b = 2.48 \times 10^{-5} \text{JK}^{-4}\text{mol}^{-1}$, and T is the temperature in Kelvin. The entropy of one mole of aluminum at such temperatures is given by the formula [JAM 2007]

(a) $S = aT + \frac{b}{3}T^3 + c$, where $c > 0$ is a constant.

(b) $S = \frac{aT}{2} + \frac{b}{4}T^3 + c$, where $c > 0$ is a constant.

(c) $S = aT + \frac{b}{3}T^3$

(d) $S = \frac{aT}{2} + \frac{b}{4}T^3$

Answer: The correct choice is (a).

Solution: $dS = \int \frac{C_V dT}{T} \Rightarrow S = \int \frac{aT + bT^3}{T} dT = aT + \frac{b}{3}T^3 + C.$

9.13   Exercise

9.13.1   Short answer type questions

1. What do you mean by a "thermodynamic process"?

2. What do you mean by a quasi-static process in thermodynamics?

3. Why is it necessary to introduce the concept of quasi-static process in thermodynamics? [Burdwan University 2021]

4. What do you mean by a reversible process in thermodynamics?

5. What is an irreversible process in thermodynamics?

6. Explain the cyclic process with an example.

7. State Kelvin–Planck and Clausius statements of the second law of thermodynamics.

8. Define entropy. What is the physical significance of entropy?

9. Explain why an adiabatic process need not be isentropic always. [Calcutta University 2021]

10. Represent the Carnot cycle in terms of T–S diagram.

11. Represent the Carnot cycle in terms of U–S diagram.

12. Represent the Carnot cycle in terms of H–S diagram.

13. Explain the principle of entropy.

14. What happens to entropy at absolute zero?

15. Explain that the perpetual motion machine is not possible according to thermodynamics. [Calcutta University 2021]

16. Explain how a refrigerator could be used for heating a room in the winter.

17. What is Gibbs paradox?

18. "Entropy is a state function"—explain.

19. What are the essential components of a heat engine?

20. Explain the significance of "source and sink" in connection with the heat engine. State their essential properties.

21. State Clausius theorem.

22. State the third law of thermodynamics.

23. "Absolute zero is unattainable"—explain.

9.13.2 Long answer type questions

1. What are the differences between a reversible and an irreversible process in thermodynamics?

2. Discuss the essential components of a heat engine with the help of a schematic diagram.

3. Prove that no engine can be more efficient than a reversible engine working between the same ranges of temperatures.

4. Prove that all reversible engines working between the same temperature ranges have the same efficiency.

5. Show that the entropy increases in natural processes. Prove also that the entropy cannot decrease for an isolated system.

6. Show that the entropy is a measure of the so-called unavailable energy.

7. Show that the entropy of the universe increases in a reversible process.

8. (a) What is Carnot's engine? Describe its operation with the help of a PV diagram and derive an expression for its efficiency.

    (b) Establish the Clausius inequality theorem.

    (c) One gram mole of a perfect gas expands isothermally to four times its initial volume. Assuming complete conversion of heat into work, calculate the change in entropy. Given: $R = 8.314$ J $(\text{mol} - \text{K})^{-1}$.
    [Delhi University 2021]

9. Calculate the change in entropy of an ideal gas in different processes.

10. Calculate the entropy of one mole of a perfect gas in terms of temperature and pressure. [WBSU 2021]

11. Show that the entropy of n mole of an ideal gas of constant heat capacity C_V at a constant temperature T and volume V is given by

$$S = C_V \ln T + nR \ln V + S_0.$$

12. Deduce an expression for the change in entropy of an ideal gas for a change of state from (P_1, V_1, T_1) to (P_2, V_2, T_2).

13. Calculate the net efficiency of a heat engine.

14. Give the Kelvin—Planck statement and Clausius statement of the second law of thermodynamics. Establish the equivalence of the above two statements. [Calcutta University 2021]

15. Show that the violation of Clausius statement of the second law of thermodynamics leads to the violation of Kelvin–Planck statement. [Burdwan University 2020]

16. Using Carnot's theorem, prove the Clausius inequality. [WBSU 2021]

17. "The perpetual motion machine of the second kind is impossible to construct"—justify this statement. [Calcutta University 2022]

18. What do you mean by Gibbs paradox? How is it resolved?

19. Show that Gibbs paradox can be removed using Sackur–Tetrode equation.

20. What is a thermodynamic scale of temperature? Prove that it is equivalent to a perfect gas scale.

21. Derive an expression for the entropy of an ideal gas.

22. Show that the work done is equal to the area of the rectangle in the T–S diagram of the Carnot cycle. [ST. Joseph's College (Autonomous), Bengaluru 2020]

23. Prove first, second, and third T–dS equations.

24. State Nernst's heat theorem and establish the equivalence of this theorem with the unattainability of absolute zero. [Burdwan University 2021]

9.13.3   Numerical problems

1. An ideal gas is heated from a temperature T_1 to a temperature T_2 by keeping its volume V constant. The gas is expanded back to its initial temperature according to the law $PV^n = $ constant, where n is a constant. If the change in entropy in the above two processes is equal, find the value of n in terms of the adiabatic index γ. $[n = \gamma + 1/2]$

2. Determine the change in entropy ΔS when 5 moles of an ideal gas are heated reversibly from a temperature 25°C to a temperature 73°C at constant pressure.

$$[9.3 \ \text{JK}^{-1}]$$

3. Show that the efficiency of the cycle ABCDA shown in Figure 9.36 is given by (Assume AC=BD)   [Calcutta University 2021]

$$\eta = \frac{2\pi \ (T_1 - T_2)}{\pi \ (T_1 - T_2) + 4 \ (T_1 + T_2)}.$$

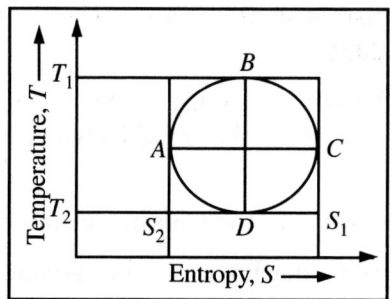

Figure 9.36 $T\text{--}S$ diagram for the cycle $ABCDA$ in problem 3.

4. A Carnot engine converts one-fifth of the heat input into work. If the sink temperature is reduced by 80°C, the efficiency gets doubled. Find the source and the sink temperature. Also, give the consequences of the Carnot cycle.

[Temperature of the source $= 400$ K, and that of the sink is 320 K]

5. An iron cube at a temperature of 400°C is dropped into an insulated bath containing 10 kg water at 25°C. The water finally reaches a temperature of 50°C at a steady state. Given that the specific heat of water is equal to $4186 \ \text{Jkg}^{-1}\text{K}^{-1}$, and the heat capacity of iron is $2990 \ \text{JK}^{-1}$. Find the entropy changes for the iron cube and the water. Is the process reversible? If so, why?

[Change in entropy of iron is $= -2195 \ \text{JK}^{-1}$, change in entropy of water is $= 3382.24 \ \text{JK}^{-1}$, and the net change in entropy is $= 1177.24 \ \text{JK}^{-1}$. Since $\Delta S > 0$, the process is irreversible.]

6. A Carnot engine operates between temperatures T_1 and T_2 with gas as the working substance whose equation of state is given by $P(V - b) = RT$. Work out the expression for the heat absorbed and the work done in each part of the cycle and show that the efficiency of the cycle is $\left(1 - \frac{T_2}{T_1}\right)$.

[Calcutta University 2021]

7. A mass "m"of fluid at temperature T_1 is mixed with an equal amount of the same fluid at temperature T_2. Prove that the resultant change of entropy of the universe is: $\frac{[2mC_V ln(T_1+T_2)/2]}{[(T1T2)1/2]}$ and also prove that it is always positive.

8. A system has heat capacity $C = \alpha T^2$ J/K with temperature 200 K, where α is a constant. Find out the change in entropy of the system when it is cooled down to the temperature of the thermal reservoir, which is at 100 K.
$$[\Delta S = -15\alpha \times 10^3 \text{ JK}^{-1}]$$ [Burdwan University 2021]

9. Assuming the relation $TdS = C_P dT - T\left(\frac{\partial V}{\partial T}\right)_P dP$, show that for an isothermal compression, $\Delta Q = -TV\alpha(P_2 - P_1)$, where ΔQ is the transfer of heat when the fluid is compressed isothermally, from a pressure P_1 to P_2, and α is the coefficient of volume expansion. [Calcutta University 2021]

10. Calculate the increase in entropy when the temperature of 1 kg of ice is raised from $-10°C$ to $+10°C$. Given that the specific heat of ice = 2.09×10^3 J kg^{-1} K^{-1}, the specific heat of water = 4.18×10^3 J kg^{-1} K^{-1}, and the latent heat of ice = 3.35×10^5 J kg^{-1}. [1455.3089 JK^{-1}] [WBSU 2021]

11. In the cycle $ABCA$ shown in Figure 9.37, the amounts of heat added to a thermodynamic system in the processes AB and BC are respectively 400 J and 100 J, and the amount of heat rejected during the process CA is 460 J. Find the efficiency of the cycle. [2/25] [Calcutta University 2022]

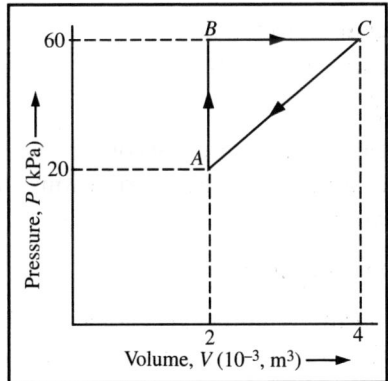

Figure 9.37 P–V diagram for the thermodynamic process in problem 11.

12. Derive an expression for the change in entropy when ice changes into steam. Find the change in entropy when 10 gm of ice melts at 0°C to 100°C.
$$[25.3168 \text{ JK}^{-1}]$$ [Vidyasagar University 2019]

CHAPTER 10

Thermodynamic Potentials and Maxwell's Relations

Just like a computer, we must remember things in the order in which entropy increases. This makes the second law of thermodynamics almost trivial. Disorder increases with time because we measure time in the direction in which disorder increases. You can't have a safer bet than that!

—Stephen Hawking

Learning Outcomes

After reading this chapter, the reader will be able to

- Understand various thermodynamic potentials such as internal energy, enthalpy, Helmholtz free energy, and Gibbs free energy and their applications

- Calculate the magnetic work done by a paramagnetic system and understand the process of creating low temperatures using the principle of adiabatic demagnetization

- Apprehend the idea of first and second-order phase transitions and Clausius–Clapeyron and Ehrenfest equations related to the phase transitions, respectively

- Derive Maxwell's thermodynamic relations

- Apply Maxwell's thermodynamic relations to derive energy equations, T–dS equations, and other thermodynamic relations connecting C_P and C_V

- Derive Joule–Kelvin coefficient for ideal and real gases like Van der Waals gas

- Describe Joule's experiment in case of adiabatic expansion of ideal and real gases

- Understand the Joule–Thomson effect for real and Van der Waals gases through porous plug experiment and the temperature of inversion

- Solve numerical problems and multiple choice questions on thermodynamic potentials, Maxwell's thermodynamic relations, and Joule–Kelvin coefficient

10.1   Introduction

The term **thermodynamic potentials** refers to a specific measure of the capacity of a thermodynamic system to perform work. It is a key concept in thermodynamics and encompasses four variables: internal energy U, Helmholtz free energy F, enthalpy H, and Gibbs free energy G. The choice of the suitable thermodynamic potential depends upon the specific conditions of the system—whether any isolated, closed, or open systems. This means each of these four potentials has its unique usage scenario and interpretation. These potentials are paramount in describing the energy changes within systems. These potentials are extensive state variables of dimensions of energy and are introduced to account for specific constraints such as isothermal, adiabatic, isochoric, and isobaric processes in a thermodynamic system. Their purpose is to allow for simple treatment of equilibrium for systems interacting with the environment. Starting from the first and second laws of thermodynamics, the differential form of four thermodynamic potentials are derived, and these are called fundamental equations. Using the thermodynamic potentials, various thermodynamic relations are derived. In this regard, knowledge of partial differentiation and properties of state functions are extensively used. The thermodynamic potentials internal energy U, Helmholtz free energy F, Gibbs free energy G, and enthalpy H assume the following interpretations:

1. Internal energy U is the capacity to do mechanical work plus the capacity to release heat. It is used when the system is isolated from its surroundings.

2. Helmholtz free energy F is the capacity to do mechanical plus nonmechanical work, that is, it is a measure of the "available work" done by the system at constant temperature and volume. The principle of minimum Helmholtz free energy maintains a system at a stable equilibrium state.

3. Gibbs free energy G is the capacity to do nonmechanical work, that is, it can be used to perform work at constant temperature and pressure. Gibbs free energy is particularly important in equilibrium because any spontaneity is governed by the decrease in Gibbs energy.

4. Enthalpy H is the capacity to do nonmechanical work plus the capacity to release heat. It is an indicator of thermal efficiency for cycles where there is work done at constant pressure.

These thermodynamic potentials are the analog of the potential energy in classical mechanics, and together with the equations of state, describe the equilibrium behavior of a thermodynamic system as a function of so-called "natural variables", such as pressure P, temperature T, volume V, and entropy S. These natural variables are a set of appropriate variables that allow us to calculate other state functions by taking partial differentiation of the thermodynamic potentials.

Maxwell's thermodynamic relations are derived using the concept of partial differentiation, statements of the laws of thermodynamics, and the definitions of thermodynamic potentials. These Maxwell's relations are used to derive other thermodynamic equations such as energy equations, T–dS equations, and the relations between specific heats C_P and C_V for ideal as well as real gases. The concept of phase transition is introduced, and the corresponding equations for the first and second-order phase transitions are derived using Maxwell's relations. The Joule–Kelvin coefficient, inversion temperature for ideal and real gases, and the porous plug experiment are described in detail. For the comprehensive presentation of the chapter, a large number of solved numerical problems and multiple choice questions with answers are also included. This allows the reader to test their understanding of thermodynamic potentials and Maxwell's thermodynamic relations.

10.2 Thermodynamic Potentials

Thermodynamic variables such as pressure P, volume V, temperature T, internal energy U, and entropy S are generally used to define the state of a thermodynamic system. The temperature T enables us in formulating the equation of state of a thermodynamic system, and the internal energy U enables us to formulate mathematically the first law of thermodynamics: $\delta Q = dU + P\,dV$, where symbols have their usual meanings. The concept of entropy S is used for the mathematical formulation of the second law of thermodynamics which is mathematically expressed as: $\delta Q = T\,dS$. This second law emphasizes that entropy increases in all natural processes. By combining these two laws of thermodynamics, we get an equation of the form

$$dU = T\,dS - P\,dV. \tag{10.1}$$

Equation (10.1) plays a pivotal role in the field of thermodynamics. The right-hand side of this equation (10.1) contains four thermodynamic variables. Any two of these four variables can be considered as independent, and with the help of the given criteria, the rest variables can be determined. For the full description of a thermodynamic phenomenon, other relations between these four thermodynamic variables are required to be found. These relations are known as **thermodynamic potentials**.

It is known from experience that every system in nature has an inherent tendency to approach equilibrium. This state of equilibrium is achieved when some of the thermodynamic potentials assume their minimum value. However, the first and the second laws of thermodynamics do not provide any information about this attainment of minimum value. This suggests that in order to get information about the condition of the thermodynamic equilibrium of a system, there is a need to supplement these laws. For example, a mechanical system is said to be in stable equilibrium when the potential energy of the system is minimal. It means that during the evolution of a process, the system must proceed in such a direction so as to minimize its potential energy. This sort of phenomenon is frequently observed in nature. Some of the examples seen are: water flows from a higher level (higher potential energy) to a lower level (lower potential energy), electric current flows from a higher potential to a lower potential, heat flows from a high-temperature source to a low-temperature object, a body falls from a higher potential to a lower potential due to gravitational field, and so on.

The behavior of the thermodynamic state functions such as internal energy U, Helmholtz free energy F, enthalpy H, and Gibbs free energy G in thermodynamics is similar to the potential energy in mechanics. These state functions acquire minimum values under different thermodynamic conditions. For example,

1. Helmholtz free energy F becomes minimum in an isothermal–isochoric process.

2. Gibbs free energy G becomes minimum in an isothermal–isobaric process.

3. Enthalpy H becomes minimum in an isobaric–adiabatic process.

These four state functions: U, F, H, and G are called thermodynamic potentials since they play the same role in thermodynamics as that played by potential energy in mechanics. Out of these four thermodynamic potentials, Helmholtz free energy F is particularly important as it provides an important connection between statistical mechanics and thermodynamics, that is, Helmholtz free energy provides a bridge between microscopic (statistical view) and macroscopic (thermodynamic view) viewpoints. The Gibbs free energy G finds wide applications in the study of phase transitions. Thus, **a thermodynamic potential can be defined as a quantity that can be used to represent the state of a system and plays a very important role for the description of chemical thermodynamics of reactions and non-cyclic processes**.

It is interesting to note that the thermodynamic potentials U, G, F, and H can be used to get Maxwell's thermodynamic relations, and in turn, these relations can be used to derive all important thermodynamic relations. The usefulness of these relations lies in the fact that they frequently relate quantities that seem to be uncorrelated. As a result, these thermodynamic relations enable us to link experimental results obtained in different ways or replace a difficult measurement with an easier one. These relations can also be used to obtain values of one property that may be straightforward from calculations or measurements of another property. Thus, it can be inferred that these thermodynamic relations are very general and extremely useful as they enormously simplify the thermodynamic properties of a system. These thermodynamic potentials are elaborately discussed in the following sections.

10.2.1   Internal Energy

The internal energy of a perfect gas depends only upon the temperature, whereas for a real gas, it is a function of both temperature and volume. For an ideal gas, the internal energy is only the kinetic energy of the constituent molecules of the gas and is independent of volume. For a real gas, it is the sum of the kinetic and potential energies of the molecules due to their mutual forces of attraction. The force of attraction between the molecules depends upon the intermolecular distance and is thus a function of the volume. As molecules in a system are moving, there is a kinetic energy. Further, due to the interaction between the molecules, there is a potential energy. The sum of these two energies makes up the internal energy of the system. This internal energy is independent of the path followed by a system in arriving at a final state f from an initial state i. Thus, the internal energy is the function of the initial and final states only. If δQ is the amount of heat absorbed by the system and δW is the amount of work by the system, then

$$U_f - U_i = dU = \delta Q - \delta W, \tag{10.2}$$

where dU is the change in internal energy of the system from an initial state i to a final state f. Using the relations for $\delta Q = T\, dS$ and $\delta W = P\, dV$, we get

$$dU = T\, dS - P\, dV. \tag{10.3}$$

Under respective suitable conditions, equation (10.3) provides, respectively, the expressions for pressure P and temperature T as

$$P = -\left(\frac{\partial U}{\partial V}\right)_S \quad \text{and} \quad T = \left(\frac{\partial U}{\partial S}\right)_V. \tag{10.4}$$

Since U is a state function, we have

$$\frac{\partial}{\partial S}\left[\left(\frac{\partial U}{\partial V}\right)_S\right]_V = \frac{\partial}{\partial V}\left[\left(\frac{\partial U}{\partial S}\right)_V\right]_S.$$

Inserting equation (10.4) in the above expression, we get

$$\left(\frac{\partial P}{\partial S}\right)_V = -\left(\frac{\partial T}{\partial V}\right)_S. \tag{10.5}$$

This equation (10.5) represents one of Maxwell's fundamental thermodynamic relations.

10.2.2 Enthalpy

Enthalpy is one of the thermodynamic potentials of a thermodynamic system and is represented by the letter H. This thermodynamic potential is equal to the internal energy U of the system plus the product of its pressure P and volume V; hence, we have $H = U + PV$. This thermodynamic potential represents the total heat content of a system and is often considered as the preferred potential in case of many chemical reactions taking place at constant pressure P. This consideration is advantageous because, at constant pressure P, the change of enthalpy dH of the system is equal to the change in its internal energy.

On differentiating the expression $H = U + PV$, the differential form of enthalpy comes out to be

$$dH = dU + P \, dV + V \, dP. \tag{10.6}$$

Using the differential form for the internal energy, $dU = T \, dS - P \, dV$ in the expression, we get

$$dH = T \, dS - P \, dV + P \, dV + V \, dP = T \, dS + V \, dP. \tag{10.7}$$

Equation (10.7) gives the differential form for enthalpy H. The same idea can be applied to the case of internal energy U to find the natural variables of enthalpy H. It is observed from equation (10.7) that $dH = 0$ when dS and dP both are zero. This indicates that the entropy S and the pressure P are the natural variables of enthalpy H. Equation (10.7) leads to the expressions for T and V, respectively, as

$$T = \left(\frac{\partial H}{\partial S}\right)_P \quad \text{and} \quad V = \left(\frac{\partial H}{\partial P}\right)_S. \tag{10.8}$$

Since H is a state function, we have $\frac{\partial}{\partial P}\left[\left(\frac{\partial H}{\partial S}\right)_P\right]_S = \frac{\partial}{\partial S}\left[\left(\frac{\partial H}{\partial P}\right)_S\right]_P$. Using this property of the state function for H and equation (10.8), we get

$$\left(\frac{\partial T}{\partial P}\right)_S = \left(\frac{\partial V}{\partial S}\right)_P. \tag{10.9}$$

Equation (10.9) represents one of Maxwell's fundamental thermodynamic relations.

10.2.3 Helmholtz free energy

The Helmholtz free energy is a thermodynamic potential that measures the useful work obtainable from a closed thermodynamic system at a constant temperature T and volume V. The concept of this free energy was introduced by the German physicist Hermann von Helmholtz in 1882. In a thermodynamic system, the mechanical equilibrium is established when a uniform pressure (for a fluid) exists throughout the system. In classical mechanics, equilibrium is defined by the condition that the sum of external forces equals to zero for the system. The link between the two is that no external forces work at equilibrium. Mechanical equilibrium is defined as the state of a thermodynamic system in which it experiences no pressure or elastic stress within it, and there is no unbalanced force or torque between the system and the surroundings. For example, a gas in a cylinder fitted with a piston is said to be in mechanical equilibrium if there is no unbalanced force on the piston. Similarly, in an isothermal–isochoric process, Helmholtz free energy F acquires its minimum value at equilibrium for a thermodynamic system.

On combining the first and second laws of thermodynamics in equilibrium, we have $dU = TdS - PdV$. If temperature T of the system remains constant, we have

$$dU = TdS - PdV = d(TS) - \delta W; \quad \Rightarrow \quad d(U - TS) = -\delta W.$$

Representing the quantity $(U - TS)$ by a function, say F, we can write the following equation

$$F = U - TS. \tag{10.10}$$

This function F is known as the Helmholtz function or the Helmholtz free energy. Thus, we have $dF = -\delta W$. It shows that in a reversible isothermal process, the work done δW by a thermodynamic system is equal to the decrease in the Helmholtz free energy dF. On account of this fact, F is also known as the work function. On differentiating equation (10.10), we get

$$dF = dU - T\, dS - S\, dT. \tag{10.11}$$

Using the relation $dU - T\, dS = -P\, dV$ in equation (10.11), we get

$$dF = -PdV - SdT. \tag{10.12}$$

Equation (10.12) provides the differential form of the Helmholtz free energy F. We can immediately see from equation (10.12) that volume V and temperature T are the natural variables of the Helmholtz free energy F. Equation (10.12) provides us entropy S and pressure P as the differential of F with respect to temperature T and volume V, respectively. These thermodynamic parameters S and P are expressed as

$$S = -\left(\frac{\partial F}{\partial T}\right)_V \quad \text{and} \quad P = -\left(\frac{\partial F}{\partial V}\right)_T. \tag{10.13}$$

Equation (10.13) shows that once Helmholtz free energy F for a system is known, we can obtain complete information about the thermal properties of the system. Further, the expression for entropy $S = -\left(\frac{\partial F}{\partial T}\right)_V$ shows that the Helmholtz free energy F decreases with the increase in temperature T at a constant volume V, since entropy S of any substance is always positive definite. The higher the entropy

S of a substance, the greater would be the rate of decrease of F. Hence, at higher temperatures, the rate of fall of Helmholtz free energy F with temperature T is maximum for gases and minimum for solids. Similarly, the expression for pressure P in equation (10.13) shows that an increase in volume V decreases the Helmholtz free energy F. Here also, the rate of fall of Helmholtz free energy F with volume V is greater at higher pressures.

One of Maxwell's relations from Helmholtz free energy F

Using equation (10.13) for S and P and the property of state function for F, that is, $\frac{\partial}{\partial T}\left[\left(\frac{\partial F}{\partial V}\right)_V\right]_T = \frac{\partial}{\partial V}\left[\left(\frac{\partial F}{\partial T}\right)_T\right]_V$, we get

$$\left(\frac{\partial S}{\partial V}\right)_T = \left(\frac{\partial P}{\partial T}\right)_V. \tag{10.14}$$

Equation (10.14) is one of the Maxwell's thermodynamic relations. From equation (10.10), we get $TS = U - F$. The product TS is known as the latent energy of the system. Thus, $U = F + TS$ can be interpreted as the internal energy of the system and is given by

internal energy = Helmholtz free energy + latent energy.

Principle of minimum Helmholtz free energy

We consider a closed system with N_j=constant, that is, $dN_j = 0$, and is in contact with a heat bath at constant temperature T, that is, $dT = 0$. Since no mechanical work is involved in the process, $dV = 0$. Under this situation, the second law of thermodynamics reduces to

$$TdS \geq dU + PdV - \mu dN = dU.$$

For constant temperature, we have $TdS = d(TS)$. Using this result in the above expression, we get

$$0 \leq TdS - dU; \quad \Rightarrow \quad 0 \leq d(TS) - dU; \quad \Rightarrow \quad 0 \leq d(TS - U);$$

$$d(U - TS) \leq 0; \quad \Rightarrow \quad dF \leq 0.$$

We see that for a system kept at constant temperature T and volume V, the Helmholtz free energy during a spontaneous change can only decrease and reaches a minimum value at equilibrium. The total amount of work that can be extracted is limited by the decrease in free energy, and some amount of work is required to be done on the system in order to increase the free energy. For all irreversible processes characterized by a constant temperature T, volume V, and particle number N in equilibrium, we have $dF < 0$, and for reversible processes, we have $dF = 0$.

10.2.4   Gibbs free energy

In thermodynamics, the Gibbs free energy is a thermodynamic potential that measures the "useful" or process-initiating work obtainable from a thermodynamic system at a constant temperature T and pressure P (isothermal–isobaric). It was developed by the American mathematician Josiah Willard Gibbs in the 1870s. It is represented by the letter G and is defined as "the energy associated with a chemical reaction that can be used to do work". This thermodynamic potential G is used to calculate the amount of work performed by a system at constant temperature T and pressure P, and is very useful in the study of phase transitions in a thermodynamic system that occur at constant temperature T and pressure P. Gibbs free energy is the capacity of a system to do nonmechanical work and the change in Gibbs free energy ΔG measures the nonmechanical work done on it. The Gibbs free energy is the maximum amount of non-expansion work that can be extracted from a closed system; this maximum can be attained only in a completely reversible process. When a system changes from a well-defined initial state to a well-defined final state, the Gibbs free energy ΔG equals the work exchanged by the system with its surroundings, minus the work of the pressure forces, during a reversible transformation of the system from the same initial state to the same final state. This free energy G of a system is equal to the sum of its enthalpy H minus the product of temperature T and entropy S of the system, and hence, is given by

$$G = H - T\,S. \qquad (10.15)$$

Gibbs free energy combines the effect of both enthalpy H and entropy S. Taking the differential of equation (10.15), we get

$$dG = dH - T\,dS - S\,dT. \qquad (10.16)$$

Substituting the differential form of enthalpy $dH = T\,dS + V\,dP$ in equation (10.16), we get

$$dG = T\,dS + V\,dP - T\,dS - S\,dT = V\,dP - S\,dT. \qquad (10.17)$$

Equation (10.17) gives the differential form of the Gibbs free energy G. It is observed from this equation (10.17) that pressure P and temperature T are the natural variables of the Gibbs free energy G. The rule of thumb for G is: "If ΔG is positive, then the reaction is not spontaneous (it requires the input of external energy to occur) and if ΔG is negative, then it is spontaneous (occurs without the input of any external energy)". The significance of the Gibbs free energy equation is that it has the ability to determine the relative importance of the enthalpy ΔH and the entropy ΔS of the system. The change in free energy ΔG of the system measures the balance between the two driving forces ΔH and ΔS. These two factors together determine whether a reaction will occur spontaneously or not.

Thus, in general, it is observed that some thermodynamic variables can be determined from the expression of a thermodynamic potential, keeping some natural variables constant during a process. This means that a given potential can be easily used to analyze a process because that thermodynamic potential remains conserved during the execution of such a process. The thermodynamic variables V and S can be determined from the Gibbs free energy G in the following way:

Equation (10.17) provides the expression for V and S as

$$V = \left(\frac{\partial G}{\partial P}\right)_T \quad \text{and} \quad S = -\left(\frac{\partial G}{\partial T}\right)_P. \tag{10.18}$$

Thus, equation (10.18) shows that entropy S and volume V of a thermodynamic system can be determined from the Gibbs free energy G.

One of Maxwell's relations from Gibbs free energy

Since the Gibbs free energy G is a state function, we must have $\frac{\partial}{\partial T}\left[\left(\frac{\partial G}{\partial P}\right)_T\right]_P = \frac{\partial}{\partial P}\left[\left(\frac{\partial G}{\partial T}\right)_P\right]_T$. Using this property of the state function of G and equation (10.18), we get

$$\left(\frac{\partial V}{\partial T}\right)_P = -\left(\frac{\partial S}{\partial P}\right)_T. \tag{10.19}$$

This equation (10.19) gives one of Maxwell's fundamental thermodynamic relations.

Principle of minimum Gibbs free energy

We consider a closed thermodynamic system at a constant pressure P. The system is in contact with a heat bath maintained at a constant temperature T. For such a system, the following thermodynamic conditions are naturally satisfied.

$$dT = 0; \quad dN_j = 0; \quad \text{and} \quad dP = 0.$$

When these conditions are used in equation (10.17), we get $dG \leq 0$. The condition $dG < 0$ holds good for an irreversible thermodynamic process, and the condition $dG = 0$ holds good for a reversible thermodynamic process in equilibrium condition.

Thus, the Gibbs energy G decreases in an irreversible process for a thermodynamic system at a constant pressure P in contact with a heat bath maintained at a constant temperature T. Further, the Gibbs free energy $G(T, P, N)$ reaches its minimum at equilibrium, which is characterized by a uniform chemical potential μ. The change in thermodynamic potentials for favorable or unfavorable reaction conditions is presented in Table 10.1.

Table 10.1 Favorable and unfavorable conditions for thermodynamic potentials

Serial No.	Favorable reaction conditions	Unfavorable reaction conditions
1	$\Delta H < 0$	$\Delta H > 0$
2	$\Delta S > 0$	$\Delta S < 0$
3	$\Delta G < 0$	$\Delta G > 0$

The following points should be noted regarding the spontaneity of reactions:

1. If $\Delta H < 0$ and $\Delta S > 0$, without even doing any calculations, one can say that the reaction will be spontaneous because $\Delta G = \Delta H - T\Delta S$ will be negative.

2. If $\Delta H > 0$ and $\Delta S < 0$, without even doing any calculations, one can say that the reaction will not be spontaneous because $\Delta G = \Delta H - T\Delta S$ will be positive.

3. Actual calculations become necessary when out of the two parameters, ΔH and ΔS, one is favorable, and the other is not. In such a case, ΔG has to be calculated to predict the spontaneity of the reaction.

10.2.5 Complete information about a system can be obtained if one of the free energies is known

We illustrate here a very important observation about thermodynamic potential. It indicates that if one of the thermodynamic potentials (free energies) is known explicitly, complete information about the thermodynamic system can be obtained. With the help of an example, this is discussed in detail below.

To express internal energy U, enthalpy H, and Gibbs free energy G in terms of Helmholtz free energy F, we have to start from the respective definitions of these thermodynamic potentials. For example, by substituting for S from equation (10.13), the internal energy U can be expressed as

$$U = F + TS = F - T\left(\frac{\partial F}{\partial T}\right)_V = -T^2\left[\frac{\partial}{\partial T}\left(\frac{F}{T}\right)\right]_V = \left[\frac{\partial\left(\frac{F}{T}\right)}{\partial\left(\frac{1}{T}\right)}\right]_V \qquad (10.20)$$

since $d\left(\frac{1}{T}\right) = -\frac{1}{T^2}dT$. Equation (10.20) is known as **Gibbs–Helmholtz equation**. It finds a wide range of applications in thermochemistry.

Similarly, on substituting for S and P from equation (10.13), the enthalpy H can be expressed as

$$H = F + TS + PV = F - T \left(\frac{\partial F}{\partial T} \right)_V - V \left(\frac{\partial F}{\partial V} \right)_T \qquad (10.21)$$

and the Gibbs free energy G can be expressed as

$$G = F + PV = F - V \left(\frac{\partial F}{\partial V} \right)_T = -V^2 \left[\frac{\partial}{\partial V} \left(\frac{F}{V} \right) \right]_T = \left[\frac{\partial \left(\frac{F}{V} \right)}{\partial \left(\frac{1}{V} \right)} \right]_T . \qquad (10.22)$$

Thus, equations (10.20), (10.21), and (10.22) clearly express that the thermodynamic potentials such as internal energy U, enthalpy H, and Gibbs free energy G can be expressed in terms of the Helmholtz free energy F. Hence, the complete information about a thermodynamic system can be obtained from the knowledge of Helmholtz free energy F. One may ask logically: Will it be said the same for other thermodynamic potentials? The answer to this question is in the affirmative.

10.2.6 Surface films and variation of surface tension with temperature

A layer of the surface of a liquid whose thickness is equal to the range of an intermolecular force is called surface film. The property of surface tension of a liquid plays an important role in studying the dynamics of a surface film. Surface film tension is the property that allows items that are not wet to float. It is also the reason why float-ant is placed on our lines and flies to prevent these objects from being wet. They will float on the surface film. For such surface films, the related thermodynamic coordinates are

1. The surface tension σ,

2. The area of the film A, and

3. The temperature T.

Using the concepts of Maxwell's thermodynamic relations, we can determine the variation of surface tension of a liquid with temperature.

Surface films

When water is exposed to air, it acts as if it were encased within an extremely thin elastic, surface membrane. This boundary is commonly known as the surface film and is interpreted as a manifestation of unbalanced molecular action. However, at the surface film, there is a surface tension due to unbalanced attraction between water molecules at the surface on one side only and upward attraction is lacking because there are no water molecules above them. Surface tension is maximum in

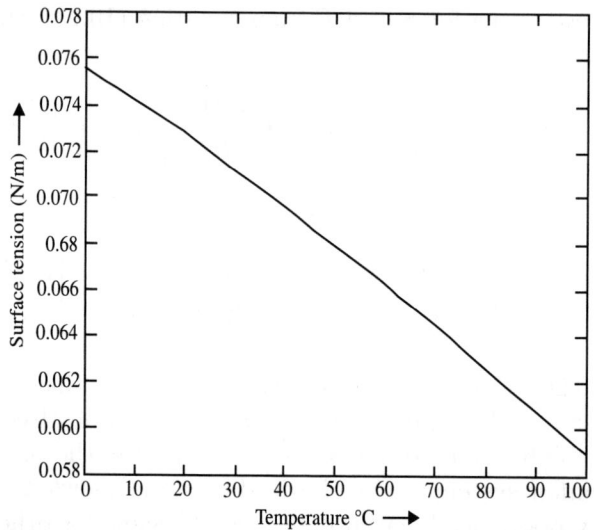

Figure 10.1 Variation of surface tension with temperature.

pure water than in any other liquid except mercury. Surface film provides support for organisms and miscellaneous particulate material; the upper as well as under surface of surface film offers mechanized support. The variation of surface tension of water in contact with air as a function of temperature is shown in Figure 10.1. From a thermodynamic point of view, the change in surface tension with temperature can be derived.

Variation of surface tension with temperature

We consider a thin rectangular liquid film of area A and movable side of length L at temperature T. Let the film be stretched adiabatically reversibly (i.e. slowly) to have the area increased by dA and temperature changed by dT. This is shown in Figure 10.2.

Here, area A of the film plays the role of volume V, and surface tension σ plays the role of (negative) pressure P. The internal energy of the film is given by

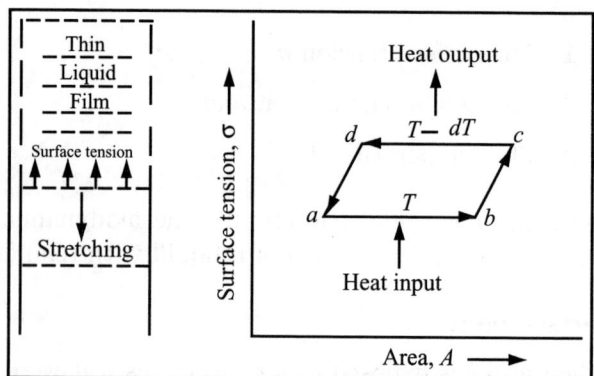

Figure 10.2 (a) Stretching of a thin liquid film and (b) Carnot cycle.

$$U = U(A, T); \quad \Rightarrow \quad dU = \left(\frac{\partial U}{\partial T}\right)_A dT + \left(\frac{\partial U}{\partial A}\right)_T dA = CdT + EdT,$$

where $C = \left(\frac{\partial U}{\partial T}\right)_A$ is the specific heat and $E = \left(\frac{\partial U}{\partial A}\right)_T$ is the surface energy (density). Since dU is a perfect differential, we have

$$\left(\frac{\partial C}{\partial A}\right)_T = \left(\frac{\partial E}{\partial T}\right)_A. \tag{10.23}$$

If heat δQ is supplied to accomplish this change, then

$$\delta Q = dU + \delta W = dU - \sigma dA,$$

where σ is the surface tension (acting as negative pressure, against which external work is done to stretch the film). Thus, we have

$$\delta Q = CdT + (E - \sigma)dA; \quad \Rightarrow \quad dS = \frac{\delta Q}{T} = \frac{C}{T}dT + \frac{E - \sigma}{T}dA.$$

Since dS is a perfect differential, we must have

$$\left[\frac{\partial}{\partial A}\left(\frac{C}{T}\right)\right]_T = \left[\frac{\partial}{\partial T}\left(\frac{E - \sigma}{T}\right)\right]_A; \quad \Rightarrow \quad \left(\frac{\partial C}{\partial A}\right)_T = T\left[\frac{\partial}{\partial T}\left(\frac{E - \sigma}{T}\right)\right]_A. \tag{10.24}$$

From equations (10.23) and (10.24), we get

$$\left(\frac{\partial E}{\partial T}\right)_A = T\left[\frac{\partial}{\partial T}\left(\frac{E - \sigma}{T}\right)\right]_A. \tag{10.25}$$

Simplifying equation (10.25), we get

$$E = \sigma - T\left(\frac{\partial \sigma}{\partial T}\right)_A. \tag{10.26}$$

This is the required expression for surface energy in terms of temperature dependence of surface tension. The surface tension decreases with the increase in temperature and hence, the surface energy increases upon stretching.

10.2.7 Magnetic work

In order to calculate the expression for magnetic work done, it is assumed that students are familiar with the basic concepts of magnetic induction $\vec{B}$, magnetic field $\vec{H}$, magnetic dipole with moment $\vec{p_m}$, and magnetization $\vec{M}$ frequently used in electricity and magnetism. It is known that in the presence of a magnetic induction field $\vec{B}$, the torque $\vec{\tau}$ acting on a magnetic dipole of moment $\vec{p_m}$ is given by $\vec{\tau} = \vec{p_m} \times \vec{B}$. The magnetic dipole moment p_m of a sample is defined as the maximum torque it experiences in the presence of a unit magnetic field $\vec{B}$. The magnetization $\vec{M}$ of a magnetic specimen is obtained from the expression: $\vec{B} = \mu \vec{H} = \mu_0(\vec{H} + \vec{M})$. Magnetization of a magnetic sample is also defined as the total magnetic moment per unit volume.

We know that the work done on a stretched string is $F\,dx$, where dx is the increase in length, and F is the tension in the elastic string. Similarly, the work done on the gas is $-P\,dV$, where P is the pressure acting on the gas and dV is the change in volume V. Following the similar procedure, the work done per unit volume on an isotropic magnetic sample in increasing its magnetization from M to $M + dM$ in a magnetic induction field $\vec{B}$ is given by $B\,dM$.

10.2.8 Cooling due to adiabatic demagnetization

Adiabatic demagnetization is a process of cooling. It occurs in magneto-caloric materials. The basic principle of adiabatic demagnetization is that these materials (especially rare-earth elements) are heated up when placed in a magnetic field and cool down when removed from the magnetic field. Thus, the temperature of certain materials is lowered when these materials are removed from the magnetic field in the adiabatic demagnetization process. This procedure was proposed by chemist Peter Debye in 1926 and independently by William Francis Giauque in 1927. It provides a very effective and useful means for cooling an already cold material, say, at about 1 K to a small fraction of 1 K.

The mechanism of adiabatic demagnetization involves a magnetic material in which some degree of disorder of its constituent particles exists at the liquid helium temperature of 4 K or below. Paramagnetic crystals like gadolinium sulfate salt $(Gd_2(SO_4)_3 : 8H_2O)$ have a certain degree of disorder in the arrangement of magnetic dipoles. When the spacing of the energy levels of the magnetic dipoles is small compared with the thermal energy, the dipoles occupy these levels equally with random orientations in space. These energy levels become separated sharply when a magnetic field is applied, that is, the corresponding energies are widely different. The lowest energy levels are occupied by the magnetic dipoles most closely aligned with the applied field. In this situation, the paramagnetic salt in contact with the liquid helium bath (maintained at a constant temperature) will transfer thermal energy to the bath. After removing contact of the paramagnetic salt with the heat bath, the magnetic field is decreased, and as a result, no heat can flow back in (an adiabatic process), and the paramagnetic sample will cool down. In this cooling process, the magnetic dipoles get trapped, that is, aligned in the lower energy states. In this process, temperatures from 0.3 K to as low as 0.0015 K can be reached.

It should be pointed out here that much lower temperatures can be achieved by an analogous procedure called adiabatic nuclear demagnetization. This process depends on the ordering (aligning) of nuclear dipoles that arise from nuclear spins. These nuclear dipoles are at least 1000 times smaller than those of the atoms. With this adiabatic nuclear demagnetization process, temperatures of the ordered nuclei have been reached as low as 16×10^{-6} K. The electronic paramagnetic refrigeration has the main disadvantages that the thermal conductivities are low, and the magnetic ordering temperatures of paramagnetic salts are high. To avoid these disadvantages, nuclear magnetic refrigeration is proposed, and is used in several specialized laboratories to refrigerate into the microkelvin temperature range.

Magneto-caloric effect (MCE)

The magneto-caloric effect (MCE) is a heating or cooling of a magnetic material when the applied magnetic field changes. At the heart of the MCEs lays coupling between the magnetic moments and external magnetic field, and in some cases the MCE involves structural transitions concomitant with magnetic transitions. MCE can be used for cooling and may offer larger efficiencies than conventional vapor-cycle refrigeration. This effect was discovered by Warburg in 1881 in iron and explained theoretically by P. Debye (1926) and independently by Giauque (1927). It consists of increasing (decreasing) the temperature of a paramagnetic specimen under adiabatic conditions in a magnitude ΔT_{ad} when an external magnetic field H is suddenly applied (removed). Since the 1930s, the magnetic cooling process has been extensively used in technology to obtain very low temperatures by adiabatic demagnetization of a paramagnetic salt. Some of the examples of paramagnetic salts are $Gd_2(SO_4)_3 : 8H_2O$. In 1997, the giant magneto-caloric effect (GMCE) has been discovered in the intermetallic ternary compound $Gd_5Ge_2Si_2$, and the first energetically efficient room-temperature magneto-caloric refrigerator based on metallic Gd as magnetic refrigerant has been developed. These discoveries have strongly stimulated both the search of new families of GMCE materials and the development of magnetic refrigeration technology with emphasis in the room-temperature range.

Pictorial demonstration of the principle of the adiabatic demagnetization

A paramagnetic specimen has randomly oriented magnetic moments, and its average magnetization is zero. This is shown at the left of Figure 10.3. When a magnetic field is

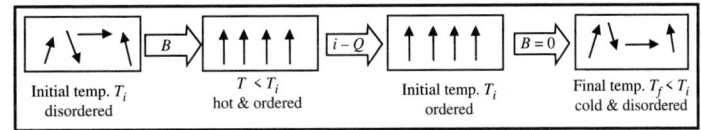

Initial temp. T_i disordered $T < T_i$ hot & ordered   Initial temp. T_i ordered   Final temp. $T_f < T_i$ cold & disordered

Figure 10.3 Variation of surface tension with temperature.

applied to the specimen, it does work on the randomly oriented magnetic moments in the paramagnetic salt and aligns them along the field. This is shown on the second left of Figure 10.3. This external work due to the magnetic field heats up the paramagnetic specimen, and its temperature rises. This heat can be taken away using a heat exchanger, leaving the specimen in an ordered state in the magnetic field, but now at the initial temperature (see the third figure from left of Figure 10.3).

Next, suddenly the magnetic field is removed (or reduced to very low values) so that the moments become disordered again by breaking away from the diminishing field and in the process doing work against it. In doing this work to become randomly oriented again, they expend internal energy, and hence, the temperature falls drastically. This is shown at the right of Figure 10.3. Thus, we have a disordered state at a temperature below the initial temperature. The cycle can be repeated again and again to attain any desirable low-temperature of the order of milli-kelvins.

The four processes of adiabatic demagnetization

The steps followed in Figure 10.3 can be summarized by the following points:

1. **Adiabatic magnetization**: Within this cycle, a paramagnetic specimen is magnetized quickly (to get an adiabatic process). As a result, its temperature increases.

2. **Isomagnetic cooling**: Within this regime, the magnetic specimen is allowed to touch a cold reservoir and is cooled down to ambient temperature in the presence of the applied magnetic field. Keeping the paramagnetic sample in thermal equilibrium with the cold reservoir, the strength of the magnetic field is increased. As a result of this process, the entropy of the magnetic sample decreases because the system becomes more ordered as the magnetic moments align themselves along the direction of the magnetic field. At this point, the temperature of the specimen still remains the same as that of the cold reservoir.

3. **Adiabatic demagnetization**: Within this cycle, the magnetic specimen is separated from the cold reservoir, and the strength of the magnetic field is lowered. The entropy of the specimen remains the same, but its temperature is reduced in reaction to the reduction in the magnetic field strength. If the specimen was already at a fairly low-temperature, this decrease in temperature can be ten-fold or greater.

4. **Isomagnetic heating**: The specimen is heated up by a cooling agent without an applied magnetic field. Inversely, the temperature of the cooling agent decreases.

This process can be repeated, permitting the sample to be cooled to very low temperatures. These steps are shown in Figure 10.3.

Mathematics of the process of adiabatic demagnetization

The method of adiabatic demagnetization is widely used for experimental purposes to obtain extremely low temperatures of the order of milli or micro kelvin temperature. The following procedure is followed in this process to create the low-temperature. First, a magnetic sample of a paramagnetic salt, say, cerium magnesium nitrate (this is already cooled to low temperatures by some other means), is magnetized isothermally. The sample is often suspended in a helium atmosphere. This can conduct away any heat that is produced in the isothermal magnetization process, and hence, the process remains isothermal. In the second step, the sample is insulated (by pumping out the helium gas) and suddenly and adiabatically demagnetized. Such a process of isothermal magnetization followed by adiabatic demagnetization is then repeated over and over again, and temperatures close to 0 K can be achieved.

We consider that δq amount of heat is added to an isotropic paramagnetic sample in the presence of a magnetic field $\vec{B}$. The work done per unit volume on this sample in increasing its magnetization (an extensive state variable) from $\vec{M}$ to $\vec{M} + d\vec{M}$ in the magnetic field $\vec{B}$ (an intensive state variable) will be given by $B \, dM$. As the sample is isotropic (we have taken a pure magnetic sample), the magnetic moment $\vec{p_m}$ and the magnetic field $\vec{B}$ are in the same direction. We assume that there is no change in the volume of the magnetic sample and hence, the increase in its internal energy per unit volume can be written as

$$dU = T \, dS + B \, dM. \tag{10.27}$$

In this magnetic context, the state function enthalpy H, Helmholtz free energy F, and Gibbs free energy G can be written as

$$H = U - B \, M, \quad F = U - T \, S, \quad and \quad G = H - T \, S = F - B \, M.$$

In differential form, the above thermodynamic potentials can be expressed as

$$dH = T \, dS - M \, dB, \quad dF = -S \, dT + B \, dM, \quad and \quad dG = -S \, dT - M \, dB.$$

Here, $\vec{M}$ is the dipole moment per unit volume and is defined as the magnetization in Am^{-1}. Using this basic information, we will now calculate the mathematical expressions $\left(\frac{\partial T}{\partial B}\right)_S$ for the process of adiabatic demagnetization. This expression is positive as T and B both increase together.

We consider the entropy S as a function of temperature T and magnetic field B. The cyclic relation among these three variables is given by

$$\left(\frac{\partial S}{\partial T}\right)_B \left(\frac{\partial T}{\partial B}\right)_S \left(\frac{\partial B}{\partial S}\right)_T = -1. \tag{10.28}$$

We want to find out the second term in equation (10.28) in terms of the measured parameters using the first and third expressions. In a reversible process, $dS = \frac{\delta Q}{T}$, and in a constant magnetic field, we know $\delta Q = C_B \, dT$. Here, S denotes the entropy per unit volume, and C_B indicates the heat capacity per unit volume in a constant magnetic field B. Thus, from these two relations, we have $\left(\frac{\partial S}{\partial T}\right)_B = \frac{C_B}{T}$.

The Maxwell relation corresponding to $\left(\frac{\partial S}{\partial P}\right)_T = (-)\left(\frac{\partial V}{\partial T}\right)_P$ becomes $\left(\frac{\partial S}{\partial B}\right)_T = \left(\frac{\partial M}{\partial T}\right)_B$ for a pure magnetic substance. Thus, equation (10.28) becomes

$$\left(\frac{\partial T}{\partial B}\right)_S = (-)\frac{T}{C_B}\left(\frac{\partial M}{\partial T}\right)_B. \tag{10.29}$$

We know that the magnetization M for a paramagnetic material is proportional to magnetic induction field B, and it falls off inversely with temperature T. These two facts give rise to the equation of state for a paramagnetic substance, which is given by $M = \frac{a\,B}{T}$. From this relation, we have $\left(\frac{\partial M}{\partial T}\right)_B = -\left(\frac{a\,B}{T^2}\right) = -\left(\frac{M}{T}\right)$. Using this result, equation (10.29) becomes

$$\left(\frac{\partial T}{\partial B}\right)_S = \frac{M}{C_B}. \tag{10.30}$$

The cooling effect is particularly effective at low temperatures when C_B is small.

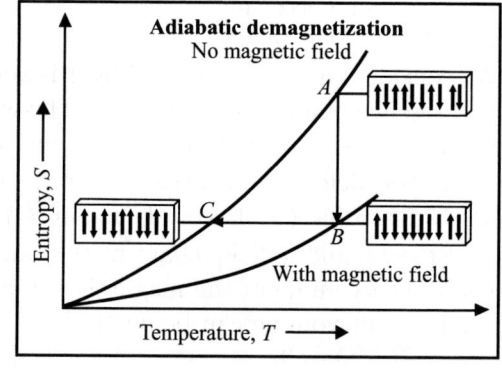

Figure 10.4 Schematic representation of isothermal magnetization and adiabatic (isoentropic) demagnetization of a paramagnetic material to achieve low-temperature.

Cooling by adiabatic demagnetization involves successive isothermal magnetization followed by adiabatic demagnetization, and this suggests that some insight into the process might be obtained by following it on an entropy: temperature $(S - T)$ diagram as in Figure 10.4.

The variation of entropy of the specimen with temperature in the absence of a magnetizing field (upper curve) and in the presence of a magnetic field (lower curve) is schematically shown in the above figure. The lower curve corresponding to the lesser entropy represents the more ordered state in the presence of a magnetic field. The process marked by "AB" represents an isothermal magnetization, and the process marked by "BC" is the corresponding adiabatic (isentropic) demagnetization. Thus, it is readily observed how this is effective in producing low temperatures. When this procedure is repeated several times, temperature is reduced to a greater extent.

Fall of temperature in adiabatic demagnetization

Rewriting equation (10.29) as $\left(\frac{\partial T}{\partial B}\right)_S = -\frac{T}{C_B}\left(\frac{\partial M}{\partial T}\right)_B$. Using $\left(\frac{\partial M}{\partial T}\right)_B = (-)\frac{a}{T^2}B$ in equation (10.29), we get the expression for fall in temperature dT as

$$T\, dT = \left(\frac{a}{C_B}\right) B\, dB. \tag{10.31}$$

If the magnetic field decreases from B_0 to 0 and T_0 and T_f are, respectively, initial and final temperatures, the change in temperature is given by

$$\int_{T_0}^{T_f} T\, dT = \left(\frac{a}{C_B}\right) \int_{B_0}^{0} B\, dB. \tag{10.32}$$

We now assume that the specific heat at a constant magnetic field does not vary with temperature. The integrals in equation (10.32) can then be readily solved to get

$$T_f{}^2 - T_0{}^2 = -\left(\frac{a}{C_B}\right) B_0{}^2 \tag{10.33}$$

or,

$$T_f - T_0 = \Delta T = -\left(\frac{a}{2\, C_B\, T_{av}}\right) B_0{}^2, \tag{10.34}$$

where $T_{av} = \left(\frac{T_f + T_0}{2}\right)$ is the average temperature of the paramagnetic specimen.

Calculation of the change in entropy at a given temperature T for a given value of B in adiabatic demagnetization

Combining the first and second laws of thermodynamics and the expression for work done in a magnetic process: $P\, dV = M\, dB$, we get

$$T\, dS = dU + M\, dB. \tag{10.35}$$

Using the differential for internal energy $dU = \left(\frac{\partial U}{\partial T}\right)_B dT + \left(\frac{\partial U}{\partial B}\right)_T dB$ in the above equation, we get

$$T\, dS = \left(\frac{\partial U}{\partial T}\right)_B dT + \left(\frac{\partial U}{\partial B}\right)_T dB + M\, dB = \left(\frac{\partial U}{\partial T}\right)_B dT + \left[\left(\frac{\partial U}{\partial B}\right)_T + M\right] dB. \tag{10.36}$$

At a constant temperature, this gives rise to $\left(\frac{\partial S}{\partial B}\right)_T = \left(\frac{1}{T}\right)\left[\left(\frac{\partial U}{\partial B}\right)_T + M\right]$. Using Maxwell's relation $\left(\frac{\partial S}{\partial B}\right)_T = \left(\frac{\partial M}{\partial T}\right)_B$, this expression reduces to

$$\left(\frac{\partial M}{\partial T}\right)_B = \left(\frac{1}{T}\right)\left[\left(\frac{\partial U}{\partial B}\right)_T + M\right]. \tag{10.37}$$

For an isothermal process along "AB" of the curve shown in Figure 10.4, we may write $dS = \left(\frac{\partial M}{\partial T}\right)_B dB$. Integrating this expression, we get

$$S = S_{B=0} + \int_0^B \left(\frac{\partial M}{\partial T}\right)_B dB. \qquad (10.38)$$

Again, for a paramagnetic material, magnetization is given by $M = \frac{[Ng^2 \mu_0 J(J+1)\mu_B^2 H]}{3K_B T} = \frac{[Ng^2 \, J(J+1)\mu_B^2 B]}{3K_B T}$ with $B = \mu_0 \, H$. This gives

$$\left(\frac{\partial M}{\partial T}\right)_B = (-)\frac{[Ng^2 \, J(J+1)\mu_B^2 B]}{3K_B T^2}. \qquad (10.39)$$

Substituting this value of $\left(\frac{\partial M}{\partial T}\right)_B$ into equation (10.38), one obtains

$$S = S_{B=0} + \int_0^B (-)\frac{[Ng^2 \, J(J+1)\mu_B^2 B]}{3K_B T^2} dB. \qquad (10.40)$$

$$\Rightarrow S = S_{B=0} - \left[\frac{Ng^2 \, J(J+1)\mu_B^2 \, B^2}{6K_B T^2}\right].$$

This leads to the change in entropy ΔS as

$$\Delta S = S - S_{B=0} = -\left[\frac{Ng^2 \, J(J+1)\mu_B^2 \, B^2}{6K_B T^2}\right]. \qquad (10.41)$$

The magneto-dielectric effect described above has been used by many researchers for the production of very low temperatures. The first experimental test of this phenomenon was carried out at temperatures down to 1.3 K. Thus, the theory was successfully tested. Some of the typical paramagnetic salts used in the experiments to attain low temperatures by this magneto-caloric effect was $FeNH_4(SO_4)_2.12H_2O$ and CeF_3.

Some comments about the process of adiabatic demagnetization

1. Electronic paramagnets are materials with net electronic magnetic moments. In the presence of a magnetic field, the electronic magnetic moments of these materials tend to align themselves with the magnetic field. Using these electronic paramagnets, the lowest temperatures that can be achieved in these methods are of the order of 1 milli-kelvin.

2. Nuclear paramagnets are the materials with a net magnetic moment caused by the individual magnetic moments of the nuclei of the material. These nuclear magnetic moments are about 1000 times smaller than the electronic magnetic moments. Hence, dipole–dipole interactions in a nuclear paramagnet are much weaker than those of an electronic paramagnet. As a result, much lower temperatures could be achieved if a nuclear paramagnet is used in the adiabatic demagnetization process. It has been possible to cool nuclear paramagnets down to 250 pico-kelvins.

The interactions between the individual particles play an important role in setting the limit of low-temperature achieved. When the applied magnetic field is very weak or absent, the individual particles tend to create their own internal magnetic field. This sets up a limit to how weak the applied magnetic field can be, because at some point, the particle interactions cancel any further cooling effects of demagnetization. As the dipole interactions between particles in nuclear paramagnetic materials are much weaker than those in electronic paramagnetic materials, the limits of the magnetic field strength and temperature are much lower in the case of nuclear paramagnetic materials.

10.2.9 Phase transition

A homogeneous part of a system having definite boundaries is defined as a phase. All physical properties within a phase of a thermodynamic system are uniform. A phase may be a chemically pure substance or may have more than one component, such as air or a mixture of two miscible liquids. The transition from one phase to another, that is, phase transition, is the process by which a thermodynamic system changes from one state to another with different physical properties. The term phase transition (or phase change) is most commonly used to describe transitions between solid, liquid, and gaseous states of matter and, in rare cases, plasma (physics). When a certain substance suffers a phase transition, certain properties of the substance change discontinuously due to the change in some external conditions. These external conditions may be temperature, pressure, or other factors. For example, the phase transition of a substance from solid to liquid is denoted by a discontinuous change in the shear strength of the substance, whereas the change to a gaseous phase is indicated by a discontinuous change in density, as shown in Figure 10.5.

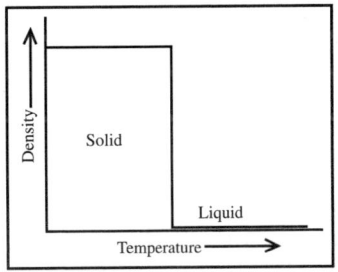

Figure 10.5 Variation of density as a function of temperature.

Phase transition normally takes place due to changes in temperature and/or pressure. The melting of ice to form water and the evaporation of water to water vapor are best-known examples of phase transition. Other examples are normal conductor to superconductivity, paramagnetism to ferromagnetism, normal helium to superfluid helium, etc. Each phase can be distinguished by the density of the constituent. For example, a frozen portion of the Arctic Ocean near the North Pole consists of ice on the top layer and liquid water beneath it. There are some water vapors in the atmosphere above the ice layer. In these three layers, the density is different because water exists separately in some combination of three phases in these layers.

According to Ehrenfest, the order of a phase transition is defined in the following way: **The order of the lowest derivative of the Gibbs enthalpy G showing a discontinuity upon crossing the coexistence curve is the order of a phase transition.** Explicitly, a phase transition between two phases, say a and b, is of order n if

$$\left(\frac{\partial^m G_a}{\partial T^m}\right)_P = \left(\frac{\partial^m G_b}{\partial T^m}\right)_P ; \qquad \left(\frac{\partial^m G_a}{\partial P^m}\right)_T = \left(\frac{\partial^m G_b}{\partial P^m}\right)_T$$

for $m = 1, 2, 3, \cdots, n-1$ and if

$$\left(\frac{\partial^n G_a}{\partial T^n}\right)_P \neq \left(\frac{\partial^n G_b}{\partial T^n}\right)_P ; \qquad \left(\frac{\partial^n G_a}{\partial P^n}\right)_T \neq \left(\frac{\partial^n G_b}{\partial P^n}\right)_T .$$

Only first- and second-order phase transitions are found to be important in physical situations.

First-order phase transitions with examples

Melting, vaporization, and sublimation are the most familiar types of phase transitions. Among the less familiar is the modification of one crystalline form to another. In these transitions, the entropy and the volume change. A first-order phase transition should satisfy the following requirements:

1. There will be changes in entropy S and volume V.

2. The Gibbs free energy $G(T, P)$ is continuous across the boundary of the phases of the system.

3. There is a discontinuous change in the first-order derivative of the Gibbs function, that is, $S = -\left(\frac{\partial G}{\partial T}\right)_P$ and $V = \left(\frac{\partial G}{\partial P}\right)_T$ are discontinuous.

4. Latent heat is associated with it.

Variations of thermodynamic parameters in first-order phase transition are shown in Figure 10.6.

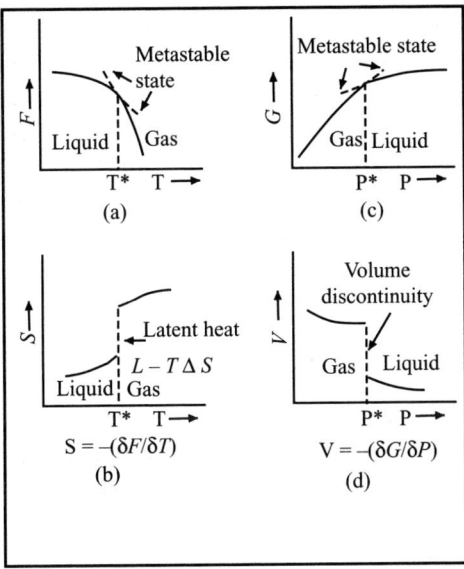

Figure 10.6 Variations of Helmholtz free energy F, Gibbs free energy G, Entropy S, and Volume V as a function of temperature T and pressure P in case of first-order phase transition.

Examples of first-order phase transition

The transformation of water into vapor at constant temperature and volume is first-order phase transition. Similarly, the transformation of ice into water at 0°C and 1 atmospheric pressure is an example of first-order phase transition.

Thermodynamics of first-order phase transition

Let us start with the so-called phase diagram of a simple substance. A phase diagram is a representation, typically in a plane, of the regions where some substance is stable in a given phase. The axes represent external control variables (intensive parameters), such as pressure, temperature, chemical potential, or an external field, or sometimes one extensive variable (volume, magnetization, etc.). Density is sometimes used in the case of fluids. The different phases are separated by lines, indicating phase transitions or regions where the system is

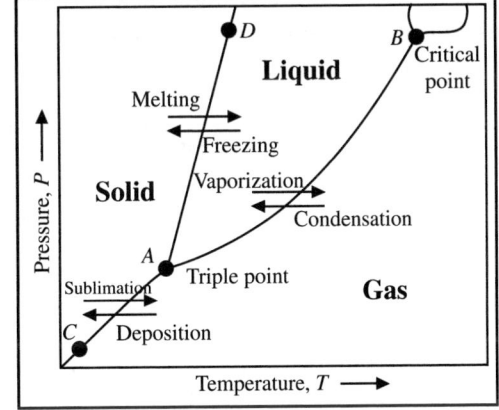

Figure 10.7 Existence of triple point and critical point in the pressure–temperature curve of a physical system.

unstable. Let us review some of the most important features of the phase diagram of a simple substance. In Figure 10.7, the phase diagram is represented in the pressure–temperature $(P\text{–}T)$ plane. The three possible phases (solid, liquid, and gas) are separated by first-order phase transition lines (continuous lines in the graph) where two phases coexist at the same time (hence the name "coexistence lines"). Phases are indicated by their names, and two special points are called T (triple point) and C (critical point). The triple point is the temperature and pressure at which solid, liquid, and vapor phases of a particular substance coexist in equilibrium.

Clausius–Clapeyron equation

First-order phase transitions take place at a particular temperature, pressure, or some other intensive variable. We could use this information to predict other temperature and pressure at which phase transitions occur. For this we must know how the changes in P and T are connected when two phases are in equilibrium. This is contained in the Clausius–Clapeyron equation.

At both a and b, the two phases are in equilibrium, so the specific Gibbs free energies of the two phases are the same: $g_1^{(a)} = g_2^{(a)}$ and $g_1^{(b)} = g_2^{(b)}$. So, if dg is the difference in Gibbs free energy between the two points, it is the same for both phases so, $dg_1 = dg_2$. But using (s and v to denote specific entropy and volume)

$$dg_1 = -s_1 \, dT + v_1 \, dP \quad \text{and} \quad dg_2 = -s_2 \, dT + v_2 \, dP.$$

$$\Rightarrow (s_1 - s_2) \, dT = (v_1 - v_2) \, dP; \quad \Rightarrow \quad \frac{dP}{dT} = \frac{s_1 - s_2}{v_1 - v_2} = \frac{s_2 - s_1}{v_2 - v_1} = \frac{\Delta s}{\Delta v}. \quad (10.42)$$

Here, $\Delta s = s_2 - s_1$ and $\Delta v = v_2 - v_1$ are the changes in specific entropy and volume of the substance, respectively. The form given by equation (10.42) doesn't look very useful as we cannot measure entropy directly. However, using $\Delta s = \frac{Q}{T}$ for an isothermal process, we can find the change in entropy during a phase transition from the concept of latent heat L. Thus, the more useful form of equation (10.42) is

$$\Rightarrow \quad \frac{dP}{dT} = \frac{Q}{T(v_2 - v_1)} = \frac{L}{T \, \Delta v}. \quad (10.43)$$

This equation (10.43) is known as the Clausius–Clapeyron equation for first-order phase transition.

Applications of Clapeyron's latent heat equation

In this section, the effects of pressure and temperature on the melting point of a solid substance are discussed in detail.

Effect of change of pressure on the melting point

It is a well-established fact that there is a change in volume when a solid is converted into a liquid. Volume may decrease or increase in the liquid state. Hence, equation

(10.43) shows that the melting point of a substance will change with the applied pressure. Accordingly, the following cases may arise:

1. If the final specific volume v_2 is greater than the initial specific volume v_1, $\Delta v = v_2 - v_1$ is positive, and hence, the derivation $\frac{dP}{dT}$ in equation (10.43) becomes a positive quantity. It indicates that under this situation, the rate of change of pressure with respect to temperature is positive. In such cases, the melting point of the substance will increase with the increase in pressure and vice versa.

2. If the final specific volume v_2 is less than the initial specific volume v_1, $\Delta v = v_2 - v_1$ becomes negative, and hence, the derivation $\frac{dP}{dT}$ in equation (10.43) becomes a negative quantity. Under this situation, the rate of change of pressure with respect to temperature becomes negative. In such cases, the melting point of the substance will decrease with the increase in pressure and vice versa.

In the case of melting of ice, the volume of water formed is less than the volume of ice taken, that is, $v_2 < v_1$. Therefore, the melting point of ice decreases with the increase in pressure. Hence, ice will melt at a temperature lower than zero degrees centigrade at a pressure higher than the normal pressure. For example, ice melts at $-9°C$ at a pressure of 100 MPa (987 atm), whereas ice melts at $0°C$ only at a pressure of 1 atm. Water behaves this way because it is one of the few known substances for which the crystalline solid is less dense than the liquid. Examples of other substances include antimony and bismuth. Increasing the pressure of ice that is in equilibrium with water at $0°C$ and 1 atm tends to push some of the molecules closer together, thus decreasing the volume of the sample. The decrease in volume and the corresponding increase in density are smaller for a solid or a liquid than for a gas, but it is sufficient to melt some of the ice.

Effect of change of pressure on the boiling point

When a liquid is converted into a gaseous system, the final volume V_2 is always greater than the initial corresponding volume V_1 of the liquid, that is, $V_2 > V_1$. Therefore, the change in volume dV is always positive. $\frac{dP}{dT}$ is a positive quantity. Hence, with the increase in pressure, the boiling point of a substance will increase and vice versa. The liquid will boil at a lower temperature at a reduced pressure.

In the case of water, the boiling point increases with an increase in pressure and vice versa. Water boils at $100°C$ only at 76 cm of Hg pressure. In the laboratories, while preparing steam, the boiling point is less than $100°C$ because the atmospheric pressure is less than 76 cm of Hg. In pressure cookers, the liquid boils at a higher temperature because the pressure inside the cooker is more than the atmospheric pressure.

Second-order phase transition with examples

The first-order phase transitions are characterized by finite changes in entropy and volume. However, there are cases where there is no change in entropy and volume, but discontinuities appear in specific heat at constant pressure and isobaric volume expansivity during a phase transition. This implies that the first-order derivatives of the Gibbs free energy are continuous, but discontinuities appear in the second-order derivatives as the substance passes from one phase to another. These types of phase transitions are called second-order phase transitions. Ehrenfest defined the order of a phase transition as the order of the lowest derivative of the Gibbs energy that shows a discontinuity while transition from one phase to another.

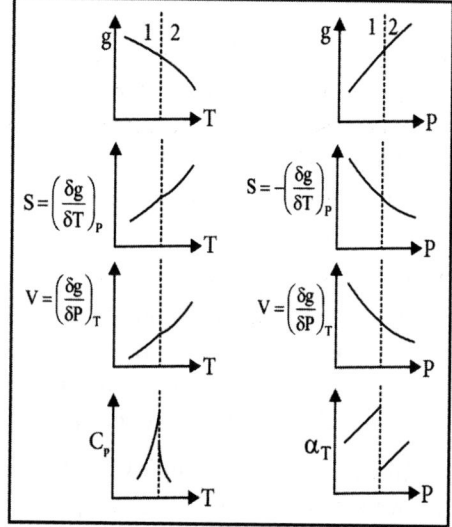

Figure 10.8 Variation of different physical parameters during the second-order phase transition.

Variations of different thermodynamic parameters during the second-order phase transition are shown in Figure 10.8.

The following are the characteristics of a second-order phase transition:

1. There are no changes in entropy and specific volume, that is, $S(T, P)$ and $V(T, P)$ are continuous.

2. Molar Gibbs function $G(T, P)$ is continuous.

3. First-order derivatives of molar Gibbs function are continuous.

4. Discontinuities arise in the second-order derivatives of Gibbs function. Thus, discontinuities are observed in the following response functions: The specific heat at constant pressure $C_P = T \left(\frac{\partial S}{\partial T} \right)_P = -T \left(\frac{\partial^2 G}{\partial T^2} \right)_P$, the isothermal compressibility $\beta_T = -\frac{1}{V} \left(\frac{\partial V}{\partial P} \right)_T = -\frac{1}{V} \left(\frac{\partial^2 G}{\partial P^2} \right)_T$, and the thermal expansion coefficient $\alpha = -\frac{1}{V} \left(\frac{\partial V}{\partial T} \right) = \frac{1}{V} \left(\frac{\partial^2 G}{\partial T \partial P} \right)$.

10.2.10 Examples of second-order phase transition

Examples of second-order phase transition are:

1. Transition from liquid helium I to liquid helium II at λ-point (2.19 K).

2. Transition from a ferromagnetic material to a paramagnetic material at the Curie point.

3. Transition of a superconducting metal into an ordinary conductor in the absence of a magnetic field.

4. Order–disorder transitions in chemical compounds and alloys.

10.2.11   Ehrenfest equations

The Clausius–Clapeyron relation does not make sense for second-order phase transitions, as both specific entropy and specific volume do not change in second-order phase transitions. Ehrenfest equations (named after Paul Ehrenfest) are equations that describe changes in specific heat capacity and derivatives of specific volume in second-order phase transitions.

The Ehrenfest equations are the equivalent of the Clausius–Clapeyron equation for second-order transitions. Consider entropy across the phase boundary. For a second-order transition, $s_1(T, P) = s_2(T, P)$ as temperature and pressure are raised infinitesimally. Then one has $s_1 + ds_1 = s_2 + ds_2$ that implies $ds_1 = ds_2$.

Considering specific entropy as a function of pressure and temperature, we can write $s(T, P)$ such that its differential can be written as $ds = \left(\frac{\partial s}{\partial T}\right)_P dT + \left(\frac{\partial s}{\partial P}\right)_T dP$.

$$\Rightarrow ds = \left(\frac{C_P}{T}\right) dT - \left(\frac{\partial v}{\partial T}\right)_P dP = \left(\frac{C_P}{T}\right) dT - \alpha \, v \, dP.$$

For phases 1 and 2 of the system, we can write

$$\left(\frac{C_{P_2}}{T}\right) dT - \alpha_2 \, v \, dP = \left(\frac{C_{P_1}}{T}\right) dT - \alpha_1 \, v \, dP.$$

This gives rise to

$$\Rightarrow \left(\frac{dP}{dT}\right) = \frac{\left(C_{P_2} - C_{P_1}\right)}{T \, v \, (\alpha_2 - \alpha_1)}. \tag{10.44}$$

Similarly, considering specific volume as a function of pressure and temperature, that is, $v(T, P)$, such that its differential can be written as

$$dv = \left(\frac{\partial v}{\partial T}\right)_P dT + \left(\frac{\partial v}{\partial P}\right)_T dP. \quad \Rightarrow \quad dv = \alpha \, v \, dT - \kappa \, v \, dP.$$

Since for a second-order phase transition, $v_1 = v_2$ leading to $dv_1 = dv_2$, we can write

$$\Rightarrow \left(\frac{dP}{dT}\right) = \frac{(\alpha_2 - \alpha_1)}{(\kappa_2 - \kappa_1)}. \tag{10.45}$$

Equations (10.44) and (10.45) are called Ehrenfest equations.

Gibbs phase rule

A phase is any physically separable material in the system. Every unique mineral is a phase (including polymorphs); igneous melts, liquids (aqueous solutions), and vapor are also considered unique phases. It is possible to have two or more phases in the same state of matter, some examples of which are solid mineral assemblages, immiscible silicate and sulfide melts, and immiscible liquids such as water and hydrocarbons. Phases may either be pure compounds or mixtures such as solid or aqueous solutions but they must "behave"as a coherent substance with fixed chemical and physical properties.

For historical reasons, geologists normally define components in terms of simple oxides (e.g. SiO_2, Al_2O_3, CaO, etc.). If two possible components always occur in the same proportions in multiple phases in a system, these can be combined into a single component (remember, we are always trying to define the minimum number of components required to make all the phases in the system).

In thermodynamics, the Gibbs phase rule provides the theoretical foundation for characterizing the chemical state of a (geologic) system. It is used to predict the equilibrium relations of the various phases present in minerals, melts, liquids, and vapors as a function of physical conditions described by variables such as pressure and temperature. Gibbs phase rule also allows us to construct phase diagrams to represent and interpret phase equilibrium in heterogeneous geologic systems. In the simplest understanding of phase diagrams, stable phase (mineral) assemblages are represented as "fields" in the "P–T diagram", and the boundaries between stable phase assemblages are defined by lines (or curves) that represent reactions between the phase assemblages. The reaction curves actually represent the condition (or the locus of points in the P–T diagram), where $\Delta G = 0$. A solid understanding of the Gibbs phase rule is required to successfully master the applications of heterogeneous phase equilibrium present in a mixture system in thermodynamics.

Thus, the Gibbs phase rule represents the condition of equilibrium between the number of coexisting phases and the components. Consider a heterogeneous system consisting of N_P phases and N_C components in equilibrium. In every phase, there are N_C different components having n_C different concentrations, but for a closed system, if the concentrations of $(N_C - 1)$ components are arbitrarily chosen, the concentration of the last component is automatically fixed. Therefore, there are $(N_C - 1)$ composition variables in each phase. For N_P phases, the number of composition variables would be $N_P (N_C - 1)$. In addition to this value, there are two other variables: pressure and temperature. Thus, the total number of variables will be $N_P (N_C - 1) + 2$.

Further, it has been shown that if the chemical potential of any one of the N_P phases is defined, then the chemical potential of the remaining $(N_P - 1)$ phases is fixed. Therefore, for N_C components, the invariant composition variables will be $N_C (N_P - 1)$. Therefore, the number of independent variables of a system will be given by

$$N_F = N_P (N_C - 1) + 2 - N_C (N_P - 1); \implies N_F = N_C - N_P + 2. \quad (10.46)$$

The Gibbs phase rule tells how many independent intensive properties denoted by N_F can be chosen to describe the equilibrium of a thermodynamic system. This will depend on the number of chemical components N_C and the number of phases N_P present in the system. According to the phase rule, the following conclusions may be drawn. For a pure substance ($N_C = 1$), the Gibbs phase rule can be applied as follows:

1. For a system with single phase, $N_P = 1$ and $N_F = N_C - N_P + 2 = 1 - 1 + 2 = 2$.

2. For a system with two phases, $N_P = 2$ and $N_F = N_C - N_P + 2 = 1 - 2 + 2 = 1$.

3. For a system with three phases, $N_P = 3$ and $N_F = N_C - N_P + 2 = 1 - 3 + 2 = 0$.

An example showing that for a single phase of a pure substance, $N_F = 2$

For a glass of liquid water, specify one of the independent intensive variables to be pressure. Choose this pressure to be 1 atmosphere. If the liquid is in the glass, the temperature can take any value between 0°C and 100°C. Within this range, the temperature can be chosen independently of the pressure. Thus, both T and P are independent. After choosing a (T, P) pair, any other property, such as volume or entropy, can be found using the steam tables or a Mollier diagram. Therefore, the remaining properties cannot be independently chosen after T and P are specified.

An example showing that for two coexistent phases of a pure substance, $N_F = 1$

For a glass of boiling water (also called saturated liquid water) in equilibrium with saturated steam, specify one of the independent intensive variables to be pressure. Choose this pressure to be 1 atmosphere. In order for the water to boil (or be saturated) at this pressure, the temperature must be 100°C. Thus, the temperature cannot be chosen independently of the pressure when both liquid and vapor water are present. Thus, only the P is independent. The temperature required at other pressures, as well as the values of all remaining thermodynamic properties, can be found in the tables for saturated steam or on a Mollier diagram.

An example showing that for three coexistent phases of a pure substance, $N_F = 0$

At the triple point, vapor, liquid, and solid all coexist. For any given substance, the triple point occurs only at one specific pair of temperature and pressure. Once it is stated the substance at the triple point, the values of this temperature and pressure pair as well as the values of all other thermodynamic properties can be found in a table or graph. Thus, no thermodynamic property can be chosen independently.

10.3   Maxwell's thermodynamic relations

Using the theory of symmetry of second derivatives and Euler's reciprocity relation, James Clark Maxwell derived a set of equations connecting thermodynamic variables in partial differential form. He used the concept of exact differentials for the derivation of these relations that connect the partial derivatives of P, V, T, and S for a simple compressible thermodynamic system. He also came up with these relations starting from the definitions of various thermodynamic potentials. These relations consist of characteristic functions of the thermodynamic potentials such as enthalpy H, Helmholtz free energy F, internal energy U, and Gibbs free energy G. These relations include thermodynamic parameters such as pressure P, entropy S, volume V, and temperature T. In thermodynamics, there are certain unknown quantities that are hard to measure in the real world. Maxwell's thermodynamic equations allow physicists to change these unknown quantities to some easily measured quantities so that the study of the subject becomes comprehensive. Here, lies the importance of Maxwell's equations, and they are thus very useful in thermodynamics.

10.3.1 Derivations of Maxwell's relations

For a (P, V, T) system undergoing an infinitesimal reversible process, the combination of the first and the second law of thermodynamics provides

$$dU = T\,dS - P\,dV. \tag{10.47}$$

This equation involves five functions of the state of the system as being determined by any two of these variables. These five state functions are U, P, V, T, and S.

We consider pressure P and temperature T as the independent variables. The state functions U, V, and S will be then functions of P and T. However, for generality, let the independent variables be represented by x and y. Thus, we have

$$U = U(x, y), \quad S = S(x, y), \quad \text{and} \quad V = V(x, y). \tag{10.48}$$

The internal energy U can now be eliminated from equation (10.47), and relations between four fundamental thermodynamic variables, P, V, T, and S, can be constructed. These relations are not merely mathematical but are alternative manifestations of the first and the second law of thermodynamics. Using the concept of differential calculus, the total differentials dU, dS, and dV can be, respectively, expressed as

$$dU = \left(\frac{\partial U}{\partial x}\right)_y dx + \left(\frac{\partial U}{\partial y}\right)_x dy; \quad dS = \left(\frac{\partial S}{\partial x}\right)_y dx + \left(\frac{\partial S}{\partial y}\right)_x dy$$

and

$$dV = \left(\frac{\partial V}{\partial x}\right)_y dx + \left(\frac{\partial V}{\partial y}\right)_x dy. \tag{10.49}$$

Substituting these values in equation (10.47), we get

$$\left(\frac{\partial U}{\partial x}\right)_y dx + \left(\frac{\partial U}{\partial y}\right)_x dy = T\left[\left(\frac{\partial S}{\partial x}\right)_y dx + \left(\frac{\partial S}{\partial y}\right)_x dy\right]$$
$$-P\left[\left(\frac{\partial V}{\partial x}\right)_y dx + \left(\frac{\partial V}{\partial y}\right)_x dy\right]. \tag{10.50}$$

Comparing coefficients of dx and dy from both sides of equation (10.50), we get

$$\left(\frac{\partial U}{\partial x}\right)_y = \left[T\left(\frac{\partial S}{\partial x}\right)_y - P\left(\frac{\partial V}{\partial x}\right)_y\right] \tag{10.51}$$

and

$$\left(\frac{\partial U}{\partial y}\right)_x = \left[T\left(\frac{\partial S}{\partial y}\right)_x - P\left(\frac{\partial V}{\partial y}\right)_x\right]. \tag{10.52}$$

As U, V, and S are exact differentials, we must have the following relations

$$\left[\frac{\partial}{\partial y}\left(\frac{\partial U}{\partial x}\right)_y\right]_x = \left[\frac{\partial}{\partial x}\left(\frac{\partial U}{\partial y}\right)_x\right]_y, \tag{10.53}$$

$$\left[\frac{\partial}{\partial y}\left(\frac{\partial V}{\partial x}\right)_y\right]_x = \left[\frac{\partial}{\partial x}\left(\frac{\partial V}{\partial y}\right)_x\right]_y, \tag{10.54}$$

and

$$\left[\frac{\partial}{\partial y}\left(\frac{\partial S}{\partial x}\right)_y\right]_x = \left[\frac{\partial}{\partial x}\left(\frac{\partial S}{\partial y}\right)_x\right]_y. \tag{10.55}$$

As the order of differentiation does not alter the value of the double derivative, we can perform this operation for the expressions given in equations (10.51) and (10.52). Differentiating equation (10.51) with respect to y at constant x, we get

$$\left[\frac{\partial}{\partial y}\left(\frac{\partial U}{\partial x}\right)_y\right]_x = \left(\frac{\partial T}{\partial y}\right)_x\left(\frac{\partial S}{\partial x}\right)_y + T\left[\frac{\partial}{\partial y}\left(\frac{\partial S}{\partial x}\right)_y\right]_x - \left(\frac{\partial P}{\partial y}\right)_x\left(\frac{\partial V}{\partial x}\right)_y$$
$$-P\left[\frac{\partial}{\partial y}\left(\frac{\partial V}{\partial x}\right)_y\right]_x. \tag{10.56}$$

And differentiating equation (10.52) with respect to x at constant y, we get

$$\left[\frac{\partial}{\partial x}\left(\frac{\partial U}{\partial y}\right)_x\right]_y = \left(\frac{\partial T}{\partial x}\right)_y\left(\frac{\partial S}{\partial y}\right)_x + T\left[\frac{\partial}{\partial x}\left(\frac{\partial S}{\partial y}\right)_x\right]_y - \left(\frac{\partial P}{\partial x}\right)_y\left(\frac{\partial V}{\partial y}\right)_x$$
$$-P\left[\frac{\partial}{\partial x}\left(\frac{\partial V}{\partial y}\right)_x\right]_y. \tag{10.57}$$

Now equating right-hand sides of equations (10.56) and (10.57) and simplifying, we get

$$\left(\frac{\partial P}{\partial x}\right)_y \left(\frac{\partial V}{\partial y}\right)_x - \left(\frac{\partial P}{\partial y}\right)_x \left(\frac{\partial V}{\partial x}\right)_y = \left(\frac{\partial T}{\partial x}\right)_y \left(\frac{\partial S}{\partial y}\right)_x - \left(\frac{\partial T}{\partial y}\right)_x \left(\frac{\partial S}{\partial x}\right)_y. \quad (10.58)$$

Here, x and y can be chosen out of P, V, T, and S. This can be done in six different ways, but only four choices give useful thermodynamic relations. These relations are derived one by one below.

Maxwell's first relation

Considering T and V as independent variables and taking $x = T$ and $y = V$, we get from equation (10.58)

$$\left(\frac{\partial S}{\partial V}\right)_T = \left(\frac{\partial P}{\partial T}\right)_V. \quad (10.59)$$

This represents Maxwell's first thermodynamic relation.

Maxwell's second relation

Considering T and P as independent variables and taking $x = T$ and $y = P$, we get from equation (10.58)

$$\left(\frac{\partial S}{\partial P}\right)_T = -\left(\frac{\partial V}{\partial T}\right)_P. \quad (10.60)$$

This represents Maxwell's second thermodynamic relation.

Maxwell's third relation

Considering S and V as independent variables and taking $x = S$ and $y = V$, we get from equation (10.58)

$$\left(\frac{\partial T}{\partial V}\right)_S = -\left(\frac{\partial P}{\partial S}\right)_V. \quad (10.61)$$

This represents Maxwell's third thermodynamic relation.

Maxwell's fourth relation

Considering S and P as independent variables and taking $x = S$ and $y = P$, we get from equation (10.58)

$$\left(\frac{\partial T}{\partial P}\right)_S = \left(\frac{\partial V}{\partial S}\right)_P. \quad (10.62)$$

This represents Maxwell's fourth thermodynamic relation.

10.4   Applications of Maxwell's thermodynamic relations

Various physical processes can be analyzed thermodynamically using Maxwell's thermodynamic relations and the equation of state obeyed by these processes. For example, the adiabatic equation of state for the process should be used when the process involved is adiabatic. It has been shown earlier that the isobaric volume expansion coefficient α_P and the isothermal compressibility β_T can be used conveniently for liquids and solids in lieu of the equation of state. Further, thermodynamic properties of such systems may often be expressed in terms of the heat capacities C_P, and C_V. Various relations between these heat capacities can be derived using Maxwell's thermodynamic relations. Using these Maxwell relations, thermodynamic potentials, and some fundamental relations, a large variety of physical processes can be dealt easily in thermodynamics. Maxwell's relations also play a very important role in the required conversion of various processes in thermodynamics. Some useful applications of Maxwell's relations in thermodynamics are discussed in detail below.

10.4.1   Clausius–Clapeyron equation

Clausius–Clapeyron equation or latent heat equation can be derived from one of Maxwell's thermodynamic relations. We consider the following Maxwell's thermodynamic relation:

$$\left(\frac{\partial S}{\partial V}\right)_T = \left(\frac{\partial P}{\partial T}\right)_V. \tag{10.63}$$

Multiplying both sides by T, we get

$$T\left(\frac{\partial S}{\partial V}\right)_T = T\left(\frac{\partial P}{\partial T}\right)_V. \tag{10.64}$$

However, we have $T\partial S = \partial Q$. Hence, equation (10.64) becomes

$$\left(\frac{\delta Q}{\partial V}\right)_T = T\left(\frac{\partial P}{\partial T}\right)_V. \tag{10.65}$$

The quantity $\left(\frac{\partial Q}{\partial V}\right)_T$ represents the amount of heat δQ absorbed or released per unit change in volume V at constant temperature T. If there is a change in volume V of a substance due to heat absorbed or released by the substance at constant temperature T, then the heat δQ represents the latent heat used when the substance changes from solid to liquid (melting) state or from liquid to vapor (boiling) state and vice versa. It should be noted that during such a change of state, the temperature of the substance remains constant.

The amount of heat δQ absorbed or released per unit mass at constant temperature is defined as the specific latent heat. Hence, we can write the specific latent heat as $\delta Q = L$. Let V_1 and V_2 be the volumes per unit mass of the substance in the first and second phase, respectively. Then, the change in volume per unit volume during the change in phase will be $\delta V = V_2 - V_1$. Hence, we have $\left(\frac{\delta Q}{\delta V}\right)_T = \frac{L}{V_2 - V_1}$. Using this result in equation (10.65), we get

$$\frac{L}{V_2 - V_1} = T\left(\frac{\partial P}{\partial T}\right)_V \; ; \quad \Rightarrow \quad \frac{dP}{dT} = \frac{L}{T(V_2 - V_1)}, \tag{10.66}$$

where dT represents the change in melting point or boiling point of the substance due to change in pressure dP. This equation is known as Clausius–Clapeyron equation or Clausius–Clapeyron latent heat equation.

10.4.2   Values of $C_P - C_V$

The specific heat capacity indicates the ability of a substance to store heat. It is a characteristic of each substance, and the specific heat capacity for the temperature change under constant pressure (C_P) and under constant volume (C_V) are frequently used to identify materials. This property is experimentally determined by the methods of **Differential Scanning Calorimetry (DSC)** and **Differential Thermal Analysis (DTA)**. These methods provide highly accurate measurement results in a short time. Generally, it is more feasible to measure C_P rather than C_V, which is easily calculated from a theoretical point of view. The difference in heat capacities $C_P - C_V$ is of particular interest to physicists for the comparison of theoretical and experimental results of a substance. Such relations connecting $C_P - C_V$ for various systems such as ideal and real gases can be easily developed with the help of Maxwell's thermodynamic relations.

The heat capacity can be defined as $C_x = T\left(\frac{\partial S}{\partial T}\right)_x$, where x may be P or V. Considering T and V as independent variables, entropy S can be written as $S = S(T, V)$. Hence, the differential increase in S can be written as

$$dS = \left(\frac{\partial S}{\partial T}\right)_V dT + \left(\frac{\partial S}{\partial V}\right)_T dV.$$

Multiplying both sides by temperature T at a constant pressure P, we can write

$$TdS_P = T\left(\frac{\partial S}{\partial T}\right)_V dT_P + T\left(\frac{\partial S}{\partial V}\right)_T dV_P.$$

$$\Rightarrow \quad T\left(\frac{\partial S}{\partial T}\right)_P = T\left(\frac{\partial S}{\partial T}\right)_V + T\left(\frac{\partial S}{\partial V}\right)_T \left(\frac{\partial V}{\partial T}\right)_P.$$

In terms of heat capacities, the above expression can be written as

$$C_P = C_V + T \left(\frac{\partial S}{\partial V}\right)_T \left(\frac{\partial V}{\partial T}\right)_P.$$

Using Maxwell's first relation $\left(\frac{\partial S}{\partial V}\right)_T = \left(\frac{\partial P}{\partial T}\right)_V$ and rearranging terms, we can write

$$C_P - C_V = T \left(\frac{\partial P}{\partial T}\right)_V \left(\frac{\partial V}{\partial T}\right)_P = T V \alpha_P \left(\frac{\partial P}{\partial T}\right)_V, \tag{10.67}$$

where $\alpha_P = \frac{1}{V} \left(\frac{\partial V}{\partial T}\right)_P$ is the volume expansion coefficient at constant pressure P.

To explain the physical significance of equation (10.67), we modify the equation by considering pressure P as a function of temperature T and volume V: $P = P(T, V)$. Hence, differential increase dP in P can be expressed as

$$dP = \left(\frac{\partial P}{\partial T}\right)_V dT + \left(\frac{\partial P}{\partial V}\right)_T dV.$$

In an isobaric process, P remains constant. So we can write

$$\left(\frac{\partial P}{\partial T}\right)_V dT_P + \left(\frac{\partial P}{\partial V}\right)_T dV_P = 0 \quad \text{so that} \quad \Rightarrow \quad \left(\frac{\partial P}{\partial T}\right)_V = - \left(\frac{\partial P}{\partial V}\right)_T \left(\frac{\partial V}{\partial T}\right)_P. \tag{10.68}$$

Using equation (10.68) into equation (10.67), we get

$$C_P - C_V = -T \left(\frac{\partial P}{\partial V}\right)_T \left[\left(\frac{\partial V}{\partial T}\right)_P\right]^2. \tag{10.69}$$

Equation (10.69), known as the **heat capacity equation,** is one of the most informative relationships in thermodynamics. The following conclusions can be drawn from this equation:

1. Since $\left(\frac{\partial P}{\partial V}\right)_T$ is negative for all known substances, and $\left[\left(\frac{\partial V}{\partial T}\right)_P\right]^2$ is always positive; the difference $(C_P - C_V)$ will always be positive. That is, C_P can never be less than C_V.

2. The factor $\left(\frac{\partial V}{\partial T}\right)_P$ is relatively small for liquids and solids, and therefore, the difference $(C_P - C_V)$ is small. Further, for water at $4°C$, we have $\left(\frac{\partial V}{\partial T}\right)_P = 0$. This indicates that for water, C_P will be equal to C_V at this temperature.

3. As $T \to 0\ K$, $C_P \to C_V$, that is, as absolute zero is approached, the heat capacities at constant pressure and at constant volume coalesce. This remarkable result is also experimentally verified.

10.4.3 $T–dS$ **Equations**

Graduate students generally consider the three $T–dS$ equations as the very "tedious equations" in thermodynamics. But frankly speaking, these equations are not at all tedious to a true lover of thermodynamics. The reasons behind this are many folds: Firstly, these equations help us to calculate the change of entropy dS in terms of either dP and dT, or dV and dT, or dV and dP during various reversible processes. Secondly, dS can also be expressed in terms of directly measurable quantities, such as the isobaric volume expansion coefficient α_P and the isothermal bulk modulus κ_T. These three $T–dS$ equations are derived below.

First $T–dS$ equation

The entropy S of a pure substance can be considered as a function of temperature T and volume V. Hence, the differential change in entropy dS can be written as

$$dS = \left(\frac{\partial S}{\partial T}\right)_V dT + \left(\frac{\partial S}{\partial V}\right)_T dV.$$

Multiplying both sides by temperature T, we get

$$TdS = T\left(\frac{\partial S}{\partial T}\right)_V dT + T\left(\frac{\partial S}{\partial V}\right)_T dV. \tag{10.70}$$

Using the definition of specific heat at constant volume $C_V = T\left(\frac{\partial S}{\partial T}\right)_V$ and one of Maxwell's relation $\left(\frac{\partial S}{\partial V}\right)_T = \left(\frac{\partial P}{\partial T}\right)_V$ in equation (10.70), we get

$$TdS = C_V dT + T\left(\frac{\partial P}{\partial T}\right)_V dV. \tag{10.71}$$

Equation (10.71) is the first $T–dS$ equation.

Second $T–dS$ equation

In order to derive the second $T–dS$ equation, we consider entropy S of a pure substance as a function of temperature T and pressure P. Hence, the differential change in entropy dS can be written as

$$dS = \left(\frac{\partial S}{\partial T}\right)_P dT + \left(\frac{\partial S}{\partial P}\right)_T dP.$$

Multiplying both sides by temperature T, we get

$$TdS = T\left(\frac{\partial S}{\partial T}\right)_P dT + T\left(\frac{\partial S}{\partial P}\right)_T dP. \tag{10.72}$$

Using the definition of specific heat at constant pressure $C_P = T\left(\frac{\partial S}{\partial T}\right)_P$ and one of Maxwell's relation $\left(\frac{\partial S}{\partial P}\right)_T = -\left(\frac{\partial V}{\partial T}\right)_P$ in equation (10.72), we get

$$TdS = C_P dT - T\left(\frac{\partial V}{\partial T}\right)_P dP. \tag{10.73}$$

Equation (10.73) is the second T–dS equation.

Third T–dS equation

In order to derive the third T–dS equation, we consider entropy S of a pure substance as a function of volume V and pressure P. Hence, the differential change in entropy dS can be written as

$$dS = \left(\frac{\partial S}{\partial P}\right)_V dP + \left(\frac{\partial S}{\partial V}\right)_P dV.$$

Multiplying both sides of this expression by temperature T, we get

$$TdS = T\left(\frac{\partial S}{\partial P}\right)_V dP + T\left(\frac{\partial S}{\partial V}\right)_P dV. \tag{10.74}$$

Further, in a constant volume process, we have $TdS = C_V dT$. Thus, we get

$$T\left(\frac{\partial S}{\partial P}\right)_V = C_V\left(\frac{\partial T}{\partial P}\right)_V. \tag{10.75}$$

And in a constant pressure process, $TdS = C_P dT$, so that

$$T\left(\frac{\partial S}{\partial V}\right)_P = C_P\left(\frac{\partial T}{\partial V}\right)_P. \tag{10.76}$$

Using equations (10.75) and (10.76) into equation (10.74), we get

$$TdS = C_V\left(\frac{\partial T}{\partial P}\right)_V dP + C_P\left(\frac{\partial T}{\partial V}\right)_P dV. \tag{10.77}$$

Equation (10.77) is the third T–dS equation.

10.4.4 Energy equations

We know that the internal energy of an ideal gas is independent of its volume but is a function of volume in the case of a real gas. In this section, we will derive some general relations for the change of internal energy dU in terms of changes in other thermodynamic variables for a pure substance.

First energy equation

In order to derive the first energy equation, we consider the internal energy U as a function of temperature T and volume V, that is, $U = U(T, V)$. Hence, the differential change in internal energy dU induced by infinitesimal changes in temperature T and volume V can be written as

$$dU = \left(\frac{\partial U}{\partial T}\right)_V dT + \left(\frac{\partial U}{\partial V}\right)_T dV. \tag{10.78}$$

For a thermodynamic system characterized by P, V, and T and undergoing an infinitesimal reversible change between two equilibrium states, combining the first and second laws of thermodynamics, the change in internal energy dU can be written as

$$dU = TdS - PdV. \tag{10.79}$$

Differentiating equation (10.79) w.r.t volume V at a constant temperature T, we get

$$\left(\frac{\partial U}{\partial V}\right)_T = T\left(\frac{\partial S}{\partial V}\right)_T - P. \tag{10.80}$$

Using equation (10.59), equation (10.80) can be written as

$$\left(\frac{\partial U}{\partial V}\right)_T = T\left(\frac{\partial P}{\partial T}\right)_V - P. \tag{10.81}$$

This is referred to as the **first energy equation**.

Using equation (10.81) in equation (10.78), we get

$$dU = C_V dT + \left[T\left(\frac{\partial P}{\partial T}\right)_V - P\right]dV. \tag{10.82}$$

Equation (10.82) can be integrated to obtain the change in internal energy associated with a given change of state for a pure substance, provided the specific heat at constant volume C_V and the equation of state for the substance are known:

$$\Delta U = \int C_V dT + \int \left[T\left(\frac{\partial P}{\partial T}\right)_V - P\right]dV. \tag{10.83}$$

Ideal gas

Let us apply equation (10.83) to the case of one mole of an ideal gas for which we have $P = \frac{RT}{V}$ so that we have

$$\left(\frac{\partial P}{\partial T}\right)_V = \frac{R}{V}$$

and the first energy equation implies that if the temperature remains constant $(dT = 0)$, we have

$$T\left(\frac{\partial P}{\partial T}\right)_V - P = \left(\frac{\partial U}{\partial V}\right)_T = \frac{RT}{V} - P = P - P = 0.$$

This indicates that the temperature remains constant, and the internal energy of an ideal gas is independent of volume. Using this result in equation (10.83), we obtain

$$\Delta U = \int_{T_1}^{T_2} C_V dT = C_V(T_2 - T_1),$$ (10.84)

where it is assumed that the heat capacity C_V at constant volume does not change over the measured range of temperature.

Real gas: using Van der Waals equation of state

Let us apply equation (10.83) to the case of a real (Van der Waals) gas for which we have $P = \frac{RT}{V-b} - \frac{a}{V^2}$. This gives rise to

$$\left(\frac{\partial P}{\partial T}\right)_V = \frac{R}{V-b}.$$

Using this result and the first energy equation at constant temperature T, we get

$$\left(\frac{\partial U}{\partial V}\right)_T = T\left(\frac{\partial P}{\partial T}\right)_V - P = \frac{RT}{V-b} - P = P + \frac{a}{V^2} - P = \frac{a}{V^2}.$$

This indicates that the temperature remains constant, and the internal energy of a real gas is a function of volume. The term $\left(\frac{\partial U}{\partial V}\right)_T$ is known as the co-pressure. It is finite, and its genesis is in intermolecular interactions.

Using this result in equation (10.83), we obtain

$$U_2 - U_1 = \int_{T_1}^{T_2} C_V dT + \int_{V_1}^{V_2} \frac{a}{V^2} dV = \int_{T_1}^{T_2} C_V dT - a\left(\frac{1}{V_2} - \frac{1}{V_1}\right).$$ (10.85)

This result shows that the internal energy of a real gas depends on volume. The change in internal energy $(U_2 - U_1)$ increases with the increase in volume V, even at fixed temperature T. This result can be understood from the fact that as intermolecular separation increases, the potential energy of interaction of the system decreases.

Real gas: using Dieterici's equation of state

To calculate the change in internal energy, let us apply equation (10.83) to the case of a real gas obeying Dieterici's equation of state that can be mathematically expressed as

$$P = \frac{RT}{V-b}\exp\left[-\left(\frac{a}{RTV}\right)\right]. \tag{10.86}$$

For a thermodynamic system described by pressure P, volume V, and temperature T and undergoing infinitesimal reversible change between two equilibrium states, the first energy equation (10.81) is

$$\left(\frac{\partial U}{\partial V}\right)_T = T\left(\frac{\partial P}{\partial T}\right)_V - P.$$

For a real gas obeying Dieterici's equation, we have

$$P = \frac{RT}{V-b}\exp\left[-\left(\frac{a}{RTV}\right)\right].$$

This gives

$$\left(\frac{\partial P}{\partial T}\right)_V = \frac{R}{V-b}\exp\left[-\left(\frac{a}{RTV}\right)\right] + \frac{RT}{V-b}\frac{\partial}{\partial T}\left[\exp\left[-\left(\frac{a}{RTV}\right)\right]\right]_V$$

$$= \frac{R}{V-b}\exp\left[-\left(\frac{a}{RTV}\right)\right] + \frac{RT}{V-b}\left[\exp\left[-\left(\frac{a}{RTV}\right)\right]\right]\left(\frac{a}{RT^2V}\right).$$

This leads to

$$T\left(\frac{\partial P}{\partial T}\right)_V = \frac{RT}{V-b}\exp\left[-\left(\frac{a}{RTV}\right)\right] + \frac{RT^2}{V-b}\left[\exp\left[-\left(\frac{a}{RTV}\right)\right]\right]\left(\frac{a}{RT^2V}\right). \tag{10.87}$$

The first term in equation (10.87) is simply equal to pressure P. Hence, we can write

$$T\left(\frac{\partial P}{\partial T}\right)_V - P = \frac{1}{V-b}\left[\exp\left[-\left(\frac{a}{RTV}\right)\right]\right]\left(\frac{a}{V}\right) = \frac{aP}{RTV}; \quad \Rightarrow \quad \left(\frac{\partial U}{\partial V}\right)_T = \frac{aP}{RTV}. \tag{10.88}$$

For an isothermal–isobaric process, we can write

$$dU = k\left(\frac{dV}{V}\right),$$

where $k = \frac{aP}{RT}$ is a constant. Integrating this expression, we get

$$\Delta U = k\ln\left(\frac{V_2}{V_1}\right) = k\ln e = k = \frac{aP}{RT}, \tag{10.89}$$

where the final volume is "e" times the initial volume.

Second energy equation

In order to investigate the dependency of internal energy U on pressure P, we write internal energy as a function of temperature T and pressure P, that is, $U = U(T, P)$. This gives rise to the increment in internal energy dU as

$$dU = \left(\frac{\partial U}{\partial T}\right)_P dT + \left(\frac{\partial U}{\partial P}\right)_T dP. \tag{10.90}$$

From the first and second law of thermodynamics, we have

$$dU = TdS - PdV$$

it follows that

$$\left(\frac{\partial U}{\partial P}\right)_T = T\left(\frac{\partial S}{\partial P}\right)_T - P\left(\frac{\partial V}{\partial P}\right)_T. \tag{10.91}$$

Using Maxwell's second relation given by equation (10.60) on the right-hand side, we get

$$\left(\frac{\partial U}{\partial P}\right)_T = -T\left(\frac{\partial V}{\partial T}\right)_P - P\left(\frac{\partial V}{\partial P}\right)_T. \tag{10.92}$$

Equation (10.92) is known as the **second energy equation** in thermodynamics.
   Using equation (10.92) into equation (10.90), we get

$$dU = \left(\frac{\partial U}{\partial T}\right)_P dT - \left[T\left(\frac{\partial V}{\partial T}\right)_P + P\left(\frac{\partial V}{\partial P}\right)_T\right] dP. \tag{10.93}$$

Thus, it may be emphasized here that the energy equations have been derived using Maxwell's thermodynamic relations. However, one can also derive these equations starting from T–dS equations as well as from the first principle. It should be noted that the reverse is also true.

10.4.5   Applications of Maxwell's relations to magnetic system

In the previous section, applications of Maxwell's relations to simple compressible substances have been discussed in detail. In this section, we will discuss T–dS equations, energy equations, and equations for the heat capacity from the combined form of the first and the second laws of thermodynamics for a **pure magnetic substance**.

First energy equation for a paramagnetic substance

We consider the internal energy U of a pure paramagnetic substance as a function of temperature T and magnetization M, that is, $U = U(T, M)$. Hence, an infinitesimal change in internal energy due to the infinitesimal changes in temperature T and magnetization M can be written as

$$dU = \left(\frac{\partial U}{\partial T}\right)_M dT + \left(\frac{\partial U}{\partial M}\right)_T dM. \tag{10.94}$$

For a pure paramagnetic material, the combination of the first and the second laws of thermodynamics can be written as

$$TdS = dU - BdM. \tag{10.95}$$

Differentiating equation (10.95) w.r.t magnetization M at constant temperature T, we get

$$\left(\frac{\partial U}{\partial M}\right)_T - B = T\left(\frac{\partial S}{\partial M}\right)_T = -T\left(\frac{\partial B}{\partial T}\right)_M. \tag{10.96}$$

Here, we have used Maxwell's relation to get $\left(\frac{\partial S}{\partial M}\right)_T = -\left(\frac{\partial B}{\partial T}\right)_M$. Equation (10.96) is the **first energy equation for a paramagnetic substance**.

First T–dS equation for a paramagnetic substance

Using equation (10.94) into equation (10.95), we get

$$TdS = \left(\frac{\partial U}{\partial T}\right)_M dT + \left(\frac{\partial U}{\partial M}\right)_T dM - BdM = \left(\frac{\partial U}{\partial T}\right)_M dT + \left[\left(\frac{\partial U}{\partial M}\right)_T - B\right] dM. \tag{10.97}$$

Using equation (10.96) and the definition of specific heat capacity at constant magnetization M, equation (10.97) can be written as

$$TdS = C_M dT - T\left(\frac{\partial B}{\partial T}\right)_M dM. \tag{10.98}$$

Equation (10.98) gives the first TdS equation for a paramagnetic substance.

Second energy equation for a paramagnetic substance

We consider the internal energy U of a pure paramagnetic substance as a function of temperature T and magnetic field B, that is, $U = U(T, B)$. Hence, an infinitesimal change in internal energy due to the infinitesimal changes in temperature T and magnetic field B can be written as

$$dU = \left(\frac{\partial U}{\partial T}\right)_B dT + \left(\frac{\partial U}{\partial B}\right)_T dB. \tag{10.99}$$

In this case, for a pure paramagnetic material, the combination of the first and the second laws of thermodynamics can be written as

$$TdS = dU - MdB. \tag{10.100}$$

Differentiating equation (10.100) w.r.t magnetic field B at constant temperature T, we get

$$\left(\frac{\partial U}{\partial B}\right)_T - M = T\left(\frac{\partial S}{\partial B}\right)_T = T\left(\frac{\partial M}{\partial T}\right)_B. \tag{10.101}$$

Here, we have used Maxwell's relation to get $\left(\frac{\partial S}{\partial B}\right)_T = \left(\frac{\partial M}{\partial T}\right)_B$. Equation (10.101) is the **second energy equation for a paramagnetic substance**.

Second T–dS equation for a paramagnetic substance

Using equation (10.99) into equation (10.100), we get

$$TdS = \left(\frac{\partial U}{\partial T}\right)_B dT + \left(\frac{\partial U}{\partial B}\right)_T dB - MdB = \left(\frac{\partial U}{\partial T}\right)_B dT + \left[\left(\frac{\partial U}{\partial B}\right)_T - M\right]dB. \tag{10.102}$$

Inserting equation (10.101) into equation (10.102) and the definition of specific heat capacity at constant magnetic field B, we get

$$TdS = \left(\frac{\partial U}{\partial T}\right)_B dT + T\left(\frac{\partial M}{\partial T}\right)_B dB = C_B dT + T\left(\frac{\partial M}{\partial T}\right)_B dB. \tag{10.103}$$

Equation (10.103) is the **second T–dS equation for a paramagnetic substance**.

These energy and T–dS equations reveal the dependence of magnetic properties on temperature T and magnetic induction field B. This effect is known as a **magneto-caloric** effect. For a paramagnetic material, the factor $\left(\frac{\partial M}{\partial T}\right)_B$ is always negative. This has the following important implications:

1. Rejection of heat occurs when the magnetic field increases in a reversible isothermal process, and vice versa; and

2. There is a drop in temperature when the magnetic field decreases in a reversible adiabatic process. This technique is widely used in experimental situations for the production of low temperatures below 1 K.

10.4.6 Change of temperature during an adiabatic process

We see that every macroscopic property of a system in equilibrium can be expressed via a set of a few measurable quantities. For example, every monoatomic or monomolecular nonmagnetic system is completely described by a set of three quantities and their temperature dependence:

1. Coefficient of thermal expansion

2. Compressibility, and

3. Heat capacity.

Introduction to Joule's experiment

When an ideal gas expands adiabatically and reversibly, it obeys the equation of state: $TV^{\gamma-1} = $ constant, where symbols have the usual meaning. This equation can be used to calculate the drop in temperature in the adiabatic process. In this process, a gas expands against an external pressure (e.g., a piston), and in pushing the piston back, the molecules of the gas perform some external work and loose their kinetic energy. Now we can think of the situation: What would happen if a gas expands into a vacuum? The result of this expansion can be realized in the following way: Consider that the gas is held inside a cylinder not by a metal piston but by a thin membrane. When the membrane breaks down, the molecules rush out into empty space (vacuum). This is certainly an irreversible expansion. It is most unlikely that all of the molecules will ever find their way back to the cylinder. In this case, the molecules will not perform any external work.

In the case of an ideal gas, there will be no internal work as there are no intermolecular forces present. There is nothing to slow down the molecules in the ideal gas in their headlong escape from the cylinder. As a result, the temperature of the gas will remain the same by the expansion. On the other hand, if the gas is a real one (say, Van der Waals type), there will be Van der Waals attractive forces acting between the molecules. Hence, during the adiabatic expansion of the gas, the molecules of the gas will slow down slightly and there will be a small drop in temperature. However, we know that at close intermolecular distances, the Van der Waals forces between the molecules are predominantly repulsive Coulomb forces. Thus, it is also possible that if the gas starts out very dense and expands irreversibly, it may initially become slightly warmer. This is due to the repulsive Coulomb forces that push the molecules apart and speed them on their way. The Joule and Joule–Thomson experiments are concerned with these scenarios.

Joule's experiment

In Joule's original experiment, there was a cylinder filled with gas at high pressure connected via a stopcock to a second cylinder with gas at a sufficiently low pressure. For the purpose of understanding the experiment, it is assumed that the second cylinder is entirely empty. These two cylinders were immersed in a water bath, and the stopcock was then opened so that the gas from the cylinder with a high pressure flowed into the empty cylinder. In this process, no heat was supplied from an external source to the system, or no heat was lost from the system to the surroundings, nor did the gas do any work. Hence, its internal energy was constant during the expansion. Joule noticed no decrease in temperature as a result of this expansion. This shows that the temperature of an ideal gas is independent of the volume if the internal energy of the gas is held constant. Hence, for an ideal gas, we have

$$\left(\frac{\partial T}{\partial V}\right)_U = 0.$$

This term $\left(\frac{\partial T}{\partial V}\right)_U$ is known as the **Joule coefficient**. On the other hand, for a real gas, one would expect a small decrease in temperature, that is, the term $\left(\frac{\partial T}{\partial V}\right)_U$ is not zero. It should be mentioned that in Joule's original experiment, the heat capacity of the water bath and the cylinders was too large to detect any fall of temperature even with real gas. However, more sensitive experiments revealed that almost all gases exhibit a fall in temperature during a Joule expansion at all temperatures investigated. The only exceptions in this series are helium and hydrogen. It was observed for helium at temperatures above about 40 K, and hydrogen at temperatures above about 200 K that

$$\left(\frac{\partial T}{\partial V}\right)_U \neq 0.$$

This shows that for such gaseous systems, the temperature is a function of volume, even if the internal energy is kept constant. Further, we state that the expression for the **Joule coefficient** of a gas can be derived if we know the equation of state for that gas. Thus, from the equation of state of an ideal gas, it can be easily shown that the Joule coefficient is zero for an ideal gas.

Free adiabatic expansion of an ideal gas

Consider a thermally insulated rigid container. The container is divided into two compartments separated by a solid partition as shown in Figure 10.9. One compartment of volume V_1 contains the gas under investigation. The other compartment is empty (vacuum). The initial temperature of the system is T_1. The solid partition is now removed, and the gas is free to expand so as to fill the entire container, whose volume is V_2. One may ask the question: What is the temperature, T_2, of the gas after the attainment of the final equilibrium

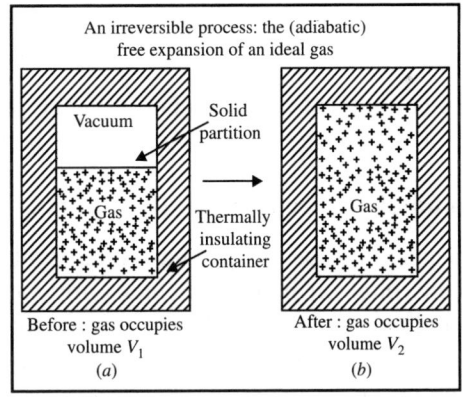

Figure 10.9 Schematic representation of free (adiabatic) expansion of an ideal gas.

state? Because the system consists of the gas, and the container is adiabatically insulated, not heat flows into the system which indicates that $Q = 0$. Furthermore, the system does not work in the expansion process, that is, $W = 0$.

An adiabatic process occurs without the transfer of heat or mass of substances between a thermodynamic system and its surroundings. In this process, energy is transferred to the surroundings only as work. The adiabatic process provides a rigorous conceptual basis for the theory used to expound the first law of thermodynamics, and as such, it is a key concept in thermodynamics. In free expansion, as there is no external pressure for the gas to expand against, the work done by or on the system is zero. Since this process does not involve any heat transfer or work, the first law of thermodynamics then implies that the net internal energy change of the system is zero. Hence, it follows from the first law of thermodynamics that the total energy of the system is conserved: that is, $\Delta U = Q - W = 0$.

Let us assume that the container itself has negligible heat capacity (which, it turns out, is not a particularly realistic assumption), so that the internal energy of the container does not change. Under these circumstances, the energy change of the system is equivalent to that of the gas. The conservation of energy, thus, reduces to

$$U(T_2, V_2) = U(T_1, V_1) \tag{10.104}$$

where $U(T, V)$ is the internal energy of the gas.

To predict the outcome of the experiment, it is only necessary to know the internal energy of the gas, $U(T, V)$, as a function of temperature T and volume V. If the initial parameters, T_1 and V_1, and the final volume, V_2 are known, then equation (10.104) yields another equation that specifies the unknown final temperature T_2.

For an ideal gas, the internal energy is independent of the volume, but it is a function of temperature only, that is, $U(T)$. In this case, the above equation (10.104) yields

$$U(T_2) = U(T_1). \tag{10.105}$$

Thus, equality of internal energy in the process implies that $T_1 = T_2$. In other words, there is no change in temperature in the free expansion of an ideal gas.

Free adiabatic expansion: ideal versus real gases

Ideal gases are assumed to have no interatomic forces among the molecules. Therefore, a free expansion neither includes any work nor there is any cooling effect (since no internal energy is spent in overcoming interatomic forces). Hence, in a free expansion of an ideal gas not only $\delta Q = 0$ but also $\delta W = 0$. In such a case, the first law of thermodynamics leads to $dU = 0$, that is, the internal energy remains constant in the free expansion of an ideal gas.

Further, it is known for ideal gases that the internal energy and enthalpy are functions of temperature only. So if internal energy U remains constant, temperature T also remains constant that means enthalpy also remains constant.

Free expansion of a real gas will cause cooling since real gases have intermolecular forces. Atoms have to spend energy in order to move farther by overcoming intermolecular forces. Thus, the expansion of a real gas occurs at the expense of internal energy that reduces the temperature. So, in case of free expansion of a real gas, there are two important consequences:

1. Free expansion causes cooling.

2. Work done in the free expansion of real gases is not zero as there is a force of attraction between the molecules of real gas.

From these concepts, it can be concluded that

1. During the free expansion of an ideal gas, both internal energy and enthalpy remain constant and there is no change in temperature.

2. During the free expansion of a real gas, none of the internal energy and enthalpy remains constant, and there is a change in temperature.

10.4.7 Free adiabatic expansion of a real gas: Porous plug experiment due to Joule–Thomson

We consider a gas at a temperature below its inversion temperature. If the gas is allowed to expand adiabatically through a porous plug, it is found that the temperature of the gas decreases. This decrease in temperature, however, is found to be very small, of the order of a few tenths of a degree per atmospheric pressure. Such a small drop in temperature is obviously not sufficient to liquefy a gas all by itself. However, in principle, this cooled gas can be used to pre-cool the incoming gas in a heat inter-exchanger before it is made to expand through the porous plug at an initial lower temperature, and this makes the fall in temperature of the incoming gas gradually steeper. If this cycle of operations is repeated several times, there is accumulated or regenerative cooling of the gas, and ultimately, the temperature of the gas may fall so much that on being subject to further Joule–Thomson expansion, it readily gets liquefied.

Linde successfully exploited the Joule–Thomson effect to produce sufficiently low temperatures required for the liquefaction of air. Later on, this process was used by Dewar and Onnes to liquefy hydrogen and helium, respectively. It should be mentioned that this method involves considerable economy over the Cascade process because in the earlier process, the gas has to be initially cooled below its inversion temperature rather than the critical temperature. The expression for the Joule–Kelvin coefficient is first derived below. The process of regenerative cooling is then discussed with a focus on the technical details of the machines used by Linde and others.

Joule–Kelvin effect: An isenthalpic process

In the porous plug experiment, the pressures on both sides of the porous plug are kept constant. Let these values be P_1 and P_2 with ($< P_1$). Further, let U_1 and V_1 be, respectively, the initial internal energy and volume of the gas. After passing through the porous plug, let the final internal energy and volume of the gas be, respectively, U_2 and V_2. Since the system is in complete thermal isolation, we have from the first law of thermodynamics that

$$\delta Q = 0 = U_2 - U_1 + P_2 V_2 - P_1 V_1.$$

This leads to

$$U_1 + P_1 V_1 = U_2 + P_2 V_2; \quad \Rightarrow \quad H_1 = H_2.$$

This expression indicates that the Joule–Kelvin expansion can be considered as a quasistatic isenthalpic process. This result clearly signifies that if a series of experiments are performed on a gas, keeping the initial temperature and pressure constant, values of the final temperature for different final pressures will be determined by the properties of the gas. On a pressure–temperature $(P–T)$ diagram, a series of discrete points are obtained, which indicate the final equilibrium states of the system with constant enthalpy. If these points are joined, a smooth curve is obtained, as shown in Figure 10.10 (a). Such a curve is known as the isenthalpic curve. It is observed from the curve that between states (1) and (4), the gas is heated up as temperature T increases with the decrease in pressure P. However, the gas cools down between the states (4) and (7), in which temperature T decreases with the decrease in pressure P. In fact, the cooling will be more if the initial state corresponds to the maximum of the isenthalpic curve (state 4).

If another series of such experiments is carried out by changing the initial temperature of the gas but keeping the initial pressure the same, another isenthalpic curve with a different value of enthalpy H will be obtained. In this way, a family of isenthalpic curves, that is, curves with $H(T, P) = $ constant for the gas under investigation will be obtained. Such a family of isenthalpic curves shown in Figure 10.10(b) is typical of most of the gases.

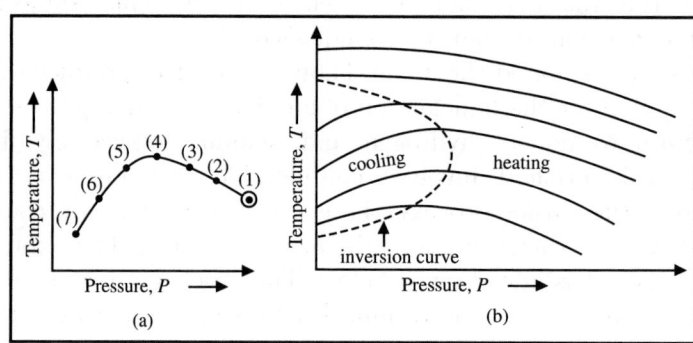

Figure 10.10 (a) Curve of an isenthalpic process, and (b) a family of such isenthalpic graphs for a typical gas.

The Joule–Kelvin coefficient μ is defined as the change in temperature associated with the corresponding change in pressure at constant enthalpy H and is denoted as $\mu = \left(\frac{\partial T}{\partial P}\right)_H$. To calculate this factor $\left(\frac{\partial T}{\partial P}\right)_H$, we express enthalpy H as a function of temperature T and pressure P, that is,

$$H = H(T, P).$$

An infinitesimal change in enthalpy H corresponding to infinitesimal changes in temperature T and pressure P can be written as

$$dH = \left(\frac{\partial H}{\partial T}\right)_P dT + \left(\frac{\partial H}{\partial P}\right)_T dP. \tag{10.106}$$

From the definition of heat capacity C_P at constant pressure P, we have

$$C_P = \left(\frac{\partial H}{\partial T}\right)_P.$$

Using this value of C_P, we can rewrite equation (10.106) as

$$dH = C_P \, dT + \left(\frac{\partial H}{\partial P}\right)_T dP. \tag{10.107}$$

Let us express this equation (10.107) in a more compact form. Using the expression for enthalpy $H = U + PV$, we get

$$dH = dU + PdV + VdP = TdS + VdP. \tag{10.108}$$

In this equation (10.108), we have used a combined form of the first and the second laws of thermodynamics. It should be mentioned that this equation (10.108) signifies an interesting result. It implies that the change in entropy in Joule–Kelvin expansion is finite, even though the process is adiabatic.

Using equation (10.108), the change in enthalpy H with pressure P can be expressed as

$$\left(\frac{\partial H}{\partial P}\right)_T = T\left(\frac{\partial S}{\partial P}\right)_T + V = -T\left(\frac{\partial V}{\partial T}\right)_P + V = V - T\left(\frac{\partial V}{\partial T}\right)_P, \tag{10.109}$$

where Maxwell's second relation given by equation (10.60) has been used. Combining equations (10.107) and (10.109), we get

$$dH = C_P \, dT + \left[V - T\left(\frac{\partial V}{\partial T}\right)_P\right] dP. \tag{10.110}$$

We know that the Joule–Kelvin expansion process is isenthalpic, that is, $dH = 0$. Under this condition of constant enthalpy, the change in temperature T with pressure P, that is, the Joule–Kelvin coefficient μ can be written as

$$\mu = \left(\frac{\partial T}{\partial P}\right)_H = \frac{1}{C_P}\left[T\left(\frac{\partial V}{\partial T}\right)_P - V\right] = \frac{T^2}{C_P}\left[\frac{\partial}{\partial T}\left(\frac{V}{T}\right)\right]_P. \tag{10.111}$$

As $P_2 < P_1$ in the Joule–Kelvin expansion process, dP will always be negative. The following three cases then arise:

1. The Joule–Kelvin coefficient μ will have a positive value only if ΔT is negative, that is, the gas cools down after passing through the porous plug.

2. A negative value of μ indicates that the heating effect is produced in the Joule–Kelvin expansion.

3. A zero value of μ indicates that there is neither cooling nor heating.

The Joule–Kelvin coefficient μ can also be expressed as $\mu = -\frac{1}{C_P}\left(\frac{\partial H}{\partial P}\right)_T$ in the following way: From the knowledge of exact differential, we have

$$\left(\frac{\partial H}{\partial P}\right)_T \left(\frac{\partial P}{\partial T}\right)_H \left(\frac{\partial T}{\partial H}\right)_P = -1; \quad \Rightarrow \quad \left(\frac{\partial T}{\partial P}\right)_H = -\left(\frac{\partial T}{\partial H}\right)_P \left(\frac{\partial H}{\partial P}\right)_T.$$

Using this expression, we get

$$\mu = \left(\frac{\partial T}{\partial P}\right)_H = -\left(\frac{\partial T}{\partial H}\right)_P \left(\frac{\partial H}{\partial P}\right)_T = -\frac{1}{C_P}\left(\frac{\partial H}{\partial P}\right)_T.$$

Further, since enthalpy H is given by

$$H = U + PV,$$

we can write

$$\left(\frac{\partial H}{\partial P}\right)_T = \left(\frac{\partial U}{\partial P}\right)_T + \left[\frac{\partial}{\partial P}(PV)\right]_T.$$

Thus, the expression for Joule–Kelvin coefficient μ becomes

$$\mu = -\frac{1}{C_P}\left(\frac{\partial H}{\partial P}\right)_T = -\frac{1}{C_P}\left[\left(\frac{\partial U}{\partial P}\right)_T + \left[\frac{\partial}{\partial P}(PV)\right]_T\right].$$

This relation can be used to calculate the Joule–Kelvin coefficient μ.

Porous plug experiment: Description

The apparatus used for the porous plug experiment is shown in Figure 10.11. It consists of a long wooden tube at the center of which there is a plug of cotton wool or compressed silk sandwiched between two perforated discs. The tube is surrounded by an insulating substance such as asbestos. The experimental gas is compressed to a known pressure, allowed to pass through coils immersed in a bath of known temperature, and then allowed to enter the tube. It comes out into the other side of the tube where the pressure is lower. The temperature of the emerging gas is measured by a sensitive thermometer. The difference in temperature on the two sides may be measured accurately using two platinum wires on either side of the tube by connecting them in opposite arms of a Wheatstone bridge.

It is found that all gases show a cooling effect over the temperature range from 4°C to 100°C, whereas hydrogen and helium gases show a heating effect after passing through the porous plug. The fall in temperature is observed to be proportional to the difference in pressure on two sides of the porous plug. Further, it is observed that the fall in temperature will be greater for a given pressure difference if the initial temperature of the gas is lower.

For every gas, there is a certain transition temperature called its inversion temperature such that if the gas is initially below its inversion temperature, there will be cooling, and if it is initially above its inversion temperature, there will be heating after passing through the porous plug. If the gas is initially at its inversion temperature, then there will be no change in temperature.

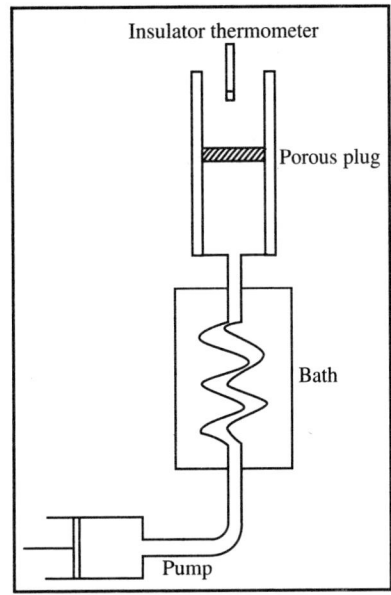

Figure 10.11 Schematic representation of the apparatus of porous plug experiment.

Theory of the porous plug experiment

Let us assume that the unit mass of the gas is enclosed on the high-pressure side of the porous plug, between the plug and a light frictionless piston, say, A. Let the volume of the gas on this side be V_1 and the pressure be P_1. Let there be another light frictionless piston B on the low-pressure side of the plug. Let B be initially in contact with the plug. As the gas in the lower part passes through the plug, it pushes the piston B forward. Let us assume that the piston B is pushed against a constant pressure P_2. Let the volume of the gas be V_2 when it has been completely expelled through the plug to the low-pressure side.

When the piston A has pushed the gas in the lower part completely through the plug, an amount of work P_1V_1 would be performed on the gas. When the gas enters the low-pressure side and pushes the piston B back, an amount of work P_2V_2 would be performed by the gas. Therefore, the net work done by the gas is

$$W = P_2V_2 - P_1V_1. \tag{10.112}$$

No heat energy is supplied to the gas from outside since the entire apparatus is thermally insulated. Also, the piston A is not pushed by any external agency. As the gas in the lower part passes through the plug, the piston A merely moves forward to keep the pressure P_1 constant. This means that no mechanical energy is supplied to the system from outside. Therefore, the energy required for doing the work W is drawn by the gas from its own internal energy content. If the internal energy of the gas below is U_1, and that above is U_2, we must have $U_2 < U_1$ so that the decrease in the internal energy $(U_1 - U_2)$ is equal to the energy used up by the gas for doing the work W. Therefore, $W = U_1 - U_2$.

Since

$$W = P_2V_2 - P_1V_1, \quad \text{we get} \quad P_2V_2 - P_1V_1 = U_1 - U_2. \quad (10.113)$$

This gives rise to

$$U_2 + P_2V_2 = U_1 + P_1V_1. \quad \text{Therefore,} \quad U + PV = \text{constant (say, } H). \quad (10.114)$$

The quantity $(U + PV)$ is called the total heat or the enthalpy of the unit mass of the gas. In the porous plug experiment, the enthalpy H of the gas remains constant so that the expansion of the gas is an isenthalpic process. Again, we know that the internal energy U consists of two parts: (a) the kinetic energy E_K, and (b) the potential energy E_P due to intermolecular forces. Therefore,

$$U = E_K + E_P. \quad (10.115)$$

The kinetic energy E_K depends only on the temperature of the gas, and the potential energy E_P depends on the separation of the molecules and, therefore, on the volume of the gas.

Let the values of E_K and E_P on the high-pressure side below the plug be E_{K_1} and E_{P_1}, and those on the low-pressure side above the plug be E_{K_2} and E_{P_2}, respectively. Therefore, the internal energy of the gas below is $U_1 = E_{K_1} + E_{P_1}$ and the internal energy of the gas above is $U_2 = E_{K_2} + E_{P_2}$. Substituting these expressions in equation (10.113), we get

$$P_2V_2 - P_1V_1 = \left[E_{K_1} + E_{P_1}\right] - \left[E_{K_2} + E_{P_2}\right] = \left[E_{K_1} - E_{K_2}\right] + \left(E_{P_1} - E_{P_2}\right).$$

This leads to

$$E_{K_1} - E_{K_2} = (P_2V_2 - P_1V_1) + \left(E_{P_2} - E_{P_1}\right). \quad (10.116)$$

If $E_{K_1} > E_{K_2}$, then there will be a decrease in kinetic energy, and there will be a decrease in the temperature of the gas. The decreases in temperature are caused by two factors:

1. $(P_2V_2 - P_1V_1)$: the increase in the value of PV

2. $(E_{P_2} - E_{P_1})$: the increase in the potential energy of the molecules. For a perfect gas, which obeys Boyle's law, $P_2V_2 = P_1V_1$, there are no intermolecular forces in a perfect gas, so $E_{P_1} - E_{P_2} = 0$. Therefore, $E_{K_1} - E_{K_2} = 0$, and there will

be no change of temperature. This is in contrast with the case of adiabatic expansion, where there will be cooling even for a perfect gas. For a real gas, in which there are intermolecular attractive forces, the potential energy increases when the molecules are pulled apart to greater distances; that is when the gas expands, so that $E_{P_2} > E_{P_1}$. There is a temperature for every real gas called the Boyle temperature. Below this temperature, PV increases as P decreases, so that $P_2 V_2 > P_1 V_1$ (provided P_1 is not too high). From equation (10.116), we see, therefore, that $E_{K_1} - E_{K_2} > 0$, so that there is a decrease in kinetic energy and a consequent fall in temperature. Above the Boyle temperature, PV decreases as P decreases, so that $P_2 V_2 < P_1 V_1$, $(P_2 V_2 - P_1 V_1)$ is therefore, negative. We can write equation (10.116) as

$$E_{K_1} - E_{K_2} = -(P_1 V_1 - P_2 V_2) + (E_{P_2} - E_{P_1}),$$

where $(P_1 V_1 - P_2 V_2)$ is positive. The contribution of $(E_{P_2} - E_{P_1})$ is therefore decreased by the amount $(P_1 V_1 - P_2 V_2)$, and the cooling becomes less. As the initial temperature of the gas is increased, the rate of increase of PV with P becomes greater, and at a particular temperature, $(P_1 V_1 - P_2 V_2)$ becomes equal to $(E_{P_2} - E_{P_1})$, so that $\left[E_{K_1} - E_{K_2}\right]$ is negative, or $E_{K_2} > E_{K_1}$, so that there is an increase in the kinetic energy of the gas, and the temperature of the gas increases when it issues through the porous plug.

Fall in temperature in the porous plug experiment

According to Van der Walls equation for one mole of a real gas, we have

$$\left(P + \frac{a}{V^2}\right)(V - b) = RT.$$

Therefore, the pressure due to the intermolecular forces of attraction is $\left(\frac{a}{V^2}\right)$. The work done against these forces when the volume increases from V_1 to V_2 is

$$W_i = \int_{V_1}^{V_2} \frac{a}{V^2} dV = \left[-\frac{a}{V}\right]_{V_1}^{V_2} = \frac{a}{V_1} - \frac{a}{V_2}.$$

This work is stored as extra internal potential energy so that the increase in potential energy is

$$E_{P_2} - E_{P_1} = W_i = \frac{a}{V_1} - \frac{a}{V_2}.$$

Substituting this in equation (10.116), we get

$$E_{K_1} - E_{K_2} = (P_2 V_2 - P_1 V_1) + \frac{a}{V_1} - \frac{a}{V_2}. \tag{10.117}$$

From Van der Waals equation of state, we have

$$PV + \frac{a}{V} - bP - \frac{ab}{V^2} = RT.$$

Therefore, from the Van der Waals equation of state, we have

$$P_1V_1 + \frac{a}{V_1} - bP_1 - \frac{ab}{V_1^2} = RT_1 \quad \text{and} \quad P_2V_2 + \frac{a}{V_2} - bP_2 - \frac{ab}{V_2^2} = RT_2.$$

Subtracting, we get

$$(P_2V_2 - P_1V_1) + \left(\frac{a}{V_2} - \frac{a}{V_1}\right) - b(P_2 - P_1) - \left(\frac{ab}{V_2^2} - \frac{ab}{V_1^2}\right) = R(T_2 - T_1).$$

Therefore,

$$P_2V_2 - P_1V_1 = R(T_2 - T_1) - \left(\frac{a}{V_2} - \frac{a}{V_1}\right) + b(P_2 - P_1) + \left(\frac{ab}{V_2^2} - \frac{ab}{V_1^2}\right).$$

Using equation (10.117), we get

$$E_{K_1} - E_{K_2} = R(T_2 - T_1) + 2\left(\frac{a}{V_1} - \frac{a}{V_2}\right) + b(P_2 - P_1) + \left(\frac{ab}{V_2^2} - \frac{ab}{V_1^2}\right). \quad (10.118)$$

We shall neglect the terms $\frac{ab}{V_2^2}$ and $\frac{ab}{V_1^2}$, which are very small compared with the other terms. To calculate the terms $\frac{a}{V_1}$ and $\frac{a}{V_2}$, we shall assume that the temperature remains approximately constant at T_1, and we shall assume that $PV \approx RT$.
Therefore,

$$P_1V_1 \approx RT_1, \text{and} P_2V_2 \approx RT_1; \quad \frac{1}{V_1} - \frac{1}{V_2} = \frac{P_1}{RT_1} - \frac{P_2}{RT_1} = \frac{P_1 - P_2}{RT_1}.$$

Let, $P_2 - P_1 = \Delta P$, and $T_2 - T_1 = \Delta T$. Therefore, we have

$$\frac{1}{V_1} - \frac{1}{V_2} = -\frac{\Delta P}{RT_1}.$$

Substituting these in equation (10.118), we get

$$E_{K_1} - E_{K_2} = R(\Delta T) + 2a\left(\frac{-\Delta P}{RT_1}\right) + b(\Delta P). \quad (10.119)$$

The left-hand side of equation (10.119) represents the decrease in kinetic energy that causes a decrease in temperature. Since we have represented the increase in temperature by ΔT, and the decrease in kinetic energy, which is equal to the decrease in heat content, is $-C_V \Delta T$, where C_V is the specific heat at constant volume. Using this relation in equation (10.119), we get

$$-C_V \Delta T = R(\Delta T) - 2a\left(\frac{\Delta P}{RT_1}\right) + b(\Delta P). \text{ Or } -(R + C_V)\Delta T = -\left(\frac{2a}{RT_1} - b\right)\Delta P.$$
$$(10.120)$$

However, $C_P = R + C_V$, where C_P is the specific heat at constant pressure. Therefore,

$$-C_P \Delta T = -\left(\frac{2a}{RT_1} - b\right)\Delta P.$$
$$(10.121)$$

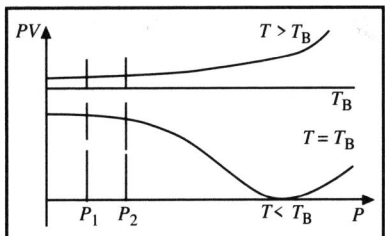

Figure 10.12 Variation of temperature in porous plug experiment.

Equation (10.121) provides the increase in temperature for a unit increase of pressure. Such changes in temperature as a function of pressure are shown in Figure 10.12.

Critical discussion on the change in temperature in porous plug experiment: Inversion temperature T_i

In the porous plug experiment, there is a decrease in pressure, so that ΔP is negative.

1. If the initial temperature of the gas is such that $\frac{2a}{RT_1} > b$, then the right-hand side of equation (10.121) will be positive. For the left-hand side also to be positive, we must have a negative value of ΔT; this means that a cooling effect is produced.

2. If $\frac{2a}{RT_1} < b$, then the right-hand side is negative. For the left-hand side to be negative, we must have a positive value of ΔT so that the heating effect is produced.

3. If $\frac{2a}{RT_1} = b$, then $\Delta T = 0$, so that the temperature corresponding to this gives the inversion temperature (T_i) of the gas, which is given by

$$\frac{2a}{RT_i} = b; \quad \Rightarrow \quad T_i = \frac{2a}{Rb}.$$
$$(10.122)$$

For a given initial temperature T, the right-hand side of equation (10.122) is constant so that

$$\frac{\Delta T}{\Delta P} = \text{constant, so } \Delta T = \text{ constant } \times \Delta P, \text{ or } \Delta T \propto \Delta P.$$

Thus, the cooling is proportional to the fall in pressure.

If the initial temperature (which is less than T_i) is made lower, the right-hand side becomes larger, so that $\frac{\Delta T}{\Delta P}$ becomes larger. Thus, for a given pressure difference ΔP, the fall in temperature (ΔT) will be greater if the initial temperature of the gas is made lower by pre-cooling it.

The quantity $\left(\frac{\Delta T}{\Delta P}\right)$ is called the Joule–Thomson "differential cooling coefficient." Its value for air calculated from equation (10.121) is 0.28° per pressure difference of 1 atmosphere per mole close to N.T.P. It agrees well with the value of 0.27° per atmosphere found experimentally.

Joule–Thomson effect for ideal gases

The Joule–Kelvin coefficient can be expressed as a sum of two terms:

$$\mu = -\frac{1}{C_P}\left[\left(\frac{\partial U}{\partial P}\right)_T + \left(\frac{\partial}{\partial P}(PV)\right)_T\right]. \qquad (10.123)$$

The first term on the RHS of equation (10.123) is a measure of deviation from Joule's law: $U = U(T)$, whereas the second term signifies deviation from Boyle's law: $PV =$ constant. For an ideal gas, both terms in equation (10.123) are zero. Therefore, the Joule–Kelvin coefficient μ is zero for an ideal gas. From this, we may conclude that if the Joule–Kelvin coefficient vanishes for a gas, it may be regarded as an ideal gas.

Joule–Thomson effect for real and Van der Waals gases

The observed characteristic features of the Joule–Kelvin coefficient μ for a real gas are the following:

1. As $\mu \neq 0$, real gases do not behave perfectly like ideal gases.

2. Except for hydrogen and helium, all gases show a cooling effect at room temperature. However, hydrogen and helium gases show a slight heating effect. Initially, this was treated as a paradox but later, it was recognized that the initial temperature of the gas must be below its inversion temperature to obtain any cooling effect.

To understand this, we start with one mole of a Van der Waals gas:

$$\left(P + \frac{a}{V^2}\right)(V - b) = RT, \qquad (10.124)$$

and the factor $\left(\frac{\partial V}{\partial T}\right)_P$ is given by

$$\left(\frac{\partial V}{\partial T}\right)_P = \frac{R}{\left(P + \frac{a}{V^2}\right) - \frac{2a}{V^3}(V - b)}. \qquad (10.125)$$

Multiplying the numerator as well as the denominator on the right-hand side of this expression by $(V-b)$ and using equation (10.124), we can write

$$\left(\frac{\partial V}{\partial T}\right)_p = \frac{R(V-b)}{RT - \frac{2a(V-b)^2}{V^3}} = \frac{(V-b)}{T\left(1 - \frac{2a(V-b)^2}{RTV^3}\right)}. \tag{10.126}$$

For $b << V$, using binomial expansion, we can write

$$\left(\frac{\partial V}{\partial T}\right)_P = \frac{V}{T}\left(1 - \frac{b}{V}\right)\left(1 - \frac{2a}{RTV}\right)^{-1},$$

$$= \frac{V}{T}\left(1 - \frac{b}{V}\right)\left(1 + \frac{2a}{RTV} + \cdots\right) = \frac{V}{T}\left(1 - \frac{b}{V} + \frac{2a}{RTV} - \frac{2ab}{RTV^2}\right), \tag{10.127}$$

where we have neglected the quadratic or higher powers of Van der Waals constants. If we now ignore the term containing the product of Van der Waals constants in equation (10.127), we can write it in a compact form as

$$T\left(\frac{\partial V}{\partial T}\right)_P - V = \frac{2a}{RT} - b. \tag{10.128}$$

On substituting this result in the expression for the Joule–Thomson coefficient $\mu = \frac{1}{C_P}\left(T\left(\frac{\partial V}{\partial T}\right)_P - V\right)$, we get

$$\mu = \frac{1}{C_P}\left(\frac{2a}{RT} - b\right). \tag{10.129}$$

In terms of inversion temperature $T_1 = (2a/Rb)$, we can rewrite it as

$$\mu = \frac{b}{C_P}\left(\frac{T_i}{T} - 1\right). \tag{10.130}$$

Comment on the Joule–Kelvin coefficient of real gas

It has been observed from equation (10.123) that the Joule–Kelvin coefficient μ can be expressed as a sum of the following two terms:

$$\mu = -\frac{1}{C_P}\left[\left(\frac{\partial U}{\partial P}\right)_T + \left(\frac{\partial}{\partial P}(PV)\right)_T\right]. \tag{10.131}$$

Equation (10.131) shows that there is always a cooling effect produced due to the first term but there may be a heating or cooling effect produced by the second term depending upon the values of temperature and pressure of the gas. Thus, the net effect of the Joule–Kelvin expansion is set by the interplay of these two terms given by equation (10.131) and can have either a positive or negative sign, that is, either heating or cooling. It will be zero when heating due to deviations from Boyle's law exactly cancels cooling due to deviations from Joule's law, that is, when

$$\left(\frac{\partial U}{\partial P}\right)_T + \left(\frac{\partial}{\partial P}(PV)\right)_T = 0.$$

Using equation (10.111), this condition can be rewritten as

$$-V + T\left(\frac{\partial V}{\partial T}\right)_P = 0. \tag{10.132}$$

This equation (10.132) can be utilized to define the inversion curve in the temperature–pressure (T, P) plane (see Figure 10.13). This curve represents the locus of points for which the Joule–Kelvin coefficient is zero, that is, $\mu = 0$. It should be mentioned that for a given initial temperature T_1 of the gas, there will be a cooling effect only if the pressure is chosen so that the initial state of the gas lies within the inversion curve, that is, $(\frac{\partial T}{\partial P})_H$ is positive. This means that if the gas is allowed to expand when it corresponds to points A or B (that is, temperatures and pressures T_1, P_1 and T_2, P_2), it will cool down. Outside the inversion curve, $(\frac{\partial T}{\partial P})_H$ is negative and the gas will warm up on expansion, that is, its temperature will increase on expansion.

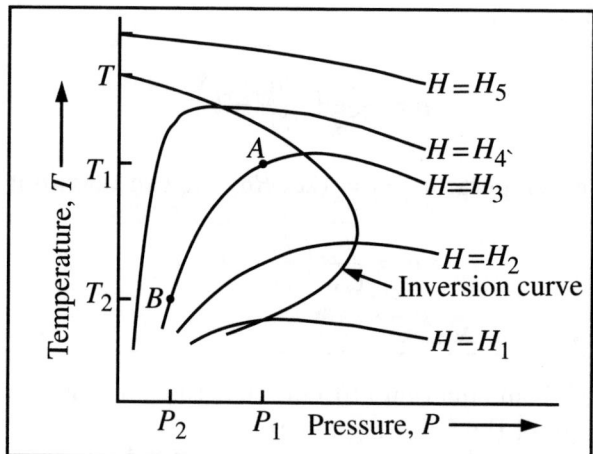

Figure 10.13 The inversion curve. Inside it, the gas cools on undergoing Joule–Thompson expansion.

Integral Joule–Thomson effect

In the above discussion, we assumed that the pressure gradient across the porous plus is gradual. However, if the pressure drop is finite, we should integrate the expression for Joule–Thomson coefficient $\mu = \frac{1}{C_P}\left(T\left(\frac{\partial V}{\partial T}\right)_P - V\right)$ to obtain the change in temperature of the gas. This gives

$$\int_{T_1}^{T_2} C_P \, dT = \int_{P_1}^{P_2}\left(T\left(\frac{\partial V}{\partial T}\right)_P - V\right) dP. \tag{10.133}$$

Using the cyclic relation $\left(\frac{\partial V}{\partial T}\right)_P = -\left(\frac{\partial V}{\partial P}\right)_T\left(\frac{\partial P}{\partial T}\right)_V$, we can write for constant T

$$\left(\frac{\partial V}{\partial T}\right)_P dP_T = -\left(\frac{\partial P}{\partial T}\right)_V dV_T. \tag{10.134}$$

Further, since

$$\int V \, dP = \int d(PV) - \int P \, dV. \tag{10.135}$$

Combining equations (10.133), (10.134), and (10.135), we obtain

$$T_2 - T_1 = \frac{1}{C_P}\left(-\int_{V_1}^{V_2} T\left(\frac{\partial P}{\partial T}\right)_V dV - \int_1^2 d(PV) + \int_{V_1}^{v_2} P \, dV\right),$$

$$= \frac{1}{C_P}\left(P_1V_1 - P_2V_2 + \int_{V_1}^{v_2} P \, dV - \int_{V_1}^{V_2} T\left(\frac{\partial P}{\partial T}\right)_V dV\right), \tag{10.136}$$

where C_P is the heat capacity over the temperature range T_1 to T_2.

For a Van der Waals gas, we have $P = \frac{RT}{V-b} - \frac{a}{V^2}$ and $\left(\frac{\partial P}{\partial T}\right)_V = \frac{R}{V-b}$ so that

$$T_2 - T_1 = \frac{1}{C_P}\left(P_1V_1 - P_2V_2 + \int_{V_1}^{V_2} P \, dV - \int_{V_1}^{V_2} \frac{RT}{V-b} \, dV\right),$$

$$= \frac{1}{C_P}\left(P_1V_1 - P_2V_2 - \int_{V_1}^{V_2} \frac{a}{V^2} \, dV\right)$$

$$= \frac{1}{C_p}\left(P_1V_1 - P_2V_2 + a\left[\frac{1}{V_2} - \frac{1}{V_1}\right]\right). \tag{10.137}$$

This equation (10.137) gives the change in temperature of a Van der Waals gas due to the Joule–Thompson effect.

10.5 Solved Problems

1. Define order parameter. Distinguish between first-order and continuous phase transition. [Calcutta University 2013]

   Answer: An **order parameter** is a measure of the degree of order of a certain physical parameter across the boundaries of a system when a phase transition in the system occurs. It normally ranges between zero in one phase and nonzero in the other. It is a measure of the changes in the physical parameter following a phase transition. For example, when a liquid becomes a gas, the order parameter is density. Its value differs greatly in the liquid and gas phases.

   In first-order phase transition, the first-order derivatives of Gibbs free energy (entropy and volume) show a discontinuity, whereas in second-order phase transition the first-order derivative of Gibbs free energy are continuous but higher order derivatives are discontinuous.

2. Deduce the Clausius–Clapeyron relation $\frac{dP}{dT} = \frac{L}{T(V_2 - V_1)}$. The symbols have their usual significance. [Calcutta University 2012, 2014, 2017]

   Answer: The two phases in both states a and b are in equilibrium. So, the specific Gibbs free energies of the two phases are the same:

   $$g_1^{(a)} = g_2^{(a)} \text{ and } g_1^{(b)} = g_2^{(b)}.$$

   So, if dg is the difference in Gibbs specific free energy between the two points, it is then the same for both phases, that is, $dg_1 = dg_2$. Writing dg in terms of specific entropy s and specific volume v, we get

   $$dg_1 = -s_1 dT + v_1 dP \quad \text{and} \quad dg_2 = -s_2 dT + v_2 dP.$$

   $$-s_1 dT + v_1 dP = -s_2 dT + v_2 dP; \quad \Rightarrow \quad (s_1 - s_2)dT = (v_1 - v_2)dP.$$

   This leads to

   $$\frac{dP}{dT} = \frac{s_2 - s_1}{v_2 - v_1} = \frac{\Delta s}{v_2 - v_1}. \tag{10.138}$$

   This form does not look very useful as we cannot measure entropy directly. However, using $\Delta s = \frac{\Delta Q}{T}$ and $\Delta Q = mL = L(m = 1)$, we have

   $$\frac{dP}{dT} = \frac{L}{T(v_2 - v_1)}. \tag{10.139}$$

3. For two-phase systems in equilibrium, pressure P is a function of temperature T only. If C_V is the specific heat at constant volume, and β_S is the adiabatic compressibility, then find the value of $\frac{C_V}{\beta_S}$.

Answer: Let us consider that the entropy S is a function of temperature T and volume V, that is, $S = S(T, V)$. Then the differential dS is given by

$$dS = \left(\frac{\partial S}{\partial T}\right)_V dT + \left(\frac{\partial S}{\partial V}\right)_T dV.$$

For adiabatic process, we get

$$\left(\frac{\partial S}{\partial T}\right)_V = -\left(\frac{\partial S}{\partial V}\right)_T \left(\frac{\partial V}{\partial T}\right)_S = -\left(\frac{\partial P}{\partial T}\right)_V \left(\frac{\partial V}{\partial T}\right)_S.$$

We have used one of Maxwell's thermodynamic relations.

Then C_V is given by

$$C_V = -T \left(\frac{\partial P}{\partial T}\right)_V \left(\frac{\partial V}{\partial T}\right)_S = -T \left(\frac{\partial P}{\partial T}\right)_V \left(\frac{\partial V}{\partial P}\right)_S \left(\frac{\partial P}{\partial T}\right)_S$$

$$= T V \beta_S \left(\frac{\partial P}{\partial T}\right)_V \left(\frac{\partial P}{\partial T}\right)_S.$$

This shows that

$$\frac{C_V}{\beta_S} = T V \left(\frac{dP}{dT}\right)^2,$$

where we have used $\beta_S = -\frac{1}{V} \left(\frac{\partial V}{\partial P}\right)_S$.

4. The vapor pressure P (in mm of mercury) of solid ammonia is given by $\log_e P = 23.03 - \frac{3754}{T}$, while that of liquid ammonia is given by $\log_e P = 19.49 - \frac{3063}{T}$, where T is in Kelvin. Calculate the triple point of ammonia.

Answer: At the triple point, the vapor pressure of the substances in each of the three states is identical. Hence, by equating the vapor pressure of solid ammonia with that of liquid ammonia, we have

$$23.03 - \frac{3754}{T} = 19.49 - \frac{3063}{T}; \quad \Rightarrow \quad \frac{1}{T}(3754 - 3063) = 23.03 - 19.49$$

$$\text{or,} \quad \frac{691}{T} = 3.54; \quad \Rightarrow \quad T = 195.2 \text{ K}.$$

5. The specific heat of water at 100°C is 1.01 and the latent heat of vaporization of water decreases with a rise in temperature at the rate of 0.64 cal/K. The latent heat of the vaporization of steam is 540 cal/g. Calculate the specific heat of saturated steam.

Answer: Here, $C_1 = 1.01$, $C_2 = ?$, $T = (273 + 100)$ K $= 373$ K, $\frac{dL}{dT} = -0.64$ cal/K. $L = 540$ cal/gm. We know that

$$C_2 - C_1 = \frac{dL}{dT} - \frac{L}{T}; \quad \Rightarrow \quad C_2 = C_1 + \frac{dL}{dT} - \frac{L}{T};$$

$$\Rightarrow \quad C_2 = 1.01 - 0.64 - \frac{540}{373}; \quad \Rightarrow \quad C_2 = -1.077 \text{ cal/gm K}.$$

6. The change in specific volume when 1 kg of water freezes is 91×10^{-6} m^3. Calculate the value of pressure when ice freezes at 272 K. The latent heat of ice is 3.36×10^5 J kg^{-1}.

Answer: We have $L = 3.36 \times 10^5$ J kg^{-1}. The freezing point of ice under normal pressure $T = 273$ K. Changing in freezing point is given by $dT = (273 - 272)$ K $= 1$ K.

Change in specific volume $V_2 - V_1 = 91 \times 10^{-6}$ m^3. According to Clapeyron's latent heat equation, we have

$$\frac{dp}{dT} = \frac{L}{T(V_2 - V_1)}; \quad \Rightarrow \quad dP = \frac{dT \times L}{T(V_2 - V_1)} = \frac{1 \times 3.36 \times 10^5}{273 \times 91 \times 10^{-6}} \text{N/m}^2$$

$$= \frac{1.352 \times 10^7}{10^5} = 135.2 \text{ atm}.$$

Hence, the pressure under which ice would freeze at 272 K is $P = (1 + 135.2)$ atm. $= 136.2$ atm.

7. Prove that in the isothermal process for an ideal gas,

(a) $G_2 - G_1 = \int RT \frac{dP}{P}$ and

(b) $G_2 - G_1 = RT \ln \frac{V_1}{V_2}$.

Answer: (a) Gibbs function for an ideal gas is defined as $G = H - TS$. For an infinitesimal reversible process,

$$dG = dH - TdS - SdT; \quad \Rightarrow \quad \text{But } dH = TdS + VdP, \quad \Rightarrow \quad dG = VdP - SdT.$$

For the isothermal process, $dT = 0$, so $dG = VdP$. For an ideal gas, $PV = RT$, then dG becomes $dG = \frac{RT}{P}dP$. Integrating this, we get

$$\int_{G_1}^{G_2} dG = \int \frac{RT}{P}dP. \quad \Rightarrow \quad G_2 - G_1 = \int \frac{RT}{P}dP.$$

(b) For an ideal gas, $PV = RT$ or $P = \frac{RT}{V}$. Differentiating it isothermally with respect to V, we get $\frac{dP}{P} = -\frac{RT}{V^2}$. Substituting this value in the above expression, we get

$$dG = V\left(-\frac{RT}{V^2}dV\right); \quad \Rightarrow \quad dG = -\frac{RT}{V}dV.$$

Integrating this expression, we get

$$\int\limits_{G_1}^{G_2} dG = -RT \int\limits_{V_1}^{V_2} \frac{dV}{V}; \quad \Rightarrow \quad G_2 - G_1 = -RT \ln\left(\frac{V_2}{V_1}\right);$$

$$\Rightarrow \quad G_2 - G_1 = RT \ln\left(\frac{V_1}{V_2}\right).$$

8. Show that in an isobaric process, the change in enthalpy is equal to the heat transferred between the system and the surroundings.

[Calcutta University 2015]

Answer: From the first law of thermodynamics, we know

$$\delta Q = dU + PdV,$$

where Q and U represent the heat exchanged and the internal energy. PdV is the work done, where P and V are the pressure and volume of the system. Integrating, we get

$$\int \delta Q = \int dU + P\int dV; \quad Or, \int \delta Q = \int d(U + PV)$$

$$= \int dH = H_{\text{final}} - H_{\text{initial}},$$

where $H = U + PV$ is the enthalpy of the system.

9. The Helmholtz free energy function F can be obtained from the internal energy U by a Legendre transformation. Using this, show that the Helmholtz free energy F is a function of the variables T and V. [Calcutta University 2012]

Answer: Considering the internal energy as a function of entropy and volume, we can write it as $U = U(S, V)$. Its differential can be written as

$$dU = \left(\frac{\partial U}{\partial S}\right)_V dS + \left(\frac{\partial U}{\partial V}\right)_S dV.$$

Combining the first and second laws of thermodynamics, we have $dU = T\,dS - P\,dV$. From this expression, we get

$$T = \left(\frac{\partial U}{\partial S}\right)_V \text{ and } P = \left(\frac{\partial U}{\partial V}\right)_S.$$

From Legendre transformation, we have

$$\phi = \left(\left(\frac{\partial U}{\partial S}\right)_V, V\right) = U(S,V) - S\left(\frac{\partial U}{\partial S}\right)_V = U - TS = F.$$

This is the Helmholtz free energy function F.

10. Starting from the expression of Helmholtz free energy $F(T,V)$ show that the specific heat at constant volume C_V is given by $C_V = -T\left(\frac{\partial^2 F}{\partial T^2}\right)$.

    [Calcutta University 2013, 2015, 2017]

Answer: Specific heat at constant volume is given by $C_V = \left(\frac{\delta Q}{dT}\right)_V = T\left(\frac{dS}{dT}\right)_V$.

The Helmholtz free energy F is given by $F = U - T\,S$. Its differential can be written as

$$dF = dU - T\,dS - S\,dT.$$

Using the first and second laws of thermodynamics, we can write

$$dF = -PdV - SdT; \quad \Rightarrow \left(\frac{\partial F}{\partial T}\right)_V = -S.$$

Differentiating this expression w.r.t temperature at constant volume, we get

$$\left(\frac{\partial S}{\partial T}\right)_V = -\left(\frac{\partial^2 F}{\partial T^2}\right)_V; \quad \Rightarrow T\left(\frac{\partial S}{\partial T}\right)_V = -T\left(\frac{\partial^2 F}{\partial T^2}\right)_V;$$

$$\Rightarrow \quad C_V = -T\left(\frac{\partial^2 F}{\partial T^2}\right)_V.$$

11. The Helmholtz free energy F of a system is given by

$$F = A + BT\,(1 - \ln T) - CT \ln V,$$

where A, B, and C are constants. Obtain expressions for pressure P, entropy S, internal energy U, enthalpy H, and Gibbs energy G.

Answer: We know that $dU = TdS - PdV$; $H = U + PV$, $F = U - TS$, and $G = H - TS$. From these relations, we can write

$$dF = -PdV - SdT.$$

Hence, the pressure P is given by

$$P = -\left(\frac{\partial F}{\partial V}\right)_T = -\left(\frac{\partial}{\partial V}(A + BT(1 - \ln T) - CT \ln V)\right)_T$$
$$= -(-CT)\frac{1}{V} = \frac{CT}{V}.$$

The entropy S of the system is given by

$$S = -\left(\frac{\partial F}{\partial T}\right)_V = -\left(\frac{\partial}{\partial T}(A + BT(1 - \ln T) - CT \ln V)\right)_T$$
$$= -\left[B(1 - \ln T) + BT\left(-\frac{1}{T}\right) - C \ln V\right] = B \ln T + C \ln V.$$

The internal energy U of the system is given by

$$U = F + TS = A + BT(1 - \ln T) - CT \ln V + T[B \ln T + C \ln V] = A + BT.$$

The enthalpy H of the system is given by

$$H = U + PV = A + BT + V \times \frac{CT}{V} = A + BT + CT.$$

Finally, the Gibbs free energy G of the system is given by

$$G = H - TS = (A + BT + CT) - T \times [B \ln T + C \ln V]$$
$$= A + BT(1 - \ln T) + CT(1 - \ln V).$$

12. For n-moles of an ideal gas, prove that

(a) $F = n \int C_V dT - nT \int \frac{C_V}{T} dT - nRT \ln V - a_1 T + a_2$

(b) $G = n \int C_P dT - nT \int \frac{C_P}{T} dT + nRT \ln P - b_1 T + b_2$

where a_1, a_2, b_1, and b_2 are constant parameters.

Answer: (a) We have the relation $F = U - TS$. This gives differential $dF = dU - d(TS)$.

Using $dU = nC_V dT$ and integrating the corresponding expression, we get

$$F = n \int C_V dT - TS + a_2.$$

Considering entropy S as a function of V and T, we get

$$dS = \left(\frac{\partial S}{\partial T}\right)_V dT + \left(\frac{\partial S}{\partial V}\right)_T dV; \quad \Rightarrow \quad TdS = nC_V dT + T\left(\frac{\partial P}{\partial T}\right)_V dV;$$
$$\Rightarrow \quad dS = n\frac{C_V}{T}dT + nR\left(\frac{dV}{V}\right); \quad \Rightarrow \quad S = n \int \frac{C_V}{T}dT + nR \ln V + a_1.$$

Using this value of S, the Helmholtz free energy becomes

$$F = n \int C_V dT - nT \int \frac{C_V}{T} dT - nRT \ln V - a_1 T + a_2.$$

(b) We have the relation $G = H - TS = U + PV - TS$. This gives differential

$$dG = dU + PdV + VdP - d(TS) = dU + nRdT - d(TS);$$
$$\Rightarrow dG = n(C_V + R)dT - d(TS); \quad \Rightarrow dG = nC_P dT - d(TS).$$

Considering entropy S as a function of P and T, we get

$$dS = \left(\frac{\partial S}{\partial T}\right)_P dT + \left(\frac{\partial S}{\partial P}\right)_T dP;$$

$$\Rightarrow TdS = nC_P dT + T\left(-\frac{\partial V}{\partial T}\right)_P dP = C_P dT - T\left(\frac{nR}{P}\right)dP;$$

$$\Rightarrow dS = n\frac{C_P}{T}dT - nR\left(\frac{dP}{P}\right); \quad \Rightarrow S = n\int\frac{C_P}{T}dT - nR\ln P + b_1.$$

Using this value of S, the Gibbs free energy G becomes

$$G = n\int C_P dT - nT\int\frac{C_P}{T}dT + nRT\ln P - b_1 T + b_2.$$

13. Starting from Gibbs energy (G), show that $C_P = -T\left(\frac{\partial^2 G}{\partial T^2}\right)$, where the symbols have their usual meaning.   [Calcutta University 2016]

Answer: Gibbs free energy G as a function of internal energy U, entropy S, and temperature T can be written as $G = U + PV - TS$.

The differential of dG can be written as $dG = dU + P\,dV + V\,dP - T\,dS - S\,dT$.

On simplification, we get $dG = V\,dP - S\,dT$.

This gives rise to

$$\Rightarrow \left(\frac{\partial G}{\partial T}\right)_P = -S. \quad \Rightarrow \left(\frac{\partial^2 G}{\partial T^2}\right)_P = -\left(\frac{\partial S}{\partial T}\right)_P = -\frac{C_P}{T}.$$

$$\text{Or, } C_P = -T\left(\frac{\partial^2 G}{\partial T^2}\right)_P.$$

14. The specific Gibbs energy of an ideal gas is given by

$$G(P,T) = RT\,\ln\left(\frac{P}{P_0}\right) - BP,$$

where B is a function of T only. Obtain the equation of state.

Answer: Here, it is given that

$$G(P,T) = RT \, \ln \left(\frac{P}{P_0} \right) - BP.$$

Writing G as a function of P and T, we get

$$dG = \left(\frac{\partial G}{\partial T} \right)_P dT + \left(\frac{\partial G}{\partial P} \right)_T dP.$$

Again $dG = dH - TdS - SdT = dU + PdV + VdP - TdS - SdT = TdS + VdP - TdS - SdT = VdP - SdT$. This leads to

$$VdP - SdT = \left(\frac{\partial G}{\partial T} \right)_P dT + \left(\frac{\partial G}{\partial P} \right)_T dP = \left[R \, \ln \left(\frac{P}{P_0} \right) - P \left(\frac{\partial B}{\partial T} \right)_P \right] dT$$

$$+ \left[RT \frac{P_0}{P} \frac{1}{P_0} - B \right] dP.$$

Comparing the coefficients of dP from both sides, we get

$$V = \frac{RT}{P} B; \quad \Rightarrow \quad P(V + B) = RT.$$

This is the equation of state of the system whose Gibbs free energy is given by $G(P,T) = RT \, \ln \left(\frac{P}{P_0} \right) - BP$.

15. A certain system is found to have Gibbs free energy $G(P,T) = RT \, \ln \left(\frac{a\,P}{(R\,T)^{\frac{5}{2}}} \right)$. Find out the specific heat at a constant pressure of the system. [Calcutta University 2013]

Answer: We know that the specific heat at constant pressure is related to the second derivative of Gibbs free energy by the expression: $C_P = -T \left(\frac{\partial^2 G}{\partial T^2} \right)_P$.

Given that, $G(P,T) = RT \, \ln \left(\frac{a\,P}{(R\,T)^{\frac{5}{2}}} \right)$. So,

$$G = RT \, \ln \left(\frac{a\,P}{(R\,T)^{\frac{5}{2}}} \right) = RT \, \ln (a\,P) - \frac{5}{2} RT \, \ln(RT)$$

$$= RT(\ln (a\,P) - \frac{5}{2} \ln\,R) - \frac{5}{2} RT \, \ln (T).$$

Differentiating this expression w.r.t temperature T at constant pressure P, we get

$$\left(\frac{\partial G}{\partial T} \right)_P = R \left[\ln (a\,P) - \frac{5}{2} \ln\,R \right] - \frac{5}{2} R \, \ln(T) - \frac{5}{2} RT \frac{1}{T}.$$

Differentiating again w.r.t temperature T at constant pressure P, we get

$$\left(\frac{\partial^2 G}{\partial T^2}\right)_P = -\frac{5}{2}\frac{R}{T}.$$

So, the expression for specific heat at constant pressure becomes

$$C_P = -T \times \left(-\frac{5}{2}\frac{R}{T}\right) = \frac{5}{2}R.$$

16. Using the Legendre transformation, obtain Gibbs free energy $G(T,P)$ from internal energy $U(S,V)$, where P, V, T, S are the usual thermodynamic parameters. [Calcutta University 2018]

    Answer: From the first and second laws of thermodynamics, we have $dU = TdS - PdV$.

    As $G = G(T,P)$ and $U = U(S,V)$, according to the knowledge of the Legendre transform, we have

    $$G = U - S\left(\frac{\partial U}{\partial S}\right)_V - \left(\frac{\partial U}{\partial V}\right)_S.$$

    Further, we have the relations

    $$\left(\frac{\partial U}{\partial S}\right)_V = T, \text{ and } \left(\frac{\partial U}{\partial V}\right)_S = -P.$$

    Using these two relations, we have the expression for Gibbs free energy as

    $$G = U - S \times T - V \times (-P) = U - TS + PV = U + PV - TS.$$

17. Starting from Gibbs energy (G), show that $H = -T^2\left[\frac{\partial}{\partial T}\left(\frac{G}{T}\right)\right]_P$, where the symbols have their usual meanings. [Calcutta University 2016]

    Answer: Gibbs free energy G as a function of internal energy U, entropy S, and temperature T can be written as $G = U + PV - TS$.

    Further, we have the relation $\frac{\partial}{\partial T}\left(\frac{G}{T}\right) = \frac{1}{T}\left(\frac{\partial G}{\partial T}\right)_P - \frac{G}{T^2}.$

    Putting the value of G in the above expression and carrying out differentiation at constant P, we get

    $$\frac{\partial}{\partial T}\left(\frac{G}{T}\right)_P = -\frac{S}{T} - \frac{G}{T^2} = -\frac{G + TS}{T^2} = -\frac{U + PV}{T^2} = -\frac{H}{T^2}.$$

This proves that

$$H = -T^2 \frac{\partial}{\partial T}\left(\frac{G}{T}\right)_P.$$

18. Using Maxwell's relation $\left(\frac{\partial S}{\partial V}\right)_T = \left(\frac{\partial P}{\partial T}\right)_V$, show that $\frac{\partial U}{\partial V} = \frac{T\alpha}{\beta_T} - P$, where α is the coefficient of volume expansion, and β_T is the isothermal compressibility.

Answer: The coefficient of volume expansion α is defined as $\alpha = \frac{1}{V}\left(\frac{\partial V}{\partial T}\right)_P$ and the isothermal compressibility $\beta_T = -\frac{1}{V}\left(\frac{\partial V}{\partial P}\right)_T$.

We consider entropy S as $S = f(T, V)$. The differential change dS in S is given by

$$dS = \left(\frac{\partial S}{\partial V}\right)_T dV + \left(\frac{\partial S}{\partial T}\right)_V dT.$$

Multiplying both sides by T, we get

$$TdS = T\left(\frac{\partial S}{\partial V}\right)_T dV + T\left(\frac{\partial S}{\partial T}\right)_V dT = T\left(\frac{\partial P}{\partial T}\right)_V dV + C_V dT,$$

where we have used the relation $\left(\frac{\partial S}{\partial V}\right)_T = \left(\frac{\partial P}{\partial T}\right)_V$.

The combination of the first and second laws of thermodynamics provides

$$dU = TdS - PdV = T\left(\frac{\partial P}{\partial T}\right)_V dV + C_V dT - PdV$$

$$= C_V dT + \left[T\left(\frac{\partial P}{\partial T}\right)_V - P\right]dV.$$

Further, we consider internal energy U as $U = f(T, V)$. Its differential change dU is given by

$$dU = \left(\frac{\partial U}{\partial V}\right)_T dV + \left(\frac{\partial U}{\partial T}\right)_V dT.$$

Comparing these two expressions for dU, we get

$$\left(\frac{\partial U}{\partial V}\right)_T = \left[T\left(\frac{\partial P}{\partial T}\right)_V - P\right] = \frac{T\alpha}{\beta_T} - P.$$

19. If the equation of state is given by

$$P = \frac{RT}{V + \frac{b}{T}}.$$

(a) Find the $\left(\frac{\partial U}{\partial V}\right)_T$.

(b) If the volume is expanded from V_1 to V_2 at a very high temperature, then what will be the change in internal energy?

Answer: (a) We have the relation

$$\left(\frac{\partial U}{\partial V}\right)_T = T\left(\frac{\partial S}{\partial V}\right)_T - P = T\left(\frac{\partial P}{\partial T}\right)_V - P.$$

As we have the expression for pressure P as

$$P = \frac{RT}{V + \frac{b}{T}}; \quad \Rightarrow \quad \left(\frac{\partial P}{\partial T}\right)_V = \frac{R}{V + \frac{b}{T}} + \frac{(-)RT}{\left(V + \frac{b}{T}\right)^2}\left(-\frac{b}{T^2}\right).$$

Using this value in the above expression, we get

$$\left(\frac{\partial U}{\partial V}\right)_T = T\left(\frac{\partial P}{\partial T}\right)_V - P = \frac{RT}{V + \frac{b}{T}} + \frac{Rb}{\left(V + \frac{b}{T}\right)^2} - P$$

$$= \frac{RT}{V + \frac{b}{T}} + \frac{Rb}{\left(V + \frac{b}{T}\right)^2} - \frac{RT}{V + \frac{b}{T}} = \frac{Rb}{\left(V + \frac{b}{T}\right)^2}.$$

(b) We have

$$\left(\frac{\partial U}{\partial V}\right)_T = \frac{Rb}{\left(V + \frac{b}{T}\right)^2}.$$

For very high temperature, that is, when $T \to \infty$, we get

$$\left(\frac{\partial U}{\partial V}\right)_T \to \frac{Rb}{V^2}.$$

If the volume is expanded from V_1 to V_2 at a very high temperature, the change in internal energy will be given by

$$U_2 - U_1 = \int_{V_1}^{V_2} \frac{Rb}{V^2}dV = Rb\left(\frac{1}{V_1} - \frac{1}{V_2}\right).$$

20. The equation of state for a certain substance is given by

$$V = \frac{RT}{P} - \frac{C}{T^3}.$$

Show that the variation of specific heat at constant pressure C_P with pressure P of such a substance varies inversely as the fourth power of temperature T.

[Calcutta University 2020]

Answer: Using Maxwell's relations, we have already shown that

$$\left(\frac{\partial C_P}{\partial P}\right)_T = -T\left(\frac{\partial^2 V}{\partial T^2}\right)_P.$$

The equation of state of the substance is given by $V = \frac{RT}{P} - \frac{C}{T^3}$. Differentiating this w.r.t temperature T at constant pressure P, we get

$$\left(\frac{\partial V}{\partial T}\right)_P = \frac{R}{P} + \frac{3C}{T^4}.$$

Differentiating the above expression w.r.t temperature T again at constant pressure P, we get

$$\left(\frac{\partial^2 V}{\partial T^2}\right)_P = -\frac{12C}{T^5}.$$

Using this, we get

$$\left(\frac{\partial C_P}{\partial P}\right)_T = -T\left(\frac{\partial^2 V}{\partial T^2}\right)_P = -T \times \left[-\frac{12C}{T^5}\right] = \frac{12C}{T^4}.$$

This proves that the variation of specific heat at constant pressure C_P as a function of pressure P of such a substance varies inversely as the fourth power of temperature T.

21. Prove that $\mu_S - \mu_H = \frac{V}{C_P}$, where $\mu_S = \left(\frac{\partial T}{\partial P}\right)_S$ and $\mu_H = \left(\frac{\partial T}{\partial P}\right)_H$.

Answer: We know that the Joule–Thomson coefficient is given by

$$\mu_H = \left(\frac{\partial T}{\partial P}\right)_H = \frac{1}{C_P}\left[T\left(\frac{\partial V}{\partial T}\right)_P - V\right].$$

Using the expression for μ_S, we can write

$$\mu_S - \mu_H = \left(\frac{\partial T}{\partial P}\right)_S - \frac{T}{C_P}\left(\frac{\partial V}{\partial T}\right)_P + \frac{V}{C_P}.$$

From Maxwell's relation, we have $\left(\frac{\partial T}{\partial P}\right)_S = \left(\frac{\partial V}{\partial S}\right)_P$. Using this relation, we get

$$\mu_S - \mu_H = \left(\frac{\partial V}{\partial S}\right)_P - \frac{T}{\left(\frac{\partial Q}{\partial T}\right)_P}\left(\frac{\partial V}{\partial T}\right)_P + \frac{V}{C_P}$$

$$= \left(\frac{\partial V}{\partial S}\right)_P - \frac{T}{T\left(\frac{\partial S}{\partial T}\right)_P}\left(\frac{\partial V}{\partial T}\right)_P + \frac{V}{C_P}$$

$$= \left(\frac{\partial V}{\partial S}\right)_P - \left(\frac{\partial T}{\partial S}\right)_P\left(\frac{\partial V}{\partial T}\right)_P + \frac{V}{C_P} = \left(\frac{\partial V}{\partial S}\right)_P - \left(\frac{\partial V}{\partial S}\right)_P + \frac{V}{C_P}.$$

Hence, $\mu_S - \mu_H = \frac{V}{C_P}$.

22. Prove that $\left(\frac{\partial U}{\partial P}\right)_T = -T\left(\frac{\partial V}{\partial T}\right)_P - P\left(\frac{\partial V}{\partial P}\right)_T$.

Answer: We know that $TdS = dU + PdV$. This gives rise to

$$T\left(\frac{\partial S}{\partial P}\right)_T = \left(\frac{\partial U}{\partial P}\right)_T + P\left(\frac{\partial V}{\partial P}\right)_T.$$

From Maxwell's relation, we have $\left(\frac{\partial S}{\partial P}\right)_T = -\left(\frac{\partial V}{\partial T}\right)_P$. Using this relation, we get

$$\left(\frac{\partial U}{\partial P}\right)_T = -T\left(\frac{\partial V}{\partial T}\right)_P - P\left(\frac{\partial V}{\partial P}\right)_T.$$

23. Starting from the thermodynamic relation

$$\left(\frac{\partial U}{\partial P}\right)_T = -T\left(\frac{\partial V}{\partial T}\right)_P - P\left(\frac{\partial V}{\partial P}\right)_T,$$

show that for a system having an equation of state of the form $PV^2 = RT$, the internal energy U is independent of pressure P at constant temperature T.

Answer: The equation of state is $PV^2 = RT$; $\Rightarrow V = \sqrt{\frac{RT}{P}}$. Differentiating this w.r.t T at constant P, we get

$$\left(\frac{\partial V}{\partial T}\right)_P = \sqrt{\frac{R}{P}}\frac{1}{2\sqrt{T}} = \frac{1}{2}\sqrt{\frac{R}{PT}}.$$

Differentiating the expression for V w.r.t P at constant T, we get

$$\left(\frac{\partial V}{\partial P}\right)_T = \sqrt{RT}\left(-\frac{1}{2}\right)P^{3/2} = \left(-\frac{1}{2}\right)\sqrt{\frac{RT}{P^3}}.$$

Using these two values in the above expression, we get

$$\left(\frac{\partial U}{\partial P}\right)_T = T\frac{1}{2}\sqrt{\frac{R}{PT}} - P\sqrt{\frac{RT}{P^3}}\left(-\frac{1}{2}\right) = -\frac{1}{2}\sqrt{\frac{RT}{P}} + \frac{1}{2}\sqrt{\frac{RT}{P}} = 0.$$

Hence, the internal energy U is independent of pressure P at constant temperature T.

24. The magnetic field on a paramagnetic salt kept at an initial temperature of 2 K is reduced from 10000 Oersted to zero. Calculate the cooling effect produced by adiabatic demagnetization of the paramagnetic salt. (Given: Curie constant per gm. mole per c.c. $= 0.042$ erg deg/gm Oersted^{-2} and $C_H = 0.42$ J g^{-1} deg^{-1}).

Answer: The cooling produced by adiabatic demagnetization of a paramagnetic salt is given by

$$\delta T = \frac{CV}{2C_H T} . H^2 .$$

Here, $T = 2$ K, $C_H = 0.42$ J g^{-1} deg^{-1} $= 0.42 \times 10^7$ erg g^{-1}deg^{-1}, $H = 10000$ Oersted. Curie constant $CV = 0.042$ erg degree/g Orested^{-2}. This expression provides

$$\delta T = -\frac{0.042 \times (10000)^2}{2 \times (0.42 \times 10^7) \times 2} = -0.25 \text{ K}.$$

25. State Gibbs phase rule. Using this rule, find the number of degrees of freedom of a system in which water is in equilibrium with saturated vapor.

[Calcutta University 2016]

Answer: Gibbs phase rule represents the condition of equilibrium between the number of coexisting phases and the components. Consider a heterogeneous system consisting of N_P phases and N_C components in equilibrium.

In every phase, there are N_C different components having N_C different concentrations but for a closed system if the concentrations of $(N_C - 1)$ components are arbitrarily chosen, the concentration of the last component is automatically fixed. Therefore, there are $(N_C - 1)$ composition variables in each phase. For N_P phases, the number of composition variables would be $N_P(N_C - 1)$. In addition to this value, there are two other variables: pressure and temperature. Thus, the total number of variables will be $N_P(N_C - 1) + 2$.

Further, it has been shown that if the chemical potential of any one of the P phases is defined, then the chemical potential of the remaining $(N_P - 1)$ phases is fixed. Therefore, for N_C components the invariant composition variables will be $N_C(N_P - 1)$. Therefore, the number of independent variables of a system will be given by

$$N_F = N_P(N_C - 1) + 2N_C(N_P - 1), \quad \Rightarrow \quad N_F = N_C N_P + 2.$$

26. Why is the slope of the fusion curve in the P–T phase diagram of water negative?

[Calcutta University 2017, 2018]

Answer: When the temperature of water increases from 0°C to approximately 4°C, its molar volume decreases and becomes minimum at this temperature (4°C). This is known as anomalous expansion of water. This indicates that the density of water increases as temperature increases from 0°C and becomes maximum at 4°C. For this reason, the slope of the fusion curve of water has a negative slope in this temperature range. Further, the value of the volume expansion coefficient of water is negative in this temperature range.

27. Draw the pressure–temperature phase diagram of H_2O indicating phases, boundaries and triple points.   [Calcutta University 2012, 2013]

Answer: The pressure–temperature phase diagram of H_2O indicating phases, boundaries, and triple points is shown in Figure 10.14.

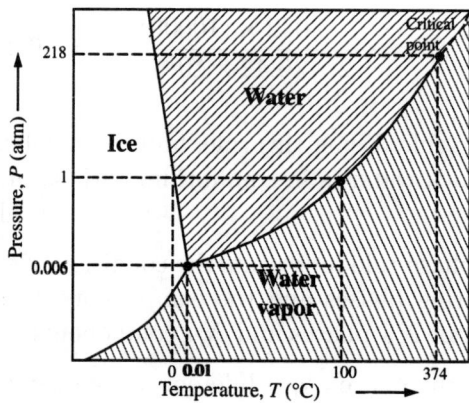

Figure 10.14 Variation of different physical parameters during the second-order phase transition for the problem 27.

28. Prove the thermodynamic relation: $\left(\frac{\partial U}{\partial V}\right)_T = T\left(\frac{\partial P}{\partial T}\right)_V - P$.
                                          [Calcutta University 2021, Delhi University 2021]

Answer: Combining the first and the second laws of thermodynamics, we have the relation: $dU = TdS - PdV$.

Differentiating this expression w.r.t volume V at constant temperature T, we get

$$\left(\frac{\partial U}{\partial V}\right)_T = T\left(\frac{\partial S}{\partial V}\right)_T - P.$$

Using Maxwell's relation $\left(\frac{\partial S}{\partial V}\right)_T = \left(\frac{\partial P}{\partial T}\right)_V$ in the above expression, we get

$$\left(\frac{\partial U}{\partial V}\right)_T = T\left(\frac{\partial P}{\partial T}\right)_V - P.$$

29. Using the thermodynamic relation: $\left(\frac{\partial U}{\partial V}\right)_T = T\left(\frac{\partial P}{\partial T}\right)_V - P$, show that the internal energy U of an ideal gas is independent of its volume V at constant temperature T.

Answer: For one mole of an ideal gas, the equation of state is

$$PV = RT; \quad \Rightarrow \quad P = \frac{RT}{V}. \quad \Rightarrow \quad \left(\frac{\partial P}{\partial T}\right)_V = \frac{R}{V}.$$

Using this value in the above expression, we get

$$\left(\frac{\partial U}{\partial V}\right)_T = T\left(\frac{\partial P}{\partial T}\right)_V - P = T \times \frac{R}{V} - P = P - P = 0.$$

This shows that the internal energy U of an ideal gas is independent of its volume V at constant temperature T.

30. Using the thermodynamic relation: $\left(\frac{\partial U}{\partial V}\right)_T = T\left(\frac{\partial P}{\partial T}\right)_V - P$, find $\left(\frac{\partial U}{\partial V}\right)_T$ for one mole of a Van der Waals gas. Using this result, show that the temperature change during free expansion of the Van der Waals gas is $\Delta T = \frac{a}{C_V} \frac{V_i - V_f}{V_i V_f}$, where V_i and V_f denote the initial and final volume, respectively.

[Calcutta University 2014]

Answer: For one mole of a real gas, the Van der Waals equation of state is given by

$$\left(P + \frac{a}{V^2}\right)(V - b) = RT; \quad \Rightarrow P = \frac{RT}{V - b} - \frac{a}{V^2}.$$

Using this relation, we get

$$\left(\frac{\partial P}{\partial T}\right)_V = \frac{R}{V - b}; \quad \Rightarrow T\left(\frac{\partial P}{\partial T}\right)_V = \frac{RT}{V - b} = P + \frac{a}{V^2};$$

$$\Rightarrow \quad T\left(\frac{\partial P}{\partial T}\right)_V - P = \frac{a}{V^2}; \quad \Rightarrow \quad \left(\frac{\partial U}{\partial V}\right)_T = \frac{a}{V^2}.$$

The Joule coefficient for a Van der Waals gas is given by

$$\mu = -\frac{a}{V^2 C_V}.$$

Thus, we see that the Joule coefficient for a Van der Waals gas is negative which implies that the temperature of the gas always decreases as it undergoes free expansion. This decrease in temperature is a consequence of the work done in overcoming the intermolecular attractive forces present in the Van der Waals gas. Over a relatively small range of temperature, $T_i < T < T_f$, we assume that the specific heat at constant volume C_V is independent of temperature T. Hence, the change in temperature of the nonideal Van der Waals gas that undergoes a free expansion can be written as

$$\Delta T = + \int_{V_i}^{V_f} \mu \, dV = + \int_{V_i}^{V_f} \left[-\frac{a}{V^2 C_V} \right] dV.$$

Here, the expression for μ has been used. This integral leads to the following result for the change in temperature

$$\Delta T = -\frac{a}{C_V} \left(\frac{1}{V_i} - \frac{1}{V_f} \right). \quad \text{Or,} \quad \Delta T = -\frac{a}{C_V} \left(\frac{V_i - V_f}{V_i V_f} \right).$$

31. If E_S and E_T are the adiabatic and isothermal elastic constants of a gas and γ is the ratio of two specific heats, C_P and C_V, then prove that $\gamma = \frac{E_S}{E_T}$.

[Calcutta University 2015, 2018]

Answer: The adiabatic and isothermal elastic coefficients are, respectively, defined by $E_S = -V \left(\frac{\partial P}{\partial V} \right)_S$ and $E_T = -V \left(\frac{\partial P}{\partial V} \right)_T$. So, the ratio of these two coefficients can be expressed as

$$\frac{E_S}{E_T} = \frac{\left(\frac{\partial P}{\partial V} \right)_S}{\left(\frac{\partial P}{\partial V} \right)_T}.$$

Considering temperature T as a function of volume V and pressure P, we can write

$$dT = \left(\frac{\partial T}{\partial V} \right)_P dV + \left(\frac{\partial T}{\partial P} \right)_V dP. \quad \text{Or,} \Rightarrow \quad \left(\frac{\partial P}{\partial V} \right)_T = -\frac{\left(\frac{\partial T}{\partial V} \right)_P}{\left(\frac{\partial T}{\partial P} \right)_V}, \quad [dT = 0].$$

We recall that

$$\delta Q = C_V dT + \left[P + \left(\frac{\partial U}{\partial V} \right)_T \right] dV.$$

So, $$\delta Q = C_V dT + \frac{C_P - C_V}{\left(\frac{\partial V}{\partial T} \right)_P} dV = C_V dT + (C_P - C_V) \left(\frac{\partial T}{\partial V} \right)_P dV$$

$$\frac{\delta Q}{C_V} = dT + (\gamma - 1) \left(\frac{\partial T}{\partial V} \right)_P dV.$$

From previous results, we can simplify and write this as

$$\frac{\delta Q}{C_V} = \left(\frac{\partial T}{\partial P} \right)_V dP + \gamma \left(\frac{\partial T}{\partial V} \right)_P dV.$$

For adiabatic process, $\delta Q = 0$ and this equation gives

$$\left(\frac{\partial P}{\partial V}\right)_S = -\gamma\frac{\left(\frac{\partial T}{\partial V}\right)_P}{\left(\frac{\partial T}{\partial P}\right)_V}. \quad \text{Or} \quad \gamma = -\left(\frac{\partial P}{\partial V}\right)_S \left(\frac{\partial T}{\partial P}\right)_V [\left(\frac{\partial T}{\partial V}\right)_P]^{-1}.$$

$$\text{Or} \quad \gamma = -\left(\frac{\partial P}{\partial V}\right)_S \left(\frac{\partial V}{\partial P}\right)_T = -\frac{\left(\frac{\partial P}{\partial V}\right)_S}{\left(\frac{\partial P}{\partial V}\right)_T}. \quad \text{Or} \quad \frac{E_S}{E_T} = -\frac{\left(\frac{\partial P}{\partial V}\right)_S}{\left(\frac{\partial P}{\partial V}\right)_T} = \gamma.$$

32. Starting from the expression $C_P - C_V = -T\left(\frac{\partial P}{\partial V}\right)_T \left(\frac{\partial V}{\partial T}\right)_P^2$ given by equation (10.69), show that for one mole of an ideal gas $C_P - C_V = R$.

Answer: For one mole of an ideal gas, we have $PV = RT$. From this, we get

$$\left(\frac{\partial P}{\partial V}\right)_T = \left(\frac{\partial \left(\frac{RT}{V}\right)}{\partial V}\right)_T = -\frac{RT}{V^2}.$$

The second factor comes out to be

$$\left[\left(\frac{\partial V}{\partial T}\right)_P\right]^2 = \left[\left(\frac{\partial \left(\frac{RT}{P}\right)}{\partial T}\right)_P\right]^2 = \left[\frac{R}{P}\right]^2 = \frac{R^2}{P^2}.$$

Using these two values in the expression $C_P - C_V = -T\left(\frac{\partial P}{\partial V}\right)_T \left(\frac{\partial V}{\partial T}\right)_P^2$, we get

$$C_P - C_V = -T\left[-\frac{RT}{V^2}\right]\left[\frac{R^2}{P^2}\right] = R.$$

33. Starting from the expression $C_P - C_V = T\left(\frac{\partial P}{\partial T}\right)_V \left(\frac{\partial V}{\partial T}\right)_P$ given by equation (10.67), show that for one mole of a real gas $C_P - C_V = R\left(1 + \frac{2a}{RTV}\right)$.

Answer: For one mole of a real gas, we have

$$\left(P + \frac{a}{V^2}\right)(V - b) = RT; \quad \Rightarrow \quad P = \frac{RT}{V - b} - \frac{a}{V^2}.$$

Differentiating the above expression w.r.t temperature T at constant volume V, we get

$$\left(\frac{\partial P}{\partial T}\right)_V = \frac{R}{V - b}.$$

Differentiating the Van der Waals equation of state for one mole w.r.t temperature T, we get

$$\left[-(V-b)\frac{2a}{V^3} + \left(P + \frac{a}{V^2}\right)\right]\left(\frac{\partial V}{\partial T}\right)_P = R;$$

$$\Rightarrow \left[-\frac{2a}{V^3} + \left(\frac{RT}{(V-b)^2}\right)\right]\left(\frac{\partial V}{\partial T}\right)_P = \frac{R}{V-b}.$$

Hence, the difference between the heat capacities comes out to be

$$C_P - C_V = T\left(\frac{\partial P}{\partial T}\right)_V \left(\frac{\partial V}{\partial T}\right)_P = \frac{\frac{R^2 T}{(V-b)^2}}{\frac{RT}{(V-b)^2} - \frac{2a}{V^3}} = \frac{R}{1 - \frac{2a}{V^3}\frac{(V-b)^2}{RT}} \text{ [as } V \gg b\text{]}$$

$$\approx \frac{R}{1 - \frac{2a}{RTV}} = R\left(1 - \frac{2a}{RTV}\right)^{-1} \approx R\left(1 + \frac{2a}{RTV}\right). \text{ [As } a \text{ is very small.]}$$

This proves that

$$C_P - C_V = R\left(1 + \frac{2a}{RTV}\right).$$

34. Prove that $\left(\frac{\partial C_V}{\partial V}\right)_T = T\left(\frac{\partial^2 P}{\partial T^2}\right)_V$ and $\left(\frac{\partial C_P}{\partial P}\right)_T = -T\left(\frac{\partial^2 V}{\partial T^2}\right)_P$.

[Calcutta University 2018]

Answer: According to the definition of C_V, we have $C_V = T\left(\frac{\partial S}{\partial T}\right)_V$. Differentiating this expression w.r.t volume V at constant temperature T, we get

$$\left(\frac{\partial C_V}{\partial V}\right)_T = T\left[\frac{\partial}{\partial V}(\frac{\partial S}{\partial T})_V\right]_T = T\left[\frac{\partial}{\partial T}(\frac{\partial S}{\partial V})_T\right]_V$$

$$= T\left[\frac{\partial}{\partial T}(\frac{\partial P}{\partial T})_V\right]_V = T\left(\frac{\partial^2 P}{\partial T^2}\right)_V.$$

From the definition of heat capacity at constant pressure, we have $C_P = T\left(\frac{\partial S}{\partial T}\right)_P$. Differentiating this expression w.r.t pressure P at constant temperature T, we get

$$\left(\frac{\partial C_P}{\partial P}\right)_T = T\left[\frac{\partial}{\partial P}(\frac{\partial S}{\partial T})_P\right]_T = T\left[\frac{\partial}{\partial T}(\frac{\partial S}{\partial P})_T\right]_P$$

$$= -T\left[\frac{\partial}{\partial T}(\frac{\partial V}{\partial T})_P\right]_P = -T\left(\frac{\partial^2 V}{\partial T^2}\right)_P.$$

35. For a thermodynamic system, show that $C_P - C_V = \alpha_P^2 \kappa_T TV$, where α_P is the isobaric volume expansion coefficient and κ_T is the isothermal bulk modulus.

[Calcutta University 2018]

Answer: From Maxwell's relation, we have $\left(\frac{\partial S}{\partial P}\right)_T = \left(\frac{\partial V}{\partial T}\right)_P$. Multiplying both sides by T, we get

$$\left(\frac{T \partial S}{\partial P}\right)_T = T \left(\frac{\partial V}{\partial T}\right)_P.$$

Considering entropy S as a function of temperature T and volume V, that is, $S = S(T, V)$, we can write the differential dS in S as

$$dS = \left(\frac{\partial S}{\partial T}\right)_V dT + \left(\frac{\partial S}{\partial V}\right)_T dV.$$

Differentiating the resulting expression w.r.t temperature T, and multiplying the resulting expression by T, we get

$$T \left(\frac{\partial S}{\partial T}\right)_P = T \left(\frac{\partial S}{\partial T}\right)_V + T \left(\frac{\partial S}{\partial V}\right)_T \left(\frac{\partial V}{\partial T}\right)_P.$$

From Maxwell's first relation, we have $\left(\frac{\partial S}{\partial V}\right)_T = \left(\frac{\partial P}{\partial T}\right)_V$. Using this relation in the above expression, we get

$$C_P - C_V = T \left(\frac{\partial P}{\partial T}\right)_V \left(\frac{\partial V}{\partial T}\right)_P.$$

From the definition, we have $\alpha_P = \frac{1}{V} \left(\frac{\partial V}{\partial T}\right)_P$ and $\kappa_T = -V \left(\frac{\partial P}{\partial V}\right)_T$. Thus, we have

$$\alpha_P^2 \kappa_T = - \left(\frac{\partial P}{\partial V}\right)_T \left(\frac{\partial V}{\partial T}\right)_P \left(\frac{\partial V}{\partial T}\right)_P \frac{1}{V} = \left(\frac{\partial P}{\partial T}\right)_V \left(\frac{\partial V}{\partial T}\right)_P \frac{1}{V}.$$

$$\Rightarrow \alpha_P^2 \kappa_T TV = \left(\frac{\partial P}{\partial T}\right)_V \left(\frac{\partial V}{\partial T}\right)_P T; \quad \Rightarrow C_P - C_V = \alpha_P^2 \kappa_T \, T \, V.$$

36. Prove the following relations where symbols have their usual significance:

(a) $\left(\frac{\partial T}{\partial V}\right)_S = - \left(\frac{\partial P}{\partial S}\right)_V$

(b) $\frac{\left(\frac{\partial P}{\partial T}\right)_S}{\left(\frac{\partial P}{\partial T}\right)_V} = \frac{\gamma}{\gamma - 1}$.

[Calcutta University 2015, 2018]

Answer:

(a) To answer the first part, see the derivation of Maxwell's third relation.

(b) The second part can be solved in the following way:

$$\frac{\left(\frac{\partial P}{\partial T}\right)_S}{\left(\frac{\partial P}{\partial T}\right)_V} = \frac{1}{\left(\frac{\partial T}{\partial P}\right)_S \left(\frac{\partial P}{\partial T}\right)_V} = \frac{1}{\left(\frac{\partial V}{\partial S}\right)_P \left(\frac{\partial P}{\partial T}\right)_V} = \frac{1}{\left(\frac{\partial V}{\partial T}\right)_P \left(\frac{\partial T}{\partial S}\right)_P \left(\frac{\partial P}{\partial T}\right)_V}$$

$$= \frac{\left(\frac{\partial S}{\partial T}\right)_P}{\left(\frac{\partial V}{\partial T}\right)_P \left(\frac{\partial P}{\partial T}\right)_V} = \frac{\frac{C_P}{T}}{\frac{C_P - C_V}{T}} = \frac{C_P}{C_P - C_V} = \frac{\gamma}{\gamma - 1}.$$

Here, we have used the expression: $C_P = C_V + T \left(\frac{\partial V}{\partial T}\right)_P \left(\frac{\partial P}{\partial T}\right)_V$.

37. Using Maxwell's thermodynamic relations, show that the ratio of adiabatic to isobaric volume expansivity is $\left(\frac{1}{1-\gamma}\right)$.

Answer: According to the definition of volume expansivity α is given by $\alpha = \frac{1}{V}\left(\frac{\partial V}{\partial T}\right)$.

Then, we have the ratio

$$\frac{\alpha_S}{\alpha_P} = \frac{V^{-1}\left(\frac{\partial V}{\partial T}\right)_S}{V^{-1}\left(\frac{\partial V}{\partial T}\right)_P} = \frac{1}{\left(\frac{\partial T}{\partial V}\right)_S \left(\frac{\partial V}{\partial T}\right)_P}$$

$$= \frac{1}{-\left(\frac{\partial P}{\partial S}\right)_V \left(\frac{\partial V}{\partial T}\right)_P} \quad \text{[using Maxwell's relation]}$$

$$= \frac{1}{-\left(\frac{\partial P}{\partial T}\right)_V \left(\frac{\partial T}{\partial S}\right)_V \left(\frac{\partial V}{\partial T}\right)_P} = \frac{T\left(\frac{\partial S}{\partial T}\right)_V}{-T\left(\frac{\partial P}{\partial T}\right)_V \left(\frac{\partial V}{\partial T}\right)_P} = \frac{C_V}{C_V - C_P} = \left(\frac{1}{1-\gamma}\right).$$

38. What is the Joule–Thomson effect? Show that the Joule–Thomson coefficient for a gas is given by $\mu = -\frac{1}{C_P}\left[T\left(\frac{\partial V}{\partial T}\right)_P - V\right]$. Using this expression, prove that the ideal gas shows no change in temperature in the Joule–Thomson effect. [Calcutta University 2016, 2018]

Answer: In thermodynamics, the Joule–Thomson effect (also known as Joule–Thomson expansion) describes the temperature change of a real gas or liquid (as differentiated from an ideal gas) when it is forced through a valve or porous plug while keeping it insulated so that no heat is exchanged with the environment. This procedure is called a throttling process or Joule–Thomson process.

The temperature of an ideal gas does not change in a throttling process. In the case of a real gas, the parameter that controls whether the gas is heated or cooled in a throttling process is

$$\mu = \left(\frac{\partial T}{\partial P}\right)_H.$$

This parameter is known as the Joule–Thompson coefficient. Given that p decreases in a throttling process, a positive Joule–Thompson coefficient implies that the temperature also decreases, and vice versa.

Let us derive a convenient expression for μ in terms of readily measured experimental parameters. Starting from the fundamental thermodynamic relation

$$dE = TdS - PdV.$$

We find that $dH = d(E + PV) = TdS + PdV$. For the case of throttling, $dH = 0$. Thus, we can write

$$0 = T\left[\left(\frac{\partial S}{\partial T}\right)_P dT + \left(\frac{\partial S}{\partial P}\right)_T dP\right] + VdP.$$

We know the equations

$$-\left(\frac{\partial S}{\partial P}\right)_T = \left(\frac{\partial V}{\partial T}\right)_P \text{ and } C_P = T\left(\frac{\partial S}{\partial T}\right)_P.$$

Using these relations, we get

$$\mu = \left(\frac{\partial T}{\partial P}\right)_H = -\frac{1}{C_P}\left[T\left(\frac{\partial V}{\partial T}\right)_P - V\right].$$

For an ideal gas (one mole), we have $PV = RT$, so that

$$\left(\frac{\partial V}{\partial T}\right)_P = \frac{R}{P} = \frac{V}{T}, \text{ and } \mu = T\frac{V}{T} - V = 0.$$

39. Prove that for an adiabatic expansion of the Van der Waals model of a real gas

$$\left(P + \frac{a}{V^2}\right)(V - b)^\gamma = \text{constant},$$

where γ is the ratio of the two specific heats, $\frac{C_P}{C_V}$.

Answer: From the first TdS equation, we have

$$TdS = C_V dT + T\left(\frac{\partial P}{\partial T}\right)_V dV. \tag{10.140}$$

Now, for an adiabatic process, $dS = 0$, so that, from above, we get

$$C_V dT = -T\left(\frac{\partial P}{\partial T}\right)_V dV. \tag{10.141}$$

For one mole of Van der Waals gas, however, we have, expressing P in terms of T and V,

$$P = \frac{RT}{V-b} - \frac{a}{V^2}; \quad \left(\frac{\partial P}{\partial T}\right)_V = \frac{R}{V-b}.$$

$$C_V dT = -\frac{RT}{V-b} dv; \quad \Rightarrow \quad C_V \frac{dT}{T} + R\frac{dV}{V-b} = 0.$$

Integrating, we get

$$C_V \ln T + R\ln(V-b) = \ln(\text{constant}); \quad \text{or,} \quad T(V-b)^{\frac{R}{C_V}} = \text{constant}.$$

Multiplying Van der Waals equation $\left(P + \frac{a}{V^2}\right)(V-b) = RT$ by $T(V-b)^{\frac{R}{C_V}} = $ constant, we get

$$\left(P + \frac{a}{V^2}\right)(V-b)^{\frac{R}{C_V}+1} = RT(V-b)^{\frac{R}{C_V}} = \text{constant}.$$

From the relation $C_P - C_V = R$, which is true only for an ideal gas, we get $\frac{R}{C_V} + 1 = \gamma$. Hence, we get

$$\left(P + \frac{a}{V^2}\right)(V-b)^{\gamma} = \text{constant}. \tag{10.142}$$

40. Show that for an isentropic transformation:

$$\left(\frac{\partial V}{\partial T}\right)_S = -\frac{C_V}{C_P - C_V}\left(\frac{\partial V}{\partial T}\right)_P \quad \text{and}$$

$$\left(\frac{\partial P}{\partial T}\right)_S = \frac{C_P}{C_P - C_V}\left(\frac{\partial P}{\partial T}\right)_V.$$

Answer: We have the relation

$$C_P - C_V = T\left(\frac{\partial P}{\partial T}\right)_V\left(\frac{\partial V}{\partial T}\right)_P. \tag{10.143}$$

Now, from the first T–dS equation, we get

$$TdS = C_V dT + T\left(\frac{\partial P}{\partial T}\right)_V dV.$$

For an isentropic process, $dS = 0$. So, we have from above

$$C_V = -T\left(\frac{\partial P}{\partial T}\right)_V\left(\frac{\partial V}{\partial T}\right)_S. \tag{10.144}$$

Again, from the second T–dS equation, we have

$$TdS = C_P dT - T\left(\frac{\partial V}{\partial T}\right)_P dP.$$

For an isentropic process, $dS = 0$, so that we obtain

$$C_P = T\left(\frac{\partial V}{\partial T}\right)_P \left(\frac{\partial P}{\partial T}\right)_S. \tag{10.145}$$

Now from equations (10.143) and (10.144), we have

$$\frac{C_V}{C_P - C_V} = \frac{-T\left(\frac{\partial P}{\partial T}\right)_V \left(\frac{\partial V}{\partial T}\right)_S}{T\left(\frac{\partial P}{\partial T}\right)_V \left(\frac{\partial V}{\partial T}\right)_P} = -\frac{\left(\frac{\partial V}{\partial T}\right)_S}{\left(\frac{\partial V}{\partial T}\right)_P}.$$

$$\left(\frac{\partial P}{\partial T}\right)_S = \frac{C_P}{C_P - C_V}\left(\frac{\partial P}{\partial T}\right)_V,$$

which is the required second relation.

41. A gas obeys the equation of state $\frac{PV}{RT} = 1 + PB(T)$, where $B(T)$ is a function of temperature only. Show that

$$C_P - C_{P_0} = -RTP\frac{d^2}{dT^2}(BT),$$

where C_{P_0} is the value of C_P at some reference temperature.

Answer: We know that

$$\left(\frac{\partial C_P}{\partial P}\right) = -T\left(\frac{\partial^2 V}{\partial T^2}\right)_P; \quad C_P = -\int T\left(\frac{\partial^2 V}{\partial T^2}\right)_P dP. \tag{10.146}$$

The equation of state is

$$\frac{PV}{RT} = 1 + PB(T); \quad \Rightarrow V = \frac{RT}{P}(1 + PB);$$

$$\left(\frac{\partial V}{\partial T}\right)_P = \frac{R}{P} + R\frac{d}{dT}(BT) : \left(\frac{\partial^2 V}{\partial T^2}\right)_P = R\frac{d^2}{dT^2}(BT).$$

From equation (10.146), we have

$$C_P = -\int RT\frac{d^2}{dT^2}(BT)dP = -RT\frac{d^2}{dT^2}(BT)\int dP;$$

$$= -RT\frac{d^2}{dT^2}(BT) + \text{constant}.$$

Now, when $P \to 0, C_P \to C_{Pp_0}$. From equation (10.146), constant $= C_{P_0}$. Therefore, we have

$$C_P = -RTP\frac{d^2}{dT^2}(BT) + C_{P_0}; \quad \Rightarrow C_P - C_{P_0} = -RTP\frac{d^2}{dT^2}(BT).$$

42. Show that for one mole of a Van der Waals gas

$$\delta Q = C_V dT + \frac{RT}{(V-b)}dV.$$

Answer: From the first T–dS equation, we have

$$TdS = C_V dT + T\left(\frac{\partial P}{\partial T}\right)_V dV. \qquad (10.147)$$

However, for one mole of a Van der Waals gas, we have

$$P = \frac{RT}{V-b} - \frac{a}{V^2}; \quad \Rightarrow \left(\frac{\partial P}{\partial T}\right)_V = \frac{R}{V-b}.$$

Using this expression and equation (10.147)

$$TdS = \delta Q = C_V dT + \frac{RT}{V-b}dV.$$

43. Show that

$$(C_P - C_V)\left(\frac{\partial^2 P}{\partial P \partial V}\right) + \left(\frac{\partial C_P}{\partial P}\right)_V\left(\frac{\partial T}{\partial V}\right)_P - \left(\frac{\partial C_V}{\partial V}\right)_P\left(\frac{\partial T}{\partial P}\right)_V = 1.$$

Answer: Let us start the problem from the left-hand side.

$$\text{L.H.S} = \left\{C_P\left(\frac{\partial^2 P}{\partial P \partial V}\right) + \left(\frac{\partial C_P}{\partial P}\right)_V\left(\frac{\partial T}{\partial V}\right)_P\right\}$$

$$- \left\{C_V\left(\frac{\partial^2 P}{\partial P \partial V}\right) + \left(\frac{\partial C_V}{\partial V}\right)_P\left(\frac{\partial T}{\partial P}\right)_V\right\}$$

$$= \frac{\partial}{\partial P}\left\{C_P\left(\frac{\partial T}{\partial V}\right)_P\right\}_V - \frac{\partial}{\partial V}\left\{C_V\left(\frac{\partial T}{\partial P}\right)_V\right\}_P$$

$$= \frac{\partial}{\partial P}\left\{T\left(\frac{\partial S}{\partial T}\right)_P\left(\frac{\partial T}{\partial V}\right)_P\right\}_V - \frac{\partial}{\partial V}\left\{\left(\frac{\partial S}{\partial T}\right)_V\left(\frac{\partial T}{\partial P}\right)_V\right\}_P$$

$$= \frac{\partial}{\partial P}\left\{T\left(\frac{\partial S}{\partial V}\right)_P\right\}_V - \frac{\partial}{\partial V}\left\{T\left(\frac{\partial S}{\partial P}\right)_V\right\}_P$$

$$= \left(\frac{\partial T}{\partial P}\right)_V\left(\frac{\partial S}{\partial V}\right)_P - \left(\frac{\partial T}{\partial V}\right)_P\left(\frac{\partial S}{\partial P}\right)_V + T\left(\frac{\partial^2 S}{\partial P\partial V}\right) - T\left(\frac{\partial^2 S}{\partial P\partial V}\right).$$

Since S is a state function, the last two terms cancel each other. Therefore, we get

$$L.H.S. = \left(\frac{\partial T}{\partial P}\right)_V\left(\frac{\partial S}{\partial V}\right)_P - \left(\frac{\partial T}{\partial V}\right)_P\left(\frac{\partial S}{\partial P}\right)_V = \left(\frac{\partial(T,S)}{\partial(P,V)}\right)_V = 1 = \text{R.H.S.}$$

44. What is the Joule–Thomson effect? Deduce an expression for J–T cooling.
[Calcutta University 2012]

Answer: The Joule–Thomson effect is the change in temperature of a real gas when forced though a valve or a porous plug while keeping it insulated so that no heat is exchanged with the surroundings.

The Joule–Thomson coefficient is a function of H,T, and P. Applying the cyclic rule, we have

$$\left(\frac{\partial H}{\partial T}\right)_P\left(\frac{\partial T}{\partial P}\right)_H\left(\frac{\partial P}{\partial H}\right)_T = -1.$$

Again, from the definitions, we have $\left(\frac{\partial H}{\partial T}\right)_P = C_P$ and $\mu_{JT} = \left(\frac{\partial T}{\partial P}\right)_H$.

Using these definitions in the cyclic rule, we get $C_P\mu_{JT}\left(\frac{\partial P}{\partial H}\right)_T = -1$.

This gives $\Rightarrow \mu_{JT} = -\frac{1}{C_P}\left(\frac{\partial H}{\partial P}\right)_T = -\frac{1}{C_P}\left(\frac{\partial(U+PV)}{\partial P}\right)_T = -\frac{1}{C_P}\left(\frac{\partial U}{\partial P} + V\right)_T$.

45. Distinguish between cooling produced by the J–T effect and adiabatic expansion.
[Calcutta University 2012]

Answer:

J–T effect	Adiabatic expansion
In case of Joule–Thomson or Joule–Kelvin effect, $dH = 0$.	In adiabatic expansion, only $\delta Q = 0$.
Joule–Thomson is highly irreversible.	However, adiabatic expansion may be reversible or irreversible.
There may be heating or cooling.	Always has a cooling effect.

46. The Joule coefficient is defined as $\eta = \left(\frac{\partial T}{\partial P}\right)_U$. Show that it is connected to the Joule–Kelvin coefficient μ by the relation.

$$\eta \left[C_P - \left(\frac{\partial (PV)}{\partial T}\right)_P \right] = \mu C_P + \left[\frac{\partial (PV)}{\partial P}\right]_T.$$

Answer: Let us take internal energy as a function of temperature and pressure, that is,

$$U = U(T, P).$$

Then, we can write

$$dU = \left(\frac{\partial U}{\partial T}\right)_P dT + \left(\frac{\partial U}{\partial P}\right)_T dP$$

or

$$\left(\frac{\partial T}{\partial P}\right)_U = \eta = -\frac{\left(\frac{\partial U}{\partial P}\right)_T}{\left(\frac{\partial U}{\partial T}\right)_P} \quad \cdots \text{(i)}$$

By definition, enthalpy $H = U + PV$, so that

$$\left(\frac{\partial U}{\partial T}\right)_P = \left(\frac{\partial H}{\partial T}\right)_P - \left[\frac{\partial}{\partial T}(PV)\right]_P = C_P - \left[\frac{\partial}{\partial T}(PV)\right]_P.$$

Substituting this result in equation (i), we obtain

$$\eta = -\frac{\left(\frac{\partial U}{\partial P}\right)_T}{C_P - \left[\frac{\partial}{\partial T}(PV)\right]_P} \quad \cdots \text{(ii)}$$

For a perfect gas, the numerator in equation (ii) is zero implying that $\eta = 0$. If we now take H as a function of T and P, we can write

$$H = H(T, P),$$

so that

$$dH = \left(\frac{\partial H}{\partial T}\right)_P dT + \left(\frac{\partial H}{\partial P}\right)_T dP.$$

For an isenthalpic process,

$$\left(\frac{\partial T}{\partial P}\right)_H = \mu = -\left(\frac{\partial H}{\partial P}\right)_T / \left(\frac{\partial H}{\partial T}\right)_P = -\frac{1}{C_P}\left(\frac{\partial H}{\partial P}\right)_T \quad \cdots \text{(iii)}$$

By definition, $H = U + PV$ and we can write

$$\left(\frac{\partial H}{\partial P}\right)_T = \left(\frac{\partial U}{\partial P}\right)_T + \left(\frac{\partial}{\partial P}(PV)\right)_T.$$

On combining this result with equation (iii), we get

$$\left(\frac{\partial U}{\partial P}\right)_T + \left(\frac{\partial}{\partial P}(PV)\right)_T = -\mu C_P;$$

$$\Rightarrow \quad \left(\frac{\partial U}{\partial P}\right)_T = -\mu C_P - \left(\frac{\partial}{\partial P}(PV)\right)_T \quad \cdots\text{(iv)}$$

Using this result in equation (ii), we get the required expression connecting the Joule coefficient with the Joule–Kelvin coefficient:

$$\eta = \frac{\mu C_P + \left(\frac{\partial}{\partial P}(PV)\right)_T}{C_P - \left(\frac{\partial}{\partial T}(PV)\right)_P}$$

or

$$\eta\left(C_P - \left(\frac{\partial}{\partial T}(PV)\right)_P\right) = \mu C_P + \left(\frac{\partial}{\partial P}(PV)\right)_T.$$

47. Prove that

$$\left[\left(\frac{\partial V}{\partial P}\right)_T - \frac{T}{C_V}\left(\frac{\partial V}{\partial T}\right)_V^2\right]\left[\left(\frac{\partial P}{\partial V}\right)_T + \frac{T}{C_P}\left(\frac{\partial p}{\partial T}\right)_V^2\right] = 1.$$

Answer: Let us start the problem from the left-hand side.

$$\text{L.H.S} = \left[-VK - \frac{T}{C_V}\alpha^2 V^2\right]\left[-\frac{1}{VK} + \frac{T}{C_p}\frac{\alpha^2 V}{K}\right].$$

Using the definition of compressibility K and the volume expansivity α, we get

$$= -VK\left[1 + \frac{T}{C_V}\alpha^2 V^2\right]\frac{1}{VK}\left[-1 + \frac{T}{C_P}\frac{\alpha^2 V}{K}\right]$$

$$= \left[1 + \frac{T}{C_V}\alpha^2 V^2\right]\left[1_{\frac{T}{C_p}}\frac{\alpha^2 V}{K}\right],$$

$$= \left[1 + \frac{C_P - C_V}{C_V}\right]\left[1 - \frac{C_P - C_V}{C_p}\right] = \frac{C_p}{C_V}\frac{C_V}{C_P} = 1 = \text{R.H.S.}$$

48. The equation of state of a thermodynamic system is given by $P = \frac{AT^3}{V}$. The internal energy of the system is

$$U = BT^n \ln \frac{V}{V_0} + f(T),$$

where B, n, and V_0 are all constant and $f(t)$ depends on the temperature only. Find the values of B and n.

Answer: From the relation $TdS = dU + PdV$, we obtain

$$
\begin{aligned}
dS &= \frac{1}{T}dU + \frac{P}{T}dV = \frac{1}{T}\left[\left(\frac{\partial U}{\partial T}\right)_V dT + \left(\frac{\partial U}{\partial V}\right)_T dV\right] + \frac{P}{T}dV; \\
&= \frac{1}{T}\left(\frac{\partial U}{\partial T}\right)_V dT + \left[\frac{P}{T} + \left(\frac{\partial U}{\partial V}\right)_T dV\right]dV; \\
&= \frac{1}{T}\left[f'(T) + nBT^{n-1}\ln\frac{V}{V_0}\right]dT + \left[\frac{1}{T}\frac{BT^n}{V} + \frac{AT^2}{V}\right]dV.
\end{aligned}
$$

Since dS is an exact differential, we obtain

$$\frac{\partial}{\partial T}\left[\frac{BT^{n-1}}{V} + \frac{AT^2}{V}\right]_V = \frac{\partial}{\partial V}\left[\frac{f'(T)}{T} + nBT^{n-2}\ln\frac{V}{V_0}\right]_T;$$

$$\Rightarrow \frac{(n-1)BT^{n-2}}{V} + \frac{2AT}{V} = nBT^{n-2}\frac{1}{V}$$

$$\Rightarrow BT^{n-2} = 2AT; \quad \Rightarrow \quad n - 2 = 1 \text{ and } B = 2A; \quad n = 3 \text{ and } B = 2A.$$

49. The internal energy U of a black body radiation enclosed in a cavity of volume V at a temperature T is given by $U = V\sigma T^4$, where the constant σ is independent of volume V and temperature T. Show that the free energy F and the entropy S of the black body radiation are, respectively, given by $F = -\frac{1}{3}V\sigma T^4 + Tf(V)$ and $S = \frac{4}{3}V\sigma T^3 - f(V)$, where $f(V)$ is some unknown function of V. Assuming $S \to 0$ as $T \to 0$, show that the radiation pressure $P = \frac{1}{3}\frac{U}{V}$.

Answer: The expression for free energy is given by $\qquad F = U - TS$.

The differential of the free energy can be written as: $\qquad dF = dU - TdS - SdT$.

Again, using the relation $TdS = dU + PdV$, we get $\qquad dF = -SdT - PdV$.

This gives the expressions for pressure P and entropy S as

$$P = -\left(\frac{\partial F}{\partial V}\right)_T, \qquad \text{and} \qquad S = -\left(\frac{\partial F}{\partial T}\right)_V.$$

From the expression $F = U - TS$ and $S = -\left(\frac{\partial F}{\partial T}\right)_V$, we get

$$U = F - T\left(\frac{\partial F}{\partial T}\right)_V = -T^2\frac{\partial}{\partial T}\left(\frac{F}{T}\right)_V.$$

Using $U = V\sigma T^4$ in this expression, we have

$$V\sigma T^4 = -T^2\frac{\partial}{\partial T}\left(\frac{F}{T}\right)_V.$$

Integrating this expression with respect to T, we get

$$\frac{F}{T} = -\frac{V\sigma T^3}{3} + f(V). \qquad \Rightarrow \qquad F = -\frac{V\sigma T^4}{3} + Tf(V).$$

Here, $f(V)$ is an unknown function of V.

Differentiating the above expression for free energy F with respect to T, we get the entropy S as

$$S = -\left(\frac{\partial F}{\partial T}\right)_V = \frac{4}{3}V\sigma T^3 - f(V). \text{ Further, as } T \to 0 \text{ and } S \to 0,$$

we have $f(v) \to 0$.

Thus, entropy $S = \frac{4}{3}V\sigma T^3$. So, $F = U - TS = V\sigma T^4 - \frac{4}{3}V\sigma T^4 = -\frac{1}{3}V\sigma T^4$.

Hence, the radiation pressure P is given by

$$P = -\left(\frac{\partial F}{\partial T}\right)_T = \frac{1}{3}\sigma T^4 = \frac{1}{3}\frac{V\sigma T^4}{V} = \frac{1}{3}\frac{U}{V} = \frac{1}{3}u.$$

Here, u is the energy density.

10.6 Multiple choice questions and answers

1. For a system at constant temperature and volume, which of the following statements is correct at equilibrium?   [GATE 2016]

   (a) The Helmholtz free energy attains a local minimum.

   (b) The Helmholtz free energy attains a local maximum.

   (c) The Gibbs free energy attains a local minimum.

   (d) The Gibbs free energy attains a local maximum.

   Answer: The correct choice is (a).

   Solution: We know that $dF = -SdT - PdV$.

2. Which among the following sets of Maxwell relations is correct? (U-internal energy, H-enthalpy, F-Helmholtz free energy, and G-Gibbs free energy).

[GATE 2010]

(a) $T = \left(\frac{\partial U}{\partial V}\right)_S$ and $P = \left(\frac{\partial U}{\partial S}\right)_V$

(c) $P = -\left(\frac{\partial G}{\partial V}\right)_T$ and $V = \left(\frac{\partial G}{\partial P}\right)_S$

(b) $V = \left(\frac{\partial H}{\partial P}\right)_S$ and $T = \left(\frac{\partial H}{\partial S}\right)_P$

(d) $P = -\left(\frac{\partial F}{\partial S}\right)_T$ and $S = \left(\frac{\partial A}{\partial P}\right)_V$

Answer: The correct choice is (b).

Solution: $dH = TdS + VdP \Rightarrow \left(\frac{\partial H}{\partial S}\right)_P = T, \quad \left(\frac{\partial H}{\partial P}\right)_S = V.$

3. Which of the following is not a Maxwell's thermodynamic relation?

(a) $\left(\frac{\partial S}{\partial V}\right)_T = \left(\frac{\partial P}{\partial T}\right)_V$

(c) $\left(\frac{\partial T}{\partial P}\right)_V = \left(\frac{\partial V}{\partial S}\right)_P$

(b) $\left(\frac{\partial T}{\partial V}\right)_S = -\left(\frac{\partial P}{\partial S}\right)_V$

(d) $\left(\frac{\partial V}{\partial P}\right)_T = \left(\frac{\partial S}{\partial T}\right)_P$

Answer: The correct choice is (d).

4. Which of the following thermodynamic relations is used for certain adiabatic changes, such as the sudden compression of a liquid or the sudden-stretching of a rod?

(a) $\left(\frac{\partial S}{\partial P}\right)_T = -\left(\frac{\partial V}{\partial T}\right)_P$

(c) $\left(\frac{\partial S}{\partial T}\right)_P = \left(\frac{\partial P}{\partial V}\right)_T$

(b) $\left(\frac{\partial S}{\partial T}\right)_P = -\left(\frac{\partial P}{\partial V}\right)_T$

(d) $\left(\frac{\partial S}{\partial P}\right)_T = -\left(\frac{\partial V}{\partial T}\right)_P$

Answer: The correct choice is (a).

Solution: The sudden stretching of a rod or sudden compression of a liquid is given by

$$\left(\frac{\partial S}{\partial P}\right)_T = -\left(\frac{\partial V}{\partial T}\right)_P.$$

5. Which one of the Maxwell's thermodynamic relations given below leads to the Clausius–Clapeyron equation?

(a) $\left(\frac{\partial S}{\partial T}\right)_P = \left(\frac{\partial P}{\partial V}\right)_T$

(c) $\left(\frac{\partial S}{\partial V}\right)_T = \left(\frac{\partial P}{\partial T}\right)_V$

(b) $\left(\frac{\partial T}{\partial P}\right)_S = -\left(\frac{\partial V}{\partial S}\right)_P$

(d) $\left(\frac{\partial S}{\partial P}\right)_T = -\left(\frac{\partial V}{\partial T}\right)_P$

Answer: The correct choice is (d).

Solution: The rate of change of temperature with pressure is given by the Clausius–Clapeyron equation, which is given as $\frac{dP}{dV} = \frac{L}{T(V_2 - V_1)}$, L is the latent heat.

This can be derived from Maxwell's first thermodynamical relation given as

$$\left(\frac{\partial S}{\partial P}\right)_T = -\left(\frac{\partial V}{\partial T}\right)_P.$$

6. Which of the following is correct if α is volume expansivity and other variables have the usual meaning in thermodynamics?

(a) $\left(\frac{\partial C_P}{\partial P}\right)_T = -TV\alpha^2$

(c) $\left(\frac{\partial C_P}{\partial P}\right)_T = -\frac{TV}{\alpha^2}$

(b) $\left(\frac{\partial C_P}{\partial P}\right)_T = TV\alpha^2$

(d) $\left(\frac{\partial C_P}{\partial P}\right)_T = \frac{TV}{\alpha^2}$

Answer: The correct choice is (a).

Solution: From Maxwell's relation, we have

$$\left(\frac{\partial S}{\partial P}\right)_T = -\left(\frac{\partial V}{\partial T}\right)_P.$$

$$\left[\frac{\partial}{\partial T}\left(\frac{\partial S}{\partial P}\right)_T\right]_P = -\left[\frac{\partial}{\partial T}\left(\frac{\partial V}{\partial T}\right)_P\right]_P \Rightarrow \left[\frac{\partial}{\partial P}\left(\frac{T\partial S}{\partial T}\right)_P\right]_T$$

$$= -T\left[\frac{\partial}{\partial T}\left(\frac{\partial V}{\partial T}\right)_P\right]_P$$

$$\Rightarrow \left(\frac{\partial C_P}{\partial P}\right)_T = -T\left(\frac{\partial^2 V}{\partial T^2}\right)_V \text{ and } \alpha = \frac{1}{V}\left(\frac{\partial V}{\partial T}\right)_P.$$

So, $\left(\frac{\partial C_P}{\partial P}\right)_T = -TV\alpha^2$.

7. A theoretical model for a real (non-ideal) gas gives the following expressions for the internal energy (U) and pressure (P),

$$U(T,V) = aV^{-2/3} + bV^{2/3}T^2 \text{ and } P(T,V) = \frac{2}{3}aV^{-5/3} + \frac{2}{3}bV^{-1/3}T^2,$$

where a and b are constants. Let V_0 and T_0 be the initial volume and initial temperature, respectively. If the gas expands adiabatically, the volume of the gas is proportional to   [JEST 2018]

(a) T (b) $T^{3/2}$ (c) $T^{-3/2}$ (d) T^{-2}

Answer: The correct choice is (c).

Solution: According to the problem, we have

$$U(T,V) = aV^{-2/3} + bV^{2/3}T^2 \text{ and } P(T,V) = \frac{2}{3}aV^{-5/3} + \frac{2}{3}bV^{-1/3}T^2.$$

Combining the first and the second laws of thermodynamics, we have

$$TdS = dU + PdV; \qquad dU = -PdV. \qquad (dS = 0)$$

Considering entropy S as a function of T and V and using $dU = -PdV$, we get

$$dU = \left(\frac{\partial U}{\partial T}\right)_V dT + \left(\frac{\partial U}{\partial V}\right)_T dV = -PdV,$$

$$\Rightarrow bV^{2/3}2TdT - \frac{2}{3}aV^{-5/3}dV + \frac{2}{3}bV^{-1/3}T^2dV = -\frac{2}{3}aV^{-\frac{5}{3}}dV - \frac{2}{3}bV^{-1/3}T^2dV;$$

$$\Rightarrow bV^{2/3}2TdT = -\frac{4}{3}bV^{-1/3}T^2dV; \quad \Rightarrow \quad -\frac{3}{2}\ln T = \ln V; \quad \Rightarrow \quad V \propto T^{-3/2}.$$

Hence, $V \propto T^{-3/2}$.

8. Which of the following can be derived by $S = S(T, V)$?

(a) $TdS = C_V dT + T\left(\frac{\partial P}{\partial T}\right)_V dV$ (c) $TdS = C_V dT - T\left(\frac{\partial P}{\partial T}\right)_V dV$

(b) $TdS = C_P dT - T\left(\frac{\partial V}{\partial T}\right)_P dV$ (d) $TdS = C_P dT + T\left(\frac{\partial V}{\partial T}\right)_V dV$

Answer: The correct choice is (a).

Solution: Writing S as a function of T and V, we get

$$dS = \left(\frac{\partial S}{\partial T}\right)_V dT + \left(\frac{\partial S}{\partial V}\right)_T dV.$$

From Maxwell's relation, we have

$$\left(\frac{\partial S}{\partial V}\right)_T = \left(\frac{\partial P}{\partial T}\right)_V \text{ and } T\left(\frac{\partial S}{\partial T}\right)_V = C_V.$$

This gives $TdS = C_V dT + T\left(\frac{\partial P}{\partial T}\right)_V dV$.

9. Which of the following can be derived by $S = S(T, P)$?

(a) $TdS = C_V dT + T(\frac{\partial P}{\partial T})_V dV$ (c) $TdS = C_P dT - T(\frac{\partial V}{\partial T})_P dP$

(b) $TdS = C_V dT - T(\frac{\partial V}{\partial T})_P dP$ (d) $TdS = C_P dT + T(\frac{\partial V}{\partial T})_P dV$

Answer: The correct choice is (c).

Solution: Considering entropy S as a function of pressure P and temperature T, we get

$$dS = \left(\frac{\partial S}{\partial T}\right)_P dT + \left(\frac{\partial S}{\partial P}\right)_T dP.$$

From Maxwell's relation, we have $\left(\frac{\partial S}{\partial P}\right)_T = -\left(\frac{\partial V}{\partial T}\right)_P$ and $T\left(\frac{\partial S}{\partial T}\right)_P = C_P$.

So, $TdS = C_P dT - T\left(\frac{\partial V}{\partial T}\right)_P dP$.

10. For an isolated thermodynamic system, P, V, T, U, S, and F represent the terms as usual. Then the following relation is true

(a) $\left(\frac{\partial F}{\partial T}\right)_V = -S$

(c) $\left(\frac{\partial U}{\partial S}\right)_V = T$

(b) $\left(\frac{\partial F}{\partial T}\right)_P = -S$

(d) $\left(\frac{\partial U}{\partial V}\right)_P = -P$

Answer: The correct choice is (a).

We know that $F = U - TS$; $\Rightarrow dF = dU - TdS - SdT = -PdV - SdT$.

11. Which of the following give Maxwell's relation $\left(\frac{\partial S}{\partial P}\right)_T = -\left(\frac{\partial V}{\partial T}\right)_P$?

(a) $dU = TdS - PdV$

(c) $dF = -SdT - PdV$

(b) $dH = TdS + VdP$

(d) $dG = -SdT + VdP$

Answer: The correct choice is (a).

Solution: We know that $dG = -SdT + VdP$. This gives us

$$\left(\frac{\partial G}{\partial T}\right)_P = -S \text{ and } \left(\frac{\partial G}{\partial P}\right)_T = V;$$

$$\Rightarrow \left(\frac{\partial}{\partial P}\left(\frac{\partial G}{\partial T}\right)_P\right)_T = \left(\frac{\partial}{\partial T}\left(\frac{\partial G}{\partial T}\right)_T\right)_P \Rightarrow \left(\frac{\partial S}{\partial P}\right)_T = -\left(\frac{\partial V}{\partial T}\right)_P.$$

12. Which among the following sets of Maxwell's relation is correct?

(a) $\left(\frac{\partial U}{\partial V}\right)_S = T$ and $\left(\frac{\partial U}{\partial S}\right)_V = P$

(c) $\left(\frac{\partial G}{\partial V}\right)_T = -P$ and $\left(\frac{\partial G}{\partial P}\right)_S = V$

(b) $\left(\frac{\partial H}{\partial P}\right)_S = V$ and $\left(\frac{\partial H}{\partial S}\right)_P = T$

(d) $\left(\frac{\partial A}{\partial S}\right)_T = -P$ and $\left(\frac{\partial A}{\partial P}\right)_V = S$

Answer: The correct choice is (b).

Solution: $dH = TdS + VdP$, $\left(\frac{\partial H}{\partial P}\right)_S = V$, and $\left(\frac{\partial H}{\partial S}\right)_P = T$.

13. Given that $H =$ the enthalpy of the system, $T =$ absolute temperature, $S =$ entropy, $G = H - TS$ is the Gibbs function for the system. In the case of reversible, isothermal and isobaric process:

(a) $G =$ constant

(c) $G < 0$ and changes with S

(b) $G > 0$ and changes with T

(d) G changes with both T and S

Answer: The correct choice is (a).

Solution: Enthalpy H is given by $H = U + PV$. From the first and second laws of thermodynamics, we can write $TdS = dU + PdV$.

Again, Gibbs free energy function G is given by $G = U + PV - TS$. This gives

$$dG = dU + d(PV) - d(TS); \implies dG = dU + VdP + PdV - SdT - TdS.$$

This gives $dG = VdP - SdT$. For an isobaric process, $dP = 0$, and for an isothermal process, $dT = 0$. Hence, we get

$$dG = 0. \text{ So, we have } G = \text{constant}.$$

14. In thermodynamics, the Gibbs function is defined as $G = H - TS$. In an isothermal, isobaric, reversible process, the Gibbs function G

(a) remains constant but not zero.   (c) varies nonlinearly.

(b) varies linearly.   (d) is zero.

Answer: The correct choice is (a).

Solution: Gibbs function G and enthalpy H are, respectively, defined by

$$G = H - TS \text{ and } H = U + PV.$$

Using H, we get the Gibbs function as

$$G = U + PV - TS; \;\Rightarrow dG = dU + d(PV) - d(TS)$$
$$= dU + P\,dV + V\,dP - T\,dS - S\,dT.$$

This gives rise to $dG = V\,dP - S\,dT$. For an isobaric process $P = $ constant, $dP = 0$, and for an isothermal process, $dT = 0$.

So, $dG = 0; \;\Rightarrow\; G = \text{constant}$.

15. A thermodynamic system is maintained at constant temperature and pressure. In thermodynamic equilibrium, its   [GATE 2008]

(a) Gibbs free energy is minimum.

(b) enthalpy is maximum.

(c) Helmholtz free energy is minimum.

(d) internal energy is zero.

Answer: The correct choice is (a).

Solution: A thermodynamic system is maintained at a constant temperature and pressure can be defined by Gibbs free energy $dG = -SdT + VdP \le 0$, that is, Gibbs free energy is minimum.

16. The internal energy E of a system is given by $E = \frac{bS^3}{VN}$, where b is a constant and other symbols have their usual meaning. The temperature of this system is equal to

(a) $\frac{bS^2}{VN}$ (b) $\frac{3bS^2}{VN}$ (c) $\frac{bS^3}{V^2N}$ (d) $(\frac{S}{N})^2$

Answer: The correct choice is (b).

Solution: $TdS = dE + PdV \Rightarrow dE = TdS - PdV \Rightarrow T = (\frac{\partial E}{\partial S})_V \Rightarrow T = \frac{3bS^2}{VN}$.

17. The free energy F of a gas consisting of N particles in a volume V and at a temperature T is $F = NK_BT \ln \left[\frac{a_0 V(K_BT)^{\frac{5}{2}}}{N}\right]$, where a_0 is constant. The internal energy U of the gas is given by

(a) $\frac{3}{2}NK_BT$

(b) $\frac{5}{2}NK_BT$

(c) $NK_BT \ln \left[\frac{a_0 V(K_BT)^{\frac{3}{2}}}{N}\right] - \frac{5}{2}NK_BT$

(d) $NK_BT \ln \left[\frac{a_0 V(K_BT)^{\frac{5}{2}}}{N}\right]$

Answer: The correct choice is (b).

Solution: We know that the free energy F is given by

$$F = U - TS; \quad \Rightarrow \quad U = F + TS.$$

Again, we have $dF = -SdT - PdV$. This leads to $\Rightarrow \left(\frac{\partial F}{\partial T}\right)_V = -S$. Hence, the internal energy U is given by $U = F - T\left(\frac{\partial F}{\partial T}\right)_V$.

Here, the free energy F is given by

$$F = NK_BT \ln \left[\frac{a_0 V(K_BT)^{\frac{5}{2}}}{N}\right] = NK_BT \ln\left(CT^{\frac{5}{2}}\right), \text{ where } C = \frac{a_0 V K_B^{\frac{5}{2}}}{N}.$$

This gives

$$\left(\frac{\partial F}{\partial T}\right)_V = NK_B \ln(CT^{\frac{5}{2}}) + NK_BT \frac{C}{CT^{\frac{5}{2}}}\frac{5}{2}T^{\frac{3}{2}};$$

$$T\left(\frac{\partial F}{\partial T}\right)_V = NK_BT \ln(CT^{\frac{5}{2}}) + \frac{5}{2}NK_BT = F + \frac{5}{2}NK_BT.$$

Using this value, we get the internal energy as

$$U = F - T\left(\frac{\partial F}{\partial T}\right)_V = F - \left(F + \frac{5}{2}NK_BT\right) = \frac{5}{2}NK_BT. \text{ (Taking magnitude.)}$$

18. A thermally insulated ideal gas of volume V_1 and temperature T expands to another enclosure of volume V_2 through a porous plug. What is the change in temperature of the gas? [JEST 2012]

(a) 0   (b) $T \ln \frac{V_1}{V_2}$ (c) $T \ln \frac{V_2}{V_1}$ (d) $T \ln \frac{V_2 - V_1}{V_2}$

Answer: The correct choice is (c).

Solution: $dH = TdS + VdP$. In the porous plug experiment due to Joule–Thompson: $dH = 0$ and $TdS = 0$, since the ideal gas is thermally insulated. Thus,

$$VdP = 0. \quad \text{Since } VdP = 0; \Rightarrow nRdT = pdV; \Rightarrow nRdT = \frac{nRTdV}{V};$$

$$dT = T\frac{dV}{V} \Rightarrow dT = T\int_{V_1}^{V_2} \frac{dV}{V} \Rightarrow dT = T\ln\frac{V_2}{V_1}.$$

19. A certain system is found to have Gibbs free energy given by $G(P,T) = RT\ln\left(\frac{aP}{(RT)^{\frac{5}{2}}}\right)$, where a and R are constants. The specific heat at constant pressure (C_P) is given by

(a) $\frac{3}{2}R$ (c) $\frac{7}{2}R$

(b) $\frac{5}{2}R$ (d) $\frac{9}{2}R$

Answer: The correct choice is (b).

Solution: We know that $S = -\left(\frac{\partial G}{\partial T}\right)_P = \frac{5}{2}R - R\ln\left(\frac{aP}{(RT^{\frac{5}{2}})}\right); \Rightarrow C_P = T\left(\frac{\partial S}{\partial T}\right) = \frac{5}{2}R$.

20. Helmholtz free energy is given by $F = -CT^4$, where C is constant and T is the temperature in Kelvin. Then, which one is the correct relation between specific heat at constant volume C_V and entropy S?

(a) $C_V = 2\,S$ (c) $C_V = 3\,S$

(b) $C_V = 4\,S$ (d) $C_V = \frac{3}{2}\,S$

Answer: The correct choice is (c).

Solution: The entropy of the system is given by

$$S = -\left(\frac{\partial F}{\partial T}\right)_V = -\left[\frac{\partial}{\partial T}\left(-CT^4\right)\right]_V = 4CT^3.$$

The specific heat at constant volume C_V is given by

$$C_V = T\left(\frac{\partial S}{\partial T}\right)_V = T \times \left[\frac{\partial}{\partial T}(4CT^3)\right]_V = T \times 12CT^2 = 3 \times 4CT^3 = 3\,S.$$

21. A thermodynamic system is described by the thermodynamic coordinates P, V, and T. Choose the valid expression(s) for the system.   [JAM 2019]

(a) $\left(\frac{\partial P}{\partial V}\right)_T \left(\frac{\partial V}{\partial T}\right)_P = -\left(\frac{\partial P}{\partial T}\right)_v$

(b) $\left(\frac{\partial P}{\partial V}\right)_T \left(\frac{\partial V}{\partial T}\right)_P = \left(\frac{\partial P}{\partial T}\right)_v$

(c) $\left(\frac{\partial V}{\partial T}\right)_P \left(\frac{\partial T}{\partial P}\right)_v = -\left(\frac{\partial V}{\partial P}\right)_T$

(d) $\left(\frac{\partial V}{\partial T}\right)_P \left(\frac{\partial T}{\partial P}\right)_v = \left(\frac{\partial V}{\partial P}\right)_T$

Answer: The correct choices are (a) and (c).

22. If the C_V is the specific heat of the ideal gas then which of the following is correct for Van der Walls gases for the same degree of freedom?

(a) $dU = C_V dT$

(b) $dU = -\frac{a}{V^2} dV$

(c) $dU = C_V dT - \frac{a}{V^2} dV$

(d) $dU = C_V dT + \frac{a}{V^2} dV$

Answer: The correct choice is (d).

Solution:

$$U = U(T,V) \Rightarrow dU = \left(\frac{\partial U}{\partial T}\right)_V dT + \left(\frac{\partial U}{\partial V}\right)_T dV \Rightarrow dU = C_V dT + \left(\frac{\partial U}{\partial V}\right)_T dV.$$

$$\left(\frac{\partial U}{\partial V}\right)_T = \frac{a}{V^2} \text{ So, } dU = C_V dT + \frac{a}{V^2} dV.$$

23. Consider a one-dimensional gas of N non-interacting particles of mass m with the Hamiltonian for a single particle is given $H = \frac{P^2}{2m} + \frac{1}{2} m \omega^2 (x^2 + 2x)$. The high-temperature specific heat in units of $R = N K_B$ (K_B is the Boltzmann constant) is [GATE 2019]

(a) 1   (b) 1.5   (c) 2   (d) 2.5

Answer: The correct choice is (a).

Solution: $< H >=< \frac{P^2}{2m} > + \frac{1}{2} m \omega^2 < x^2 > + \frac{1}{2} m \omega^2 < 2x >= \frac{N K_B T}{2} + \frac{N K_B T}{2} + U_0.$

$\Rightarrow < H >= N K_B T + U_0; \quad \Rightarrow \quad C_v = \frac{\partial H}{\partial T} = N K_B.$

24. For a Van der Waals gas, the equation of the adiabatic curve in the variables T and V is given by

(a) $T(V - b)^{\frac{R}{C_P}} = \text{constant}$

(b) $T(V - b)^{\frac{R}{C_V}} = \text{constant}$

(c) $T(V - b)^{-\frac{R}{C_P}} = \text{constant}$

(d) $T(V - b)^{-\frac{R}{C_V}} = \text{constant}$

Answer: The correct choice is (b).

Solution: $(P + \frac{a}{V^2})(V - b) = RT$ and $dU = C_V dT + \frac{a}{V^2} dV.$

For adiabatic process: $dQ = 0 = dU + P dV; \quad \Rightarrow \quad -dU = P dV.$

This leads to

$$-C_V dT - \frac{a}{V^2} dV = \left(\frac{RT}{V-b} - \frac{a}{V^2} \right) dV; \quad \Rightarrow -C_V dT = \left(\frac{RT}{V-b} \right) dV;$$

$$\Rightarrow -\int \frac{C_V dT}{RT} = \int \frac{dV}{V-b} \quad \Rightarrow \frac{C_V}{R} \ln Tk = \ln(V-b);$$

$$V-b = (Tk)^{-\frac{C_V}{R}}; \quad \Rightarrow V-b = (T)^{-\frac{C_V}{R}} \times (k)^{-\frac{C_V}{R}}; \quad \Rightarrow (V-b)T^{-\frac{C_V}{R}} = (k)^{-\frac{C_V}{R}}$$

Hence, we have $(V-b)^{\frac{R}{C_V}} T = $ Constant.

25. The Clausius–Clapeyron equation indicates that the increase of pressure increases the melting point:

    (a) in the case of all substances.
    (b) in the case of all substances that expand on solidification.
    (c) in the case of all substances contract on solidification.
    (d) in the case of all substances that neither expand nor contract on solidification.

    Answer: The correct choice is (c).

    Solution: Clausius–Clapeyron equation is given as

    $$\frac{dP}{dT} = \frac{L}{T(V_2 - V_1)} \quad \Rightarrow \text{ if } V_2 > V_1, \frac{dP}{dT},$$

    is positive. If volume V increases, the pressure P increases with temperature T.

26. If there is a 10% decrease in the atmospheric pressure at a hill compared to the pressure at sea level, then the change in the boiling point of water is ...° C.

    (Take latent heat of vaporization of water as 2270 kJ/kg and the change in specific volume at the boiling point to be 1.2 m³/kg.)

    Answer: 2

    Solution: $\frac{dP}{dT} = \frac{1}{T}\frac{L}{V_2 - V_1} \rightarrow dT = dP \times T \times \frac{V_2 - V_1}{L} = \frac{0.1 \times 1.01 \times 10^5 \times 373 \times 1.2}{2270 \times 10^3} = 0.02 \times 10^2 = 2°C.$

27. Ice of density ρ_1 melts at pressure P and absolute temperature T to form water of density ρ_2. The latent heat of melting of 1 gram of ice is L. What is the change in the internal energy ΔU resulting from the melting of 1 gram of ice?

    [JEST 2014]

    (a) $L + P(\frac{1}{\rho_2} - \frac{1}{\rho_1})$ (c) $L - P(\frac{1}{\rho_1} - \frac{1}{\rho_2})$

    (b) $L - P(\frac{1}{\rho_2} - \frac{1}{\rho_1})$ (d) $L + P(\frac{1}{\rho_1} - \frac{1}{\rho_2})$

Answer: The correct choice is (d).

Solution: From the first law of thermodynamics, we have $dU = \delta Q - dW = \delta Q - PdV$. This gives rise to

$$dU = mL - PdV \quad \Rightarrow \quad dU = L - P\int_{\rho_1}^{\rho_2}\left(-\frac{1}{\rho^2}\right)d\rho = L + P\left[\frac{1}{\rho_1} - \frac{1}{\rho_2}\right].$$

Hence, $V = \frac{1}{\rho} \Rightarrow dV = (-\frac{1}{\rho^2})d\rho$.

28. Clausius–Clapeyron equation is

(a) $\frac{dP}{dT} = \frac{\Delta S}{\Delta V}$ (b) $\frac{dP}{dT} = \frac{1}{\Delta S \Delta V}$ (c) $\frac{dP}{dT} = \frac{T\Delta S}{\Delta V}$ (d) $\frac{dP}{dT} = \frac{\Delta V}{T\Delta S}$

when ΔS = change in entropy during the change of state and dV is change in volume in two phases.

Answer: The correct choice is (a).

29. If the ratio of isothermal and adiabatic elasticities is $\frac{E_S}{E_T}$, then which of the following is true:

(a) $\frac{E_S}{E_T} = \frac{C_P}{C_V}$

(b) $\frac{E_S}{E_T} = \frac{C_V}{C_P}$

(c) $\frac{E_S}{E_T} = C_P.C_V$

(d) $\frac{E_S}{E_T} = \sqrt{\left(\frac{C_P}{C_V}\right)}$

Answer: The correct choice is (a).

30. The relation between C_P and C_V for a real Van der Waals gas is

(a) $C_P - C_V = R$

(b) $C_P - C_V = R^{-1}$

(c) $C_P - C_V = R(1 + \frac{2a}{RTV})$

(d) $C_P - C_V = R(1 + \frac{2a}{RbV})$

Answer: The correct choice is (c).

31. In an experiment, a certain quantity of an ideal gas at temperature T_0, pressure P_0, and volume V_0 is heated by a current flowing through a wire for a duration of t seconds. The volume is kept constant, and the pressure changes to P_1. If the experiment is performed at constant pressure starting with the same initial conditions, the volume changes from V_0 to V_1. The ratio of the specific heats at constant pressure and constant volume is   [JEST 2018]

(a) $\frac{P_1-P_0}{V_1-V_0}\frac{V_0}{P_0}$ (b) $\frac{P_1-P_0}{V_1-V_0}\frac{V_1}{P_1}$ (c) $\frac{P_1 V_1}{P_0 V_0}$ (d) $\frac{P_0 V_0}{P_1 V_1}$

Answer: The correct choice is (a).

Solution: (I) Heating at constant volume provides

$$\frac{P_0}{T_0} = \frac{P_1}{T_1} \Rightarrow T_1 = \frac{P_1}{P_0}T_0; \quad \Rightarrow \quad Q = C_V(T_1 - T_0) = C_V\left(\frac{P_1}{P_0} - 1\right)T_0.$$

(II) Heating at constant pressure provides

$$\frac{V_0}{T_0} = \frac{V_1}{T_1} \Rightarrow T_1' = \frac{V_1}{V_0}T_0; \quad \Rightarrow \quad Q' = C_p(T_1' - T_0) = C_pT_0\left(\frac{V_1}{V_0} - 1\right).$$

Using the ideal gas equation, we get $PdV + VdP = RdT$. At constant pressure, this becomes $PdV = RdT$. Hence,

$$dT_p = \frac{P}{R}dV = \frac{P_0}{R} \times (V_1 - V_0); \quad \text{and} \quad dT_v = \frac{V}{R}dP = \frac{V_0}{R}(V_1 - V_0).$$

$$C_v \times \frac{V_0}{R}(P_1 - P_0) = C_p \times \frac{P_0}{R}(V_1 - V_0); \quad \Rightarrow \quad \frac{C_P}{C_V} = \frac{V_0(P_1 - P_0)}{P_0(V_1 - V_0)} = \left(\frac{P_1 - P_0}{V_1 - V_0}\right) \times \frac{V_0}{P_0}.$$

32. A frictionless heat-conducting piston of negligible mass and heat capacity divides a vertical, insulated cylinder of height $2H$ and cross-sectional area A into two halves. Each half contains one mole of an ideal gas at temperature T_0 and pressure P_0 corresponding to STP. The heat capacity ratio $\gamma = C_P/C_V$ is given. A load of weight w is tied to the piston and suddenly released. After the system comes to equilibrium, the piston is at rest, and the temperatures of the gases in the two compartments are equal. What is the final displacement y of the piston from its initial position, assuming $yW \gg T_0C_V$?   [JEST 2018]

(a) $\frac{2H}{\sqrt{\gamma}}$ (c) $\frac{H}{\sqrt{\gamma}}$

(b) $H\gamma$ (d) $\frac{2H}{\gamma}$

Answer: The correct choice is (c).

Solution: $\frac{P_0V_0}{T_0} = \frac{P_2V_2}{T_2}; \Rightarrow \frac{P_0A \times H}{T_0} = \frac{P_2(A(H-y))}{T_2}; \Rightarrow P_2 = \frac{T_2 \times P_0H}{T_0(H-y)}$ ...(i)

$\Rightarrow \frac{P_0A \times H}{T_0} = P_1\frac{A(H+y)}{T_2} \Rightarrow P_1 = \frac{T_2}{T_0} \times \frac{P_0H}{(H+y)}$ ...(ii)

Total change in internal energy of the system = net energy input = wy. Hence, $2nC_V(T_2 - T_0) = wy$. As $wy \gg C_VT_0$ and $n = 1$ mol, we have

$$T_2 = \frac{wy}{2C_v}...(iii) \text{ and } C_v = \frac{R}{\gamma - 1}.$$

Also, at equilibrium, we have $P_2 - P_1 = \frac{w}{A}$. Using the value of T_2 in (i) and (iii), and substituting (ii) from (i), we get

$$\frac{wy}{2C_V}\frac{P_0H}{T_0(H-y)} - \frac{wy}{2C_V}\frac{P_0H}{T_0(H+y)} = \frac{w}{A} \Rightarrow \frac{HP_0y}{2C_VT_0}\left(\frac{1}{H-y} - \frac{1}{H+y}\right) = \frac{1}{A};$$

$$\Rightarrow \frac{A \times H \times P_0 y}{2 \times \frac{R}{\gamma - 1} T_0} \left[\frac{H + y - H + y}{H^2 - y^2} \right] = 1 \quad \Rightarrow \quad [\text{use } AH = V_0];$$

$$\Rightarrow \frac{P_0 V_0}{T_0} \times \frac{y}{2R} \times \frac{2y \times \gamma^{-1}}{[H^2 - y^2]} = 1; \quad \Rightarrow \quad \frac{R \times y^2 \times \gamma^{-1}}{R(H^2 - y^2)} = 1;$$

$$\Rightarrow y^2 \gamma - y^2 = H^2 - y^2; \Rightarrow; \quad \Rightarrow \quad y = \sqrt{\frac{H}{\gamma}}.$$

33. An aluminum plate of mass 1 kg at 95°C is immersed in 0.5 l of water at 20°C kept inside an insulating container, and is then removed. If the temperature of water is found to be 23°C, then the temperature of the aluminum plate is ...° C (the specific heat of water and aluminum are 4200 J/kg − K and 900 J/kg-K, respectively; the density of water is 1000 kg/m^3) [JAM 2016]

Answer: 94.36

Solution: $-M_a S_a (T_{a_f} - T_{a_i}) = M_w\, S_w (T_{w_f} - T_{w_i}) \quad \Rightarrow -0.1 \times 4200(T_{a_f} - 368) = 0.5 \times 900(296 - 293);$

$\Rightarrow -2100(T_{a_f} - 368) = 450 \times 3 \Rightarrow (T_{a_f} - 368) = \frac{450}{700} = -0.64;$

$\Rightarrow = 368 - 0.64 = 367.36 = 367.36 - 273 = 94.36.$

34. A rigid triangular molecule consists of three noncolinear atoms joined by rigid rods. The constant pressure molar specific heat (C_p) of an ideal gas consisting of such molecules is [JAM 2015]

(a) 6R   (b) 5R   (c) 4R   (d) 3R

Answer: The correct choice is (c).

Solution: The degrees of freedom of the rigid triangular molecule is $= 6$.

The internal energy is given by $U = \frac{6RT}{2}$. So, the specific heat at a constant volume is $\Rightarrow C_V = (\frac{\partial U}{\partial T})_V = 3R \Rightarrow C_P = C_V + R = 4R.$

35. Joule–Thomson effect depends upon

(a) initial temperature only   (c) both (a) and (b)

(b) pressure difference only   (d) neither (a) nor (b)

Answer: The correct choice is (c).

36. A solid melts into a liquid via first-order phase transition. The relationship between the pressure P and the temperature T of the phase transition is $P = -2T + P_0$, where P_0 is a constant. The entropy change associated with the phase transition is 1.0 Jmole$^{-1}K^{-1}$. The Clausius–Clapeyron equation for the latent heat is $L = T(\frac{dP}{dT})\Delta v$. Here, $\Delta v = v_{\text{liquid}} - v_{\text{solid}}$ is a change in molar volume at the phase transition. The correct statement relating the values of the volume is   [JAM 2006]

(a) $v_{\text{liquid}} = v_{\text{solid}}$

(b) $v_{\text{liquid}} = v_{\text{solid}} - \frac{1}{2}$

(c) $v_{\text{liquid}} = v_{\text{solid}} - 1$

(d) $v_{\text{liquid}} = v_{\text{solid}} + 2$

Answer: The correct choice is (b).

Solution: Since $P = -2T + P_0 \Rightarrow \frac{dP}{dT} = -2$.

It is given $L = T(\frac{dP}{dT})\Delta v \Rightarrow L = -2T\Delta v \Rightarrow \frac{dL}{dT} = -2\Delta v$.

Since $dS = 1.0$ Jmole$^{-1}K^{-1}$, $dS = \frac{dQ}{T} = \frac{mdL}{dT} = 1 \Rightarrow 1 = -2\Delta v \Rightarrow \Delta v = -\frac{1}{2}$.

37. Joule–Thomson coefficient is

(a) $(\frac{\partial T}{\partial P})_H = \frac{1}{C_P}(\frac{2a}{RT} - b)$

(b) $(\frac{\partial T}{\partial P})_H = \frac{1}{C_P}(\frac{2b}{RT} - a)$

(c) $(\frac{\partial T}{\partial P})_H = C_P(\frac{2a}{Rb} - T)$

(d) $(\frac{\partial T}{\partial P})_H = \frac{1}{C_V}(\frac{2a}{Rb} - T)$

Answer: The correct choice is (a).

38. Across a first-order phase transition, the free energy is   [GATE 2013]

(a) proportional to the temperature

(b) a discontinuous function of temperature

(c) a continuous function of the temperature but its first derivative is discontinuous

(d) such that the first derivative with respect to temperature is continuous

Answer: The correct choice is (c).

39. If r is the number of existing phases, f the degrees of freedom, and n the number of components in a system, then the Gibbs phase rule is

(a) $r + f = n$

(b) $r + f = n + 1$

(c) $r + f = n + 2$

(d) $r + f = n - 1$

Answer: The correct choice is (c).

10.7 Exercise

10.7.1 Short answer type questions

1. What do you mean by a thermodynamic relation?

2. What do you mean by thermodynamic potential?

3. Define internal energy, enthalpy, Helmholtz free energy, and Gibbs free energy.

4. State favorable conditions for various thermodynamic potentials.

5. Write down Maxwell's four thermodynamic relations.

6. What do you mean by first and second-order phase transitions?

7. What is a phase diagram?

8. Write down the first T–dS and first energy equations for a paramagnetic substance.

9. What do you mean by first and second-order phase transition.

10. Write down any two important differences between the first and the second-order phase transition. [Burdwan University 2021]

11. Draw the pressure–temperature diagram of H_2O indicating the phases, boundaries, and the triple point. [Calcutta University 2021, Delhi University 2021]

12. What is an adiabatic system?

13. State Gibbs phase rule.

14. What is meant by adiabatic demagnetization?

15. What is magneto-caloric effect?

16. What is Joule–Thomson coefficient?

17. Show that enthalpy remains constant during Joule–Thomson experiment. [Delhi University 2021]

18. What do you mean by inversion line and inversion point in case of Joule–Thomson effect? [Burdwan University 2021]

10.7.2 Long answer type questions

1. Explain the concept of thermodynamic potentials.

2. Write and explain the four thermodynamic potentials. [St. Joseph's College (Autonomous), Bengaluru 2020]

3. Using thermodynamic potential, derive Maxwell's thermodynamic relations.

4. Using the Clausius–Clapeyron equation, explain the rise/fall of the boiling point of a liquid with increase/decrease of pressure and also the rise/fall of melting point with increase of pressure corresponding to increase/decrease of volume upon phase transition to liquid state.

5. Using the Jacobian concept on thermodynamic variables, prove Maxwell's four thermodynamic relations.

6. Define four thermodynamic potentials. Using these potentials, derive the four Maxwell's thermodynamic relations. [Delhi University 2021]

7. Obtain any two Maxwell's thermodynamic relations from thermodynamic potentials and give their significance.
[St. Joseph's College (Autonomous), Bengaluru 2019]

8. Prove that the change in enthalpy in an isobaric process is equal to the heat transferred.

9. Prove that the change (decrease) in the Helmholtz function in an isothermal process is equal to the work done by the system.

10. Establish that for a system of constant volume and in contact with a heat reservoir, the equilibrium corresponds to the minimum of Helmholtz free energy.

11. Prove that for a reversible isothermal and isobaric process, Gibbs function remains constant.

12. Starting from the second law of thermodynamics, show that for a mechanically isolated system, the Helmholtz free energy never increases.
[Calcutta University 2022]

13. Derive the energy equation [Delhi University 2021, Calcutta University 2021]

$$\left(\frac{\partial U}{\partial V}\right)_T = T\left(\frac{\partial P}{\partial T}\right)_V - P.$$

14. Using the fact that dS is an exact differential, derive the following relation:

$$\left(\frac{\partial U}{\partial V}\right)_T = T\left(\frac{\partial P}{\partial T}\right)_V - P.$$

Hence, show that for a Van der Waals gas, the internal energy is not a function of temperature alone. [Calcutta University 2020]

15. Prove that [Delhi University 2021]

$$G = H + T \left(\frac{\partial G}{\partial T} \right)_P \quad \text{and} \quad F = U + T \left(\frac{\partial F}{\partial T} \right)_V.$$

16. What is free expansion? Show that for an ideal gas, internal energy depends only on temperature and is independent of pressure and volume.
[Burdwan University 2021]

17. Describe the Joule–Thomson porous plug experiment.

18. Calculate the change in temperature in the porous plug experiment.

19. Discuss the behavior of the Joule–Thomson coefficient below, above, and at the inversion temperature.

20. Discuss the Joule–Thomson effect for an ideal gas.

21. Discuss the Joule–Thomson effect for a real gas obeying Van der Waals equation of state.

22. Discuss the cases when "Joule–Thomson coefficient" is negative, positive, and zero. Obtain the expression for a temperature of inversion of the gas. Explain why hydrogen and helium show a heating effect at ordinary temperatures while other gases show a cooling effect. [Delhi University 2021]

23. Differentiate between the cooling process through adiabatic expansion and the Joule–Thomson effect. [ST. Joseph's College (Autonomous), Bengaluru 2019, Burdwan University 2020, Calcutta University 2022]

24. State the principle of production of low-temperature using adiabatic demagnetization. [Burdwan University 2020]

25. Prove that [Burdwan University 2021]

    (a) Heat is generated under compression for a substance that expands on heating, and

    (b) Cooling takes place for a substance that contracts on heating.

26. What is adiabatic demagnetization? Explain how it could be utilized to achieve ultra-low temperatures?

27. Discuss the process of adiabatic cooling with the help of the entropy—temperature behavior of paramagnetic specimen.

28. What is phase transition? Discuss the classification of phase transition with examples.

29. Show that, during first order phase transition, the entropy of the entire system is a linear function of the total volume. [Burdwan University 2020]

30. With the help of the necessary diagram, distinguish between first and second-order phase transitions. Derive Clausius–Clapeyron equation of latent heat.
   [Delhi University 2021]

31. Derive Ehrenfest's equations for second-order phase transition.
   [Delhi University 2021]

32. Write down the characteristics of second-order phase transition with a suitable example.   [Calcutta University 2022]

33. Represent the behavior of specific Gibbs function, entropy, volume, and heat capacity with temperature by drawing plots for first and second-order phase transitions.

34. Derive the Clausius–Clapeyron equation from T–dS equation.
   [Calcutta University 2022]

35. What is the significance of the Clausius–Clapeyron equation relating to first-order phase transition?

36. Why is an isentropic process not necessarily an adiabatic process?
   [Burdwan University 2020]

10.7.3   Numerical problems

1. Show that $C_P = T \left(\frac{\partial V}{\partial T}\right)_P \left(\frac{\partial P}{\partial T}\right)_S$ and $C_V = -T \left(\frac{\partial P}{\partial T}\right)_V \left(\frac{\partial V}{\partial T}\right)_S$.
   [Burdwan University 2021]

2. Prove that

   (a) $\left(\frac{\partial T}{\partial P}\right)_V \left(\frac{\partial S}{\partial V}\right)_P - \left(\frac{\partial T}{\partial V}\right)_P \left(\frac{\partial S}{\partial P}\right)_V = 1.$

   (b) $\left(\frac{\partial P}{\partial T}\right)_S \left(\frac{\partial V}{\partial S}\right)_P - \left(\frac{\partial P}{\partial S}\right)_T \left(\frac{\partial V}{\partial T}\right)_S = 1.$

3. Using thermodynamic potentials deduce the Gibbs–Helmoholtz relation

$$U = -T^2 \left[\frac{d}{dT}\left(\frac{F}{T}\right)\right]_V = F - T \left(\frac{\partial F}{\partial T}\right)_V.$$

4. Show that enthalpy and Gibbs potential are related by

$$H = -T^2 \left[\frac{d}{dT}\left(\frac{G}{T}\right)\right]_P = G - T \left(\frac{\partial G}{\partial T}\right)_P.$$

5. Prove that Gibbs potential and Helmholtz free energy are related by

$$G = -V^2 \left[\frac{d}{dT}\left(\frac{F}{V}\right)\right]_T = F - V \left(\frac{\partial F}{\partial V}\right)_T.$$

6. Prove that

$$C_P - C_V = -T \left(\frac{\partial P}{\partial T} \right)_V \left[\left(\frac{\partial V}{\partial T} \right)_P \right]^2 .$$

7. Show that for a real Van der Waals gas:

$$C_P - C_V = R \left[1 + \frac{2a}{RTV^3} (v - b)^2 \right] . \qquad \text{[Calcutta University 2021]}$$

8. Show that for an isentropic transformation:

$$\left(\frac{\partial V}{\partial T} \right)_S = - \frac{C_V}{C_P - C_V} \left(\frac{\partial V}{\partial T} \right)_P . \qquad \text{[Calcutta University 2021]}$$

9. Prove that

$$P = T \left(\frac{\partial S}{\partial V} \right)_T - T \left(\frac{\partial U}{\partial V} \right)_T .$$

10. Show that

(a) $dH - dU = dG - dF = d(PV)$ and

(b) $dU - dF = dH - dG = d(TS)$.

11. Derive the Euler relation: $U = TS - PV + \sum_i \mu_i N_i$.

And hence, the relations: $H = TS + \sum_i \mu_i N_i$; $\quad F = -PV + \sum_i \mu_i N_i$; $\quad G = \sum_i \mu_i N_i$.

And also the Gibbs—Duhem relation: $0 = SdT - VdP + \sum_i N_i d\mu_i$.

12. A gas obeys the relation $P(V - b) = RT$, where b is a constant. The specific heat at constant volume C_V of the gas is constant. Show that

(a) The internal energy U of the gas is a function of temperature T only.

(b) During the adiabatic process, the gas follows the relation $P(V - b)^\gamma = $ constant, where γ is a constant.

13. One mole of an ideal gas expands from volume V_0 to $2V_0$ under Joule expansion. What is the change of entropy of the gas, the surroundings, and the universe during this Joule expansion? [Change of entropy of the gas $= $ NA KB ln (2); Change of entropy of the surroundings $= -$ NA KB ln (2); and the change of entropy of the universe $= 0$] [Burdwan University 2021]

14. An iron cube at a temperature of 400°C is dropped into an insulated bath containing 10 kg water at 25°C. The water finally reaches a temperature of 50°C at a steady state. Given that the specific heat of water is equal to 4186 $\text{Jkg}^{-1}\text{K}^{-1}$. Find the entropy changes for the iron cube and the water. Is the process reversible? If so, why?

$$[1177.24 \,\text{JK}^{-1}. \text{ As } \Delta S > 0, \text{ the process is irreversible}]$$

15. Instead of using the concept of thermodynamic potentials, use the idea of a Carnot cycle to derive equation (10.26) for the surface film having the following steps:

    (a) isothermal stretching at temperature T and surface tension σ,

    (b) adiabatic expansion to cool to temperature T–dT with rise of surface tension to $\sigma + d\sigma$,

    (c) isothermal contraction at T–dT, and

    (d) adiabatic contraction to regain initial temperature T and surface tension σ.

16. Show that the difference between specific heats at constant magnetic field and at constant magnetization is given by

$$C_B - C_M = \frac{\chi V B^2}{T}.$$

Hint: Use the analogous formula for this difference from the formula:

$$C_P - C_V = -T \left(\frac{\partial P}{\partial T}\right)_V \left[\left(\frac{\partial V}{\partial T}\right)_P\right]^2 \text{ to get } C_B - C_M = -T \left(\frac{\partial B}{\partial T}\right)_M \left(\frac{\partial M}{\partial T}\right)_B.$$

17. Using the first energy relation: $\left(\frac{\partial U}{\partial V}\right)_T = T \left(\frac{\partial P}{\partial T}\right)_V - P$, show that the internal energy of a Van der Waals gas is given by $U = \int C_V dT - \frac{a}{V^2} + U_0$, where U_0 is a constant energy.

18. Consider N spin-$\frac{1}{2}$ spins in a magnetic field B. The spin system is in thermal contact with an ideal gas of N particles in a volume V. Initially, the two systems have a temperature T. Assume $g\mu_B B \gg K_B T$. If we slowly reduce the magnetic field to zero, what becomes the temperature of the gas?

19. Calculate the change in temperature of boiling water when the pressure is increased by 27.12 mm of Hg. The normal boiling point of water at atmospheric pressure is 100°C.

    Given: Latent heat of steam $= 537$ cal/gm and, a specific volume of steam $= 1674$ cm^3.   [1°C or 1 K]

20. Calculate the pressure and temperature of the triple-point of water. Given that the lowering of melting point of ice per atmosphere increase of pressure is 0.0072°C and the saturated vapor pressure at 0°C is 4.60 mm of Hg, while at 1°C, it is 4.94 mm of Hg.　　　　$[p = 4.6024$ mm of Hg, and $t = 0.00715$°C$]$

21. Calculate the specific heat of saturated steam at 100°C from the following data: the latent heat L at 90°C $= 545.25$ cal, L at 100°C $= 539.30$ cal, L at 110°C $= 533.17$ cal, and the specific heat of water at 100°C $= 1.013$ cal/g.
$$[-1.036 \text{ cal/g}]$$

22. Calculate the decrease in the melting point of ice when the pressure changes by 1 atmosphere, a specific volume of ice at 273 K is 1.091×10^{-3} m^3 kg^{-1} and that of water at 273 K is 10^{-3} m^3 kg^{-1}. The latent heat of ice is $L = 3.36 \times 10^5$ J kg^{-1} and 1 atmosphere $= 10^5$ N m^{-2}.
$$[-7.46 \times 10^{-3} \text{ K}] \text{ [BMS College Bengaluru 2017]}$$

23. What is the temperature change when one Joule expands 1 mole of Helium enclosed in a $(0.1 \text{ m})^3$ container into a $(0.2 \text{ m})^3$ container? $[\Delta T = -0.243 \text{ K}]$

Use the Van der Waals equation of state with $a = 0.00346$ J m^3 mol^2 and the molar heat capacity of Helium, $C_V = 12.48$ J K^{-1} mol^{-1}.

CHAPTER 11

Classical and Quantum Theory of Black Body Radiation

My futile attempts to fit the elementary quantum of action somehow into the classical theory continued for a number of years and they cost me a great deal of effort. Many of my colleagues saw in this something bordering on tragedy. But I feel differently about it. For the thorough enlightenment I thus received was all the more valuable. I now knew for a fact that the elementary quantum of action played a far more significant part in physics than I had originally been inclined to suspect and this recognition made me see clearly the need for the introduction of totally new methods of analysis and reasoning in the treatment of atomic problems.

—Max Planck

Learning Outcomes

After reading this chapter, the reader will be able to

- Understand the distribution of energy density in a black body radiation as a function of wavelength and temperature

- Derive classical laws of black body radiation such as Wien distribution law and Rayleigh–Jeans law

- Get an idea about the development of quantum theory of radiation

- Understand Planck's quanta postulates and explain the black body radiation spectrum

- Derive Planck's law of black body radiation

- Verify Planck's law of black body radiation experimentally

- Derive Wien distribution law and Rayleigh–Jeans law and explain ultraviolet catastrophe from Planck's law of black body radiation

- Determine the temperature of cosmic microwave background radiation using Planck's law of black body radiation

- Solve numerical problems and multiple choice questions on black body radiation

11.1 Introduction

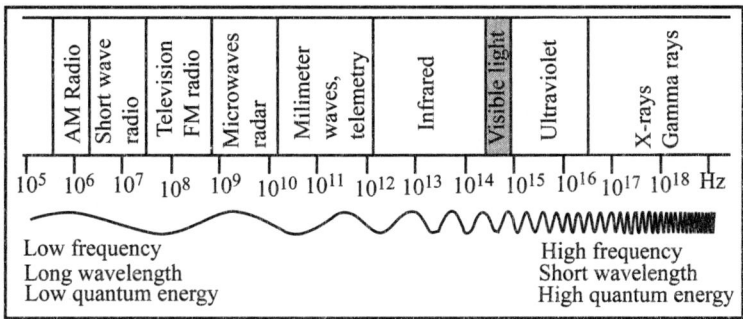

Figure 11.1 The whole electromagnetic spectrum. Thermal radiation ranges in frequency from the shortest infrared rays through the visible-light spectrum to the longest ultraviolet rays.

Radiation emitted from the surface of a heated source is known as **thermal radiation**. In this process, thermal energy is spread out in all directions in the form of electromagnetic radiation and travels directly to its point of absorption at the speed of light. It does not require an intervening medium for its propagation. The wavelength of thermal radiation ranges from the longest infrared rays through the visible-light spectrum to the shortest ultraviolet rays. Such electromagnetic spectrum is shown in Figure 11.1 as a function of frequency. The distribution of radiant energy with their corresponding intensities within various ranges of wavelengths is governed by the temperature of the emitting surface. Further, it is observed that the total radiant heat energy emitted from a surface is proportional to the fourth power of its absolute temperature. This is referred to as **Stefan's law**.

Each and every material object radiates (or absorbs) thermal radiation. The rate at which this radiation (or absorption) occurs depends upon the nature of the surface of the material object. Objects that are good emitters are also good absorbers. This is referred to as **Kirchhoff's radiation law**. For example, a blackened surface is an excellent emitter as well as an excellent absorber. If the same surface is silvered, its emissivity as well as absorptive power becomes poor. A black body is one that absorbs all the radiant energy that falls on it. Such a perfect absorber is also a perfect emitter.

Energy is also transferred by the process of radiation. For example, the transfer of energy from the Sun to the earth takes place by this process. Another example of this process is the heating of a room with an open-hearth fireplace. The flames, coals, and hot bricks radiate heat directly to the objects in the room with a small portion of this heat being absorbed by the intervening air. Most of the air drawn from the room and heated in the fireplace does not reenter the room by the process of convection but is carried up the chimney together with the products of combustion. As a rule of thumb, approximately 35 percent of the total energy created due to airburst is emitted as thermal radiation and the emitted light and heat in this process are capable of causing skin burns, eye injuries, and starting fires of combustible material at considerable distances.

Radiation has wave-like behavior. It was suggested in 1900 that radiation might also have particle-like properties. This idea on heat radiation came in apparently as an innocuous work. From everyday experience, we know that it is the radiation that burns the toast in an oven, that warms our hands and body in front of a fire, and that provides the intense glare of a furnace. It is experimentally found by physicists that the distribution of energy in each of the different frequencies (i.e. colors) varies with the temperature of the radiation. When a body is gradually heated from low to high temperature, it emits radiation from red to orange to white heat, and the frequencies with the greatest energy of this radiation change correspondingly.

The characteristic features of electromagnetic radiation and various properties of thermal radiation are discussed in detail in this chapter. The spectrum of black body radiation and its pure temperature dependence are elaborated for completeness of the description of radiation. The development of classical electromagnetic theory in explaining the features of black body radiation is mentioned with its limitations. Various laws proposed by Kirchhoff, Stefan–Boltzmann, Wien, and Rayleigh–Jeans are thoroughly discussed in the context of black body radiation. The drawbacks of classical theory of black body radiation are mentioned with reference to the ultraviolet catastrophe. The concept of radiation pressure and Saha's ionization formula are presented in detail in this chapter.

11.2 Spectral distribution of black body radiation

A black body completely absorbs all radiations incident on it and emits radiations in all directions and at all wavelengths with the maximum possible monochromatic intensity. The intensity of such emitted radiation varies continuously as a function of wavelength. The characteristic features of the variation of intensity is that at a specified temperature, the intensity of such black body radiation increases with the increase in wavelength, reaches to a maximum at a given wavelength, and then decreases steadily with the increase in wavelength. Further, it is observed that at a given wavelength, the amount of emitted radiation increases with an increase in temperature. The intensity curves shift to the left toward the shorter-wavelength region with the increase in temperature. This indicates that a larger fraction of the intensity of radiation is emitted at shorter wavelengths at higher temperatures.

The definition of a black body, terms related to black body radiation, the nature of intensity, and its variation as a function of temperature and wavelength of the emitted radiation are thoroughly described in the following sections.

11.2.1 What is a black body?

In general, radiation that is incident on an object is partially absorbed and partially reflected. In thermodynamic equilibrium, the rate at which an object absorbs radiation is the same as the rate at which it emits. Therefore, a good absorber of radiation (any object that absorbs radiation) is also a good emitter. A perfect absorber absorbs all electromagnetic radiation incidents on it. Such an object is called a **black body**.

Further, the definition of a black body may be viewed in the following way: Anybody in thermal equilibrium with the surroundings

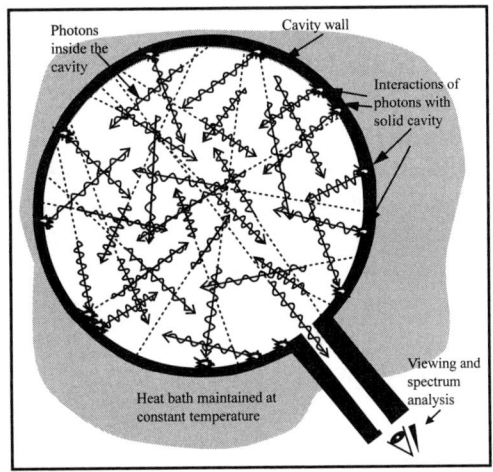

Figure 11.2 The black body is a cavity filled with photons bouncing back and forth.

satisfying Planck's radiation formula is known as a **black body**. However, it should be mentioned that a perfect black body does not exist in nature. It is a highly idealized theoretical object that is described as something "absorbing all incident radiation". It is commonly pictured as a cavity or empty bottle/box in which waves/photons are bouncing back and forth between walls at a certain temperature of the cavity. The cavity enclosing the black body radiation has a little peephole through which escaping radiation can be observed and can be experimentally tested. A black body is supposed to contain all essential aspects of the radiation from a real body, such as the visible glow from a lump of iron at a temperature 1000°C, the Sun at a temperature 6000°C, or the invisible infrared faint glow of a human body at a temperature 27°C. But one may ask the question: Why is a lump of iron, the Sun, or a human body thought of as an empty bottle with a peephole (which is supposed to be an ideal example of a black body)?

Thus, it is recognized that a good physical approximation to a black body is a tiny hole made in a cavity with no other openings. Such a representation is schematically shown in Figure 11.2. In such a cavity, any ray of radiation entering through the hole will be reflected by the walls of the cavity and will be eventually absorbed by them. On the other hand, due to the thermal chaotic motion of the molecules within the walls, some radiation is emitted in addition to the absorption of some part of it.

After a "sufficiently long time", a state of equilibrium between the walls of the cavity and the radiation inside the cavity is reached. In this state of equilibrium, a detailed balance between the amount of emitted radiation of a given wavelength, polarization, and direction equals, on the average, the amount of absorbed radiation having the same properties. This indicates that a thermodynamic equilibrium exists between the walls of the cavity and the radiation enclosed within the cavity. Hence, radiation from such a cavity with a tiny hole, in practice, can be approximated very closely as the radiation of a black body. For example, a building with windows can be considered as a familiar representative of a black body radiation cavity with a hole. It is observed that the windows appear darker than the outside walls. This is because most of the light that gets into a window never reflects back to the observer. Regardless of the wavelength, the radiation is almost completely absorbed. Further, this description is independent of the nature of the material of the inner walls of the cavity. Hence, it implies that no matter whatever be the material, a tiny hole in a cavity will radiate approximately as **an ideal black body**.

It should be mentioned that it is next to impossible to construct a **perfect absorber** in the laboratory. But in 1859, Kirchhoff devised a very beautiful method to implement a black body in reality. He made a small hole in the side of a large box and it served as an excellent absorber since any radiation that goes through the hole bounces back and forth inside the box, and a major portion of it gets absorbed on each bounce (see Figure 11.2). A very small portion of the incident radiation has little chance of ever getting out of it again. The reverse arrangement of this fact can be utilized to have a good emitter of black body radiation. One can take an oven with a tiny hole in one side, and the radiation coming out of the hole is presumably as good as a representation of a **perfect emitter**.

Some terms related to spectral distribution of black body radiation

The emitted radiation from a black body is quantitatively characterized by the **total energy density** u at any point that denotes the total radiant energy for all wavelengths ranging from 0 to ∞ per unit volume around that point. Its unit is J m^{-3}. The **spectral energy density** u_λ for the wavelength λ is a measure of the radiated energy per unit wavelength per unit volume. Hence, the quantity $u_\lambda d\lambda$ denotes the energy per unit volume in the wavelength range between λ and $\lambda + d\lambda$. This spectral energy density u_λ and the total energy density u are related through the expression:

$$u = \int_0^\infty u_\lambda d\lambda.$$

A heated black body radiates energy that depends on the wavelength as well as temperature. The amount of radiant energy from a body at a given wavelength λ per second per unit surface area per unit range of wavelength defines the total emission power E_λ of that body for the wavelength λ. Therefore, $E_\lambda d\lambda$ denotes radiant energy per unit area per second in the wavelength range between λ and $\lambda + d\lambda$. The total emission power $E(T)$ as a function of temperature T can be obtained by integrating the parameter $E_\lambda d\lambda$ over all wavelength range from 0 to ∞, that is, $E(T)$ is given by

$$E(T) = \int\limits_0^\infty E_\lambda d\lambda.$$

Emissivity e of a given surface is defined as the ratio of the radiation emitted by the surface of a given body at a given temperature to the radiation emitted from the surface of a black body at the same temperature. Hence, emissivity e is given by

$$e = \frac{E(T)}{E_B(T)},$$

where $E_B(T)$ is the emissivity for a black body at a given temperature T. The value of e lies in the range $0 \le e \le 1$. The emissivity of a real surface is a function of temperature T, wavelength λ, and the direction of radiation.

Sometimes, the term **radiance** is used to express the distribution of intensity of the emitted radiation from a hot body. It is defined as the power radiated per unit area per unit wavelength interval at a given wavelength λ and a given temperature T. It is expressed in the unit of watt per square meter per meter.

Characteristic features of a black body

A black body possesses various characteristic features, some of which are stated below:

1. It can absorb radiation of all wavelengths falling on it. So, its absorption coefficient is unity: $a_\lambda = 1$ at all wavelengths λ.

2. It cannot reflect radiation falling on it. So, its reflection coefficient is zero: $r_\lambda = 0$ at all wavelengths λ.

3. It cannot transmit radiation falling on it. So, its transmission coefficient is zero: $t_\lambda = 0$ at all wavelengths λ.

4. As the body can neither reflect nor transmit radiation of all wavelengths falling on it, it appears black and is termed as black body.

5. The perfect black body is an ideal concept. There is no surface that can be regarded as perfectly black.

6. Examples of nearly perfect black bodies are: platinum black that can absorb about 98% of the visible light radiation incident on it, Ferry's black body and lamp black that can absorb about 96% of the visible radiation incident on it.

Sources of black body radiation

The black body radiators are used for a variety of commercial applications. Such black body radiators almost approach an ideal black body and emit visible light or radiation used for other practical purposes. Some of the black bodies include stoves, incandescent light bulbs, electric heaters, night vision equipment, burglar alarms, warm-blooded animals, the Sun, the stars, etc. Some typical examples of black body radiators are the following:

1. The electrical energy is converted into light energy in a filament bulb. When the switch in a circuit containing the bulb is made "ON", the filament of the bulb is initially not fully heated. In the beginning, it radiates energy in the infrared regions and goes to red, then yellow, until it reaches an almost white light when fully energized. Thus, a filament bulb can be used as a black body radiator.

2. A metal piece during welding emits radiation of various wavelengths. As the welding temperature increases, initially the metal piece glows red, then orange, and the color goes on changing until it becomes a bright, bluish cast. The glow is very intense at very high temperatures, and painful to look at using the naked eye. For this reason, welders routinely use dark goggles to avoid damage to their eyes.

3. There are certain objects such as ice cube and hot charcoal that emit radiation in the infrared region. Such objects are utilized in night vision equipment to detect infrared radiation and convert this into a visible image. These equipment are used for the detection of people and warm-blooded animals at night.

4. Infrared radiation is also emitted from animals and cannot be visible with the naked eye. A thermal camera sensitive to infrared, however, can be used to observe such thermal radiation from animals. The formed image appears to be a glowing object due to the black body radiation, unlike during the day when the person reflects the light falling on them.

Since the emitted radiation also depends upon the temperature, sources of black body radiation are classified into several categories according to the temperature range. The hot and warmer objects emit more energetic radiation and cool faster than the cooler objects. In the absence of a heat source, the objects will sooner or later reach the same temperature as the surroundings and are said to be in thermal equilibrium. The temperature corresponding to this thermal equilibrium depends on the object under test. For example, a low-temperature black body is reasonable for applications such as calibrating infrared (IR) sensor that looks at buildings, vehicles, or human bodies. Based on the temperature range, the sources of black body radiation are available in three main categories:

1. Low-temperature black body in the temperature range from −40°C to +150°C.

2. High-temperature extended area black body in the temperature range from ambient up to +600°C.

3. High-temperature black body cavity in the temperature range from ambient up to +1200°C.

11.2.2 Variation of the energy density of black body radiation

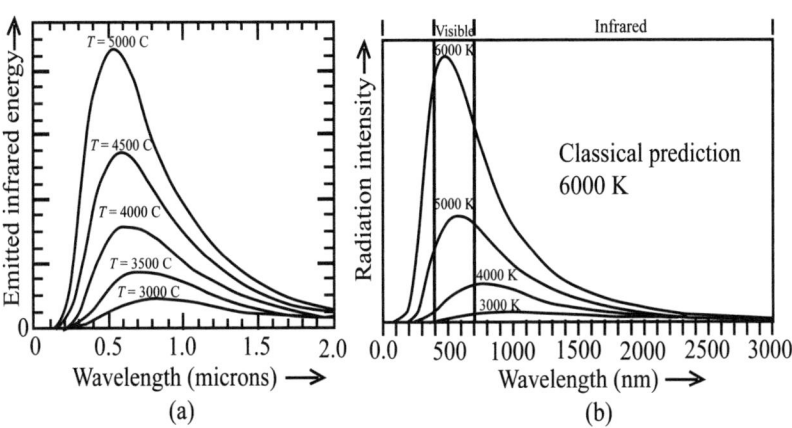

Figure 11.3 Variation of the energy density of emitted radiation from a black body as a function of wavelength at selective temperatures marked against each curve. The figure on the right shows the same features, but the visible part of the electromagnetic radiation is highlighted.

By the 1890s, experimental techniques had improved sufficiently that it was possible to make fairly precise measurements of the energy distribution of the black body radiation. At the University of Berlin in 1895, Wien and Lummer punched a small hole in the side of an otherwise completely closed oven, and began to measure the radiation coming out of it. The radiation beam coming out of the hole was passed through a diffraction grating that scattered the different wavelengths/ frequencies in different directions toward a screen. A detector was moved up and down along the screen to find the amount of radiant energy emitted in each frequency range. Wien and Lummer observed a radiation intensity versus wavelength curve close to the distribution shown in Figure 11.3.

The experimental results of the variation of energy density of black body radiation as a function of wavelength at selective temperatures are shown in Figure 11.3. Each curve in the figure corresponds to a particular temperature. It is observed from the figure that at a fixed wavelength, the intensity of radiation increases with the increase in temperature. Additionally, as the temperature of an object increases, the peak intensity moves to shorter wavelengths. The peak intensity and the corresponding temperature provide very useful physical information about the radiating system. For example, the temperature on the surface of the Sun is about 6000°C, and the peak in the spectrum of the Sun appears in the yellow region of the visible portion of the spectrum. This phenomenon can also be observed by slowly heating a piece of metal. At room temperature, the metal does not emit any light visible to the naked human eye. However, if it is heated to approximately 500°C, it begins to glow red. At 500°C, the metal emits visible light primarily in the red portion of the visible spectrum. As the temperature of the metal increases to approximately 1500°C, it begins to glow white. At 1500°C, the metal emits radiation at all visible wavelengths. As the combination of all visible colors produces white light, the metal looks white at this high temperature.

Figure 11.3 shows experimental curves of black body radiation between the wavelength λ and the intensity of radiation E_λ (energy radiated per second per unit area at a wavelength λ) emitted by a black body at selective temperatures. The figure on the right shows a similar characteristic spectrum highlighting the visible part of the electromagnetic radiation. From the figure, the following characteristic features are to be noted:

1. The black body radiation depends on the temperature of the enclosure only and is independent of its size, shape, or the nature of the material of the walls of the enclosure. An iron horseshoe, a ceramic vase, and a piece of charcoal — all exhibit the same spectrum of black body radiation if their temperatures are the same. Hence, the black body radiation curves are universal.

2. At a particular temperature T, the energy emitted from the black body increases with the increase in wavelength, reaches a maximum at a particular value of wavelength, say λ_m, and then decreases steadily with a further increase in wavelength λ. This energy approaches toward zero as $\lambda \to \infty$.

3. The wavelength λ_m of maximum emission shifts toward the lower wavelength side as the temperature of the black body increases. Further, it is observed that $\lambda_m T$ = a constant. This is known as Wien's displacement law.

4. The spectrum of the energy distribution of a black body is continuous, that is, a plot of energy density E_λ of black body radiation at wavelength λ against wavelength λ at a particular temperature T is a continuous curve. This is shown in Figure 11.3.

5. The wavelength of emitted radiation varies continuously from $\lambda = 0$ to $\lambda = \infty$. The emitted energy from a black body is not confined to a single wavelength λ but is distributed amongst various wavelengths ranging from zero to infinity.

6. With the increase in temperature, the radiant energy per second per unit surface area per unit range of wavelength of the black body radiation also increases, that is, E_λ for T_2 is larger than E_λ for T_1 when $T_2 > T_1$.

7. The area under the curve represents the total energy (E) emitted by a perfect black body per second per unit area over the complete wavelength range at that temperature. This area is found to increase according to Stefan–Boltzmann law that states that the total intensity of radiation or total emissive power is proportional to the fourth power of the absolute temperature of the black body. Thus, if a black body is at absolute temperature T, then the total intensity of radiation is given by $E = \sigma T^4$, where σ is a constant, called Stefan's constant. This constant σ has a unit $\mathrm{Wm}^{-2}\mathrm{K}^{-4}$ and a value 5.67×10^{-8}.

11.2.3   Similarity between thermal radiation and ideal gas

No. of observations	Thermal radiation	Ideal gas
1	The radiation in an enclosure at a constant temperature has all possible wavelengths and proceeds in all directions.	The molecules in an ideal gas are in random motion in all directions with all possible values of velocity ranging from 0 to ∞.
2	The black body radiation exerts pressure on the walls of the enclosure.	The ideal gas also exerts pressure on the walls of the container in which the gas is kept.
3	The quantum of black body radiation contains energy ranging from 0 to ∞.	The molecules of an ideal gas contain velocities ranging from 0 to ∞ and hence possess kinetic energy ranging from 0 to ∞.
4	The energy density plays a similar role to that of the internal energy of an ideal gas.	The internal energy of an ideal gas plays a similar role to that of the energy density of the black body radiation.

11.3   Kirchhoff's law

Before the formal derivation of Kirchhoff's law, it was experimentally established that a good absorber is also a good emitter, and a poor absorber, at the same time, is also a poor emitter. This indicates that a good reflector must be a poor absorber. For the same reason, reflective metallic coatings are made on lightweight emergency thermal blankets. These blankets lose very little amount of heat by radiation. Kirchhoff related the emissive and absorptive power of such reflectors and absorbers and recognized the universality and uniqueness of the function that describes the emissive power of a black body. But the precise form of this universal function was not clear to him. Lord Rayleigh and Sir James Jeans made attempts to describe this function in classical terms, resulting in **Rayleigh–Jeans law**. This law turned out to be inconsistent with experimental observations and resulted in the ultraviolet catastrophe. In 1900, the correct form of the function was discovered by Max Planck, assuming the quantized character of the black body radiation. This functional form is termed as **Planck's law of black body radiation**.

Kirchhoff's law states that **at a given wavelength, the ratio of the emissive power to the absorptive power for radiation is the same (constant) for all bodies at the same temperature and is equal to the emissive power of a perfectly black body at that temperature**.

This law can be mathematically expressed as

$$\frac{e_\lambda}{a_\lambda} = \text{constant} = E_\lambda, \tag{11.1}$$

where e_λ and a_λ are the emissive and absorptive powers of the body at a given temperature T and at a given wavelength λ, and E_λ is the emissive power of a perfectly black body.

11.3.1   Derivation of Kirchhoff's law

We consider an enclosure at a uniform temperature T and completely insulated from the surroundings. The wall of the enclosure is opaque to the thermal radiation of all wavelengths. The enclosure is filled with thermal radiation of all wavelengths emitted by its walls. We place another body M of emissive power e_λ and absorptive power a_λ inside the enclosure at a larger distance from the wall of the enclosure. The thermal equilibrium between the body M and the enclosure is achieved through the following two facts:

1. If the temperature difference between the body and the enclosure exists, a Carnot engine can transfer heat reversibly from one to the other till the temperatures of both of them are the same. During such a process, some amount of heat is converted into work. Such conversion of heat into work will continue as long as there is a temperature difference between the body and the enclosure. Thus, the system acts as the continual source of work by conversion of heat of a single body without maintaining another body at a lower temperature. This is in contradiction to the second law of thermodynamics.

2. Also, the idea of entropy demands that the temperature difference between the body and the enclosure should vanish at equilibrium. It is a well-known fact that the entropy becomes maximum at equilibrium, and the maximum value of entropy demands that the temperature difference should vanish.

These arguments point to the fact that the body and the enclosure will be at the same temperature when they attain an equilibrium state.

Thus, in the equilibrium condition, the body and the enclosure attain the same temperature. This is true if a composite body with different emissive and absorptive powers is placed at different parts inside the enclosure. Under this situation, the total energy emitted will be equal to the total energy absorbed at equilibrium. This is independent of the positions or orientations of the body with respect to the walls of the enclosure. This indicates the isotropic character of the radiation inside the enclosure.

The amount of energy emitted by an elemental area dA of the body A in the direction between θ, $\theta + d\theta$ and ϕ, $\phi + d\phi$ is $e_\lambda d\lambda dA \cos \theta \sin \theta d\theta d\phi$. Therefore, the amount of energy emitted by dA for all wavelengths and all possible θ and ϕ is given by

$$dA \int_0^\infty e_\lambda d\lambda \int_0^{\pi/2} \cos \theta \sin \theta d\theta \int_0^{2\pi} d\phi = \pi dA \int_0^\infty e_\lambda d\lambda. \tag{11.2}$$

Thus, for the whole body, the emission is given by

$$\pi \left(\sum dA \right) \int_0^\infty e_\lambda d\lambda. \tag{11.3}$$

But the amount of energy emitted by an elemental area dA' of the enclosure in the direction of dA of the body is given by

$$dQ_\lambda = E_\lambda d\lambda dA' \cos \theta' \frac{dA \cos \theta}{r^2}, \tag{11.4}$$

where r is the distance between dA' of the enclosure and dA of the body, E_λ is the emissive power of the surface of the enclosure, and θ' is the angle made by the normal dA' with the direction of emission.

So, the amount of energy absorbed by dA of the body is given by

$$a_\lambda dQ_\lambda = a_\lambda E_\lambda d\lambda dA \cos \theta \frac{dA' \cos \theta'}{r^2} = a_\lambda E_\lambda d\lambda dA \cos \theta d\omega$$

$$= a_\lambda E_\lambda d\lambda dA \cos \theta (\sin \theta d\theta d\phi), \tag{11.5}$$

where $d\omega = \dfrac{dA' \cos \theta'}{r^2} =$ solid angle subtended by dA' at $dA = \sin \theta d\theta d\phi$.

So, the total energy absorbed by dA from the total surface of the enclosure for all wavelengths is given by

$$dA \int_0^\infty a_\lambda E_\lambda d\lambda \int_0^{\pi/2} \cos\theta \sin\theta d\theta \int_0^{2\pi} d\phi = dA \int_0^\infty a_\lambda E_\lambda d\lambda \times \frac{1}{2} \times 2\pi = \pi dA \int_0^\infty a_\lambda E_\lambda d\lambda.$$

(11.6)

So, for the whole body, the absorption is given by

$$\pi \left[\sum dA\right] \int_0^\infty a_\lambda E_\lambda d\lambda,$$

In the equilibrium condition at a particular temperature T, we must have

$$\pi \left(\sum dA\right) \int_0^\infty e_\lambda d\lambda = \pi \left[\sum dA\right] \int_0^\infty a_\lambda E_\lambda d\lambda,$$

This indicates that $\int_0^\infty e_\lambda d\lambda = \int_0^\infty a_\lambda E_\lambda d\lambda$.

This is valid for any value of wavelength λ, so we must have $e_\lambda = a_\lambda E_\lambda$.

If the same body A is placed in another enclosure at the same temperature T, and with E'_λ as the emissive power of its surface but having a different shape and nature of the walls,

$$e_\lambda = a_\lambda E_\lambda. \tag{11.7}$$

Since e_λ and a_λ depend on the nature of the body and its temperature only and not upon the surroundings, we have $E_\lambda = E'_\lambda$.

For a black body, we have $a_\lambda = 1$, and if it is placed inside an enclosure, we have $e_\lambda = E_\lambda$. This shows that E_λ is equal to the emissive power of a black body. Thus, radiation in any hollow enclosure is independent of the nature and shape of the walls and is identical to the black body radiation at the same temperature. Hence, Kirchhoff's law can be stated as: **At any temperature, the ratio of the emissive power to the absorptive power of a substance is constant, equal to the emissive power of a perfectly black body.**

11.4 Radiation pressure

When electromagnetic radiation is incident upon the surface of a substance, interaction between electromagnetic radiation and the molecules of the substance takes place. Such substances may be of various types, including clouds of particles or gases. This interaction can be absorption, reflection, or some of both (the common case) and generates pressure exerted upon any surface exposed to electromagnetic radiation. This pressure due to the electromagnetic radiation is known as the **Radiation pressure**. It should also be mentioned that various types of substances emit radiation and thereby experience a resulting pressure. The concept of radiation pressure was first put forward by Johannes Kepler way back in 1619. He used the concept of radiation pressure to explain the observation that the tail of a comet always points away from the Sun.

It is experimentally found that the thermal energy radiated from objects depends upon the temperature. This emitted electromagnetic radiation has both energy and momentum. Hence, the emission of such radiation reduces the temperature of the objects. The momentum of the emitted radiation also causes a reactive force that provides pressure across the radiating surface when expressed per unit area. The forces generated by the radiation pressure are generally too small to be detected under everyday circumstances. However, these forces do play a crucial role in some branches of physics, such as astronomy and astrodynamics. For example, had the effects of the Sun's radiation pressure on the spacecraft of the Viking program been ignored, the spacecraft would have missed Mars orbit by about 15000 km.

Further, it is known that the momentum of a classical electromagnetic field or the momenta of photons (particles of light) provide the existence of radiation pressure. The interaction of electromagnetic waves or photons with matter involves an exchange of momentum. From the law of conservation of momentum, it is understood that any change in the total momentum of the waves or photons must involve an equal and opposite change in the momentum of the matter it interacted with (Newton's third law of motion). Such an interaction results in the transfer of momentum provides a general explanation for the existence of radiation pressure.

In this section, the expressions for radiation pressure are calculated in two different situations, as mentioned below:

1. When the radiation falls normally on the surface and

2. When the radiation is diffuse.

11.4.1 Radiation pressure due to normal radiation

The energy of a photon is given by

$$E = mc^2 = h\nu. \tag{11.8}$$

This provides the mass-equivalent of the photon as $m = \dfrac{h\nu}{c^2}$. Also, the momentum of a photon is given by

$$mc = \frac{h\nu}{c^2} \times c = \frac{h\nu}{c} = \frac{E}{c}. \tag{11.9}$$

We consider the case of normal incidence of the photons on a surface. Under this situation, the momentum imparted to the surface per unit area per second (i.e. pressure p) is given by

$$p = \sum \frac{E}{c} = \frac{E_T}{c}, \tag{11.10}$$

where E_T is the total energy incident per unit area per second on the surface.

If u is the energy density, the total energy passing per second through any area A of the surface normal to the radiation is given by $E_T = u \times (Ac)$.

Therefore, the energy radiated per unit area per second will be

$$E_T = \frac{uAc}{A} = uc; \quad \Rightarrow \quad u = \frac{E_T}{c}. \tag{11.11}$$

Equation (11.11) is a very important result in thermodynamics. Further, comparing equations (11.10) and (11.11), we get $p = u$. If the momentum of the electromagnetic radiation per unit area per second is absorbed by the surface, it will then represent the time-rate of change of momentum per unit area, that is, the pressure of the radiation. Thus, the pressure of radiation for normal incidence is equal to the energy density of the electromagnetic radiation.

11.4.2 Radiation pressure due to diffuse radiation

We consider a beam of photons incident at an angle θ_z to the surface OB. Surface BC is taken to be normal to the beam so that energy incident on BC per second $EA = u\,c\,A$, where A is the surface area of BC ($A = A' \cos\theta_z$), and A' is the surface area of the surface OB. It is clear from the figure that the total energy crossing the surface BC is equal to the energy incident on the surface OB per second. Therefore, energy incident on the surface OB per unit area per second

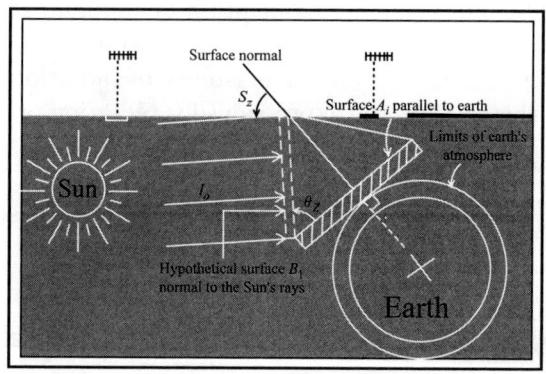

Figure 11.4 Schematic diagram for the calculation of the radiation pressure exerted by electromagnetic radiation from the Sun as it strikes the Earth's surface.

$$\frac{ucA}{A'} = \frac{uc\,A' \cos\theta_z}{A'} = uc\,\cos\theta_z. \tag{11.12}$$

Therefore, the momentum incident on the surface OB per unit area per second [see Figure 11.4]

$$\frac{uc \, \cos \theta_z}{c} = u \cos \theta_z$$

and the component of this momentum in the direction normal to OB will be

$$= u \cos \theta_z \, (\cos \theta_z) = u \cos^2 \theta_z. \tag{11.13}$$

If all the incident photons are absorbed, then this would represent the rate of change of momentum per unit area per second normal to the surface OB, that is, the pressure of the radiation on the surface OB will be

$$p = u \cos^2 \theta_z. \tag{11.14}$$

In the case of diffuse radiation, the radiation is incident from all possible directions with equal probability. Therefore, the above term is to be summed over all beams. Suppose there are N beams of radiation meeting on a surface; then, we can write total radiation pressure as

$$p = \sum u \cos^2 \theta_z. \tag{11.15}$$

To calculate the $\sum u \cos^2 \theta_z$, let us imagine a hemisphere of radius r. If N beams are considered to be uniformly distributed over the surface of the hemisphere, then

$$\text{beams per unit area} = \frac{N}{2\pi r^2}.$$

Therefore, beams through a ring area confined between θ and $\theta_z + d\theta_z$ will be

$$dN = \text{ area of the ring } \times \frac{N}{2\pi r^2} = 2\pi r^2 \sin \theta_z d\theta_z \times \frac{N}{2\pi r^2} = N \sin \theta_z d\theta_z. \tag{11.16}$$

Hence, the expression for pressure becomes

$$p = \sum dN \times u \cos^2 \theta_z = \sum N \sin \theta_z d\theta_z \times u \cos^2 \theta_z = \sum Nu \cos^2 \theta_z \sin \theta_z d\theta_z. \tag{11.17}$$

The ring area will assume the surface of the hemisphere if θ_z is varied from 0 to $\pi/2$. Therefore, for pressure at a point, the center, due to the beams through a hemispherical surface, will be

$$p = \int_0^{\pi/2} Nu \cos^2 \theta_z \sin \theta_z d\theta_z = U \int_0^{\pi/2} \cos^2 \theta_z \sin \theta_z d\theta_z = \frac{U}{3}. \tag{11.18}$$

Here, $U = Nu$. Therefore, the pressure of the diffuse radiation is equal to one-third of the total energy density.

Radiation pressure from sunlight: What is the magnitude of the radiation pressure produced by sunlight at the earth's distance from the Sun? How does it compare with atmospheric pressure?

The power per unit area R of sunlight arriving at the top of earth's atmosphere, called the solar constant, is 1.36×10^3 Wm^{-2}. It is related to the energy density U by

$$R = \frac{cU}{4}; \quad \Rightarrow \quad U = \frac{4R}{c}. \tag{11.19}$$

Again, we know that

$$P_{\text{rad}} = \frac{1}{3}U = \frac{4R}{3c} = \frac{4 \times \left(1.36 \times 10^3 \text{ Wm}^{-2}\right)}{3 \times (3.00 \times 10^8 \text{ ms}^{-1})} = 6.04 \times 10^{-6} \text{ Pa}.$$

For comparison, about 99% of the earth's atmosphere lies below 35 km. At that altitude, the atmospheric pressure is about 10^3 Pa. Hence, it is observed that the radiation pressure of sunlight is about nine orders of magnitude smaller than the pressure at an altitude just below 35 km.

11.5 Temperature dependence of the black body radiation spectrum

The black body radiation spectrum shown in Figure 11.3 shows that at a fixed temperature, the energy density of the black body radiation varies typically with the wavelength of the radiation. Therefore, to determine the temperature dependence of the energy density of the black body radiation, the total energy density over the entire wavelength range, that is, from 0 to ∞ is considered, not at a particular value of wavelength. Such temperature dependence of the black body radiation spectrum is discussed below.

11.5.1 Stefan–Boltzmann law

According to Stefan's law, the total amount of heat energy E radiated by a perfectly black body per second per unit area is directly proportional to the fourth power of the absolute temperature T of the black body.

Mathematically, this law can be expressed as

$$E \propto T^4 \text{ or } E = \sigma T^4, \tag{11.20}$$

where σ is a constant, known as Stefan's constant. Sometimes, this law is referred as Stefan's fourth power law. The law, in this form, refers only to the emission of black body radiation and not to the net loss. If a black body A at an absolute temperature T is surrounded by another black body B at an absolute temperature T_0, then the amount of heat lost by the black body A is σT^4, and the amount of heat absorbed by the black body A from the black body B is σT_0^4. So, the net amount of heat lost by the black body A per unit area per second is given by

$$E = \sigma \left(T^4 - T_0^4\right). \tag{11.21}$$

Equation (11.21) is known as Stefan–Boltzmann law. The total energy radiated per unit surface area of a black body per unit time is shown in Figure 11.5 as a function of temperature T. The solid line is fit to the experimental data with equation (11.20). It is observed from the figure that the data-points are nicely fitted with equation (11.20), which confirms that Stefan's fourth power law is obeyed by such type of radiation.

Experimental verification of Stefan's law

Experimental arrangement due to Lummer and Pringsheim for the verification of Stefan's law is shown in Figure 11.6. The hollow vessel A in the figure is kept at 100°C by immersing it in boiling water. It acts as a standard source of radiation for calibrating the bolometer G from time to time. The black body B, employed for the range of temperatures 200°C and 600°C, consisted of a mixture of sodium and potassium nitrates. The melting point of this nitrate is 219°C. This salt bath could be maintained at any desired temperature by placing it in contact with an iron cylinder heated to 1300°C by a gas furnace. The temperature could be measured with a thermo element.

The measuring instrument shown at G was the Lummer–Kurlbaum surface bolometer. Besides, there are a number of water-cooled shutters so that radiation can be stopped or allowed to fall at will. To measure the spectrum of the radiation, the radiation from B was passed first through a fluor spar prism, and the spectral energy distribution was measured by a linear bolometer. The bolometer was kept at different distances from the black body, and the inverse square law was verified.

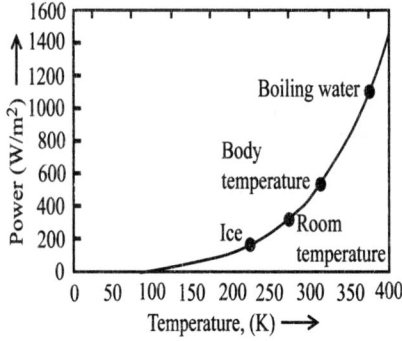

Figure 11.5 Variation of total energy radiated per unit time per unit surface area of a black body as a function of temperature T. The solid line is a fit to the data according to the Stefan–Boltzmann law.

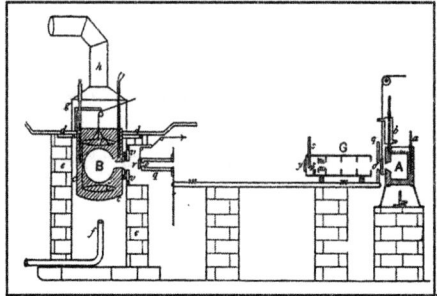

Figure 11.6 Experimental setup due to Lummer and Pringsheim for the verification of Stefan's law.

Thermodynamic proof of Stefan's law

Let us suppose that the black body radiation is enclosed in an evacuated chamber with perfectly reflecting walls and a perfectly reflecting moving piston. The purpose of assuming perfectly reflecting walls is to avoid heat exchange between the walls and the radiation; otherwise, the thermal capacity of the walls will appear in the calculation.

We consider a cylinder of volume V filled with black body radiation at a temperature T. Let u be the energy per unit volume (energy density) of the radiation inside the cylinder, and P be the pressure of the radiation. The total energy U of the radiation is then given by the product of energy per unit volume and the volume of the cylinder. Hence, we have

$$U = u \times V. \tag{11.22}$$

According to Maxwell's theory of electromagnetic radiation, the pressure P exerted by radiation is given by

$$P = \frac{1}{3}u. \tag{11.23}$$

Let us suppose that a small amount of heat δQ is supplied to the cylinder filled with radiation and at the same time, the volume of the cylinder is changed by an amount dV. If dU is the change in internal energy of the radiation and δW is the corresponding external work done, we have from the first law of thermodynamics

$$\delta Q = dU + \delta W = dU + PdV. \tag{11.24}$$

Substituting the values of U and P, respectively, from equations (11.22) and (11.23) into equation (11.24), we have

$$\delta Q = d(uV) + \frac{1}{3}udV = udV + Vdu + \frac{1}{3}udV = \frac{4}{3}udV + Vdu. \tag{11.25}$$

According to the second law of thermodynamics, we have $\delta Q = TdS$. Using this in equation (11.25), we get

$$TdS = \frac{4}{3}udV + Vdu; \quad \Rightarrow \quad dS = \frac{4}{3}\frac{u}{T}dV + \frac{V}{T}du. \tag{11.26}$$

Considering S to be a function of V and u, that is, $S = f(V, u)$, we have

$$dS = \left(\frac{\partial S}{\partial V}\right)dV + \left(\frac{\partial S}{\partial u}\right)du. \tag{11.27}$$

Comparing equations (11.26) and (11.27), we get

$$\left(\frac{\partial S}{\partial V}\right) = \frac{4}{3}\frac{u}{T} \text{ and } \left(\frac{\partial S}{\partial u}\right) = \frac{V}{T}.$$

Again, we have

$$\left(\frac{\partial^2 S}{\partial u \partial V}\right) = \left(\frac{\partial^2 S}{\partial V \partial u}\right), \text{ that is, } \frac{\partial}{\partial u}\left(\frac{\partial S}{\partial V}\right) = \frac{\partial}{\partial V}\left(\frac{\partial S}{\partial u}\right).$$

This leads to

$$\frac{\partial}{\partial u}\left(\frac{4}{3}\frac{u}{T}\right) = \frac{\partial}{\partial V}\left(\frac{V}{T}\right).$$

Or,

$$\frac{4}{3}\frac{1}{T} - \frac{4}{3}\frac{u}{T^2}\frac{\partial T}{\partial u} = \frac{1}{T}; \quad \Rightarrow \quad \frac{1}{3}\frac{1}{T} = \frac{4}{3}\frac{u}{T^2}\frac{\partial T}{\partial u}; \quad \Rightarrow \quad \frac{\partial u}{u} = 4\frac{\partial T}{T}. \qquad (11.28)$$

Integrating equation (11.28), we get

$$\ln u = 4\ln T + a, \text{ or } u = aT^4,$$

where a is the constant of integration. We know that the energy E radiated per unit area per second from a perfectly black body at absolute temperature T and the energy of radiation u inside an enclosure at the same temperature are related by

$$E = \frac{1}{4}u\ c, \qquad (11.29)$$

where c is the velocity of light. Substituting the value of u in this expression, we get

$$E = \frac{1}{4}a\ c\ T^4 = \sigma T^4, \qquad (11.30)$$

where $\sigma = \frac{1}{4}a\ c$ is called *Stefan's constant*. This is known as *Stefan's law*.

Application of Stefan's law

1. Measurement of high temperature: Stefan's law is widely used for measurement of very high temperature with the help of instruments called radiation pyrometers. Such pyrometers need not be place in contact with the body whose temperature is being measured; hence, there is no upper limit to the rank of temperature measurable with such pyrometers. However, in such measurements, we assume both to be perfectly black, which is not always so. Thus, the measured temperature will be less than the actual one.

   Ferry's radiation pyrometer: In this pyrometer, the radiation from the source entering from the right is focused by a concave mirror M through the diaphragm D on a blackened plate P, known as the receiver. The position of the mirror can be adjusted to achieve accurate focusing, which is judged by viewing the image of the source through an eyepiece E placed behind a hole in the mirror M. The receiver is attached to one junction (hot) of a thermocouple. The other junction (cold) is shielded from the radiation by T. A milli-voltmeter mV is connected to the thermocouple. The incident radiation is absorbed by the receiver heats it up, whence the milli-voltmeter shows a deflection.

As the electromotive force E in the thermocouple is proportional to the intensity of the incident radiation, we have from Stefan's law,

$$E = a\left(T_1^b - T_2^b\right), \tag{11.31}$$

where T_1 and T_2 are, respectively, the temperatures of the source and the receiver and a is constant. The value of b varies between 3.8 and 4.2, being constant for a particular instrument. Hence, this pyrometer should be calibrated in comparison with a standardized thermocouple before use.

2. Temperature of the Sun

   (a) **Solar constant:** The solar constant is the average amount of energy that will fall in one minute on one square centimeter of a perfectly black surface held normal to the Sun's rays and placed at the mean distance of the earth from the Sun provided there were no absorption in the earth's atmosphere.

      This constant can be measured with an instrument known as a pyrheliometer, and its value is found to be nearly 1.937 cal/cm^{2}min. In SI, the value is nearly 1356 J/m^{2}s.

   (b) **Temperature of the Sun:** Assuming the Sun to be a perfectly black body of radius r and temperature T, then from Stefan's law, heat radiated by it per second is given by $Q = 4\pi r^2 \sigma T^4$. This radiant heat is distributed uniformly over the entire surface area $4\pi R^2$ of an imaginary sphere of radius R concentric with the Sun, where R is the mean distance of the earth from the Sun. Hence, from the definition, the solar constant S is given by

$$S = \frac{4\pi r^2 \sigma T^4}{4\pi R^2} \times 60; \quad \Rightarrow T^4 = \left(\frac{R}{r}\right)^2 \frac{S}{\sigma} \times \frac{1}{60}.$$

      Substitution of known values R, r, S, and σ gives $T = 5723$ K.

11.5.2   Newton's law of cooling

If the temperature T of a black body radiator is slightly higher than the temperature of the surroundings T_0, an interesting conclusion can be made from Stefan–Boltzmann law of black body radiation given by equation (11.21). Let us put $T = T_0 + x$, where the difference in temperature $x = T - T_0$ is assumed to be very small. Under this situation, we can write equation (11.21) in the following form:

$$E = \sigma \left(T^4 - T_0^4\right) = \sigma \left[\left(T_0 + x\right)^4 - T_0^4\right]$$
$$= \sigma \left(T_0^4 + 4T_0^3 x + 6T_0^2 x^2 + 4T_0^3 x^3 + x^4 - T_0^4\right).$$

Neglecting the higher power of x, we get

$$E = 4\sigma T_0^3 x; \quad \Rightarrow \quad E = K_N \left(T - T_0\right), \tag{11.32}$$

where $K_N = 4\sigma T_0^3$ is a constant. Thus, for the small difference in temperature, heat lost per unit area per second by a black body due to radiation is proportional to the difference in temperature of the body with the surroundings. This is known as **Newton's law of cooling**. Like Stefan's law, Newton's law of cooling also holds true not only for a black body but also for any other body. However, for any radiating body, the proportionality constant is different and depends on the nature of the surface of the body.

Time of cooling

Let us consider a body of mass m and specific heat c placed in a surrounding of temperature θ_0. Let the body be cooled from the temperature θ_1 to θ_2 $(\theta_1 > \theta_2)$ in time t. If the body cools from a temperature θ by $d\theta$ in time dt, then from Newton's law of cooling, the net rate of loss of heat by the body is

$$\frac{dQ}{dt} = b(\theta - \theta_0), \tag{11.33}$$

where b is a constant, which, among other quantities, also involves the nature and the area of the surface of the body. The rate of heat absorbed by the body is given by

$$\frac{dQ}{dt} = -m \, c \frac{d\theta}{dt}. \tag{11.34}$$

The negative sign indicates that the temperature θ is decreasing with time t. Equating equations (11.33) and (11.34), we get

$$-m \, c\frac{d\theta}{dt} = b(\theta - \theta_0); \quad \text{or,} \quad -\frac{d\theta}{dt} = \frac{b}{m \, c}(\theta - \theta_0) = K(\theta - \theta_0).$$

Here, $K = \frac{b}{m \, c}$ is a constant. So, the rate of decrease in temperature, that is, the rate of cooling of a body is proportional to its difference in temperature with the surroundings. This is an alternative form of Newton's law of cooling.

From the above expression, we get $-\frac{d\theta}{(\theta - \theta_0)} = K \, dt$. Integrating this expression within proper limits, we get

$$-\int_{\theta_1}^{\theta_2} \frac{d\theta}{\theta - \theta_0} = K \int_0^t dt; \quad \Rightarrow t = \frac{1}{K} \ln \frac{\theta_1 - \theta_0}{\theta_2 - \theta_0} = \frac{m\,c}{b} \ln \frac{\theta_1 - \theta_0}{\theta_2 - \theta_0}. \quad (11.35)$$

This equation (11.35) gives the time required for cooling of a black body from a temperature θ_1 to θ_2 when the surrounding temperature is θ_0.

Cooling curve

The graphical variation of the temperature of a heated body with time is called its cooling curve. Such a cooling curve is schematically shown in Figure 11.7. The slope of the tangent drawn at a point of the curve gives the rate of decreased temperature of the body at that particular time. The plot of the rate of decrease of temperature with the difference in temperature is found to be a straight line passing through the origin. Such a cooling curve is used to find the thermal conductivity of a bad conductor in graduate-level experiments.

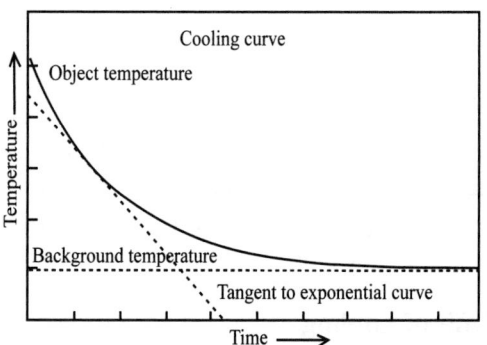

Figure 11.7 Curve from Newton's law of cooling.

11.5.3   Wien's displacement law

When radiation from a black body is passed through a prism, a continuous spectrum is obtained in which the intensity of radiation is different in different parts of the spectrum. The energy of the spectrum is distributed in various wavelengths varying from zero to infinity. The law that connects the intensity with wavelength is known as the law of distribution of intensity of the black body radiation. At a particular temperature, it is observed that the intensity of the black body radiation increases with the increase in wavelength, becomes maximum at a particular wavelength, and finally decreases with a further increase in wavelength. Thermodynamic analysis shows that with the increase in temperature, the wavelength at which the intensity of the black body radiation becomes maximum shifts toward a lower value.

Mathematically, it can be represented by

$$\lambda_m T = \text{constant}, \quad (11.36)$$

where λ_m is the wavelength at which the intensity of the radiation is the maximum corresponding to a given temperature T. This law is known as *Wien's displacement law* and was derived by *Wilhelm Wien* in 1893 using the thermodynamic argument. Wien considered adiabatic expansion of a cavity containing waves of light in thermal equilibrium and showed that, under slow expansion or contraction, the energy of electromagnetic radiation reflecting off the walls changes in exactly the same way as the frequency or wavelength due to Doppler's effect. From the principle of the quasi-static process in thermodynamics, it is known that a thermal equilibrium state stays in a thermal equilibrium when the expansion or contraction process is carried out very slowly. The principle of adiabatic expansion or contraction allowed Wien to conclude that for each mode, the adiabatic invariant energy/frequency is only a function of the other adiabatic invariant, the frequency/temperature.

11.6 Classical theory of black body radiation

The characteristic features of the variation of energy density of black body radiation as a function of wavelength are shown in Figure 11.3 at some selective temperatures. Several attempts based on classical and quantum theory have been made to explain this variation. An overview of the necessary classical theory of black body radiation is presented below.

Light is an electromagnetic wave. It is produced when an electric charge vibrates. The word "vibrates" specifically means any change in how the charge moves—speeding up, slowing down, or changing direction. It should be mentioned that heat is just the kinetic energy of random motion of the constituent particles of the system. Electrons in a hot object vibrate in random directions, and as a result, light is produced. A hotter object means more energetic vibrations, and so more light is emitted by a hotter object, that is, it glows brighter. It would be shown that the classical physics could not explain the shape of the spectrum of black body radiation.

The electrons in a hot object can vibrate with a range of frequencies, ranging from very few vibrations per second to a huge number of vibrations per second. In fact, there is no limit to how great the frequency can be. According to classical physics, each frequency of vibration should have the same energy. Since there is no limit to how great the frequency can be, there is no limit to the energy of the vibrating electrons at high frequencies. This means that, according to classical physics, there should be no limit to the energy of the light produced by the electrons vibrating at high frequencies. But it is **WRONG!!** Experimentally, it is found that the energy density of the black body radiation spectrum is always small at shorter wavelengths, that is, at higher frequencies.

Stefan and Boltzmann analyzed the characteristic features of the black body radiation and found that the energy radiated per unit area per unit time is directly proportional to the fourth power of temperature of the radiator on an absolute scale. Later, this law was experimentally verified and was also derived from Planck law of black body radiation. Wien proposed a phenomenological law known as Wien's distribution law that predicted correctly zero energy density for $\lambda = 0$ and $\lambda = \infty$. But, its prediction of finite energy density for $T = \infty$ contradicted Stefan's T^4 law.

Considering the equipartition theorem and the number of modes in the frequency range between ν and $\nu + d\nu$, Rayleigh and Jeans calculated the energy density of the black body radiation as a function of wavelength λ and temperature T. This law is mathematically expressed as

$$dp(\nu, T) = \rho\nu(T)d\nu = \frac{8\pi K_B T \nu^2 d\nu}{c^3},$$

where symbols have their usual meaning. It predicted the variation of intensity of the black body radiation well at lower frequencies and higher temperatures. However, the experimental data performed on the black body radiation showed slightly different results than what was expected by this law. The experimental results showed a bell type of curve, but according to the Rayleigh–Jeans law, the total energy of the black body radiation becomes infinite when integrated over the whole range of frequency. Ehrenfest later dubbed this phenomenon as the "ultraviolet catastrophe".

In 1900, Max Planck introduced the idea of quantization of energy (this is a clear deviation from the classical concept of electromagnetic radiation) and deduced the relationship between the energy density of the black body radiation spectrum as a function of wavelength at various temperatures. This law successfully explained the variation of the energy density of black body radiation over the whole range of wavelengths at various temperatures.

The classical laws of black body radiation due to Wien, and Rayleigh and Jeans are described in detail below.

11.6.1 Wien's radiation formula

The radiation from a black body is conveniently expressed in terms of the spectral energy density $u(\lambda, T)$, or *energy per unit volume per unit frequency (wavelength) of the radiation within the black body cavity*. For radiation in equilibrium with the walls, the power emitted per square centimeter of the opening is simply proportional to the energy density of the radiation in the cavity. Because the cavity radiation is isotropic and unpolarized, one can average over direction to show that the relation between the power radiated per unit area per unit frequency (wavelength) $J(\lambda, T)$ and $u(\lambda, T)$ is $J(\lambda, T) = u(\lambda, T) \frac{c}{4}$, where c is the speed of the radiation.

In 1896, *Wilhelm Wien* made some arbitrary assumptions regarding the mechanism of emission and absorption of black body radiation inside an enclosure. These assumptions are:

1. The radiation inside a hollow enclosure may be supposed to be produced by a resonator of molecular dimensions.

2. The frequency of the wave emitted is proportional to the kinetic energy of the resonator, that is,

$$\frac{1}{2}mv^2 = \alpha\nu = \alpha\frac{c}{\lambda},$$

   where α is a constant, and ν is the frequency of the light emitted.

3. The intensity of radiation of any particular wave is proportional to the number of resonators that have got the requisite energy, that is,

$$E_\lambda = e^{-\frac{\frac{1}{2}mv^2}{K_BT}} \times \psi(v) = e^{-\frac{\alpha\nu}{K_BT}} \times \psi(v),$$

   where $\psi(v)$ is a function of the energy of the molecular resonator and, hence, of its velocity. It is therefore a function of the wavelength.

Wilhelm Carl Franz Wien

Wilhelm Carl Werner Otto Fritz Franz Wien (January 13, 1864 and August 30, 1928) was a German physicist who, in 1893, used theories about heat and electromagnetism to deduce Wien's displacement law. This law calculates the emission of a black body at any temperature from the emission at any one reference temperature. He also formulated an expression for the black body radiation, which is correct in the photon-gas limit. His arguments were based on the notion of adiabatic invariance and were instrumental for the formulation of quantum mechanics. Wien received the 1911 Nobel Prize for his work on heat radiation. He was a cousin of Max Wien, inventor of the Wien Bridge.

Comparing this with Wien's displacement law, we get

$$E_\lambda d\lambda = \frac{A}{\lambda^5} e^{-\frac{c_2}{\lambda K_BT}} d\lambda, \tag{11.37}$$

where c_2 is found to vary considerably with wavelength λ. Equation (11.37) is known as Wien's radiation formula[1] for a black body.

---

[1] W. Wien, On the Division of Energy in the Emission-spectrum of a Black Body, *Philosophical Magazine* 43, no. 262 (1897): 214–220.

Wien's explanation of the distribution law

For the short values of wavelength λ, the exponential factor becomes very large and contributes more than that of the other factor λ^{-5}. This means that at shorter wavelengths, E_λ increases with λ. On the other hand, the exponential factor becomes very small at higher values of λ. In this range, the factor λ^{-5} dominates mostly and hence, E_λ should decrease at higher λ.

At first sight, it is observed that Wien's law is apparently good to explain the black body radiation curve. However, the comparison of the curve obtained from Wien's distribution law with the experimental one indicates that in the shorter wavelength range, Wien's law fits very well, but a marked difference between these curves is observed at the higher wavelength range. This implies an error in the theoretical distribution law, which is too large to ascribe to experimental uncertainties and indicates a flaw in the theory. Wien could neither explain the failure of this relation nor supply a better theory for this.

Although Wien's law does not hold good for a complete explanation of the black body radiation curve, but one can deduce the maximum spectral emissive power dependence on temperature using this law in the following way:

At $\lambda = \lambda_m$, Wien's displacement law indicates that $\lambda_m T = b$. Using this expression in Wien's distribution law, we get

$$E_{\lambda_m} = A\, \lambda_m^{-5}\, e^{-\left(\frac{a}{\lambda_m T}\right)} = A\, \lambda_m^{-5}\, e^{-\left(\frac{a}{b}\right)}. \tag{11.38}$$

This leads to the result that

$$E_{\lambda_m} \propto \lambda_m^{-5}; \quad \Rightarrow \quad E_{\lambda_m} \propto \left(\frac{T}{b}\right)^5; \quad \Rightarrow \quad E_{\lambda_m} \propto T^5. \tag{11.39}$$

Experimental verification of Wien's distribution law

Within a year of the discovery of Wien's distribution law of black body radiation (see equation 11.37), the great German spectroscopist Friedrich Paschen had confirmed Wien's prediction by working in the then difficult infrared range of 1 to 4 µm and at temperatures of 400 to 1600 K. It is observed from Figure 11.8 that Paschen had made most of his measurements in the maximum energy region of a body heated to 1500 K and found good agreement with Wien's exponential law.

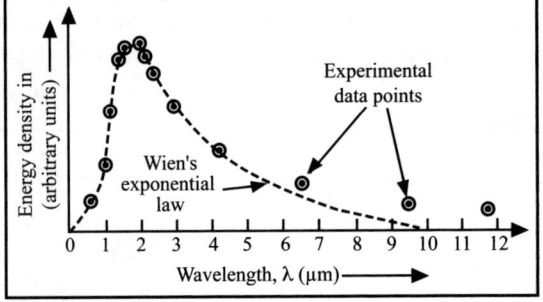

Figure 11.8 Discrepancy between Wien's law and experimental data for a black body at 1500 K.

However, in 1900, Lummer and Pringsheim extended the measurements to 18 µm, and Rubens and Kurlbaum went even farther—to 60 µm. Both these experimental groups concluded that Wien's law failed miserably in this region. This is clearly evident from Figure 11.8. At longer wavelengths, the experimental data points lie well above Wien's theoretical curve. Thus, we see that Wien's distribution law is successful in explaining the black body radiation spectrum up to wavelength $\lambda = 4$ µm, that is, it can explain the increase in energy density with wavelength and the maximum of energy density at a particular value of λ, but it fails to explain the spectrum at higher wavelengths. Thus, it provided a hint for a better theory to explain the radiation spectrum of a black body for all wavelengths.

Proof of Wien's displacement law from thermodynamic consideration

It can be proved that the spectral distribution is independent of the shape of the enclosure, but for simplicity, we shall consider the spherical enclosure, which is capable of expanding radially like a football bladder. Let some black body radiation be enclosed inside the sphere, and u be the energy per unit volume (energy density) of the radiation at a temperature T. The pressure of the radiation is P. If V is the volume of the spherical enclosure, its total energy is then given by

$$U = u \times V. \tag{11.40}$$

We consider that the enclosure is expanding adiabatically with a uniform velocity v. From the first law of thermodynamics, we have

$$\delta Q = dU + \delta W = dU + PdV = 0, \tag{11.41}$$

as the process is adiabatic, $\delta Q = 0$.

According to Maxwell electromagnetic theory of radiation, the expression for pressure P exerted by radiation on the walls of the enclosure is given by

$$P = \frac{1}{3}u. \tag{11.42}$$

Substituting the values of U and P in equation (11.41), we get

$$d\left(uV\right) + \frac{1}{3}udV = 0; \quad \Rightarrow \quad udV + Vdu + \frac{1}{3}udV = 0$$

$$\frac{4}{3}udV + Vdu = 0; \quad \Rightarrow \quad \frac{4}{3}\frac{dV}{V} + \frac{du}{u} = 0.$$

Integrating the above expression, we get

$$\frac{4}{3}\ln V + \ln u = \text{constant}. \tag{11.43}$$

This leads to

$$V^{4/3}u = \text{constant.} \tag{11.44}$$

But, from Stefan's law, we have $u = aT^4$, where a is constant. Using this relation in equation (11.44), we get

$$V^{4/3}T^4 = \text{constant.} \qquad \Rightarrow \quad V^{1/3}T = \text{constant.} \tag{11.45}$$

During the adiabatic expansion of the enclosure, it produces a change in the wavelength of radiations reflected from the moving walls due to Doppler's effect. Let us calculate the change in wavelength due to Doppler's effect.

Let OA be a ray of wavelength λ incident at an angle θ on the wall of the enclosure in position S_1S_2. The particular wave crest strikes the wall at A and is reflected in the direction AM. When the next crest reaches A, the wall has moved a distance $AC(= vT)$ where v is the velocity of the expansion of the wall, and the crest is reflected from point B of the new position $S_1'S_2'$ of the wall. The wave is reflected in the direction BN. If λ is the wavelength when reflection takes place from the wall S_1S_2 and λ_1 is the new wavelength when reflection takes place from the wall at $S_1'S_2'$, we then have

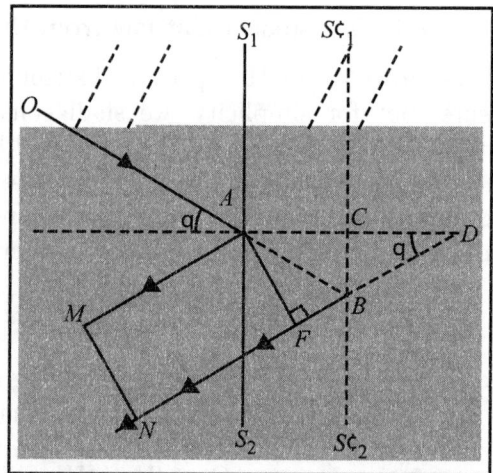

Figure 11.9 Diagram for the proof of Wien's displacement law.

$$\lambda = AM \qquad \text{and} \qquad \lambda_1 = AB + BN. \tag{11.46}$$

Change in wavelength $d\lambda$ is then given by

$$d\lambda = \lambda_1 - \lambda = (AB + BN) - AM.$$

If AF is the perpendicular drawn from A on BN, then we have $AM = NF$. This gives rise to $d\lambda = (AB + BF)$. From the Figure 11.9, we have $AB = BD$. The change in wavelength $d\lambda$ takes the form

$$d\lambda = DB + BF = FD = AD\cos\theta = 2AC\cos\theta = 2vT\cos\theta = \frac{2v\lambda\cos\theta}{c}. \tag{11.47}$$

Every ray inside the spherical enclosure undergoes repeated reflections. The path of a single ray is shown in Figure 11.9. It is clear from the figure that the angle of incidence θ at every reflection remains the same. The distance traveled by radiation between two consecutive reflections is given by $2r\cos\theta$.

The time between two consecutive reflections is $\dfrac{2r\cos\theta}{c}$, so the number of reflections per second is $\dfrac{c}{2r\cos\theta}$ and the number of reflections in time dt will be $\dfrac{cdt}{2r\cos\theta}$.

If r_0 be the initial radius of the enclosure, then after a time t, its radius will be $r = r_0 + vt$ so that $dr = vdt$ and $dt = \dfrac{dr}{v}$.

So, the number of reflections in time dt will be

$$\frac{c}{2r\cos\theta}\frac{dr}{v}.$$

The change in wavelength in time dt corresponding to dr is equal to the change in wavelength at one reflection times the number of reflections in time dt. Hence, we get the expression for $d\lambda$ as

$$d\lambda = \frac{2v\lambda\cos\theta}{c} \times \frac{c}{2r\cos\theta} \times \frac{dr}{v} = \lambda\frac{dr}{v}. \qquad (11.48)$$

This gives

$$\frac{d\lambda}{\lambda} = \frac{dr}{r}. \qquad (11.49)$$

Integrating both sides, we have $\ln\lambda = \ln r = \ln k$, where k is any constant. This gives rise to

$$\frac{\lambda}{r} = \text{constant}. \qquad (11.50)$$

From equation (11.45), we have $V^{1/3}T = $ constant. This leads to $\left(\frac{4\pi r^3}{3}\right)^{1/3} T = $ constant . Hence, we conclude that

$$rT = \text{constant}. \qquad (11.51)$$

Combining equations (11.50) and (11.51), we get

$$\lambda T = \text{constant}. \qquad (11.52)$$

Thus, if radiation of a particular wavelength at a certain temperature is adiabatically altered to a shorter wavelength, then the temperature changes in the inverse ratio. This is the usual statement of Wien's displacement law.

11.6.2   Rayleigh–Jeans law of black body radiation

In order to explain the spectra of black body radiation, Rayleigh–Jeans used the classical ideas of electromagnetic theory and made certain assumptions. These assumptions can be stated as follows:

1. The radiating system is composed of a collection of linear harmonic oscillators. These atomic oscillators in the walls of the enclosure continuously exchange energy in any amount with electromagnetic radiation. This radiation has a wavelength varying from zero to infinity. Ultimately an equilibrium condition is established when the energy density of radiation assumes an equilibrium value determined by the temperature of the black body radiation.

2. Electromagnetic radiations (waves) are reflected from the walls of the container. The incident and the reflected waves interfere to form stationary waves. The principles of superposition for the formation of stationary waves are employed to determine the number of possible independent vibrations between the frequency ranges ν and $\nu + d\nu$ per unit volume.

3. The law of equipartition of energy is then utilized to find the energy of each independent mode of vibration.

Using these assumptions, Rayleigh–Jeans derived the following expression for the energy density of black body radiation within the wavelength range between λ and $\lambda + d\lambda$ at an absolute temperature T

$$E_\lambda d\lambda = \frac{8\pi K_B T}{\lambda^4} d\lambda. \qquad (11.53)$$

Here, K_B is the Boltzmann constant with a value of $1.38 \times 10^{-23} \mathrm{JK}^{-1}$.

Derivation of Rayleigh–Jean's distribution law of black body radiation

In order to derive the energy distribution law as a function of wavelength, Rayleigh–Jean considered the following facts regarding black body radiation:

1. The radiation enclosed in a hollow cubic enclosure with perfectly reflecting walls is supposed to consist of a number of waves. These waves travel in all possible directions in the enclosure and undergo multiple reflections from different regions of the wall. Stationary vibrations are formed inside the enclosure with nodal points at the wall. These nodal points are formed due to the interference of the reflected waves with the corresponding incident waves. They calculated the number of such modes of vibration per unit volume within the frequency ranges between ν and $\nu + d\nu$ at a particular temperature T.

2. The average energy of such mode of vibration assigned to the black body radiation was calculated using the principle of equipartition of energy.

3. The total energy of the radiation was then calculated as a function of wavelength at a particular temperature T.

Following these three steps, Rayleigh–Jean's formula for the variation of energy of the black body radiation is calculated below:

Calculation of the number of modes of vibration per unit volume within the frequency ranges between ν and $\nu + d\nu$

We consider that the black body radiation is enclosed in a cubical box with perfectly reflecting walls of length L at a temperature T.

Inside the cubical box of side L, the waves bounce back and forth. The walls of the box are rigidly fixed and perfectly conducting. Therefore, the electric field of the electromagnetic wave *must be zero at the walls of the box*, and so we can only fit waves into the box that are multiples of half a wavelength. The first few examples of such waves are shown in Figure 11.10.

In the x-direction, the wavelengths of the waves that can be fitted into the box are those for which $\dfrac{n_x \lambda_x}{2} = L$, where n_x takes any positive integral value, $1, 2, 3, \cdots$. Similarly, for the y and z directions, we have $\dfrac{n_y \lambda_y}{2} = L$ and $\dfrac{n_z \lambda_z}{2} = L$, where n_y and n_z are also positive integers.

The expression for the waves that fit into the box in the x-direction is given by

$$A(x) = A_0 \sin(k_x x),$$

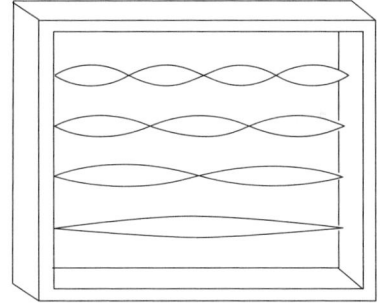

Figure 11.10 Standing waves formed inside the box of length L.

where k_x is the component of the wave-vector of the mode of oscillation in the x-direction.

Now, $k_x = \frac{2\pi}{\lambda_x}$. Hence, the values of k_x that fit into the box are those for which $\lambda_x = \frac{2L}{n_x}$. This gives the value of k_x as

$$k_x = \frac{2\pi}{2L} \times n_x = \frac{\pi n_x}{L}.$$

Similarly, the values of k_y and k_z along the y and z directions are, respectively, given by

$$k_y = \frac{2\pi}{2L} \times n_y = \frac{\pi n_y}{L}, \qquad \text{and} \qquad k_z = \frac{2\pi}{2L} \times n_z = \frac{\pi n_z}{L}.$$

Let us now plot a three-dimensional diagram with axes k_x, k_y, and k_z showing the allowed values of k_x, k_y, and k_z. These form a regular cubical array of points, each of them defined by the three integers, n_x, n_y, and n_z. This is shown in Figure 11.11. This is exactly the same as the velocity, or momentum space which is introduced for the particles in the kinetic theory of gases (KTG).

The waves can oscillate in three dimensions, but the components of their k-vectors, k_x, k_y, and k_z, must be such that they are associated with one of the points of the lattice in k-space. A wave oscillating in three dimensions with any of the allowed values of n_x, n_y, and n_z satisfies the boundary conditions, and so every point in the lattice represents a possible mode of oscillation of the waves within the box, consistent with the boundary conditions. The equation for the wave in three dimensions is given by

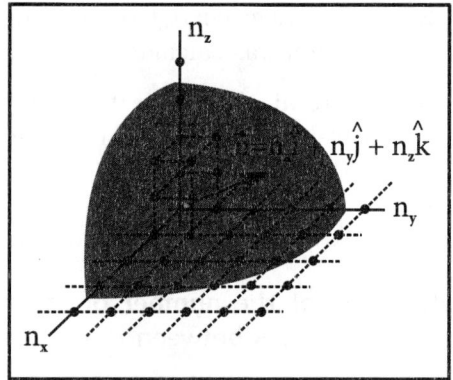

Figure 11.11 The positive values of n_x, n_y, and n_z determine the wave vectors and the number of modes of vibrations.

$$\frac{\partial^2 A(x,y,z,t)}{\partial x^2} + \frac{\partial^2 A(x,y,z,t)}{\partial y^2} + \frac{\partial^2 A(x,y,z,t)}{\partial z^2} = \frac{1}{c^2}\frac{\partial^2 A(x,y,z,t)}{\partial t^2}. \quad (11.54)$$

The trial solution describing the modes of oscillation of the wave in equation (11.54) can be written as

$$A(x,y,z,t) = A_0 \sin(k_x x)\sin(k_y y)\sin(k_z z)\sin(\omega t). \quad (11.55)$$

This contains both the space and the time part of the solution. To find the dispersion relation for the waves, that is, the relation between ω and k_x, k_y, and k_z, we insert the trial solution given by equation (11.55) into the wave equation (11.54). This procedure provides the following result

$$|k|^2 = k_x^2 + k_y^2 + k_z^2 = \frac{\omega^2}{c^2},$$

where k is the 3D wave-vector. Using the values of $k_x = \frac{\pi n_x}{L}$, $k_y = \frac{\pi n_y}{L}$, and $k_z = \frac{\pi n_z}{L}$, we get

$$\frac{\omega^2}{c^2} = |k|^2 = k_x^2 + k_y^2 + k_z^2 = \frac{\pi^2}{L^2}(n_x^2 + n_y^2 + n_z^2).$$

This leads to

$$n_x^2 + n_y^2 + n_z^2 = \frac{\omega^2 L^2}{\pi^2 c^2} = \frac{4L^2 \nu^2}{c^2} = \left(\frac{2L\nu}{c}\right)^2. \tag{11.56}$$

This equation (11.56) shows that in 3D cases, the wavelengths and frequencies are determined by three integers n_x, n_y, and n_z. Each choice of n_x, n_y, and n_z corresponds to a particular mode of vibration (frequency). The total number of frequencies is the total number of possible sets of n_x, n_y, and n_z. The number of modes of oscillation within the frequency ranges between ν and $\nu + d\nu$ can be calculated using equation (11.56).

We take the help of statistical physics to calculate the number of modes of oscillation $g(\nu)d\nu$ in the frequency interval between ν and $\nu + d\nu$. This is a straightforward calculation since we need only to count up the number of lattice points in the interval between k and $k + dk$ in the k-space corresponding to ν and $\nu + d\nu$, as shown in Figure 11.12.

Thus, the number of modes of oscillation within the frequency range between ν and $\nu + d\nu$ is the volume of the first octant $\left(\frac{1}{8} - th\right)$ of the spherical shell with radii equal to $\left(\frac{2L\nu}{c}\right)$ and $\left(\frac{2L(\nu + d\nu)}{c}\right)$. Hence, the number of modes in the given frequency range is given by

$$g(\nu)d\nu = \frac{1}{8} \times \frac{4}{3}\pi \left[\left(\frac{2L}{c}(\nu + d\nu)\right)^3 - \left(\frac{2L}{c}\nu\right)^3\right] = \frac{4\pi V}{c^3}\nu^2 d\nu,$$

where $V = L^3$ is the volume of the black body radiation enclosure. Since the electromagnetic waves are transverse in nature, there are two possible states of polarization for each mode. Therefore, for such radiation, the total number of modes of oscillation $g(\nu)d\nu$ within the frequency range between ν and $\nu + d\nu$ is

$$g(\nu)d\nu = 2 \times \frac{4\pi V}{c^3}\nu^2 d\nu = \frac{8\pi V}{c^3}\nu^2 d\nu. \tag{11.57}$$

So, the number of modes of oscillation per unit volume is given by

$$g(\nu)d\nu = \frac{8\pi}{c^3}\nu^2 d\nu. \tag{11.58}$$

Calculation of the average energy of such modes of vibration

We know that the radiation enclosed in a hollow cubic enclosure with perfectly reflecting walls consists of a number of waves. These waves form modes of vibration due to the interference of incident and reflecting waves. In thermodynamic equilibrium at a temperature T, these modes are treated as degrees of freedom with energy equal to $\frac{1}{2}K_B T$ per degree of freedom. This is because, if we wait for a long enough time, there are processes that enable energy to be exchanged between the apparently independent modes of oscillation. Thus, after a long time, each mode of oscillation will be in thermodynamic equilibrium and will attain the same average energy $\langle E \rangle$.

Each mode has two degrees of freedom: one for the kinetic energy and the other for the potential energy. Hence, the average energy per mode of vibration is given by

$$\langle E \rangle = \frac{1}{2}K_B T + \frac{1}{2}K_B T = K_B T. \quad (11.59)$$

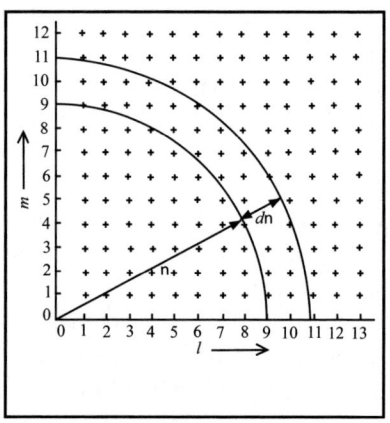

Figure 11.12 The allowed number of modes between ν and $\nu + d\nu$.

Calculation of the total energy per unit volume, that is, the energy density as a function of temperature

At a particular temperature T, the emitted energy of the black body radiation within the frequency ranges between ν and $\nu + d\nu$ is obtained by taking the product of equations (11.58) and (11.59), and the result is found to be

$$E_\nu d\nu = \frac{8\pi\nu^2}{c^3}d\nu \times \langle E \rangle = 8\pi\frac{\nu^2}{c^3}d\nu \times K_B T = \frac{8\pi K_B T}{c^3}\nu^2 d\nu. \quad (11.60)$$

Equation (11.60) represents Rayleigh–Jean's law for the distribution of energy of the black body radiation in terms of frequency. This energy density E_ν is proportional to the temperature T and to the square of the frequency ν.

Rayleigh–Jean's formula for black body radiation in terms of wavelength λ

The Rayleigh–Jean's law (see equation 11.60) for the distribution of energy of the black body radiation in terms of frequency is given by

$$E_\nu d\nu = \frac{8\pi K_B T}{c^3}\nu^2 d\nu.$$

Using the relation $\nu = \dfrac{c}{\lambda}$, we get $\quad \Rightarrow \quad d\nu = -\frac{c}{\lambda^2}d\lambda$.

Taking the magnitude and inserting it in the above expression, we get

$$E_\lambda d\lambda = 8\pi K_B T \frac{\left(\frac{c}{\lambda}\right)^2}{c^3}\frac{c}{\lambda^2}d\lambda = \frac{8\pi K_B T}{\lambda^4}d\lambda. \quad (11.61)$$

This equation (11.61) gives the energy density of black body radiation in terms of wavelength λ and temperature T according to Rayleigh–Jeans formula.

Experimental verification of Rayleigh–Jean's formula

An experimental plot of the black body radiation spectrum, together with the theoretical prediction of the Rayleigh–Jeans law, is shown in Figure 11.13. At long wavelengths, the Rayleigh–Jeans law is in reasonable agreement with experimental data, but at short wavelengths, major disagreement is apparent.

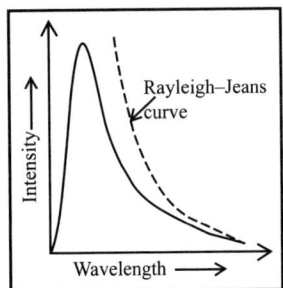

As λ approaches zero, the function $E_\lambda(T)$ given by equation (11.61) approaches infinity. Hence, according to classical theory, not only should short wavelengths predominate in a black body spectrum, but also the energy emitted by any black body should become infinite in the limit of zero wavelengths. In contrast to this prediction, the experimental data plotted in Figure 11.13 shows that as λ approaches zero, $E_\lambda(T)$ also approaches zero. This mismatch of the theory and experiment was so disconcerting that

Figure 11.13 Comparison between experimental results and the curve predicted by Rayleigh–Jeans law for the distribution of energy density of black body radiation.

scientists called it the *ultraviolet catastrophe*. This "catastrophe"—infinite energy— occurs as the wavelength approaches zero. The word *ultraviolet* was applied because ultraviolet wavelengths are short. Thus, Rayleigh–Jeans formula is not successful in explaining the full black body radiation spectrum and provides a hint for a better theory. In the next section, the ultraviolet catastrophe is discussed in detail.

11.6.3 Ultraviolet catastrophe

In the 19th century, physicists attempted to calculate how much power was radiated by a black body. They turned to two tools: classical harmonic oscillators and the equipartition theorem. The classical harmonic oscillator theory of radiation provided that the number of oscillation modes in a 3D box is proportional to the square of frequency of the wave and that the power of a wave was dependent on the frequency of it. The amount of energy of each mode of vibration was determined by the equipartition theorem and was found to be dependent on the temperature. These two things result that as the number of vibrational modes increases, there is a massive increase in degrees of freedom, and therefore, energy increases all the way up to infinity. Hence, these calculations indicate that any object above absolute zero emits an infinite amount of radiation, most of it in ultrahigh energy gamma rays. This blowup of energy as the frequency increases is termed the *ultraviolet catastrophe*. This is clearly unphysical as the total radiated power of a cavity is not observed to be infinite, a point that was made independently by Einstein and by Lord Rayleigh and Sir James Jeans in 1905. This divergence of energy of black body radiation leading to ultraviolet catastrophe is shown in Figure 11.14.

The phrase "ultraviolet catastrophe" refers to the fact that the Rayleigh–Jeans law accurately predicts experimental results at radiative frequencies below 10^5 GHz

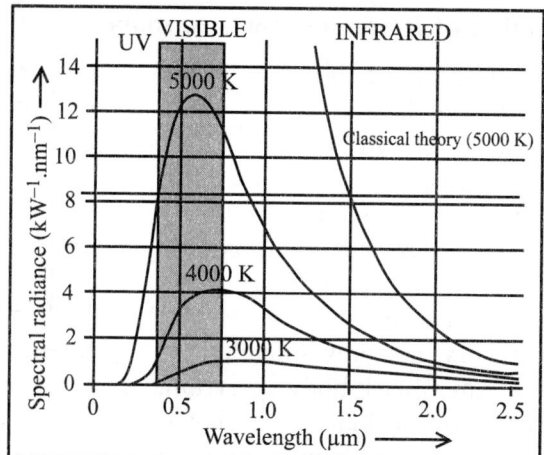

Figure 11.14 The ultraviolet catastrophe is the error in the "Rayleigh–Jeans" law (depicted as "classical theory" in the graph) for the energy emitted by an ideal black body at short wavelengths. The error, much more pronounced for shorter wavelengths, is the difference between the black curve, as classically predicted by the "Rayleigh–Jeans" law, and the measured curve as predicted by Planck's law.

but begins to diverge with experimental observations as these frequencies reach the ultraviolet region of the electromagnetic spectrum.

Hence, the ultraviolet catastrophe, also called the Rayleigh–Jeans catastrophe, is the prediction of late 19th-century/early 20th-century classical physics that an ideal black body at thermal equilibrium will emit radiation in all frequency ranges, emitting more energy as the frequency increases. Calculation of the total amount of radiated energy over all frequency ranges shows that a black body would release an infinite amount of energy, contradicting the principles of conservation of energy, and this fact indicates that a new model for the behavior of the black body was needed.

Mathematics of the catastrophe

Applying the law of classical theory of electromagnetic theory, Rayleigh–Jeans derived an expression for the energy density of the black body radiation as a function of wavelength λ and temperature T. This is recast as

$$E_\lambda d\lambda = \frac{8\,\pi\,K_B\,T}{\lambda^4}d\lambda. \tag{11.62}$$

This law states that at a particular temperature T, the energy density E_λ of the black body radiation is inversely proportional to the fourth power of wavelength λ. This law poses a severe problem while calculating the total energy radiated by the black body over all frequency (wavelength) ranges.

The classical expression for the total energy density of electromagnetic radiation can be obtained by integrating equation (11.62) over all wavelength limits and is found to

$$U = \int_0^\infty E_\lambda d\lambda = \int_0^\infty \frac{8\pi K_B T}{\lambda^4} d\lambda.$$

This expression leads to

$$U = \left[-\frac{8\pi K_B T}{3\lambda^3} \right]_0^\infty \to \infty. \tag{11.63}$$

Thus, according to this result, the total energy density of electromagnetic radiation inside an enclosed cavity is infinite! This is clearly an absurd result. If the treatment is correct, we would get an infinite amount of energy on opening the cover of the cavity. This prediction is known as the ultraviolet catastrophe because the Rayleigh–Jeans law diverges defectively from experimental observations in the ultraviolet region of the spectrum.

It should be mentioned that although the Rayleigh–Jeans law diverges at high frequencies, it is in excellent agreement with the measured spectrum at low frequencies and at high temperatures. This result was derived by Lord Rayleigh in 1900. This was one of the key problems in classical physics at the end of the 19th century. Let us think—what has gone wrong with it?

Failure of classical physics in explaining the spectrum of black body radiation

In a hot object, electrons can vibrate with a range of frequencies, ranging from very few vibrations per second to a huge number of vibrations per second. In fact, there is no limit to how great the frequency can be. According to classical physics, each

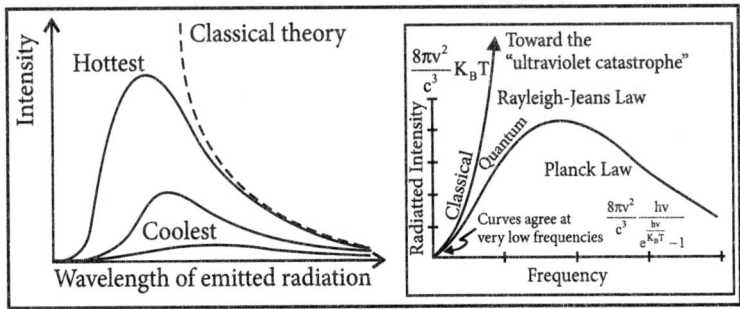

Figure 11.15 Variation of the energy density of emitted radiation from a black body as a function of wavelength at selective temperatures, but the dotted line shows the theoretical prediction using classical theory. The figure on the right shows the different portions of the curves relating to classical and quantum behavior.

frequency of vibration should have the same energy. Since there is no limit to how great the frequency can be, there is no limit to the energy of the vibrating electrons at high frequencies. This means that, according to classical physics, there should be no limit to the energy of the light produced by the electrons vibrating at high frequencies. However, the exact spectrum of black body radiation depends upon the properties of the material and the temperature, and this experimental phenomenon of black body radiation could not be adequately explained by classical physics. It was found that the observed intensity of black body radiation as a function of wavelength varies with temperature. It was observed further that attempts to explain or calculate this spectral distribution from classical theory were complete failures. A theory developed by Rayleigh and Jeans predicted that the intensity should go to infinity at short wavelengths. Since the intensity actually drops to zero at short wavelengths, the Rayleigh–Jeans result was called the "ultraviolet catastrophe". There was no agreement between theory and experiment in the ultraviolet region of the black body spectrum. This created a deadlock in explaining the shape of the black body spectrum and remained as a mystery to the physicists for a long time.

Unfortunately, the Rayleigh–Jeans formula fails horribly to predict the actual results of the experiments. Notice that the radiancy in this equation is inversely proportional to the fourth power of the wavelength, which indicates that at short wavelength (i.e., near 0), the radiancy will approach infinity. (The Rayleigh–Jeans formula is the region to the right of the graph.) The data (the other three curves in the graph) actually show a maximum radiancy, and below the λ_{max} at this point, the radiancy falls off, approaching 0 as λ approaches 0. This failure is called the *ultraviolet catastrophe*, and by 1900, it had created serious problems for classical physics because it called into question the basic concepts of thermodynamics and electromagnetic that were involved in reaching that equation.

11.7 Quantum theory of black body radiation

By the late 19th century, most of the physicists thought that the physics discipline was well equipped with the theories to explain the observed natural phenomena in physics. The universe appeared to the physicists as a simple and orderly place containing matter. These matters consisted of massive particles with accurately defined location and motion, and electromagnetic radiation whose exact position in space could not be fixed and not be assigned with mass. Matter and energy were thus considered distinct and unrelated phenomena. However, scientists faced some inconvenient observations that could not be explained by the theories available at that time. These observations were the black body radiation spectrum, Balmer's formula for the line spectrum of the hydrogen atom, and Michelson–Morley's null results and challenged to the very foundations of Newtonian concepts of **space, time, and mass** on one hand and determination of motion and position according to Newton's laws of motion on the other. The resolutions to these "three dark clouds", according to Lord Kelvin, unfolded a new era in Physics at the beginning of the 20th century. The generic term for subsequent developments is modern physics, and in contrast, physics up to the 19th century is called **classical physics**.

The mystery of black body radiation triggered the birth of modern physics in 1900, when Max Planck in an "act of despair" invented the idea of the smallest **quantum of energy**. Planck viewed the **concept of introducing quantum** to be merely a mathematical trick to resolve the scientific deadlock of classical physics in explaining the variation of the energy density of black body radiation as a function of wavelength and temperature. However, Planck adopted this trick without any real physical meaning to it. Nevertheless, it worked well in explaining the observed features of the black body radiation spectrum. Later, Einstein used a similar idea of "quanta of light" called **photons** to come up with a simple formula for the photoelectric effect, for which he was awarded the Nobel Prize in Physics in 1921. It is to be mentioned here that Einstein won the Nobel Prize for the formula but not for its derivation based on the concept of quanta because Swedish scientists did not believe in any reality of light quanta or light particles. In later years, Einstein confessed that he did not believe in light quanta, but the reservations of the inventors were overwhelmed by the snowball of quantum mechanics starting to roll in the 1920s. A hundred years later, black body radiation was back at the center of discussion as the cornerstone of climate alarmism based on the idea of atmospheric "back-radiation" from the so-called "greenhouse gases" that causes "global warming".

Max Karl Ernst Ludwig Planck

Max Karl Ernst Ludwig Planck was born in Kiel, Germany, on April 23, 1858. His father, J.W. Planck, was a Professor of Constitutional Law at the University of Kiel and later in Gottingen. Planck studied at the Universities of Munich and Berlin, where his teachers were Kirchhoff and Helmholtz, and received his doctorate of philosophy at Munich in 1879. He was Privatdozent in Munich from 1880 to 1885, then associate professor of theoretical physics at Kiel until 1889, in which year he succeeded Kirchhoff as professor at Berlin University. He remained in the post until retirement in 1926. Planck became president of the Kaiser Wilhelm Society for the Promotion of Science, a post he held until 1937. The Prussian Academy of Sciences appointed him a member in 1894 and a permanent secretary in 1912.

Planck's earliest work was on the subject of thermodynamics. He got interest in the subject from his studies under Kirchhoff and R. Clausius and published papers on entropy, thermoelectricity, and the theory of dilute solutions. At the same time, he paid attention to the problem of variation of energy of black body radiation as a function of temperature that was at variance with the predictions of classical physics. In 1900, Planck was able to deduce the relationship based on the revolutionary idea that the energy emitted by a resonator could only take on discrete values or quanta. The energy for a resonator of frequency ν is $h\nu$, where h is a universal constant called Planck's constant. This marked a turning point in the history of physics. The importance of the discovery, with its far-reaching effect on classical physics, was not appreciated at first. Albert Einstein used this idea to explain the photoelectric effect. Planck's work was published in the Annalen der Physik and summarized in two books: Thermodynamik (Thermodynamics) (1897) and Theorie der Wärmestrahlung (Theory of heat radiation) (1906).

Planck was elected to Foreign Membership of the Royal Society in 1926 and was awarded the Society's Copley Medal in 1928. He faced a troubled and tragic period in his life during the Nazi government in Germany. He felt it is his duty to remain in his country but openly opposed some of the Government's policies, particularly the persecution of the Jews. Planck suffered great hardship after his home was destroyed by bombing during the war. He was revered by his colleagues not only for the importance of his discoveries but for his great personal qualities. He was also a gifted pianist and is said to have at one time considered music as a career. In 1885, he was appointed as Associate Professor in his native town, Kiel, and he married a friend of his childhood, Marie Merck, who died in 1909. He remarried her cousin Marga von Hoesslin. Three of his children died young, leaving him with two sons. He suffered a personal tragedy when one of them was executed for his part in an unsuccessful attempt to assassinate Hitler in 1944. Such a miserable life Planck had! Professor Planck died at Gottingen on October 4, 1947.

According to modern quantum theory, electromagnetic radiation is considered to be the flow of photons through space. These photons are called light quanta and are packets of energy $h\nu$, where the symbol h is Planck's constant, and ν is the same as that of the frequency of the electromagnetic wave of classical theory. Photons having the same energy $h\nu$ are all alike, and their number density corresponds to the intensity of the radiation. These photons always move with the universal speed of light.

11.7.1 Planck's quantum hypothesis: A rescue from the dead-lock

In this section, the experimental results of the variation of the energy density of black body radiation as a function of wavelength at selective constant temperatures are discussed in detail. The failure of classical physics to explain the observed features is highlighted. Planck quantum postulates are presented as a solution for the explanation of the variation of the spectral distribution of black body radiation. Planck law of black body radiation is derived using the historical quantum assumption, and the experimental verification of the law is described. With the help of Planck law, already deduced laws of black body radiation such as Wien's distribution law, Rayleigh–Jeans law, Stefan–Boltzmann law, and Wien's displacement law are derived. Planck units, such as Planck length, Planck time, Planck mass, and Planck temperature, are mentioned to satisfy the needs of the interested readers. The average energy of a possible frequency of radiation in an absolute vacuum is calculated and found to be correct with the prediction of quantum electrodynamics. In the end, a huge number of solved numerical problems and multiple choice questions with answers are given for the benefit of the students.

One of the major assumptions of classical physics was that energy increased or decreased in a smooth, continuous manner. For example, classical physics predicted that as wavelength decreased, the intensity of the radiation an object emits should increase in a smooth curve without limit at all temperatures (ultraviolet catastrophe). Thus, classical physics could not explain the sharp decrease in the intensity of radiation emitted at shorter wavelengths (primarily in the ultraviolet region of the spectrum). In 1900, however, the German physicist Max Planck (1858–1947) explained the ultraviolet catastrophe by proposing (in what he called "an act of despair") that the energy of electromagnetic waves is quantized rather than continuous. This means that for each temperature, there is a maximum intensity of radiation that is emitted in a black body object, corresponding to the peaks, so the intensity does not follow a smooth curve as the temperature increases, as predicted by classical physics. Thus, energy could be gained or lost only in integral multiples of some smallest unit of energy, **a quantum**.

It is worth mentioning that both Planck and Einstein introduced discrete quanta of energy as a "mathematical trick" without physical reality in order to avoid the ultraviolet catastrophe long before the quantum mechanics of atoms was formulated in the 1920s in the form of Schrodinger's wave equation.

Experimental evidences in support of quantization of energy

The concept of quantization of energy put forward by Max Planck helped to resolve a number of previously unexplained natural phenomena that are briefly mentioned below:

1. The dead-lock situation in black body radiation (infinite energy at lower wavelengths) was resolved successfully.

2. Using the idea of the quantization of energy, Einstein explained the anomaly in the specific heat of solids at low temperatures. Debye improved the Einstein model and explained the experimentally observed T^3 dependence of the specific heat at low temperatures.

3. The scattering of X-rays from a graphite block was found to exhibit a spectrum with intensity peaks at two wavelengths: one at the same wavelength λ as the incident radiation and the other at a longer wavelength λ'. This wavelength λ' was found to depend on the direction of scattering, whereas the wavelength λ was independent of the direction. This effect known as the **Compton effect** could not be explained by the classical theory of radiation. This effect was successfully explained by considering the special theory of relativity and electromagnetic waves consisting of photons that are particles of light in the same sense that electrons or other massive particles are particles of matter.

4. Einstein used the idea of the quantization of energy of electromagnetic radiation and explained the characteristic features of the photoelectric effect successfully.

5. In 1913, Bohr used the concept of quantization of energy and developed an atomic model with certain assumptions and finally explained the line spectrum of hydrogen atoms. In 1914, Franck and Hertz demonstrated experimentally the existence of discrete energy states in atoms.

During this period, several fundamental unresolved problems occupied some of the greatest minds of that time. For example, the KTG and the equipartition theorem put forth by Clausius, Maxwell, and Boltzmann was not widely accepted. The atomic and molecular theory of the structure of matter came under attack, as the origin of the "resonances" in molecules, which were assumed to be the origin of spectral lines, was unknown. It should be pointed out that there was no direct evidence for fundamental particles until 1897, when J. J. Thomson discovered the electron. There are other experimental observations that led to the development of the quantum theory of radiation. These observations are briefly described below.

Planck did not realize how radical and far-reaching his proposals were. He viewed his strange assumptions as mathematical constructions to provide a formula that fits the experimental data. It was not until later, when Einstein used very similar ideas to explain the Photoelectric Effect in 1905, that it was realized that these assumptions described "real Physics" and were much more than mathematical constructions to provide the right formula.

Planck's revolutionary hypothesis: a gateway of modern physics

Modern physics in the form of quantum mechanics and relativity theory was born at the beginning of the 20th century from an apparent collapse of classical deterministic physics. The collapse resulted from a couple of scientific paradoxes, which appeared unsolvable using classical physics, both connected to light as electromagnetic waves described by Maxwell equations. These paradoxes are:

1. Ultraviolet catastrophe of black body radiation: infinite energy and

2. Nonexistence of ether as a medium carrying electromagnetic waves.

Other scientists, such as Albert Einstein, Niels Bohr, Louis de Broglie, Erwin Schrodinger, and Paul M. Dirac, advanced Planck theory and made possible the development of quantum mechanics—a mathematical application of the quantum theory that maintains that energy is both matter and a wave, depending on certain variables. Quantum mechanics thus takes a probabilistic view of nature, sharply contrasting with classical mechanics, in which all precise properties of objects are, in principle, calculable. Today, the combination of quantum mechanics with Einstein's theory of relativity is the basis of modern physics.

Planck thus returned to Newton's corpuscular theory of light, which had been replaced by Maxwell's wave theory in the late 19th century. The new particle statistics of thermodynamics developed by Ludwig Boltzmann was also in favor of the corpuscular theory of particles. In an "act of despair' Planck gave up deterministic continuum physics for statistics of particles and thus opened the door to modern physics with wave–particle duality viewed as a resolution of the inescapable contradiction between wave and particle. Einstein picked up Planck's quanta as a patent clerk in one of his five articles during his "annus mirabilis" in 1905 and suggested an explanation of a law of photo-electricity which had been discovered experimentally. This gave Planck's quanta a boost, and in 1923, Einstein won the Nobel Prize in Physics, not for his explanation based on light as particles, which the Nobel Committee did not buy, but for the "discovery" of a law that had already been discovered experimentally.

Both Planck and Einstein introduced discrete quanta of energy as a "mathematical trick" without physical reality in order to avoid the ultraviolet catastrophe long before the quantum mechanics of atoms was formulated in the 1920s in the form of Schrodinger's wave equation, even before the existence of atoms had been experimentally confirmed. Planck, Einstein, and Schrodinger refused to embrace the new quantum mechanics with the wave function as the solution of Schrodinger's wave equation being interpreted as a probability distribution of discrete particles. They were, therefore, left behind as modern physics took off on a mantra of wave–particle duality into a new era of atomic physics, with the atomic bomb as evidence that the direction was correct. The inventors of quantum mechanics were thus expelled from the new world they had created, but the question remains today: Is light waves or particles? What is really wave–particle duality?

There is ample evidence that light is a wave, well described by Maxwell's equations. There are some aspects of light connected to the interaction of light and matter in the emission and absorption of light that are viewed to be difficult to describe as wave mechanics, with black body radiation as the basic problem. If black body radiation captured in Planck Law of Radiation can be derived by wave mechanics, then the main motivation of particle statistics disappears, and a return to rational determinism may be possible. And after all, Schrodinger's equation is a wave equation, and Schrodinger firmly believed that there are no particles, only waves as solutions of his wave equation.

11.7.2 Planck's law of black body radiation

On the very day when I formulated this law, I began to devote myself to the task of investing it with a true physical meaning. This quest automatically led me to study the interrelation of entropy and probability - in other words, to pursue the line of though inaugurated by Boltzmann.

—Max Planck

Planck's quantum postulates

A black body in thermal equilibrium emits electromagnetic radiation called black body radiation. The radiation has a specific spectrum and intensity that depends only on the temperature of the body. Max Planck, in 1901, accurately described the radiation by assuming that electromagnetic radiation was emitted in discrete packets (or quanta). Planck quantum hypothesis is a pioneering work, heralding the advent of a new era of modern physics and quantum theory. Planck postulate is one of the fundamental principles of quantum mechanics. The postulate was introduced by Max Planck in 1900 for the derivation of the law of black body radiation. The postulates are the following:

1. A black body radiation chamber is filled up not only with radiation but also with simple oscillators or resonators of molecular dimensions. These oscillators are termed as Planck oscillators or resonators and can vibrate with all possible frequencies.

2. The oscillators cannot radiate or absorb energy continuously, but the energy of the oscillators in a black body is emitted or absorbed in the form of packets or quanta called photons. The energy of such photons is quantized and is given by $E_n = nh\nu$, where n is an integer $(1, 2, 3, \cdots)$, h is Planck constant, and nu is the frequency of the oscillator. Photon is not a material body but is considered to be a mass-less packet of energy.

3. Whenever a black body emits or absorbs energy, it does so in whole number multiples of photons, that is, $nh\nu$, where $n = 1, 2, 3, \cdots$ and never $1.2, 2.5, 3.7,$ etc.

This assumption is the most revolutionary in character and states that the exchange of energy between radiation and matter cannot take place continuously but only in certain multiples of the fundamental frequency of the oscillators. These assumptions allowed Planck to derive a formula for the entire spectrum of the radiation emitted by a black body. Planck was unable to justify this assumption based on classical physics; he considered quantization as being purely a mathematical trick rather than (as is now known) a fundamental change in the understanding of the world.

Qualitative understanding of the postulate

One may ask the question: what is the origin of energy quantization of electromagnetic waves within the cavity of a black body radiation? The answer to this question is related to the emission processes taking place at the walls of the cavity. Radiation is emitted by the atoms in the walls that undergo harmonic oscillations. A simple calculation of quantum mechanics shows that the spectrum of the harmonic oscillator (see Figure 11.16) is characterized by evenly spaced energy levels with energy differences equal to $h\nu$.

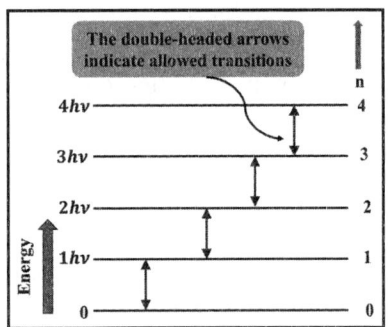

Figure 11.16 Allowed energy levels for an oscillator with frequency ν.

Significance of Planck's quantum postulate

This is to highlight the fact that Planck's quantum postulate was applied to several physical observations with grand success. In 1905, in one of his three most important papers, Albert Einstein adapted Planck's quantum postulate to explain the photoelectric effect, but Einstein proposed that the energy of photons themselves was quantized and that quantization was not merely a feature of microscopic oscillators. Planck's postulate was further applied to understand the Compton effect and was applied by Niels Bohr to explain the emission spectrum of hydrogen atoms, and the correct value of the Rydberg constant was derived. These physical observations led to the foundation of Planck's quantum postulate strong.

Physicists realized later that quantization forms a basis for all the fundamental laws of nature. The laws of mechanics discovered by Isaac Newton worked very well in the household world because the effect of quantization is negligible in this case. The effect of quantization becomes dominant in the world of electrons, atoms, and molecules, and the character of natural laws changes in a drastic way. Various mathematical techniques have been developed to represent these laws in the theory of quantum mechanics, which incorporates Newton's laws of classical mechanics as a special case under the conditions of the household macroscopic world. In quantum mechanics, Planck constant h plays the role of quantum of action. This is to mention that h has a unit of angular momentum.

Derivation of Planck's law of black body radiation

Max Planck in 1900 derived an empirical formula to explain the experimentally observed distribution of energy in the spectrum of black body radiation. The formula is deduced using the following assumptions that are called Planck hypothesis:

1. A radiation chamber of volume V and at temperature T is filled up with black body radiation. The walls of the chamber are made up of molecules that can vibrate with all possible frequencies and these molecules are called resonators and can exchange energy with the radiation. The vibration of the resonator entails one degree of freedom only.

2. The oscillators cannot radiate or absorb energy continuously. An oscillator of frequency ν can only radiate or absorb energy in units of quanta of magnitude $h\nu$, where h is a universal constant, called Planck constant. This assumption is revolutionary in character. In other words, this assumption states that the exchange of energy between radiation and matter cannot take place continuously but is limited to a discrete set of values $0, h\nu, 2h\nu, 3h\nu, 4h\nu, \cdots, nh\nu$, that is, in multiples of some small unit, called quantum.

The following steps are to be executed sequentially in order to derive Planck law of black body radiation:

1. The radiation enclosed in a hollow cubic enclosure with perfectly reflecting walls is supposed to consist of a number of waves. These waves travel in all possible directions and undergo multiple reflections from different regions of the wall. Stationary vibrations are formed inside the enclosure with nodal points at the walls. These nodal points are formed due to the interference of the reflected waves with the corresponding incident waves. The number of such modes of vibration per unit volume $g(\nu)d\nu$ within the frequency ranges between ν and $\nu + d\nu$ at a particular temperature T is then calculated.

2. Planck assumed that each mode of vibration behaves as an oscillator, and these oscillators are distributed in various energy states with energies 0, $h\nu$, $2h\nu$, $3h\nu$, $4h\nu$, $\cdots$, $nh\nu$ according to Maxwell–Boltzmann distribution law. The average energy $\langle E \rangle$ of these oscillators corresponding to each mode of vibration is calculated using Planck quantum hypothesis.

3. Finally, the energy of the black body radiation within the frequency ranges between ν and $\nu+d\nu$ is calculated at a particular temperature T by multiplying $g(\nu)d\nu$ with $\langle E \rangle$.

Calculation of the number of modes of oscillation $g(\nu)d\nu$ within the frequency range between ν and $\nu + d\nu$

It is worth mentioning that the procedure for calculating the number of modes of oscillation $g(\nu)d\nu$ within the frequency range between ν and $\nu + d\nu$ is identical for both cases of Rayleigh–Jeans and Planck. So, the results given by equation (11.57) are used to get the number of modes of oscillation $g(\nu)d\nu$ in the given frequency range in the volume V. This is recast as

$$g(\nu)d\nu = 2 \times \frac{4\pi V \nu^2 d\nu}{c^3} = \frac{8\pi V \nu^2 d\nu}{c^3}. \tag{11.64}$$

Calculation of the average energy of Planck quantum oscillator

If N is the total number of Planck resonators and E is their total energy, then the average energy of a Planck oscillator is given by $\langle E \rangle = \frac{E}{N}$. According to Maxwell–Boltzmann law, the number of oscillators having energies $0, h\nu, 2h\nu, 3h\nu, \cdots, Nh\nu$ are in the ratio

$$1 : e^{-\frac{h\nu}{K_B T}} : e^{-\frac{2h\nu}{K_B T}} : e^{-\frac{3h\nu}{K_B T}} : \cdots : e^{-\frac{Nh\nu}{K_B T}}.$$

If N_0 is the number of oscillators having energy zero, then the number of oscillators N_1 with energy $E_1 = h\nu$ is given by $N_1 = N_0 e^{-\frac{h\nu}{K_B T}}$. Similarly, the number of oscillators N_2 with energy $E_2 = 2h\nu$ is given by $N_2 = N_0 e^{-\frac{2h\nu}{K_B T}}$ and, in general, the number of oscillators N_n with energy $E_n = nh\nu$ is given by $N_n = N_0 e^{-\frac{nh\nu}{K_B T}}$ and so on.

Thus, the total number of oscillators N is given by

$$N = N_0 + N_0 e^{-\frac{h\nu}{K_B T}} + N_0 e^{-\frac{2h\nu}{K_B T}} + \cdots + N_0 e^{-\frac{nh\nu}{K_B T}} + \cdots.$$

$$\Longrightarrow N = N_0 \left(1 + e^{-\frac{h\nu}{K_B T}} + e^{-\frac{2h\nu}{K_B T}} + \cdots + e^{-\frac{nh\nu}{K_B T}} + \cdots\right) = \frac{N_0}{1 - e^{-\frac{h\nu}{K_B T}}}. \tag{11.65}$$

Equation (11.65) gives the total number of oscillators N with total energy E.

The expression for total energy E is calculated in the following way:

$$E = N_0 \times E_0 + N_1 \times E_1 + N_2 \times E_2 + N_3 \times E_3 + \cdots + N_n \times E_n + \cdots$$

$$= N_0 \times 0 + N_0\, e^{-\frac{h\nu}{K_B T}} \times h\nu + N_0\, e^{-\frac{2h\nu}{K_B T}} \times 2h\nu + N_0\, e^{-\frac{3h\nu}{K_B T}} \times 3h\nu$$

$$+ \cdots + N_0\, e^{-\frac{nh\nu}{K_B T}} \times nh\nu + \cdots .$$

This leads to

$$E = N_0\, h\nu \left[e^{-\frac{h\nu}{K_B T}} + 2\, e^{-\frac{2h\nu}{K_B T}} + 3\, e^{-\frac{3h\nu}{K_B T}} + \cdots + n\, e^{-\frac{nh\nu}{K_B T}} + \cdots \right]$$

$$= N_0\, h\nu\, \frac{e^{-\frac{h\nu}{K_B T}}}{\left(1 - e^{-\frac{h\nu}{K_B T}}\right)^2}. \tag{11.66}$$

Dividing this equation (11.66) by equation (11.65), we get the average energy of Planck's oscillator as

$$\langle E \rangle = \frac{E}{N} = N_0\, h\nu\, \frac{e^{-\frac{h\nu}{K_B T}}}{\left(1 - e^{-\frac{h\nu}{K_B T}}\right)^2} \times \frac{1}{\dfrac{N_0}{1 - e^{-\frac{h\nu}{K_B T}}}} = \frac{h\nu}{e^{\frac{h\nu}{K_B T}} - 1}. \tag{11.67}$$

Equation (11.67) is a very important result in black body radiation. It also provides the average energy of a photon.

Calculation of total energy of the black body radiation within frequency ranges between ν and $\nu + d\nu$

The energy $E(\nu)d\nu$ of the black body radiation contained within the frequency ranges between ν and $\nu + d\nu$ in volume V of the radiation chamber at a temperature T can be obtained by multiplying equation (11.67) with equation (11.64). This provides

$$E(\nu)d\nu = g(\nu)d\nu) \times \langle E \rangle = \frac{8\pi V \nu^2 d\nu}{c^3} \times \frac{h\nu}{\left(e^{\frac{h\nu}{K_B T}} - 1\right)}.$$

Hence, the energy density $u(\nu)d\nu$ of the black body radiation within the frequency ranges between ν and $\nu + d\nu$ at a temperature T is given by

$$u(\nu)d\nu = \frac{E(\nu)d\nu}{V} = \frac{8\pi \nu^2 d\nu}{c^3} \times \frac{h\nu}{\left(e^{\frac{h\nu}{K_B T}} - 1\right)} = \frac{8\pi h}{c^3} \times \frac{\nu^3}{\left(e^{\frac{h\nu}{K_B T}} - 1\right)} d\nu. \tag{11.68}$$

Equation (11.68) provides the famous Planck law of black body radiation expressed in terms of frequency ν.

Planck's law of black body radiation in terms of wavelength λ

Planck law of black body radiation in terms of wavelength λ can be obtained from equation (11.68) by changing frequency ν to wavelength λ using the relation $\nu = \frac{c}{\lambda}$. This can be achieved in the following ways:

We know that, $\nu = \frac{c}{\lambda}$. This gives $|d\nu| = \left|-\frac{c}{\lambda^2}d\lambda\right| = \frac{c}{\lambda^2}d\lambda$. Using these results in equation (11.68), the energy per unit volume $u(\lambda)d\lambda$ of the chamber of the black body radiation having wavelengths in the ranges between λ and $\lambda + d\lambda$ is given by

$$u(\lambda)d\lambda = \frac{8\pi h}{c^3} \times \frac{\left(\frac{c}{\lambda}\right)^3}{\left(e^{\frac{hc}{\lambda K_B T}} - 1\right)} \times \frac{c}{\lambda^2}d\lambda = \frac{8\pi hc}{\lambda^5} \frac{d\lambda}{\left(e^{\frac{hc}{\lambda K_B T}} - 1\right)}. \qquad (11.69)$$

This is Planck law of black body radiation expressed in terms of wavelength λ.

Conceptual understanding of Planck's law of black body radiation

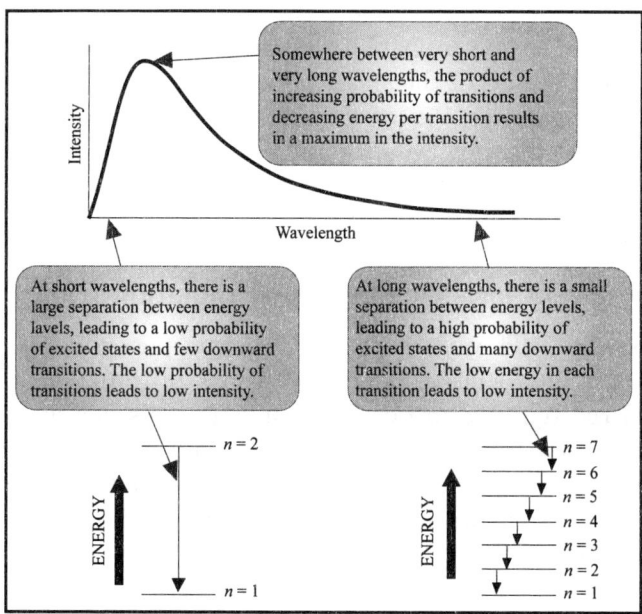

Figure 11.17 In Planck model, the average energy associated with a given wavelength is the product of the energy of a transition and a factor related to the probability of the transition occurring.

A schematic diagram of the variation in intensity of black body radiation, according to Planck law, is shown in Figure 11.17 as a function of wavelength λ. It is observed from the figure that for a black body at temperatures up to several hundred degrees, the majority of the radiation is in the infrared radiation of the electromagnetic spectrum. At higher temperatures, the total radiated energy increases, and the intensity peak of the emitted spectrum shifts to shorter wavelengths so that a significant portion is radiated as visible light. The figure contains three distinct regions: one at short wavelengths, one at somewhere between very short and very long wavelengths, and the other at very long wavelengths. According to Planck, the average energy of the radiation is the average energy difference between levels of the oscillator, weighted according to the probability of the wave being emitted. This weighting is based on the occupation of higher-energy states as described by the Boltzmann distribution that indicates that the probability of a state being occupied is proportional to the factor $e^{-\left(\frac{E}{K_B T}\right)}$, where E is the energy of the state (with reference to the energy of the ground state).

At low frequencies, the energy levels are close together (see on the right of Figure 11.17), and many of the energy states are excited because the Boltzmann factor $e^{-\left(\frac{E}{K_B T}\right)}$ is relatively large for these states. Therefore, there are many contributions to the outgoing radiation, although each contribution has very low energy.

For the high-frequency radiation, the allowed energies are very far apart (see on the left of Figure 11.17). The probability of thermal agitation exciting these high energy levels is small because of the small value of the Boltzmann factor for a large value of E. At high frequencies, the low probability of excitation results in very little contribution to the total energy, even though each quantum is of large energy. This low probability "turns the curve over" and brings it down to zero again at short wavelengths.

11.7.3   Deduction of laws of black body radiation from Planck's law

Planck's law describes the variation of the power per unit area (intensity, I) emitted from a thermal source (black body) as a function of the emission wavelength λ and the temperature of the source T. This law is originated from a first-principles derivation and agrees very well with experimental observations. Other laws of black body radiation, such as Wien's distribution law, Rayleigh–Jeans formula, and Stefan–Boltzmann law (all these laws are derived from the classical concepts), can be deduced from Planck's law of black body radiation under specific conditions. These laws of black body radiation are derived from Planck's radiation law mentioned below.

Deduction of Wien's distribution law from Planck's law

Wien's law describes the dependence of wavelength corresponding to maximum emission intensity λ_{max} of black body radiation as a function of the temperature of the source of black body radiation. This law is empirical; that is, it is originated from experimental observations only, not from first principles. This law does not predict the detailed shape of the intensity of the emission spectrum as a function of wavelength, while Planck's law does. The graphical comparison between Planck's law and Wien's distribution law is shown in Figure 11.18, where the solid curves give the emission spectra predicted by Planck's law as a function of frequency, and the short-dashed line is due to Wien. In the same graph, the intensity variation is plotted by a dotted line due to Rayleigh–Jeans law. The existence of frequency corresponding to the maximum emission intensity ν_{max} is also shown in the figure.

In 1893, Wien showed that the spectral distribution of energy emitted from a black body at a temperature T can be expressed as

$$u_\lambda d\lambda = C\frac{1}{\lambda^5}f(\lambda T)d\lambda,$$

where $u_\lambda d\lambda$ is the energy density of radiation between wavelengths λ and $\lambda + d\lambda$, C is a constant and $f(\lambda T)$ is a function of the product λT.

In terms of wavelength λ, Planck's law of black body radiation is expressed as

$$u(\lambda)d\lambda = \frac{8\pi hc}{\lambda^5}\frac{d\lambda}{\left(e^{\frac{hc}{\lambda K_B T}} - 1\right)}.$$

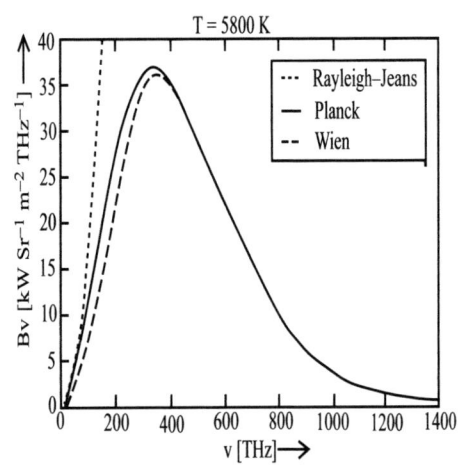

Figure 11.18 Relative comparison of the variation of intensity of black body radiation at a temperature 5800 K due to Rayleigh–Jeans law with Wien's approximation and Planck's law.

In the lower wavelength region, we have $e^{\frac{hc}{\lambda K_B T}} \gg 1$, and therefore, 1 in the denominator on the right-hand side of Planck's radiation law can be neglected. Thus, we get

$$u(\lambda)d\lambda = \frac{8\pi hc}{\lambda^5}\frac{d\lambda}{\left(e^{\frac{hc}{\lambda K_B T}}\right)} = \frac{8\pi hc}{\lambda^5}e^{-\frac{hc}{\lambda K_B T}}d\lambda = \frac{C_1}{\lambda^5}e^{-\frac{C_2}{\lambda}}d\lambda,$$

where $C_1 = 8\pi hc$ and $C_2 = \frac{hc}{K_B T}$ are constants. This expression possesses a similar form to that of Wien's distribution law. Hence, Wien's distribution law of black body radiation can be derived from Planck's law.

Deduction of Wien's displacement law from Planck's law

When radiation from a black body is passed through a prism, a continuous spectrum is obtained in which the intensity of radiation is different in different parts of the spectrum. The energy of the spectrum is distributed in various wavelengths varying from zero to infinity. The law that connects the intensity with wavelength is known as the law of distribution of intensity of black body radiation. At a particular temperature, it is observed that the intensity of the black body radiation increases with the increase in wavelength becomes maximum at a particular wavelength and finally decreases with a further increase in wavelength. Thermodynamic analysis shows that with the increase in temperature of the black body, the wavelength at which the intensity of the black body radiation is at its maximum shifts toward the lower value.

Wien's displacement law states that **the black body radiation curve for different temperatures peaks at a wavelength inversely proportional to the temperature**, that is, the wavelength at which the peak in the radiation curve occurs shifts toward a lower value with the increase in temperature. This shift in the peak is a direct consequence of the Planck radiation law. Mathematically, Wien's displacement law can be expressed as

$$\lambda_m\, T = \text{constant}, \tag{11.70}$$

where λ_m is the wavelength at which the intensity of the black body radiation is maximum corresponding to a given temperature T.

From Planck's law of black body radiation, we have

$$u(\lambda)d\lambda = \frac{8\pi hc}{\lambda^5}\, \frac{d\lambda}{\left(e^{\frac{hc}{\lambda K_B T}} - 1\right)} = \frac{8\pi K_B^5 h T^5}{h^4\, c^4}\, \frac{x^5}{e^x - 1} \ \text{with}\ x = \frac{hc}{\lambda K_B T}.$$

Differentiating the above expression with respect to x, we get

$$\frac{\partial u(\lambda)}{\partial x} = \frac{8\pi K_B^5\, T^5}{h^4\, c^4}\, \frac{x^4}{e^x - 1} \left[5 - \frac{xe^x}{e^x - 1}\right].$$

For optimum value of $u(\lambda)$ at $\lambda = \lambda_m$, we have

$$5 - \frac{xe^x}{e^x - 1} = 0 \quad \Longrightarrow \quad 5e^x - 5 - xe^x = 0; \quad \Longrightarrow \quad 5 = (5 - x)e^x. \tag{11.71}$$

Equation (11.71) is a transcendental equation. The solution of this transcendental equation yields the value of wavelength λ corresponding to the maximum intensity of $u(\lambda)$, that is, it gives λ_m. This transcendental equation has two solutions. The trivial solution, $x = \frac{hc}{\lambda K_B T} = 0$, is not a general one since it is only valid in the high-temperature limit: $T \to \infty$. The general solution can be obtained

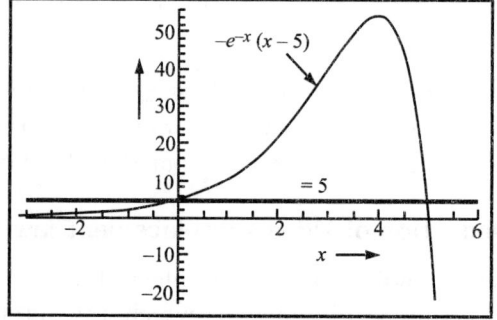

Figure 11.19 Graphical solutions of the transcendental equation (11.71).

graphically by plotting the two equations $y = 5$ and $y = -(x - 5)e^x$ on the same graph and looking for the (nonzero) value of x where the two plots intersect. Such a plot is shown in Figure 11.19.

It is observed from the graph that the non-trivial solution corresponds to $x = \frac{hc}{\lambda K_B T} \approx 5$. The actual value is $x = 4.96511$. Putting this value of $x = 4.96511$ in the expression $x = \frac{hc}{\lambda K_B T}$ and using $\lambda = \lambda_m$, we get

$$4.96511 = \frac{hc}{\lambda_m K_B T}; \implies \lambda_m T = \frac{hc}{K_B \times 4.96511} = 2.899 \times 10^{-3} \text{ mK} = \text{constant}.$$

In this expression, values of the universal constants h, c, and K_B are inserted to get the numerical value in the right-hand side. **The relation $\lambda_m T = $ constant is known as Wien's displacement law**. Thus, it is clearly observed that Wien's displacement law can be successfully derived from Planck's law of black body radiation. Table 11.1 summarizes the black body temperatures necessary to provide a peak for emitted radiation in various regions of the spectrum. The third column gives the energy of the photon corresponding to the wavelength given by Wien's displacement law.

Rayleigh–Jean's distribution law from Planck's law

In order to explain the variation of emission spectra of black body radiation as a function of wavelength λ, Rayleigh–Jeans derived a theoretical formula using the classical ideas of electromagnetic theory. This formula is given by

$$u(\lambda)d\lambda = \frac{8\pi K_B T}{\lambda^4}d\lambda. \tag{11.72}$$

Table 11.1 Values of wavelength and temperature corresponding to the peak of the black body radiation in the whole range of spectrum according to Wien's law.

Region	Wavelength (cm)	Energy (eV)	Temperature of black body (K)
Radio	> 10	$< 10^{-5}$	< 0.03
Microwave	$10 - 0.01$	$10^{-5} - 0.01$	$0.03 - 30$
Infrared	$0.01 - 7 \times 10^{-5}$	$0.01 - 2$	$30 - 4100$
Visible	$7 \times 10^{-5} - 4 \times 10^{-5}$	$2 - 3$	$4100 - 7300$
Ultraviolet	$4 \times 10^{-5} - 1 \times 10^{-7}$	$3 - 10^3$	$7300 - 3 \times 10^5$
X-rays	$1 \times 10^{-7} - 1 \times 10^{-9}$	$10^3 - 10^5$	$3 \times 10^5 - 3 \times 10^8$
Gamma rays	$< 10^{-9}$	$> 10^5$	$> 3 \times 10^8$

Here, symbols have their usual meaning.

In Planck's radiation law (see equation 11.69), we have a term involving exponential: $\dfrac{1}{\left(e^{\frac{hc}{\lambda K_B T}}\right)}$. Further, in the long wavelength region, the term $\left(\frac{hc}{\lambda K_B T}\right)$ is small at a finite temperature T. Expanding the exponential term of equation (11.69), Planck's law can be expressed to the following form

$$u(\lambda)d\lambda = \frac{8\pi hc}{\lambda^5} \frac{d\lambda}{\left[1 + \left(\frac{hc}{\lambda K_B T}\right) + \cdots - 1\right]} = \frac{8\pi hc}{\lambda^5} \times \frac{\lambda K_B T}{hc} d\lambda = \frac{8\pi K_B T}{\lambda^4} d\lambda. \quad (11.73)$$

This equation (11.73) is exactly the same as that given by equation (11.72), which is Rayleigh–Jeans law for black body radiation. This shows that the Rayleigh–Jeans law for black body radiation can be derived from Planck's law.

Stefan–Boltzmann law from Planck's law

According to Stefan's law, the total amount of heat energy E radiated from a perfectly black body per second per unit area is directly proportional to the fourth power of its absolute temperature T.

Mathematically, this law can be expressed as

$$E \propto T^4 \qquad \Longrightarrow \qquad E = \sigma T^4 \qquad (11.74)$$

where σ is a constant, called Stefan's constant. This law is sometimes referred to as Stefan's fourth power law. This law can be derived from Planck's law of black body radiation in the following way:

The total energy density inside the black body chamber across the entire wavelength limit can be determined by integrating Planck's law of black body radiation as

$$U = \int_0^\infty u(\lambda)d\lambda = \int_0^\infty \frac{8\pi hc}{\lambda^5} \frac{1}{\left(e^{\frac{hc}{\lambda K_B T}} - 1\right)} d\lambda.$$

Let $y = \frac{hc}{\lambda K_B T}$, and this gives $d\lambda = -\frac{hc \, dy}{K_B T y^2}$. Using these two substitutions, the above expression becomes

$$U = \int_\infty^0 \frac{8\pi K_B^5 T^5 \, y^5}{h^4 \, c^4 \, (e^y - 1)} \left(-\frac{hc}{y^2 \, K_B T}\right) dy = \int_0^\infty \frac{8\pi K_B^4 T^4}{h^3 \, c^3} \times \frac{y^3}{(e^y - 1)} dy$$

$$= \frac{8\pi K_B^4 T^4}{h^3 \, c^3} \int_0^\infty \frac{y^3}{(e^y - 1)} dy.$$

Using the value of the standard definite integral, $\int\limits_{0}^{\infty} \frac{y^3}{(e^y-1)} dy = \frac{\pi^4}{15}$, we have $U = \frac{8\pi K_B^4 T^4}{h^3 c^3} \frac{\pi^4}{15}$.

So, the total energy emitted from the unit surface area of the black body per second is

$$E = \frac{c}{4}\frac{U}{4} = \frac{c}{4}\frac{8\pi K_B^4 T^4}{h^3}\frac{\pi^4}{c^3}\frac{\pi^4}{15} = \sigma\,T^4. \tag{11.75}$$

Here, $\sigma = \frac{2\pi^5 K_B^4}{15\,h^3\,c^3} = 5.654 \times 10^{-8}$ J s^{-1} m^{-2} K^{-4} is known as Stefan's constant. This equation (11.75) represents the Stefan–Boltzmann law for black body radiation. Hence, it is evident that the Stefan–Boltzmann law can be derived from Planck's law of black body radiation.

An important comment about Planck's law

According to the quantum theory of Max Planck, the average energy of a mode with frequency ν is given by [see equation (11.67)]

$$\langle E \rangle = \frac{h\nu}{e^{\frac{h\nu}{K_B T}} - 1}. \tag{11.76}$$

This equation determines the average number of photons in a single mode of frequency ν in thermal equilibrium

$$\langle n_\nu \rangle = \frac{\langle E \rangle}{h\,\nu} = \frac{1}{e^{\frac{h\nu}{K_B T}} - 1}. \tag{11.77}$$

This is called the photon occupation number in thermal equilibrium. We see that at high frequencies and low temperatures, $\frac{h\nu}{K_B T} \gg 1$, the occupation number is $e^{-\frac{h\nu}{K_B T}}$, which is just the standard Boltzmann factor. At low frequencies and high temperatures, however, the occupation number becomes $\frac{K_B T}{h\,\nu}$. Evidently, there is much more to photon statistics than has been apparent from this elementary treatment.

Let us reorganize the expression for the number density of photons of different frequencies in light of our considerations of the form of the Boltzmann distribution. We recall that the Boltzmann distribution has two parts, the Boltzmann factor and the degeneracy of the energy state. Let us rewrite Planck's distribution, given by equation (11.68), in terms of the number density of photons in the frequency interval between ν and $\nu + d\nu$.

$$N(\nu)d\nu = \frac{u(\nu)d\nu}{h\nu} = \frac{8\pi\nu^2 d\nu}{c^3}\frac{1}{e^{\frac{h\nu}{K_B T}} - 1}.$$

Multiplying numerator and denominator by h^3, we get

$$N(\nu)d\nu = \frac{8\pi}{h^3}\left(\frac{h\nu}{c}\right)^2 d\left(\frac{h\nu}{c}\right) \frac{1}{e^{\frac{h\nu}{K_B T}} - 1} = \frac{4\pi p^2 dp}{h^3} \times 2 \times \frac{1}{e^{\frac{h\nu}{K_B T}} - 1}, \qquad (11.78)$$

where $p = \frac{h\nu}{c}$ is the momentum of the photon. Equation (11.78) has the following important implications:

1. The term $4\pi p^2 dp$ is the differential volume of momentum space for the photons that have energies in the range $h\nu$ to $h(\nu + d\nu)$.

2. The factor "2" corresponds to the two polarization states of the photon (or electromagnetic wave).

3. The term h^3 is the elementary volume of phase space, and hence, the term $\frac{4\pi p^2 dp}{h^3}$ tells us how many states there are available in the frequency interval between ν and $\nu + d\nu$.

4. The final term $\frac{1}{e^{\frac{h\nu}{K_B T}} - 1}$ is the photon occupation number that has been derived above.

11.7.4   What happens to Planck's law in the limit $h \to 0$?

Having postulated quantization and derived a fine radiation law (it fitted the data spectacularly well and avoided the UV catastrophe), Planck let the quantization constant, h, slide toward zero as he had intended all along. Immediately, he was back in trouble with his equations: the catastrophe reappeared, and he found himself where his predecessors had failed. When h becomes zero, Planck's law is blurred into the classical laws that were so extravagantly wrong. Nothing he could do would keep matters satisfactory if h vanished. The constant h had to remain finite, or all that he had gained was lost. Furthermore, a unique value of the constant was indicated to give the best fit to experimental data. It bore upon Planck – although he fought against the conclusion with all his might – that the quantization assumption was essential, not just a trick to manipulate the equations. Mathematically, we see that as $h \to 0$,

$$\langle E \rangle = K_B T,$$

which is the energy corresponding to the law of **equipartition of energy**.

Let us now try to understand what happens to the total energy contained in a box filled with black body radiation of all frequencies at temperature T if Planck's constant h were reduced to zero. Under the equilibrium condition, the total energy of the black body radiation inside an enclosure of volume V at a temperature T is given by

$$E(T) = \frac{8\pi h V}{c^3} \int\limits_0^\infty \frac{\nu^3 d\nu}{e^{\frac{h\nu}{K_B T}} - 1}.$$

If we take the limit $h \to 0$, that is, in the classical limit, we have

$$\lim_{h \to 0} E(T) = \lim_{h \to 0} \frac{8\pi h V}{c^3} \int\limits_0^\infty \frac{\nu^3 d\nu}{e^{\frac{h\nu}{K_B T}} - 1} = \lim_{h \to 0} \frac{8\pi V}{c^3} \int\limits_0^\infty \frac{h\nu^3 d\nu}{e^{\frac{h\nu}{K_B T}} - 1}$$

$$= \frac{8\pi V}{c^3} \int\limits_0^\infty \frac{h\nu^3 d\nu}{1 + \frac{h\nu}{K_B T} - 1} \quad (h \to 0 \text{ neglecting higher order terms of } \frac{h\nu}{K_B T})$$

$$= \frac{8\pi K_B T V}{c^3} \int\limits_0^\infty \nu^2 d\nu \to \infty.$$

This result shows that the total energy turns out to be infinite when h approaches zero. This exactly happens in the classical picture using Rayleigh–Jeans formula, which is referred to as **ultraviolet catastrophe**.

What Planck Did Not Do

Remarkably, Planck did not set $K_B T$, which would have resulted in the Rayleigh–Jeans law. Probably, he did not do this because he had already rejected the statistical point of view of Maxwell and Boltzmann. Instead, in his words, "I had no alternative than to tackle the problem once again – this time from the opposite side – namely from the side of thermodynamics, my own home territory where I felt myself to be on safer ground. In fact my previous studies of the Second Law of Thermodynamics came to stand me in good stead now, for at the very outset I hit upon the idea of correlating not the temperature of the oscillator but its entropy with its energy ⋯. Nobody paid any attention to the method which I adopted and I could work out my calculations completely at my leisure, with absolute thoroughness, without fear of interference or competition."

11.7.5 Experimental verification of Planck's law

Toward the end of the 19th century, physicists carried out experimental investigations to record the spectral intensity of radiation emitted from a black body as a function of wavelength at various temperatures. At that time, it was a real-world challenge to the physicists due to the unavailability of high-sensitive instruments, and they were forced to upgrade the instruments for better measurements of the intensity of black body radiation. Planck thus began his research paper with the acknowledgment of the beautiful works of several researchers at the Physikalisch-Technische-Reichsanstalt (Imperial Physico-Technical Institute, the forerunner of modern Germany's National Bureau of Standards). The literature says that at this institute, the physicists were conducting and improving the most advanced experiments in infrared Bolometry of the time, and Planck used these experimental results in support of the theoretical basis for the formula of black body radiation he invented. In fact, Planck could not have made his discovery prior to October 1900, even though he had been working on the problem of black body radiation for several years.

Historically, the properties of radiation emitted from a heated body were known from experiments well before an all-inclusive theory was developed. Some of these experimental details are mentioned briefly below.

In the month of July 1900, Otto Richard Lummer and Ernst Georg Pringsheim published their experimental results of the black body radiation spectrum. In this experiment, they used an apparatus that Lummer and Ferdinand Kurlbaum had built in 1898. The experimental setup is shown in Figure 11.20. In the experimental setup, they punched a small hole in the side of an otherwise completely closed oven. A ceramic tube was placed in the oven. A platinum cylinder sheet was kept within the ceramic tube and heated to obtain a temperature in the range between $-188°C$ and

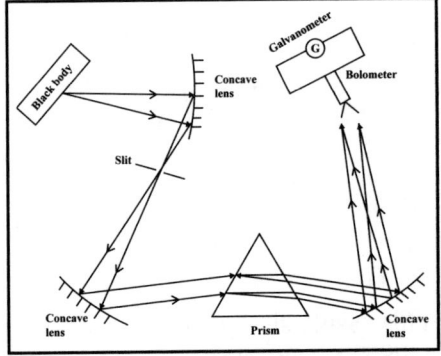

Figure 11.20 A schematic diagram of the experimental setup on the black body radiation, conducted by Otto Lummer and Ferdinand Kurlbaum in 1898.

$1650°C$. Using a bolometer, Lummer and Pringsheim measured the power distribution in the infrared radiation coming out from the hole in their cavity radiator. It should be pointed out that the design of the bolometer was inspired by that of the bolometer that Samuel P. Langley had used for astronomical measurements since 1880. Lummer and Pringsheim measured the amount of heat in wavelengths as long as 8.3 microns (red light has a wavelength of a little less than one micron) by varying the temperature of the platinum tube from $-188°C$ (-306.4 Fahrenheit) to $+1650°C$ ($+3002$ Fahrenheit).

It is estimated from Wien's displacement law that at $+1650°C$, a black body radiates most intensely at a wavelength of 1.486 microns. Lummer and Pringsheim carried out their measurements at wavelengths longer than this and extended the realm of their measurements into that part of the infrared spectrum where black body radiation conforms less to the law that Wilhelm Wien had proposed in 1896 and more to the Rayleigh–Jeans law. However, Lummer and Pringsheim did not obtain a discrepancy between their observations and Wien's law large enough to give them certainty. Later in the year, Heinrich Leopold Rubens and Ferdinand Kurlbaum extended the range of measurable wavelengths to 50 microns and obtained results that removed all doubt. On October 07, 1900, Rubens revealed his experimental results of black body radiation to Planck over afternoon tea, and Planck solved the problem of black body radiation in that evening.

Before closing this section, a brief introduction is presented about the instrument **Bolometer**, which is used to measure the temperature of the black body radiator.

Bolometer

The name bolo-meter is formed from the Greek word "bolo" that means shot. This instrument was used by experimental physicists to discover new atomic and molecular absorption lines in the invisible infrared portion of the electromagnetic spectrum. Langley used this instrument particularly to measure the spectrum emitted from the Sun and interpreted his experimental results as a demonstration that "heat" and "light" from the Sun is just two different names of the same physical phenomena.

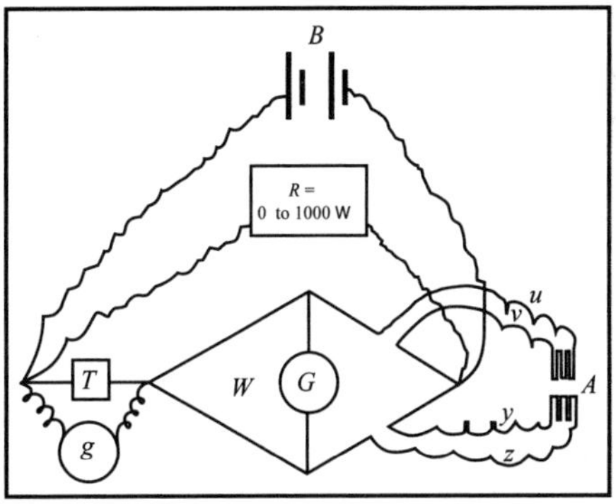

The instrument "bolometer" is specially constructed with an element of temperature-sensitive material. A semiconductor bead is used as an active material and is supported between two pigtail leads. When a radio frequency power is applied to the element of

Figure 11.21 Circuit diagram of Langley's bolometer. The bolometer "A" was connected to the Wheatstone bridge by the insulated copper wires "u, v, y, z".

the bolometer, the element is heated up by the absorbed power, and a change in its electrical resistance takes place. This bolometer is used in a bridge circuit so that the small changes in resistance can be easily detected and the corresponding power can be measured by the substitution method (i.e., the substitution of dc or low-frequency power to produce an equivalent heating effect). A D'Arsonval meter movement is usually used as the null indicator.

For the measurement of temperature using a bolometer, two principles are followed. According to one principle of measurement (the principle used in the balanced bridge), the bridge is initially balanced with low-frequency bias power. Radio frequency power is then applied to the bolometer, and the bias power is gradually removed until the bridge is again balanced. The actual radio frequency power is then equated to the bias power removed. According to another principle of measurement (the principle used in the unbalanced bridge), the bridge is not rebalanced after the radio frequency power is applied. Rather, the indicator reading is converted directly into power by calibration previously performed.

In the beginning, the emitted infrared radiation was measured by bolometers. The bolometer used was developed based on a similar device by the American astronomer Samuel Pierpont Langley and consisted of two platinum strips covered with lampblack. One strip was shielded from radiation, and one was exposed to it. The strips formed two branches of a Wheatstone bridge that was fitted with a sensitive galvanometer and connected to a battery, as seen in Langley's drawing of the circuit diagram in Figure 11.21. Electromagnetic radiation falling on the exposed strip (at "A") would heat it and change its resistance. The bolometer was enclosed in a cylindrical holder made of nonconducting material. It had the capability to detect radiation from a cow standing 400 m away and was sensitive to differences in temperature of 0.001 K (Langley 1881).

Microwave background radiation—an excellent experimental proof of Planck's law

The most famous black body radiation spectrum is cosmological in origin. Just after the "Big Bang", the Universe was essentially a "fireball". In the Big Bang model, the Universe starts out from a hot, dense initial state and subsequently expands and cools. It had been noted by Gamow and collaborators that the Big Bang model predicts a background of cosmic radiation, a relic from the hot early phase. Its temperature had been estimated to be of the order of a few degrees Kelvin (values from 5 K to 50 K can be found during that time). During the time of the "Big Bang", the energy associated with the radiation was completely dominating that associated with matter. The early Universe was also

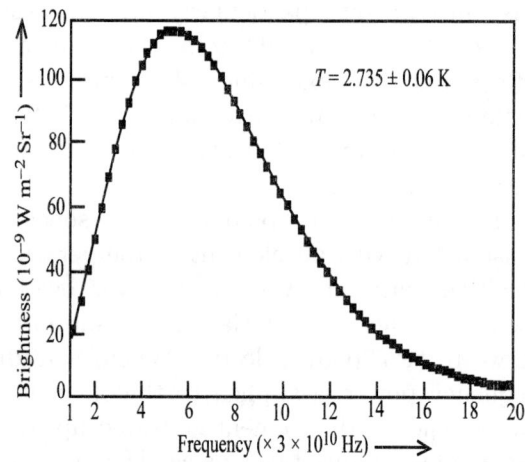

Figure 11.22 Cosmic background radiation spectrum measured by the Far Infrared Absolute Spectrometer (FIRAS) aboard the Cosmic Background Explorer satellite (COBE).

pretty well described by equilibrium statistical thermodynamics, which means that the radiation had a black body spectrum. The photons making up the cosmic background radiation have rather low energies. The spectrum peaks at $\epsilon = 2.82\,K_B T = 6.6 \times 10^{-4}\,eV$ that corresponds to wavelengths of about a millimeter in the infrared. These wavelengths belong to the long-wavelength tail of the spectrum, and do not penetrate the atmosphere. Further, these wavelengths can be detected in the microwave region of a few centimeters without much difficulty. As the Universe expanded, the radiation was gradually Doppler shifted to ever larger wavelengths (in other words, the radiation did work against the expansion of the Universe and, thereby, lost energy), but its spectrum remained invariant. This primordial radiation is still detectable as a faint microwave background that pervades the whole universe.

This microwave background radiation was discovered accidentally by Penzias and Wilson in 1965[2]. The history of Arno Penzias and Robert Wilson is quite amusing. These two young radio astronomers employed by Bell Laboratories at Holmdel, New Jersey, were observing the sky with a radio-telescope that had been built to investigate radio transmissions from communication satellites. They had the most advanced radio receiver of the time, a so-called horn antenna, with a "cold load" cooled with liquid Helium to suppress interference with the detector heat. But despite this, they found a persistent, isotropic receiver noise that was significantly larger than what they had expected. Also, after checking their equipment thoroughly and removing a "white dielectric" (pigeon droppings), they found that this mysterious background noise corresponded to an antenna temperature of about 3.5 K at 7.35 cm and was always present in the signal. This was the cosmic background microwave radiation.

Until recently, it was difficult to measure the full spectrum with any degree of precision because of strong microwave absorption and scattering by the Earth's atmosphere. However, all of this changed when the COBE (Cosmic Background Explorer) satellite was launched in 1989. It took precisely nine minutes to measure the perfect black body spectrum, which is reproduced in Figure 11.22. The solid line through the data points is a fit according to the famous Planck's law of black body radiation. This fit provided a characteristic temperature 2.735 K of the cosmic background radiation. This temperature was found to match well with that of the prediction of the "Big Bang" theory. In a real sense, this temperature of the cosmic background radiation can be regarded as the "temperature of the Universe". After the discovery of the CMB, the Big Bang model of cosmology was established, and in 1978, Penzias and Wilson were rewarded with the Nobel Prize in Physics for this wonderful discovery.

11.8 Infrared radiation

Infrared radiation (IR) is a region of the electromagnetic radiation spectrum where wavelengths of such radiation range from about 700 nanometers (nm) to 1 millimeter (mm). Wavelengths of infrared waves are longer than those of visible light but shorter than those of radio waves. IR is classified into three categories: short wave (or near-infrared, or IR-A), medium wave (or medium or middle infrared, or IR-B), and long wave infrared (or far infrared, or IR-C).

---

[2]A. A. Penzias and R. W. Wilson, A Measurement of Excess Antenna Temperature at 4080 Mc/s, *Astrophysical Journal* 142 (1965): 419–421.

Like ordinary light, infrared light can also be focused, reflected, and polarized. Infrared radiation plays an important role in taking images at night and in detecting various objects under suitable conditions.

11.8.1 Herschel experiment: Discovery of infrared radiation

Sir Frederick William Herschel (1738–1822) was born in Hanover, Germany, and became well-known as both a musician and an astronomer. He moved to England in 1757 and, with his sister Caroline, constructed telescopes to survey the night sky. Their work resulted in several

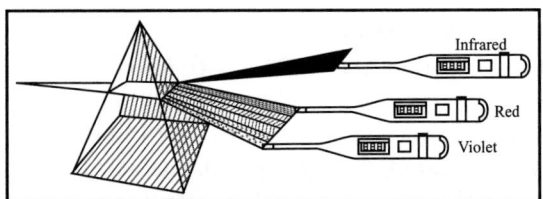

Figure 11.23 Herschel experimental set up that used a prism and thermometers. This simple set up eventually led to the discovery of the infrared region of the electromagnetic spectrum.

catalogs of double stars and nebulae. Herschel is famous for his discovery of the planet Uranus in 1781, the first new planet found since antiquity. Herschel made another dramatic discovery in 1800. He wanted to know how much heat was passed through the different colored filters he used to observe sunlight. He noted that filters of different colors seemed to pass different amounts of heat. Herschel thought that the colors themselves might be of varying temperatures, and so he devised a clever experiment to investigate his hypothesis.

In Herschel's experiment, shown in Figure 11.23, sunlight was directed through a glass prism to create a spectrum. The emergent light was divided into its colors, and a "rainbow" was created. He measured the temperature of each color in the rainbow. He used three thermometers with blackened bulbs (to better absorb the heat) and placed one bulb against each color while the other two were placed beyond the spectrum as control samples. As he measured the temperatures of the violet, blue, green, yellow, orange, and red light, he noticed that all of the colors had temperatures higher than the controls and that the temperature of the colors increased from the violet to the red part of the spectrum.

After noticing this pattern, Herschel decided to measure the temperature just beyond the red portion of the spectrum in a region apparently devoid of sunlight. In order to achieve the ambient air temperature in the room, Herschel placed a thermometer beyond the red end of the visible spectrum. To his surprise, he observed that this region of the spectrum had the highest temperature of all. Therefore, Herschel made the conclusion that there must be some invisible form of light beyond the visible spectrum that he called "calorific rays" (the Latin word calor means heat). These experimental results were reported to the Royal Society. In 1800, Herschel expressed this discovery in his own language, "In this case, radiant heat will at least partly, if not chiefly, consist, if I may be permitted the expression, of invisible light; that is to say, of rays coming from the Sun, that have such a momentum as to be unfit for vision".

Herschel performed additional experiments on what he called calorific rays (derived from the Latin word for heat) beyond the red portion of the spectrum. He found that they were reflected, refracted, absorbed, and transmitted in a manner similar to visible light. What Sir William had discovered was a form of light (or radiation) beyond red light, now known as infrared radiation. (The prefix infra means below.) Herschel's experiment was important because it marked the first time that someone demonstrated that there were types of light that we cannot see with our eyes. At the National Museum of Science and Industry in London, Herschel's experiment is on display so that people can see the instrument and sense its historical importance.

We know that the refractive index n of a glass varies nonlinearly with wavelength, and as a result, the sunlight gets dispersed into its various components while passing through a glass prism. In Herschel's experiment, it was also observed that the dispersed wavelengths were not uniformly spaced; there was a larger concentration toward the infrared. This indicated an increase in the observed "temperature" toward the longer wavelengths.

11.8.2 Infrared radiation from human body

Using late 19th century laboratory physics and recent biophysics, we may find out why the human body radiates most at a wavelength of about 10 μm. The specific intensity of the human skin can be approximated with that of a modified black body as $I_\lambda = \epsilon(\lambda_m) B_\lambda(T)$ where B_λ is some function of wavelength λ and temperature T and $\epsilon(\lambda)$ is the emissivity of human skin in the wavelength range of $1\,\mu m < \lambda < 14\,\mu m$ and is nearly unity, $\epsilon(\lambda) = 0.98 \pm 0.01$. We may actually say it radiates like a black body. Wien's displacement law may be used to find out in which wavelength range it radiates most. Assuming a skin temperature of $300\,\mathrm{K} < T < 310\,\mathrm{K}$, we arrive at $9/3\,\mu m < \lambda_{\max}(T) < 9.7\,\mu m$.

The total surface area of the human body can be approximated by the depth-weight formula as $area = 71.85 \times W^{-0.425} \times H^{0.725}$. Here, W is the weight in kg, and H is the depth in cm. The total surface area for a typical adult male is $1.73\,\mathrm{m}^2$. The total radiated power at a skin temperature of $T = 306\,\mathrm{K}$ may be calculated from Stefan's law as $P_{\mathrm{rad}} = area \times \sigma \times T^4 = 860\,\mathrm{mW}$. Although only a $\frac{5}{7}$ fraction of the total skin surface is considered as effective radiator; the uncovered skin is a bright infrared source. We shine bright in the infrared while the human eye detects only 1% of the light at 0.69 μm, and 0.01% at 0.75 μm, and so we effectively cannot see wavelengths longer than about 0.75 μm.

It is not only the radiation of the human body that we cannot see by our eyes. In space, there are many regions that are hidden from optical telescopes because they are embedded in dense regions of gas and dust. However, infrared radiation, having wavelengths that are much longer than visible light, can pass through dusty regions of space without being scattered. This means that we can study objects hidden by gas and dust in the infrared, which we cannot see in visible light, such as the center of our galaxy and regions of newly forming stars.

11.9   Derivation of Planck's law of black body radiation from various approaches

In this section, we will derive Planck's law of black body radiation following the approach of Prof. S. N. Bose. This law was derived by Prof. Bose in 1924 by considering the equilibrium properties of the black body radiation. In fact, the derivation of Planck's law is one of the most important applications of Bose–Einstein statistics.

11.9.1   Planck's alternative derivation of black body radiation law

We consider a cavity of volume V filled with black body radiation at a temperature T. This system of black body radiation has been looked at historically from two, practically identical but conceptually different points of view. These are:

1. The radiation cavity is considered as an assembly of harmonic oscillators with quantized energies $\left(n_i + \frac{1}{2}\right)\hbar\omega_i$, where $n_i = 0, 1, 2, 3, \cdots$, and ω_i is the (angular) frequency of an oscillator, or

2. The radiation cavity is considered as a gas of identical and indistinguishable quanta–the so-called photons–the energy of a photon of frequency ω_i of the radiation mode being equal to $\hbar\omega_i$.

The first point of view is essentially the one adopted by Planck (1900), except that we have also included here the zero-point energy of the oscillator. For the thermodynamics of the radiation, this zero-point energy is of no great consequence and may be dropped altogether. Planck considered that the oscillators are distinguishable from one another as their angular frequencies ω_i are different. Hence, he applied Maxwell–Boltzmann statistics to these oscillators. However, Planck pointed out that the expression for the single-oscillator partition function $Z_1(V, T)$ would be different from the classical expression because now the energies accessible to the oscillators are discrete rather than continuous. In quantum mechanical situation, the single-oscillator partition function is given by

$$Z_1(\beta) = \frac{1}{2\sinh\left(\frac{\hbar\omega_i}{2K_BT}\right)}.$$

For an assembly of N-oscillators, the partition function is given by

$$Z_N(\beta) = \left(Z_1(\beta)\right)^N = \left(\frac{1}{2\sinh\left(\frac{\hbar\omega_i}{2K_BT}\right)}\right)^N = e^{-\frac{N}{2}\frac{\hbar\omega_i}{K_BT}}\left(1 - e^{\frac{\hbar\omega_i}{K_BT}}\right)^{-N}.$$

Taking the natural logarithm of the above expression, we get

$$\ln Z_N(\beta) = \ln\left(e^{-\frac{N}{2}\frac{\hbar\omega_i}{K_BT}}\left(1 - e^{\frac{\hbar\omega_i}{K_BT}}\right)^{-N}\right) = -\frac{N}{2}\frac{\hbar\omega_i}{K_BT} - N\ln\left(1 - e^{\frac{\hbar\omega_i}{K_BT}}\right).$$

The expectation value of the energy of a Planck oscillator of frequency ω is then given by

$$U = -\frac{\partial \ln Z_N(\beta)}{\partial \beta} = -\frac{\partial}{\partial \beta}\left(-\frac{N}{2}\frac{\hbar\omega_i}{K_BT} - N\ln\left(1 - e^{\frac{\hbar\omega_i}{K_BT}}\right)\right) = N\left[\frac{\hbar\omega_i}{2} + \frac{\hbar\omega_i}{e^{\frac{\hbar\omega_i}{K_BT}} - 1}\right].$$

Excluding the zero-point term $\frac{1}{2}\hbar\omega_i$, the average energy of these N-oscillators can be written as

$$\langle E_i \rangle = \frac{\hbar\omega_i}{e^{\frac{\hbar\omega_i}{K_BT}} - 1}. \tag{11.79}$$

Now, the number of normal modes of vibration per unit volume of the cavity in the frequency range ω and $\omega + d\omega$ is given by the Rayleigh expression

$$g(\omega)d\omega = 2 \times 4\pi \times \left(\frac{1}{\lambda}\right)^2 \times d\left(\frac{1}{\lambda}\right) = \frac{\omega^2 d\omega}{\pi^2 c^3}, \tag{11.80}$$

where factor 2 has been included to take into account the duplicity of the transverse modes; the symbol c denotes the speed of light. By equations (11.79) and (11.80), the energy per unit volume, that is, energy density associated with the angular frequency range ω and $\omega + d\omega$ is given by

$$u(\omega)d\omega = \langle E \rangle\, g(\omega)d\omega = \frac{\hbar}{\pi^2 c^3}\frac{\omega^3 d\omega}{e^{\frac{\hbar\omega}{K_BT}} - 1}. \tag{11.81}$$

In terms of frequency ν, equation (11.81) can be expressed as

$$u(\nu)d\nu = \frac{8\pi h}{c^3} \times \frac{\nu^3}{\left(e^{\frac{h\nu}{K_BT}} - 1\right)}d\nu. \tag{11.82}$$

This is Planck's formula for the distribution of energy over the black body spectrum. Integrating equation (11.82) over all values of frequency ν, the total energy density of the black body radiation in the cavity can be obtained.

11.9.2 Bose's derivation of Planck's law of black body radiation

The second point of view originated with Bose (1924) and Einstein (1924, 1925). Bose investigated the problem of the "distribution of photons over the various energy levels" in the system. However, instead of worrying about the allocation of the various photons to the various energy levels, Prof. Bose concentrated on the statistics of the energy levels themselves. He examined questions such as

1. What is the "probability" of an energy level $E_i = \hbar\omega_i$ being occupied by n_i photons at a time?

2. What is the mean value of n_i?

3. What is the mean value of E_i? And so on.

These energy levels are distinguishable. So, he applied Maxwell–Boltzmann statistics to these energy levels. The mean values of n_i and E_i, then, turn out to be

$$\langle n_i \rangle = \frac{\sum_{n_i=0}^{\infty} n_i e^{-\left(\frac{n_i \hbar\omega_i}{K_B T}\right)}}{\sum_{n_i=0}^{\infty} e^{-\left(\frac{n_i \hbar\omega_i}{K_B T}\right)}} = \frac{1}{e^{\frac{\hbar\omega_i}{K_B T}} - 1}$$

$$\langle E_i \rangle = \hbar\omega_i \langle n_i \rangle = \frac{\hbar\omega_i}{e^{\frac{\hbar\omega_i}{K_B T}} - 1}. \tag{11.83}$$

This result is identical to the earlier result derived in equation (11.79). To obtain the number of photon states with momenta lying between $\frac{\hbar\omega}{c}$ and $\frac{\hbar(\omega+d\omega)}{c}$, Bose made use of the connection between this number and the "volume of the relevant region of the phase space". This gives the number of photon states in volume V with momenta lying between $\frac{\hbar\omega}{c}$ and $\frac{\hbar(\omega+d\omega)}{c}$ as

$$g(\omega)d\omega = 2 \times \left(\frac{V}{h^3}\right) \left[4\pi \times \left(\frac{\hbar\omega}{c}\right)^2 \times \left(\frac{\hbar \, d\omega}{c}\right)\right] = \frac{V\omega^2 d\omega}{\pi^2 c^3}. \tag{11.84}$$

This is also identical to the earlier result shown in equation (11.80).

By combining equations (11.83) and (11.84), the energy per unit volume, that is, energy density associated with the angular frequency range ω and $\omega + d\omega$ is given by

$$u(\omega)d\omega = \langle E \rangle \, g(\omega)d\omega = \frac{\hbar\omega}{e^{\frac{\hbar\omega}{K_B T}} - 1} \frac{\omega^2 d\omega}{\pi^2 c^3} = \frac{\hbar}{\pi^2 \, c^3} \frac{\omega^3 d\omega}{e^{\frac{\hbar\omega}{K_B T}} - 1}. \tag{11.85}$$

In terms of frequency ν, equation (11.85) can be expressed as

$$u(\nu)d\nu = \frac{8\pi h}{c^3} \times \frac{\nu^3}{\left(e^{\frac{h\nu}{K_B T}} - 1\right)} d\nu. \qquad (11.86)$$

This is exactly the distribution formula of Planck. It must be noted that, although the emphasis lay elsewhere, the mathematical steps that led Bose to his final result went literally parallel to the ones occurring in the oscillator approach!

11.9.3 Einstein's derivation of Planck's law of black body radiation

Albert Einstein went deeper into the problem of derivation of Planck's law of black body radiation. He considered the statistics of both the photons and the energy levels taken together. During the process of distributing photons over the various energy levels, he highlighted the basic fact that the photons are indistinguishable – this fact was also implicitly taken care of in Bose's treatment. Einstein's derivation of the desired distribution was essentially the same as given in last section, with one important difference. This difference indicates that the total number of photons in any given volume was indefinite, and as a result, the constraint of a fixed N was no longer present. As a result, the Lagrange multiplier α did not enter into the discussion, and to that extent, the final formula for $\langle n_i \rangle$ was simpler:

$$\langle n_i \rangle = \frac{1}{e^{\frac{E}{K_B T}} - 1}. \qquad (11.87)$$

This result is identical to that given in equation (11.83), with $E = \hbar\omega$. The subsequent steps in Einstein's derivation are basically the same to those of due to S. N. Bose. Factor 2 in this expression arises essentially from the same reason as in the expression due to Rayleigh–Jeans. But in the present context, it would be more appropriate to consider factor 2 as the representation of the two states of polarization of the photon spin.

11.10 Radiation as a photon gas

Black body radiation is an electromagnetic radiation given off by a warmer object. At a given temperature, the plot of the intensity of such black body radiation as a function of wavelength indicates that the radiation intensity approaches zero in a smooth and continuous manner as the wavelength is decreased. But the classical physics predicts that as wavelength approaches toward lower value, the intensity of the radiation should approach smoothly towards infinity at all temperatures. Thus, classical physics could not explain the sharp decrease in the intensity of radiation emitted at shorter wavelengths, that is, in the ultraviolet region of the spectrum.

This discrepancy between experimental results and the classical prediction is referred to as the "ultraviolet catastrophe". The German physicist Max Planck resolved this ultraviolet catastrophe in 1900 by proposing (in what he called "an act of despair") that the energy of electromagnetic waves is quantized rather than continuous. Thus, energy could be gained or lost only in integral multiples of some smallest unit of energy. This is known as **quantum**. This quantum of energy has been given the name **"photon"**. This means that for each temperature, there is a maximum intensity of radiation that is emitted in a black body object, so the intensity does not follow a smooth curve as the temperature increases, as predicted by classical physics.

The electromagnetic radiation inside a black body chamber exists as patterns of standing waves or modes. A single mode is like a standing wave on a guitar string and is characterized by a frequency ν. Planck made a bold hypothesis that the energy of such a mode is quantized; that is, only certain discrete values of energy of these oscillation modes are allowed. The energy E_n of nth mode is given by

$$E_n = \left(n + \frac{1}{2}\right)\hbar\omega = \left(n + \frac{1}{2}\right)h\nu,$$

where ν is the oscillation frequency, h is Planck's constant, and n is an integer starting with 0. This shows that each oscillation mode can exist in any one of an infinite number of energy states whose energies are equally separated by the energy $\hbar\omega$. To discuss the properties of the photon gas, for the time being, we ignore the factor $\frac{1}{2}$ in the expression for E_n as it has no effect on the results we look for. Hence, we take $E_n = n\hbar\omega$ as the energy of the nth mode whose (angular) frequency is ω. When the energy of a mode is E_n, we say that there are n photons in the mode. Each photon has energy equal to $\hbar\omega$.

11.10.1 Photon picture of black body radiation

A photon is the quantum of light and is considered as an elementary particle. It does not possess rest mass and electric charge. Albert Einstein used the concept of photons to explain the experimental observations of the photoelectric effect, which was not successfully interpreted by the classical wave model of light. In particular, the photon model was very successful in accounting for the frequency dependence of energy of the emitted radiation in the photoelectric effect. Max Planck explained the variation of energy density of black body radiation as a function of wavelength using the semi-classical models, in which light is still described by Maxwell's equations, but the material objects that emit and absorb light do so in amounts of energy that are quantized.

The emission of photons occurs in many natural processes. The natural sources for photon emission are light sources such as floor lamps or lasers. Photons are also emitted from an accelerated charged particle. This phenomenon is known as synchrotron radiation. Further, photons of various energies are seen to be emitted or absorbed when a molecular, atomic, or nuclear transition occurs from higher to lower energy levels or from lower to higher energy levels, respectively. In the annihilation process between a particle and its corresponding antiparticle, photons are also found to be emitted. It should be mentioned that in all these processes, photons carry both energy and momentum.

Particle nature of photon

The experimental observations on the photoelectric effect clearly demonstrate that when light interacts with matter, it behaves as if it was made of discrete packets of energy called quanta. Each quantum of light has energy $h\nu$ and momentum $\frac{h\nu}{c}$. The fact that a light quantum has definite energy as well as momentum allows us to associate a particle with it. This particle was called a photon. The particle nature of electromagnetic radiation was further confirmed in 1924 by the experiment of A. H. Compton on scattering of X-rays by electrons. The phenomenon of an increase in the wavelength of X-ray photons scattered by the striking electrons is called the Compton effect.

Photon number is not conserved

We are familiar with the general notion that the number of particles N is a fixed number in a given system. This is a reasonable assumption in the nonrelativistic case if the particles possess nonzero mass. However, this is not true in the case of relativistic systems. This assumption also breaks down for the case of photons, which are *zero-mass* bosons. In fact, photons enclosed in a container of volume V at a temperature T can readily be absorbed or emitted by the walls of the container, resulting in a change in the number of photons. When a photon collides with an electron in the wall, the electron gets excited to a higher energy state, and a photon is removed from the photon gas. This electron comes back to the lower energy level in a series of steps, each one of which releases an individual photon back into the system of photon gas. It should be kept in mind that the sum of the photon energies of the emitted photons is the same as that of the absorbed photon but the number of emitted photons in this process will vary. As a result of this lack of constraint on the number of photons in the system, it can be shown that **the chemical potential of the photons must be zero**. It readily follows that photons obey a simplified form of Bose–Einstein statistics in which there is an unspecified total number of particles. This type of statistics is called *photon statistics*.

Characteristic features of the photon picture of electromagnetic radiation

So far, we have described various properties of photons. Some of the properties of photons are listed below:

1. A photon is a "quantum" of electromagnetic energy in physics. Sometimes, a photon is considered as a "bundle" of electromagnetic energy. Photons are a basic unit of nature called an "elementary particle". They are not thought to be made up of smaller particles.

2. The photon gas is an ideal gas of particles of zero rest mass. A photon has no electric charge and is a stable particle.

3. Photons do not interact with each other. Therefore, there is no need to consider interparticle interactions. In the interaction of radiation with matter, radiation behaves as if it is made of particles called photons.

4. In the collision of a photon with a particle (e.g. photon–electron collision), the total energy and total momentum remain conserved. However, the number of photons may not be conserved in a collision. A new photon may be created, or a photon may be absorbed in a collision.

5. Spin quantum number of each photon is 1 (in units of $\frac{h}{2\pi}$). Therefore, the Bose–Einstein distribution is applicable to a photon gas.

6. The spin degeneracy of a photon is two (corresponding to two independent directions of polarizations). The degeneracy is not three, as in the case of particles of spin 1 and non zero rest mass.

7. For electromagnetism, the photon is the gauge boson. Therefore, all other quantum numbers of the photon, such as lepton number, baryon number, and flavor quantum numbers, are zero.

8. All photons of electromagnetic radiation of a particular frequency ν, or wavelength λ have the same energy E and momentum $\frac{h\nu}{c}$. c is the speed of light in a vacuum. Such energy and momentum are independent of the intensity of the radiation.

9. Photons are electrically neutral and are not deflected by electric and magnetic fields.

10. Photons do not interact among themselves. Hence, an assembly of photons cannot reach equilibrium by exchanging energy among them. However, photons interact with other particles, like charged particles. So, photons can exchange energy and momentum in the presence of charged particles and can attain equilibrium.

11. The chemical potential of a photon gas is zero.

12. The fugacity (z) for a photon gas is 1.

13. Photons are mass-less particles, and obey the dispersion relation $E = p\,c$.

14. Thermodynamic quantities like pressure, number density, energy density, and entropy of a photon gas depend only on temperature.

11.10.2   A photon near a black hole

A photon has energy $E = h\nu$. According to Einstein's famous mass–energy relation $E = mc^2$, an inertial mass m can be assigned to a photon. This mass m is given by

$$m = \frac{E}{c^2} = \frac{h\nu}{c^2}.$$

Thus, when a light beam passes near a heavy star, its trajectory has to be deflected due to the force of gravity of the heavy star. Such deflection of a light beam coming from a distant star is indeed experimentally observed when the beam passes near the Sun.

We may also expect that when a photon leaves a star, its energy should decrease because of the gravitational field of the star. This indeed happens and manifests itself in a decrease in frequency that is usually referred to as the gravitational red shift. One can experimentally calculate the gravitational red shift by noting down the potential energy of a photon on the surface of the star. Using the above expression for mass m, the gravitational potential energy of the star can be written as

$$\text{Gravitational potential energy} \approx -\frac{GMm}{R} = -\frac{GM}{R} \times \frac{h\nu}{c^2},$$

where M is the mass of the star, R is its radius, and G is the universal gravitational constant. Thus, when a light beam reaches the earth, its frequency would be determined from the following energy equation:

$$h\nu' = h\nu - \frac{GM}{R} \times \frac{h\nu}{c^2}.$$

This leads to the relative change in frequency as

$$\frac{\Delta\nu}{\nu} = \frac{\nu - \nu'}{\nu} = \frac{GM}{R\,c^2}. \tag{11.88}$$

In deriving equation (11.88), the effect of the earth's gravitational field has been neglected. Equation (11.88) shows that if the mass of the star is very large and radius R is small so that the RHS of equation (11.88) exceeds unity, then the light beam will not be able to escape from the star. The star for which this condition is satisfied is termed **a black hole**. It is to be noted that the general theory of relativity (GTR) should be used for the theoretical description of black holes. According to the theory of GTR, the limiting radius of a star, under this situation, comes out to be

$$R_s = \frac{2GM}{c^2}.$$

This radius R_s is known as the **Schwarschild radius**. This result indicates that if the mass of the star is contained inside a sphere of radius $R < R_s$, a light beam will never leave the star, and the star will be termed as a black hole.

For a star of mass $M \approx 10M_s \approx 1.989 \times 10^{34}$ g, where $M_s = 1.989 \times 10^{33}$ g is the mass of the Sun, this radius R_s comes out to be

$$R_s \approx \frac{2GM}{c^2} \approx \frac{2 \times (6.67 \times 10^{-8}) \times (1.989 \times 10^{34})}{(3 \times 10^{10})^2} \text{ cm} \approx 29.481 \text{ km} \approx 30 \text{ km}.$$

Experimentally, the existence of such black holes with radius $R_s \approx 10$ km has been detected in a real situation.

11.10.3   Planck's formula and quantum electrodynamics

We conclude this chapter with a curious historical fact. In 1913, Planck published the second edition of his lectures on the **Theory of Thermal Radiation**, where he gave a new derivation of his own formula derived earlier. Conceptually, both derivations were similar to one another, but he made a modification that slightly changed the final result. In the first paper published in October 1900, he assumed that the energy of the black body radiation could be subdivided into discrete values (which later turned out to be the true quanta of light): 0, $E(= h\nu)$, $2E$, $3E$, $\cdots$, etc. In 1913, he realized that the energy of the mode could lie in the intervals $(0, E)$, $(E, 2E)$, $(2E, 3E)$, $(3E, 4E)$, $\cdots$, etc. and that the equilibrium state of the radiation was reached by transitions among those intervals. It should be noted that this is almost like having no discreteness at all! With this modification, Planck's earlier derivation goes through with only a little change: the average energy of the radiation in the interval $(0, E)$ is $\frac{1}{2}E$, in $(E, 2E)$ is $E + \frac{1}{2}E$, in $(2E, 3E)$ is $2E + \frac{1}{2}E$, etc. Thus, there is a shift in the average energy of the radiation mode by $\frac{1}{2}E$.

Hence, the average energy of the oscillator will be given by

$$\langle E \rangle = \frac{\hbar\omega}{e^{\frac{\hbar\omega}{K_B T}} - 1} + \frac{1}{2}\hbar\omega. \tag{11.89}$$

The term added on the right-hand side of equation (11.89) is small, but it turned out to have observable implications in the theory of specific heats of solids. And remarkably, it was detected by experiments of that time that the $\frac{1}{2}\hbar\omega$-term indeed needed to be there!

Equation (11.89) has another highly nontrivial implication. The average number of photons (i.e., the quanta of light) per mode is:

$$n_\omega = \frac{\text{average energy per mode}}{\text{energy of a single photon}} = \frac{\langle E \rangle}{\hbar\omega} = \frac{1}{e^{\frac{\hbar\omega}{K_B T}} - 1} + \frac{1}{2}.$$

In an absolute vacuum, the temperature is zero (since there are no particles in a vacuum, there is no thermal motion, and hence, by definition, the temperature must vanish). Again, in the $lim_{T \to 0}$, we have $lim_{T \to 0} e^{\frac{\hbar\omega}{K_B T}} \to e^{\infty} = \infty$ for all $\omega \neq 0$. Then, from the above expression, we get $(n_{\omega})_{T \to 0} = 0 + \frac{1}{2}$.

This says that in a vacuum, there is, on average, $\frac{1}{2}$ photons for each possible frequency of radiation, and hence, the vacuum is not really empty! This result was proved several decades later in quantum electrodynamics. Thus, Planck's derivation given by equation (11.89) predicted a correct result in Quantum Electrodynamics.

11.11 Applications of black body radiation

Radiations from black bodies are widely used in various practical applications. Some of these applications are mentioned below:

1. Radiations from black bodies are used for lighting, heating, security purposes, and thermal imaging. Such radiations are also used for testing and measurement purposes.

2. This radiation can be used to measure the temperature of astrophysical objects. For example, the temperature of cosmic microwave background radiation (CMBR) has been determined using this radiation.

3. A source of black body radiation with a known temperature is effectively used for calibrating and testing the radiation thermometers.

4. The color temperature of a source is defined as the temperature (in kelvin) at which the heated black body radiator matches the color of the light source. Thus, color temperature is a characteristic of visible light and has several important applications in various fields, such as photography, videography, and publishing. The color temperature of a light source is determined by comparing its chromaticity with a theoretical, heated black body radiator.

5. The black body radiation (in the infrared region) emitted from objects has been used to design infrared thermometers. Such thermometers are widely used to measure temperature in the appropriate range. Laser thermometers are such thermometers in which a laser beam is used to design the thermometer. Similarly, there are noncontact thermometers to describe the device's ability to measure temperature from a distance. By knowing the amount of infrared energy emitted from the object and its emissivity, the temperature of the object can be determined.

6. This radiation can be used for two-color pirometry.

11.12 Solved numerical problems

1. The emitted energy from a black body at a temperature 1646 K is maximum at a wavelength 1.78 μm. Calculate the temperature of a planet (assume this planet to be a black body) that emits radiation of maximum energy at a wavelength 14.0 μm.

   Answer: According to Wien's displacement law, we have

   $$\lambda_m\, T = \text{constant} = 2.898 \times 10^{-3} \text{ m K.}$$

   Using this expression for the given black body and the planet, we get

   $$(\lambda_m)_{BB} T_{BB} = (\lambda_m)_{\text{Planet}} T_{\text{Planet}}; \implies 1.78 \text{ μm} \times 1646 \text{ K} = 14.0 \text{ μm} \times T_{\text{Planet}}$$

   $$T_{\text{Planet}} = \frac{1.78 \times 1646}{14} \text{ K} = 209.27 \text{ K.}$$

   Hence, the temperature of the planet is 209.27 K.

2. The temperature of skin of a human body is approximately 35°C. What is the wavelength of the emitted radiation from the skin of a human body at which the radiation peak occurs at the temperature of the skin?

   Answer: According to Wien's displacement law, we have

   $$\lambda_m\, T = \text{constant} = 2.898 \times 10^{-3} \text{ m K.}$$

   The factor λ_m corresponds to the wavelength at which maximum power is emitted from a black body at a temperature T. Using this expression, we get

   $$\lambda_m\, T = 2.898 \times 10^{-3} \text{ m K}; \implies \lambda_m = \frac{2.898 \times 10^{-3} \text{ m K}}{308 \text{ K}} = 9.40 \text{ μm.}$$

   Hence, the wavelength of the emitted radiation from the skin of a human body at which the radiation peak occurs is 9.40 μm.

3. The temperature of the surface of the Sun is 5450°C. Calculate the rate at which energy is radiated from the area of 1×10^{-4} m^2 lying on its surface. Assume that the temperature of the surrounding space is 0 K. Given, Stefan's constant $= 5.71 \times 10^{-8}$ W m^{2}K^{-4}.

   Answer: The temperature of the surface of the Sun is $(5450+273)$ K $= 5723$ K.

   From Stefan's law of black body radiation, we have $E = \sigma \left(T^4 - T_0^4\right)$. Using this law, rate of energy radiated from the area of 1×10^{-4} m^2 lying on the surface of the Sun is given by

   $$E = 5.7 \times 10^{-8} \times 1 \times 10^{-4} \times \left(5723^4 - 0^4\right) \text{ W} = 6114.6 \text{ W.}$$

4. A spherical black body with a radius of 0.10 m radiates power of 500 W at a temperature 600 K. If its radius becomes one-third of its previous value and the temperature tripled, find the power radiated from the spherical black body in watt.

   Answer: The surface area of a sphere is $4\pi R^2$, where R is its radius. According to Stefan's law of black body radiation, the power P radiated from a spherical black body at a temperature T is proportional to $T^4 R^2$. According to the problem, for the two spherical black bodies, we have

   $$\frac{p'}{p} = \left(\frac{T'}{T}\right)^4 \left(\frac{R'}{R}\right)^2 = (3)^4 \times \left(\frac{1}{3}\right)^2 = 3^2 = 9.$$
   $$\Rightarrow P' = 9P = 9 \times 500 \text{ W} = 4500 \text{ W}.$$

   Thus, in the second case, the radiation from the spherical black body will be 4500 W.

5. A metal ball of mass 5 kg, diameter 0.20 m, and temperature 127°C is suspended in a box whose walls are at a temperature 27°C. Calculate the maximum rate of fall in temperature of the metal ball. Given that the specific heat of the metal = 0.2 and Stefan's constant = 5.7×10^{-8} in S.I. unit.

   Answer: As the metal ball is inside a closed box, it will behave like a black body, and its temperature will fall at the maximum rate. From Stefan's law of black body radiation, the net rate of loss of thermal energy from the metal ball is

   $$\frac{dQ}{dt} = \sigma A \left(T^4 - T_0^4\right) = 5.7 \times 10^{-8} \times 4\pi \left(\frac{0.2}{2}\right)^2 \times \left[(400)^4 - (300)^4\right]$$
   $$= 125.286 \text{ W}.$$

   If the temperature of the metal ball falls by $d\theta$ in time dt, we have

   $$\frac{dQ}{dt} = ms\frac{d\theta}{dt} \quad \text{or,} \quad \frac{d\theta}{dt} = \frac{1}{ms}\frac{dQ}{dt};$$
   $$\Rightarrow \frac{d\theta}{dt} = \frac{125.268}{5 \times 0.2 \times 4200}°\text{C s}^{-1} = 0.02983°\text{C s}^{-1}.$$

6. A thin rectangular sheet of brass of sides 15.0×10^{-2} m and 12.0×10^{-2} m is heated in a furnace up to a temperature 600 °C and then taken out. How much electric power is needed to maintain the sheet at this temperature? Given that the emissivity of the brass is 0.250 and Stefan–Boltzmann constant is $\sigma = 5.7 \times 10^{-8}$ Wm^{-2}K^{-4}. Further neglect the loss of heat due to convection.

Answer: The sheet radiates heat from both sides. So, the amount of heat radiated per second by the sheet, from Stefan–Boltzmann law, is given by

$$Q = 2\epsilon\sigma AT^4. \tag{11.90}$$

Here, the emissivity is $\epsilon = 0.250$, Stefan–Boltzmann constant is $\sigma = 5.67 \times 10^{-8} \text{Wm}^{-2}\text{K}^{-4}$. The cross sectional area of the rectangular sheet is given by

$$A = 15 \times \times 10^{-2} \text{ m} \times 12 \times 10^{-2} \text{ m} = 180 \times 10^{-4} \text{ m}^2,$$

and the temperature is given by $T = (600 + 273)\text{K} = 873\text{K}$. So, the amount of electric power that needs to be supplied to the sheet to maintain its temperature constant is given by

$$Q = 2 \times 0.25 \times 5.67 \times 10^{-4} \times 180 \times 10^{-4} \times (873)^4 \text{ W} = 296.4 \text{ W}.$$

7. The rate at which the earth receives radiation at its surface from the Sun is 1400 Wm^{-3}. The distance of the center of the Sun from the surface of the earth is 15×10^{11} m, and the radius of the Sun is 7×10^8 m. Calculate the temperature of the surface of the Sun from the given data. Assume that the Sun behaves as a black body.

Answer: Let R be the radius of the Sun. Its surface area will be then given by $4\pi R^2$. Assuming the Sun to be a black body at a surface temperature T, and applying Stefan's law of black body radiation, the rate of emitted radiation from the Sun comes out to be

$$Q = 4\pi R^2 \, \sigma T^4, \tag{11.91}$$

where σ is the *Stefan's constant*.

If the distance of the center of the Sun from the surface of the earth is d, then the rate of incidence of radiant energy on a unit area of the earth's surface is given by

$$= \frac{Q}{4\pi d^2} = \frac{4\pi R^2 \, \sigma T^4}{4\pi d^2} = \frac{R^2 \, \sigma T^4}{d^2}. \tag{11.92}$$

According to the problem, this is equal to

$$\frac{R^2 \sigma T^4}{d^2} = 1400 \text{ Wm}^{-2}, \tag{11.93}$$

where $R = 7 \times 10^8$ m, $d = 1.5 \times 10^{11}$ m, and $\sigma = 5.67 \times 10^{-8}$ Wm^{-2}K^{-4}. Hence, we get

$$T^4 = \frac{1400 \times (1.5 \times 10^{11})^2}{(7 \times 10^8)^2 \times 5.67 \times 10^{-8}}; \qquad \Rightarrow \quad T = 5803 \text{ K}.$$

So, the temperature of the surface of the Sun is 5803 K.

8. Assume that a planet radiates heat at a rate proportional to the fourth power of its surface temperature T, and that the planet attains such a steady temperature that the loss of heat is exactly compensated by the heat gained from the Sun. Show that the surface temperature of the planet will vary inversely as the square root of its distance from the Sun. Assume that other things remain the same.

Answer: Let us assume that the Sun is a perfectly black body of radius R and surface temperature T_1. Then, from *Stefan's law*, heat radiated by the Sun per second is

$$Q = 4\pi R^2 \sigma T_1^4, \tag{11.94}$$

where σ is *Stefan's constant*.

So, the amount of radiant energy incident on unit area per second on a planet situated at a distance d from the Sun is given by

$$= \frac{Q}{4\pi d^2} = \frac{4\pi R^2 \sigma T_1^4}{4\pi d^2} = \frac{R^2 \sigma T_1^4}{d^2}. \tag{11.95}$$

Hence, the total energy incident upon the planet per second is

$$= \frac{R^2 \sigma T_1^4}{d^2} \times \text{ cross-section of the planet}$$

$$= \frac{R^2 \sigma T_1^4}{d^2} \times \pi r^2 \quad [r = \text{radius of the planet}]$$

Let T be the surface temperature of the planet in the steady state. Then, the energy radiated from it per second is given by $4\pi r^2 \sigma T^4$. So, in the steady state, that is, at thermal equilibrium, we have

$$\frac{R^2 \sigma T_1^4}{d^2} \times \pi r^2 = 4\pi r^2 \sigma T^2 \quad \text{or,} \quad \Rightarrow \quad T^4 = \frac{R^2 T_1^4}{4d^2} \quad \Rightarrow \quad T = T_1 \sqrt{\frac{R}{2d}}.$$

As R and T_1 are constants, we have $T \propto \frac{1}{\sqrt{d}}$.

This shows that the surface temperature of the planet will vary inversely to the square root of its distance from the Sun.

9. The wavelength of the cosmic microwave background radiation is 1.1 mm. Calculate the temperature of the cosmic microwave background radiation.

   Answer: According to Wien's displacement law, $\lambda_m T = \text{constant} = 2.898 \times 10^{-3}$ m K.

   So we have,

   $$T = \frac{2.898 \times 10^{-3}}{\lambda_m} K = \frac{2.898 \times 10^{-3}}{1.1 \times 10^{-3}} K = 2.7 \text{ K}.$$

   This temperature was recorded by Penzius and Wilson in 1964 in a famous experiment and was identified as the temperature of the cosmic microwave background radiation. It should be mentioned here that this temperature is very close to the prediction of the Big Bang theory of the universe and is considered as the experimental verification of the Big Bang theory.

10. A spherical black body of radius 1 cm is enclosed in an evacuated chamber. If the chamber is at a temperature of 300 K, find out the amount of heat that must be supplied per second to the black body to keep it at a temperature 1000 K. Neglect conduction of heat. (Given: Stefan's constant, $\sigma = 5.67 \times 10^{-8}$ Wm^{-2}K^{-4}.)   [Calcutta University 2017]

    Answer: According to Stefan–Boltzmann law, the amount of radiation emitted per sec per unit area (which is equal to the supplied heat in these cases) is given by

    $$Q = \sigma(T^4 - T_0^4),$$

    where T_0 and T are, respectively, the initial and final temperatures and σ is Stefan's constant. Here, $T = 1000$ K and $T_0 = 300$ K. So,

    $$Q = (5.67 \times 10^{-8})(1000^4 - 300^4) = 56240.73 \text{ J/sec.m}^2.$$

    As the radius of the body is 1 cm, its surface area will be $= 4\pi(0.01)^2$ m$^2 = 1.26 \times 10^{-3}$ m^2.

    So, the heat supplied per second will be $= 70.86$ J/sec.

11. A wire of length 1 m and radius 1 mm is heated by passing an electric current through it to produce 1 kW of radiant power. Calculate the temperature of the wire. Treat the wire as a perfect black body and ignore any end effects.

    Answer: In this problem, we have to use the following concept:

    The electric power supplied by the current, $W =$ power radiated from the wire, $A\sigma T^4$, where σ is Stefan's constant with a value of 5.67×10^{-8} Js^{-1}m^{-2}K^{-4},

A is the surface area of the radiating surface of the wire, and T is its temperature. Thus, we have

$$W = A\sigma T^4; \quad \Rightarrow \quad T = \left(\frac{W}{A\sigma}\right)^{1/4}.$$

The surface area of the radiating surface of the wire is

$$A = 2\pi r \, l = 2\pi \times 10^{-3} \times 1.0 = 6.283 \times 10^{-3} \text{ m}^2.$$

Hence, the temperature T of the wire is

$$T = \left(\frac{W}{A\sigma}\right)^{1/4} = \left(\frac{1000}{6.283 \times 10^{-3} \times 5.67 \times 10^{-8}}\right)^{1/4} = 1294 \text{ K}.$$

12. At a temperature of 500°C, a black body A radiates thermal energy at the rate of 1.02×10^4 Js^{-1}. Calculate Stefan's constant σ if the surface area of the black body is 0.5 m^2.

    Answer: According to Stefan's law, the total energy radiated from a black body with surface area A per second at a temperature T is given by $E_T = A \sigma T^4$.

    From this expression, we have Stefan's constant σ as

    $$\sigma = \frac{E_T}{AT^4} = \frac{1.02 \times 10^4}{0.5 \times (273 + 500)^4} = 5.7 \times 10^{-8} \text{ Jm}^{-2}\text{s}^{-1}\text{K}^{-4}.$$

13. A perfect black body A at an initial temperature 300°C is allowed to cool inside an evacuated chamber surrounded by melting ice at the rate of 0.35°C s^{-1}. The mass, specific heat, and surface area of the black body are, respectively, 32 g, 0.10 calg^{-1}C^{-1}, and 8 cm^2. Calculate Stefan's constant.

    Answer: According to Stefan's law, the energy of the black body radiation per unit area per second from a black body at a temperature T surrounded by a chamber at temperature T_0 is given by $E = \sigma(T^4 - T_0^4)$.

    The total thermal energy radiated per second is given by

    $$E_T = A\sigma(T^4 - T_0^4) \text{ erg s}^{-1} = \frac{\sigma A(T^4 - T_0^4)}{J} \text{ cal s}^{-1}.$$

    The heat energy supplied to the black body is given by $ms\Delta T$. Equating these two expressions, we get

    $$ms\Delta T = \frac{\sigma A(T^4 - T_0^4)}{J}.$$

This gives

$$\sigma = \frac{ms\Delta T J}{A(T^4 - T_0^4)} = \frac{32 \times 0.10 \times 0.35 \times (4.2 \times 10^7)}{8 \times ((573)^4 - (273)^4)}$$

$$= 5.74 \times 10^{-5} \text{ erg } \text{cm}^{-2}\text{s}^{-1}\text{K}^{-4}.$$

14. The radius of the Sun and the temperature of the surface of the Sun are, respectively, given by 7×10^8 m and 6000 K. Assuming the Sun to radiate as a black body, answer the following questions:

(a) What is the total amount of radiation emitted per second from the surface of the Sun?

(b) If the distance between the earth and the Sun is 1.5×10^{11} m, how much energy falls on an area of 1 m^2 at the earth's surface perpendicular to the direction of the Sun?

Answer: (a) Just as the energy density is related to the emissive power, the amount of energy radiated by a surface is given by

$$E(T) = \frac{c}{4}U(T) = \frac{c}{4}\sigma T^4$$

$$= \frac{1}{4}(3 \times 10^8 \text{ m/s}) \left(7.57 \times 10^{-16} \text{ J/(m}^3\text{K}^4)\right)(6000 \text{ K})^4$$

$$= 7.36 \times 10^7 \text{ W m}^{-2}.$$

So, the total radiated power is given by

$$P = E(T) \times A = E(T) \times 4\pi R^2 = 7.36 \times 10^7 \text{ Wm}^{-2} \times 4\pi \times (7 \times 10^8 \text{ m}^2)$$

$$= 4.5 \times 10^{25} \text{ W}.$$

(b) The total radiation falls on the surface of a sphere of radius 1.5×10^{11} m, so that a fraction $(1/4\pi D^2) = 1/\left(4\pi \times (1.5 \times 10^{11} \text{ m})^2\right)$ is intercepted by a single square meter.

Hence, the power falls on an area of 1 m^2 on the surface of the earth is given by

$$p = \frac{P}{4\pi D^2} = \frac{4.5 \times 10^{25} \text{ W}}{4\pi \times (1.5 \times 10^{11} \text{ m})^2} = 1.6 \times 10^3 \text{ W}.$$

It should be borne in mind that the solar constant is 1.36×10^3 W.

15. The earth receives an amount of heat energy 2.0 cal cm^{-2}min^{-1} from the Sun. If the angular diameter of the Sun is 32' and is treated as a black body, find the surface temperature of the Sun. Given: $\sigma = 5.7 \times 10^{-12}$ W cm^{-2}K^{-4}.

Answer: Combining Stefan's law and the definition of solar constant, we have $T^4 = \dfrac{S}{\sigma}\left(\dfrac{R}{r}\right)^2$, where σ is the Stefan's constant, and S is the solar constant. Here, it is given by

$$S = \frac{2}{60}\ \text{calcm}^{-2}\text{s}^{-1} = \frac{2 \times 4.2}{60} \times \left(10^2\right)^2 \text{J} \times \text{m}^{-2} \times \text{s}^{-1} = 14 \times 10^2\ \text{J}\,\text{m}^{-2}\text{s}^{-1}.$$

Again the angular radius is given by $\left(\dfrac{r}{R}\right) = \dfrac{\text{radius of the Sun}}{\text{distance of the earth from the Sun}}.$

The angular diameter can be expressed in radians in the following way:

As angular diameter $= 32'$, the angular radius

$$= \frac{r}{R} = \frac{32'}{2} = 16' = \left(\frac{16}{60}\right)^{\circ} = \frac{16}{60} \times \frac{\pi}{180}\ \text{radian.}$$

This provides $\implies \dfrac{R}{r} = \dfrac{180 \times 60}{16 \times \pi}.$

Thus, $T^4 = \dfrac{14 \times 10^2}{\sigma} \times \left(\dfrac{180 \times 60}{16\pi}\right)^2.$ [Since $\sigma = 5.7 \times 10^{-12}\ \text{W cm}^{-2}\text{K}^{-4}$.]

This leads to $T = 5830$ K.

16. When the Sun is directly overhead, the thermal energy incident on the earth is 1.4 kW m^{-2}. Assuming that the Sun behaves like a perfect black body, show that the total intensity of radiation emitted from the Sun is 6.43×10^7 Wm^{-2} and hence, estimate the temperature of the Sun. The radius of the Sun is 7×10^5 km, and the Sun is at a distance of 1.5×10^8 km from the earth.

Answer: Solar constant S is the heat energy received by $1\,\text{m}^2$ of the earth's surface per second. If R is the radius of the Sun and r is the distance between the Sun and the earth, the total intensity of radiation emitted from the Sun will be $\sigma T^4\ Wm^{-2}$ and that from the Sun's surface will be $\sigma T^4 4\pi R^2$.

The radiation received per second per m^2 of earth's surface will be $S = \sigma T^4 \dfrac{4\pi R^2}{4\pi r^2}.$

Solving,

$$\sigma T^4 = S\frac{r^2}{R^2} = 1400 \left(\frac{1.5 \times 10^8}{7 \times 10^5}\right)^2 = 6.43 \times 10^7\ \text{Wm}^{-2}.$$

$$\text{This leads to} \quad T = \left(\frac{6.43 \times 10^7}{5.67 \times 10^{-8}}\right)^{1/4} = 5880\ \text{K}.$$

So, the temperature of the Sun is 5800 K.

17. The mass of the Sun is 2×10^{30} kg, its radius is 7×10^8 m, and its effective surface temperature is 5700 K. Calculate the mass of the Sun lost per second by radiation and the time required for the mass of the Sun to diminish by 1%.

Answer: The power radiated from the Sun

$$P = \sigma T^4 4\pi R^2 = 5.67 \times 10^{-8}(5700)^4 \times 4\pi \times (7 \times 10^8)^2 = 3.68 \times 10^{26} \text{ W.}$$

So, the mass loss per second $m = \dfrac{P}{c^2} = \dfrac{3.68 \times 10^{26}}{(3 \times 10^8)^2} = 4.1 \times 10^9$ kg/s.

Time taken for the mass of the Sun to decrease by 1% is

$$\frac{M}{100} \times \frac{1}{m} = \frac{2 \times 10^{30}}{100 \times 4.1 \times 10^9} = 4.88 \times 10^{18}\text{s}$$

$$= \left[\frac{4.88 \times 10^{18}}{365 \times 24 \times 3600} \right] \text{ year} = 1.55 \times 10^{11} \text{ year.}$$

18. A 10 g body heated at a temperature of 127°C is suspended inside an evacuated chamber maintained at a temperature of 27°C. The specific heat of the material of the body is 0.10 cal $g^{-1}C^{-1}$, and its surface area is 10 cm^2. Find the rate of cooling of the body if Stefan's constant is 5.72×10^{-5} erg cm^{-2} s^{-1} K^{-4}.

Answer: According to Stefan's law, the total thermal energy (in heat unit) of the black body radiation with surface area A per second from a black body at a temperature T surrounded by a chamber at temperature T_0 is given by

$$\frac{\sigma A(T^4 - T_0^4)}{J} \text{cal/s.}$$

If the ΔT be the cooling rate of the body with mass m and specific heat s, we can write

$$ms\Delta T = \frac{\sigma A(T^4 - T_0^4)}{J}.$$

This gives

$$\Delta T = \frac{\sigma A(T^4 - T_0^4)}{msJ} = \frac{(5.72 \times 10^{-5}) \times 10 \times ((400)^4 - (300)^4)}{10 \times 0.1 \times (4.20 \times 10^7)} = 0.24°\text{C/s.}$$

19. The operating temperature of the filament of a bulb is 3000 K. The emissivity and the surface area of the bulb are, respectively, 0.35 and 0.25 cm^2. Calculate the wattage of the bulb if Stefan's constant is 5.67×10^{-5} erg cm^{-2} s^{-1} deg^{-4}.

Answer: According to Stefan's law, the total thermal energy of a black body (here, the bulb is considered to be a black body) with surface area A per second is given by

$$E_T = e\sigma AT^4 = 0.35 \times 5.67 \times 10^{-5} \times 0.25 \times (3000)^4 = 40 \text{ W}.$$

20. Consider a photon gas enclosed in volume V and at temperature T in equilibrium. The photon is a mass-less particle. Show that the number of photons in the enclosure of volume V is proportional to the cube of temperature on an absolute scale. [Calcutta University 2013]

Answer: According to Plank's law of black body radiation, the number of photons in a volume V is given by

$$N = \int_0^\infty f(E)g(\nu)d\nu = \int_0^\infty \frac{1}{e^{\frac{h\nu}{K_BT}} - 1} \frac{8\pi\nu^2}{c^3} d\nu = \int_0^\infty \frac{8\pi}{c^3} \frac{\nu^2 \, d\nu}{e^{\frac{h\nu}{K_BT}} - 1}.$$

Let us put $\frac{h\nu}{K_BT} = x$; $\Rightarrow$ $d\nu = \frac{K_BT}{h} dx$. Using these substitutions in the above expression, we get the number of photons as

$$N = \frac{8\pi(K_BT)^3}{h^3c^3} \int_0^\infty \frac{x^2 dx}{e^x - 1} = \frac{8\pi(K_BT)^3}{h^3c^3} \times 2.405.$$

This shows that the number of photons in a cavity of volume V filled with black body radiation is proportional to the cube of temperature on an absolute scale.

21. A photon is a mass-less particle. Show that the average energy of the photons enclosed in a volume V and in equilibrium at temperature T is given by $\langle\epsilon\rangle = 2.70K_BT$.

Answer : According to Planck's law of black body radiation, the total number of photons per unit volume at a particular temperature T is given by

$$N = \frac{\int_0^\infty u_\nu \, d\nu}{h\nu} = \frac{\int_0^\infty \frac{8\pi h\nu^3}{c^3} \times \frac{1}{\left(e^{\frac{h\nu}{K_BT}} - 1\right)} d\nu}{h\nu} = \int_0^\infty \frac{8\pi}{c^3} \times \frac{\nu^2 \, d\nu}{e^{\frac{h\nu}{K_BT}} - 1}.$$

Let $\frac{h\nu}{K_BT} = x$; $\Rightarrow d\nu = \frac{K_BT}{h} dx$. The above integral then becomes

$$N = \frac{8\pi(K_BT)^3}{h^3c^3} \int_0^\infty \frac{x^2 dx}{e^x - 1} = \frac{8\pi(K_BT)^3}{h^3c^3} \times 2.405.$$

This is the total number of photons per unit volume of the cavity filled with black body radiation at a temperature T.

The average energy $\langle \epsilon \rangle$ of the photons at a temperature T can be calculated by dividing the total energy per unit volume by the total number of photons per unit volume. Hence, the average energy is

$$\langle \epsilon \rangle = \frac{\int\limits_0^\infty u_\nu \, d\nu}{N} = \frac{\int\limits_0^\infty \frac{8\pi h}{c^3} \nu^3 \times \frac{1}{\left(e^{\frac{h\nu}{K_B T}} - 1\right)} d\nu}{\frac{8\pi (K_B T)^3}{h^3 c^3} \times 2.405} = \frac{\frac{8\pi K_B^4 T^4}{h^3 c^3} \frac{\pi^4}{15}}{\frac{8\pi (K_B T)^3}{h^3 c^3} \times 2.405}$$

$$= \frac{K_B T \pi^4}{15 \times 2.405} = 2.70 \times K_B T.$$

Hence, the average energy of the photons at a temperature T of the radiation enclosed in the cavity is given by

$$\langle \epsilon \rangle = 2.70 \times K_B T.$$

So, the total number of photons emitted per unit area per sec by a black body at temperature T is given by

$$\frac{1}{4} \times N \times c = \frac{1}{4} \times \frac{8\pi (K_B T)^3}{h^3 c^3} \times 2.405 \times c = \frac{2\pi (K_B T)^3}{h^3 c^2} \times 2.405.$$

22. Calculate the average energy of an oscillator if its frequency and temperature are, respectively, 5×10^{14} Hz and 5000 K.

Answer: Here, we have $h = 6.63 \times 10^{-34}$ Js and $\nu = 5 \times 10^{14}$ Hz. Hence, the energy of the oscillator is given by

$$h\nu = 6.63 \times 10^{-34} \times 5 \times 10^{14} \text{ J} = 3.32 \times 10^{-19} \text{ J}$$

and the energy corresponding to the temperature $T = 5000$ K is given by

$$K_B T = 1.38 \times 10^{-23} \times 5000 \text{ J} = 6.90 \times 10^{-20} \text{ J}.$$

Assuming that these oscillators obey Planck's law of black body radiation, their average energy is given by

$$\langle E \rangle = \frac{h\nu}{e^{\frac{h\nu}{K_B T}} - 1} = \frac{3.32 \times 10^{-19}}{e^{\frac{3.32 \times 10^{-19}}{6.90 \times 10^{-20}}} - 1} = 2.7 \times 10^{-21} \text{ J}.$$

Note that this energy is smaller than $K_B T$.

23. Calculate the number of modes of vibration at a 100 cc chamber in the wavelength range from (i) 5000 to 5002 Å and (ii) 8000 to 8005 Å.

Answer: The number of modes within wavelength range λ and $\lambda + d\lambda$ in a chamber of volume V is given by

$$N_\lambda d\lambda \times V = \frac{\frac{8\pi K_B T}{\lambda^4} d\lambda \times V}{K_B T} = \frac{8\pi V}{\lambda^4} d\lambda = \frac{8 \times 3.14 \times 100}{(5000 \times 10^{-8})^4} \times (2 \times 10^{-8})$$

$$= 8.03 \times 10^{12}.$$

For the second case, the number of modes in the given volume V is given by

$$N_\lambda d\lambda \times V = \frac{\frac{8\pi K_B T}{\lambda^4} d\lambda \times V}{K_B T} = \frac{8\pi V}{\lambda^4} d\lambda = \frac{8 \times 3.14 \times 100}{(8000 \times 10^{-8})^4} \times (5 \times 10^{-8})$$

$$= 3.066 \times 10^{12}.$$

24. The Sun has an average temperature of $T = 6000°C$ on its surface. Calculate the radiation pressure on the surface of the Sun.

Answer: The pressure exerted by a photon gas, that is, radiation pressure P is given

$$P = \frac{8\pi^5 (K_B)^4}{45 h^3 c^3} T^4.$$

Substituting the values of the constants and $T = (6000 + 273)$ K, we get

$$P = \frac{8\pi^5 (K_B)^4}{45 h^3 c^3} T^4 = \frac{8\pi^5}{45} \times \frac{\left(6273 \times 1.38 \times 10^{-23} \text{ JK}^{-1}\right)^4}{\left(3 \times 10^8 \text{ ms}^{-1} \times 6273 \text{ K} \times 6.626 \times 10^{-34} \text{ Js}\right)^4}$$

$$= 0.39 \text{ Nm}^{-2}.$$

25. An FM radio transmitter has a power output 100 kW and operates at a frequency of 94 MHz. What is the number of photons emitted from the transmitter per second?

Answer: A photon of frequency 94 MHz has energy

$$E = h\nu = \left(6.62618 \times 10^{-34} \text{ J.s}\right) \times 94 \times 10^6 \text{ Hz} = 6.23 \times 10^{-26} \text{ J}.$$

The radio transmitter emits energy at a rate of 100 kJs^{-1}. Hence, the number of photons N_{ph} emitted from the transmitter per second is given by

$$N_{ph} = \frac{\text{power}}{\text{energy/photon}} = \frac{100 \times 10^3 \text{ W}}{6.23 \times 10^{-26} \text{ J}} = 1.605 \times 10^{30} \text{ s}^{-1}.$$

26. How many photons are emitted from a 100 W sodium lamp (550 nm) in one second? Assume that the lamp is 100% efficient in converting electrical energy into light.

Answer: The energy of a photon of wavelength $\lambda = 550$ nm is given by

$$E_{ph} = \frac{h\,c}{\lambda} = \frac{(2.998 \times 10^8 \text{ ms}^{-1})\,(6.62618 \times 10^{-34} \text{ J.s})}{550 \times 10^{-9} \text{ m}} = 3.61 \times 10^{-19} \frac{\text{J}}{\text{photon}}.$$

Using the 100-watt lamp, the number of photons emitted per second is given by

$$\frac{P}{E_{ph}} = \frac{100 \text{ W}}{3.61 \times 10^{-19} \frac{\text{J}}{\text{photon}}} = 2.77 \times 10^{20} \frac{\text{photons}}{\text{s}}.$$

27. Consider 1 cc of black body radiation at 727°C. How many photons are available in this black body radiation? Also find the average value of energy of the photon. (Given: $\int_{0}^{\infty} \frac{x^2}{e^x - 1} dx = 2.405$.)

Answer: According to Planck's law of black body radiation, the number of photons per unit volume is given by

$$N = \int_{0}^{\infty} \frac{8\pi}{c^3} \frac{\nu^2}{e^{\frac{h\nu}{K_B T}} - 1} d\nu,$$

where symbols have their usual meaning. Let us substitute

$$\frac{h\nu}{K_B T} = x; \quad \Longrightarrow \quad d\nu = \frac{K_B T}{h} dx.$$

Using these substitutions, we get

$$N = \frac{8\pi (K_B T)^3}{h^3 c^3} \int_{0}^{\infty} \frac{x^2 dx}{e^x - 1} = \frac{8\pi (K_B T)^3}{h^3 c^3} \times 2.405.$$

Putting $K_B = 1.38 \times 10^{-16}$, $T = 1000$ K, $c = 3 \times 10^{10}$ cm/s, and $h = 6.63 \times 10^{-27}$ in CGS unit, we get the number of photons per unit volume as

$$N = 2.017 \times 10^{10}.$$

The expression for total energy of the radiation per unit volume is given by
$E = \frac{8\pi^5 (K_B T)^4}{15 h^3 c^3}$.

Hence, the average energy per photon is

$$\frac{E}{N} = \frac{\frac{8\pi^5 (K_B T)^4}{15 h^3 c^3}}{\frac{8\pi (K_B T)^3}{h^3 c^3} \times 2.405} = \frac{\pi^4 (K_B T)}{15 \times 2.405} = 8.96 \times 10^{-13} \mathrm{erg} = \frac{8.96 \times 10^{-13}}{10^7} \mathrm{J}$$
$$= 8.96 \times 10^{-19} \text{ J.}$$

The energy corresponding to the room temperature is given by

$$K_B T = 1.38 \times 10^{-23} \times 300 \text{ J} = 4.14 \times 10^{-21} \text{ J.}$$

This shows that the average energy of the oscillator, under given conditions, is much larger than the energy corresponding to room temperature.

28. The cosmic microwave background radiation (CMBR) has a temperature of approximately 2.7 K.

   (a) What is the value of wavelength λ_{max} (in m) that corresponds to the maximum spectral density $u(\lambda, T)$ of CMBR?

   (b) What is the value of frequency ν_{max} (in Hz) that corresponds to the maximum spectral density $u(\nu, T)$ of CMBR?

   (c) Do the maxima $u(\lambda, T)$ and $u(\nu, T)$ correspond to the same photon energy? If not, why?

Answer:

(a) From Wien's displacement law, we have $\lambda_{max} = \frac{hc}{5 K_B T}$. This gives

$$\lambda_{max} = \frac{hc}{5 K_B T} = \left(\frac{6.6 \times 10^{-34} \times 3 \times 10^8}{5 \times 1.38 \times 10^{-23} \times 2.7} \right) m = 1.1 \text{ mm.}$$

(b) The frequency ν_{max} is given by

$$\nu_{max} = 2.8 \frac{K_B T}{h} = \left(\frac{2.8 \times 1.38 \times 10^{-23} \times 2.7}{6.6 \times 10^{-34}} \right) \text{Hz} = 1.58 \times 10^{11} \text{ Hz.}$$

(c) The maxima $u(\lambda, T)$ and $u(\nu, T)$ do not correspond to the same photon energy. The reason of that is

$$u(\lambda, T)d\lambda = -u(\nu, T)d\nu \text{ and } \frac{d\nu}{d\lambda} = -\frac{c}{\lambda^2}.$$

It can be easily checked using the expression: $u(\lambda, T) = \frac{8\pi hc}{\lambda^5} \frac{1}{\exp\left(\frac{hc}{\lambda K_B T}\right) - 1}$.

29. What is the approximate number of CMBR photons hitting the earth per second per square meter?

Answer: The energy flux is given by

$$J = \sigma\left(T_{CMBR}\right)^4 = 5.7 \times 10^{-8}\left(\frac{W}{K^4\,m^2}\right) \times 2.7^4\,K^4 = 3 \times 10^{-6}\left(\frac{W}{m^2}\right).$$

The average energy of a photon is given by

$$\langle\epsilon\rangle = \frac{u(T)}{N} = \frac{\frac{8\pi^5 K_B^4 T^4}{15h^3 c^3}}{\frac{8\pi K_B^3 T^3 \times 2.404}{h^3 c^3}} = \frac{\pi^4}{15 \times 2.404}K_B T \approx 2.7 K_B T.$$

So, the approximate number of CMBR photon hitting the earth per second per square meter is given by

$$N = \frac{J}{\langle\epsilon\rangle} \approx \frac{3 \times 10^{-6}}{2.7 \times 1.38 \times 10^{-23} \times 2.7} \approx 3 \times 10^{16}\frac{\text{photons}}{m^2\,s}.$$

30. Two parallel black planes are, respectively, at the temperature, T_1 and T_2. The energy flux between these planes in vacuum is due to the black body radiation. A third black plane is inserted between the other two and is allowed to come to an equilibrium temperature T_3. Find this temperature T_3, and show that the energy flux between planes 1 and 2 is cut in half due to the presence of the third plane.

Answer: The configuration of the plates is shown below:

Without the third plane, the energy flux per unit area is

$$J^0 = \sigma\left(T_1^4 - T_2^4\right).$$

The equilibrium temperature of the third plane can be found from the energy balance given by

$$\sigma\left(T_1^4 - T_3^4\right) = \sigma\left(T_3^4 - T_2^4\right); \quad \Rightarrow \quad T_1^4 + T_2^4 = 2T_3^4; \quad \Rightarrow \quad T_3 = \left(\frac{T_1^4 + T_2^4}{2}\right)^{1/4}.$$

The energy flux between the first and second planes in the presence of the third plane is given by

$$J = \sigma\left(T_1^4 - T_3^4\right) = \sigma\left(T_1^4 - \frac{T_1^4 + T_2^4}{2}\right) = \frac{1}{2}\sigma\left(T_1^4 - T_2^4\right) = \frac{1}{2}J^0.$$

Thus, the third plane cuts the energy flux into half.

Super insulation: Many layers of aluminized Mylar foil loosely wrapped around the helium bath (in a vacuum space between the walls of a liquid helium cryostat) are used for creating super insulation. If there are N layers of heat shield, the energy flux reduction for such N heat shields is given by

$$J_n = \frac{J^0}{N+1}.$$

31. The internal energy U of a black body radiation enclosed in a cavity of volume V at a temperature T is given by $U = V\sigma T^4$, where the constant σ is independent of volume V and temperature T. Show that the free energy F and the entropy S of the black body radiation are, respectively, given by $F = -\frac{1}{3}V\sigma T^4 + Tf(V)$ and $S = \frac{4}{3}V\sigma T^3 - f(V)$, where $f(V)$ is some unknown function of V. Assuming $S \to 0$ as $T \to 0$, show that the radiation pressure $P = \frac{1}{3}\frac{U}{V}$.

    Answer: The expression for free energy is given by $\qquad F = U - TS.$

    The differential of the free energy can be written as: $\qquad dF = dU - TdS - SdT.$

    Again using the relation, $TdS = dU + PdV$, we get $\qquad dF = -SdT - PdV.$

    This gives the expressions for pressure P and entropy S as

    $$P = -\left(\frac{\partial F}{\partial V}\right)_T; \qquad \text{and} \qquad S = -\left(\frac{\partial F}{\partial T}\right)_V.$$

    From the expression $F = U - TS$ and $S = -\left(\frac{\partial F}{\partial T}\right)_V$, we get

    $$U = F - T\left(\frac{\partial F}{\partial T}\right)_V = -T^2\frac{\partial}{\partial T}\left(\frac{F}{T}\right)_V.$$

Using $U = V\sigma T^4$ in this expression, we have

$$V\sigma T^4 = -T^2 v \frac{\partial}{\partial T}\left(\frac{F}{T}\right)_V.$$

Integrating this expression with respect to T, we get

$$\frac{F}{T} = -\frac{V\sigma T^3}{3} + f(V). \quad \Rightarrow \quad F = -\frac{V\sigma T^4}{3} + Tf(V).$$

Here, $f(V)$ is an unknown function of V. Differentiating the above expression for free energy F with respect to T, we get the entropy S as

$$S = -\left(\frac{\partial F}{\partial T}\right)_V = \frac{4}{3}V\sigma T^3 - f(V). \text{ Further, as } T \to 0, S \to 0, \text{ we have } f(v) \to 0.$$

Thus, entropy $S = \frac{4}{3}V\sigma T^3$. So, $F = U - TS = V\sigma T^4 - \frac{4}{3}V\sigma T^4 = -\frac{1}{3}V\sigma T^4$.

Hence, the radiation pressure P is given by

$$P = -\left(\frac{\partial F}{\partial T}\right)_T = \frac{1}{3}\sigma T^4 = \frac{1}{3}\frac{V\sigma T^4}{V} = \frac{1}{3}\frac{U}{V} = \frac{1}{3}u.$$

Here, u is the energy density.

11.13 Multiple choice questions and answers

1. The absorptive power of a perfectly black body is equal to

   (a) 0.5

   (b) 0

   (c) 1

   (d) ∞

   Answer: The correct choice is (c).

2. The spectrum of a black body radiation is/are

   (a) band

   (b) continuous

   (c) line

   (d) absorption

   Answer: The correct choices are (a) and (c).

3. The average energy of a Planck's oscillator is

   (a) $E = h\nu$

   (b) $E = nh\nu$

   (c) $E = mc^2$

   (d) $E = \dfrac{h\nu}{e^{\frac{h\nu}{K_B T}} - 1}$

   Answer: The correct choice is (d).

4. Two stars A and B radiate maximum energy at wavelengths 360 nm and 480 nm, respectively. The ratio of their absolute temperatures is

   (a) 3:4

   (b) 4:3

   (c) 256:81

   (d) 81:256

   Answer: The correct choice is (c).

5. A black body has a maximum wavelength λ_m at 2000 K. Its corresponding wavelength at 3000 K will be

   (a) $\dfrac{3}{2}\lambda_m$

   (b) $\dfrac{2}{3}\lambda_m$

   (c) $\dfrac{16}{81}\lambda_m$

   (d) $\dfrac{81}{16}\lambda_m$

   Answer: The correct choice is (b).

6. The radiant energy incident normally on the surface of the earth from the Sun is $8.4 \times$ J/m^2 $-$ min. What would have been the radiant energy incident normally on the earth if the Sun had a temperature twice the present value?

   (a) 16.8×10^4 J/m^2 $-$ min

   (b) 33.6×10^4 J/m^2 $-$ min

   (c) 62.7×10^4 J/m^2 $-$ min

   (d) 134.4×10^4 J/m^2 $-$ min

   Answer: The correct choice is (d).

7. A sphere is at a temperature 600 K. Its cooling rate is R in an external environment with temperature 200 K. If temperature falls to 400 K, then cooling rate R' will be

   (a) $\dfrac{3R}{16}$

   (b) $\dfrac{16R}{3}$

   (c) $\dfrac{9R}{27}$

   (d) None of these.

   Answer: The correct choice is (a).

8. The rectangular surface of area 8×10^{-2} m $\times$ 4×10^{-2} m of a black body at 127°C, emits energy at the rate of E per second. If the length and breadth of the surface are reduced to half of its initial value and the temperature raised to 327°C, the rate of emission of energy will be

(a) $\dfrac{3}{8}E$

(c) $\dfrac{9}{16}E$

(b) $\dfrac{81}{16}E$

(d) $\dfrac{81}{64}E$

Answer: The correct choice is (d).

9. The intensity of radiation emitted by the Sun has its maximum value at a wavelength of 510 nm and that emitted by the North Star has the maximum value at 350 nm. If these stars behave like black bodies, then the ratio of the surface temperatures of the Sun and the North Star is

(a) 1.46

(c) 1.21

(b) 0.69

(d) 0.83

Answer: The correct choice is (c).

10. A black body is at a temperature of 2880 K. The energy of radiation emitted by this object with wavelength between 499 nm and 500 nm is U_1, between 999 nm and 1000 nm is U_2, and between 1499 nm and 1500 nm is U_3. If Wien's constant is $b = 288 \times 10^6$ nm K, then

(a) $U_1 = 0$

(c) $U_1 > U_2$

(b) $U_3 = 0$

(d) $U_2 > U_1$

Answer: The correct choice is (d).

11. Consider black body radiation contained in a cavity maintained at 2000 K. If the volume of the cavity is reversibly and adiabatically increased from 10 cm^3 to 640 cm^3, the temperature of the cavity changes to   [GATE 2004]

(a) 800 K

(c) 600 K

(b) 700 K

(d) 500 K

Answer: The correct choice is (d).

According to Wien's displacement law, we have $V^{1/3}T=$ constant. Using this law, we get

$$V_1^{1/3}T_1 = V_2^{1/3}T_2; \quad \Rightarrow \quad (10)^{1/3} \times 2000 = (640)^{1/3} \times T_2; \quad \Rightarrow \quad T_2 = 500 \text{ K}.$$

12. The temperature of a cavity of fixed volume is doubled. Which of the following is true for the black body radiation inside the cavity?   [GATE 2003]

(a) Its energy and the number of photons both increase 8 times.

(b) Its energy increases 8 times and the number of photons increases 16 times.

(c) Its energy increases 16 times and the number of photons increases 8 times.

(d) Its energy and the number of photons both increase 16 times.

Answer: The correct choice is (c).

For black body radiation inside a cavity with fixed volume, we have $E \propto T^4$. So, the total energy increases by 16 times when the temperature of the cavity becomes double at a fixed volume. Energy of each photon is K_BT when the temperature of the cavity becomes double at a fixed volume, energy of each photon becomes $2K_BT$.

So, the number of photons is given by

$$= \frac{\text{Total energy}}{\text{Energy of each photon}} = \frac{16 \times \text{Total initial energy}}{2 \times \text{Energy of each photon}}$$

$$= 8 \times \text{initial number of photons.}$$

Therefore, the number of photons increases by 8 times.

13. The plots of intensity versus wavelengths for three black bodies at temperatures T_1, T_2, and T_3, respectively, show peaks at wavelengths λ_1, λ_2, and λ_3 such that $\lambda_1 > \lambda_2 > \lambda_3$. Their temperatures are such that

(a) $T_1 > T_2 > T_3$ (c) $T_2 > T_3 > T_1$

(b) $T_1 > T_3 > T_2$ (d) $T_3 > T_2 > T_1$

Answer: The correct choice is (d).

14. The energy spectrum of a black body exhibits a maximum around a wavelength λ_0. The temperature of the black body is now changed such that to energy is maximum around a wavelength $\dfrac{3\lambda_0}{4}$. The power radiated by the black body will now increase by a factor of

(a) $\dfrac{4}{3}$ (c) $\dfrac{64}{27}$

(b) $\dfrac{16}{9}$ (d) $\dfrac{256}{81}$

Answer: The correct choice is (d).

15. The power incident on a detector of lights 100 nW. Determine the number of photons per second incident on the detector if the wavelength is 800 nm.

(a) 4×10^{11} photons   (c) 4×10^{10} photons

(b) 2×10^{11} photons   (d) 1×10^{11} photons

Answer: The correct choice is (a).

We know that energy incident per second= number of incident photons per second $\times \frac{hc}{\lambda}$. From this relation, we get

$$100 \times 10^{-9} = N \times \frac{hc}{\lambda}; \quad \Rightarrow \quad N = \frac{100 \times 10^{-9} \times 800 \times 10^{-9}}{6.627 \times 10^{-34} \times 3 \times 10^8} = 4 \times 10^{11} \text{ photons.}$$

16. What is the energy of a typical visible photon? About how many photons enter the eye per second when one looks at a weak source of light such as the moon, which produced light of intensity of about 3×10^{-4} W/m^2?

(a) 2.0 eV, 2.5×10^{11}

(b) 2.3 eV, 2.5×10^{10}

(c) 2.3 eV, 5.2×10^{10}

(d) 2.3 eV, 2.2×10^{11}

Answer: The correct choice is (b).

The wavelength of visible light is between 400 and 700 nm. We take a typical value of 550 nm. So, the energy of this light of wavelength 550 nm is given by

$$E = \frac{hc}{\lambda} \text{ J} = \frac{6.627 \times 10^{-34} \times 3 \times 10^{8}}{800 \times 10^{-9}} \text{ J} = \frac{6.627 \times 10^{-34} \times 3 \times 10^{8}}{1.602 \times 10^{-19} \times 800 \times 10^{-9}} \text{eV} = 2.3 \text{ eV}.$$

On the atomic level, this energy is significant, but it is extremely small to the everyday standards. When we look at the moon, the energy entering our eye per second is given by IA, where I is the intensity and A is the area of the pupil.

We know that the diameter of the pupil is about 6 mm, so its area will be of the order of $A \approx 3 \times 10^{-5}$m^2. So, the number of photons entering our eye per second is

$$= \frac{IA}{E} = \frac{3 \times 10^{-4} \times 3 \times 10^{-5}}{2.3 \times 1.602 \times 10^{-9}} \approx 2.5 \times 10^{10} \text{ photons/s.}$$

17. A spherical black body with radius R and at the temperature T (K) emits an energy E J/s. Another spherical black body with radius $2R$ and at temperature $2T$ (K) emits an energy equal to

(a) 2 E J/s

(b) 16 E J/s

(c) 4 E J/s

(d) 64 E J/s

Answer: The correct choice is (d).

According to Stefan's law of black body radiation, we have the energy radiated per unit area per unit time is directly proportional to the fourth power of absolute temperature T.

This gives $E = \sigma T^4$. Then we have $E = \sigma T^4 \times$ area $\propto R^2 T^4$.

So, $\frac{E_2}{R_2^2 T_2^4} = \frac{E_1}{R_1^2 T_1^4} \Rightarrow \frac{E_1}{R_1^2 T_1^4} \times R_2^2 T_2^4 = \left(\frac{R_2}{R_1}\right)^2 \times \left(\frac{T_2}{T_1}\right)^4 \Rightarrow E = 4 \times 16 \times E = 64E.$

18. Consider a black body radiation enclosed in a cavity whose walls are at temperature T. The radiation is in equilibrium with the walls of the cavity. If the temperature of the walls is increased to 27 and the radiation is allowed to come to equilibrium at the new temperature, the entropy of the radiation increases by a factor of

(a) 2   (c) 8

(b) 4   (d) 16

Answer: The correct choice is (c).

For black body radiation, the energy is given by

$$F = \frac{-8\pi^5 K_B^4 T^4}{45\hbar^2 C^3} V \text{ and, } S = -\left(\frac{\partial F}{\partial T}\right)_V = \left(\frac{32\pi^5 K_B^4}{45\hbar^2 C^3}\right) V T^3.$$

If the temperature increased from T to $2T$ then entropy will increase from S to $8S$.

19. Two bodies A and B have thermal emissivity of 0.01 and 0.81, respectively. The outer surface areas of both the bodies are the same. The rate of emission of total radiant power of the two bodies is also the same. The wavelength of B corresponding to maximum spectral radiant energy in the radiation from B is shifted from the wavelength corresponding to maximum spectral radiant energy in the radiation from A by 1.00 µm. If the temperature of A is 5802 K, then the temperature of B is

(a) 1934 K   (c) 2901 K

(b) 11604 K   (d) None of the above

Answer: Hence, the correct choice is (a).

According to Wien's displacement law, $\lambda_m T = \text{constant} \Rightarrow \lambda_A T_A = \lambda_B T_B$.

We know that the power radiated by a body with emissive power ε at a temperature T is given by

$$P = \varepsilon \sigma T^4 A. \quad \text{This provides} \quad \varepsilon_a T_A^4 = \varepsilon_B T_B^4$$

$$T_B = \left(\frac{\varepsilon_A}{\varepsilon_B}\right)^{\frac{1}{4}} T_A = \left(\frac{0.01}{0.81}\right)^{\frac{1}{4}} \times 5802 = \frac{1}{3} \times 5802 = 1394 \, \text{K}$$

20. A black body is at the temperature of 527°C. Which one of the following values of the temperature makes it to radiate twice as much energy per second?

(a) 951 K   (c) 1032 K

(b) 638 K   (d) 1227 K

Answer: The correct choice is (a).

According to Stefan's law of black body radiation,

$$E = \sigma T^4 \qquad \Rightarrow \qquad \frac{E_2}{E_1} = \left(\frac{T_2}{T_1}\right)^{\frac{1}{4}}$$

$$\Rightarrow \quad (\frac{T_2}{T_1})^4 = 2 \qquad \Rightarrow \qquad T_2 = (2)^{\frac{1}{4}} \times 800 \text{ K} = 951 \text{ K}$$

21. For a system of independent noninteracting 1D oscillators, the value of the free energy per oscillator, in the limit $T \to 0$, is

(a) $\frac{1}{2}\hbar\omega$ (c) $\frac{3}{2}\hbar\omega$

(b) $\hbar\omega$ (d) 0

Answer: The correct choice is (a).

For the given system $z_N = \left[2\sinh\frac{\hbar\omega}{2K_BT}\right]^{-N}$

$$F = -K_BT \ln z_N = NK_BT \ln\left[2\sinh\left(\frac{\hbar\omega}{2K_BT}\right)\right]$$

$$= NK_BT \ln\left[\frac{2(e^{\frac{\hbar\omega}{2K_BT}} - e^{-\frac{\hbar\omega}{2K_BT}})}{2}\right]$$

$$= NK_BT \ln e^{\frac{\hbar\omega}{2K_BT}}\left(1 - e^{-\frac{\hbar\omega}{K_BT}}\right)$$

$$= NK_BT \ln e^{\frac{\hbar\omega}{2K_BT}} + NK_BT \ln\left(1 - e^{-\frac{\hbar\omega}{K_BT}}\right)$$

$$\frac{F}{N} = \frac{\hbar\omega}{2} + K_BT \ln\left(1 - e^{-\frac{\hbar\omega}{K_BT}}\right)$$

$$\Rightarrow \quad \frac{F}{N} = \frac{\hbar\omega}{2} + 0 = \frac{\hbar\omega}{2} \quad (K_BT \to 0).$$

22. For a black body radiation in a cavity, photons are created and annihilated freely as a result of emission and absorption by the walls of the cavity. This is because

(a) The chemical potential of the photons is zero.
(b) Photons obey Pauli Exclusion Principle.
(c) Photons are spin 1 particles.
(d) The entropy of the photons is very large.

Answer: The correct choice is (a).

Since photon has zero chemical potential, the number of photons cannot be conserved. Thus, photons can be created and annihilated.

23. If E is the total energy emitted by a body at a temperature T Kelvin and $E_{\max}$ is the maximum energy emitted by it at the same temperature, then

(a) $E \propto T^4, E_{\max} \propto T^5$

(b) $E \propto T^4, E_{\max} \propto T^{-5}$

(c) $E \propto T^{-4}, E_{\max} \propto T^4$

(d) $E \propto T^4, E_{\max} \propto T^4$

Answer: The correct choice is (a).

24. Star A has the peak of its black body radiation at λ_A. Star B has its peak at λ_B, which is one-fourth that of λ_A. If Star A's surface temperature is T_A, how does the surface temperature T_B of Star B compare?

(a) $T_B = 16 T_A$

(b) $T_B = 4 T_A$

(c) $T_B = T_A/4$

(d) $T_B = T_A/16$

Answer: The correct choice is (b).

25. The average number of photons in equilibrium inside a cavity is proportional to [GATE 2006]

(a) T

(b) T^2

(c) T^3

(d) T^4

Answer: The correct choice is (c).

26. The average number of energy quanta of the oscillators is given by

(a) $<n> = \frac{1}{e^{\beta \hbar \omega} - 1}$

(b) $<n> = \frac{e^{-\beta \hbar \omega}}{e^{-\beta \hbar \omega} - 1}$

(c) $<n> = \frac{1}{e^{\beta \hbar \omega} + 1}$

(d) $<n> = \frac{e^{-\beta \hbar \omega}}{e^{-\beta \hbar \omega} + 1}$

Answer:

$$<n> = \frac{\sum n_i e^{-\beta \varepsilon_i}}{Z} = \frac{0.e^{-\frac{\beta \hbar \omega}{2}} + 1.e^{-\frac{3\beta \hbar \omega}{2}} + 2.e^{-\frac{5\beta \hbar \omega}{2}}}{Z} = \frac{1.e^{-\frac{3\beta \hbar \omega}{2}} + 2.e^{-\frac{5\beta \hbar \omega}{3}} + 3.e^{-\frac{7\beta \hbar \omega}{2}} + ...}{\frac{e^{-\frac{\beta \hbar \omega}{2}}}{(1 - e^{-\beta \hbar \omega})}}$$

$$= \frac{(1.e^{-\frac{3\beta \hbar \omega}{2}} + 2.e^{-\frac{5\beta \hbar \omega}{2}} + 3.e^{-\frac{7\beta \hbar \omega}{2}})(1 - e^{-\beta \hbar \omega})}{e^{-\frac{\beta \hbar \omega}{2}}} = (1.e^{-\beta \hbar \omega} + 2.e^{-2\beta \hbar \omega} + 3.e^{-3\beta \hbar \omega} + ...)(1 - e^{-\beta \hbar \omega})$$

$$= (1.e^{-\beta \hbar \omega} + 2.e^{-2\beta \hbar \omega} + 3.e^{-3\beta \hbar \omega} + ...) - (1.e^{-2\beta \hbar \omega} + 2.e^{-3\beta \hbar \omega} + 3.e^{-4\beta \hbar \omega} + ...) =$$

$$e^{-\beta \hbar \omega} + e^{-2\beta \hbar \omega} + e^{-3\beta \hbar \omega} + ... = \frac{e^{-\beta \hbar \omega}}{1 - e^{-\beta \hbar \omega}} = \frac{1}{e^{\beta \hbar \omega} - 1}.$$

Hence, the correct choice is (a).

11.14 Exercise

11.14.1 Short answer type questions

1. What is thermal radiation?

2. State some characteristics of thermal radiation.

3. What do you mean by radiant energy?

4. What is a black body? How can a black body be realized in practice?

5. State some characteristics of a black body.

6. What do you mean by diffuse radiation? Write an expression for it.

7. Explain why does radiation exert pressure.

8. Define emissive and absorptive powers.

9. State and explain Kirchhoff's law of black body radiation.

10. State Stefan's law of black body radiation.

11. State Wien's law of black body radiation.

12. State Wien's displacement law of black body radiation.

13. State Rayleigh–Jeans of black body radiation.

14. State Planck's law of black body radiation.

15. What is photon? Write down some properties of photons.

16. What is black body radiation? Deduce an expression for energy density and pressure of diffuse radiation.

17. What do you mean by chemical potential? Show that the chemical potential of a photon is zero.

18. What do you mean by diffuse radiation? Write an expression for it.

19. State Rayleigh–Jeans law of black body radiation and derive it from Planck's law of black body radiation.

20. What is the temperature of microwave background radiation?

21. What do you mean by infrared radiation?

22. What is the quantum theory of radiation?

23. What are the experimental evidences which led to birth of quantum theory of radiation?

24. What is the ultraviolet catastrophe?

25. What are the energy and momentum of a photon?

26. Photon number is not conserved – explain.

27. Explain the particle nature of the photon.

28. Explain – radiation exerts pressure.

29. State Planck's law of black body radiation.

30. What is a photon? Write down some properties of photons.

31. What do you mean by chemical potential? Show that the chemical potential of a photon gas is zero.

32. State Rayleigh–Jeans law of black body radiation and derive it from Planck's law of black body radiation.

33. What is the temperature of microwave background radiation?

34. What do you mean by infrared radiation?

11.14.2   Long answer type questions

1. Define the emissive and absorptive power of a substance. Show that at any temperature, the ratio of emissive power to the absorptive power of a substance is constant and is equal to the emissive power of a perfectly black body.

2. Deduce an expression for energy density and pressure of diffuse radiation.

3. What are the drawbacks of classical theory in explaining the observed features of black body radiation?

4. Calculate the number of modes of Planck's oscillator within frequency ranges between ν and $\nu + d\nu$ in a cavity filled with black body radiation per unit volume at a temperature T.

5. Calculate the average energy of Planck's oscillator within frequency ranges between ν and $\nu + d\nu$ in a cavity filled with black body radiation per unit volume at a temperature T.

6. Derive Planck's law of black body radiation.

7. Plot graphically the energy density of black body radiation according to Planck's law at two different temperatures.

8. Describe microwave background radiation in support of Planck's law of black body radiation.

9. Following the approach of Prof. S. N. Bose, derive Planck's law of black body radiation.

10. Following the approach of Prof. A. Einstein, derive Planck's law of black body radiation.

11. Write a short note on the experimental verification of Planck's law of black body radiation.

12. Deduce Wien's distribution law using Planck's law of black body radiation.

13. Deduce Stefan's law of black body radiation using Planck's law of black body radiation.

14. Write a short note on Planck's unit of mass, length, and time.

15. Explain – how does the temperature of cosmic background radiation determined using Planck's law of black body radiation.

16. What is a bolometer? What is its function?

17. Describe Herschel's experiment for the detection of infrared radiation.

18. (i) Define the emission power of a body. State Stefan–Boltzmann's law.
(ii) Derive Planck's radiation formula from Bose–Einstein statistics. Hence, show that Wien's distribution law and Rayleigh–Jeans law can be obtained from Planck's law of black body radiation. [CU 2010, 2014]

19. What are the drawbacks of classical theory in explaining the observed features of black body radiation?

20. Write a short note on the "quantum theory of radiation".

21. Why is the chemical potential μ of photons in a box, and also acoustic phonon in a crystal, taken to be zero?

22. What are the characteristic features of photon picture of electromagnetic radiation?

23. Write down the postulates of Planck's quantum theory of radiation.

24. Derive Planck's law of black body radiation.

25. Plot graphically the energy density of black body radiation according to Planck's law at two different temperatures and explain.

26. Describe microwave background radiation in support of Planck's law of black body radiation.

27. Write a short note on Planck's unit of mass, length, and time.

28. What are the significances of Planck's unit?

29. Explain how the temperature of cosmic background radiation determined using Planck's law of black body radiation.

11.14.3   Numerical problems

1. Show that in a black body radiation chamber of volume V at a temperature T, the number of modes of vibration $g(\nu)d\nu$ within the frequency range between ν and $\nu + d\nu$ is given by

$$g(\nu)d\nu = 8\pi V \frac{\nu^2}{c^3} \, d\nu.$$

2. Calculate the average energy of Planck's oscillator within frequency ranges between ν and $\nu + d\nu$ in a cavity filled with black body radiation per unit volume at a temperature T.

3. Derive Planck's law of black body radiation in terms of wavelength.

4. Deduce Wien's displacement law from Planck's law of black body radiation.

5. Calculate the density of states of a photon gas.

6. Calculate the energy density of a photon gas.

7. Calculate the pressure of a photon gas.

8. Determine the energy of a photon and the number of photons emitted per second by a $P = 2$ mW He-Ne laser that operates on the wavelength $\lambda = 632.8$ nm. Interpret the results.   [1.96 eV, 6.37×10^{15} photons/s].

9. The temperature of the skin of a person is $\theta_{skin} = 35°$C. Determine the wavelength at which the radiation emitted from the skin reaches its peak.
    [$\lambda m = 9.40 \times 10^{-6}$ m]

10. Estimate the net loss of energy during one day. Express the result in calories by use of the conversion relation 1 cal $= 4.184$ J. [9.5 μm, 155 W, 3207 kcal].

11. Estimate the net loss of power by the human body in an environment of temperature $\theta_{env} = 20°$C. Assume that the emittance of the human skin is $\epsilon = 0.98$ for the infrared region and the typical surface area of the human body is estimated to be $A = 2$ m^2 and its temperature is $33°$C.
    [155 W, 3207 kcal]

12. Calculate the number density of photons in the following cases:

   (a) For the entire observable universe with a temperature of $T = 2.73$ K.

   (b) At room temperature $T = 300$ K.

   (c) At $T = 1500$ K.

   (d) On the surface of the Sun with $T = 6000$ K.

   $[n = 4.128 \times 10^8 \ m^{-3}, \ n = 5.478 \times 10^{14} \ m^{-3}, \ n = 6.847 \times 10^{16} \ m^{-3}$ and $n = 4.382 \times 10^{18} \ m^{-3}]$

13. Show that the internal energy of a photon gas is $U = 2.701 \ N \ K_B T$ whereas that for a classical ideals gas is $U = 1.50 \ N \ K_B T$.

Calculation of the Number of Accessible States to an Ideal Gas

1A Calculation of the number of accessible states to an ideal gas

We consider an ideal gas enclosed in a container of volume V at a temperature T. The gas consists of N number of molecules, each of mass m. Suppose the total energy of the system lies in a narrow range from $(E - dE)$ to E. Any molecule of the ideal gas lying within this energy range is described by a state having an elementary volume

$$\Delta\Gamma_{\text{elementary}} = dq_1 \, dq_2 \, dq_3 \, \cdots \, dq_{3N} \, dp_1 \, dp_2 \, dp_3 \, \cdots \, dp_{3N}, \tag{1}$$

where $q's$ and $p's$ are, respectively, the position and momentum coordinates of the molecules of the gaseous system. Hence, the total volume available to the system in the $6N$-dimensional phase space Γ_V is given by

$$\Gamma_V = \int \Delta\Gamma_{\text{elementary}} = \int dq_1 \, dq_2 \, dq_3 \, \cdots \, dq_{3N} \, dp_1 \, dp_2 \, dp_3 \, \cdots \, dp_{3N}.$$

This leads to

$$\Gamma_V = \int dq_1 \, dq_2 \, dq_3 \, \cdots \, dq_{3N} \times \int dp_1 \, dp_2 \, dp_3 \, \cdots \, dp_{3N} = V_{\text{position}} \times V_{\text{mom}}, \tag{2}$$

where V_{position} and V_{mom} are the respective total volume in position and momentum space. The corresponding total number of quantum states (micro states) in the phase space is given by

$$\Omega = \frac{\Gamma_V}{h^{3N}} = \frac{1}{h^{3N}} \int dq_1 \, dq_2 \, dq_3 \, \cdots \, dq_{3N} \, dp_1 \, dp_2 \, dp_3 \, \cdots \, dp_{3N} = \frac{V_{\text{position}} \times V_{\text{mom}}}{h^{3N}}. \tag{3}$$

In the case of a perfect gas (i.e. an ideal gas) considered here, the constituent molecules do not interact with one another, that is, the potential energy of the gas is zero, and thus, the total energy E of the gas is wholly kinetic and is given by

$$E = \sum_{i=1}^{3N} \frac{p_i^2}{2m}. \tag{4}$$

Since the energy does not depend on the position coordinates of the molecules, and the coordinates of a molecule are independent of the coordinates of other molecules, the integral over the molecular position coordinates just gives the volume in position space. Hence, the position integral provides

$$\int dq_1 \, dq_2 \, dq_3 \, \cdots \, dq_{3N} = V^N. \tag{5}$$

Using equation (5) in equation (3), we get

$$\Omega = \frac{V^N}{h^{3N}} \int dp_1 \, dp_2 \, dp_3 \, \cdots \, dp_{3N} = \frac{V^N}{h^{3N}} \times V_{\text{mom}}, \tag{6}$$

where $V_{\text{mom}} = \int dp_1 \, dp_2 \, dp_3 \, \cdots \, dp_{3N}$. The value of this volume V_{mom} in momentum space occupied by the molecules of the gaseous system can be calculated in the following way: Since the energy of the system is confined in the range $(E - dE)$ to E, we have the following condition for the gas

$$(E - dE) \leq \sum_{i=1}^{3N} \frac{p_i^2}{2m} \leq E; \quad \Rightarrow \quad 2m(E - dE) \leq \sum_{i=1}^{3N} p_i^2 \leq 2mE.$$

This expression can be written as

$$\left(\sqrt{2m(E - dE)}\right)^2 \leq (p_1^2 + p_2^2 + \cdots + p_{3N}^2) \leq \left(\sqrt{2mE}\right)^2. \tag{7}$$

Hence, the momentum integral in equation (6) is the volume of a hyper-shell having inner and outer radii $\sqrt{2m(E - dE)}$ and $\sqrt{2mE}$, respectively, in the $3N$-dimensional momentum space. One such shell in the energy range between $(E - dE)$ and E in energy space is shown in Figure 1A.1. Here, the momentum coordinates of one molecule cannot be separated from those of the others.

Let us first consider this case in a 3D space. In this space, the volume of a spherical shell having inner and outer radii, respectively, as $\sqrt{2m(E - dE)}$ and $\sqrt{2mE}$, is given by

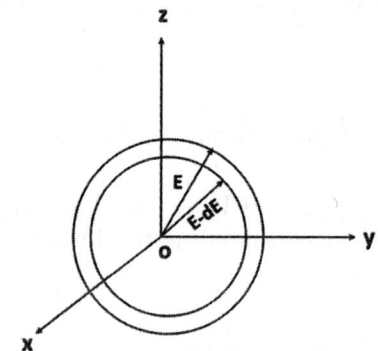

Figure 1A.1 The region between $(E - dE)$ and E shows the number of states $g(E)dE$ within the energy ranges between $(E - dE)$ and E for the particles.

$$\int dp_1 \, dp_2 \, dp_3 = \frac{4\pi}{3} \left[\left(\sqrt{2mE}\right)^3 - \left(\sqrt{2m(E - dE)}\right)^3 \right].$$

This leads to

$$\int dp_1 \, dp_2 \, dp_3 = \frac{\pi^{\frac{3}{2}}}{\Gamma\left(\frac{3+2}{2}\right)} \left[\left(\sqrt{2mE}\right)^3 - \left(\sqrt{2m(E-dE)}\right)^3\right]. \tag{8}$$

Following this analogy, we can write the volume of a hypershell having inner and outer radii, respectively, $\sqrt{2m(E-dE)}$ and $\sqrt{2mE}$ in the $3N$-dimensional momentum space as

$$\int dp_1 \, dp_2 \, dp_3 \, \cdots \, dp_{3N} = \frac{\pi^{\frac{3N}{2}}}{\Gamma\left(\frac{3N+2}{2}\right)} \left[\left(\sqrt{2mE}\right)^{3N} - \left(\sqrt{2m(E-dE)}\right)^{3N}\right]$$

$$= \frac{(2\pi mE)^{\frac{3N}{2}}}{\left(\frac{3N}{2}\right)!} \left\{1 - \left(1 - \frac{dE}{E}\right)^{\frac{3N}{2}}\right\} = \frac{(2\pi mE)^{\frac{3N}{2}}}{\left(\frac{3N}{2}\right)!} \left[1 - \left\{1 - \frac{3N}{2}\frac{dE}{E}\right.\right.$$

$$+\frac{1}{2!}\left(\frac{3N}{2}\right)\left(\frac{3N}{2}-1\right)\left(\frac{dE}{E}\right)^2 - \frac{1}{3!}\left(\frac{3N}{2}\right)\left(\frac{3N}{2}-1\right)\left(\frac{3N}{2}-2\right)\left(\frac{dE}{E}\right)^3 + \cdots\left.\left.\right\}\right].$$

Since N is very large, we will have $\left(\frac{3N}{2}-1\right) \approx \left(\frac{3N}{2}\right)$, $\left(\frac{3N}{2}-2\right) \approx \left(\frac{3N}{2}\right)$, and so on. Using these results, the volume in momentum space can be written as

$$V_{\text{mom}} = \frac{(2\pi mE)^{\frac{3N}{2}}}{\left(\frac{3N}{2}\right)!} \left[1 - \left\{1 - \frac{3N}{2}\frac{dE}{E} + \frac{1}{2!}\left(\frac{3N}{2}\frac{dE}{E}\right)^2 - \frac{1}{3!}\left(\frac{3N}{2}\frac{dE}{E}\right)^3 + \cdots\right\}\right]$$

$$= \frac{(2\pi mE)^{\frac{3N}{2}}}{\left(\frac{3N}{2}\right)!} \left[1 - \exp\left(-\frac{3N}{2}\frac{dE}{E}\right)\right].$$

If N is very large $\sim 10^{23}$, and ΔE, say, $0.0001 \, eV$ and $E = 1 \, eV$, then $e^{-\frac{3N}{2}\frac{\Delta E}{E}} \sim e^{-18} \to 0$.

This leads to

$$V_{\text{mom}} = \int dp_1 \, dp_2 \, dp_3 \, \cdots \, dp_{3N} = \frac{(2\pi mE)^{\frac{3N}{2}}}{\left(\frac{3N}{2}\right)!}. \tag{9}$$

Using equation (9) into equation (6), we have the number of states (microstates) within the volume of the hypershell as

$$\Omega = \frac{V^N V_{\text{mom}}}{h^{3N}} = \frac{V^N (2\pi mE)^{\frac{3N}{2}}}{h^{3N}\left(\frac{3N}{2}\right)!} = \frac{V^N}{\left(\frac{3N}{2}\right)!}\left[\frac{2\pi mE}{h^2}\right]^{\frac{3N}{2}}. \tag{10}$$

Equation (10) gives the total number of accessible micro states to the ideal gas of volume V with total energy E.

Alternative approach to calculate the number of accessible states Ω

We consider a system consisting of classical non relativistic free particles with the same mass m. This type of gas is popularly known as Boltzmann gas. The Hamiltonian H of such a gaseous system is given by $H = \sum\limits_{i=1}^{3N} \frac{p_i{}^2}{2m}$, where p_i is the momentum of the ith particle in one direction. The expression for Hamiltonian can be rearranged to write

$$\sum_{i=1}^{3N} p_i{}^2 = 2m \sum_{i=1}^{3N} \epsilon_i = 2mE = R^2.$$

We consider the situation in a momentum space, and R is the effective radius of N particles in this momentum space. These N particles occupy a $3N$-dimensional momentum space and a $3N$-dimensional spatial space. In the momentum space, the particles will be occupying a sphere of radius R. Since entropy is related to the logarithm of this number of states, it can be shown that the logarithm of the number of states contained in a volume with radius R and surface area Σ are nearly equal. Hence, for convenience, we will choose the number of states within a volume of radius R. Thus, the total volume available to the N particles will be a product of the volume of spatial and momentum space. A particle occupies a spatial volume V, and hence, N particles will occupy volume V^N. Now we have to find the volume of the particles in momentum space. The volume of a $3N$-dimensional sphere of radius R is given by

$$V_{3N}(R) = \frac{\pi^{\frac{3N}{2}} R^{3N}}{\frac{3N}{2}!}.$$

Hence, the total volume of the particles in momentum and spatial space is given by $V^N \left(\frac{\pi^{\frac{3N}{2}} R^{3N}}{\frac{3N}{2}!} \right)$. Substituting the value of R, the expression for the total volume occupied by the molecules comes out to be

$$V^N \left(\frac{\pi^{\frac{3N}{2}} (2mE)^{\frac{3N}{2}}}{\frac{3N}{2}!} \right).$$

However, the minimum volume required for the occupation of a microstate in this $6N$-dimensional space is $h^3 N$. Hence, the number of states available to the particle is

$$\Omega = \left(\frac{V}{h^3} \right)^N \frac{(2\pi mE)^{\frac{3N}{2}}}{\frac{3N}{2}!}.$$

This leads to the expression of the number of accessible states to the system of N particles as

$$\Omega = \left(\frac{V}{h^3}\right)^N \frac{(2\pi mE)^{\frac{3N}{2}}}{\frac{3N}{2}!} = \frac{V^N}{\left(\frac{3N}{2}\right)!}\left[\frac{2\pi mE}{h^2}\right]^{\frac{3N}{2}}. \tag{11}$$

This is exactly the same as given by equation (10).

Useful Thermodynamic Relations

2A Useful Thermodynamic Relations

2A.1 Maxwell's thermodynamic relations

1. $\left(\frac{\partial S}{\partial V}\right)_T = \left(\frac{\partial P}{\partial T}\right)_V$.

2. $\left(\frac{\partial S}{\partial P}\right)_T = -\left(\frac{\partial V}{\partial T}\right)_P$.

3. $\left(\frac{\partial T}{\partial V}\right)_S = -\left(\frac{\partial P}{\partial S}\right)_V$.

4. $\left(\frac{\partial T}{\partial P}\right)_S = \left(\frac{\partial V}{\partial S}\right)_P$.

2A.2 Expressions for isobaric volume expansion coefficient α_P, isothermal compressibility β_T, and isothermal elasticity κ_T

1. $\alpha_P = \frac{1}{V}\left(\frac{\partial V}{\partial T}\right)_P$.

2. $\beta_T = -\frac{1}{V}\left(\frac{\partial V}{\partial P}\right)_T$.

3. $\kappa_T = \frac{1}{\beta_T} = -V\left(\frac{\partial P}{\partial V}\right)_T$.

2A.3 Various expressions for C_V, C_P, and $C_P - C_V$

1. $C_P - C_V = R$;   for one mole of an ideal gas.

2. $C_P - C_V = T\left(\frac{\partial P}{\partial T}\right)_V \left(\frac{\partial V}{\partial T}\right)_P = T\,V\,\alpha_P\left(\frac{\partial P}{\partial T}\right)_V$.

3. $C_P - C_V = -T\left(\frac{\partial P}{\partial V}\right)_T \left[\left(\frac{\partial V}{\partial T}\right)_P\right]^2 = T\,V\,\kappa_T\,\alpha_P^2$.

4. $C_V = -T\left(\frac{\partial^2 F}{\partial T^2}\right)$.

5. $C_P = -T\left(\frac{\partial^2 G}{\partial T^2}\right)$.

6. $\left(\frac{\partial C_V}{\partial V}\right)_T = T\left(\frac{\partial^2 P}{\partial T^2}\right)_V$.

7. $\left(\frac{\partial C_P}{\partial P}\right)_T = -T\left(\frac{\partial^2 V}{\partial T^2}\right)_P$.

8. $C_P - C_V = R\left(1 + \frac{2a}{RTV}\right)$;   for one mole of a Van der Waals gas when $V \gg b$.

2A.4 Thermodynamic relations from combination of first and second law of thermodynamics

The combination of the first and the second laws of thermodynamics leads to the following relation

$$dU = TdS - PdV. \tag{12}$$

1. $T = \left(\frac{\partial U}{\partial S}\right)_V.$

2. $\left(\frac{\partial U}{\partial S}\right)_P = T - P\left(\frac{\partial V}{\partial S}\right)_T = T - \frac{P}{\left(\frac{\partial S}{\partial V}\right)_T} = T - \frac{P}{\left(\frac{\partial P}{\partial T}\right)_V} = T - \frac{P\,\beta_T}{\alpha_P}.$

3. $\left(\frac{\partial U}{\partial S}\right)_P = T - P\left(\frac{\partial V}{\partial S}\right)_P = T - P\left(\frac{\partial V}{\partial T}\right)_P\left(\frac{\partial T}{\partial S}\right)_P = T - \frac{P\,V\,\alpha_P\,T}{C_P}.$

4. $-P = \left(\frac{\partial U}{\partial V}\right)_S.$

5. $\left(\frac{\partial U}{\partial S}\right)_P = T\left(\frac{\partial S}{\partial V}\right)_P - P = T\left(\frac{\partial S}{\partial T}\right)_P\left(\frac{\partial T}{\partial V}\right)_P - P = \frac{C_P}{V\,\alpha_P} - P.$

6. $\left(\frac{\partial U}{\partial V}\right)_T = T\left(\frac{\partial S}{\partial V}\right)_T - P = T\left(\frac{\partial P}{\partial T}\right)_V - P.$

7. $\left(\frac{\partial U}{\partial P}\right)_V = T\left(\frac{\partial S}{\partial P}\right)_V = T\left(\frac{\partial S}{\partial T}\right)_V\left(\frac{\partial T}{\partial P}\right)_V = \frac{C_V\,\beta_T}{\alpha_P}.$

8. $\left(\frac{\partial U}{\partial P}\right)_T = T\left(\frac{\partial S}{\partial P}\right)_T - P\left(\frac{\partial V}{\partial P}\right)_T = -T\left(\frac{\partial V}{\partial T}\right)_P + P\,V\,\beta_T = -T\,V\,\alpha_P + P\,V\,\beta_T.$

9. $\left(\frac{\partial U}{\partial P}\right)_S = -P\left(\frac{\partial V}{\partial P}\right)_S = P\,V\,\beta_S.$

10. $\left(\frac{\partial U}{\partial T}\right)_S = -P\left(\frac{\partial V}{\partial T}\right)_S = -\frac{P}{\left(\frac{\partial T}{\partial V}\right)_S} = \frac{P}{\left(\frac{\partial P}{\partial S}\right)_V} = \frac{P}{\left(\frac{\partial P}{\partial T}\right)_V\left(\frac{\partial T}{\partial S}\right)_V} = \frac{P\,C_V\,\beta_T}{T\,\alpha_P}.$

11. $\left(\frac{\partial U}{\partial T}\right)_V = T\left(\frac{\partial S}{\partial T}\right)_V = C_V.$

12. $\left(\frac{\partial U}{\partial T}\right)_P = T\left(\frac{\partial S}{\partial T}\right)_P - P\left(\frac{\partial V}{\partial T}\right)_P = C_P - P\,V\,\alpha_P.$

2A.5 T–dS Equations

1. $TdS = C_V dT + T\left(\frac{\partial P}{\partial T}\right)_V dV.$

2. $TdS = C_P dT - T\left(\frac{\partial V}{\partial T}\right)_P dP.$

3. $TdS = C_V\left(\frac{\partial T}{\partial P}\right)_V dP + C_P\left(\frac{\partial T}{\partial V}\right)_P dV.$

2A.6 Energy equations

1. $\left(\frac{\partial U}{\partial V}\right)_T = T\left(\frac{\partial P}{\partial T}\right)_V - P.$

2. $\left(\frac{\partial U}{\partial P}\right)_T = -T\left(\frac{\partial V}{\partial T}\right)_P - P\left(\frac{\partial V}{\partial P}\right)_T.$

2A.7 Thermodynamic relations between Helmholtz free energy F, pressure P, volume V, temperature T, and entropy S

We have the relation

$$dF = -SdT - PdV. \tag{13}$$

1. $S = -\left(\frac{\partial F}{\partial T}\right)_V$.

2. $\left(\frac{\partial F}{\partial T}\right)_P = S - P\left(\frac{\partial V}{\partial T}\right)_P = -S - PV\alpha_P$.

3. $\left(\frac{\partial F}{\partial T}\right)_S = -S - P\left(\frac{\partial V}{\partial T}\right)_S = -S - \frac{P}{\left(\frac{\partial T}{\partial V}\right)_S} = -S + \frac{P}{\left(\frac{\partial P}{\partial T}\right)_V\left(\frac{\partial T}{\partial S}\right)_V} = -S + \frac{C_V P \beta_T}{\alpha_P T}$.

4. $-P = \left(\frac{\partial F}{\partial V}\right)_T$.

5. $\left(\frac{\partial F}{\partial V}\right)_P = -S\left(\frac{\partial T}{\partial V}\right)_P - P = -\frac{S}{V\alpha_P} - P$.

6. $\left(\frac{\partial F}{\partial V}\right)_S = -S\left(\frac{\partial T}{\partial V}\right)_S - P = +S\left(\frac{\partial P}{\partial S}\right)_V - P = S\left(\frac{\partial P}{\partial T}\right)_V\left(\frac{\partial T}{\partial S}\right)_V - P = \frac{S\alpha_P T}{C_V \beta_T} - P$.

7. $\left(\frac{\partial F}{\partial P}\right)_T = -P\left(\frac{\partial V}{\partial P}\right)_T = PV\beta_T$.

8. $\left(\frac{\partial F}{\partial P}\right)_V = -S\left(\frac{\partial T}{\partial P}\right)_S - P\left(\frac{\partial V}{\partial P}\right)_S = -S\left(\frac{\partial V}{\partial S}\right)_P + PV\beta_S = -\frac{SV\alpha_P T}{C_P} + PV\beta_S$.

9. $\left(\frac{\partial F}{\partial S}\right)_T = -P\left(\frac{\partial V}{\partial S}\right)_T = -\frac{P}{\left(\frac{\partial S}{\partial V}\right)_T} = -\frac{P}{\left(\frac{\partial P}{\partial T}\right)_V} = -\frac{P\beta_T}{\alpha_P}$.

10. $\left(\frac{\partial F}{\partial P}\right)_V = -S\left(\frac{\partial T}{\partial P}\right)_V = -\frac{S\beta_T}{\alpha_P}$.

11. $\left(\frac{\partial F}{\partial S}\right)_V = -S\left(\frac{\partial T}{\partial S}\right)_V = -\frac{ST}{C_V}$.

12. $\left(\frac{\partial F}{\partial S}\right)_P = -S\left(\frac{\partial T}{\partial S}\right)_P - P\left(\frac{\partial V}{\partial S}\right)_P = -\frac{TS}{C_P} - \frac{PV\alpha_P T}{C_P}$.

2A.8 Thermodynamic relations between enthalpy H, pressure P, volume V, temperature T, and entropy S

We have the relation

$$dH = TdS + VdP. \tag{14}$$

1. $T = \left(\frac{\partial H}{\partial S}\right)_P$.

2. $\left(\frac{\partial H}{\partial S}\right)_V = T + V\left(\frac{\partial P}{\partial S}\right)_V = T + V\left(\frac{\partial P}{\partial T}\right)_V\left(\frac{\partial T}{\partial S}\right)_V = T + \frac{V\alpha_P T}{\beta_T C_V}$.

3. $\left(\frac{\partial H}{\partial S}\right)_T = T + V\left(\frac{\partial P}{\partial S}\right)_T = T - \frac{V}{\left(\frac{\partial V}{\partial T}\right)_P} = T - \frac{1}{\alpha_P}$.

4. $V = \left(\frac{\partial H}{\partial P}\right)_S$.

5. $\left(\frac{\partial H}{\partial P}\right)_T = T\left(\frac{\partial S}{\partial P}\right)_T + V = -T\left(\frac{\partial V}{\partial T}\right)_P + V = -TV\alpha_P + V$.

6. $\left(\frac{\partial H}{\partial P}\right)_V = T\left(\frac{\partial S}{\partial P}\right)_V + V = T\left(\frac{\partial S}{\partial T}\right)_V\left(\frac{\partial T}{\partial P}\right)_V + V = \frac{C_V\,\beta_T}{\alpha_P} + V.$

7. $\left(\frac{\partial H}{\partial V}\right)_S = V\left(\frac{\partial P}{\partial V}\right)_S = -\kappa_S.$

8. $\left(\frac{\partial H}{\partial T}\right)_P = T\left(\frac{\partial S}{\partial T}\right)_P = C_P.$

9. $\left(\frac{\partial H}{\partial P}\right)_V = T\left(\frac{\partial S}{\partial V}\right)_P = \frac{T}{\left(\frac{\partial V}{\partial S}\right)_P} = \frac{T}{\left(\frac{\partial V}{\partial T}\right)_P\left(\frac{\partial T}{\partial S}\right)_P} = \frac{C_P}{V\,\alpha_P}.$

10. $\left(\frac{\partial H}{\partial V}\right)_T = T\left(\frac{\partial S}{\partial V}\right)_T + V\left(\frac{\partial P}{\partial V}\right)_T = T\left(\frac{\partial P}{\partial T}\right)_V - \kappa_T = \kappa_T\left(T\,\alpha_P - 1\right).$

11. $\left(\frac{\partial H}{\partial T}\right)_V = T\left(\frac{\partial S}{\partial T}\right)_V + V\left(\frac{\partial P}{\partial T}\right)_V = C_V + \frac{V\,\alpha_P}{\beta_T}.$

12. $\left(\frac{\partial H}{\partial T}\right)_S = V\left(\frac{\partial P}{\partial T}\right)_S = \frac{V}{\left(\frac{\partial T}{\partial P}\right)_S} = \frac{V}{\left(\frac{\partial V}{\partial S}\right)_P} = \frac{V}{\left(\frac{\partial V}{\partial T}\right)_P\left(\frac{\partial T}{\partial S}\right)_P} = \frac{T}{\alpha_P\,C_P}.$

2A.9   Thermodynamic relations between Gibbs free energy G, pressure P, volume V, temperature T, and entropy S

We have the relation

$$dG = -SdT + VdP. \tag{15}$$

1. $V = \left(\frac{\partial G}{\partial P}\right)_T.$

2. $\left(\frac{\partial G}{\partial P}\right)_V = -S\left(\frac{\partial T}{\partial P}\right)_V + V = -\frac{S\,\beta_T}{\alpha_P} + V.$

3. $\left(\frac{\partial G}{\partial P}\right)_S = -S\left(\frac{\partial T}{\partial P}\right)_S + V = -S\left(\frac{\partial V}{\partial S}\right)_P + V = -\frac{S\,T\,V\,\alpha_P}{C_P} + V.$

4. $-S = \left(\frac{\partial G}{\partial T}\right)_P.$

5. $\left(\frac{\partial G}{\partial T}\right)_V = -S + V\left(\frac{\partial P}{\partial T}\right)_V = -S + \frac{V\,\alpha_P}{\beta_T}.$

6. $\left(\frac{\partial G}{\partial T}\right)_S = -S + V\left(\frac{\partial P}{\partial T}\right)_S = -S + \frac{C_P}{\alpha_P\,T}.$

7. $\left(\frac{\partial G}{\partial V}\right)_P = -S\left(\frac{\partial T}{\partial V}\right)_P = -\frac{S}{V\,\alpha_P}.$

8. $\left(\frac{\partial G}{\partial V}\right)_T = V\left(\frac{\partial P}{\partial V}\right)_T = -\kappa_T.$

9. $\left(\frac{\partial G}{\partial V}\right)_S = -S\left(\frac{\partial T}{\partial V}\right)_S + V\left(\frac{\partial P}{\partial V}\right)_S = S\left(\frac{\partial P}{\partial S}\right)_V + V\left(\frac{\partial P}{\partial V}\right)_S = \frac{S\,T\,\alpha_P}{\beta_T\,C_V} - \kappa_S.$

10. $\left(\frac{\partial G}{\partial S}\right)_V = -S\left(\frac{\partial T}{\partial S}\right)_V + V\left(\frac{\partial P}{\partial S}\right)_V = -\frac{S\,T}{C_V} + \frac{S\,\alpha_P\,T}{\beta_T\,C_V}.$

11. $\left(\frac{\partial G}{\partial S}\right)_P = -S\left(\frac{\partial T}{\partial S}\right)_P = -\frac{S\,T}{C_P}.$

12. $\left(\frac{\partial G}{\partial S}\right)_T = V\left(\left(\frac{\partial P}{\partial S}\right)_T\right) = \frac{V}{\left(\frac{\partial S}{\partial P}\right)_T} = -\frac{V}{\left(\frac{\partial V}{\partial T}\right)_P} = -\frac{1}{\alpha_P}.$

Concept of Negative Temperature

3A Temperature in Thermodynamics

For an infinitesimal reversible process, a combination of first and second laws of thermodynamics results

$$dS = \frac{dU}{T} + Y\,dX, \tag{16}$$

where $Y\,dX$ denotes the generalized expression for work done by the system, dS is the change in entropy, and dU is the change in internal energy of the system. Equation (16) leads to the definition of temperature T as

$$\frac{1}{T} = \left(\frac{dS}{dU}\right)_X; \quad \Rightarrow \quad T = \frac{1}{\left(\frac{dS}{dU}\right)_X}. \tag{17}$$

Thus, equation (17) indicates that the temperature at any point depends on the slope of the $S - U$ curve. If the slope of this curve (point A in Figure 3A.1) is positive, the temperature will be positive. On the other hand, the temperature will be negative for the negative slope of the curve (point C in Figure 3A.1).

In general, the entropy S of a thermodynamic system is an increasing function of the internal energy U, resulting in a positive temperature of the system. But in thermodynamics, we may have model systems for which entropy S is not restricted to be a monotonically increasing function of the internal energy U. The elements of such systems would have only two energy states in thermal equilibrium. The lowest

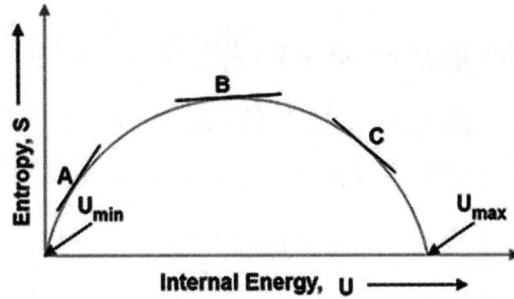

Figure 3A.1 Energy levels of a paramagnetic system with number of dipoles in the presence of an external magnetic field $\vec{B}$.

possible energy $U_{\min}$ (see Figure 3A.1) of the system is achieved when all the elements of the system are in their lowest energy state at the lowest possible temperature. This corresponds to a highly ordered state of the thermodynamic system with entropy

772

$S = 0$. Similarly, when all the elements are in the highest energy state at a higher temperature, the energy of the system becomes maximum $(U_{\max})$ (see Figure 3A.1), and this state again corresponds to a highly ordered state of the system with entropy $S = 0$. At the intermediate energies, elements of the system are distributed in these two states with their corresponding probabilities depending upon the values of energy of these two states, and the disorder of the system increases, thereby increasing the entropy. Therefore, the entropy of the system passes through a maximum (point B in Figure 3A.1) between the lowest $U_{\min}$ and the highest energy $(U_{\max})$ of the system and then diminishes with increasing U. This is shown in Figure 3A.1. The maximum of the entropy corresponds to

$$\left(\frac{dS}{dU}\right)_X = 0, \tag{18}$$

and according to equation (17), this corresponds to $T \to \infty$.

Hence, in cooling the system from negative to positive temperature, the system passes through ∞ K instead of 0 K. This indicates that negative temperatures are not **colder** than 0 K but are **hotter** than infinite temperature. Thus, thermodynamic principles and theorems can be applied to the negative as well as positive temperature states with suitable modifications.

This principle of variation in temperature from positive to negative through infinity can be demonstrated with the help of a paramagnetic system. This is described in detail below.

3A.1 Paramagnets

Materials that exhibit a small positive magnetic susceptibility in the presence of an external magnetic field are called paramagnetic, and the effect is termed as paramagnetism. In the absence of an external magnetic field, the orientations of atomic magnetic moments are random, leading to net magnetization zero. When an

Figure 3A.2 Energy levels of a paramagnetic system with number of dipoles in the presence of an external magnetic field $\vec{B}$.

external magnetic field is applied, dipoles line-up with the direction of the field, resulting in a positive magnetization. In this type of magnetic materials, the dipoles do not interact, and extremely large magnetic fields are required to align all of the dipoles. In addition, the effect is lost as soon as the magnetic field is removed. The thermal agitation randomizes the directions of the magnetic dipoles. Thus, an increase in temperature decreases the paramagnetic effect. The magnetic susceptibility of these materials is slightly positive and lies in the range $+10^{-5}$ to

$+10^{-2}$. Some examples of paramagnetism materials are aluminum, calcium, titanium, alloys of copper, etc.

We consider a paramagnetic substance having N magnetic atoms per unit volume placed in an external magnetic field $\vec{B}$. Let us assume that each atom has a spin $\frac{1}{2}$ and an intrinsic magnetic moment μ_B. In the presence of the magnetic field $\vec{B}$, the potential energy of such a magnetic dipole is given by

$$E = -\vec{\mu_B}.\vec{B} = -\mu_B \, B \cos \theta, \tag{19}$$

where θ is the angle between the direction of the magnetic moment $\vec{\mu_B}$ and the magnetic field $\vec{B}$. Classically, the magnetic moment can point along any direction in space, that is, θ can take any value between 0 and 2π. However, owing to the restrictions imposed by the principles of quantum mechanics, θ can have only two values: 0° and 180°, but there is no restriction on the number of atoms per state. This is a tailor made situation for a two-level system shown in Figure 3A.2. The characteristic features of such a tailor-made system under consideration are mentioned below:

1. As the spins are oriented either along or opposite to the direction of the magnetic field $\vec{B}$, only two energy levels with energies $-\mu_B \, B$ and $+\mu_B \, B$ will be available to the system. Thus, we have two energy states with energies $E_1 = -\mu_B \, B$ and $E_2 = +\mu_B \, B$. This is shown in Figure 3A.2.

2. Moreover, the energy levels are non-degenerate. That is, there is only one state at each level.

3. A state may contain any number of atoms.

For such a paramagnetic two-level system, we will calculate energy and entropy using the thermodynamic principles and comment on the variation of temperature of the system.

If n number of dipoles (each paramagnetic atom behaves as a dipole in the presence of magnetic field) are oriented along the direction of the magnetic field $\vec{B}$ and the remaining $(N - n)$ dipoles are oriented anti parallel to the magnetic field $\vec{B}$ (see Figure 3A.3), the total energy of the system in thermal equilibrium at an absolute temperature T can be written as

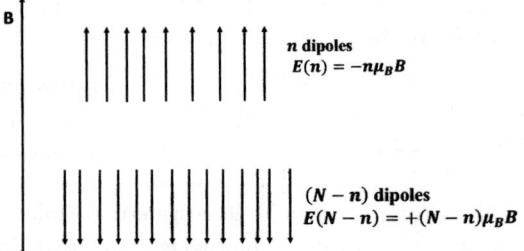

Figure 3A.3 Schematic arrangements of dipoles of a paramagnetic material in the presence of a magnetic field $\vec{B}$.

$$E(n) = n \times (-\mu_B \, B) + (N - n) \times (\mu_B \, B). \tag{20}$$

This equation (20) gives the total energy of this two-level system in the presence of the external magnetic field $\vec{B}$. Such energy levels are shown in Figure 3A.3.

The number of configurations with n dipoles oriented parallel to the magnetic field and $(N - n)$ dipoles anti parallel is given by

$$\Omega(n) = \frac{N!}{n!(N - n)!}. \tag{21}$$

This is the corresponding statistical weight of the above distribution. Once we know the number of configuration (number of states), the entropy $S(n)$ of the system can be obtained from the Boltzmann entropy relation in the following way:

$$S(n) = K_B \ln \left[\Omega(n)\right] = K_B \ln \left[\frac{N!}{n!(N - n)!}\right] = K_B \left[\ln(N!) - \ln(n!) - \ln(N - n)!\right]. \tag{22}$$

The numbers N and n are very large, so we can apply Stirling's approximation: $\ln(N!) = N \ln N - N$ to simplify equation (22). Using this approximation, equation (22) can be written as

$$S(n) = K_B \ln \left[N \ln N - N - n \ln n + n - (N - n)\ln(N - n) + (N - n)\right].$$

This leads to

$$S(n) = K_B \left[N \ln N - n \ln n - (N - n)\ln(N - n)\right]. \tag{23}$$

Equation (17) relates the temperature T of the system to the entropy S by

$$\frac{1}{T} = \left(\frac{\partial S(n)}{\partial E(n)}\right)_{B,\mu}. \tag{24}$$

Further, this equation (24) can be expressed as

$$\frac{1}{T} = \left(\frac{\partial S(n)}{\partial n}\right)\left(\frac{\partial n}{\partial E(n)}\right). \tag{25}$$

Differentiating equation (20) w.r.t n, we get

$$\frac{\partial E(n)}{\partial n} = -2\mu_B B \tag{26}$$

and differentiating equation (23) w.r.t n, we get

$$\frac{\partial S(n)}{\partial n} = K_B \left[\ln \left(\frac{N - n}{n}\right)\right]. \tag{27}$$

Using the results of equations (26) and (27) in equation (25), we have

$$\frac{1}{T} = \left[K_B \ln \left(\frac{N-n}{n} \right) \right] \left[\frac{-1}{2\mu_B B} \right] = \frac{K_B}{2\mu_B B} \ln \left(\frac{n}{N-n} \right). \qquad (28)$$

This equation (28) provides the temperature of the system in terms of the number of dipoles in the two energy states: $-\mu_B B$ and $+\mu_B B$ in the presence of an external magnetic field $\vec{B}$.

3A.2   Concept of negative temperature

Equation (28) has two very important implications:

1. If the number of dipoles n oriented along the direction of the magnetic field $\vec{B}$ is greater than half the total number of dipoles N, that is, $n > \frac{N}{2}$, the temperature T of the system is positive $(T > 0)$. [Check this by taking $n = 0.6N$]

2. On the other hand, if the number of dipoles n oriented along the direction of the magnetic field $\vec{B}$ is less than half the total number of dipoles N, that is, $n < \frac{N}{2}$, the temperature T of the system is negative $(T < 0)$. [Check this by taking $n = 0.3N$]

Thus, it is physically observed that in a negative temperature state, more than half the dipoles are aligned anti parallel to the direction of the magnetic field $\vec{B}$. At absolute zero temperature, all the dipoles will be in the lower-energy state. As the temperature is increased gradually, dipoles move to the higher energy state. When $T \to \infty$, both the states are equally populated with spin-up and spin-down dipoles. If the number of dipoles in the higher energy is more than that in the lower energy state, one is tempted to say that the temperature is hotter than infinity, even though the configuration is not energetically favorable. This signifies that population inversion corresponds statistically to a state of negative temperatures, implying that a negative temperature is hotter than $T = \infty$. And a system in a negative temperature state has more energy than it has at a positive temperature state! This is a paradoxical result.

In paramagnetic substance, the interaction between the ionic magnets and the lattice are so strong that the substance cannot exist in a state of population inversion for an appreciable time. However, in 1951, Pound Purcell and Ramsay established the existence of negative absolute temperature by using nuclear magnetic resonance techniques. It was found that the nuclear magnetic moments of the Li atoms in LiF interact extremely slowly with the lattice and elapse several minutes before the equilibrium condition is achieved with the lattice. As a result, the system exists in the state of population inversion, that is, the state of negative temperature for quite sometimes. Researchers at the Max-Planck Institute and the Ludwig-Maximilians University, Munich, worked with a Bose–Einstein Condensate of about 112000 very cold potassium atoms (39 K). A Bose–Einstein condensate was made up of atoms that occupied the same quantum state. They put these atoms in an optical dipole trap that confines the atoms and keeps them together. Then, they used a tunable 3D optical lattice that acts like a crystalline structure for the potassium atoms to reside in and creates an associated energy band structure. The optical lattice consists of a bunch of laser beams set to create a standing wave that acts like a lattice structure for the potassium atoms to be in. The potassium atoms in this lattice structure remain in their lowest internal energy state, but they have an associated band structure. The lattice confines the motional states of the atoms to this lowest energy band. However, this band consists of many states (momenta states) for each atom.

Chemical Potential

4A   Chemical Potential

Very often, the term "chemical potential" is not well understood by the students. After studying thermal physics and statistical mechanics for several times, students are still in a lot of confusion about the meaning of the term "chemical potential". This quantity is represented by the letter μ. Typically, students learn the definition of μ, its properties, its derivation in some simple cases, and its consequences, and work out numerical problems on it. Still, students ask the question: "What is the chemical potential?" and "What does it actually mean?" Attempts are made in this appendix to clarify the meaning of this physical quantity μ with some simple examples.

The concept of chemical potential has appeared first in the classical works of J. W. Gibbs. Since then, it has become actually a subtle concept in thermodynamics and statistical mechanics. It is not easy to grasp the meaning and significance of chemical potential μ, like thermodynamic concepts such as temperature T, internal energy E, or even entropy S. In fact, chemical potential μ has acquired a reputation as a concept not easy to grasp even for the experienced physicist. Chemical potential was introduced by Gibbs within the context of an extensive exposition on the foundations of statistical mechanics. In his exposition, Gibbs considered a grand canonical ensemble of systems in which the exchange of particles occurs with the surroundings. In this description, the chemical potential μ appears as a constant required for a necessary closure to the corresponding set of equations. Thus, a fundamental connection with thermodynamics is achieved by observing that the unknown constant μ is indeed related to standard thermodynamic functions like the Helmholtz free energy $F = U - TS$ or the Gibbs thermodynamic potential $G = F + PV$ through their first derivatives. μ, in fact, appeared as a conjugate variable to volume V.

4A.1   Comments about chemical potential

We are familiar with the term potential used in mechanical and electrical system. A capacity factor is associated with each potential term. For example, in a mechanical system, mass is the capacity factor associated with the gravitational potential $g(h_2 - h_1)$, where h_1 and h_2 are the corresponding heights, and the gravitational work done is given by $mg(h_2 - h_1)$. In the case of an electrostatic system, the work done in moving a charge q from a position of electrostatic potential V_1 to another position of electrostatic potential V_2 is $q(V_2 - V_1)$. Here, q plays the role of capacity factor. In a similar way, the capacity factor, and potential can be defined for a chemical system. **Mole** is considered to be the capacity factor and ***molar free energy*** as the potential (known as the **chemical potential**). Then, the amount of work done to transfer n moles of a substance from a state of molar free energy F_1 to another state of molar free energy F_2 will be given by $n(F_2 - F_1)$. **A chemical substance that is free to move from one place to another will move continuously from a state of higher chemical potential to a state of lower chemical potential. In a state of equilibrium, the chemical potential is the same throughout the system.**

Let us have a look at the chemical potential from a statistical point of view. In a micro-canonical ensemble, total energy U, volume V, and number of particles N are constant. In a canonical ensemble, the possibility of exchange of energy among the systems (by placing the systems in contact with a large heat bath) is taken into account. But the number of particles is held constant. But there are many systems in which the number of particles is not constant. For example, growing of a crystal, melting of ice, and some chemical reactions. Naturally, the question arises: **How can one extend the methods developed for a constant number of particles to handle non constant particle numbers?**

We have seen that If two systems have the same temperature, there is no net flow of thermal energy between the systems. Again, for two systems having the same pressure, there will be no net change of volumes. Now we ask the question: **If there is no net flow of particles between two systems, which quantity would have the same value?** It is the chemical potential μ. If two systems are at the same temperature and only have a single chemical species and the same value of the chemical potential, there will be no net flux of particles from one side of the system to another. But if the chemical potential is different, there will be a net flux of particle numbers. This is observed when we look at two gases with different particle numbers in the same volume.

We know that temperature and pressure are thermodynamically defined by maximizing entropy with respect to variations in energy and volume, respectively. The interpretation of temperature is that it remains the same for two systems in thermal equilibrium. Similarly, the pressure remains the same for two systems in mechanical equilibrium. Again it is known that the entropy is also a function of the number of particles (or oscillators, or magnetic moments, $\cdots$) making up the system. Two systems which can exchange particles are in diffusive contact and can come to diffusive equilibrium. We expect that two systems will adjust the concentration of particles until equilibrium is obtained; presumably, the subsystem with the most number of particles will tend to give up particles to the other subsystem(s). In this case, an observable, known as the chemical potential, becomes the same for two systems in diffusive equilibrium and plays the analogous role as those of temperature and pressure in thermal and mechanical equilibrium, respectively. Thus, **the chemical potential is the observable that controls the state of a system in diffusive equilibrium**.

Consider a thermodynamic system that can exchange energy and particles with a reservoir. The volume of the system can change. When a system gets energy through heating, its internal energy U increases by an amount TdS. If the system expands by a volume dV at a pressure P, the work done by the system is PdV. If the system gains dN particles of the same type, then internal energy U increases by μdN, where μ is the chemical potential. The total change in internal energy can then be written as

$$dU = TdS - PdV + \mu dN. \tag{29}$$

This relation is commonly called the thermodynamic identity, valid for infinitesimal, reversible processes. At constant volume and entropy, it follows from equation (29) that the parameter μ can be expressed as

$$\mu = \left(\frac{\partial U}{\partial N} \right)_{S,V}. \tag{30}$$

This mathematical expression can be used to find chemical potential for thermodynamic systems.

4A.2 Chemical potential in terms of other thermodynamic functions

Chemical potential is related to various thermodynamic functions such as internal energy, Helmholtz free energy, enthalpy, and Gibbs free energy. These mathematical relations are briefly discussed below.

Chemical potential in terms of internal energy U

We consider a homogeneous system consisting of independent components in several coexisting phases. Further, we assume that a given phase (ith) of this composite (mixture) system is characterized by the number of moles N_i, entropy S and volume V. Thus, the physical properties of the phase can be considered as a function of the independent variables V, S and the number of moles of the constituents $1, 2, 3, \cdots, i, \cdots$. As for example, the internal energy of the system can be written as

$$U = U(S, V, N_1, N_2, N_3, \cdots, N_i, \cdots, N_n),$$

where N_i is the number of particles of the ith component of the system that contains n components. So the infinitesimal change in internal energy can be written as

$$dU = \left(\frac{\partial U}{\partial S}\right)_{V,N_1,N_2,N_3,\cdots} dS + \left(\frac{\partial U}{\partial V}\right)_{S,N_1,N_2,N_3,\cdots} dV + \left(\frac{\partial U}{\partial N_1}\right)_{S,V,N_2,N_3,\cdots} dN_1$$

$$+ \left(\frac{\partial U}{\partial N_2}\right)_{S,V,N_1,N_3,\cdots} dN_2 + \cdots.$$

$$= \left(\frac{\partial U}{\partial S}\right)_{V,N} dS + \left(\frac{\partial U}{\partial V}\right)_{S,N} dV + \sum_{i=1}^{n} \left(\frac{\partial U}{\partial N_i}\right)_{S,V,N_i} dN_i. \tag{31}$$

Where the subscript N means that the number of particles of each component is held fixed, and the subscript N_i means that the number of particles of each component except for the ith component is held constant. Let us introduce a new variable μ_i to the partial derivative $\left(\dfrac{\partial U}{\partial N_i}\right)_{S,V,N_i}$ so that

$$\mu_i = \left(\frac{\partial U}{\partial N_i}\right)_{S,V,N_i}.$$

With this notation, equation (29) can be written as

$$dU = TdS - PdV + \sum_{i=1}^{n} \mu_i dN_i. \tag{32}$$

This is the mathematical form of first law of thermodynamics for a system that contains n species.

Chemical potential in terms of Helmholtz free energy F

The Helmholtz free energy F may be written as a function of temperature T, volume V, and n independent variables $N_1, N_2, N_3, \cdots, N_i, \cdots, N_n$, that is,

$$F = F(T, V, N_1, N_2, N_3, \cdots, N_i, \cdots, N_n). \tag{33}$$

The corresponding infinitesimal change in the Helmholtz free energy can be written as

$$dF = \left(\frac{\partial F}{\partial T}\right)_{V,N} dT + \left(\frac{\partial F}{\partial V}\right)_{T,N} dV + \sum_{i=1}^{n} \left(\frac{\partial F}{\partial N_i}\right)_{T,V,N_j} dN_i$$

$$= -SdT - PdV + \sum_{i=1}^{n} \mu_i dN_i = -SdT - PdV + \mu_1 dN_1 + \mu_2 dN_2 + \mu_3 dN_3 + \cdots,$$

where

$$\mu_1 = \left(\frac{\partial F}{\partial N_1}\right)_{T,V,N_2,\cdots} \quad \text{and} \quad \mu_2 = \left(\frac{\partial F}{\partial N_2}\right)_{T,V,N_1,\cdots}. \tag{34}$$

Thus, the chemical potentials are the rate of change of free energy per mole at constant volume and temperature, keeping other elements also constant.

Chemical potential in terms of enthalpy H

Following a similar procedure, the chemical potential μ can be expressed in terms of the thermodynamic function enthalpy H in the following way: Enthalpy is given by $H = U + PV$ and is a function of

$$H = H(S, P, N_1, N_2, N_3, \cdots, N_i, \cdots, N_n).$$

The corresponding infinitesimal change in the enthalpy can be written as

$$dH = TdS + VdP + \sum_{i=1}^{n} \mu_i dN_i, \quad \text{where} \quad \mu_i = \left(\frac{\partial H}{\partial N_i}\right)_{S,P,N_j,\cdots}. \tag{35}$$

This equation (35) provides the chemical potential μ in terms of the thermodynamic function enthalpy H.

Chemical potential in terms of Gibbs free energy G

The chemical potential μ can be obtained from the Gibbs free energy G in the following way. The Gibbs free energy function G is a function of several variables like

$$G = G(T, P, N_1, N_2, N_3, \cdots, N_i, \cdots, N_n).$$

The corresponding infinitesimal change in the Gibbs free energy function can be written as

$$dG = -SdT + VdP + \sum_{i=1}^{n} \mu_i dN_i,$$

where

$$\mu_i = \left(\frac{\partial G}{\partial N_i}\right)_{T,P,N_j,\cdots}. \tag{36}$$

These derivations show that the chemical potential can be written as

$$\mu_i = \left(\frac{\partial U}{\partial N_i}\right)_{S,V,N_j,\cdots} = \left(\frac{\partial H}{\partial N_i}\right)_{T,V,N_j,\cdots} = \left(\frac{\partial H}{\partial N_i}\right)_{S,P,N_j,\cdots} = \left(\frac{\partial G}{\partial N_i}\right)_{T,P,N_j,\cdots}$$

Thus, the chemical potential of any component in a mixture system can be obtained by differentiating any of the thermodynamic functions $U, F, H,$ and G with respect to the corresponding number of particles, keeping other thermodynamic variables fixed.

In the following section, we would like to look at what equation (30) tells us for some simple systems.

4A.3 Illustrating the idea of chemical potential with the help of an example

To illustrate the idea of chemical potential, we consider a simple idealized system. The single-particle energy eigenvalues of this system are quantized in integer multiples of energy ϵ as shown in Figure 4A.1. In general, the energy ϵ depends on the volume of the system, but in this example, we will insist on holding V constant. Further, we assume that the energy eigenstates are nondegenerate. The system is like that of a collection of simple harmonic oscillators, but with the zero of the single-particle energy scale shifted.

Further, we assume that the system consists of two distinguishable particles labeled as R and B, and the total initial energy of the system is $U = 2\epsilon$. We then ask the question: How many microstates are available to the system consisting of these two particles? The corresponding distribution of microstates for the two particles subject to the condition $U = 2\epsilon$ is shown in Table I.

Table I

R	B
2ϵ	0
0	2ϵ
ϵ	ϵ

Table II

R	B	W
2ϵ	0	0
0	2ϵ	0
0	0	2ϵ
ϵ	ϵ	0
ϵ	0	ϵ
0	ϵ	ϵ

Table III

	R	B	W
	ϵ	0	0
	0	ϵ	0
	0	0	ϵ

This shows that there are three microstates, that is, $\Omega = 3$. The entropy of the system is then given by

$$S = K_B \ln \Omega = K_B \ln 3, \qquad (37)$$

where K_B is the Boltzmann constant and Ω is the number of microstates.

In this system, let us add another distinguishable particle, labeled W, with zero energy, keeping the volume of the system fixed. The addition is done in such a way that the total energy is still $U = 2\epsilon$. Then, the available microstates for the system are shown in Table II. So, the number of microstates in this case will be six, that is, $\Omega = 6$ and the new entropy becomes $S = K_B \ln 6$. Thus, entropy increases from $S = K_B \ln 3$ to $S = K_B \ln 6$. But, to get the chemical potential of the system, we have to keep entropy constant. Let us calculate the chemical potential of the system under this situation. According to equation (37), it is calculated as the change in internal energy of

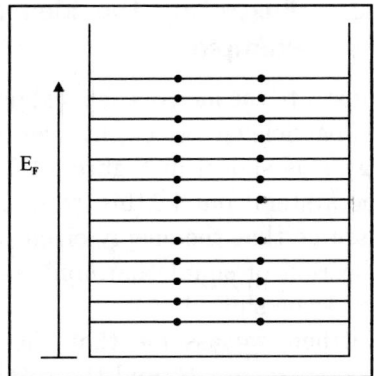

Figure 4A.1 Single-particle energy states are filled up to the Fermi level. The spacing of energy levels is shown schematically only.

the system when one more particle is added to it, holding the volume and the entropy constant. The only way this can be done for this simple system is to add the particle W with energy $-\epsilon$, or, physically, to add the particle while allowing the internal energy of the system to decrease by ϵ. This makes the total energy of the new system $U = \epsilon$, and the number of microstates again becomes $\Omega = 3$. These microstates are shown in Table III.

As the number of microstates remains constant in this case, the entropy also remains constant at $S = K_B \ln 3$. So the energy of the added particle is $\Delta U = -\epsilon$. Hence, according to equation (30), the chemical potential becomes

$$\mu = -\epsilon. \qquad (38)$$

The negative sign in equation (38) simply indicates that the energy of the system must decrease as the particle is added to the system so that the entropy remains constant.

4A.4   Chemical potential of an ideal gas

In statistical physics, it is a standard practice to calculate first the partition function of a system; other thermodynamic parameters, such as internal energy, pressure, entropy, and chemical potential, are then determined from the expressions involving the derivatives of the logarithm of the partition function. This procedure is adapted to get the expression for the chemical potential of a classical ideal system consisting of N identical but indistinguishable particles. The partition function Z_N for such a system is given by

$$Z_N = \frac{1}{N!}(Z_S)^N = \frac{1}{N!}\left(\sum_i e^{-\beta E_i}\right)^N = \frac{1}{N!}\left[\left(\frac{m}{2\beta\pi\hbar^2}\right)^{\frac{3}{2}}gV\right]^N, \qquad (39)$$

where Z_S is the single-particle partition function given by

$$Z_S = \left[\left(\frac{m}{2\beta\pi\hbar^2}\right)^{\frac{3}{2}}gV\right].$$

Taking logarithms on both sides of equation (39), we get

$$\ln Z_N = N\left[\ln(gV) + \frac{3}{2}\ln(K_BT) + \frac{3}{2}\ln\left(\frac{m}{2\pi\hbar^2}\right)\right] - \ln(N!)$$

$$= N\left[\ln(gV) + \frac{3}{2}\ln(K_BT) + \frac{3}{2}\ln\left(\frac{m}{2\pi\hbar^2}\right)\right] - N\ln N + N.$$

This leads to

$$\ln Z_N = N\left[\ln\left(\frac{gV}{N}\right) - \frac{3}{2}\ln\beta + \frac{3}{2}\ln\left(\frac{m}{2\pi\hbar^2}\right) + 1\right]. \qquad (40)$$

The entropy is related to the partition function Z by

$$S = K_B\ln Z + \frac{U}{T}.$$

Using this relation, we get the expression for the entropy of the ideal gas as

$$S = K_B\ln Z + \frac{U}{T} = K_B\left(N\left[\ln\left(\frac{gV}{N}\right) - \frac{3}{2}\ln\beta + \frac{3}{2}\ln\left(\frac{m}{2\pi\hbar^2}\right) + 1\right]\right) + \frac{U}{T},$$

$$= NK_B\left(\ln\left(\frac{gV}{N}\right) + \frac{3}{2}\ln(K_BT) + \frac{3}{2}\ln\left(\frac{m}{2\pi\hbar^2}\right) + 1\right) + \frac{1}{T}\times\frac{3}{2}NK_BT.$$

This leads to

$$S = NK_B\left[\ln\left(\frac{gV}{N}\right) + \frac{3}{2}\ln\left(\frac{mK_BT}{2\pi\hbar^2}\right) + \frac{5}{2}\right]$$

$$= NK_B\left[\ln\left(\frac{gV}{N}\right) + \frac{3}{2}\ln\left(\frac{mU}{3N\pi\hbar^2}\right) + \frac{5}{2}\right]. \qquad (41)$$

Equation (41) provides the expression for the entropy of the ideal system and is known as the **Sackur–Tetrode equation**.

Calculation of the chemical potential of the ideal gaseous system

We know that the chemical potential μ is defined by the relation

$$\mu = \left(\frac{\partial F}{\partial N}\right)_{T,V},$$

where F is the Helmholtz free energy given by $F = -K_B T \ln Z_N$. Using this expression for Helmholtz free energy, we get the chemical potential μ of a classical system as

$$\mu = \left(\frac{\partial F}{\partial N}\right)_{T,V} = \left[\frac{\partial}{\partial N}(-K_B T \ln Z_N)\right]_{T,V} = -K_B T \left[\frac{\partial}{\partial N}(\ln Z_N)\right]_{T,V},$$

$$= -K_B T \left(\frac{\partial}{\partial N}\left(N\left[\ln\left(\frac{gV}{N}\right) - \frac{3}{2}\ln\beta + \frac{3}{2}\ln\left(\frac{m}{2\pi\hbar^2}\right) + 1\right]\right)\right)_{T,V},$$

$$= -K_B T \left(\left[\ln\left(\frac{gV}{N}\right) - \frac{3}{2}\ln\beta + \frac{3}{2}\ln\left(\frac{m}{2\pi\hbar^2}\right) + 1\right] + N\frac{N}{gV}\left(-\frac{gV}{N^2}\right)\right).$$

This leads to

$$\mu = -K_B T \ln\left[\left(\frac{gV}{N}\right)\left(\frac{mK_B T}{2\pi\hbar^2}\right)^{3/2}\right], \tag{42}$$

where g is the spin-degeneracy of the particles, given by $g = 2s + 1$, and s is the spin of the particle.

The interesting feature of equation (42) is that the chemical potential μ is negative. One may ask the question: "Why is the chemical potential of a classical system negative?" Actually, in the classical limit, the quantity in square brackets, in equation (42), is large and very much greater than 1, making μ a negative number. This condition will be satisfied whenever T is large, and the volume per particle $\frac{V}{N}$ is large compared to the cube of the thermal de Broglie wavelength, $\lambda = \frac{h}{p}$, where $p^2 = 3mK_B T$ (which follows from the equipartition theorem). In fact, μ must be negative because in order to add a particle while keeping the entropy and volume constant, the particle must carry "negative energy" — or rather, it must be added while the internal energy of the ideal gas is allowed to decrease by cooling.

The internal energy of a classical system is given by $U = \frac{3}{2}NK_B T$. Let us add a single-particle to the system so that N becomes $(N + 1)$, and the internal energy of the system changes by μ so that the new internal energy becomes $(U + \mu)$. Then, from equation (41), the expression for entropy S for this new system becomes

$$S' = (N+1)K_B \left[\ln \left(g \frac{V}{(N+1)} \right) + \frac{3}{2} \ln \left(\frac{m(U+\mu)}{3(N+1)\pi\hbar^2} \right) + \frac{5}{2} \right]. \tag{43}$$

We wish to find the value of μ that makes S in equation (41) and S' in equation (43) equal in keeping with $\mu = \left(\dfrac{\partial U}{\partial N} \right)_{S,V}$ as per equation (30). In the classical limit, N is large, and we have $\dfrac{1}{N} << 1$, and $|\mu| << U$. Expanding the logarithms in equation (43) to first order and simplifying the expression, we see that the value of μ that makes the entropy $S' = S$ is precisely that given by equation (41).

Checking of equality of entropy in these two cases

Expanding the logarithm and keeping only the first order term, we get

$$S' = (N+1)K_B \left[1 + \left(\frac{gV}{(N+1)} \right) + \frac{3}{2} \left(1 + \frac{m(U+\mu)}{3(N+1)\pi\hbar^2} \right) + \frac{5}{2} \right].$$

Further, $(N+1) \approx N$, and $|\mu| << U$. Using these approximations, we get

$$S' = NK_B \left[1 + \left(\frac{gV}{(N)} \right) + \frac{3}{2} \left(1 + \frac{m(U)}{3(N)\pi\hbar^2} \right) + \frac{5}{2} \right].$$

This leads to

$$S' = NK_B \left[\ln \left(\frac{gV}{N} \right) + \frac{3}{2} \ln \left(\frac{m(U)}{3N\pi\hbar^2} \right) + \frac{5}{2} \right] = . \tag{44}$$

Stated differently, adding a particle to the ideal gas would cause the number of accessible microstates and hence entropy to increase, unless at the same time U is forced to decrease. This indicates that the only way to hold the entropy of the system constant is to reduce the internal energy U, which suppresses the number of microstates. Hence, μ is negative. Thus, it is observed that if we wish to add a particle physically while holding the entropy constant, the internal energy must be forced to decrease by cooling the system.

4A.5 Chemical potential of an ideal Fermi gas

The chemical potential μ for a classical gas is negative, that is, $\mu < 0$. In the classical limit, that is, at high-temperature T and for large value of N, both Fermi and Bose gases should behave classically; in particular, their chemical potentials should also be negative. Keeping this feature in knowledge, we look at the form of the chemical potential μ for an ideal Fermi gas. The variation of chemical potential for a fermion gas is shown in Figure 4A.2.

At high temperatures, the form of μ does indeed approach that of an ideal classical gas. But we see that at low temperatures, $\mu > 0$ which is a very interesting result for the fermion gas. Let us explain this variation. At $T = 0$, all the low-lying single-particle states up to the Fermi energy ϵ_F are filled in accordance with the Pauli exclusion principle. These states are shown in Figure 4A.1, and each state is filled with two spin 1/2 fermions, one with spin up and another with spin down. All states above the Fermi level ϵ_F are empty. There is only one microstate available to the system, and the entropy $S = K_B \ln \Omega = K_B \ln 1 = K_B \times 0 = 0$ in accordance with the third law of thermodynamics.

Let another fermion be added to the system at $T = 0$. It must go into a single-particle state at or just above the Fermi level. So the increase in energy of the system is $\Delta U = \epsilon_F$, and thus, $\mu = \epsilon_F$. It should be noted that this is a positive quantity. There is still only one available microstate, so the entropy of the Fermi gas is still zero. Equation (30) shows that this energy is really the chemical potential for the Fermi gas at this zero temperature.

As the temperature rises, the total internal energy of the system increases, and some of the fermions begin to occupy excited states. The entropy of the system increases due to an increase in temperature because more

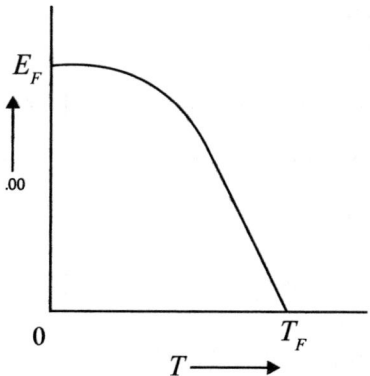

Figure 4A.2 Variation of chemical potential μ of an ideal Fermi gas with temperature T.

microstates become available to the system. We consider further the gas at a very low but nonzero temperature, and add one more particle to it. To satisfy equation (30), the entropy of the system must not increase when the particle is added. The new particle must go into one of the states close to the Fermi level ϵ_F, since fermions leave these states first, when excited into higher states. In fact, the new particle must go into a low-lying, vacant single-particle state, which will be a little below ϵ_F. The gas must also be cooled a little, to avoid increasing the number of accessible microstates. Usually, adding a particle to a system causes an increase in the number of accessible microstates, even if the internal energy of the system remains constant. The reason is that the number of ways of distributing the total energy among the particles increases. The change in internal energy of the Fermi gas, $\Delta U = \mu$, must therefore be positive but a little smaller than ϵ_F.

At slightly higher temperatures, more fermions are excited into higher single-particle states. More of the low-lying single-particle states become vacant, and the energy of the lowest lying single-particle states gets smaller. To add a new particle without increasing the entropy requires the new particle to go into a low-lying single-particle state, which is considerably below ϵ_F. Once again, the gas is slightly cooled to avoid an increase in the number of microstates and, therefore, entropy. Hence, the chemical potential μ will be below ϵ_F. Thus, μ decreases from ϵ_F, to smaller and smaller values, until μ becomes zero just below the temperature $T_F = \dfrac{\epsilon_F}{K_B}$, at which even the single-particle ground state is unlikely to be occupied.

After this point, μ becomes negative. If a new particle is further added to the system, the internal energy must decrease to comply with the constraint spelled out in equation (44), that the entropy remains constant. As the temperature rises, the gas eventually begins to mimic classical behavior, and the chemical potential decreases and becomes increasingly negative. In this way, the variation of μ of an ideal Fermi gas with temperature T can be completely understood. This variation is shown in Figure 4A.2. A simple calculation for the chemical potential of a free Fermi gas is presented below.

Calculation of chemical potential μ for a fermion gas in three-dimensions

The density of single-particle states of non-interacting, (i.e., free) electrons in three dimensions is given by

$$g(\epsilon) = \frac{V}{2\pi^2}\left(\frac{2m}{\hbar^2}\right)^{\frac{3}{2}} \epsilon^{\frac{1}{2}}. \tag{45}$$

At $T = 0$, all states are occupied up to the "Fermi energy", which is the zero-temperature limit of the chemical potential. This is determined by the condition that

$$N = \frac{V}{2\pi^2}\left(\frac{2m}{\hbar^2}\right)^{\frac{3}{2}} \int_0^{\epsilon_F} \epsilon^{\frac{1}{2}} d\epsilon = \frac{V}{2\pi^2}\left(\frac{2m}{\hbar^2}\right)^{\frac{3}{2}} \frac{2}{3}\epsilon_F^{3/2}, \tag{46}$$

where N is the total number of electrons. Equation (46) gives the expression for Fermi energy at zero temperature as

$$\epsilon_F = \frac{\hbar^2}{2m}(3\pi^2 n)^{\frac{2}{3}}, \tag{47}$$

where $n = \dfrac{N}{V}$. The corresponding Fermi temperature $\left(T_F = \dfrac{\epsilon_F}{K_B}\right)$ associated with this Fermi energy is given by

$$T_F = \frac{\hbar^2}{mK_B}\frac{1}{2}(3\pi^2)^{\frac{2}{3}}(n)^{\frac{2}{3}}. \tag{48}$$

At finite temperature, each single-particle state has a mean number of particles given by the Fermi–Dirac distribution. Hence, the total number of particles N can be written as

$$N = \frac{V}{2\pi^2}\left(\frac{2m}{\hbar^2}\right)^{\frac{3}{2}} \int_0^{\infty} \frac{\epsilon^{\frac{1}{2}}}{e^{\beta(\epsilon-\mu)} + 1} d\epsilon. \tag{49}$$

Equating equations (46) and (49), we get

$$\frac{3}{2}\epsilon_F^{3/2} = \int\limits_0^\infty \frac{\epsilon^{\frac{1}{2}}}{e^{\beta(\epsilon-\mu)}+1}d\epsilon. \tag{50}$$

From equation (50), we can determine $\mu(T)$ as a function of T. However, it is not possible to give a closed form of the analytical expression for $\mu(T)$. Therefore, it is determined numerically. Let us write equation (50) in dimensionless form by defining

$$\tilde{\mu} = \frac{\mu}{\epsilon_F}, \quad \tilde{T} = \frac{T}{T_F}. \tag{51}$$

In terms of these dimensionless "tilde" variables, equation (50) can be written as

$$\frac{3}{2} = \int\limits_0^\infty \frac{x^{\frac{1}{2}}}{e^{(x-\tilde{\mu})/\tilde{T}}+1}dx. \tag{52}$$

Equation (52) has been used to determine $\tilde{\mu} \equiv \mu/\epsilon_F$ numerically as a function of $\tilde{T} \equiv T/T_F$, and the result is shown in Figure 4A.3. It is found that $\mu(T)$ changes sign at $T = T^*$, where

$$\frac{T^*}{T_F} = 0.9887. \tag{53}$$

Incorporating the spin degeneracy of 2, the classical expression for the chemical potential can be written as

$$\mu_{\text{class}}(T) = -\frac{3}{2}K_BT\ln\left(\frac{T}{T_Q}\right), \tag{54}$$

where

$$T_Q = \frac{\hbar^2}{mK_B}2\pi\left(\frac{n}{2}\right)^{\frac{2}{3}}. \tag{55}$$

The spin degeneracy appears through the replacement of n in the expression for T_Q for spin-less particles by $\frac{n}{2}$, that is, the (number) density of particles per spin species.

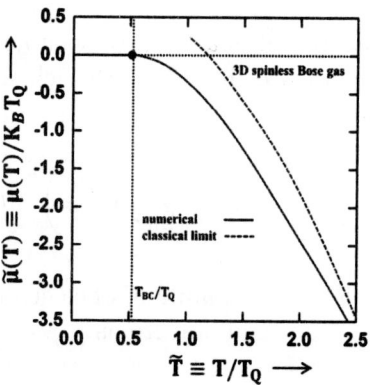

Figure 4A.3 The solid curve shows $\tilde{\mu}(T) \equiv \mu(T)/\epsilon_F$ against $\tilde{T} \equiv T/T_F$. The value of $\mu(T)$ is found to vanish at $T = T^*$ where $T^*/T_F = 0.9887$. This is indicated by one of the dashed vertical lines. The classical (high-temperature) limit, $\tilde{\mu}(T)$, is shown by the dotted curve. The classical result vanishes at $T = T_Q$. At high temperatures, the numerically calculated curve and the curve for the classical limit approach each other.

Equation (54) can also be obtained from equation (49) by neglecting the factor of +1 in the denominator. In this limit, it is justified to neglect +1 because μ is quite large and negative; as a result, the exponential dominates.

One can then extract $e^{\beta\mu}$ out of the integral. Evaluating the integral over ϵ using

$$\int_0^\infty x^{1/2} e^{-x} dx = \Gamma(3/2) = \frac{\sqrt{\pi}}{2} \text{ gives equation (37) with } T_Q \text{ given by equation (55).}$$

Comparing equation (55) with equation (48), we see that

$$T_Q = \frac{4\pi}{(6\pi^2)^{2/3}} T_F = 0.8271 T_F,$$

$$\Rightarrow \quad \tilde{T}_Q \equiv \frac{T_Q}{T_F} = 0.8271. \tag{56}$$

Writing $\tilde{\mu}_{\text{class}} = \dfrac{\mu_{\text{class}}}{\epsilon_F}$, we can express equation (54) as

$$\tilde{\mu}_{\text{class}} = -\frac{3}{2}\tilde{T} \ln\left(\frac{\tilde{T}}{\tilde{T}_Q}\right), \tag{57}$$

which is shown by the dotted line in Figure 4A.3.

For electrons at metallic densities, T_F is typically several tens of thousands of Kelvin, far higher than room temperature. In this limit, $T \ll T_F$, so μ is extremely close to its low-temperature limit ϵ_F (which is, of course, positive). But in the case of the classical ideal gas, we have observed opposite behavior at very high-temperature limit in which μ, given by equation (57), is large in magnitude but negative.

Another purpose is to point out that the different temperature scales, T_F, T^*, and T_Q, all depend on parameters of the system and fundamental constants in the same way, that is,

$$T_i = c_i \frac{\hbar^2}{m K_B} n^{2/3}, \tag{58}$$

where i refers to F, Q, or c^*, and c_i is a (dimensionless) numerical constant with values

$$c_F = \frac{1}{2}(3\pi^2)^{2/3} = 4.7854, \quad c^* = 0.9887 \times c_F = 4.7312, \quad c_Q = 2^{1/3}\pi = 3.9582, \tag{59}$$

obtained, respectively, from equations (48), (53), and (55).

4A.6   Chemical potential of a Bose–Einstein gas

The density of single-particle states of non-interacting spin-less boson in three dimensions is

$$\rho(\epsilon)d\epsilon = 2\pi V \left(\frac{2m}{h^2}\right)^{\frac{3}{2}} \epsilon^{\frac{1}{2}} d\epsilon. \tag{60}$$

The mean occupancy of a single-particle state is given by the Bose–Einstein distribution, so the mean total number of particles is given by

$$N = \int_0^\infty 2\pi V \left(\frac{2m}{h^2}\right)^{\frac{3}{2}} \frac{\epsilon^{\frac{1}{2}}}{e^{\frac{\epsilon-\mu}{K_B T}} - 1} d\epsilon = 2\pi V \left(\frac{2m}{h^2}\right)^{\frac{3}{2}} \int_0^\infty \frac{\epsilon^{\frac{1}{2}}}{e^{\frac{\epsilon-\mu}{K_B T}} - 1} d\epsilon. \tag{61}$$

Equation (61) is used to compute the chemical potential $\mu(T)$ as a function of temperature. It is an implicit relation. In general, we cannot take $\mu(T)$ outside the integral to get an explicit expression for $\mu(T)$. However, in the classical limit (at very high temperatures), $\mu(T)$ can be obtained explicitly. In this case, the factor (-1) in the denominator of equation (61) can be neglected (which is justified in this limit since μ is large and negative, so the exponential dominates). Thus, the factor (-1) can be neglected if $e^{\beta(\epsilon-\mu)} \gg 1$. Under this situation, the Bose–Einstein distribution function $\langle n \rangle = \frac{1}{e^{\beta\epsilon} e^{-\beta\mu} - 1}$ takes the form $\langle n \rangle = e^{\beta\mu_{\text{class}}} e^{-\beta\epsilon}$, which is the Maxwell–Boltzmann statistics. Here, μ is replaced by its classical value μ_{class}.

Neglecting (-1) in the denominator in equation (61), we get the total number of particles per unit volume as

$$n = \frac{N}{V} = 2\pi \left(\frac{2m}{h^2}\right)^{\frac{3}{2}} e^{\beta\mu_{\text{class}}} \int_0^\infty \epsilon^{\frac{1}{2}} e^{-\beta\epsilon} d\epsilon. \tag{62}$$

in which the factor $e^{\beta\mu_{\text{class}}}$ has been pulled out of the integral. Making substitution $x = \beta\epsilon$ and using the result that $\int_0^\infty x^{\frac{1}{2}} e^{-x} dx = \Gamma\left(\frac{3}{2}\right) = \frac{\sqrt{\pi}}{2}$, we get

$$e^{-\beta\mu_{\text{class}}} = \left(\frac{2\pi m K_B T}{h^2}\right)^{\frac{3}{2}} \frac{1}{n} = \left(\frac{2\pi m K_B T}{h^2}\right)^{\frac{3}{2}} \frac{V}{N}. \tag{63}$$

This can be expressed as

$$\mu_{\text{class}}(T) = -\frac{3}{2} K_B T \ln\left(\frac{T}{T_Q}\right), \tag{64}$$

where T_Q is given by

$$T_Q = \frac{h^2}{2\pi m K_B} \left(\frac{N}{V}\right)^{\frac{2}{3}}. \tag{65}$$

It is to be noted that at high-temperature limit, $T >> T_Q$ and it is evident from equation (64) that the chemical potential is negative.

Taking equation (65) to the 3/2-power and slightly rearranging the factors, we get

$$\left(K_B T_Q\right)^{\frac{3}{2}} = \left(\frac{h^2}{2\pi m}\right)^{\frac{3}{2}} \frac{N}{V}. \tag{66}$$

Substituting equation (66) into equation (61), we get

$$\left(K_B T_Q\right)^{\frac{3}{2}} = \frac{2}{\sqrt{\pi}} \int_0^\infty \frac{\epsilon^{\frac{1}{2}}}{e^{\frac{\epsilon-\mu}{K_B T}} - 1} d\epsilon. \tag{67}$$

Let us define $\mu_S = \frac{\mu}{K_B T_Q}$, $T_S = \frac{T}{T_Q}$, and $x = \frac{\epsilon}{K_B T}$. Using these substitutions, equation (67) can be written in dimensionless form, as mentioned below.

$$\left(K_B T_Q\right)^{\frac{3}{2}} = \frac{2}{\sqrt{\pi}} \left(K_B T\right)^{\frac{3}{2}} \int_0^\infty \frac{x^{\frac{1}{2}}}{e^{\left(x - \frac{\mu}{K_B T}\right)} - 1} dx; \Rightarrow \left(\frac{T_Q}{T}\right)^{\frac{3}{2}} = \frac{2}{\sqrt{\pi}} \int_0^\infty \frac{x^{\frac{1}{2}}}{e^x e^{-\frac{\mu}{K_B T}} - 1} dx;$$

$$\Rightarrow 1 = \frac{2}{\sqrt{\pi}} \int_0^\infty \frac{(T_S x)^{\frac{1}{2}}}{e^x e^{-\frac{\mu_S K_B T_Q}{K_B T}} - 1} T_S dx; \Rightarrow 1 = \frac{2}{\sqrt{\pi}} \int_0^\infty \frac{(T_S x)^{\frac{1}{2}}}{e^x e^{-\frac{\mu_S}{T_S}} - 1} T_S dx.$$

Let us further assume that $T_S x = x_0$. Using this substitution, the above expression becomes

$$1 = \frac{2}{\sqrt{\pi}} \int_0^\infty \frac{x_0^{\frac{1}{2}}}{e^{\left(\frac{x_0-\mu_S}{T_S}\right)} - 1} dx_0. \tag{68}$$

Equation (68) is the dimensionless form of equation (61) and can be used to determine μ_S numerically as a function of T_S. Such numerical results are shown by the solid line in Figure 4A.3. The classical (high-T) expression in equation (64) can also be put in dimensionless form as

$$\mu_{S(class)}(T) = \frac{\mu_{class}(T)}{K_B T_Q} = -\frac{3}{2} T_S \ln T_S, \tag{69}$$

and this is shown by the dotted curve in Figure 4A.3. At high temperatures, the curve numerically calculated from equation (68) approaches the curve obtained from the classical (high-T) limit given by equation (69).

It is observed from equation (68) that $\mu(T)$ becomes less negative with the decrease in temperature, and it is equal to zero at the Bose–Einstein condensation temperature T_{BC}. The value of T_{BC} can be obtained from equation (68) by setting μ_S equal to zero. Defining a new integration variable y by $y = \frac{x_0}{T_S}$, equation (68) becomes

$$1 = \frac{2}{\sqrt{\pi}} \left(T_{BC}^S\right)^{\frac{3}{2}} \int\limits_0^\infty \frac{y^{\frac{1}{2}}}{e^y - 1} dy, \tag{70}$$

where $T_{BC}^S = \frac{T_{BC}}{T_Q}$. The integral is evaluated as follows:

$$\int\limits_0^\infty \frac{y^{\frac{1}{2}}}{e^y - 1} dy = \int\limits_0^\infty \frac{e^{-y} y^{\frac{1}{2}}}{1 - e^{-y}} dy = \int\limits_0^\infty e^{-y} y^{\frac{1}{2}} \left[1 + e^{-y} + e^{-2y} + e^{-3y} + \cdots\right] dy,$$

$$= \int\limits_0^\infty e^{-y} y^{\frac{1}{2}} \left[1 + \frac{1}{2^{\frac{3}{2}}} + \frac{1}{3^{\frac{3}{2}}} + \frac{1}{4^{\frac{3}{2}}} + \cdots\right] dy. = \Gamma\left(\frac{3}{2}\right) \varsigma\left(\frac{3}{2}\right),$$

where $\Gamma(x) = \int\limits_0^\infty t^{x-1} e^{-t} dt$ is the Gamma function and $\varsigma(x) = 1 + \frac{1}{2^x} + \frac{1}{3^x} + \frac{1}{4^x} + \cdots$ is the zeta function.

We know that $\Gamma\left(\frac{3}{2}\right) = \frac{\sqrt{\pi}}{2}$. The value of $\varsigma\left(\frac{3}{2}\right)$ is not known exactly, but its numerical value is 2.612. Using these values in equation (70), we get

$$T_{BC}^S = \frac{T_{BC}}{T_Q} = \left[\frac{1}{\varsigma\left(\frac{3}{2}\right)}\right]^{\frac{2}{3}} = 0.5272. \tag{71}$$

This temperature is marked in Figure 4A.4. From equations (65) and (71), we get

$$T_{BC} = T_Q \times 0.5272 = \frac{h^2}{2\pi m K_B} \left(\frac{N}{V}\right)^{\frac{2}{3}} \times 0.5272. \tag{72}$$

This is the expression for Bose–Einstein condensation temperature T_{BC}. At temperatures below T_{BC}, a finite fraction of the particles is in the lowest quantum state (with energy 0). It should be mentioned that, in this region, the chemical potential is actually not quite zero but of order $\frac{1}{N}$. Figure 4A.3 plots $\mu(T)$ in the thermodynamic limit, $N \to \infty$, and so shows $\mu(T)$ equal to zero for $T \leq T_{BC}$.

4A.7 Variation of chemical potential μ of a Bose–Einstein gas with degeneracy g due to spin

For a system consisting of non-interacting bosons with degeneracy $g(= 2s + 1)$ due to spin s, the factor $e^{-\beta\mu}$ becomes

$$e^{-\beta\mu} = \frac{V}{N}\left(\frac{2\pi m K_B T\, g^{\frac{2}{3}}}{h^2}\right)^{\frac{3}{2}}.$$

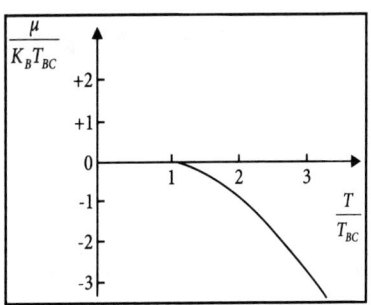

Figure 4A.4 Variation of scaled chemical potential $\frac{\mu}{K_B T_{BC}}$ of a Bose–Einstein gas as a function of scaled temperature $\frac{T}{T_{BC}}$.

Thus, we see that for a fixed value of degeneracy g, this condition $e^{-\beta\mu} \gg 1$ will be satisfied if temperature T is very high, mass m is very high, and the number density $n = \frac{N}{V}$ is very low. The chemical potential μ for such a system is always negative. With the decrease in temperature, the factor $\frac{V}{N}\left(\frac{2\pi m K_B T\, g^{\frac{2}{3}}}{h^2}\right)^{\frac{3}{2}}$ decreases, and hence, the chemical potential μ also decreases. Now, one may ask a simple question: At what temperature will μ be equal to zero? To have an answer to this question, we put $\mu = 0$ in equation (68) and denote the corresponding temperature by T_{BC}, say. The expression for this temperature comes out to be

$$T_{BC} = \frac{h^2}{2\pi m K_B}\left(\frac{N}{2.612\, g\, V}\right)^{\frac{2}{3}} = \frac{h^2}{2\pi m K_B}\left(\frac{N}{2.612\, (2s+1)\, V}\right)^{\frac{2}{3}}. \tag{73}$$

This temperature T_{BC} is known as the Bose condensation temperature. If the temperature is further reduced from T_{BC}, μ cannot be positive, nor can it become negative again. The only possibility is that μ will remain equal to zero after it has attained zero value. The variation of the chemical potential with scaled temperature is shown in Figure 4A.4.

Thermal Ionization: Saha Ionization Equation

5A   Saha ionization equation

At a high enough temperature and/or density, the atoms in a gas suffer collisions due to their high thermal energy, and some of the atoms get ionized, making an ionized gas. In this process, a number of electrons that are normally bound to the atom in orbits around the atomic nucleus become free and thus form an independent electron gas cloud coexisting with the surrounding gas of atomic ions and neutral atoms. These ionized atoms and electrons generate an electric field that causes motion of the charges, and a current is generated in the gaseous medium. This current produces a localized magnetic field. The state of matter thus created is called **plasma**. In thermal equilibrium, the ionization state of such a gaseous system is related to the ionization potential, temperature, and pressure of the system. Thus, the **Saha ionization equation**. expresses how the state of ionization of any particular element in a star changes with varying temperatures and pressures. This equation takes into account the combined ideas of quantum mechanics and statistical mechanics for its derivation and is used to explain the spectral classification of stars. This equation was developed by the Indian astrophysicist **Prof. Meghnad Saha** in 1920.

5A.1   Derivation of Saha ionization equation

According to Prof. M. N. Saha, the temperatures in the interior of stars are extremely high, and the elements present there are mostly in the atomic state. Saha argued that under the prevailing conditions inside the stars, atoms move very rapidly and undergo frequent collisions. In the process of such collisions, valence electrons are stripped off from their orbits. This is referred to as **thermal ionization** and is accompanied by electron recapture to form neutral atoms. The degree of such thermal ionization depends on the temperature of the star. Using the Saha ionization equation, a general relation between the degree of thermal ionization and the temperature can be obtained from the statistical description of plasma in **thermodynamic equilibrium**.

The Saha ionization equation relates the temperature, pressure, and the ionization potential of atoms to their degree of ionization. This equation holds only for weakly ionized plasmas for which the Debye length is large. This means that the screening of the Coulomb interaction of ions and electrons by other ions and electrons is negligible. In order to derive the Saha ionization equation, the following assumptions are made:

1. Only single ionization occurs in the interior of the star.

2. The contribution of nuclear spin to the degree of ionization is negligible.

3. The dynamical equilibrium between ionization and recombination (due to electron capture) is achieved at a given temperature and pressure.

4. Multiple ionization, atoms in the excited states, and the presence of impurities are not taken into account.

5. Interaction of the gas with the wall of the container is not considered. Such interaction can result in surface ionization and in the ionization of the gas by the electrons ejected from a hot wall.

The thermal ionization of the atoms inside the star can be looked upon as some kind of chemical reaction given by

$$A \rightleftharpoons A^+ + e. \tag{74}$$

where A stands for the atom under consideration, A^+ is a singly ionized positive ion, and e is the electron. From the knowledge of thermodynamics of a chemical reaction involving different species at equilibrium, we can write

$$\xi_A \mu_A + \xi_{A^+} \mu_{A^+} + \xi_e \mu_e = 0. \quad \text{or,} \quad \Rightarrow \quad \sum_k \xi_k \mu_k = 0, \tag{75}$$

where ξ_A, ξ_{A^+}, and ξ_e are the respective stoichiometric coefficients for A, A^+, and e and μ_A, μ_{A^+}, and μ_e are, respectively, the chemical potentials for A, A^+, and e. For the reaction given by equation (74), we have $\xi_A = 1$, $\xi_{A^+} = -1$, and $\xi_e = -1$.

Let N_a, N_i, and N_e be, respectively, the numbers of neutral atoms, ions, and electrons of the gaseous system enclosed in a box of volume V at a temperature T. We can then write for equation (74) as

$$\frac{N_i N_e}{N_a} = S(T, P), \tag{76}$$

where $S(T, P)$ is the **Bose function**. It is a function of temperature and pressure. It should be noted that high temperature favors ionization, but high pressure favors recombination. This equation provides the relative number of three types of particles in equilibrium when the number of ionizations per second equals the number of recombinations per second. We know that the number of particles in a given energy level is proportional to the Boltzmann factor for that level, and the total number of particles is proportional to the sum of Boltzmann factors for all the energy levels, that is, to the partition function. In terms of the partition functions for the involved species, equation (76) can be written as

$$\frac{N_i\, N_e}{N_a} = \frac{Z_i\, Z_e}{Z_a}. \tag{77}$$

The partition function Z_k (for kth component) is the sum of Boltzmann factors over all the states that include both translation and internal (rotational, vibrational, and electronic) energy states. As the total energy of a particle is the sum of its translational and internal energies, the total partition function Z_k will be the product of its translational Z_{tran}^k and internal partition functions Z_{int}^k. Hence, we have

$$Z_k = Z_{\text{tran}}^k\, Z_{\text{int}}^k. \tag{78}$$

It is to be noted that the respective internal partition functions are functions of T only but not of pressure. Assuming these entities as indistinguishable, equation (77), in terms of the partition functions, can be written as

$$\frac{N_i\, N_e}{N_a} = \frac{Z_i\, Z_e}{Z_a} = \frac{\dfrac{\left(Z_{\text{tran}}^i\, Z_{\text{int}}^i\right)^{N_i}}{N_i!}\, \dfrac{\left(Z_{\text{tran}}^e\, Z_{\text{int}}^e\right)^{N_e}}{N_e!}}{\dfrac{\left(Z_{\text{tran}}^a\, Z_{\text{int}}^a\right)^{N_a}}{N_a!}}. \tag{79}$$

Using the expression for the single-particle translational partition function for the kth component: $Z_k = V\left(\frac{2\pi m_k K_B T}{h^2}\right)^{\frac{3}{2}}$, the total partition functions for neutral atoms, positive ions, and electrons can be, respectively, written as

$$Z_a = \frac{1}{N_a!}\left[V\left(\frac{2\pi m_a K_B T}{h^2}\right)^{\frac{3}{2}} Z_{\text{int}}^a\right]^{N_a}; Z_i = \frac{1}{N_i!}\left[V\left(\frac{2\pi m_i K_B T}{h^2}\right)^{\frac{3}{2}} Z_{\text{int}}^i\right]^{N_i}; \text{ and }$$

$$Z_e = \frac{1}{N_e!}\left[V\left(\frac{2\pi m_e K_B T}{h^2}\right)^{\frac{3}{2}} Z_{\text{int}}^e\right]^{N_e}. \tag{80}$$

Here m_a, m_i, and m_e are the masses of an atom, an ion, and an electron, respectively. We know that the chemical potential μ_k, in terms of the partition function Z_k, is given by

$$\mu_k = -\frac{1}{\beta}\left(\frac{\partial\, \ln Z_k}{\partial N_k}\right)_{T,V}. \tag{81}$$

Using this definition, the chemical potential μ_k for the N_a neutral atoms can obtained in the following way: Taking the natural logarithm of the partition function given by equation (81) for N_a neutral atoms, we get

$$\ln Z_a = N_a \ln \left[V \left(\frac{2\pi m_a K_B T}{h^2} \right)^{\frac{3}{2}} Z_{int}^a \right] - \ln \left(N_a! \right)$$

$$\Rightarrow \quad \ln Z_a = N_a \ln \left[V \left(\frac{2\pi m_a K_B T}{h^2} \right)^{\frac{3}{2}} Z_{int}^a \right] - N_a \ln N_a + N_a. \qquad (82)$$

Differentiating this equation (82) with respect to N_a at constant T and V, we get the chemical potential μ_a for these N_a neutral atoms as

$$\mu_a = -\frac{1}{\beta} \left(\frac{\partial}{\partial N_a} \left[N_a \ln \left[V \left(\frac{2\pi m_a K_B T}{h^2} \right)^{\frac{3}{2}} Z_{int}^a \right] - N_a \ln N_a + N_a \right] \right)_{T,V}.$$

This leads to

$$\mu_a = -K_B T \ln \left[\frac{V}{N_a} \left(\frac{2\pi m_a K_B T}{h^2} \right)^{\frac{3}{2}} Z_{int}^a \right]. \qquad (83)$$

Following equation (83), the chemical potentials μ_i and μ_e, respectively, for N_i ions and N_e electrons can be written as

$$\mu_i = -K_B T \ln \left[\frac{V}{N_i} \left(\frac{2\pi m_i K_B T}{h^2} \right)^{\frac{3}{2}} Z_{int}^i \right] \quad \text{and}$$

$$\mu_e = -K_B T \ln \left[\frac{V}{N_e} \left(\frac{2\pi m_e K_B T}{h^2} \right)^{\frac{3}{2}} Z_{int}^e \right]. \qquad (84)$$

Using the values of respective chemical potentials from equation (84) and the values of respective stoichiometric coefficients: $\xi_a = 1$, $\xi_i = -1$, and $\xi_e = -1$ in equation (75), we get

$$\xi_A \mu_A + \xi_{A^+} \mu_{A^+} + \xi_e \mu_e = 0; \qquad \Rightarrow \qquad \xi_a \mu_a + \xi_i \mu_i + \xi_e \mu_e = 0.$$

$$\Rightarrow \quad (+1) \left\{ -K_B T \ln \left[\frac{V}{N_a} \left(\frac{2\pi m_a K_B T}{h^2} \right)^{\frac{3}{2}} Z_{int}^a \right] \right\}$$

$$+ (-1) \left\{ -K_B T \ln \left[\frac{V}{N_i} \left(\frac{2\pi m_i K_B T}{h^2} \right)^{\frac{3}{2}} Z_{int}^i \right] \right\}$$

$$+ (-1) \left\{ -K_B T \ln \left[\frac{V}{N_e} \left(\frac{2\pi m_e K_B T}{h^2} \right)^{\frac{3}{2}} Z_{int}^e \right] \right\} = 0.$$

Using the relation: $PV = n_k RT = \frac{N_k}{N_A} RT = \frac{N_k}{N_A} N_A K_B T = N_k K_B T. \Rightarrow \frac{V}{N_k} = \frac{K_B T}{P}$, we get

$$\Rightarrow \quad -\ln\left[\frac{K_B T}{P_a}\left(\frac{2\pi m_a K_B T}{h^2}\right)^{\frac{3}{2}} Z^a_{\text{int}}\right] + \ln\left[\frac{K_B T}{P_i}\left(\frac{2\pi m_i K_B T}{h^2}\right)^{\frac{3}{2}} Z^i_{\text{int}}\right]$$

$$+ \ln\left[\frac{K_B T}{P_e}\left(\frac{2\pi m_e K_B T}{h^2}\right)^{\frac{3}{2}} Z^e_{\text{int}}\right] = 0$$

$$\Rightarrow \quad \frac{P_i P_e}{P_a} = \left(\frac{2\pi m_i m_e}{m_a h^2}\right)^{\frac{3}{2}} (K_B T)^{\frac{5}{2}} \left(\frac{Z^i_{\text{int}} Z^e_{\text{int}}}{Z^a_{\text{int}}}\right). \tag{85}$$

Here P_a, P_i, and P_e are the partial pressures exerted by the neutral atoms, the ions, and the electrons, respectively. It is important to point out that for most reactions of the type $A \rightleftharpoons A^+ + e$, we measure energies from the same reference level. For this particular case, we arbitrarily assign zero energy to the ground state. The internal partition functions can then be written as

$$Z^a_{\text{int}} = Z_a; \quad Z^e_{\text{int}} = Z_e, \quad and \quad Z^i_{\text{int}} = Z_i e^{-\left(\frac{\epsilon^*}{K_B T}\right)}, \tag{86}$$

where ϵ^* is the ionization energy.

The ratio $\frac{m_i}{m_a} \approx 1$ as the mass of a hydrogen atom is 10^4 times the mass of an electron. Further, for free atoms and ions, the possible internal states are electronic states. Hence, for all practical purposes, these internal partition functions may be taken to be the ground state degeneracies g_a, g_i, and g_e. But for an electron, we have $g_e = 2$.

Using these values of degeneracy and the expressions for the internal partition functions given by equation (86), equation (85) can be written as

$$\frac{P_i P_e}{P_a} = \left(\frac{2\pi m_e}{h^2}\right)^{\frac{3}{2}} (K_B T)^{\frac{5}{2}} \left(\frac{2g_i}{g_a}\right) e^{-\left(\frac{\epsilon^*}{K_B T}\right)}. \tag{87}$$

A system of ions and electrons will be electrically neutral when the number of ions is equal to the number of electrons. So $P_i = P_e$. Also, the total pressure is the sum of the partial pressures, and we can write $P = P_a + P_i + P_e$, so that $P_a = P - 2P_e$. Using these results, equation (87) can be written as

$$\frac{P_i P_e}{P_a} = \frac{P_e^2}{P - 2P_e} = \frac{P\left(\frac{P_e}{P}\right)^2}{1 - 2\frac{P_e}{P}} = \frac{Px^2}{1 - 2x}, \tag{88}$$

where $x = \frac{P_e}{P}$ denotes the mole fraction of electrons, that is, the fraction of electrons in the system. Using equation (88) into equation (87), we get

$$\frac{x^2}{1-2x} = \left(\frac{2\pi m_e}{h^2}\right)^{\frac{3}{2}} \frac{(K_B T)^{\frac{5}{2}}}{P} \left(\frac{2g_i}{g_a}\right) e^{-\left(\frac{\epsilon^*}{K_B T}\right)}. \tag{89}$$

This equation (89) expresses the degree of ionization as a function of temperature T, pressure P, and the ionization potential of the atoms present in the interior of a star. This is commonly referred to as **Saha's ionization equation**. This equation (89) implies that the degree of ionization will be more if the temperature is high, the pressure is low, and the ionization potential is also low. This formula finds a wide range of applications in astrophysics. It has been successfully used to explain the Fraunhofer spectrum of the Sun, and the spectrum of the stars, and to estimate the temperature of the atmosphere. Now a days, it finds wide applications in plasma physics as well as in the ionization processes in semiconductor physics.

5A.2 Discussion on Saha's ionization equation

The Saha equation is used to relate the ionization state of an element to the temperature and pressure. This equation provides physical information about the ionization state of an element in terms of easily measurable quantities like temperature and pressure. Further, this equation provides guidelines to the spectral classification of stars and to find the ratio of particle densities for two different ionization levels of the element.

This can be illustrated with the help of an example. Let us consider a carbon atom in an ionization state r. A carbon atom has four electrons in the outer most orbit in its neutral state, but a carbon ion has three electrons. In its ground state, two electrons would occupy its $n = 1$ state, and the third one would occupy its $n = 2$ state. The atom can be raised to the next higher ionization state with the ejection of the latter electron if a photon provides the appropriate energy known as the ionization potential, ϵ^*. Such a transition is shown in Figure 5A.1. It takes the atom from the ground state of the rth ionization state to the ground state of the $(r + 1)$th ionization state. The transition can initiate in any energy state of the original ion and terminate in any energy state of the final ion, as shown by transition 2 in the figure.

The Saha equation can be seen as a restatement of the equilibrium condition for the chemical potentials and simply states that the potential for an atom of ionization state r to ionize is the same as the potential for an electron and an atom of ionization state $(r + 1)$; the potentials are equal. Therefore, the system

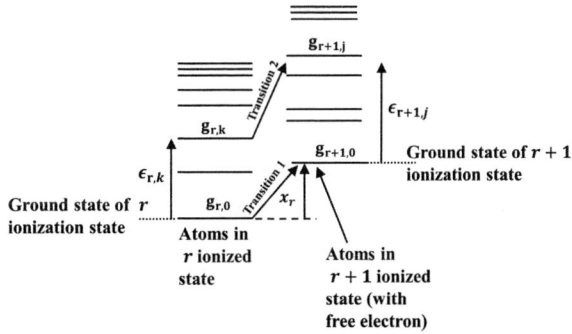

Figure 5A.1 Schematic energy band diagram of an atom in two ionization states. The $(r+1)$ state has one bound electron less than the r state.

is in equilibrium, and no net change of ionization occurs. The equilibrium condition prevails when the plasma is in local thermodynamic equilibrium. But this does not happen in the optically-thin corona. In this case, equilibrium ionization states are estimated by detailed statistical calculation of collision and recombination rates.

The Saha equation can be written as

$$\frac{n_{r+1}n_e}{n_r} = \frac{Z_{r+1}g_e}{Z_r} \left(\frac{2\pi m_e K_B T}{h^2}\right)^{\frac{3}{2}} e^{-\left(\frac{\epsilon^*}{K_B T}\right)}, \tag{90}$$

where n_{r+1} is the density of states in the ionized $(r+1)$th state (m^{-3});

n_r is the density of states in the ionized (r)th state (m^{-3});

n_e is the density of electrons (m^{-3});

Z_{r+1} is the internal partition function of the ionized $(r+1)$th state;

Z_r is the internal partition function of the ionized (r)th state;

$g_e = 2$ is the statistical weight of the electrons;

ϵ^* is the ionization potential of state r reach to $(r+1)$ state, ground level to ground level transition.

The following features are to be noted from the Saha equation given by equation (90):

1. The Saha equation (90) yields the value of $\frac{n_{r+1}n_e}{n_r}$, where n_{r+1}, n_e, and n_r are the densities of the atoms in the higher ionization states $(r+1)$, of the free electrons, and of the atoms in the lower ionization state (r), respectively.

2. This equation is derived from the principle of detailed balance and using the statistical distributions such as the Planck function, Boltzmann formula, and Maxwell–Boltzmann distribution law.

3. At a fixed temperature T, it is apparent from equation (90) that the electron density n_e greatly affects the ionization ratio $\frac{n_{r+1}}{n_r}$. A large electron density results in more collisions and, hence more recombination and thus lowers the ratio $\frac{n_{r+1}}{n_r}$. At a fixed n_e, the ionization ratio $\frac{n_{r+1}}{n_r}$ is enhanced by the larger partition function Z_{r+1} and higher temperature T. This parameter favors the more highly ionized state. Higher temperature also makes more available phase space for the electrons released during ionization.

4. **Saha equation for hydrogen plasma**: The degree of ionization $\frac{n_{r+1}}{n_r}$ can be calculated using this Saha ionization equation (90). We consider hydrogen atoms in an optically thick gas in thermal equilibrium at a temperature $T = 6400$ K, the temperature of the photosphere of the Sun. Further, we consider that only hydrogen is present in the photosphere, and neutral hydrogen is in the rth state. Then, the partition function for this case can be written as

$$Z_r(T) = g_{r,0} + g_{r,1}e^{-\left(\frac{\epsilon_{r,1}}{K_B T}\right)} + g_{r,2}e^{-\left(\frac{\epsilon_{r,2}}{K_B T}\right)} + \cdots = \sum_k g_{r,k}e^{-\left(\frac{\epsilon_{r,k}}{K_B T}\right)}. \quad (91)$$

The first term of equation (91) is simply the ground-state statistical weight $g_{r,0} = 2$ because the electron can have two spin states (up and down) relative to the proton. We consider the second term for neutral hydrogen: $g_{r,1}e^{-\left(\frac{\epsilon_{r,1}}{K_B T}\right)}$. The atom requires $\epsilon_{r,1} = 10.2$ eV to reach the $n = 2$ state from the ground state. For $T = 6400$ K, one has $K_B T = 0.56$ eV. The statistical weight of the $n = 2$ state for the neutral hydrogen is $g_{r,1} = 8$; (six p states and two s states).

The second term thus becomes $g_{r,1}e^{-\left(\frac{\epsilon_{r,1}}{K_B T}\right)} = 8 \times e^{-\left(\frac{10.2}{0.56}\right)} = 9.8 \times e^{-8}$. This is small compared to $g_{r,0} = 2$ and can be safely ignored. Similarly, other higher-order terms can be safely ignored. Thus, for most astrophysical applications involving hydrogen, the ground-state statistical is a satisfactory approximation for $Z_r(T)$. Thus, we have $Z_r(T) = g_{r,0} = 2$. As the proton has no bound electrons in the ionized state, we have $Z_{r+1}(T) = g_{r+1,0} = 1$. The statistical weight for the electron is $g_e = 2$.

In the pure hydrogen plasma, there is one free electron for each ionized proton. So, we will have $n_{r+1} = n_e$. Then, from equation (90), we have

$$\frac{n_{r+1}^2}{n_r} = \frac{Z_{r+1}g_e}{Z_r}\left(\frac{2\pi m_e K_B T}{h^2}\right)^{\frac{3}{2}} e^{-\left(\frac{\epsilon^*}{K_B T}\right)}.$$

This leads to

$$\frac{n_{r+1}^2}{n_r} = 2.41 \times 10^{21} \times \frac{1 \times 2}{2} \times (6400)^{\frac{3}{2}} \times e^{-\left(\frac{13.6}{0.56}\right)} = 3.5 \times 10^{16}\text{m}^{-3}. \quad (92)$$

When the plasma is exactly 50% ionized, we will have $n_{r,1} = n_r$. This provides $n_{r+1} = 3.5 \times 10^{16}\text{m}^{-3}$. Thus, at 6400 K, the total hydrogen (HI and HII) number density, or equivalently the total proton number density, is $n = n_{r+1} + n_r = 7 \times 10^{16}\text{m}^{-3}$.

At lower densities n, still at the same temperature, one would expect their fewer collisions leading to recombination and thus the plasma to be more than 50% ionized, that is, $\frac{n_{r+1}}{n_r} > 1$. This follows from the constancy of $\frac{n_{r+1}^2}{n_r} = \frac{n_{r+1}}{n_r}n_{r+1}$ that tells us that an increase in the ionization ratio $\frac{n_{r+1}}{n_r}$ above unity implies a reduced n_{r+1}. In turn, n_r is less than the reduced n_{r+1} because $\frac{n_{r+1}}{n_r} > 1$, the total density $n = n_{r+1} + n_r$ is thus reduced. Conversely, if the density is greater, the hydrogen will be then less 50% ionized. It will be mostly neutral for only moderately higher values.

5. Ionization in the Sun and universe: In the photosphere of the Sun, the mass density is ($\sim 3 \times 10^{-4}$ kg m^{-3}) that corresponds to 2×10^{23} protons m^{-3}. Since this is substantially larger than 50% density at $T = 6400$ K, the plasma must be more than 50% neutral. Using $n = n_{r+1} + n_r = 2 \times 10^{23}$ m^{-3} at $T = 6400$ K, we get $\frac{n_{r+1}}{n_r} = 4 \times 10^{-4}$. It is mostly neutral. At a somewhat higher temperature, $T \approx 23000$ K, the ration comes out to be $\frac{n_{r+1}}{n_r} = 50$ that indicates that the gas would be very much ionized. The temperatures for reaching an ionized state are much less than the value one might naively guess, namely that obtained from $K_B T = 13.6$ eV, i.e. $T = 160000$ K. In the dilute gases often found in Astrophysics, the temperature required for ionization is quite low. Once an atom is ionized, it has few opportunities to become neutral again. This is due to less chance of interaction with a free electron in a dilute gas.

6. The Saha equation can also give insight into the state of ionization of hydrogen at the early universe. Standard models indicate that, as the expanding universe cools, the ionized plasma recombines to neutral hydrogen. The universe then becomes transparent to optical phonon. At this time ($\sim 10^5$ years), the density is ($n \sim 10^9$ m^{-3}). The curves for this total density of electron and proton cross the 50% axis at the temperature of 4000 K. Thus, the hydrogen becomes neutral at about this temperature. The relatively cool photons can no longer ionize the hydrogen so they are free to travel freely through the universe. According to Saha's ansatz, when the universe had expanded and cooled such that the temperature reached about 3000 K, electrons recombined with protons forming H atoms. Suddenly, the universe became transparent to most electromagnetic radiation. That 3000 K surface, red shift by a factor of about 1000 generates the 3 K "cosmic background" of temperature 3 K. This cosmic radiation fills the universe today.

5A.3   Application of Saha's ionization equation

Let us test the applicability of Saha's equation (90) in some simple cases. We assume that the stellar atmosphere consists of pure hydrogen, and we are interested in determining the degree of ionization of hydrogen in such a stellar atmosphere. A hydrogen atom has only one bound electron; hence, there are only two possible stages of ionization. When the electron is not removed from the hydrogen atom, we get the neutral hydrogen atom. When the electron is removed from the hydrogen atom, we get a free proton. Let us denote such neutral hydrogen atoms and protons by I and II, respectively. The numbers of the hydrogen atoms and ions are, respectively, say, N_I and N_{II}. The degree of ionization of hydrogen is measured by the ratio $\frac{N_{II}}{N_{II}+N_I}$. This ratio $\frac{N_{II}}{N_{II}+N_I}$ is plotted as a function of temperature T in Figure 5A.2.

The characteristic features of the curves shown in Figure 5A.2 are the following:

1. When the temperature is low (up to ~ 8500 K), the fraction $\frac{N_{II}}{N_{II}+N_I}$ is zero. Here, the ionization of hydrogen atoms does not take place.

2. With further increase in T, $\frac{N_{II}}{N_{II}+N_I}$ starts increasing from zero and over a range of temperature (7500 K to 12000 K), $\frac{N_{II}}{N_{II}+N_I}$ increases at a rapid rate approaching toward a saturation value. In this temperature range, ionization starts, and with an increase in T, a large number of hydrogen atoms get ionized, indicating a huge increase in N_{II}.

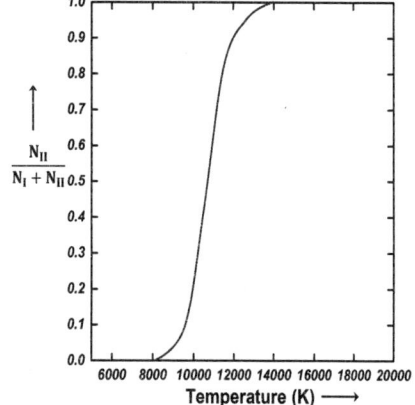

Figure 5A.2 Temperature variation of $\frac{N_{II}}{N_{II}+N_I}$ for Hydrogen

3. With further increase in T, $\frac{N_{II}}{N_{II}+N_I}$ saturates to 1, implying complete ionization of the hydrogen gas. Therefore, the ratio $\frac{N_{II}}{N_{II}+N_I}$ becomes 1.

Figure 5A.3 shows the graphical variation of Saha's equation (90). It is a log–log plot of the ratio of number density in $(r+1)^{th}$ state and r^{th} state as a function of temperature at different fixed values of the total densities n. A value of zero of the ratio $\frac{n_{r+1}}{n_r}$ indicates that 50% of the atoms are ionized. Further, it is observed from the figure that the gas in the photosphere of the Sun remains in quite a neutral state as $\frac{n_{r+1}}{n_r} = 4 \times 10^{-4}$ at $T = 6400$ K. This is shown in Figure 5A.3 by the solid circles.

The figure covers a wide range of densities from a very low value of 10^5 nucleus m^{-3} to 10^{25} m^{-3}. At this higher value of density (10^{25} m^{-3}), the physical spacing of the atoms begins to approach the size of the atoms and the ideal gas laws are no longer applicable. With further increase in density, of the order of $\sim 10^{32}$ m^{-3} (at the center of the Sun), the spacing between the atoms becomes less than the size of the ground-state orbital of the electron. As a result, the wave functions representing the electrons start overlapping each other. Hence, the electrons can no longer be associated with a given atom, and Saha's ionization equation is not applicable in this case.

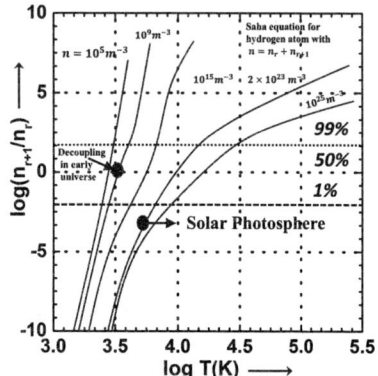

Figure 5A.3 Log–log plot of ratio $\frac{n_{r+1}}{n_r}$ for various fixed total densities, $n = n_{r+1} + n_r$ as a function of temperature T.

Thermodynamic Functions of a Photon Gas

6A   Radiation as a photon gas

In the nineteenth century, physicists applied classical electromagnetic theory to explain the experimental results of black body radiation but were unable to provide an adequate explanation. This was a major problem to the physicists as classical theory predicted an infinite amount of energy of the radiation emitted from a black body. This gross disagreement was called the "ultraviolet catastrophe". Max Planck presented a paper on December 14, 1900, in which he guessed the answer to this problem of black body radiation. This guess marked the very beginning of quantum mechanics.

The electromagnetic radiation inside a black body chamber exists as patterns of standing waves or modes. A single mode is like a standing wave on a guitar string and is characterized by a frequency. Planck made a bold hypothesis that the energy of such a mode is quantized, ithat is, only certain energies of these oscillation modes are allowed. Thus, the energy E_n of nth mode is given by $E_n = \left(n + \frac{1}{2}\right) \hbar\omega = \left(n + \frac{1}{2}\right) h\nu$, where ν is the oscillation frequency, h is Planck's constant, and n is an integer starting with 0. Thus, each oscillation mode can exist in any one of an infinite number of energy states whose energies are equally separated by the energy $\hbar\omega$. In the discussion of the properties of the photon gas, for the time being, we shall ignore $\frac{1}{2}$ in the expression for E_n as it has no effect on the results we seek. Therefore, we take $E_n = n\hbar\omega$ as the energy of the nth mode whose (angular) frequency is ω. When the energy of a mode is E_n, we say that there are n photons in the mode. Each photon has energy equal to $\hbar\omega$. Thus, according to the prescription of Max Planck, a black body radiation chamber consists of a number of photons in various energy states with different amounts of energy.

6A.1   Properties of a photon gas

The properties of a photon gas are the following:

1. In physics, a photon is a bundle of electromagnetic energy. The photon is sometimes referred to as a "quantum" of electromagnetic energy. Photons are

not thought to be made up of smaller particles. They are a basic unit of nature called an elementary particle.

2. The photon gas is an ideal gas of particles of zero rest mass. A photon has no electric charge and is a stable particle.

3. Photons do not interact with each other; therefore, there is no need to consider inter-particle interactions.

4. The number of particles in the isolated photon gas is not conserved.

5. Photon has spin 1 (in a unit of $\frac{h}{2\pi}$); therefore, Bose–Einstein distribution is applicable to a photon gas.

6. The spin degeneracy is two (corresponding to two independent directions of polarizations). The degeneracy is not three, as in the case of particles of spin 1 and non-zero rest mass.

7. The photon is the gauge boson for electromagnetism, and therefore all other quantum numbers of the photon (such as lepton number, baryon number, and flavor quantum numbers) are zero.

6A.2   Thermodynamic functions of photon gas

The energy of the massive particles in a classical gas at equilibrium is distributed according to the Maxwell–Boltzmann distribution. This distribution in equilibrium is established as the massive particles collide with each other and exchange energy (and momentum) in the process. There is also an equilibrium distribution in a photon gas. But there exists a sharp difference—photons do not collide with each other (except under very extreme conditions). Hence, the equilibrium distribution in a photon gas must be established by some other means. The most common way that leads to such an equilibrium distribution in a photon gas is by the interaction of the photons with matter. If the photons are absorbed and emitted by the walls of the system containing the photon gas, the equilibrium distribution for the photons will be similar to a black body distribution at a temperature equal to that of the walls of the system containing the photon gas.

Number of photons is not conserved, but energy is conserved: A very important difference between a gas of massive particles and a photon gas with a black body distribution is that the number of photons in the system is not conserved. A photon may collide with an electron in the wall, exciting it to a higher energy state, removing a photon from the photon gas. This electron may drop back to its lower level in a series of steps, each one of which releases an individual photon back into the system of photon gas. Although the sum of the photon energies of the emitted photons is the same as that of the absorbed photon, the number of emitted photons will vary. As a result of this lack of constraint on the number of photons in the system, it can be shown that **the chemical potential of the photons must be zero**.

In the following sections, the thermodynamic properties of a photon gas, such as distribution function, density of states, internal energy, pressure, chemical potential, entropy, enthalpy, Helmholtz free energy, and Gibbs free energy, are calculated in detail.

6A.3   Statistics of a photon gas

Thermal radiation can be considered to be a photon gas consisting of photons that have no rest mass but possess momenta. The number of photons, though treated as particles, is not conserved. The total energy of photon is, however, constant. There is no restriction on the number of photons occupying the same quantum number or a compartment in a cell of phase space. Thus, the photons follow the Bose–Einstein distribution, the thermodynamics probability of which is given by

$$\Omega = \prod_{i=1}^{s} \frac{(n_i + g_i - 1)!}{n_i!(g_i - 1)!}.$$

The condition of maximum thermodynamic probability gives

$$\sum_{i=1}^{s} \ln \frac{g_i + n_i}{n_i} \delta n_i = 0. \tag{93}$$

We know that for a gas consisting of photons, the total energy of the photon gas is conserved, but the number of photons in such a system is not conserved. This constraint of the total energy provides

$$\epsilon = \sum_{i=1}^{s} n_i \epsilon_i = \text{constant}; \qquad \Rightarrow \delta \epsilon = \sum_{i=1}^{s} \epsilon_i \delta n_i = 0. \tag{94}$$

We shall apply the Lagrangian method of undetermined multipliers to find the most probable distribution subject to the condition that the total energy of the photon gas is conserved. For achieving the most probable distribution, we multiply equation (94) by $-\beta$ and add the resulting expression to equation (93), which finally yields

$$\sum_{i=1}^{s} \left[\ln \left(\frac{n_i + g_i}{n_i} \right) - \beta \epsilon_i \right] \delta n_i = 0. \tag{95}$$

As the variations δn_i are independent of each other, we get from equation (95)

$$\ln \left(\frac{n_i + g_i}{n_i} \right) - \beta \epsilon_i = 0; \qquad \Rightarrow \quad \frac{n_i + g_i}{n_i} = e^{(\beta \epsilon_i)};$$

$$\Rightarrow \quad 1 + \frac{g_i}{n_i} = e^{(\beta \epsilon_i)}; \qquad \Rightarrow \quad \frac{g_i}{n_i} = e^{\beta \epsilon_i} - 1.$$

This leads to

$$n_i = \frac{g_i}{e^{\beta \epsilon_i} - 1}. \tag{96}$$

This equation represents the most probable distribution of the photon number among various energy levels of a photon gas system obeying Bose–Einstein statistics. Here, the parameter β is given by $\beta = \frac{1}{K_B T}$. Hence, the number of photons per state is given by

$$f(\epsilon_i) = \frac{n_i}{g_i} = \frac{1}{e^{\beta \epsilon_i} - 1}.$$

The expression for energy of a photon is given by $\epsilon = h\nu$. Using this expression for $\epsilon = \epsilon_i$, we get

$$f(\epsilon) = \frac{1}{e^{\beta \epsilon} - 1} = \frac{1}{e^{\frac{h\nu}{K_B T}} - 1}. \tag{97}$$

This is the distribution function for a photon gas. Here, we have used $\beta = \frac{1}{K_B T}$. Comparing with Bose–Einstein distribution function, it is observed that the chemical potential of a photon gas is zero. This greatly simplifies the calculation of thermodynamic parameters of the photon gas.

6A.4 Density of states of a photon gas

We consider a radiation chamber of volume V filled with black body radiation at a temperature T. We know that this black body radiation inside the chamber can be considered as a photon gas. The density of states for such a photon gas can be calculated in the following way.

The energy of a photon of frequency ν is $\epsilon = h\nu$, and its momentum is $p = \frac{h}{\lambda} = \frac{h\nu}{c} = \frac{\epsilon}{c}$ where c is the velocity of light. The elemental volume in phase space (the volume of the radiation chamber is considered as the phase space) is given by

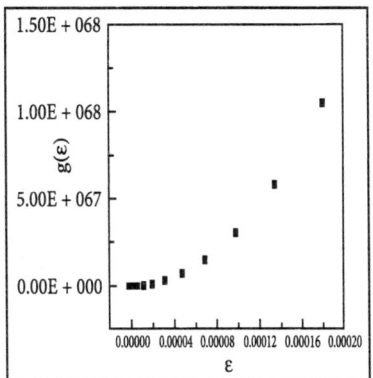

Figure 6A.1 Variation of the density of states of a photon gas, calculated according to equation (100), as a function of energy.

$$dp_x dp_y dp_z dx dy dz = dp_x dx dp_y dy dp_z dz$$
$$= h \times h \times h = h^3.$$

The total volume of the phase space available to the photon gas is given by

$$2 \times \int \int \int \int \int \int dp_x dp_y dp_z dx dy dz = 2 \times \int \int \int dx dy dz \int \int \int dp_x dp_y dp_z.$$

Factor 2 arises because the photons of the electromagnetic field (light) can be both right-handed and left-handed polarized. But we have $\int \int \int dx dy dz = V$, the volume of the system in ordinary position space, that is, the volume occupied by the photons in the black body chamber.

Hence, the total volume of the phase space corresponding to the momentum interval between p and $p + dp$ and position space can be written as

$$V_{phase} = 2 \times V \int \int \int dp_x dp_y dp_z = 2V \times \text{ volume between } p \text{ and } p$$

$+ dp$ in the momentum space.

The quantity $\int \int \int dp_x dp_y dp_z$ represents the volume of the system in the momentum space corresponding to the momentum interval between p and $p + dp$. This volume comes out to be $= 4\pi p^2 dp$. Hence, the volume of the phase space between the momentum interval p and $p + dp$ is given by

$$V_{p,p+dp} = 2V \times 4\pi p^2 dp = 8\pi V p^2 dp.$$

The number of states within this volume of the phase space is obtained by dividing this volume by the elementary volume of the phase space (h^3). So, the number of states within this momentum interval per unit volume is

$$g(p)dp = \frac{V_{p,p+dp}}{V h^3} = \frac{8\pi V p^2 dp}{V h^3} = \frac{8\pi p^2 dp}{h^3}. \tag{98}$$

This is the density of states for a photon gas within the momentum interval between p and $p + dp$.

Density of states of a photon gas in terms of energy

The density of states of a photon gas in terms of energy can be calculated in the following way:

We know from equation (98) that the number of photon states within the momentum interval between p and $p + dp$ per unit volume is given by

$$g(p)dp = \frac{8\pi p^2 dp}{h^3}.$$

We know that the energy and momentum of a photon are related by $p = \frac{\epsilon}{c}$. Using this relation, we get the number of states per unit volume within the energy interval between ϵ and $\epsilon + d\epsilon$ as

$$g(\epsilon)d\epsilon = \frac{8\pi p^2 dp}{h^3} = \frac{8\pi \left(\frac{\epsilon}{c}\right)^2 \frac{d\epsilon}{c}}{h^3} = \frac{8\pi \epsilon^2 d\epsilon}{c^3 h^3}. \tag{99}$$

This gives the density of states of photons at energy ϵ as

$$g(\epsilon) = \frac{8\pi\epsilon^2}{c^3 h^3}. \tag{100}$$

The variation of the density of states of a photon gas as a function of energy is shown in Figure 6A.1. The value of density of states is calculated according to equation (100).

6A.5 Energy density u_ν of a photon gas

The number of photons $n(p)dp$ per unit volume within the momentum interval between p and $p + dp$ can be expressed as

$$n(p)dp = f(p)g(p)dp = \frac{1}{e^{\frac{pc}{K_B T}} - 1} \times \frac{8\pi p^2 dp}{h^3} = \frac{8\pi}{h^3} \frac{1}{e^{\frac{pc}{K_B T}} - 1} p^2 dp. \tag{101}$$

Hence, the density of photons (number of photons per unit volume in the momentum space) in the momentum space is given by

$$n'(p)dp = \frac{n(p)dp}{4\pi p^2 dp} = \frac{2}{h^3} \frac{1}{e^{\frac{pc}{K_B T}} - 1}.$$

The number of photons $n(\epsilon)d\epsilon$ per unit volume within the energy interval between ϵ and $\epsilon + d\epsilon$ is then given by

$$n(\epsilon)d\epsilon = f(\epsilon)g(\epsilon)d\epsilon = \frac{1}{e^{\frac{\epsilon}{K_B T}} - 1} \times \frac{8\pi\epsilon^2}{c^3 h^3} d\epsilon = \frac{8\pi}{c^3 h^3} \frac{1}{e^{\frac{\epsilon}{K_B T}} - 1} \epsilon^2 d\epsilon.$$

Since each photon has energy $\epsilon(= h\nu)$, the energy density of photons within the energy interval ϵ and $\epsilon + d\epsilon$ can be written as

$$u_\epsilon d\epsilon = \epsilon n(\epsilon)d\epsilon = \frac{8\pi}{c^3 h^3} \frac{1}{e^{\frac{\epsilon}{K_B T}} - 1} \epsilon^3 d\epsilon.$$

In terms of frequency ν, this energy density within the frequency interval ν and $\nu + d\nu$ can be expressed as

$$u_\nu d\nu = \frac{8\pi}{c^3 h^3} \frac{1}{e^{\frac{h\nu}{K_B T}} - 1} (h\nu)^3 h d\nu = \frac{8\pi h}{c^3} \frac{\nu^3}{e^{\frac{h\nu}{K_B T}} - 1} d\nu.$$

This provides the energy density of the photon gas at a frequency ν and at a temperature T as

$$u_\nu = \frac{8\pi h}{c^3} \frac{\nu^3}{e^{\frac{h\nu}{K_B T}} - 1}. \tag{102}$$

Equation (102) gives the energy density of a photon gas at a temperature T.

6A.6 Number of photons in a photon gas

The expression for the energy density of a photon gas within the frequency interval between ν and $\nu + d\nu$ is given by

$$u_\nu = \frac{8\pi h}{c^3} \frac{\nu^3}{e^{\frac{h\nu}{K_B T}} - 1} d\nu.$$

The energy of a single photon is $\epsilon = h\nu$. Hence, the number of photons per unit volume within the frequency interval ν and $\nu + d\nu$ is given by

$$dN_\nu = \frac{u_\nu d\nu}{h\nu} = \frac{1}{h\nu} \times \frac{8\pi h}{c^3} \frac{\nu^3}{e^{\frac{h\nu}{K_B T}} - 1} d\nu = \frac{8\pi}{c^3} \frac{\nu^2 d\nu}{e^{\frac{h\nu}{K_B T}} - 1}.$$

So, the total number of photons in a volume V at a temperature T is given by

$$N = \int_0^\infty dN_\nu \times V = \int_0^\infty \frac{8\pi V}{c^3} \frac{\nu^2 d\nu}{e^{\frac{h\nu}{K_B T}} - 1} = \frac{8\pi V}{c^3} \int_0^\infty \frac{\nu^2 d\nu}{e^{\frac{h\nu}{K_B T}} - 1}.$$

Let us put $\frac{h\nu}{K_B T} = x$; therefore, $\frac{h}{K_B T} d\nu = dx$. Using these substitutions, we get

$$N = \frac{8\pi V}{c^3} \int_0^\infty \frac{\nu^2 d\nu}{e^{\frac{h\nu}{K_B T}} - 1} = \frac{8\pi V}{c^3} \int_0^\infty \frac{\left(\frac{K_B T x}{h}\right)^2 \frac{K_B T dx}{h}}{e^x - 1} = \frac{8\pi V}{c^3} \left(\frac{K_B T}{h}\right)^3 \int_0^\infty \frac{x^2}{e^x - 1} dx.$$

This leads to

$$N = \frac{8\pi V}{c^3} \left(\frac{K_B T}{h}\right)^3 \int_0^\infty \frac{x^{3-1} dx}{e^x - 1} = \frac{8\pi V}{c^3} \left(\frac{K_B T}{h}\right)^3 \Gamma(3)\zeta(3). \tag{103}$$

Here $\zeta(3)$ is the Riemann Zeta function. Using the values of $\Gamma(3)$ and $\zeta(3)$, we get the expression for N as

$$N = \frac{8\pi V}{c^3} \left(\frac{K_B T}{h}\right)^3 \times 2.404 = \frac{8\pi K_B^3}{c^3 h^3} \times 2.404 \times VT^3 = 2.0206 \times 10^7 VT^3. \tag{104}$$

Equation (104) gives the number of photons in a photon gas at a temperature T. Further, this equation (104) shows that the number of photons in a photon gas is directly proportional to the cube of absolute temperature T and is directly proportional to the volume V. Figure 6A.2 shows the variation of the number of photons calculated using equation (104) with temperature T in an enclosure of volume $V = 10^{-4}$ m^3. It should be mentioned that this is a very important result in radiation physics.

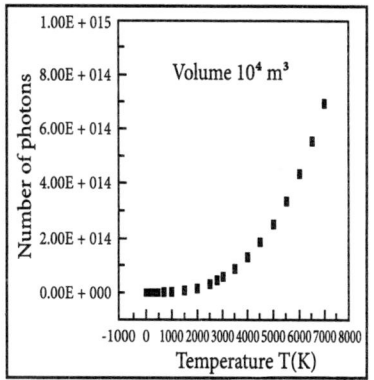

Figure 6A.2 Variation of the number of photons, calculated according to equation (104), as a function of temperature T.

6A.7 Internal energy of a photon gas

According to equation (102), the energy density u_ν of a photon gas in a black body radiation chamber of volume V at a temperature T within the frequency range between ν and $\nu + d\nu$ is given by

$$u_\nu d\nu = \frac{8\pi h}{c^3} \frac{\nu^3}{e^{\frac{h\nu}{K_B T}} - 1} d\nu.$$

Here, the symbols have their usual meaning. So, the total internal energy U of the photon gas in the volume V is given by

$$U = \int_0^\infty V u_\nu d\nu = \frac{8\pi h V}{c^3} \int_0^\infty \frac{\nu^3}{e^{\frac{h\nu}{K_B T}} - 1} d\nu.$$

Let us put $\frac{h\nu}{K_B T} = x$; therefore, $\frac{h}{K_B T} d\nu = dx$. Using these substitutions in the above expression, the expression for total internal energy U becomes

$$U = \frac{8\pi h V}{c^3} \int_0^\infty \frac{\left(\frac{x K_B T}{h}\right)^3}{e^x - 1} \left(\frac{K_B T}{h}\right) dx = \frac{8\pi h V}{c^3} \left(\frac{K_B T}{h}\right)^4 \int_0^\infty \frac{x^3}{e^x - 1} dx.$$

This gives

$$U = \frac{8\pi h V}{c^3} \left(\frac{K_B T}{h}\right)^4 \Gamma(4)\zeta(4) = \frac{8\pi h V}{c^3} \left(\frac{K_B T}{h}\right)^4 \times 6 \times \frac{\pi^4}{90}$$

$$= \frac{8\pi h V}{c^3} \left(\frac{K_B T}{h}\right)^4 \times \frac{\pi^4}{15}.$$

After simplification, we get the total internal energy as

$$U = \frac{8\pi^5(K_B)^4}{15h^3c^3} \times VT^4 = 7.532 \times 10^{-16} \times VT^4. \tag{105}$$

Equation (105) provides the expression for the total internal energy of a photon gas kept in a volume V at a temperature T. Further, equation (105) shows that the total internal energy of the photon gas is directly proportional to the fourth power of absolute temperature T and is directly proportional to the volume V of the black body radiation chamber.

6A.8   Average energy of a black body photon

The average energy of a black body photon can be obtained by dividing equation (105) by equation (104), and this comes out to be

$$\langle E_0 \rangle = \frac{U}{N} = \frac{\frac{8\pi^5(K_B)^4}{15h^3C^3}}{VT^4} \frac{8\pi K_B^3}{c^3h^3} \times 2.404 \times VT^3$$

$$= \frac{\pi^4}{15 \times 2.404}(K_BT) \; J = 2.33 \times 10^{-4}T \text{ eV}. \tag{106}$$

At room temperature ($T = 300$ K), the average energy of a black body photon comes out to be

$$\langle E_0 \rangle = 2.33 \times 10^{-4}T\text{eV} = 2.33 \times 10^{-4} \times 300 \; eV = 69.9 \text{ meV}.$$

It is to be noted that the value of energy (classical) corresponding to room temperature is

$$K_BT = \frac{1.38 \times 10^{-23} \times 300}{1.602 \times 10^{-19}} \text{ eV} = 0.0258 \text{ eV} = 25.8 \text{ meV}.$$

This shows that the average energy of a photon at room temperature is much larger than the corresponding classical energy K_BT.

We consider a macroscopic harmonic oscillator of mass $m = 0.01$kg moving with a velocity $V_{\max} = 0.1$ ms^{-1} (maximum) and amplitude $A = 0.01$ m. The frequency of this oscillator is $v = \frac{V_{\max}}{2\pi A} = 1.6 \; Hz$; its time period and energy are, respectively, given by

$$T = v^{-1} = 0.63 \text{ s and } E = \frac{1}{2}mV_{\max}^2 = 5 \times 10^{-5} \text{ J}.$$

Further, the product of energy and time period has the dimension of an "action", and in this case, the value of the action is 3.14×10^{-5} Js. This is about 5×10^{28}

times larger than Planck's constant h. At this frequency of oscillation $\nu = 1.6$Hz, the quantum of energy is given by

$$E_0 = h\nu = 6.627 \times 10^{-34} \times 1.6 \text{ J} = 1.06 \times 10^{-33} \text{ J}.$$

Hence, the ratio $\frac{E_0}{E}$ comes out to be

$$\frac{E_0}{E} = \frac{1.06 \times 10^{-33} \text{ J}}{5 \times 10^{-5} \text{ J}} = 2.12 \times 10^{-29}.$$

This is a very small quantity, and hence, the quantum effects can be neglected. On the other hand, quantum effects become important for high- frequency electromagnetic waves in black body radiation. This example clearly indicates the regime of validity of quantum theory. In black body radiation, there is a clear deviation of the classical theories from the experimental results at higher frequencies, leading to ultraviolet catastrophe.

6A.9   Pressure of a photon gas

The expression for pressure of an ideal gas consisting of N molecules, each of mass m and enclosed in a volume V, is given by

$$P = \frac{1}{3}m\frac{N}{V}\langle v^2 \rangle,$$

where $\sqrt{\langle v^2 \rangle}$ is the root mean square speed of the particles. If $M(= mN)$ be the mass of the molecules of the gas, then the expression for pressure becomes

$$P = \frac{1}{3}\frac{M}{V}\langle v^2 \rangle.$$

It is known that in a photon gas, all photons have the same velocity c. So, the expression for pressure exerted by a photon gas will be given by

$$P = \frac{1}{3}\frac{M}{V}\langle v^2 \rangle = \frac{1}{3}\frac{M}{V}c^2 = \frac{1}{3}\frac{U}{V},$$

where U is the total internal energy of the photon gas.

The equation of state for an ultra-relativistic quantum gas (which inherently describes photons) is given by $U = 3PV$.

Now, from equation (105), we get the expression for total internal energy U of a photon gas as

$$U = \frac{8\pi^5 (K_B)^4}{15h^3 c^3} V T^4.$$

Hence, the expression for pressure P of a photon gas becomes

$$P = \frac{1}{3}\frac{U}{V} = \frac{1}{3}\frac{1}{V}\frac{8\pi^5(K_B)^4}{15h^3c^3}VT^4 = \frac{8\pi^5(K_B)^4}{45h^3c^3}T^4. \tag{107}$$

This equation (107) shows that the pressure of a photon gas is proportional to the fourth power of the temperature T and is independent of volume V.

6A.10   Entropy of a photon gas

We know that the change in entropy dS is given by

$$dS = \frac{\delta Q}{T} = \frac{1}{T}(dU + PdV) = \frac{1}{T}\left(d(uV) + \frac{u}{3}dV\right) = \frac{1}{T}\frac{4}{3}udV.$$

Integrating this expression, we get

$$S = \frac{1}{T}\frac{4}{3}uV = \frac{1}{T}\frac{4}{3}\frac{U}{V}V = \frac{4U}{3T}.$$

Again, from equation (105), we know that the total internal energy U of a photon gas is given by

$$U = \frac{8\pi^5(K_B)^4}{15h^3c^3}VT^4.$$

This leads to the expression for entropy as

$$S = \frac{4}{3}\frac{U}{T} = \frac{4}{3}\frac{1}{T}\frac{8\pi^5(K_B)^4}{15h^3c^3}VT^4 = \frac{32\pi^5(K_B)^4}{45h^3c^3}VT^3 = \frac{4\pi^2(K_B)^4}{45h^3c^3}VT^3. \tag{108}$$

This equation (108) shows that the entropy of the photon gas is proportional to the volume V of the enclosure and is proportional to the cube of absolute temperature T.

6A.11   Enthalpy of a photon gas

We know that enthalpy is given by

$$H = U + PV.$$

For a photon gas, the pressure P is related to the energy density by $P = \frac{1}{3}\frac{U}{V}$. Using this relation, the expression for enthalpy H becomes

$$H = U + PV = U + \frac{1}{3}\frac{U}{V} \times V = U + \frac{U}{3} = \frac{4}{3}U.$$

Again, the expression for the total internal energy of a photon gas is given by $U = \frac{8\pi^5 (K_B)^4}{15 h^3 c^3} V T^4$. Using this value of U, we have the expression for enthalpy H as

$$H = \frac{4}{3} U = \frac{4}{3} \frac{8\pi^5 (K_B)^4}{15 h^3 c^3} V T^4 = \frac{32\pi^5 (K_B)^4}{45 h^3 c^3} V T^4 = \frac{4\pi^2 (K_B)^4}{45 \hbar^3 c^3} V T^4. \tag{109}$$

This equation (109) shows that the enthalpy of the photon gas is proportional to the volume V of the enclosure and is proportional to the fourth power of absolute temperature T.

6A.12   Helmholtz free energy of a photon gas

We know that the Helmholtz free energy F is given by $F = U - TS$.

For a photon gas, the entropy S is given by $S = \frac{4}{3} \frac{U}{T}$. Using this expression for entropy, the Helmholtz free energy F becomes

$$F = U - TS = U - T \times \frac{4}{3} \frac{U}{T} = U - \frac{4}{3} U = -\frac{U}{3}.$$

Again, the expression for the total internal energy U of a photon gas is given by

$$U = \frac{8\pi^5 (K_B)^4}{15 h^3 c^3} V T^4.$$

Using this value of U, we have the expression for the Helmholtz free energy F as

$$F = -\frac{1}{3} U = -\frac{1}{3} \frac{8\pi^5 (K_B)^4}{15 h^3 c^3} V T^4 = \frac{8\pi^5 (K_B)^4}{45 h^3 c^3} V T^4. \tag{110}$$

This equation (110) shows that the Helmholtz free energy F of the photon gas is directly proportional to the volume V of the enclosure and is also directly proportional to the fourth power of absolute temperature T.

6A.13   Gibbs free energy of a photon gas

We know that the Gibbs free energy G is given by

$$G = U + PV - TS.$$

For a photon gas, entropy S and pressure P are, respectively, given by $S = \frac{4}{3} \frac{U}{T}$ and $P = \frac{1}{3} \frac{U}{V}$. Using these expressions, Gibbs free energy G takes the form

$$G = U + PV - TS = U + \frac{1}{3} \frac{U}{V} \times V - T \times \frac{4}{3} \frac{U}{T} = U + \frac{1}{3} U - \frac{4}{3} U = 0. \tag{111}$$

This equation (111) shows that the Gibbs free energy G of a photon gas is zero.

6A.14   Chemical potential of a photon gas

Photons are bosons with a unique property. The chemical potential of a photon gas in equilibrium in a volume V and at temperature T is formally given by $\mu = 0$.

The physical reason for setting $\mu = 0$ is that the number of photons in the volume cannot be arbitrary; rather, the number of photons is constantly and automatically being adjusted so that the photon gas is in thermal equilibrium with the walls of the container at a constant temperature T. That is, the walls of the container constantly absorb and reemit photons. It should be mentioned that even a gas of photons far out in space does not contain a fixed number of photons since photons can be annihilated or created in collisions (although the scattering cross-section for this process is very small, and photons would have to be annihilated or created in pairs to conserve charge conjugation). Therefore, when writing the thermodynamic identity for a photon gas, the term μdN in the first law of thermodynamics should be omitted since N cannot be held fixed anyway. This is formally consistent with the setting of $\mu = 0$.

Further, it is easy to see why μ should be set equal to 0 by considering the distribution function for the photons. Let us consider a single photon state of energy ϵ of a photon gas. The photon gas is in thermal equilibrium at a temperature T with its chemical potential μ. Photons are spin 1 (in units of $\hbar$) bosons and should, therefore, obey Bose–Einstein statistics. The average occupancy of the state of energy ϵ is therefore given by

$$\langle n(\epsilon)\rangle = \frac{1}{e^{\beta(\epsilon-\mu)} - 1},$$

where $\beta = \frac{1}{K_B T}$. This formula is valid for any Bose system. If we now set $\mu = 0$, we get

$$\langle n(\epsilon)\rangle = \frac{1}{e^{\beta\epsilon} - 1}.$$

This is the standard form of Planck's distribution law of black body radiation. All thermodynamic quantities can be derived from this equation for the average occupancy of a single photon state of energy ϵ.

For example, the total internal energy of a Bose–Einstein gas is

$$U = \int_0^\infty \epsilon\langle n(\epsilon)\rangle g(\epsilon)d\epsilon = \int_0^\infty \frac{\epsilon g(\epsilon)d\epsilon}{e^{\frac{(\epsilon-\mu)}{K_B T}} - 1}, \tag{112}$$

where $g(\epsilon)d\epsilon$ is the density of states, that is, the number of single-particle states with energy between ϵ and $\epsilon + d\epsilon$.

We set $\mu = 0$ and use the density of states factor for a photon gas in a box, $g(\epsilon)d\epsilon = \frac{V\omega^2 d\omega}{\pi^2 c^3}$, where a single photon of angular frequency ω has energy $\epsilon = \hbar\omega$. Equation (112) then leads to

$$U = \int_0^\infty \hbar\omega\langle n(\epsilon)\rangle g(\epsilon)d\epsilon = \frac{V}{\pi^2 c^3} \int_0^\infty \frac{\hbar\omega^3 d\omega}{e^{\frac{\hbar\omega}{KT}} - 1}. \tag{113}$$

This equation (113) gives the same expression for the total energy of a photon gas kept inside a box of volume V. This integral can be done analytically, of course. The result is the familiar one for black body radiation with the internal energy $U = V\left(\frac{8\pi^5 K^4}{15h^3 c^3}\right)T^4$. Therefore, the photon gas corresponds to a Bose gas for which the chemical potential μ is zero.

Planck's Units

7A   Planck's units

In 1899, Max Planck first proposed his radical theory of energy quantization. He proposed to build a system of "natural units" from a few of the more important constants in physics. These important constants include the speed of light, the universal gravitational constant, the Planck constant, and the Boltzmann constant. These quantities have great significance in various fields in physics. Combining these fundamental constant quantities, Planck generated various expressions with units of mass, length, time, and temperature. These units are known as Planck units. These four units signify us something different about the nature of reality. Before describing these units, a brief description about the four fundamental physical constants is presented below.

7A.1   The speed of light c

The speed of light c in a vacuum is a natural unit for speed and has magnitude $c = 299,792,458$ ms^{-1}. According to the special theory of relativity, this speed is the upper limit for the speed at which conventional matter or energy can travel through space. It is the universal "speed limit". It was recognized during the information age that the photons of electromagnetic radiation and material objects are used as the carriers of information. The speed of light is then a restriction on the speed at which information may travel. The speed of light can be used in time of flight measurements to measure large distances to extremely high precision. In a paper published in 1865, James Clerk Maxwell proposed that light is an electromagnetic wave and travels at speed c. In 1905, Albert Einstein postulated that the speed of light c with respect to any inertial frame of reference is a constant and is independent of the motion of the light source.

7A.2   Universal gravitational constant G

The gravitational constant, denoted by the capital letter G, is an empirical physical constant with magnitude $G = 6.67428 \times 10^{-11}$ N m^2 kg^{-2}. This universal gravitational constant is involved in the calculation of gravitational effects in Newton's law of universal gravitation and in Einstein's theory of general relativity.

It quantifies the relation between the geometry of space–time and the energy–momentum tensor. It contains the natural units of the fundamental quantities of mechanics, such as length, mass, and time. Gravity is obviously an essential characteristic of the universe, which makes the gravitational constant an obvious candidate for one of the fundamental descriptors of reality.

7A.3 Planck's constant h

The Planck constant h was first postulated by Max Planck in 1900. It is a fundamental physical constant characteristic of the mathematical formulations of quantum mechanics and has magnitude $h = 6.626 \times 10^{-34}$ Js. Planck constant describes the behavior of particles and waves on the atomic scale, including the **particle aspect of light**. In its traditional form, this constant h is the proportionality constant that relates the frequency and energy of a photon, a quantum of the electromagnetic radiation. It can be seen as a subatomic-scale constant. The dimension of Planck constant is the product of energy multiplied by time, a quantity called **action**. Hence, h plays two major important roles: it is sometimes called the **quantum of action**, and in its reduced form, $\hbar \left(= \frac{h}{2\pi} \right)$, it is known as the **quantum of angular momentum**. For example, the angular momentum of an electron bound to an atomic nucleus is quantized and can only be a multiple of $\hbar$. The second form is considered by many to be the more fundamental of the two, but it did not appear until 1930.

7A.4 Boltzmann's constant K_B

The Boltzmann constant, symbolized by K_B, is a fundamental constant of physics occurring in nearly every statistical formulation of both classical and quantum physics. It is named after Ludwig Boltzmann and has magnitude $K_B = 1.3806504 \times 10^{-23}$ JK^{-1}. The physical significance of K_B is that it provides a measure of the amount of energy corresponding to the random thermal motions of the particles making up a substance. The average energy per degree of freedom for a classical system at equilibrium at a temperature T is $\frac{1}{2} K_B T$. This corresponds very well with experimental data. It has the same unit as that of entropy and determines the quantum of this quantity. Entropy and information are related. The smallest amount of information is the bit – a choice between one of two things (1 or 0, yes or no, true or false, etc). The quantum of entropy is thus the entropy of a bit $S = K_B \ln 2$. Surprisingly, Boltzmann himself never tried to determine the constant that now bears his name. This constant plays an important role in determining the probability of occupation of a state with energy E at a temperature T in equilibrium. With the help of these four fundamental constants: c, G, h, and K_B, expressions for the Planck length l_{Planck}, the Planck time t_{Planck}, the Planck mass m_{Planck}, and the Planck temperature T_{Planck} are derived below.

7A.5   Planck's length l_{Planck}

These four fundamental constants are combined in a way that gives the Planck units. If the unit corresponds to the desired quantity, it has just made a Planck unit. For example, the Planck length is obtained if the combination of the constants ends in meters. Thus, the Planck length l_{Planck} can be obtained as

$$l_{Planck} = \sqrt{\frac{\hbar G}{c^3}} = \sqrt{\frac{1.055 \times 10^{-34} \text{ Js} \times 6.67426 \times 10^{-11} \text{ N m}^2 \text{ kg}^{-2}}{(299,792,458 \text{ ms}^{-1})^3}}$$
$$= 1.616 \times 10^{-35} \text{ m.} \tag{114}$$

This is small beyond comprehension. The next biggest material thing is a proton, the diameter of which is of the order of 10^{-15} m. This is a full 20 orders of magnitude bigger. Think of something that's about 10^5 m across (100 km). The big island of Hawaii comes to mind. If a proton was blown up to the size of the island of Hawaii, the Planck length would be as big as the original proton.

7A.6   Planck's time t_{Planck}

The Planck's time t_{Planck} can be obtained in the following way:

$$t_{Planck} = \sqrt{\frac{\hbar G}{c^5}} = \sqrt{\frac{1.055 \times 10^{-34} \text{ Js} \times 6.67426 \times 10^{-11} \text{ N m}^2 \text{ kg}^{-2}}{(299,792,458 \text{ ms}^{-1})^5}}$$
$$= 5.391 \times 10^{-44} \text{ s.} \tag{115}$$

Let us try to understand this time scale. We know that the speed of light (photon) is the greatest speed with which a material particle can move. The diameter of a proton $(10^{-15}$ m$)$ is the smallest length so far in nature. Now we ask the question: How long does it take a photon to cross the diameter (D) of a proton? The time taken by the photon is given by

$$t = \frac{D}{c} = \frac{1.0 \times 10^{-15} \text{ m}}{299,792,458 \text{ ms}^{-1}} = 3 \times 10^{-24} \text{ s.} \tag{116}$$

We see here that we are 20 orders of magnitude short. Again it is known that the universe is 13.8 billion years old. This time, when expressed in seconds, it becomes

$$t = 13.8 \times 10^9 \times 365.24 \times 24 \times 60 \times 60 \; s = 4.35 \times 10^{17} \text{ s.} \tag{117}$$

This time is twenty orders of magnitude larger than a millisecond. If the time it took a photon to cross the diameter of a proton was slowed to the point where the photon needed the entirety of time itself to complete its task, the Planck time t_{Planck} would last a thousandth of a second.

7A.7 Planck's mass m_{Planck}

The Planck's mass m_{Planck} can be obtained in the following way:

$$m_{\text{Planck}} = \sqrt{\frac{\hbar c}{G}} = \sqrt{\frac{1.055 \times 10^{-34} \text{ Js} \times 299792458 \text{ ms}^{-1}}{\left(6.67426 \times 10^{-11} \text{ N m}^2 \text{ kg}^{-2}\right)}} = 2.716 \times 10^{-8} \text{ kg}. \quad (118)$$

This mass is nearly equal to 22 g. This is the mass like that of a speck of dust. Let us compare this mass to that of a uranium atom, the heaviest naturally occurring atom. The mass of this uranium atom is given by $m = 238 \, u = 4.0 \times 10^{-25}$ kg. Again, we know that the heaviest known subatomic particle is the top quark, and its mass is given by $m = 173 \frac{GeV}{c^2} = 3.1 \times 10^{-25}$ kg.

Both of these values are about 17 orders of magnitude smaller than the Planck mass. Whereas the Planck length l_{Planck} and Planck time t_{Planck} seem to represent some lower limit on how finely space and time can be divided, the Planck mass m_{Planck} seems to be an upper limit on how big the small things in nature can be. No elementary particle will ever be more massive than the Planck mass m_{Planck}.

7A.8 Planck's temperature T_{Planck}

The Planck's temperature T_{Planck} can be obtained in the following way:

$$T_{\text{Planck}} = \sqrt{\frac{\hbar c^5}{G K_B^2}} = \sqrt{\frac{1.055 \times 10^{-34} \text{ Js} \times (299,792,458 \text{ ms}^{-1})^5}{\left(6.67426 \times 10^{-11} \text{ N m}^2 \text{ kg}^{-2}\right) \times \left(1.38 \times 10^{-23} \text{ JK}^{-1}\right)^2}}$$
$$= 8.903 \times 10^{32} \text{ K}. \quad (119)$$

How hot is this? No human beings or nature has approached such temperature. The interiors of the hottest stars are close to a billion kelvin (10^9 K) – 24 orders of magnitude short. In recent times, the hottest laboratory experiments are being carried out inside very large particle accelerators, such as the Tevatron at Fermilab near Chicago and the Large Hadron Collider (LHC) at CERN near Geneva. The temperatures generally observed in these cases are of the order of quadrillions of kelvins (10^{15} K). This temperature is still 18 orders of magnitude short. In contrast, the coldest temperatures ever achieved in the laboratory are a few hundred picokelvins (10^{-10} K). The entire range of temperatures achieved so far is an astounding 25 orders of magnitude, but we're still short 8 additional zeros. The Planck temperature T_{Planck} is so hot that it seems to be meaningless. As we shall soon see, that's the point.

For the next 50 years or so, Planck's notion of a natural unit system—one derived from physical laws, not accidents of human history—was considered an interesting diversion with little or no meaning. The primary reason for this was probably that quantum theory and general relativity were just too new and unfamiliar. (Relativity did not even exist at the time of Planck's publication.) The physics of the modern era was a strange world that few understood at first.

The Planck units have no practical application. No car odometer will be calibrated in Planck lengths, no stopwatch will tick off Planck times, and no thermometer will ever give temperatures as a teeny, tiny fraction of the Planck value. These numbers tell us the limits of physics as we currently know it and maybe even the limit of physics as it could ever be known. That's why it's an important theory.

Index